PROPERTY OF

Physical Science Today

Advisory Board

Matthew Sands, Ph.D.,
University of California, Santa Cruz

Robert Kolenkow, Ph.D.,
Massachusetts Institute of Technology

George Wetherill, Ph.D.,
University of California, Los Angeles

Gregory Choppin, Ph.D.,
Florida State University

Contributors

Isaac Asimov, Ph.D.,
Boston University

Adolph Baker, Ph.D.,
Lowell Technological Institute

Marion Bickford, Ph.D.,
University of Kansas

Geoffrey Burbidge, Ph.D.,
University of California, San Diego

Gregory Choppin, Ph.D.,
Florida State University

Michael Chriss, M.S.,
College of San Mateo

David Eisenberg, Ph.D.,
University of California, Los Angeles

John Fowler, Ph.D.,
University of Maryland

Harry Gray, Ph.D.,
California Institute of Technology

Alan Holden, B.S. (Retired),
Bell Telephone Research Laboratories

Lester Ingber, Ph.D.,
University of California, San Diego

Robert Kolenkow, Ph.D.,
Massachusetts Institute of Technology

Cindy Lee, Ph.D. Candidate,
Scripps Institution of Oceanography

Peter Lonsdale, Ph.D. Candidate,
Scripps Institution of Oceanography

Granger Morgan, Ph.D.,
University of California, San Diego

Manuel Rotenberg, Ph.D.,
University of California, San Diego

Matthew Sands, Ph.D.,
University of California, Santa Cruz

Arthur Schawlow, Ph.D.,
Stanford University

Victor Weisskopf, Ph.D.,
Massachusetts Institute of Technology

George Wetherill, Ph.D.,
University of California, Los Angeles

Physical Science Today

CRM BOOKS
Del Mar, California

Copyright © 1973 by Communications Research
Machines, Inc. All rights reserved. No portions of this
book may be reproduced in any form without
permission from the publisher,
Communications Research Machines, Inc.
Del Mar, California 92014
Library of Congress Catalog Card Number: 72-85255
Manufactured in the United States of America
First Printing

Preface

What happens when the nonscientist plunges into the study of physical science? Too often he discovers that he has entered a different and seemingly alien world with very little preparation for coping with it. Even the language is new and strange: *inclined planes, vectors, valence . . . ORBITAL quantum numbers(!) ;* He approaches the whole thing with trepidation because he KNOWS it is going to be awful.

That is why the people who joined together to work on *Physical Science Today* decided to approach the subject in a somewhat unconventional manner. The book does not begin abruptly with a classification of all the facts and figures of science, nor does it start out with a jolting panorama of idealized situations. It begins instead with simple pictures of matter, of atoms, of light—pictures that are ways of organizing our thoughts about the physical world. Through everyday examples and analogies, bridges are built leading into—and back from—that world to show how closely connected it is to the familiar territory of our daily experience. The world of physical science is not an alien one. It is the world we all know, expressed in ideas that are sprinkled liberally with imaginative insight.

It is true that these ideas may be understood on many levels. But how can the nonscientist appreciate the deeper and more intellectually satisfying levels without some background and preparation? To meet this problem, the book takes a cyclic approach to important topics such as the structure of matter, the atom, chemical change, and light. The introductory treatments of these topics in the early chapters, while complete in themselves, are followed and reinforced by more thorough discussions later on.

Although the actual sequence of topics is novel, the ordering is a deliberate attempt to pace the presentation of new material; hence, the recent discoveries of planetary and earth science—topics that are intrinsically interesting to the nonscientist—unfold after the abstractions of physical laws and, accordingly, may be understood on a deeper level. The topic of energy, as another example, is broadly described in early chapters, later analyzed in terms of the laws of thermodynamics, and finally treated concretely in terms of the energy resources of our environment. The book reflects the unique way that scientific knowledge builds upon itself. Once a few concepts are understood and an intellectual framework has been erected, it becomes increasingly easy to discover connecting points where new ideas may be attached to the framework. Each time the book returns to a topic, the degree of sophistication is greater to keep pace with growing abilities.

It is also true that physical science often appears to be synonymous with the formidable realm of mathematics. But mathematics merely is a device for working out ideas. And it is the *ideas* of physical science that the book stresses—the thoughts that people have had about the world, and the reasons why they had them. Of course there are numbers, and some arithmetic, because having a numerical estimate of, say, the size of an atom and the energy resources of a country helps us think more clearly about such things. There are even a few simple equations, but most of them, with the exception of the optional chapter on relativity, are written out to make them easy to understand. Problems are not presented in the traditional manner. Instead they appear in context with the subject being discussed, or they are worked out fully on illustrated problems pages in the belief that the process of discovery, rather than the answer itself, is infinitely more interesting.

There is not, however, an explicit discussion of scientific process or of the nature of physical laws. In a book that is intended to be a survey of the field, these topics are best understood by looking at the actual achievements of physical science in understanding the world. For example, the development of atomism serves as a case study for illustrating the interplay of thought, observation, and experiment that characterizes scientific endeavor. The history of atomism shows clearly how models in physical science can be modified or discarded when their predictive power is tested by new experience, and how new principles are gradually generalized as they are extended to new situations.

Physical Science Today is designed to portray physical science as it is—a perhaps surprisingly humanistic endeavor to explore and understand the world around us. It is a book that is shaped as it is by scientists who are, more than scientists, artists and philosophers and friends.

Cecie Starr
Associate Publisher, Sciences

Overview

How Science Views the World
- The World Around Us
  1. Thinking About Things
  2. Seeing Things
  3. Measurements and Numbers
  4. The Atom Hunters
  5. Stuff and Substance

Graphic Pause

Concepts in Physics
  6. Pushes and Pulls
  7. Motion
  8. Energy
  9. Relativity

Wishing Won't Make It So

Space Science
  10. Exploring Outer Space
  11. Origin and Evolution of the Universe
  12. Early History of the Solar System
  13. The Third Planet

Earth Science
  14. Materials of the Earth
  15. The Changing Atmosphere
  16. Waters of the Earth
  17. The Restless Earth
  18. A Nice Place to Visit...

The Beginning and the End

The Physical and Chemical Atom
  19. The Quantum Atom
  20. Bonds Between Atoms
  21. States of Matter
  22. Chemical Change
  23. Giant Molecules

Graphic Pause

Electricity and Magnetism
  24. The Original Charge Account
  25. Magnetism and Electromagnetism
  26. Electromagnetic Radiation
  27. Lasers: The Light Fantastic

Graphic Pause

High-Energy Physics
  28. Inside the Nucleus
  29. Fission, Fusion, and Nuclear Energy
  30. Elementary (?) Particles and Symmetry

Contents

The World Around Us	1

1. Thinking About Things — 5
Ways of Thinking — 6
The Search for Order — 7
The Atomic Model — 9
The Behavior of Matter — 11
The Chemical Elements — 13
Chemical Change — 14
Inside the Atom — 15

2. Seeing Things — 19
Light and Dark — 20
Shadows and How Light Travels — 21
Diffraction — 23
Absorption — 25
Refraction — 26
The Color and Energy of Photons — 31
The Quantum Photon — 33

3. Measurements and Numbers — 37
Units of Measurement — 38
Approximation and Error — 40
Distance and Length — 41
Measuring Surfaces — 42
Measuring Volumes — 44
Proportions — 45
How Much Time — 45
Other Quantities — 47
Writing Large and Small Numbers — 48

4. The Atom Hunters — 55
Early Pictures of Matter — 55
The Chemical Elements and Atomic Theory — 57
The Chemical Atom — 60
Combining Weight and Atomic Weight — 60
The Number of Atoms and Avogadro's Number — 63
The Size of Atoms — 65
Seeing Atoms — 66

5. Stuff and Substance — 69
Density and the Spacing of Atoms — 69
Temperature and the Motion of Atoms — 71
Pressure and the Collisions of Atoms — 75
Change of State — 76
Carrying Energy — 76
Chemical Change — 78
   Valence — 78
   The Periodic Table — 80
Atomic Spectra — 84
The Structure of the Atom — 84

Graphic Pause — 87

6. Pushes and Pulls — 89
Measuring Forces — 90
Force Is a Vector — 91
Combining Forces — 93
Taking Forces Apart — 94
Forces Always Come in Pairs — 96
The Laws of Statics — 98
Gravity — 100
The Idea of a Field of Force — 104
The Electric Force — 105
Nuclear Force — 109

7. Motion — 111
Describing Motion — 111
Change in Position — 113
Change in Time — 114
Speed and Velocity — 114
Uniform Motion — 116
Momentum — 119
Force and Momentum — 125
Motion With a Constant Force — 127
Changing Forces — 129
Circular Motion — 131

8. Energy — 135
Mechanical Energy—Kinetic and Potential — 137
Force and Work — 138
Heat Energy — 143
Heat Energy to Mechanical Energy — 144
Chemical Energy — 145
Other Forms of Energy — 145
Energy Runs Downhill — 146

9. Relativity — 149
The Michelson-Morley Experiment — 152
Fitzgerald and the Contraction of Motion — 154
The Lorentz Transformations — 155
The Special Theory of Relativity — 156
The General Theory of Relativity — 165

Wishing Won't Make It So	168

10. Exploring Outer Space 173
The Solar System	175
Motions Through the Zodiac	176
Earth's Neighbors	180
Other Stars, Other Planets	184
Star Populations	186
Galaxies	188

11. Origin and Evolution of the Universe 193
The Expanding Universe	193
The Existence of Radioactive Isotopes	195
The Origin and Evolution of Stars	198
The Conversion of Hydrogen Into Helium	200
The Formation of Heavier Elements	207

12. Early History of the Solar System 213
The Record of the Meteorites	215
The Age of the Solar System	218
Chemical Composition of the Solar Nebula	221
Evidence From the Moon	224

13. The Third Planet 233
The Early Earth	233
The Earth Today	234
The Earth's Composition	238
Formulation of an Earth Model	239
The Nature of the Earth's Crust	241
Continents and Ocean Basins	245
Why Study the Earth?	248

14. Materials of the Earth 251
Crystalline and Amorphous Solids	251
Important Mineral Groups	254
The Silicates	254
The Carbonates	254
Formation of Magmas and Igneous Rocks	256
Formation of Sedimentary Rocks	259
Weathering	259
Erosion	261
Transportation	261
Deposition	264
Lithification	265
Chemical Sedimentation	266
Formation of Metamorphic Rocks	268
The Rock Cycle	272

15. The Changing Atmosphere 275
The Atmosphere—A "Still Shot"	277
Atmospheric Pressure	277
Regions of the Atmosphere	279
Components of the Atmosphere	280
Atmospheric Stability	280
Dynamics of the Atmosphere	281
The Balance of Heat Energy	281
The Distribution of Heat Energy	284
Fronts, Highs, Lows, and Tonight's Weather	286
A Model for the Atmosphere	288
Controlling the Weather	289

16. Waters of the Earth 293
The World Ocean	293
Oceanic Currents	295
Waves and Tides	298
Upwelling and the Distribution of Nutrients	299
Man's Use and Misuse of the Ocean	300
River, Lake, and Ground Water	301
Solid Water	305
A History of the Hydrosphere	306

17. The Restless Earth 311
Continental Drift	312
Evidence From Oceanography	313
Evidence From Geophysics	314
Paleomagnetism	314
Sea Floor Spreading	317
Plate Tectonics	319
Mountain Building	323

18. A Nice Place to Visit... 329
The Study for Survival	329
Technological Innovation and Substitution	330
A Case Study: Our Use of Energy	331
The Study for Benefits and Use	338
Geothermal Power	339
Solar Power	341
Nuclear Power	342
The Study for Challenge	344

The Beginning and the End: An Imaginative Essay	346

19. The Quantum Atom 357
Electron, Proton, Neutron: Atomic Building Blocks	357
Thomson and the Search for Electrons	357
Millikan's Oil Drops	359
Rutherford Finds the Nucleus	361
The Energies of Atoms and Photons	363
The Photoelectric Effect	364
Energy Levels and the Balmer Formula	365
Ground Levels and Excited Levels	367
The Periodic Table and Atomic Structure	368
Bohr's Quantum Atom	368
The Pauli Principle and the Shell Model	370
Periodic Shells, Periodic Table	371
Waves and Particles	372
Waves and Uncertainties	375
Electrons Clouds	376

20. Bonds Between Atoms 279
Ionic Bonds	379
Covalent Bonds	382
Intermediate Polar Bonds	384
Dispersion Forces	385
Polyatomic Molecules and Bonding	386
Explanation of the Attraction in Covalent Bonds	386
Molecular Bonding Geometry	387
Hydrocarbons	389
Names for Organic Compounds	391
Right Hand, Left Hand, and Life	392
Unsaturated Hydrocarbons	395
Metals	397

21. States of Matter 399
Restlessness	399
Crystallinity	399
Diffraction	401
Solid, Liquid, and Gas	404
Allotropy	405
Imperfections	408
Dislocations	409
How Solids Are Held Together	410
Electrons in Metals	413
Electrons in a Crystal	414
Conductors and Insulators	415
Plasmas	416

22. Chemical Change — 419
- Balanced Equations — 419
- Chemical Structure and Reactivity — 421
- Energy Changes in Chemical Reactions — 422
- Reaction Rates — 424
- Types of Reactions — 426
  - Substitution Reactions — 427
  - Acid-Based Reactions: Proton Transfer — 428
  - Addition and Elimination — 430
  - Isomerization — 430
  - Oxidation-Reduction — 431
- Chemical Changes in the Environment — 432
  - Mercury Reactions — 432
  - The Chemistry of Smog — 435

23. Giant Molecules — 439
- Polymers From Monomers — 439
- The Sizes of Giant Molecules — 441
- DNA: A Polymer for Storage and Replication of Information — 443
- Fibrous Proteins — 444
- Enzymes: Our Protein Catalysts — 445
- The Structure and Function of Something Called Chymotrypsin — 446
- Viruses as Giant Molecules — 448

Graphic Pause — 453

24. The Original Charge Account — 455
- Electric Charge — 455
- Electricity and the Structure of Matter — 456
  - Electrons in Matter: Conductors and Insulators — 457
  - Simple Electric Phenomena — 458
- Electric Current — 459
- Potential Energy and Voltage — 462
- The Connection Between Force and Potential — 462
- Charges, Conductors, and Capacitors — 465
- Electric and Mechanical Power — 467
- Electric Resistance and Ohm's Law — 468
- Electromotive "Force" — 469
- Electric Power Transmission — 471

25. Magnetism and Electromagnetism — 477
- Electric and Magentic Fields — 480
- Magnetic Forces on Charges — 481
- The Magnetic Force on Currents — 484
- Units of Current and Magnetic Field — 487
- Magnetism and Matter — 487
  - Paramagnetism — 489
  - Ferromagnetism — 490
  - Diamagnetism — 490
- Magnetism and Electricity — 492
  - Generators and Transformers — 492
  - Electromagnetic Induction — 493
- Magnetism and Relativity — 497

26. Electromagnetic Radiation — 499
- Waves — 499
- Electricity, Magnetism, and Electromagnetic Waves — 503
- Electromagnetic Waves and Energy — 505
- The Electromagnetic Spectrum — 505
  - Radio Waves — 508
  - The Microwave Region — 511
  - The Infrared Region — 513
  - Visible Light — 515
  - Ultraviolet Light, X-Rays, and Gamma Rays — 515
  - Light Rays — 517
- Interference — 519
- Polarized Light — 520

27. Lasers: The Light Fantastic — 522
- Kinds of Lasers — 525
- Properties and Uses of Laser Light — 528
  - Beam Divergence — 528
  - Power Density — 529
  - Scattering and the Raman Effect — 531
  - Coherence — 533
- Lasers and Lensless Photography — 534
- Laser Safety — 538

Graphic Pause — 541

28. Inside the Nucleus — 543
- Nuclear Stability and the Neutron/Proton Ratio — 544
- Nuclear Size — 544
- Nuclear Energy — 545
- Binding Energy — 546
- Nuclear Forces — 548
- Types of Radioactive Decay — 549
- Half-Life of Radioactive Decay — 551
- Detection of Radiation — 552
- Nuclear Reactions — 555
- Accelerators — 556
- The Nuclear Shell Model — 556

29. Fission, Fusion, and Nuclear Energy — 559
- Discovery of Fission — 559
- The Liquid Drop Model of the Nucleus — 562
- Mass Distribution in Fission — 562
- Energy of Fission and the Chain Process — 563
- The Manhattan Project — 566
- Nuclear Reactors — 566
- Nuclear Energy — 568
- Nuclear Fusion — 571

30. Elementary (?) Particles and Symmetry — 575
- Conservation Laws — 576
- Angular Momentum — 577
- Hadrons and Leptons — 579
- Quantum Theory of Forces — 581
- Particles and Antiparticles — 582
- Reactions of Elementary Particles — 583
- Isotopic Spin and Hypercharge — 585
- Conservation Laws and Symmetry — 587

Appendix 1. Constants, Conversions and Astronomical Data — 592
Appendix 2. Star Maps — 594
Contributing Consultants — 598
Selected Bibliography — 604
Glossary of Scientific Terms — 610
Index — 624
Credits and Acknowledgments — 634

The World Around Us

In Switzerland the thick haze that forms over so many industrialized parts of the world is still nonexistent, and the air is unbelievably clear. During my summer visits to Geneva I find great pleasure in going out to the nearby meadows and contemplating the stars, which are suspended like radiant crystals in that clear night air. I pick out the constellations, those ancient perceptions by which we first imposed order on the heavens, and I search for individual stars—Polaris, Arcturus, Antares—those permanent, silent points of reference in the northern skies. And I observe with great satisfaction that in our chaotic times the stars, at least, provide reliable assurance of order in the universe. This is an order of the grandest scale, and much of it can be explained by physical laws that have been painstakingly carved out of centuries.

I thought of this one night as I gazed upward, and after my mind had moved across the skies, musing on our accumulated wisdom, it happened that I slipped into a comfortable sleep and I began to dream. Strangely, in my dream I continued to scan the skies and the constellations above me for what seemed to be a long, unhurried time. Then, almost imperceptibly, a few stars began to move—to move in this fixed universe of ours!

Imprisoned in that dream I watched more and more stars move slowly apart, gathering speed and then rushing into vast, erratic orbits. The skies became laced with their fiery paths and I, beneath that terrifying canopy, felt all reason vanish. How could they move so incredibly fast? Considering their distance from earth, they would be moving faster than light—and didn't Einstein say that couldn't be so? My mind suddenly had become an unknown quantity in the midst of chaos. There were no laws, there was no order; my frames of reference did not exist, nor did my explanations of them.

I awakened drenched with sweat. Feeling somewhat foolish—yet, I must admit, undeniably worried—I glanced upward. The familiar array of stars was there, silent and ordered as I have always known it to be. It was with incredible relief that I welcomed the sight of those tangible affirmations of our understanding of the universe.

In that moment I sensed the fear and wonder sweeping through past ages. I thought of the disbelief of the Chinese who witnessed the supernova of the year 1054, the explosion that left behind the vast remnants we call the Crab Nebula. Here was an event that totally contradicted the first meticulous scratchings into the workings of the physical universe—but those observers could not awaken and call it a dream. They could only record this phenomenon, and others, and thus lay the foundations for the pyramid of human experience that would begin its slow, upward thrust from fear and uncertainty.

In this century the process has accelerated. Building on the cumulative record of past impressions—from Chinese astronomers to Newton to Einstein—physical science is reaching new and often unexpected plateaus. And the view today sweeps out from DNA to the expanding universe, from matter to the stars.

Consider the distant probes of astrophysics, a science that has taken the laws we use to explain the natural world and has carried them into the unknown—that elusive outer limit of knowledge which turns

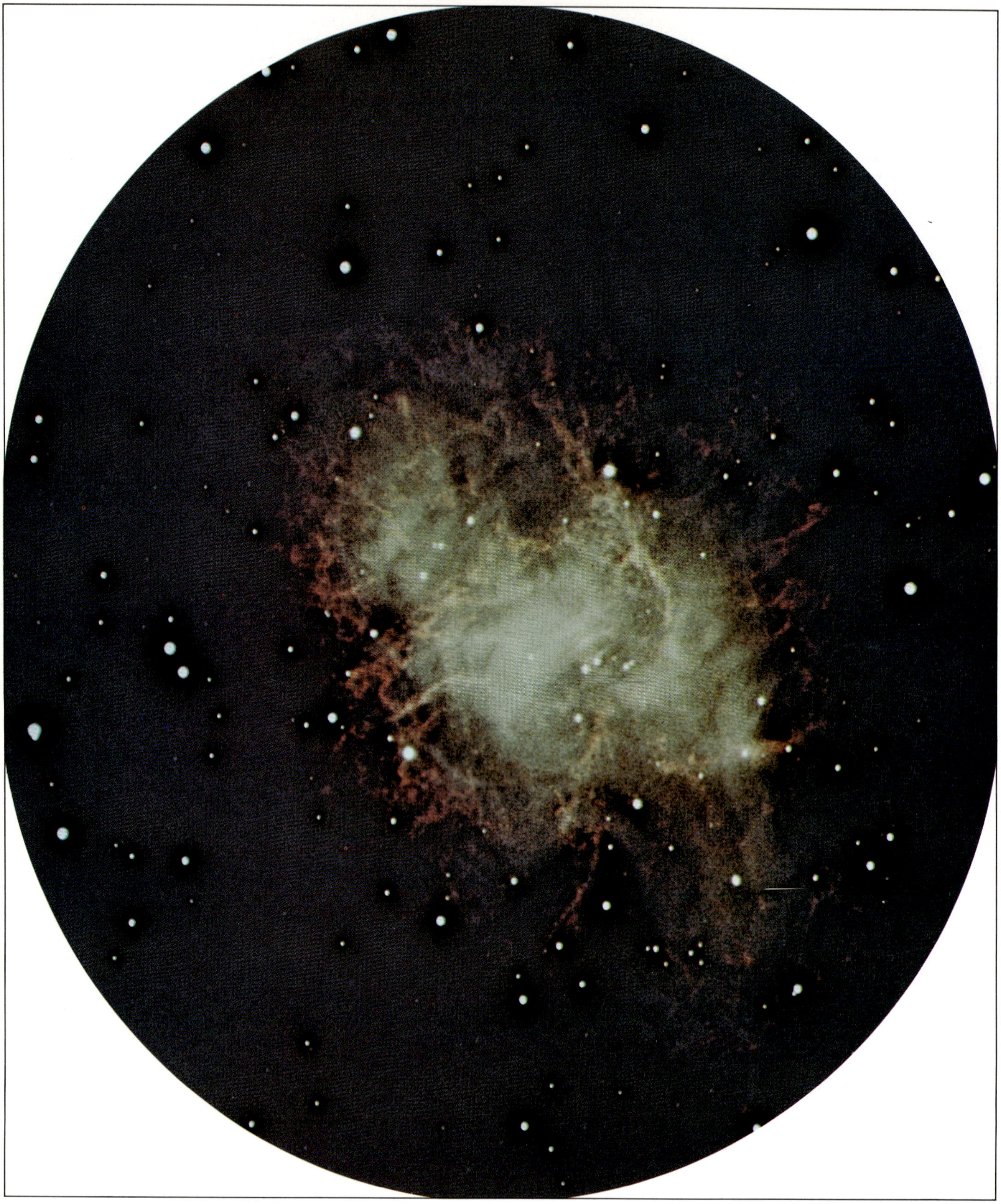

up in newer forms to haunt our curious minds. It emerges in red giants and white dwarfs; it tantalizes us with collapsing stars and exploding galaxies. Magellan in the sixteenth century traced out the finite surface of his world; explorers in this century identify a more distant and seemingly finite boundary—one they cannot see directly but are beginning to comprehend. It is more than *10 billion light years* away, beyond the billions of stars in our galaxy, beyond the many millions of galaxies equal in size to our own. And all the explanations of astrophysics cannot penetrate through it. From beyond this limit, light—the only messenger between worlds—never can reach our part of the universe.

Nor does the search for knowledge lead only to the limits of an expanding universe. One of the most remarkable achievements of our time is the discovery of an inner form of that boundary, the world of atoms. Ultimately, all the regularities of form and structure we see in the natural world—from snowflakes to stars—reside in the symmetry of atomic patterns and the forces that bind the elementary units of matter together. But this boundary, too, is elusive. What, after all, is the "elementary" unit, the irreducible particle of matter? Not very long ago in human history the irreducibles were Fire, Air, Earth, and Water; in Newton's time the elementary units were considered to be all the myriad kinds of matter—Salt, for instance, and Water and Iron and Emeralds. And now, even the historic isolation of the one-hundred-odd species of atoms appears to be a boundary to still another world, the world inside the atom. It is populated not only with protons, neutrons, electrons, photons, and neutrinos, but with an enormous host of such bewildering oddities as mesons and baryons. Who knows where these challenging frontiers will lead?

The incredible thing, however, is not so much the expansion of our outer and inner horizons but the tremendously reassuring knowledge that basic natural laws span the full range from one extreme boundary to the other. With these laws we can find stability and order even in the apparent chaos of the unknown.

And this is what physical science is about. All the far-reaching probes into the formation of the earth, the birth of our sun, the development of the basic elements of life, the nature of matter—all these things help us know our place in the universe. In physical science we find the certainty that nature works according to exact laws, that humankind is not helpless before the whims of a capricious universe. These edifying and stabilizing laws allow us to fall asleep at night and trust that the world, as we know it, will still exist for the day that follows.

Victor F. Weisskopf

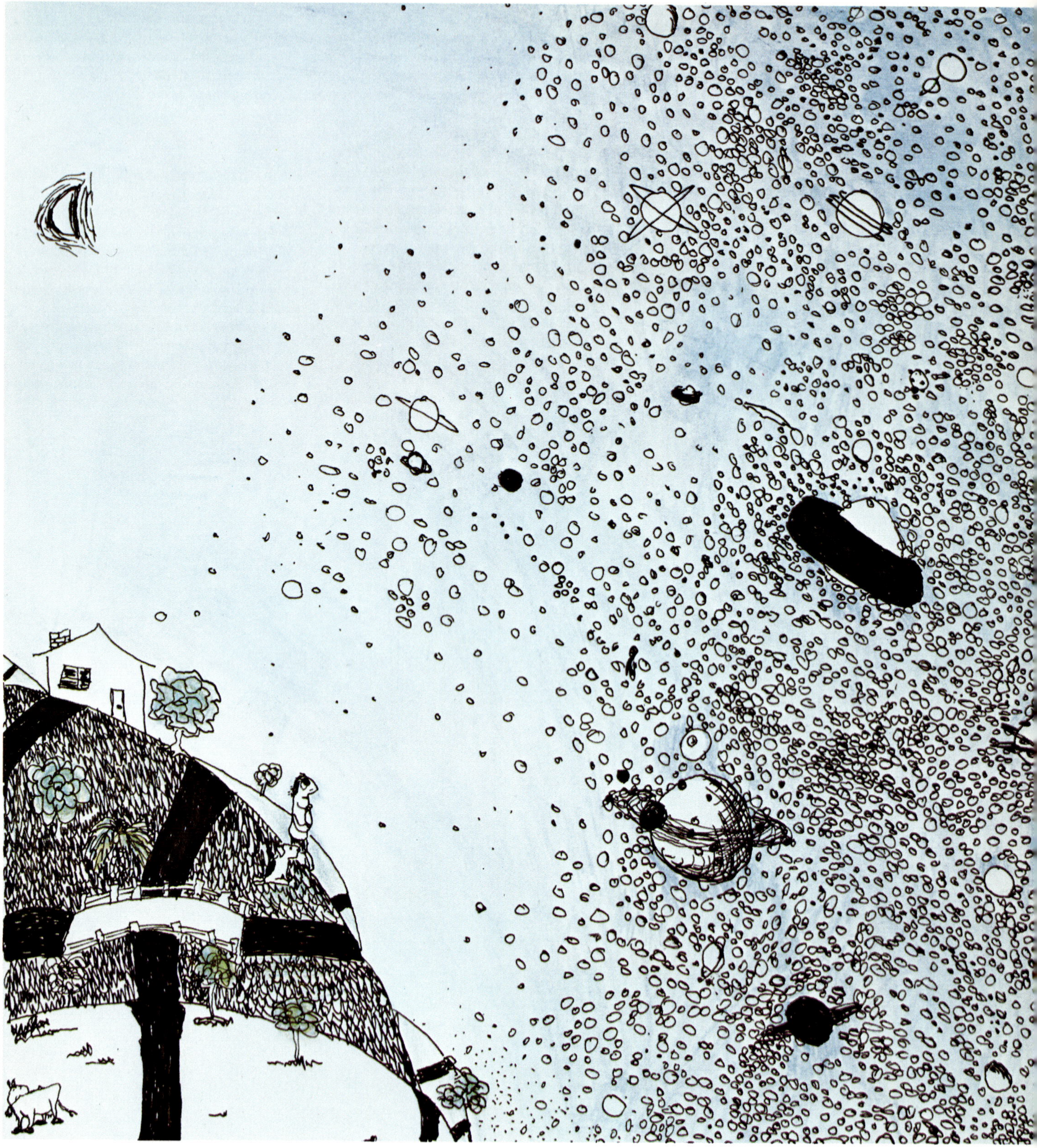

1
Thinking About Things

The universe through the eyes of a twentieth-century child.

The world is a wonderfully complex mosaic of people and things, earth and sky, objects and materials. And the patterns of that mosaic are constantly moving and changing about us. Each day we are bombarded with an incredible variety of impressions, sensations, and images, all entering into our consciousness. There are things to be dealt with, to be sorted out into coherent bits of knowledge that will influence how we will react as we are faced with a multitude of daily decisions—where we step, which roads we follow; what we choose to do or choose not to do or even want to talk about.

How is it we manage to put together all that we know so that it makes some sense? Very little is known about how information is processed in the brain. Somehow the brain receives an impression of the changing world and, somehow, goes to work on it. Although we do not yet know how all the mechanisms work together, our own experience gives us some general notion of the processes that must be going on.

We know, for example, that we are able to find similarity in things that are really quite different. We see a chunk of sandstone, of granite, and of marble and each time we think "rock"—even though they are all different things. Somehow our mind holds an idea of what a "rock" is. Marble fits that idea; a carnation does not. In our consciousness we have a *model* of what a rock is. And when something we see has enough of the characteristics that have been built into our model we decide it is *like* the model. We decide it's a rock.

By the time we reach adulthood we have thousands of models. Nearly every word we know stands for a model that is stored away in our mind. The word "elephant," for instance, evokes an unambiguous model; the ideas associated with it are relatively precise. But other models are not as clear-cut. Say "human dignity" or "time" or "universe" and see what images these words evoke.

What good are models? Why do we make them? One reason is to communicate with other people. If you shout, "Look out for the elephant!" you must have identified something you saw as fitting your model of "elephant." But more than this, you assume other people have a similar model of "elephant" and will know what you are talking about and will move out of its way. And the word "elephant" was invented—as a symbol for a shared model—so that you may exchange your thoughts with others.

But models have another and perhaps more important use, for we would need them even if we never chose to speak. Models provide clues about what is likely to happen, because our models include *behavior* as well as appearance. By using them we are able to make predictions about the future, and to decide how to behave in the hope that the future may come out more to our liking.

No model is perfect, yet we have a tendency to think models are more accurate than they really are. This occasionally gets us in trouble, because often what a model tells us to expect is not what actually happens. The real world may differ from a model, and a prediction may not necessarily come true. Anyone who has predicted the weather or planned his weekend by weather forecasts knows that. Not remembering the limits of a model can lead us into difficulty, and in trying to keep life simple we often find that we have pushed a model further than we should. So we correct the model, hoping to predict better next time.

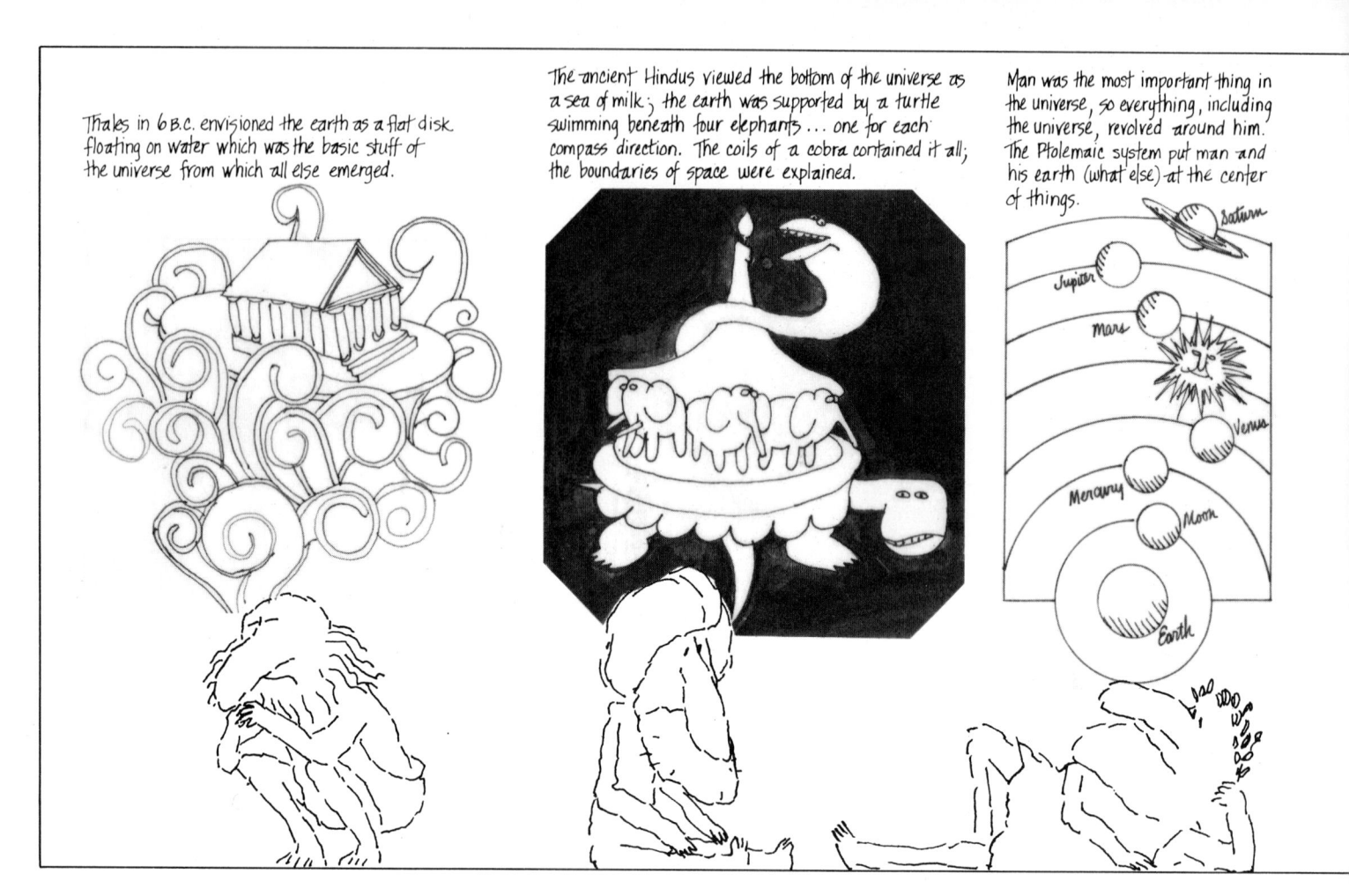

Thales in 6 B.C. envisioned the earth as a flat disk floating on water which was the basic stuff of the universe from which all else emerged.

The ancient Hindus viewed the bottom of the universe as a sea of milk; the earth was supported by a turtle swimming beneath four elephants... one for each compass direction. The coils of a cobra contained it all; the boundaries of space were explained.

Man was the most important thing in the universe, so everything, including the universe, revolved around him. The Ptolemaic system put man and his earth (what else) at the center of things.

WAYS OF THINKING

A large part of our intellectual activity consists of building and refining models of the world and our place in it. From our earliest years, our culture provides us with models that introduce order into the seeming chaos of the world. Often these models are simplified so that things will *seem* to be simpler—so that they can be thought of in manageable ways. These models are not final answers, but they do suggest ways in which things and events may be connected.

Our collections of models are many and varied. Some are worked out in great detail; others are less structured. Some have been given special names: *Sociology*—a way of thinking about how people relate to each other in groups in terms of kinship, class, culture and subculture, neighborhood, and city. *Psychology*—a way of thinking about what goes on in the mind of an individual, alone and in relation to others. *Biology*—a way of thinking about living things. The myriad forms of life; their growth and functioning, their development and death, their evolution through time. *Physical science*—a way of thinking about the nature and working of all material things. The universe and stars; the earth and its rocks, waters, and atmosphere. Sound and light. All things we make and use; what we ourselves are made of. And how it all works.

But all these different ways of searching for order in a seemingly chaotic world are approximate and incomplete. Even their separation into "subjects" may mask their interdependence. These subjects must be

Physical Science Today

In the fifteenth century, Copernicus decided that the sun, not the earth, was the center of the universe. His model didn't fit in with the models of his contemporaries, so nobody listened to him for a long time.

Today our models of the solar system are based on models of a much grander scale—the boundaries of which are now recognized to be in the origin of the universe itself (but what is the universe?)

interrelated to be understood. How can sociology understand the functioning of communities without understanding the psychology of its individual members? How can psychology understand the way the mind works without concern for the biological processes going on in the brain? How, for that matter, can biology understand the working of the brain without thinking about the chemistry of its cells or the physics of the electric impulses in the nerve fibers? And how can physical science understand its own view of the material world without taking into account the working of the human mind? And so we come full circle, for we need sociology and psychology to understand how we know and communicate about the physical world.

All knowledge is connected; all knowledge is incomplete. Sometimes we struggle with the parts and, on occasion, we try to come to terms with the whole. This book presents one of those parts, the view of the world that has been constructed by the physical scientist. It is the story of how men have tried, and failed or succeeded, to find some understanding about the material world and its relation to man.

Figure 2. Throughout history men have created models for explaining the motion of the sun, the moon, the planets. What was the sky? Where did it begin and end? The desire to build a satisfying model of space—to give boundaries to it so that it could be thought of in manageable ways—persists even today. Rather than being amused by the naïve elaborations made on ancient models, perhaps we may view them as analogs to our own present-day models of our solar system: all seek to show the connectability of things, of objects, and of events that we know about from experience and from the patterns of thinking in our culture. One is no less a part of the human experience than another; only the levels of understanding change as we continue to build and to refine our models of the natural world and our place in it.

THE SEARCH FOR ORDER

It all begins with curiosity. In the United States everyone has an idea that summers are hotter than winters. Most people even have some idea of what the hottest and coldest temperatures usually are. In Arizona, for instance, temperatures reach 120 degrees Fahrenheit in the summer and

1 | Thinking About Things

SEPTEMBER Temperature				
Day	Max.	Min.	Avg.
1	101	71	86
2	101	71	86
3	101	71	86
4	101	70	86
5	101	70	86
6	101	70	86
7	100	70	85
8	100	69	85
9	100	69	85
10	100	69	85
11	100	69	85
12	100	68	84
13	99	68	84
14	99	68	84
15	99	68	84
16	99	67	83
17	98	67	83
18	98	67	83
19	98	67	83
20	97	66	82
21	97	66	82
22	97	66	82
23	96	65	81
24	96	65	81
25	96	65	81
26	95	64	80
27	95	64	80
28	94	63	79
29	94	63	79
30	93	62	78

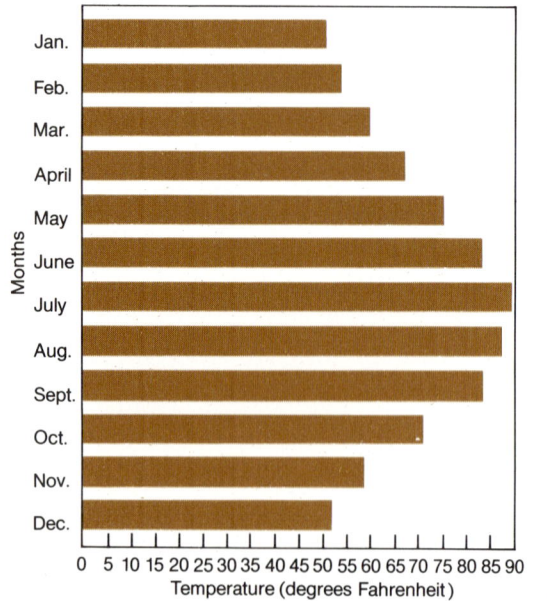

drop to 30 degrees Fahrenheit in the winter. Suppose you just happened to be curious about how the weather in a certain place changes through the year. Maybe you have reason to want to know—you're interested in planning a vacation—or maybe you just like to think about the weather. You might start by reading the thermometer once each month for a year. But on some days you notice the reading happens to be hotter or colder than usual. If you decide to take the whole thing seriously, you begin reading the temperature every day. You notice, of course, that the temperature varies from sunrise to sunset, so you decide to make your reading at noon each day.

Now you are starting to get a lot of numbers, or *data*. What are you going to do with it? Say you decide that what you really want to know is what the "typical" noonday temperature is for each month. You take all your readings for each day and calculate the monthly average. You then have twelve numbers, each one the average noonday temperature (also called the *mean* noonday temperature) for each month. At the end of your observations you may decide to plot a graph and put it on the wall of your room. If your curiosity continues you might decide to do the same thing for another year. The results you get will be similar but not exactly the same, and from your graphs you begin to see that the temperature is a little different from year to year.

The sequence just described is one kind of scientific inquiry. Beginning with curiosity you move on to careful observations, keeping track of things by recording numbers, worrying about how to improve the measurements, and organizing the data in some convenient way. In the end you have a relationship between temperature and month of the year. Such a relationship is described by saying that the temperature is known *as a function of* the time of year. If this relationship is the same from year to year, you may have established a *natural law*—which merely means you have discovered a repeating pattern in the behavior of nature.

Your work also shows some of the typical problems of scientific work. For one thing, the results are never perfect. One year you may find the average noonday temperature in April is 65 degrees Fahrenheit; the next year you may discover it is 63 degrees Fahrenheit. If you wanted to predict the average midday temperature for *next* April you could make a pretty good guess, but your prediction would have some uncertainty. The law you have found is not precise; it has some *error*.

Perhaps, in the course of your work, you begin to wonder: "Why are the summers hotter?" Noticing that the sun appears higher in the sky during the summer, you wonder whether the difference has some connection with temperature. So each day at noon you measure the angle of the sun above the horizon as well—finding, eventually, the sun is in fact higher in the summer sky. You are beginning to find another relationship, another connection, between different observations. Then you begin to think: "Of course. When the sun is directly overhead it gives more heat." Perhaps you try to figure out in detail how much more heat arrives from the summer sun than from the winter sun and, from that, how much hotter the surface of the earth should get. By now things are getting complicated, and you have to draw on other knowledge—the properties of the sun's rays, what heat is, how things get hot and cold, and so on.

You have again been following a typical course of scientific study: you are beginning to make more complicated connections and you are beginning to make a *theory*. Your course is leading you toward an explanation of *why* something happens. In your mind you are actually building a model of how something works, and of how different things are connected. Your theory, or your model, goes something like this: *Heat comes from the sun. More heat means a higher temperature. In the summer the*

Physical Science Today

sun gives more heat. So summers have higher temperatures. If you can even determine *how much* hotter the summer will be, then your theory will be that much better.

If your theory is going to be interesting to anyone else, however, it must work in places other than your home town. Say you have a friend in Chile, where the hottest months occur when we have our winter. Will your theory work there?

This example is typical of the way of science—curiosity, observation, searching for relationships, thinking through and working out ideas, making a model that ties it all together. But there is another point. Science generally is asking "why" but always with a special meaning for the word: *How can I understand this in relation to something else I know?* With this question, science weaves an ever expanding web of connections between the known and the unknown. It is a cumulative process, for as the web enlarges, more and more questions—and connections—are made possible.

There are other questions as well—questions of values and goals. These we define for ourselves out of the conditions of our lives and on the basis of what is known, to be sure. But we also define them on the basis of what we imagine to be desirable, and worth attaining. Science helps us to understand the consequences of our actions, but for help in deciding what consequences we *should* wish to achieve we must look outside of science—to such human endeavors as ethics, religion, and philosophy.

Figure 3. (Left, above) This chart shows the average daily temperature for a 1-month period in Phoenix, Arizona. (Below) The temperatures are plotted for a 1-year period as well. These data were provided by the US Weather Bureau in Phoenix.

THE ATOMIC MODEL

Over the past century or so the activities of many physical scientists—chemists, physicists, astronomers, and others—have been aimed at discovering some understanding of the nature of the materials of the world. All of this work has led to our present view of the structure of matter. It is a remarkable view, or model, that has far-reaching and profound consequences for our ability to understand the nature of the physical world.

All the world is made of atoms. Everything we see, touch, and know about consists of basic, minute particles of matter—atoms—millions, billions, billions of billions of billions of tiny specks of matter. Each one behaves in its own way, according to its own special rules. It sticks to other atoms, it moves back and forth, it flies through space. Everything that exists, everything that happens is, at the base, what atoms are and what atoms do.

Yet in all the world there are only 100 or so different kinds of atoms, tiny "fuzzy" balls of basic matter governed by relatively simple rules of behavior. Together they comprise the variety and complexity of the world. Materials that are heavy or light, hard or soft, hot or cold, and everything that happens—motion, change, walking, flying, breathing, flowing, living, dying—stem from the actions of atoms, from their motion, from their behavior.

When you stand near the ocean, you see wind-driven waves rippling the water's surface and breaking on the shore. Here myriad atoms that make up the water are hugging together, sliding tightly over one another like a milling crowd and giving the water its smoothness and flow. Atoms push against each other, surging forward with the wave. And some gather into small groups to make droplets of spray as the wave breaks.

Under your feet you can feel the solid earth. Here atoms of a different kind are locked tightly together in a rigid pattern, each holding its place despite the pressure of your foot. But sometimes their interlocking is not strong enough to resist: they break apart and the earth crumbles into

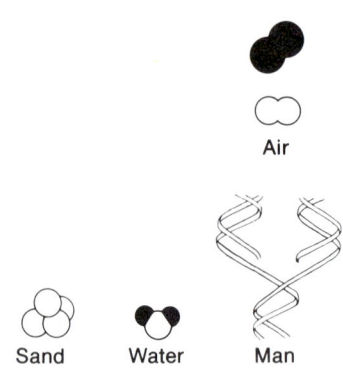

Figure 4. If you *could* magnify our familiar world by a factor of millions, you would discover complex but repetitive arrangements of molecules.

pieces. Sometimes they break again and again under the beating of the waves, and rock turns to sand. Each grain of sand in its turn is a small colony of atoms that has managed to hold together.

Perhaps there is a wind moving against your face. Here again are different atoms—this time spread thinly and flying helter-skelter, knocking against one another and being pushed against your face by the wind.

Your own body is an immense collection of specially selected atoms, each in its place with its own part to play. The atoms of your flesh tend to hold to their relative places, yielding when touched but bound together enough not to break easily. The atoms of your blood are special groups traveling together, carrying the energy for life. The atoms of your muscles are grouped in special ways that can pull, or not pull, to swing your arm or perhaps bend your leg. Atoms of your eyes sense the presence of light and send the images of the world through other atoms to your brain. The atoms of your brain hold together in fantastic patterns, delicately tracing out thought and consciousness.

All of these atoms are tiny balls of matter, more or less spherical—and all are of similar size even though some are larger or smaller relative to one another. These specks of matter come in about 100 different species. Take any piece of matter apart and you will find only these same 100 or so different kinds of atoms. And all of the most common materials are made up of selections from only 20 or 30 different kinds. Each kind of

10   Physical Science Today

atom has its own special way of behaving that permits its identification. All atoms of a particular kind, however, are identical: you can't tell one from another. Each atom of oxygen, for example, is precisely the same as every other atom of oxygen throughout the world.

The different species of atoms are similar to one another in certain important respects. Place any two atoms near each other and they will try to pull closer together. In some cases they pull together strongly; in others, the pull is weaker. Place an oxygen atom near a hydrogen atom and they will attract each other and draw together. But they will do this only to a point. Once they get close enough to each other the attraction diminishes and they will fiercely resist any attempt to push them closer together. In fact, if you try to do so they will push apart with a strong repulsive force.

Touch the cold metal of an ice cube tray that is still in a freezer and your flesh will stick to it. The atoms of the metal and the atoms of your flesh are pulled toward each other. Push hard against the metal, however, and it resists: its atoms will not allow the atoms of your flesh to come any closer. Things hold together because their atoms attract one another, but they keep their shape because each atom occupies its own segment of space and won't let another atom enter that space.

THE BEHAVIOR OF MATTER

The atomic view of matter helps us understand not only how matter is constructed but also how matter behaves—how the world about us moves and changes. Put a drop of water on your finger and imagine yourself looking into it with a microscope that magnifies 100 million times. If you could do this, you would see an incredible jumble of spheres of matter, each appearing to be about 1 centimeter in diameter under magnification. Two different kinds of atoms are present: smaller ones called "hydrogen" and larger ones called "oxygen." As they move about, two of the smaller hydrogen atoms always stay attached to each oxygen atom. Each triplet of hydrogen-oxygen-hydrogen is a *molecule* of water. The atomic parts of a molecule stick together in a more-or-less rigid shape; the attraction those parts have for one another is stronger than the attraction they have for the atoms of other molecules. Rather than just sticking together, the water molecules slide over and around one another. This motion prevents the water droplet from taking on a definite form, and it results in what we call a *liquid.*

The molecules of water slide easily over one another because they are constantly in motion, moving back and forth like so many dancers

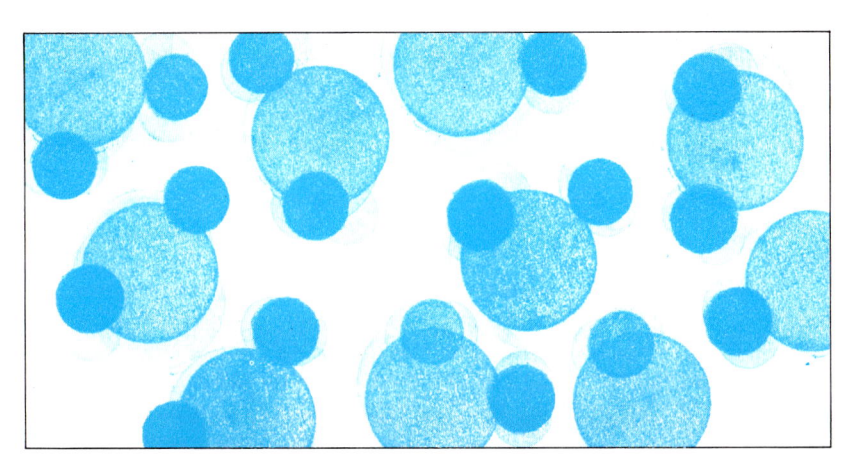

Figure 5. This drawing shows what you would see if you could peer into a drop of water magnified 100 million times larger than real life. Two hydrogen atoms and one oxygen atom are fastened together to make a molecule of water. The molecules are continually moving, sliding over one another. The lightly shaded areas depict the same scene an instant later, when all the molecules have moved to new positions.

Figure 6. What you would see if you looked at ice magnified about 100 million times. All of the atoms are connected together in a pattern of hexagons. The atoms do not change positions, but they do jiggle around a fixed place.

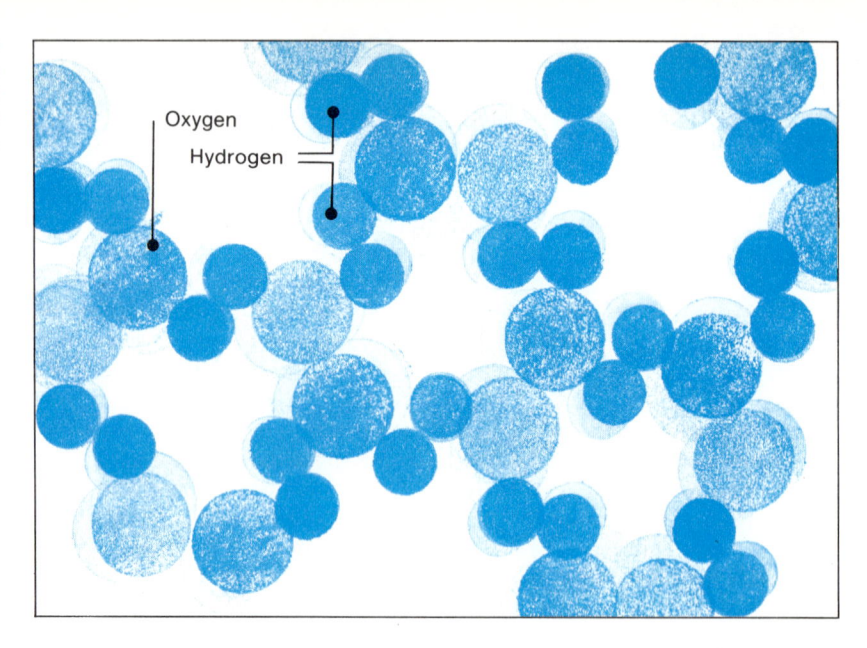

crowded together on a small dance floor. Each molecule has an *energy* of motion. Bumping one against the other, each shares the common energy of the water droplet in roughly the same amount. This shared energy of motion is the *heat* in the liquid.

If water is heated to a higher temperature, more heat energy is added to it; and with more heat there is more motion. When we submerge our fingers in hot water the atoms of the water move violently against our fingers, and we feel this as "hot." If the water is cooled, the energy of motion is diminished and the atoms move about more slowly. But if energy continues to be taken from the drop of water, at some point the character of the motion changes abruptly. Instead of slithering about all over the place, each molecule settles down in a particular place and jiggles back and forth rapidly without ever going anywhere. The water ceases to flow; it becomes ice. It is now a *solid.*

When water becomes solid, when it freezes into ice, the decreased energy of motion is no longer great enough to overcome the forces attracting each molecule to the other molecules nearby. Each atom stays close to its proper place, pulling on nearby atoms but also holding them off at a proper distance. Together they create a pattern of hexagons that expresses the inner nature of the hydrogen and oxygen atoms—a pattern we call a crystal of ice.

But if a piece of ice is heated, the motion of the molecules increases until the ties between them are broken by the violent movement. The solid ice returns to liquid water.

And if still more heat energy is added to the water droplet, it disappears; it evaporates away. The motion of the molecules becomes more violent and the molecules at the surface of the droplet are no longer held back by their attraction to the others. One by one they tear free to sail off through space. The droplet disappears as the water loses its atoms to the air.

Now look at ordinary air through the imaginary microscope that can magnify things 100 million times. Here you see different groupings of atoms. This time there are oxygen atoms and nitrogen atoms. They are nearly always joined in pairs of the same kind: an oxygen atom with an oxygen atom, a nitrogen atom with a nitrogen atom. You are seeing molecules of oxygen and molecules of nitrogen. The atoms of a single mole-

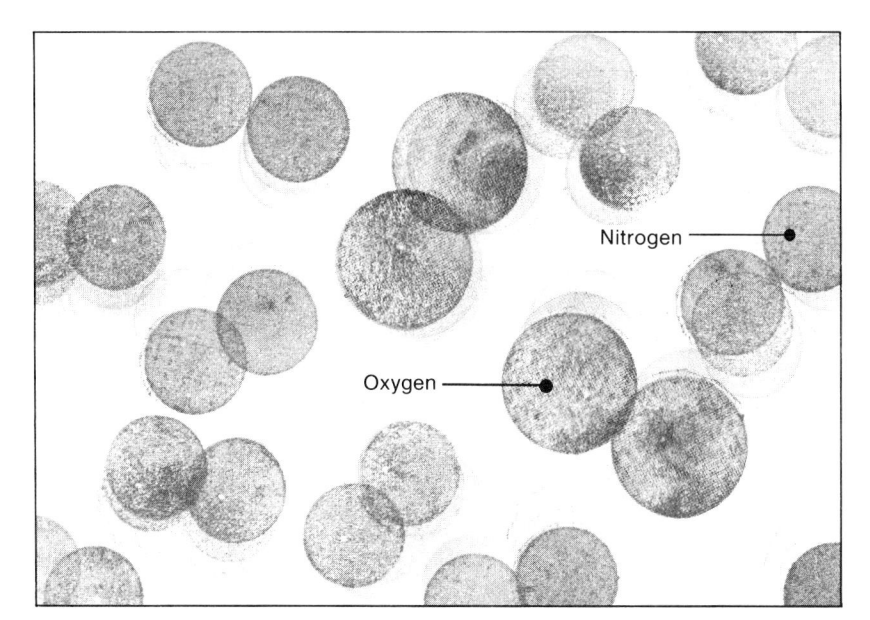

Figure 7. What you would see if you looked at air magnified about 100 million times larger than real life. Molecules of oxygen (two oxygen atoms stuck together) and molecules of nitrogen (two nitrogen atoms stuck together) fly about, punctuating empty space.

cule hug together, but the attraction of each molecule to the others is rather weak, so the energy of motion of the molecules does not permit them to stick together. They fly about in space, bumping against one another and bouncing from one encounter to the next. When matter consists of atoms that behave in this way, it is called a *gas.*

What would you observe if you could focus that same microscope on parts of your body? In a bone you would see such atoms as oxygen, calcium, and others holding tightly together to make a solid structure. In muscle or flesh you would find atoms of oxygen, hydrogen, carbon, nitrogen, phosphorus, and now and then some others. The atoms are linked together in the long chains and complicated blobs that make fibers and tissues. Although the atoms hold together tightly, the chains and blobs are flexible and can bend. And between and among the tissues and fibers you would see the moving atoms of the liquids—the blood and lymph fluids, complicated jumbles of atoms in large and small groups carrying the energy and the materials of life.

What images do you now have in your mind? You "saw" atoms sticking together and keeping their distance from other atoms; you "saw" the insides of a liquid, a solid, a gas. This is what a scientist "sees"—this is his own idea of how atoms behave. His years of theorizing, research, and experimentation tell him that is the way things *seem* to be at the atomic level. But actually there is no way to build a microscope having the magnification of your imaginary microscope. It is impossible to watch individual atoms as they move about in matter. Nevertheless, there are ways of "seeing" that indicate atoms do exist and behave in certain ways. From these observations a picture has been worked out that shows how matter is built and how it behaves on a microscopic level. This is the atomic view of the world.

THE CHEMICAL ELEMENTS

Hundreds of years of scientific study were needed to discover that all the materials in the world can be broken apart into basic substances that cannot, by any sort of chemical method, be broken apart into simpler substances. As researchers discovered these basic substances—*chemical elements,* as they came to be called—they gave each one a descriptive

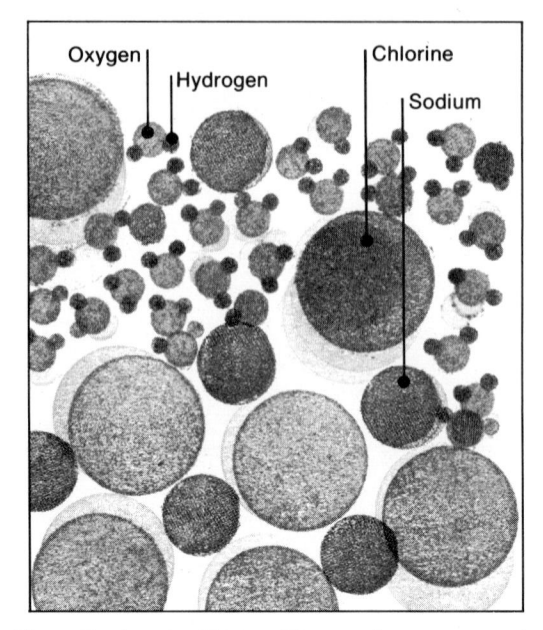

Figure 8. A grain of table salt is a regular arrangement of chlorine atoms and sodium atoms in a crystal. Each water molecule contains two hydrogen atoms and one oxygen atom, which are attracted to the chlorine and sodium atoms. They tug at and jostle the atoms of the crystal, pulling them off one by one and carrying them into the liquid. Table salt dissolves in water because the external pull of the water atoms is greater than the force that holds the chlorine and sodium atoms to the crystal. Substances such as wood and aluminum do not dissolve in water because the atoms making them up pull more tightly to their neighbors and the water atoms cannot pull them off.

name: oxygen, hydrogen, carbon, nitrogen, and so forth. They also worked out a system of abbreviations—O, H, C, N, and so forth—based on those names. Today we know that all the material of the earth is made of approximately 100 chemical elements.

These chemical elements are now known to be substances that contain only one kind of atom. And each element contains atoms only of its own kind. Now that atoms are known to exist, the names of the elements are used to label each atom of that kind. Hydrogen atoms are the ones found in the chemical element hydrogen; oxygen atoms are found in the chemical element oxygen; and so on. Abbreviations for each kind of element are also used to signify single atoms. You can, for instance, write H—O—H to represent two atoms of hydrogen and one of oxygen bound together to make a molecule of water.

CHEMICAL CHANGE

When table salt is poured into water it gradually disappears. It dissolves into the water. What is going on? Molecules of the water rub against the grains of salt and find the sodium atoms and chlorine atoms that make up the salt crystal. The sodium and chlorine atoms are attracted more to the atoms of the water than to themselves. The water molecules pick up the atoms of the salt one by one, making new combinations of atoms, new molecules that drift off among the molecules of water. This is one example of *chemical change,* the formation of new kinds of molecules from the atomic parts of other molecules.

Another form of chemical change is burning. Consider how charcoal burns in a barbecue. For the most part, charcoal is made up of carbon atoms that are holding together tightly to make solid pieces of the charcoal. But carbon atoms are also attracted strongly to oxygen atoms—they readily form molecules of carbon and oxygen atoms. Two such kinds of molecules can be formed: carbon monoxide (CO), which contains one atom each of carbon and oxygen; and carbon dioxide (CO_2), which contains two atoms of oxygen and one of carbon. Both kinds of molecules make a gas at ordinary room temperatures. So the burning of charcoal is a chemical process in which the carbon atoms found in the charcoal combine with oxygen atoms found in the air, making a gas of carbon monoxide and carbon dioxide molecules.

The flame of the charcoal fire is the hot gas of the newly formed molecules. The flame is hot because the new molecules are created with rapid motion, a motion caused by the strong attraction of the carbon and oxygen atoms. When the new molecules of carbon and oxygen are formed, the component atoms "snap" together sharply. This rapid snapping together sets up such a commotion that molecules in the neighborhood end up acquiring tremendous speeds. The chemical reaction has produced energy that appears as the heat of the flame.

Why does the charcoal have to be heated before it will burn? Because the carbon atoms normally ignore the oxygen molecules of the air and keep to themselves. They must first be moved a little bit away from one another. When a fire is started, the heat energy gets the carbon atoms of the charcoal moving about rapidly, so they are more easily separated from one another. At the same time, the oxygen atoms of the air are likewise moved apart. The various atoms then come together in new combinations to make up new molecules composed of carbon and oxygen. Once that starts to happen, more heat is made to keep the fire going.

Another kind of fire is going on all the time in our bodies, but it is much more gentle and controlled. The food we eat contains quantities of carbon, hydrogen, and oxygen atoms as well as others, bonded to-

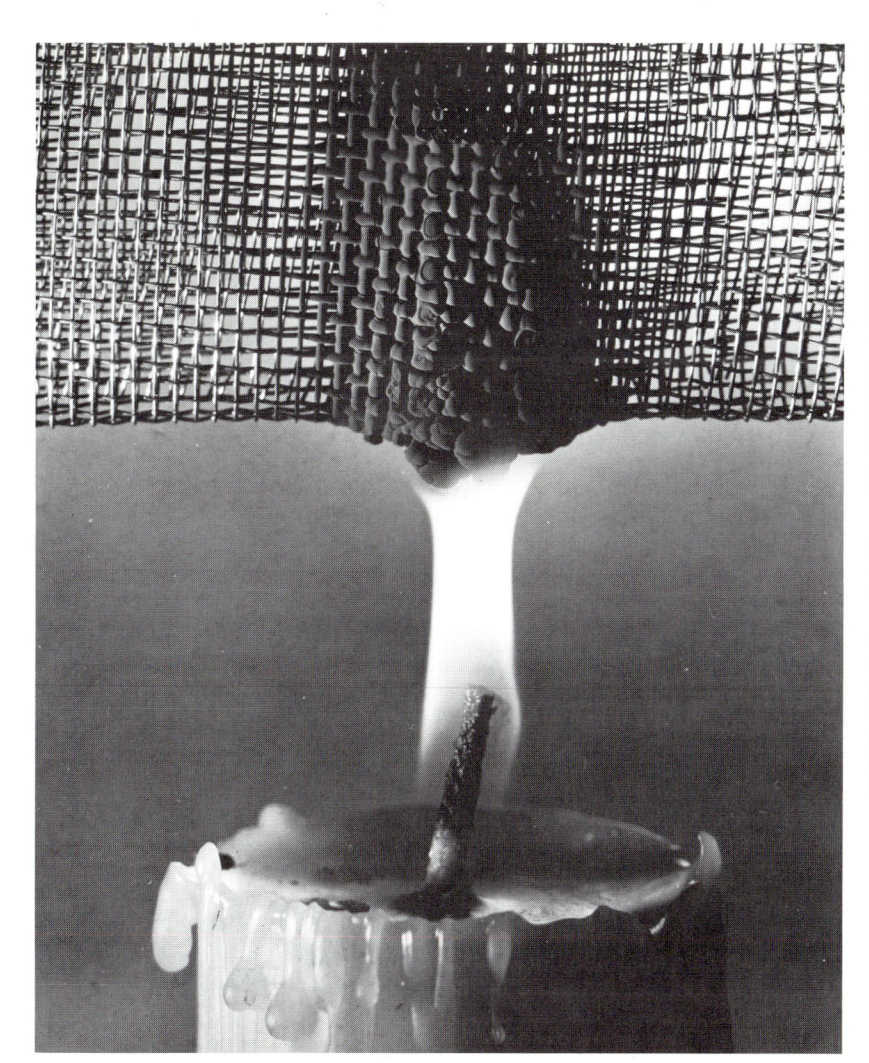

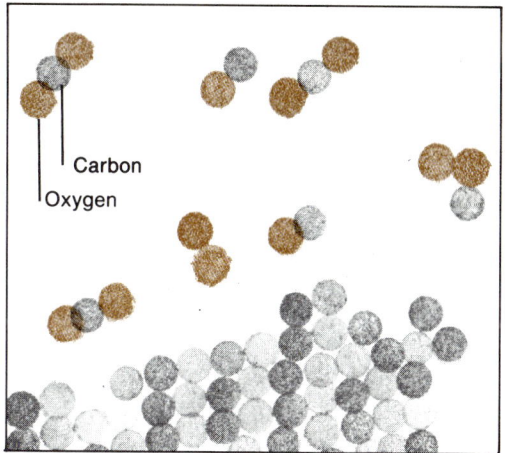

Figure 9. (Left) What happens when heat from the candle flame reaches the carbon atoms in this screen? When carbon burns, molecules of oxygen (two atoms) are colliding with carbon atoms and knocking them loose. One loosened carbon atom may combine with one oxygen atom to form carbon monoxide (CO) or with two oxygen atoms to form carbon dioxide (CO_2). The nitrogen atoms do not do anything except get in the way of the oxygen atoms and so slow down the burning. In a gas of pure oxygen, the burning is much faster.

gether in many ways to create complicated molecules. Our digestive system takes in various kinds of molecules of our food and transports them through our blood to the tissues. Our lungs take oxygen atoms from the air. These atoms are picked up by red blood cells and transported to the cells where the oxygen is combined with carbon and hydrogen atoms to produce molecules of water, H_2O, and carbon dioxide, CO_2. When these chemicals are formed, energy is given up—heat energy that keeps us warm and acts as the fuel for our movements.

INSIDE THE ATOM

What else can be said about the tiny, basic specks of matter called atoms? A great deal can be said about their internal structure, which determines how each kind of atom behaves.

Each of the 100 different kinds of atoms has a different *internal structure*. This structure determines whether the matter made up of atoms of a particular kind will be a solid or a gas; it also determines what kind of molecules it will make with other atoms as well as what kind of chemical change will occur with the molecules.

At the center of each atom is its *nucleus*, a fantastically small core that is 100,000 times smaller than the atom itself. If a baseball represented

Figure 10. The *electron "cloud"* is a fuzzy, nearly spherical ball made of a number of electrons. Depending on the kind of atom, there can be from one to several hundred electrons in it. Even though this spherical ball is referred to as a "cloud," it does have substance and rigidity, and that is why atoms cannot be pushed too closely together. The *nucleus* is a tiny space too small to be depicted even at this "magnification," for it is only 1/100,000 the diameter of the atom itself. But it contains nearly all of the atom's weight—about 99.98 percent of it!

an atom, the nucleus would be a tiny spot much smaller than a period that is printed on this page—in fact, it would be smaller than the sharp point of a pin. Although the nucleus occupies almost no space at all, it carries nearly all of the atom's weight. The weight of *any* object is essentially the combined weight of all its atomic nuclei.

Each kind of atom has a different weight, as determined by the weight of its nucleus. The lightest atom is hydrogen; one of the heaviest, uranium, weighs about 240 times more. The weights of almost all other atoms lie somewhere in between those two weights.

The nucleus of the atom also carries a charge of electricity. An interesting fact about electricity is that the *amount* of electrification can exist only in combinations of what is called the *basic electric charge.* And each atomic nucleus carries a certain number of these basic electric charges. The nucleus of a hydrogen atom carries 1 basic electric charge, an oxygen atom carries 8, a uranium atom carries 92.

The amount of electric charge on the nucleus of an atom determines the nature of the atom. So it is customary to distinguish one kind of atom from another by giving its *atomic number* instead of, or in addition to, its name. The atomic number of an atom is just the number of basic electric charges carried by the nucleus of the atom. Hydrogen has atomic number 1; oxygen, atomic number 8; uranium, atomic number 92.

It also happens that the more units of electric charge a nucleus carries, the more weight it has. Atoms of higher atomic number are the heavier atoms: hydrogen is very light, uranium is heavy.

The "body" of an atom—the ball that surrounds the pointlike nucleus—consists of an *electron cloud,* made up of a group of smeared-out objects called *electrons.* A hydrogen atom has only one electron in its cloud; a uranium atom has ninety-two. Although the electron cloud makes up nearly all of the volume of an atom, it has only about 1/4,000 of the weight of the atom; the rest of the weight is contained in the nucleus. But it is this cloud that gives the atom its form and determines its behavior.

An atom is held together by electric forces. Each electron in the atom carries exactly one basic electric charge—but the nature of the charge is *opposite* that of the nucleus. The nucleus is said to consist of a number of units of *positive* electric charge, and each electron has 1 unit of *negative* electric charge. It is the electric force of attraction between the positive charge of the nucleus and the negative charge of each electron that pulls the atom together.

Suppose you were to start out with a bare nucleus of oxygen, which has a positive electric charge of 8 units. This nucleus would pull toward itself eight electrons, each with 1 unit of negative electricity. When these eight electrons have been accumulated, the negative electricity of the electrons cancels the effect of the positive electricity of the nucleus. The atom is then said to be electrically *neutral.* It will no longer have a strong attraction for more electrons.

For most purposes, an atom—or an atomic species—may be described by giving the weight of its nucleus as well as its atomic number, which is the electric charge of the nucleus. You then know that the number of electrons in the electron cloud of a neutral atom necessarily corresponds to the electric charge of the nucleus.

The nature of each kind of atom is determined completely by the behavior of its electron cloud. And that behavior depends only on the *number* of electrons present in the cloud. An atom of oxygen has a nucleus with 8 units of positive charge that attracts eight electrons. By the nature of the electric forces and of the electrons themselves, the eight electrons will always gather into a cloud of a particular size and shape surrounding the nucleus to make an atom of oxygen. Because all electrons behave according to the same laws of nature, all oxygen atoms will be the same.

Again, the weight of a nucleus increases roughly in proportion to the amount of its electric charge. So, also, does the number of electrons in an atom increase with the weight of that atom. Finally, the *size* of the atom also increases with its atomic number and its weight. The more electrons residing in the body of the atom, the larger it will be.

The most common atom in the entire universe is the hydrogen atom. In fact, it is the same basic atom from which the entire universe was created. It has a nucleus with 1 unit of positive electricity, so it has only one electron in its cloud. It is the lightest atom of all—atomic number 1. Other common atoms of the world are carbon, atomic number 6; nitrogen, atomic number 7; and oxygen, atomic number 8. The atoms that occur naturally in the world fill out nearly all atomic numbers to uranium, atomic number 92. Beyond uranium there are a few more atoms—up to about atomic number 105—that appear rarely in nature but have been made artificially by man.

Why these 100-odd atomic varieties exist—indeed, why the laws governing the natural world *allow* only these kinds of atoms to be built—is the subject of modern research in nuclear physics, and the subject of later chapters in this book.

But for now, you have a picture of the most recent model of the atom. Each atom is visualized as having a heavy core, giving the atom its weight and providing an attraction that can pull in just the right number of electrons. Surrounding the core, a cloud of that same number of electrons fills out the space of the atom, giving the atom its size and its shape. And the behavior of the atom—the way it interacts with other atoms to make the molecules of the material world—derives from the behavior of the electron cloud.

This is the atom that makes up all of the world. Everywhere atoms stick together, everywhere they move in a perpetual dance. They let go of some to combine with others, forming one substance out of another in the process. They give up energy or take up energy. All we know as life exists because atoms are at work. A substance such as water is simple—merely oxygen atoms and hydrogen atoms in the combination H_2O. Yet its actions are so varied: water flows in a stream, waves break on a shore, droplets fly in the air to make clouds. Take many more kinds of atoms and put them together in many, many ways—never the same, all organized into an infinite number of shapes and relationships. How incredibly complex it becomes—sometimes even complex enough to become a living man or woman with the capacity to think about an atom, and what it does.

The universe through the eyes of a medieval child of science.

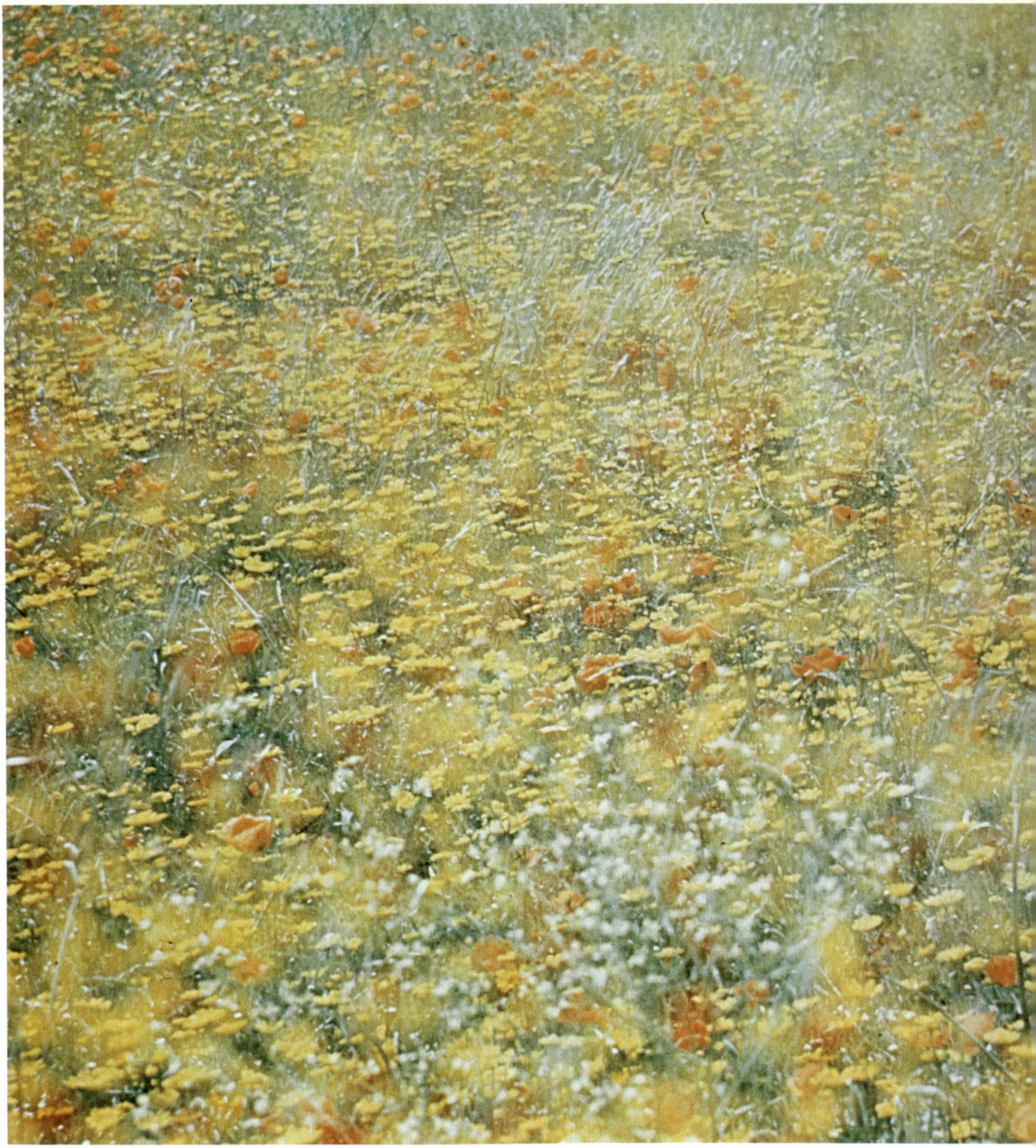

All we know and will ever hope to learn about the world comes to us through our senses. We taste things and we smell them, we hear and feel and see them with specialized parts of the body that receive information from the external world and transmit it to the brain. Each of the senses provides us with important knowledge about our surroundings, knowledge that makes life possible and meaningful. Of all the senses, however, the sense of sight gives us by far the richest and most varied information about the world. During nearly every waking moment our eyes cast about, picking up views of moving and stationary patterns — from sunrises and stars to the earth and its inhabitants. Each one of us may register, through our eyes, billions of sights during a lifetime.

All of this is possible because there is in nature a special phenomenon called *light,* and because organs of sight developed during the long span of biological evolution so that light could be used for contact with the outside world.

But how do we "see" things? What is happening, for instance, as your eyes move across the page of this book? You see this page because something travels from the page to your eye. That "something" is light. But what *is* light? It is a pure form of energy that travels through space. In the history of science there have been many arguments about the precise form taken by this energy. Is the energy carried in small specks or particles, like miniature droplets sprayed out from a hose? Or is it carried in waves, like ripples on a lake? Today the energy of light is known to be electric in nature, and it turns out that *both* views are partly correct. Light energy behaves as if it were a stream of tiny droplets. But it also has properties that make it seem at times to be wavelike. This discussion of how light works begins by considering light as a flow of tiny particles of electric energy, which scientists call *photons* — a word that's derived from the Greek word for "light."

You see the page of this book because each part of it is emitting photons in all directions. The photons arriving at your eyes tell you the page is there. You decide something is visible when it informs you of its presence by *radiating* streams of photons, which your eyes intercept.

Photons originate in the atoms of things. The atoms and molecules of every visible object vibrate continually and every now and then the atomic vibrations shake off tiny bits of their energy of vibration as photons of light. Now if the atoms of something visible are losing their energy of vibration by emitting photons, this energy must somehow be restored, otherwise it would soon run down. In a *self-luminous* object the energy is supplied from within. The white-hot wire inside an electric light bulb is self-luminous; so is a candle, and so is the sun. In each of these instances, the object is being heated by some source of energy — electricity, chemical burning, or nuclear energy. But the heat itself is nothing more than the violent motion of the atoms of the material. During their motion the atoms jostle one another, setting up the internal vibrations that become the source of all visible photons.

Atoms or molecules may also begin vibrating internally by *receiving* light from an external source. When photons from one object arrive at the surface of another object, the energy they carry may set the receiving atoms into vibration. A waterfall, for instance, isn't self-luminous; neither is a horse or a motorcycle. You can see these objects only when they are

2
Seeing Things

Light from the sun interacts with the atoms in a field of flowers.

Figure 2. (Right) Photons of light radiate in all directions from a flower. Those photons reaching our eyes stimulate our visual sensory mechanism, which sends a message to the brain. There the message is decoded as "flower."

Figure 3. (Below) Place a playing card 1 foot away from a small lamp. Now move the card so that it is 2 feet away from the lamp. How much light arrives at the card now? The same light arriving at the card that is *1 foot* away spreads out so that it illuminates *four cards* of the same size at a distance of *2 feet*. So each card must be receiving only one-fourth the original amount of light. The amount of light received from a small lamp decreases in proportion to the *square of the distance* from the lamp.

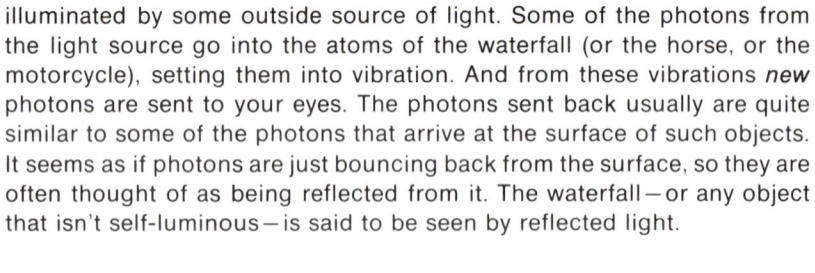

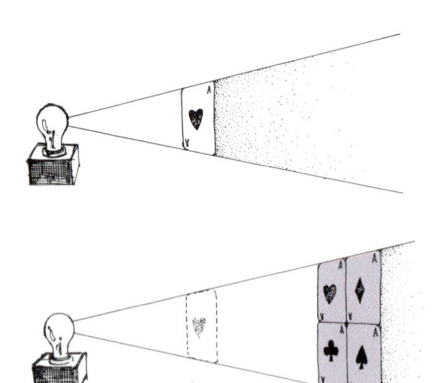

illuminated by some outside source of light. Some of the photons from the light source go into the atoms of the waterfall (or the horse, or the motorcycle), setting them into vibration. And from these vibrations *new* photons are sent to your eyes. The photons sent back usually are quite similar to some of the photons that arrive at the surface of such objects. It seems as if photons are just bouncing back from the surface, so they are often thought of as being reflected from it. The waterfall—or any object that isn't self-luminous—is said to be seen by reflected light.

LIGHT AND DARK

Imagine you are in a room at night with a friend. Only one lamp lights the room. As your friend moves about, his face appears lighter or darker, depending on how close he is to the lamp. Why is this so?

Think of the photons streaming out in all directions from the lamp. During a given second of time the lamp emits a certain quantity of photons and, during the next second, another equal quantity. It does this for *every* second the lamp is on. Now think of your friend's face when it is close to the lamp. Of the total amount of photons leaving the lamp, a certain fraction arrives at his face. When he moves away from the lamp, a smaller fraction of the light gets to his face. Fewer photons arrive at his face in each second, so fewer atoms of his face are stimulated enough to send photons back to your eyes. *Lighter* means more light—more photons in each second; *darker* means less light—less photons in each second. And *blackness* means no light at all.

But wait a moment—you "see" black all the time. You don't see a black object by means of reflected light, however. You know it is there be-

Figure 4. (Above) We see the white cat against the dark background because the white cat is *reflecting* most of the photons that reach it; the background is *absorbing* the photons. (Below) This time the black cat is absorbing most of the photons and the background is reflecting them.

cause you can't see what's behind it. Either you are seeing its outline—its silhouette—or you are distinguishing parts of it that are not totally black. When it is illuminated, a black object doesn't send light to your eyes because the light that reaches it is gobbled up: none is sent back. New photons of visible light are not created. The light is completely absorbed.

The atoms of an object that appears black are of a kind that does not vibrate in a way that will send off photons of visible light. Objects that appear white or brightly colored have different kinds of atoms, which vibrate in a way that sends off many photons. The atoms of dark or gray colors are in-between: sometimes they send back photons and sometimes not. But remember that the amount of light reaching your eyes also depends on how much light is getting to the object in the first place. Even a white or brightly colored object that is not getting much light will not send much back. At dusk, for example, the sunlight fades away rapidly and it becomes increasingly difficult to distinguish one color from the other. Eventually everything turns gray, then black.

SHADOWS AND HOW LIGHT TRAVELS

The richly varied patterns of the world exist partly because the surfaces of all objects react differently to the light shining on them by sending back more or less light, and partly because some surfaces of the same thing are receiving more or less light than others. And some variation in light and dark comes about because everything is casting shadows.

When you position your hand between a light source and a page of this book, a shadow appears on the page because your hand is preventing some of the photons of light from reaching it. The shadow, however, is more than just a dark area; it is a recognizable outline that follows the

2 | Seeing Things 21

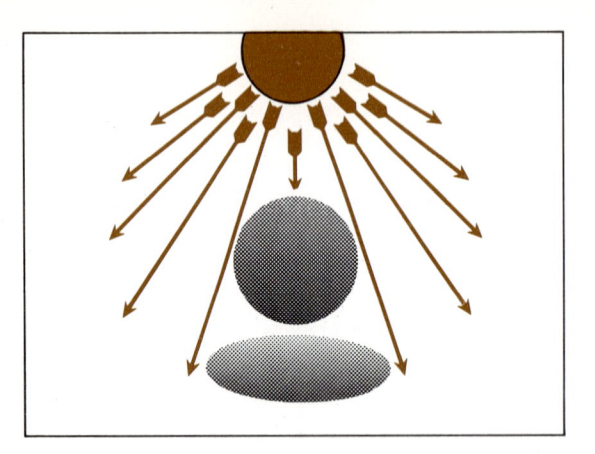

Figure 5. Photons from a light source travel in a straight line. The photons that just pass by the edge of an object illuminate the edge of the shadow it casts.

shape of your hand. The fact that the shadow is some kind of image of your hand is an important clue about the nature of light:

The photons of light travel through the air in a straight line.

The photons of light emitted by a lamp don't just wander about aimlessly. Each photon travels in a straight line until it hits something. The photons that just miss the edge of your hand continue on in a straight line to the book, thereby producing the edge of the shadow.

In the diagrams accompanying this chapter, lines have been drawn that attempt to show how the photons are traveling through space. You can't draw what a photon "looks like" while it's traveling. All that can really be shown in a diagram is a *representation* of the *path* taken by a photon, the straight line it moves along. Wherever such a path is shown, imagine that many photons—one after another—run down the same path. And it is the paths taken by light that determine the geometric shapes of things we see. These paths, with their streaming photons, are called *light rays*. Light rays are a convenient abstraction for thinking about the behavior of the streams of photons and for understanding how we see.

Imagine that from every point on a luminous object there are innumerable rays of light radiating straight out in all directions. It is clearly impossible to think about *all* of these rays—there are more than can be imagined—but by selecting a few to think about and by following these rays to their end, it is possible to trace out what is happening. In the diagrams in this chapter, only some of the rays are shown to illustrate what is going on or some particular rays are selected that have special significance. There are always uncountable others that are not shown.

The straight lines made by rays of light are seen most clearly in a *beam* of light from such sources as hand flashlights, aerial searchlights, and theater spotlights. All the rays of such a beam lie in nearly the same direction. Anything put in the path of the beam will be illuminated only in a small area. If the air is clear the beam itself is not visible, and the light will show up only where the beam happens to hit something. If the air is dusty or smoky, however, the particles in the air will receive some of the light and become visible along the length of the beam. This effect can be seen when a searchlight probes into shrouds of fog at night, or when the sun's light penetrates the openings in a dense forest canopy and shines through the mists that linger over the forest floor.

The idea of light rays can be used to think more precisely about shadows. Have you ever stopped to think about how fast a shadow moves? Could you, for instance, ever move fast enough to leave your shadow behind? As silly as the notion sounds, it is not a total impossibility. Because your shadow is created by light rays streaming past you toward some surface, it will take some time for them to reach their destination. If you *were* able to move fast enough, you could be somewhere else by the time the light around the edge of the shadow reached that surface. It wouldn't be easy, of course, because *light travels at the speed of 300 million meters per second.*

Suppose you are standing in the sunlight, waving your hand back and forth. The photons making the shadow of your hand take a few billionths of a second to get from your hand to the ground. But in that short time your hand has moved slightly, so the shadow is not following the motion of your hand exactly. The discrepancy is so small, however, that you would never notice it.

How, then, do we know how fast the photons of light travel? Some scientists have actually measured how much a shadow falls behind when the object casting it is moving rapidly. The speed of light was first measured this way in 1849. Armand Fizeau made a "toothed wheel" by cutting fine slots all around the edge of a metal disk. He then arranged to spin the wheel and compare the position of the shadows of the teeth when they were moving to the position of the shadows formed by the stationary wheel. He was thus able to measure how far behind the shadow lagged when the wheel was rotated and, from his measurements, to determine the speed of light.

DIFFRACTION

If you look closely at a shadow you will see that its outline is never perfectly sharp and, in fact, may be quite fuzzy. You might expect the shadow to be perfectly sharp if the light source producing it were a point, but that is impossible. Even a small flashlight bulb, for instance, has some size, and every point on the bulb—or whatever is emitting light—is the starting point for rays of light. Rays from each of these points illuminate the edge of an object and produce their own shadow, but all the shadows are not in the same place. The total shadow is the effect of all the shadows produced by the different parts of the source of light.

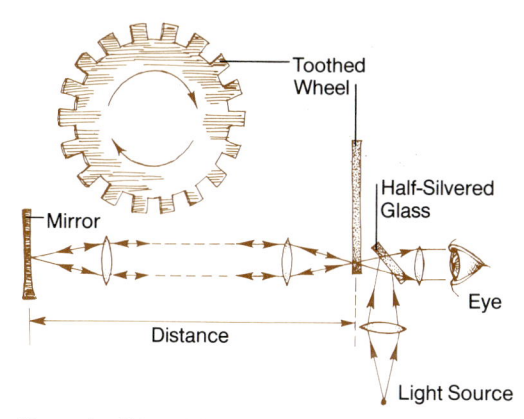

The completely dark region of a shadow occurs where none of the light rays from the source can arrive. Other regions are partly dark be-

Figure 6. (Above) Armand Fizeau used a toothed wheel apparatus to calculate the speed of light. He measured the number of revolutions per second that were required to eclipse the light rays sent through the wheel and reflected back from a mirror positioned a distance away.

Figure 7. (Left) Beams of light filtering through a redwood forest illustrate that light travels in a straight line.

Figure 8. Light waves from a dim light source cause interference bands as they pass by the edges of a paper clip. The image is a shadow recorded directly on film. This photograph was made without any lens in the camera. Focused images through a lense would destroy the interference bands.
(Fundamental Photographs)

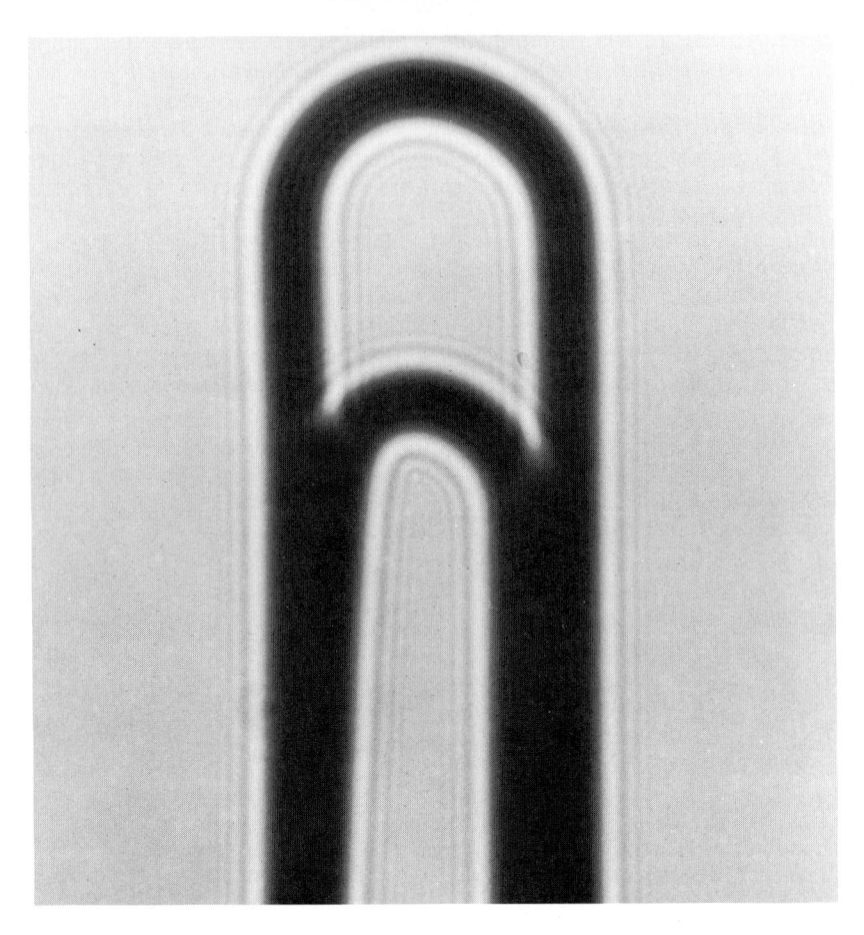

cause they are illuminated by only part of the source, so less light is arriving. The "fuzzy" region comes to an end where light rays are not obstructed by the object casting the shadow.

If the source of light is large or, more commonly, if there are several sources of light, shadows are never completely dark. Light from somewhere gets behind the object and the object casts only a partial, or "soft," shadow. Nor do all of the sources of the light arriving at the object have to be self-luminous. When another white or light-colored object in the neighborhood receives light it sends much of its light back out again, and this reflected light, too, may illuminate a place that would otherwise be in shadow.

Shadows can never be *perfectly* sharp for still another reason that has to do with the basic nature of light. Suppose you make a very small source of light—as you can do, for example, by shining light through a small hole—and then put some object with sharp edges in the path of the light to make a shadow. When you look closely at the shadow that is created, you see that it does not have a crisp edge and is in fact fuzzier than you would think it should be from the size of the hole. Rather, the transition from dark to light occurs gradually, and a series of light and dark bands appears near the edge of the shadow. The effect is most striking when the light source isn't too close to the object casting the shadow.

How do the photons from a light source make such a shadow? If you think only in terms of light rays you will be unable to understand what happens. A model that describes light only in terms of photons traveling in straight lines—along light rays—is incomplete. In part, light has the properties of a wave: an undulating vibration in pure space. And it is this

Physical Science Today

Figure 9. When waves arrive at an obstruction, as illustrated by this photograph of a harbor entrance, the water behind the obstruction is not disturbed; the obstruction casts a "shadow."

wavelike aspect of light that controls the behavior of the photons. Earlier you read that the emission of photons occurs because of the internal vibrations of atoms. Perhaps it is not surprising, then, that the light carries with it something of the same kind of vibration.

It is not possible to make a complete and correct description of light either in terms of particle-like photons or in terms of vibrations or waves; both aspects exist in the light. For most everyday experiences, you can make an adequate description of light in terms of photons moving along straight lines—along rays of light. But from time to time you will come across effects that can only be understood by taking into account the vibrations carried along with the light—when, for example, you look closely at shadows that seem sharp, or when you see colored films of oil on water.

What is happening when light goes past the edge of an object—and some of the light does not travel along the straight paths we would expect—is called the *diffraction* of light. It comes about because of the wavelike character of light. A discussion of the details of how the wave nature of light produces the diffraction effects at the edge of a shadow is too complicated to go into now. You will, however, come back later to the vibrations associated with light when you read about the nature of color, which has been ignored until now.

ABSORPTION

Before leaving the subject of shadows, go back for a moment and think about what happens to light that is blocked when a shadow is formed.

When you hold a sheet of heavy, black paper between yourself and a light source, you see that light will not penetrate through it. The paper stops all of the photons of light. A sheet of thin, white paper, on the other hand, will let some of the photons through; even the shadow it casts will not be as dark. When the photons of light arrive at a white surface, the atoms there are set into vibration by the incoming energy and in turn become new sources of photons. But only some of the energy is sent *backward* as reflected light. Some of it is also sent *forward* into the paper and is absorbed. And some of it is lost: some energy goes into vibrations that heat up the paper rather than sending off light.

When light can't get through paper—or any object such as a table, a wall, and a book—the paper (or whatever) is said to be *opaque*. However, some objects, such as a thin sheet of white paper, let light through but in such a way that the paths taken by the photons are scrambled up. You can see light coming out the other side but you can't make out the shape of the things that sent out the light—you can't "see" through the paper. Such objects are called *translucent,* and they are said to *diffuse* the light going through. Many lampshades, for instance, are made of translucent materials. When light rays pass through a substance almost as if it weren't there, so that you can "see" clearly through it, the substance is *transparent.* Water and other liquids, many plastics, and, of course, air are familiar transparent materials. So is glass.

Yet glass, like other solids, is packed tightly with atoms. How is it, then, that the photons of light can pass right through it? One reason is found in the way the atoms of glass respond to the light: they do not absorb any of its energy. When light arrives the atoms vibrate in a regular way and immediately send the energy out again. Another reason is found in the arrangement of these atoms. The *surface* of glass is quite smooth, and the *interior* of glass is packed uniformly with atoms. The properties of the atoms and their smooth arrangement mean that large numbers of atoms act *in cooperation* when light photons arrive. In other words, the photons set many atoms into *similar vibrations,* and the atoms work together in their effects on the photons, sending them on their way with hardly any disturbance at all.

Now consider what happens when you take a piece of clear plastic and scratch the entire surface by sandpapering it. You will no longer be able to see through it. When the surface of plastic is roughened, the atoms do not all cooperate in the same way. And when they vibrate they send off their photons in all directions. The surface is no longer transparent.

Or consider a jar full of tiny glass beads. Because of the spaces between the beads, the glass in the jar is not smooth throughout and the light that comes in is sent out in all directions. So you can't see through it, either.

Or take a glass of clear water. Put several drops of ink in it and the water becomes opaque, or nearly so. The molecules of the ink spread throughout the water, and they are of a kind that do not help the light on its way. Any light that reaches an ink molecule is immediately gobbled up. The light is absorbed.

Finally, take a glass jar of water and shake it violently so that many tiny bubbles of air are formed in the water. The water becomes "milky" and you can no longer see through it. The water is no longer uniform; the cooperation of the atoms is interfered with.

REFRACTION

When you shine a narrow beam of light into, say, a fish tank, the beam goes through the water and continues on its way out the other side. But

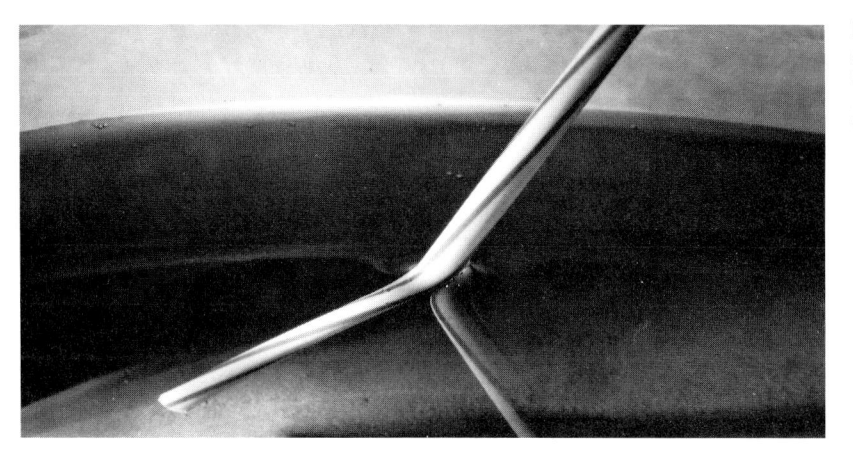

Figure 10. A drinking straw appears to bend at a sharp angle where it enters the water. This *refraction* occurs because the speed of light is less in water than it is in air. The *reflected* image of the straw is also visible. (Fundamental Photographs)

as it enters the water the beam of light is *bent.* The light rays, the paths taken by the photons of light, change direction suddenly at the edge of the water because the *speed* of light is somewhat *slower* in the water. When light arrives at an angle to the water's surface, one "edge" of the light slows down and the other "edge" moves slightly ahead of it. By the time the light penetrates the water it has swerved from its original course, much like a car swerves when its right wheels accidentally go off onto a soft shoulder and are slowed down. If a beam of light is directed straight down into a container of water, both "edges" are slowed together and the light doesn't bend. The most bending occurs when the angle of incidence is greatest. Similar bending occurs for the same reasons when light leaves the water and goes back out into the air.

The bending of light rays on entering or leaving a transparent material is called *refraction.* Exactly how much bending will occur depends on the angle of incidence and on the nature of the intercepting material — the kind of atoms it has and the way its atoms are packed together.

By carefully measuring what happens when light enters a transparent material at one angle, and then at another, and at another, and by repeating such measurements over and over again with many different transparent materials, scientists found that the change in angle when light is refracted always happens in a predictable way. It is most easily described by saying there is a special relationship between the angle the light makes when it *arrives* at a surface and the angle it makes when it *leaves* the surface. This relationship may be expressed in mathematical form, and it may also be depicted by a graph, as shown in the simple illustration in the right margin. Once you know the mathematical formula, you can draw the graph that represents the same relationship for any particular kind of transparent material. All you need to know is one number — called the *index of refraction* — that tells you what the refracting property of the material has been determined to be.

When we look at something through a transparent substance, then, is the object really where it appears to be? We are accustomed to thinking that light comes straight to our eyes, so when light is bent by refraction we still think it is coming straight to our eyes and we perceive it to be in a place where it is not. Suppose you are looking at a small frog in a tank. The light sent from the frog to your eyes is bent on leaving the water. You are not used to thinking about that, so you imagine the light comes straight to your eye. You "see" the position of the frog straight ahead of you. But if you tried to reach in and grab the frog where it appeared to be you would find nothing there (much to the relief of the frog).

Normally we look at things through air, rather than water, so the effect of air on light is important to us. Does the air also make objects ap-

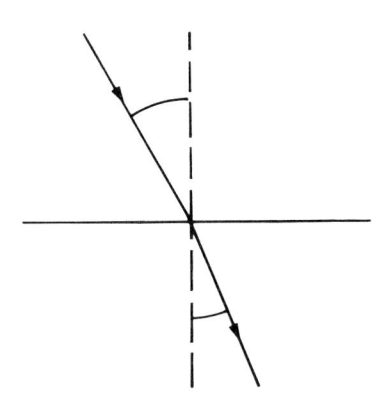

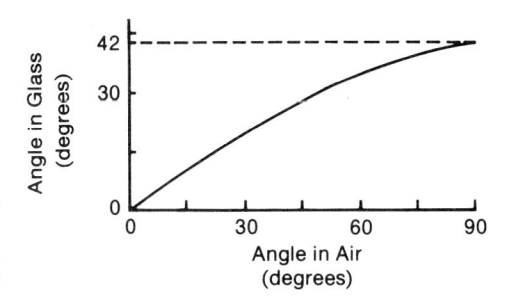

Figure 11. The bending of light from air into glass with index of refraction equal to 1.5.

2 | Seeing Things

Extending Our Sense of Vision

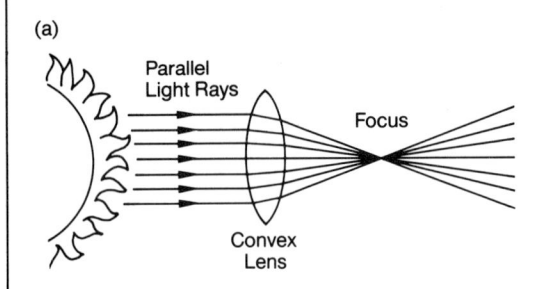

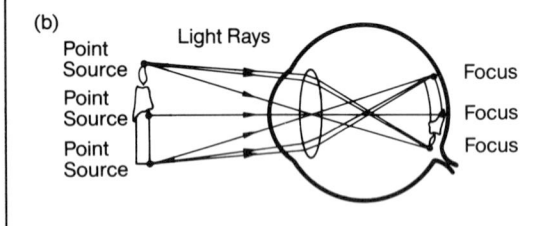

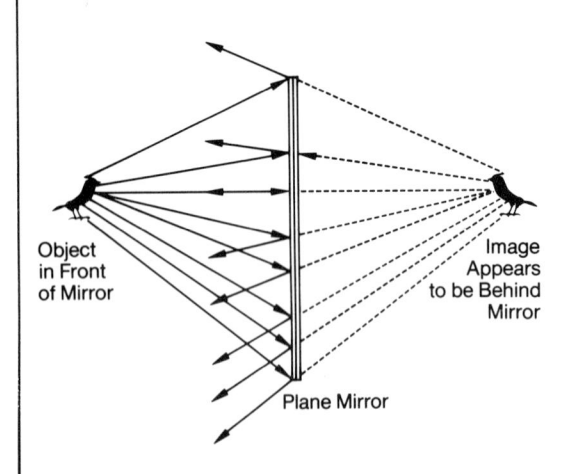

What good are the ideas of refraction and reflection? We can put them to use to see worlds other than our own. *Lenses* are pieces of glass or similar transparent materials that have been ground so that their surfaces are curved. When parallel light rays go through different parts of a convex lens, as in (a), they are bent, or refracted, by different amounts. This kind of lens sends the rays on their way to a *focus point,* which is a place where rays emanating from a single point on the object sending out light converge at a single point to help form the *image* of the object; the light from each different point on the object is brought to a different point to form the image.

Our eyes contain a convex lens that works to focus light rays on the back of the eye, the retina. In (b), a candle is sending light rays to an eye. (Only seven rays are shown but you know there are countless others.) Three rays are shown emanating from the same point in the candle flame. One passes through the center of the lens and is sent in a straight line onto the retina; the other two arrive at the lens at a different angle and are refracted. All the rays emanating from that point in the candle flame will be bent in a way that makes them converge in the same place on the retina. The same thing happens for every light ray that every part of the candle sends out. As shown in (b), the result is a retinal image that is upside down; but somehow through the processes of the brain the image is reverted and we see things right side up. In nearsighted vision, the focal points do not extend all the way back to the retina; in farsighted vision, the focal points converge "behind" the retina. Both faults can be corrected with still another kind of lens —eyeglasses.

We also use the light reflected from a mirror to help us in extending our vision. Consider how a *plane mirror* works. Light rays from an object strike the mirror's surface and are reflected back at an angle that is exactly the same as the angle at which they arrived. If you extend these rays backward, as in (c), they intersect at a place that is the same distance *behind* the mirror as the object is in *front* of the mirror, forming what we call the object's *virtual image.* Virtual images are reversed: your left ear appears to be a right ear, printed words appear to be printed backward. If you look at a *curved mirror,* however, you see that different parts of the mirror are putting different parts of the image in different places. The size and shape of the image changes, in much the same way that it changes when an object is seen through a lens. Such curved mirrors are used in astronomical telescopes for focusing the light of distant stars onto photographic film.

The Magnifying Lens and How It Works

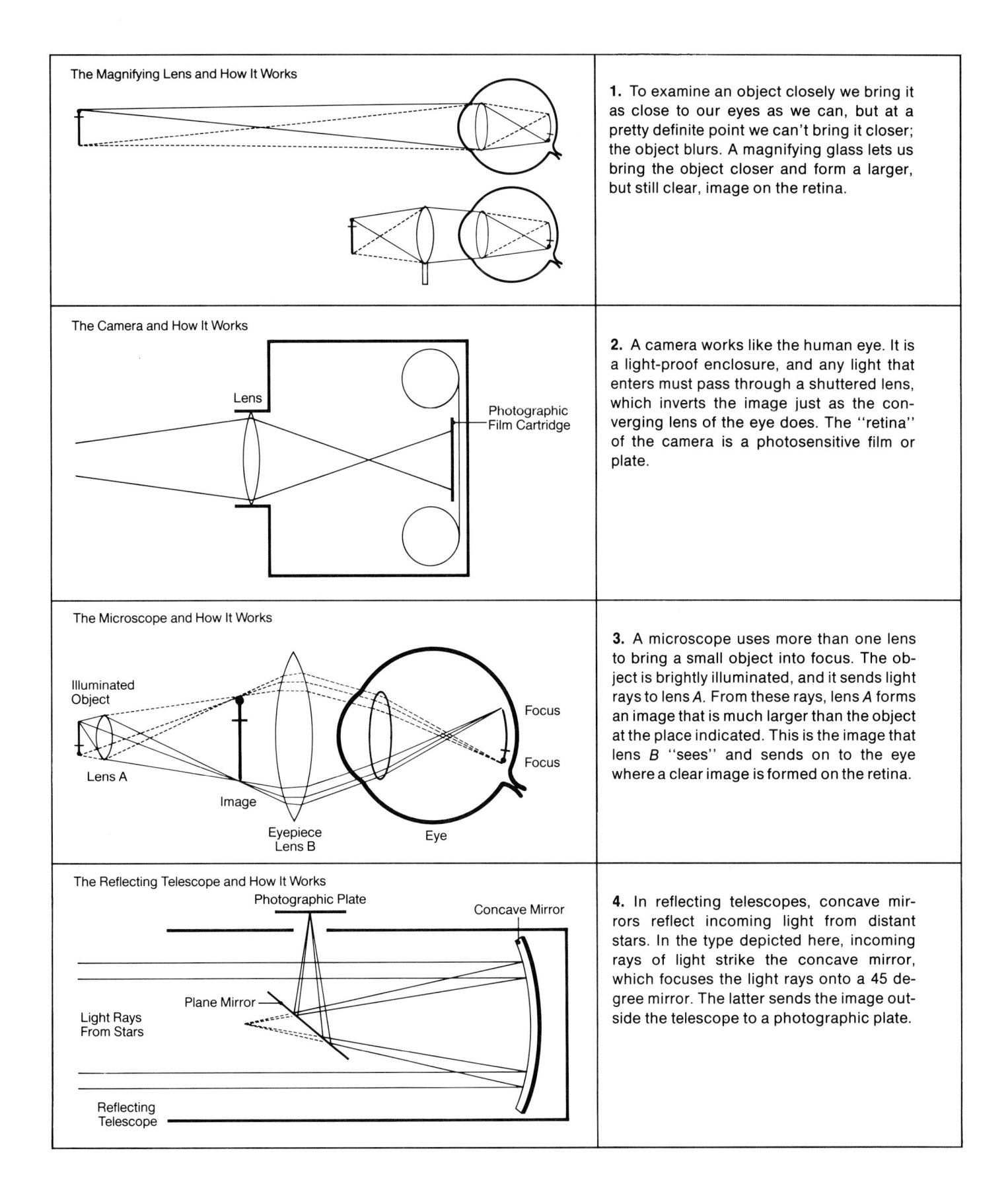

1. To examine an object closely we bring it as close to our eyes as we can, but at a pretty definite point we can't bring it closer; the object blurs. A magnifying glass lets us bring the object closer and form a larger, but still clear, image on the retina.

The Camera and How It Works

Lens
Photographic Film Cartridge

2. A camera works like the human eye. It is a light-proof enclosure, and any light that enters must pass through a shuttered lens, which inverts the image just as the converging lens of the eye does. The "retina" of the camera is a photosensitive film or plate.

The Microscope and How It Works

Illuminated Object
Focus
Focus
Lens A
Image
Eyepiece Lens B
Eye

3. A microscope uses more than one lens to bring a small object into focus. The object is brightly illuminated, and it sends light rays to lens A. From these rays, lens A forms an image that is much larger than the object at the place indicated. This is the image that lens B "sees" and sends on to the eye where a clear image is formed on the retina.

The Reflecting Telescope and How It Works

Photographic Plate
Concave Mirror
Plane Mirror
Light Rays From Stars
Reflecting Telescope

4. In reflecting telescopes, concave mirrors reflect incoming light from distant stars. In the type depicted here, incoming rays of light strike the concave mirror, which focuses the light rays onto a 45 degree mirror. The latter sends the image outside the telescope to a photographic plate.

Images in Mirrors and Light Reflection

When you look into a mirror you see what appears to be a whole world behind the mirror—although you know that it is not really there at all. What is going on when you see something in a mirror? How is that false world created?

Step 1. Hold a daisy in front of a mirror. Looking toward the mirror, you see a copy of the daisy, except that now the right side and the left side are reversed. What causes this reflection? As light strikes the daisy, many of the photons are not absorbed but instead are reflected off in all directions. Some of them hit the mirror, where they bounce off, and some of these reflected photons reach your eye. When they bounce, they obey the law of reflection, so they come back in very specific directions (see Figure A).

You know that light generally travels in straight lines, so it appears that the photons have come from some place behind the mirror (see Figure B). When the photons *appear* to be coming from something that is not really there, you see an *image* of the real object.

If a mirror is flat, the image of the daisy—or of anything—appears to be exactly the same distance *behind* the mirror as the real daisy is in front of the mirror, as you can see from the geometry of Figure B. Also the size of the image of the daisy is the same as the size of the real daisy. Do you understand why this is so?

Step 2. Suppose now that the mirror is not flat, but is *convex*—that is, it is curved toward you. The paths taken by the photons are shown in Figure C. To find the location of the image behind the mirror, extend the paths of the photons bouncing off the mirror. Is the image of the daisy closer or farther away than it was for a flat mirror? Does the daisy now appear larger or smaller than it really is?

Step 3. Next think about what happens if the mirror is curved the other way. A mirror having a hollow side toward you is *concave*. The paths taken by the photons are shown in Figure D. Is the image closer or farther away than it was for a flat mirror? Does it appear larger or smaller than it really is?

pear in wrong places? Hardly at all. The atoms of the air do affect light in the same way all transparent materials affect light. But air is a very thin substance. Its atoms are spread so far apart that their effect on light is much less than the effect of the atoms in water—in fact, roughly a thousand times less.

Early in the history of science, some people thought that air or something similar was necessary to "carry" light from one place to another. Those who preferred to think of the wave nature of light insisted that the vibrations of light must be carried along by some vibrating substance. But then, light was known to travel along just as well in a vacuum —a region of space from which air and all other matter has been removed. Light from the sun, for example, reaches the earth after traveling through a long distance of empty space. For a time, most scientists still believed there was *something* left in the vacuum, and they called it the "ether." Finally, during the early 1900s, they came to the conclusion that "something" did not have to be present in empty space before light could go through it.

Air is so thin that it is nearly like empty space in its effect on light. But it does have an effect on light that makes a big difference in our daily experience. When you look up at the sky at night you see the blackness of cosmic space. Except for the tiny points of starlight, nothing is out there to send light to your eyes. What a difference in the daytime! You look up at the same empty space, yet the daytime sky is bright blue.

What is this "blue" sky? It is the air above us. As photons from the sun reach the air, their vibrations are felt by the atoms of the air, which do their own vibrations and become sources of new photons of light. Just as they do in other transparent substances, the atoms of the air generally act in cooperation to send the photons along in their original direction. But the atoms of the air are spread far apart, and not uniformly at that. So their cooperation is not always perfect. The spaces between the atoms are like "bubbles" of vacuum, much like the tiny bubbles of air in a glass of water that has been agitated. Because of this, the air is not perfectly transparent. When you look at the blue sky, then, what you see are atoms of the air, signaling their presence by sending you photons they have received from the sun.

THE COLOR AND ENERGY OF PHOTONS

Why is the sky blue? Why not green, or brown, or yellow? Why, for that matter, is a rose red and not blue? The color you see is related to a characteristic of photons that is carried with them—*their energy*. Each photon is distinguished by the quantity of energy it carries with it. This energy is given up each time a photon is created by a vibrating atom; it is received by your eyes when photons are absorbed there. One photon may have a different quantity of energy than another, and photons moving along a ray of light may have many different energies. When you say, "That rose is red," you are saying that your eyes have told you something about the kinds of photons you are receiving from the rose.

When you look at a rainbow, you see bands of color ranging from red through orange, yellow, and green to blue. *These colors are spread out according to the energies of the photons of the light.* The photons carrying the color "red" have less energy, and the photons carrying the color "blue" have more energy—about twice as much energy in each blue photon as in each red one.

But how do your eyes know that the photons from the blue end of the rainbow have more energy? In the light-sensitive surface of the retina

Figure 12. The relation between color and the energy of photons is illustrated in this graph.

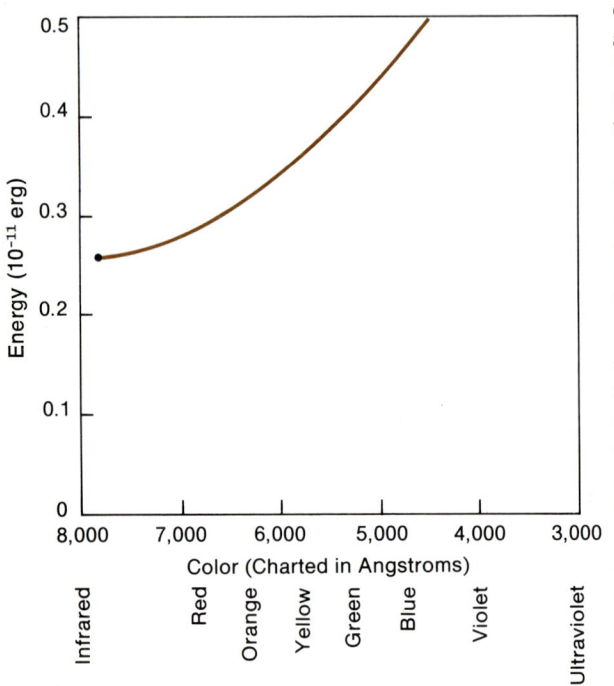

Figure 13. (Below) How do we see color? Red photons arrive at the eye, where they are focused at different places. Red signals are sent to the brain only through red-sensitive cells; blue-signals are sent only by blue-sensitive cells; and green signals are sent only by green-sensitive cells.

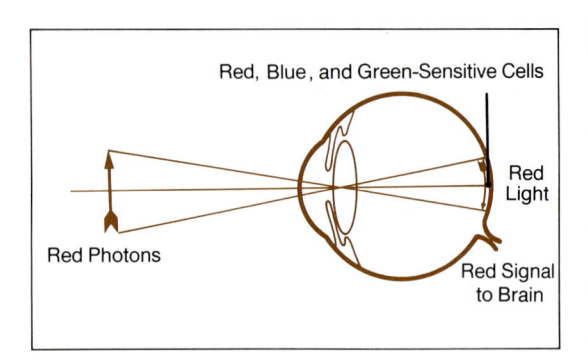

of the eye, several different kinds of cells absorb the light. Some have atoms that are put together in molecules in such a way that the energy carried by photons of red light is just the right amount to rearrange the atoms to new positions and cause them to send a signal to the brain—a "red" signal. Other cells in the eye have a different, more closely knit arrangement of atoms, and only photons with the larger energy of blue light will rearrange them and signal the brain.

Although the exact details of color vision are not known, it seems there are only three kinds of cells in the eye that tell us about color. Each responds to many different energies of photons, but one responds most to colors near red, another to colors near green, and the third to colors near blue. We see a color when messages from one kind of cell are stronger than those from the other two.

With only three kinds of light-sensitive cells, the eye sorts through an uncountable number of different gradations of color. If both red and green signals are reasonably strong and only the blue signal is weak, you see "orange." Similarly, if both the green and the blue signals are strong, you see "greenish-blue," and so on. But each color merely represents a *different proportion* of the three basic signals transmitted from the eye to the brain.

The rainbow colors are called the "pure" colors. Each one is carried by photons of a particular energy and affects the eye in a particular way. But all possible *combinations* of rainbow colors are also seen by us not as the combinations but as new colors. For example, the last color at the blue end of the rainbow is a blue-violet. In combination with some photons of pure red light, it appears royal purple.

Most of the colors you see are not pure colors. Any one color of light contains photons of many different energies in some particular proportions. The eye sorts out the amount of each of the pure colors present and sends its signals to the brain, which in the end registers what is called a single color—taupe, turquoise, or whatever.

And what about the things that send signals to the eye in the first place? You have been reading about the color of "light," but what about, say, the red color of a rose? A red rose is seen by the red light it sends to the eye. The atoms of a red petal respond in their own way to sunlight. When photons carrying blue light from the sun reach it, their energy causes the atoms to vibrate in a way that absorbs the photon's energy. But when photons carrying red light arrive, the atoms of the petals vibrate in a way that sends those photons back out again. All the incoming colors are absorbed except the red and the near-red; they alone carry the signal of the rose's "color."

Once photons of light are thought of as interacting with matter, it becomes clear that color is not an intrinsic property of things or of only the light that illuminates them. Rather, *color depends on both matter and light.* Not that we would necessarily want to eliminate all the intrinsic associations. Shakespeare's Luv would be less like a Red, Red Rose if we had to think of her only as absorbing the blue or the green photons.

Are there "colors" you can't see? Because color is something that you *see,* perhaps the question doesn't make sense. But if you think of color in terms of the energy of the photons of light you may certainly ask, "Is there *light* I can't see? Are there photons that don't give the sensation of light to the human eye?" The answer is yes.

The sun emits many kinds of photons that are not visible to us. Some of these photons have energies lower than that of red light, and the light is called *infrared.* Some photons have energies higher than the energy of violet light and their characteristic light is called *ultraviolet.* Although infrared and ultraviolet light extend beyond the visible spectrum, their energy has been captured by special photographic films.

The properties of "invisible" light are similar in all respects to visible light. It, too, is a form of electric energy. It travels in particles called photons. And it travels at the same speed: 300 million meters per second. All the forms this pure energy comes in have been given seemingly unrelated names—cosmic ray photons; gamma rays; x-rays; ultraviolet light, visible light, and infrared light; microwaves; and radio waves—because at the time they were being discovered people were not yet aware of the "laws" that tied them together. But taken as a whole they represent what is now called the *electromagnetic spectrum,* the subject of a later chapter.

Figure 14. The white light from the sun is spread out by a prism of glass into a spectrum of the "pure" colors of the rainbow. The photons of each color carry a different energy and are refracted in a slightly different way by the glass. The red photons have less energy than the blue ones and are bent more by the prism. The visible portion is depicted in the context of the entire electromagnetic spectrum, which is the subject of a later chapter.

THE QUANTUM PHOTON

Before about 1860, the study of light was a completely independent part of science. Then it was discovered that electric and magnetic forces could travel through space in waves—something like waves on the surface of water or sound waves through the air. The predicted properties of these "electromagnetic" waves were found to explain many of the properties of light. So it was concluded that light was composed of just such waves. But then in the early 1900s it was discovered that light also behaved in many ways like a stream of particles—the photons that have just been described. Today we know that a complete description of light must include both aspects—waves and particles.

This new description of light is called the quantum theory of light, or quantum electrodynamics. According to this theory, light consists of particle-like photons that carry electric and magnetic forces. The motion of the photons is determined by the spreading out of the waves, and the greater the magnitude of the waves, the more photons there are. Thus, in a sense, the electromagnetic waves may be thought of as "probability waves," because they determine the likelihood of finding a photon at a particular place. The nature of electromagnetic waves will be discussed in greater detail in a later chapter. Before leaving the discussion of light, however, you may take a look at the basic connection between the photon idea of light and the wavelike properties of light.

As a photon travels along carrying its vibrating electric force, these vibrations are spread out into a wave, somewhat reminiscent of the rip-

Figure 15. The waves of light are something like the waves of water that radiate outward when a pebble is dropped into a pond. But the waves are not confined to a surface as they are on a pond; light waves radiate outward in all directions.

ples on the surface of a lake. The spacing between the ripples—the distance from the crest of one ripple to the crest of the next—is always the same for each kind of photon. Depending on the vibration each photon carries, the ripples may be close together, or they may be far apart. This spacing is called *wavelength.* When the wavelength of the vibrations of photons is measured, it turns out that each color of photon has its own wavelength.

The fact that photons of different colors have different wavelengths is responsible for the band of colors reflected by some objects. If you take a phonograph record and hold the edge of it next to your eyes, a halo of rainbow colors will appear when light strikes the record. Photons of different wavelengths are being sent in different directions by the rippled surface of the record.

This spreading out of colors by a finely rippled surface is called the "diffraction grating effect." It is the effect that scientists use to measure the wavelength of the color carried by different photons. Using a grating made with a precise number of ripples in each centimeter, they send white light onto the grating and measure how the different colors come out in different directions. When the spacing between the ripples of the grating is compared with the angles taken by the different colors, it's possible to calculate the wavelength. In this way, the wavelength for photons of green light, for example, has been calculated to be one-half of one millionth of a meter.

The wave nature of light shows up in another familiar way when, as a popular song has it, "The oil in the puddles makes taffeta patterns that run down the drains in colored arrangements . . ."[©] These colors emerge because a thin film of oil has a thickness close to the wavelength of light. But the oil may be deeper in places. Where the thickness is equal to the wavelength of green light, you see the color green; where it is somewhat thicker—equal to the wavelength of red light—you see the color red.

Each color photon has its own vibration spacing (wavelength). But you have also seen that the color of light is determined by the *energy* carried by the photons. The relation between the wavelength and the energy of the photons of each color is an important law of quantum physics. This law may be stated simply:

The wavelength of any color is equal to a special number divided by the energy carried by the photon of that color.

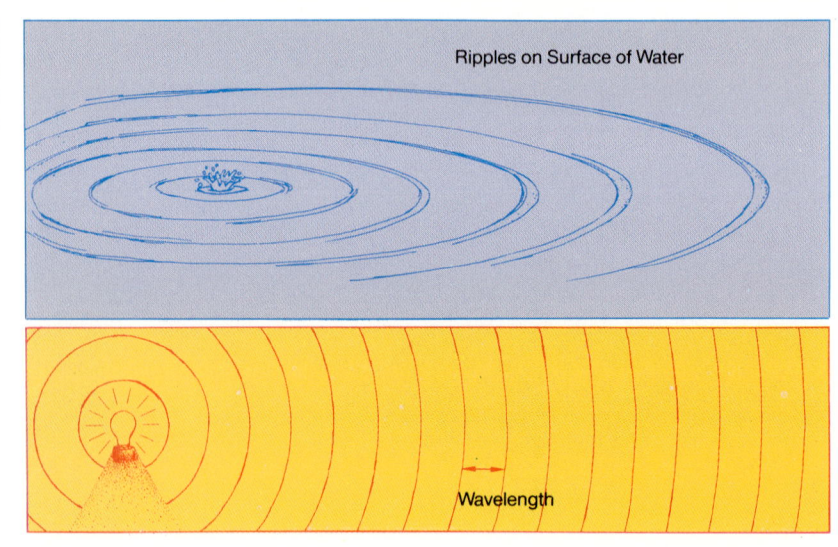

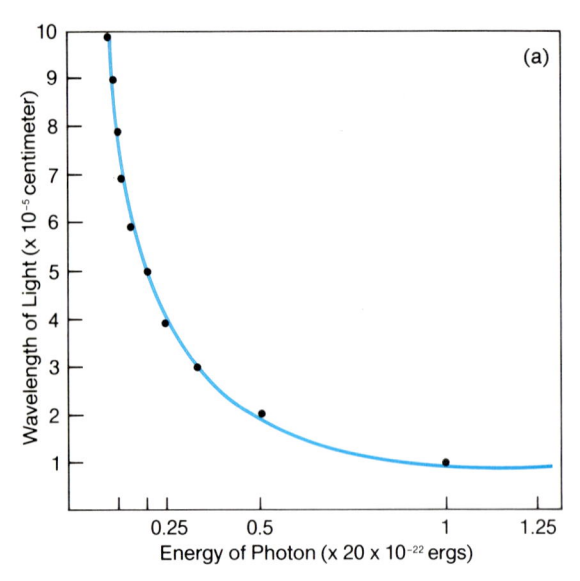

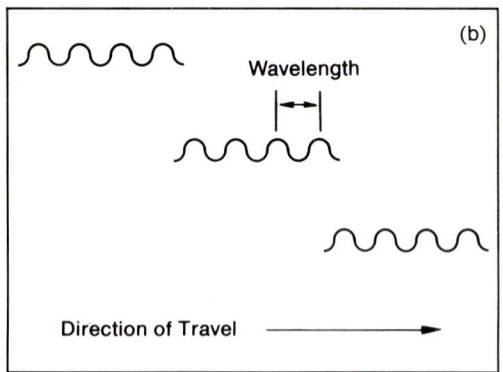

Figure 16. (a) The law of quantum physics relates the energy of a photon of light to the wavelength of its vibration for each color. (b) A visual model of a photon is sketched here. The whole photon carries the energy.

34　　　　　　　　　　　　　　　　　　　　　　　　　　Physical Science Today

It is also possible to write an equation that says the same thing. For each color,

$$\text{Wavelength} = \frac{\text{special number}}{\text{energy of photon}}$$

This basic law of nature tells us that the higher energy photons have the shorter wavelengths. It was first understood through the work of Max Planck and Albert Einstein. The special number is related to a number first discovered by Planck and called "Planck's constant." These two scientists pointed out that several experiments into the nature of light that seemed strange at the time were understandable only by relating the energy of a photon with the wavelength of its vibration.

The wave nature of light, together with the fact that light travels at a definite speed, leads to another characteristic of photons: the rapidity of their electric vibrations. The number of times these vibrations make a complete wave in each second of time is called its *frequency*. Because each photon is characterized by a definite spacing—a definite wavelength—and because the whole photon is traveling along at the unique speed always traveled by light, the speed of vibration of the light arriving at some place has a particular frequency. The closer together the waves of light, the more rapid the vibration will be when the light arrives at some destination. Again, you have an elegant law of nature that can be expressed simply:

Frequency (the rapidity of the vibrations of light) is equal to the speed of light divided by the wavelength of the vibration.

or, in a shorthand way,

$$\text{Frequency} = \frac{\text{speed of light}}{\text{wavelength}}$$

Here, then, is the scientist's model of how light works. You see things because little packages of energy—photons—are sent to your eyes. Each photon travels in a more-or-less straight line, and always at the same fantastically high speed. Each photon is a vibrating electric force, and the quantum of energy carried by the vibration is a certain amount for each color of photon. And the wavelength and the frequency have particular values for each color—and energy—carried by the light. It is through this pulsating energy of light that the world can be explored.

THESE SIGNS MAINTAINED BY THE WATSON LAKE LIONS CLUB
MOUNTAIN VIEW CITY LIMIT

One of the important features of science is that it deals not only with the qualities of things—that something is big, or hot, or blue—it also concerns itself with quantities—that something is *so* big, or *so* hot, or *so* blue. It says, for example, not merely that the moon is "far away," but that it is 400 million meters away from earth. Not merely that atoms are "small," but that if you could count the atoms along the edge of a table you would find 10 billion atoms in each meter length.

Why are numbers so important? Usually there are several possible ways of thinking about things and there may be several possible models for describing them. But if numbers are used in the descriptions, the models must also give the *right* numbers. With numbers we have a way of carefully checking whether an idea is right or wrong; we have a way of improving our models. It is one thing to say that hydrogen gas burns with oxygen to produce water. It means much more to say that for each kilogram of hydrogen burned, 9 kilograms of water will be produced. Indeed, only after observing that chemical materials always combine in weights that are related by small, simple numbers did scientists begin to understand that matter probably was made up of atoms.

Our understanding of nature is based on models we hold in our minds, models of how we think things work. In science, models are much more useful when they are expressed in quantitative terms. The idea of an atom as a permanent, unchanging entity came about when researchers saw that even though dissolving, burning, and boiling could change the *form* of materials, the total *weight* of the materials being analyzed would not change.

The power of science depends on the ability to make statements about the quantities of things, and much effort has been given to developing ways of discovering what those quantities are. How big is something? How often, or how fast, does it happen? A scientific observation involves not only seeing what exists or what happens; it also involves making *measurements*, which are the results of definite procedures for finding out some number.

Take an example from probability that involves the simplest kind of measurement: *counting*. If you toss a penny often enough, it will land "heads up" about half the time. If you toss a penny, say, 30 times and count the number of times it lands heads up, will the number be exactly 15? Don't bet on it. You are about 6 times more likely to get some other number. If you do the experiment of 30 tosses over and over again, you will discover that more often than not (about 60 percent of the time, to be more specific) the number of heads will be between 13 and 17. That is something you can discover about nature simply by counting things.

If you want to describe the speed of a rotating wheel, you can put a chalk mark on its rim and then count the number of times the mark makes a complete revolution in 1 minute as the wheel is turned. If the number turns out to be 60, you would say the speed is 60 revolutions per minute (often abbreviated 60 rpm).

Perhaps you are familiar with the Geiger counter, or the "scintillation counter." This instrument is used to measure atomic radiation, which consists of fast-flying atomic particles. When a fast-flying particle goes through the counter, an electric signal is produced that can cause an audible "click" in an earphone or a loudspeaker. If there is radioactive material nearby, some of the radiation it emits will go through the counter and

3

Measurements and Numbers

In Western cultures, at least, man is incredibly compulsive when it comes to numbers. In 1942 a homesick highway worker in British Columbia posted a sign stating the mileage to his home town; tourists still add to this collection.

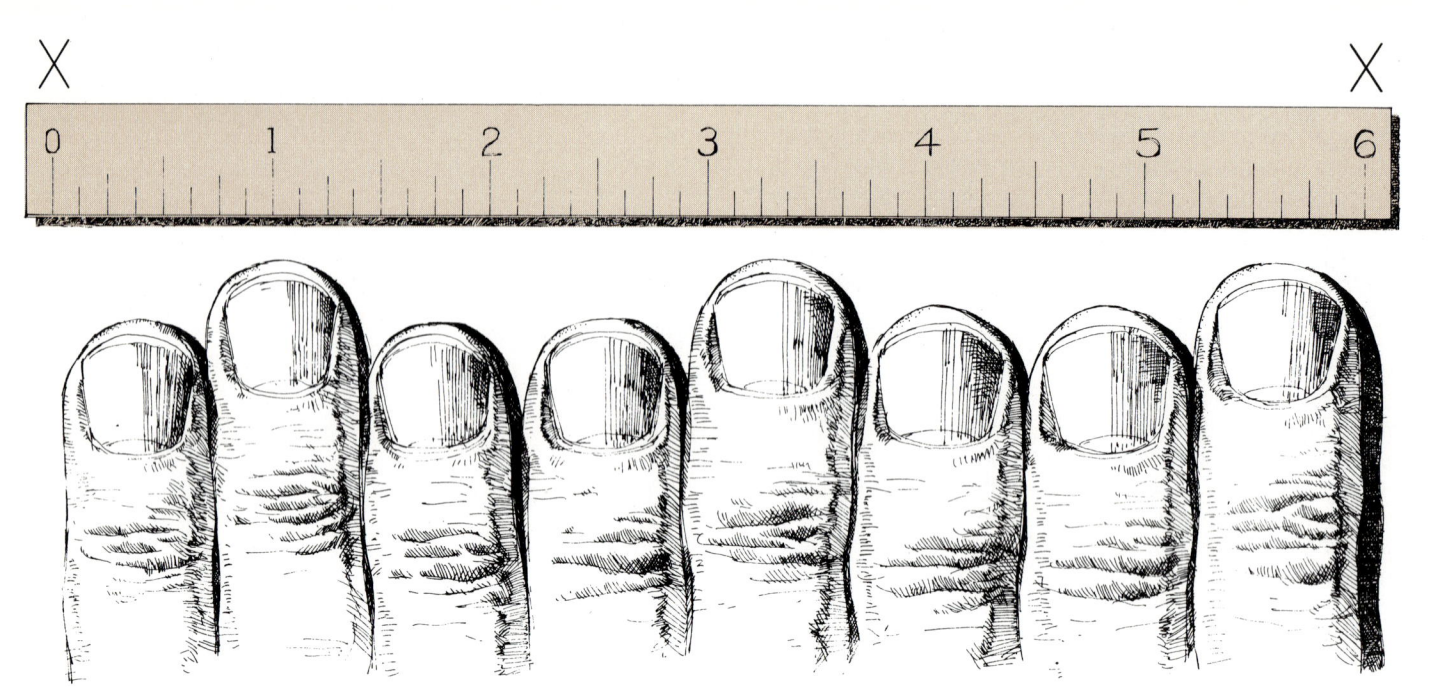

Figure 2. How can you measure the distance between the two crosses on this page? First you decide on some unit of measurement. Here, it is a unit of distance. You may choose an "inch" or a "fingerwidth." Then you relate what you are measuring to that standard unit. In this case you count off how many times an "inch" will fit into the space, then you say that 6 inches, or six of the chosen units, will fit. Or you say that "The measured distance is 6 times larger than 1 inch, the unit of measure I chose to use." If you use the other unit of measure you would say, "I measure the distance to be 8 times larger than my chosen unit—8 fingerwidths."

the quantity of radiation can be measured by counting the number of clicks that occur in each minute. The amount, or the *intensity,* of the radiation arriving at any particular place is expressed as so many counts per minute.

UNITS OF MEASUREMENT

Many kinds of measurement are useful in science: Measurements of the sizes of things—the length, the area, the volume. Measurements of distances, of temperatures, of times, of forces, of speeds, of colors. And many others. All such measurements are used to determine a number, or several numbers, that can be associated with the situation being observed. Because all numbers are related in some way to the process of counting, all measurements are related directly or indirectly to the idea of counting something.

There are two crosses positioned along this page (Figure 2). What is the distance from one cross to the other? If you happen to have a ruler with you perhaps you might use it to find the distance is 6 inches. If you don't have a ruler, perhaps you might count off how many times the width of your index finger fits into the space between the crosses. Then the distance might be "8 finger-widths." Either way, you will have made a reasonable measurement; both involve basically the same process.

Every measurement is a process of comparing something to a *unit of measurement.* In the example given above, a number is compared to a unit of distance. Such a comparative number is often called a *ratio*, or a *proportion.* When you say a distance is 6 inches, you are indicating the distance measured is 6 times larger than the unit of measurement (inch). Or you may say the ratio of the distance to the standard unit of distance is 6 to 1; or, more simply, the ratio of the distance to the unit of measurement is 6.

When the distance being measured is *smaller* than the unit of measurement, a different sort of comparison may be used. For example, a small distance may be expressed as one-fifth (1/5) of an inch. In other words, if that small distance is put together 5 times, it will add up to 1 inch. Another, and perhaps more common, way of thinking is to start out with a different unit for the measurement. Suppose you choose to divide 1 inch into 10 equal parts. "The tenth of an inch" would be the new unit of measurement. You might then say that the small distance is 2 of these units, or two-tenths (2/10) of an inch. The two ways of thinking are equivalent: a choice between them is a matter of convenience.

What units of measurement should you choose? Any choice is arbitrary. Throughout history, thousands of different units have been used for various kinds of measurements. And thousands are still in use: The inch, the meter, the mile, the parsec, the bushel, the carload. And on and on. And it can get incredibly confusing.

But all units of measurement are equally "valid." Science can make no choice between one and another because the choice is always a matter of convenience. Perhaps one unit represents convenience in making the measurement, or convenience in remembering the result, or convenience in doing calculations. More importantly, perhaps, it may represent convenience in being able to tell other people about the measurement. For all these reasons (particularly the last), scientists from all over the world congregate every year or so to talk over units of measurement and to agree, for the sake of convenience, on a standard set of units that everyone will use. Then if people wish to compare measurements they can do so without a long discussion of how big is a finger-width compared to an inch.

Scientists are not alone in wanting to agree on common units. With the growing closeness of the world community, trade among different countries can benefit from a common set of units of measurement. Without that agreement, there is a problem of thinking about selling "pounds" of a product to someone who thinks about "grams"; or worse, of trying to use a 1/2-inch wrench to repair a car that has its bolts and nuts cut to 1-centimeter size. So all the nations of the world except one—the United States—have agreed to begin using a common system of measurement, called the *metric system.* This set of units uses the *meter* as the unit of distance.

Science is an international activity and scientists everywhere have been using some version of the metric system of measurement for more than a century. The United States, too, will undoubtedly join the rest of the world in adopting the metric system and will gradually abandon the "English system" now used in everyday life. Because the metric system is much more convenient for scientific work—and because it is the system that will be used in the future—it is used most often in this book.

What is a meter? Originally it was intended to be a length that was a certain small fraction (a forty-millionth) of the distance around the world. But that turned out to be an inconvenient definition because nobody could say just how far that was. So somebody decided on an even more arbitrary definition. Two scratches were made on a metal bar and the distance between them was called "1 meter." The bar was placed in a safe place (in France) and copies of the bar were made for anyone who wanted to know exactly how long a meter really was.

But then scientists became interested in making more and more precise measurements. And they found, of course, that the distance between the two scratches on the standard meter was not perfectly precise. When they looked at the scratches with a microscope, the scratches looked wide and fuzzy. No one could tell exactly *where* the center of the scratch was. Recently the "meter" was given a new definition. It was chosen to be as close to the old definition as anyone could tell but it is

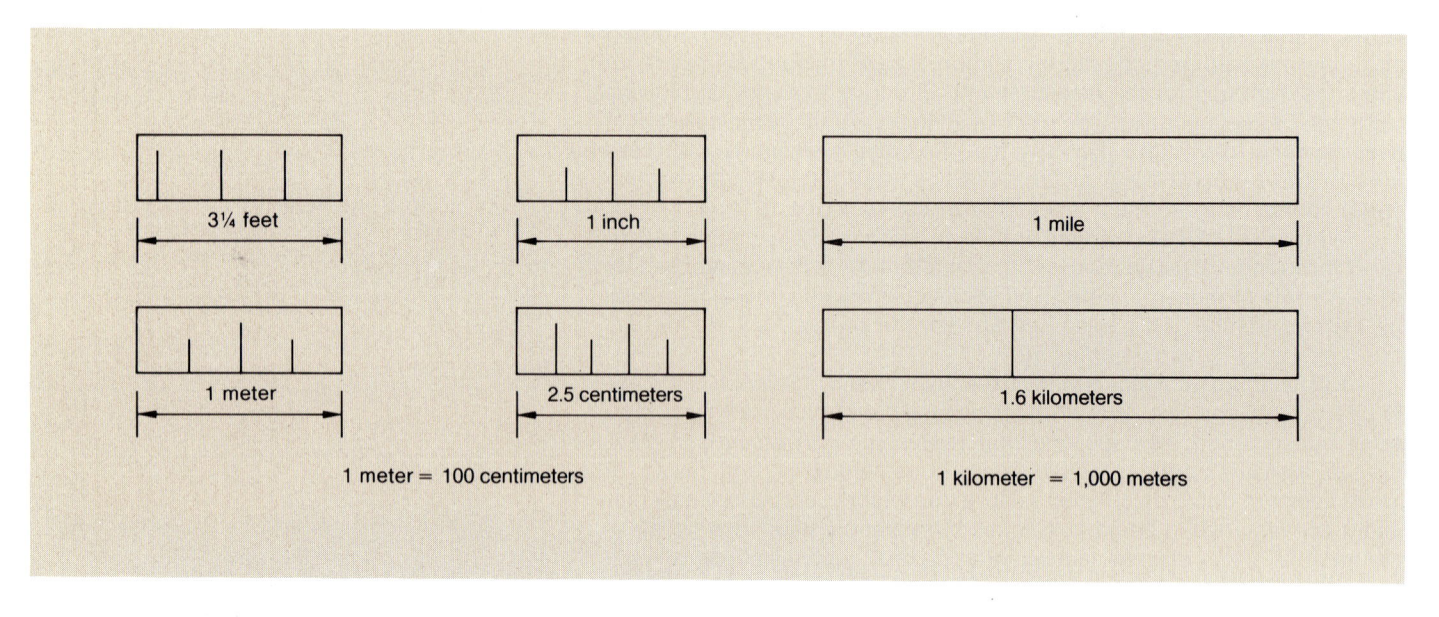

Figure 3. The physical object being measured doesn't change in size, but the *units* used to measure it can vary. When depicting an object, it may be more convenient to *scale* the object to make it easier to view.

more precise. The meter is now *defined* to be a certain distance that is 1,650,763.73 times larger than the wavelength of the orange-red light that comes out of the vibrations of krypton atoms.

APPROXIMATION AND ERROR

Because science often deals with quantitative statements, many people believe that science is an "exact" subject. The term "exact" usually implies the numbers of science are absolutely precise or correct, as when someone says, "I have exactly $3.27 in my wallet." Some scientific observations can give similarly precise numbers. When you say you have five toes on each foot, you mean exactly five; when a scientist says an oxygen atom has eight electrons he means exactly eight. But most of the quantitative statements about the world are not exact. In fact, except for instances of simple counting *they cannot be exact.* Every scientific measurement of size gives a number that is only approximate. So every numerical statement about the sizes of things in the world is only approximately right.

Why is it impossible to measure size perfectly? First, there are bound to be some imperfections in the measuring process. Second, we never are sure we know exactly what we mean by what we are measuring. If you used a ruler to measure the distance between the two crosses, you probably had some difficulty in placing one end of the ruler exactly on one of the crosses. Moreover, the ruler itself was not "perfect"—it was not made perfectly, and it probably has changed since it was made. And then, what does "the distance between the crosses" mean? Between the "centers" of the crosses? But where are the centers, exactly? Under a magnifying lens the crosses look fuzzy, so how is it possible to say where the center is? Do you think you could ever say *exactly* what the distance is between the crosses?

Every measurement is in some sense a rough measurement; it will have some *error.* This word is used to express how far off from some ideal number a result might be. Science usually tries to indicate how imprecise or inexact a measurement is by giving an estimate of how large the error might be. For example, if a scientist says the distance between the crosses is 6 inches, he always means it is "about" 6 inches. If his measurement is very rough — possibly just a quick guess — he might decide it could be anywhere between 5-1/2 inches and 6-1/2 inches. He might say the distance is "6 inches plus or minus half an inch," which may be written 6 ± 1/2 inches. What if he measures the distance more precisely and finds it is 6.02 inches but isn't really *sure* that it is 6.02 inches and not 6.01 inches or 6.03 inches? Then he would say the distance is "6.02 inches with a probable error of one-hundredth of an inch either way," or, compactly, 6.02 ±0.01 inches.

DISTANCE AND LENGTH

Usually distance is thought of as the number of times a basic unit of distance is counted off in going from one point to the other. Distance may be measured along a road, a path, or some curve that goes from one point to another. You may speak of the distance from New York to San Francisco along some particular route; you may talk about the distance around a circle. If you speak of the distance from A to B without specifying how you go from A to B, you generally mean the shortest possible distance. If A and B are on the surface of the earth, the shortest distance over the surface is the so-called "great circle route." Scientifically speaking, you would in general consider the shortest distance to be along a straight line from A to B.

Often when you think of some particular object you want to refer to some particular distance associated with it, so you use the name given to it — height, for instance, or length, or width, or thickness, or diameter. When you use such ideas, you mean the shortest distance from some edge of an object to an opposite edge. At other times you may require a different idea; when, for example, you refer to girth or circumference. Then you mean the distance around the outside of the object.

In the English system, distances (and therefore lengths) are measured with reference to such units as the inch, the foot, the yard, and the mile, all of which are related in complicated but undoubtedly familiar ways. In the metric system, distance is measured with reference to the meter or to some related unit of measurement such as the *centimeter*, which is defined to be exactly one-hundredth (1/100) of a meter. The *kilometer* is also used, and it is defined to be exactly equal to 1,000 meters. Various units of distance are depicted in Figure 3.

When the distance between two points or two objects is considered to be, say, 30 meters, we normally think it would be possible to place a 1-meter stick exactly 30 times between the two points. But many situations do not really lend themselves to that sort of procedure. Perhaps the Empire State Building sits right between the two points; perhaps the two points are on opposite sides of the Amazon River. More often than not we only think we are using the ideal way of measuring, but in fact we are employing an *indirect way of measuring* based on some model of how we believe the world works.

Suppose there are two trees A and B on opposite sides of a river. What is the distance between them? One way to find out is shown in Figure 4. If you measure the lengths of the sides $\overline{AB}$ and $\overline{BC}$ of a triangle you have two numbers — say, 3 and 5. Now divide the first number by the second. When 3 is divided by 5 the result is 3/5, which is the proportion (or ratio) of the two lengths. Now you may use a fact that was discovered by the science (or mathematics) of geometry: The *proportions* of the sides of small triangles are the same as the *proportions* of the sides of large triangles, as long as the angles between the sides are the same. This

Figure 4. The method of *triangulation* may be used to estimate the distance to an inaccessible object. The similarity of triangles with equal angles is used to estimate the distance across the river by drawing a model of the actual triangle, and then *scaling* it according to the length between two of the accessible objects.

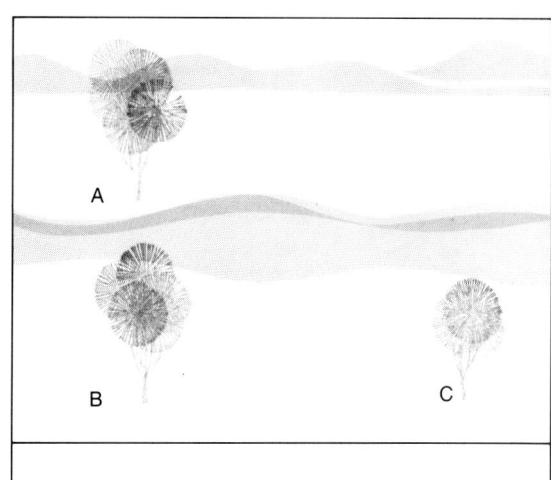

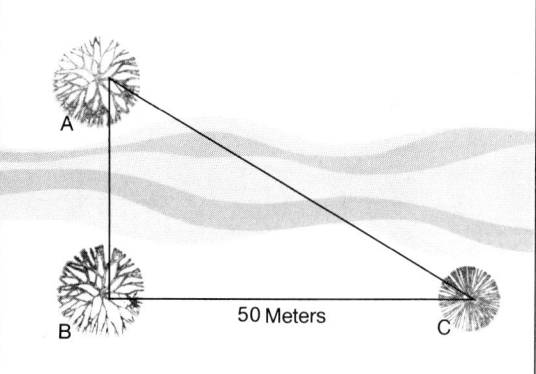

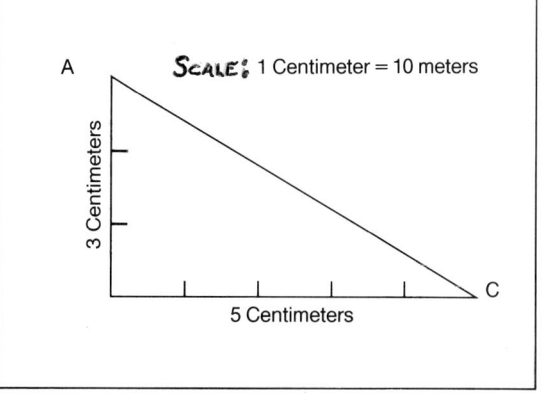

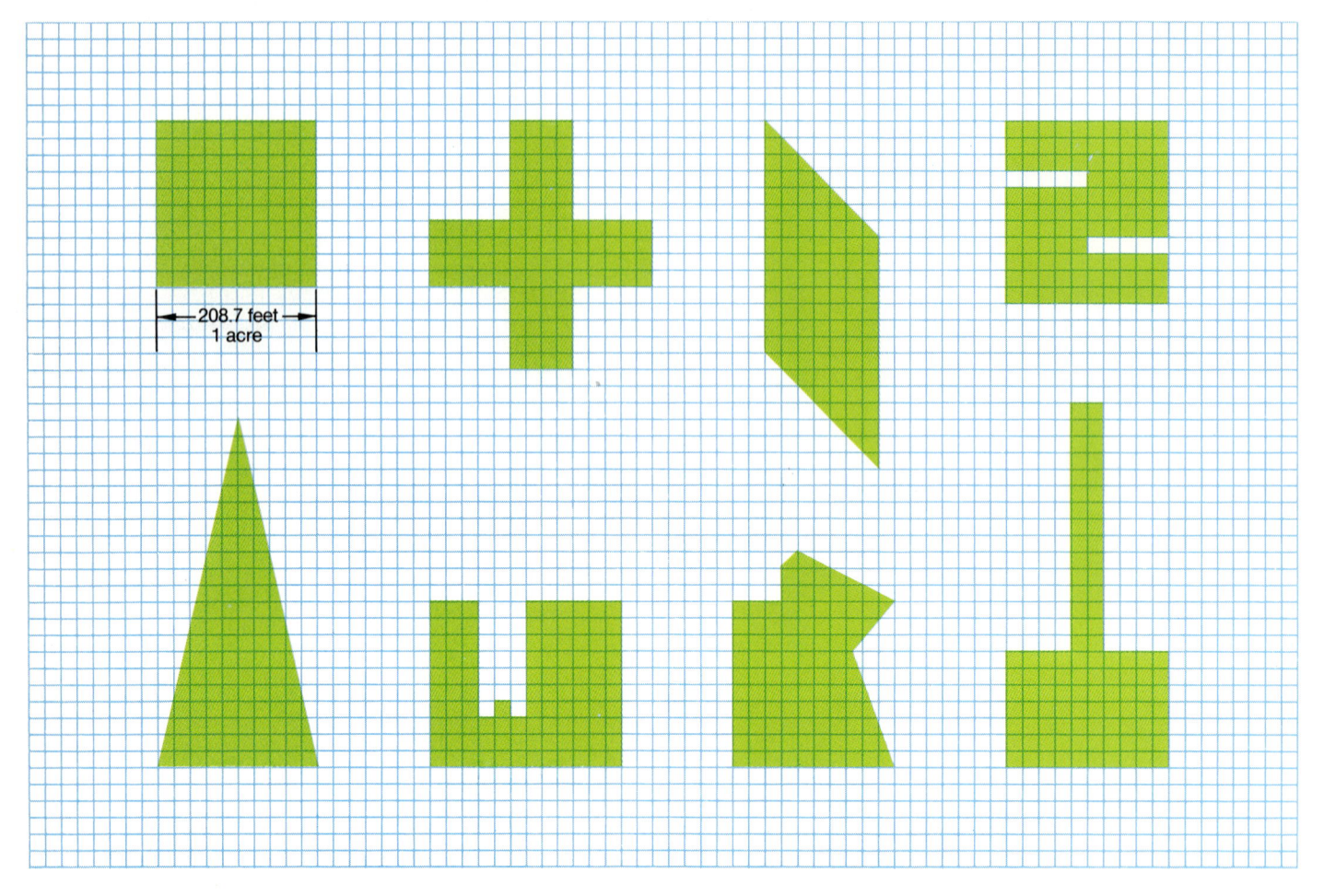

Figure 5. If you were to purchase some land, you would want to know how large it is. How would you know that the areas depicted here are equal?

important fact is used extensively in surveying and in scientific measurements. How far away from the earth is the moon? How far above the earth is a man-made satellite that you see orbiting in the sky overhead? Geometry can be used to measure such distances indirectly.

MEASURING SURFACES

Suppose you wish to paint the walls of a room. How much paint you will need depends on how much *surface* the walls have. The word "area" is used to describe a measure of the amount of surface. To determine the area of a surface, you must first choose a *unit of area* and then calculate how many of those units will completely cover the surface.

You could, for example, choose the "hand" as a unit of area. You would have to find out how much of the surface a "hand" will cover. If the "hand" is a useful unit, you should also be able to use it to find out the area of, say, the surface of a page in a book. How many "hands" would cover that surface? The measurement would be difficult because a "hand" is a complicated shape. Measurements of surfaces are easier when the unit of area is chosen to be a *square* of some convenient size.

The acre is the most common unit of area for measuring the surface of parcels of land. It is defined as the area of a square that is 208.7 feet along each edge. However, a 1-acre parcel of land does not have to be

Physical Science Today

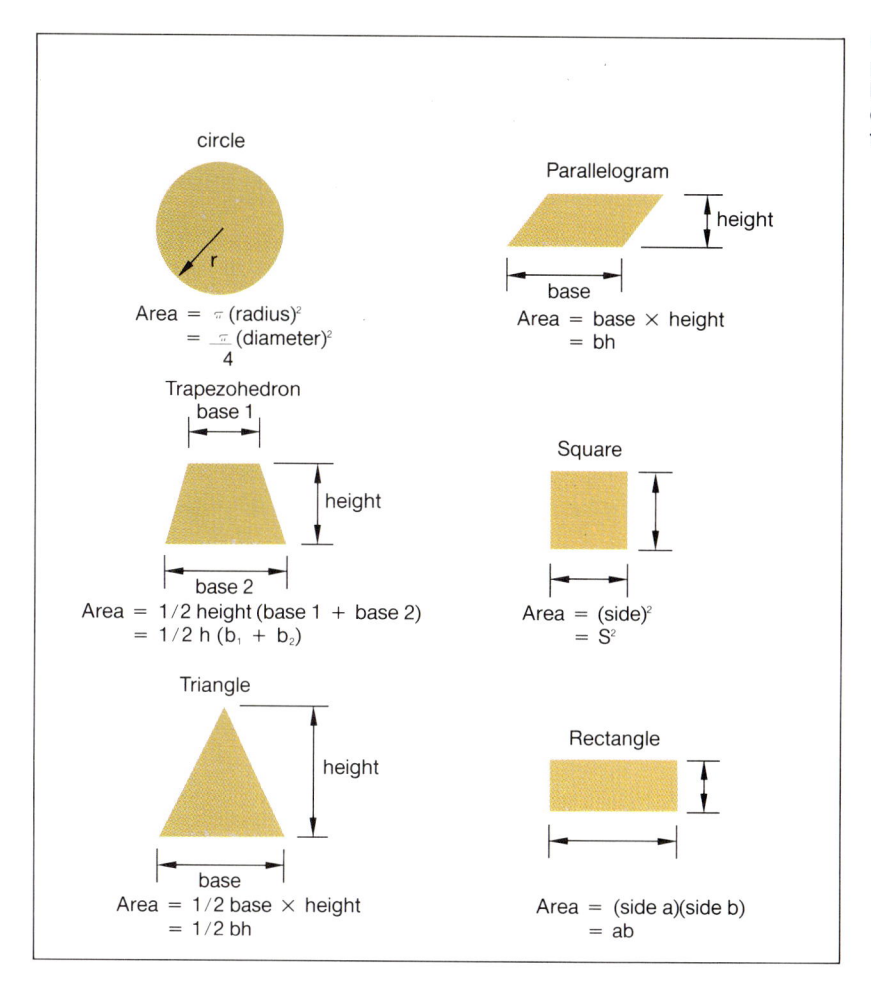

Figure 6. You can estimate the area of a surface by pasting down cards of small, known areas within its boundaries. Or, if the shape of the area is simple enough, you can use one of these convenient formulas to calculate the exact area.

circle

Area = π (radius)^2
     = $\frac{\pi}{4}$ (diameter)^2

Parallelogram

Area = base × height
     = bh

Trapezohedron

Area = 1/2 height (base 1 + base 2)
     = 1/2 h (b$_1$ + b$_2$)

Square

Area = (side)^2
     = S^2

Triangle

Area = 1/2 base × height
     = 1/2 bh

Rectangle

Area = (side a)(side b)
     = ab

square: it may be a parcel that has the same amount of surface as the standard square. Figure 5 illustrates scale drawings of several parcels of land, all of which have an area of 1 acre. How can you determine they are equivalent in area? You can choose a smaller unit of area—say, a small square that represents 1/100 of an acre—and then show that about 100 of those smaller squares will cover each of the odd-shaped parcels of land.

Most commonly a unit of area is chosen that is simply described in relation to the unit of length. The "square inch," for example, means the unit of area is a square piece of surface that measures 1 inch along each side. What is the area of this page, measured in square inches?

In the metric system of measurement, the unit of area is the *square meter.* It is the area covered by a square with sides 1 meter long. Sometimes related units of area, such as the *square centimeter* and the *square kilometer,* are more convenient. If a table top has an area of 2 square meters, what is its surface area measured in square centimeters? If a parcel of land has a surface area of 3 square kilometers, what is its area measured in square meters?

The science, or mathematics, of geometry studies how the surface areas of various objects can be determined if some of the dimensions of the object—the length, the width, the height, the diameter, and so on—are known. For example, the area of a wall can be figured out by multiplying the length of the wall and its height. As another example, the surface area of an orange is about 3 times as large as a square that has

3 | Measurements and Numbers

Box
Volume = Length x Width x Height

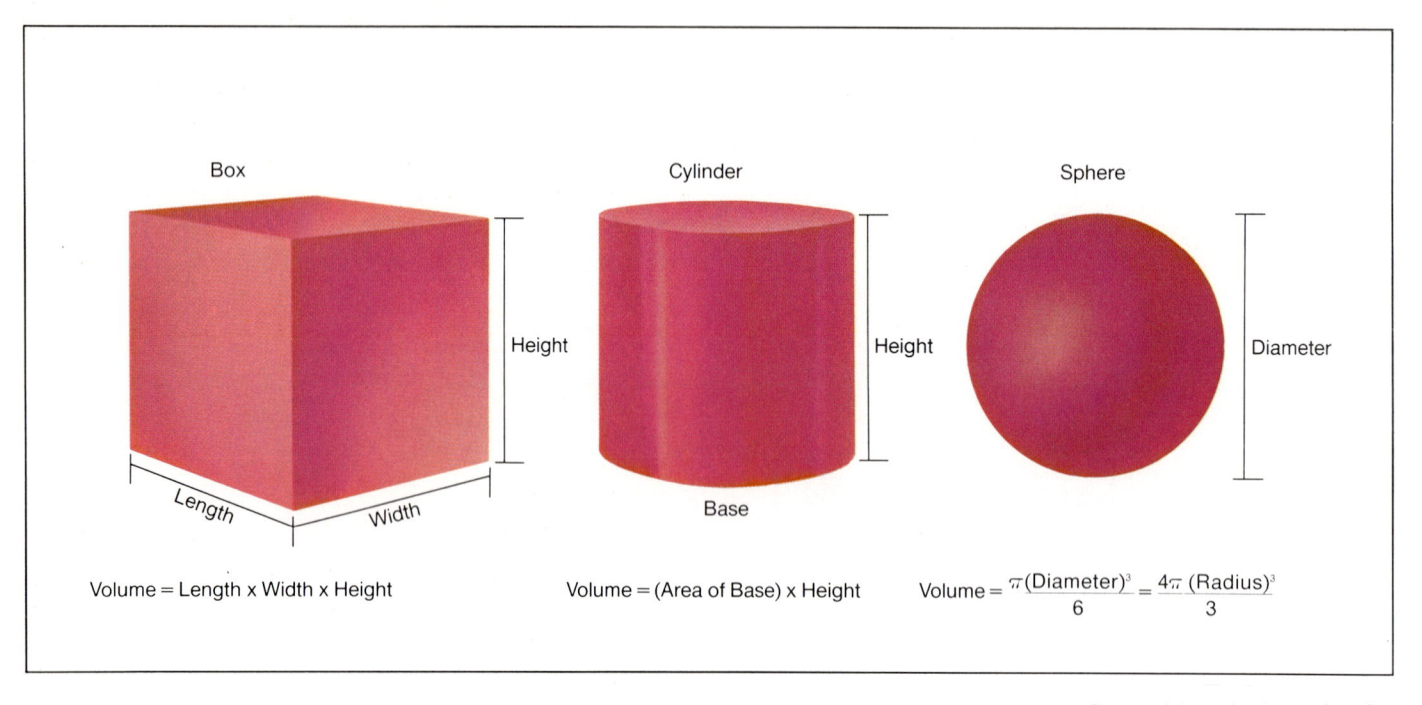

Cylinder
Volume = (Area of Base) x Height

Sphere
$$\text{Volume} = \frac{\pi(\text{Diameter})^3}{6} = \frac{4\pi(\text{Radius})^3}{3}$$

Figure 7. Because volume is simply a measure of the amount of *space* occupied by something, any volume can be calculated by multiplying its three linear dimensions.

sides equal to the diameter of the orange. Several handy formulas for determining areas are given in Figure 6.

MEASURING VOLUMES

Quantities of milk are commonly measured in *quarts*. When you buy a quart of milk you get the amount of milk that fills up a certain definite quantity of space. An "amount of space" is called a *volume*. A quart is a unit of measure of volume; a container that holds 1 quart of milk has a volume equal to 1 quart. A gallon container has inside it a volume of space equal to the unit of volume. A gallon, as it happens, has a volume 4 times as large as a quart.

In measuring the volume of any space, you first choose your unit of volume and then figure out how many of those units will fit inside the space. You may find out *experimentally* by pouring a liquid such as water into the space and counting the number of quarts of water that can be poured in. Or you may figure out a volume *theoretically* by doing a calculation based on other measurements. For instance, if you know some of the dimensions of the space, you may figure out the volume by using the science of geometry.

In the metric system, the basic unit of volume is the *cubic meter*. It is defined to be the space occupied by a cube that is 1 meter along each edge. Sometimes the *cubic centimeter* is used, which is the volume of a cube that is 1 centimeter along each edge. Another unit of volume, the *liter,* is almost the same as a quart; it has the same volume as 1,000 cubic centimeters.

The cubic meter and the cubic centimeter are chosen as units of volume because they make it easier to figure out how many cubes of a particular size will fit into a space, once the linear dimensions of the space are known. If the floor of a room measures 10 meters by 5 meters, and if its ceiling is 3 meters high, it is easy to figure out that 150 cubic meters, each measuring 1 meter on a side, will fit into the room. The volume of the room is equal to 150 cubic meters. Figure 7 shows how the volumes of various kinds of spaces can be worked out from their dimensions.

PROPORTIONS

If someone says: "This parcel of land is twice as big as that one," what does he mean? Twice as long on each side or twice as much area? It makes a difference. If it is twice as long on each *side,* it has *4 times* as much surface *area.* Or if it is 3 times as long on each side, it has 9 times as much area. *For a surface of a given shape,* area increases in proportion to the *square* of the linear dimension. (The word "square" is used here with its meaning in algebra. The "square of a number is that number multiplied by itself: "three squared," written as 3^2, is the same thing as 3×3, or 9.) If you make an accurate "scale model" car that is only 1/10 as long as a real car, how much less than the real car would it weigh?

If you wish to compare *spaces* of the same shape but of different sizes, then the *volume* of the space is proportional to the *cube* of the linear dimension. A box that is 4 times larger in each dimension (4 times longer, 4 times wider, 4 times higher) has a volume that is larger by 4^3, or $4 \times 4 \times 4$, or 64.

HOW MUCH TIME

So far, you have been reading about measures of space—distance, area, and volume. Now take a look at a completely different kind of measurement—the measurement of time. It is difficult to think of a way of saying what "time" is in words, but we all seem to get the idea as we grow up.

Begin with the idea of an event, or an instantaneous "happening." It could be the striking of a bell, or a flash of lightning, or any occurrence such as the arrival of a car at an intersection. When we "measure" time, we measure the *interval of time between two events.* If someone says, "It took 5 hours to fly from San Francisco to Boston," he is thinking of a measurement of the interval between a first event (leaving San Francisco) and a second event (arriving at Boston). And he may have measured the time interval by looking at his watch when he departed and when he arrived.

But a watch is nothing more than a special gadget for *counting* small pieces of time. Inside the watch a small, oscillating wheel continually swings back and forth, and a small set of gears counts the number of times the wheel oscillates. The count shows up in the position of the hands of the watch.

Many kinds of "clocks" have been invented for measuring time. Most of them, particularly the more accurate and convenient ones, use the same basic principle as a wristwatch. They all have an *oscillator,* which is a device that swings back and forth in a regular way. It may be a mechanical oscillator, such as the balance wheel in a watch or the pendulum on a grandfather's clock. Or it may be an electric oscillator, in which an electric current swings back and forth.

The swinging motion of an oscillator marks off equal pieces of time —just as the increments on a ruler mark off equal pieces of distance. You may, for example, think of the "upswing" of the oscillator as the time mark and the "downswing" as the space between marks. When we measure the time interval between two events, we merely are counting the number of upswings that occur between the first event and the second event.

The oscillation of light and dark between day and night was probably the first time marker used. One oscillation—from light to dark and back to light again—was chosen as the *unit of time* and was called the

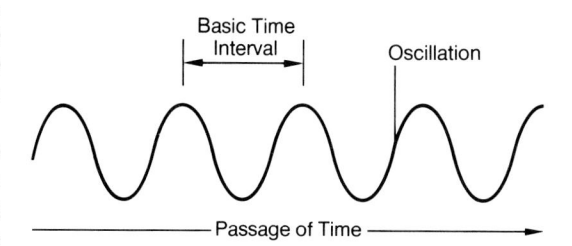

Figure 8. Clocks are *oscillators.* The time interval between each peak of an oscillation is always the same. To mark time you simply count the number of peaks that have passed and multiply that number by the time interval.

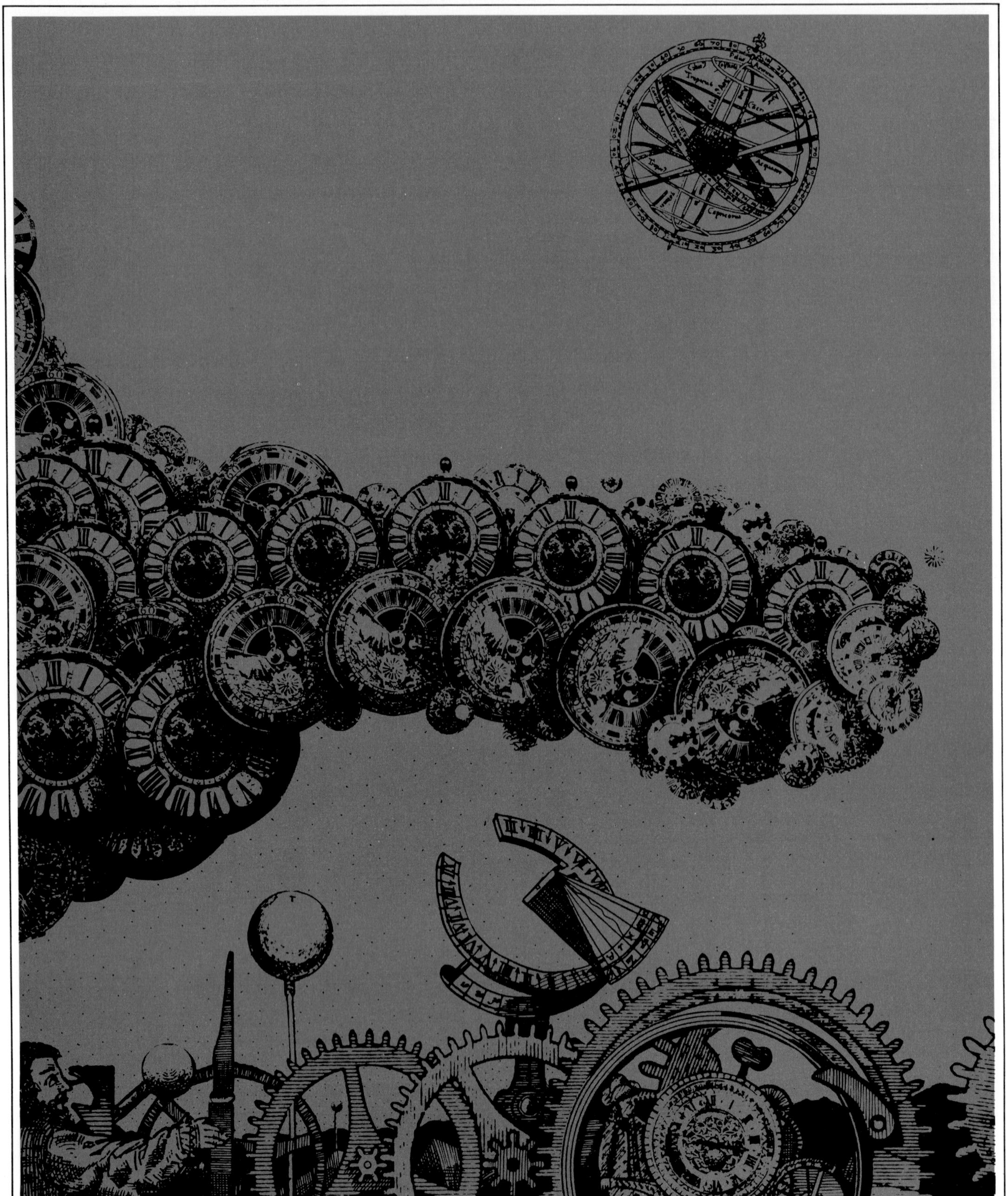

"day." Later, people felt the need for a smaller unit, so the "hour" was invented as a new unit. The hour was defined to be 1/24 of a day (a strange choice, when you think about it). Then people wanted a still smaller unit, so a segment of time only 1/60 as big as an hour was chosen and called a "minute (tiny) piece of an hour," or "minute" for short. Then even that was too big for some purposes, so someone invented a "second minute fraction" of an hour—equal to 1/60 of a minute—and it became known as the "second." The second is now the most common unit of time for scientific measurements. It was once defined to be 1/60 of 1/60 of 1/24 of a day—or 1/86,400 of the duration of a day.

But then people discovered that apparently accurate clocks did not always count the same number of seconds in each day—because days are not all the same length. So for a while the official, standard second was defined to be a certain small fraction of a year. But then it was found that each "year" appears to be of slightly different duration, so there was a lot of discussion about making a new definition of the second.

A few years ago people noticed that certain atomic vibrations are affected very little by outside influences. Clearly those vibrations could be used to make a truly regular oscillator. Today, "atomic clocks" are used as the most precise way of measuring time. They work in a way that is similar to the principle of the watch. A set of electronic devices counts up the vibrations of certain atoms, much as the gears of a watch count up the oscillations of the balance wheel in a watch. Because of the better reliability of atomic clocks, the second finally was defined to be a time interval that is a certain number of times larger than the time for one atomic vibration of some particular atom to occur. Specifically, a "second" is now officially the time required for 9,192,637,770 oscillations of a particular kind of vibration of the atom cesium 133.

OTHER QUANTITIES

In addition to space and time, many other ideas in science have been refined enough to be described in terms of quantitative measures. Temperature, mass, speed, loudness, weight, electric charge, color, hardness, force, and density are a few examples; there are many more. All of these quantities are defined in terms of a procedure for making a measurement, and a unit of the measure is defined. An example is temperature; its unit, the "degree Centigrade." Many units, however, are not defined directly but are defined indirectly in terms of units of other measures. Consider *speed,* which is defined to be the ratio of the distance traveled to the time interval that passes during the trip. In other words, speed is distance divided by time:

$$\text{Speed} = \frac{\text{distance}}{\text{time}}$$

Because of this definition of speed, it is most convenient to define the *unit* of speed as the unit of length divided by the unit of time. For example, if a car travels 35 meters in 5 seconds, you would write

$$\text{Speed} = \frac{35 \text{ meters}}{5 \text{ seconds}} = 7 \frac{\text{meters}}{\text{seconds}}$$

When speed is calculated in this way, its unit of measurement comes out as "meters-divided-by-seconds." That unit could be given a name of its own; the "swift," for instance. But that has not been done. Instead, the unit is named the "meter per second," which can be written in several ways:

$$\frac{\text{Meters}}{\text{Seconds}} = \text{meter per second} = \text{meter/second} = \text{m/sec}$$

Figure 9. Our early ancestors used the alternation of day and night as a clock; today we use the vibrations of a certain kind of atom as a measure of the passage of time. Historically we have devised elaborate ways of dividing time. The devices we have come up with are in a sense a measure of our culture's increasing concern with technical complexity.

You will encounter many such *derived units of measurement* in this book. Sometimes they will keep their derived unit names and sometimes they will be given special invented names. For example, "kilograms per cubic meter" is a unit for density. But the unit of power, which is the "joule per second," is also called the "watt."

WRITING LARGE AND SMALL NUMBERS

Suppose you want to make an estimate of the earth's "weight"—or, more properly, its total mass expressed in terms of the standard kilogram. (A standard kilogram is the mass of a standard block of material that weighs a little over 2 pounds.) How could you go about it? You could do it in the following way, which is the same way the scientist Isaac Newton did it about 250 years ago.

You might guess that most of the earth is made up of rock or similar stuff. The rocks you see around you weigh about 3 times as much as water. Each cubic meter of water has a mass of approximately 1,000 kilograms. Now if you could figure out how many cubic meters there are in the whole earth, you would be able to estimate its mass. It would be the number of cubic meters multiplied by 3,000 kilograms, the approximate mass of each cubic meter of rock.

Try calculating the number of cubic meters—in other words, the volume—of the earth. The radius of the earth is about 4,000 miles. Now a mile is about as long as 1,600 meters, so the radius of the earth is 4,000 times larger than 1,600 meters, or

Radius of earth = 4,000 × 1,600 meters = 6,400,000 meters

What is the *volume* of the earth? The volume of a sphere is about 4 times the cube of its radius. More precisely, it is equal to $(4\pi/3) \times R^3$ but π is approximately equal to 3. So,

Volume of earth = 4 × (6,400,000)^3
= 4 × 6,400,000 × 6,400,000 × 6,400,000

How big is all that? If you multiply it all out, you will get

Volume of earth = 1,047,376,000,000,000,000,000 cubic meters

That's obviously an inconvenient number to work with. First, it takes a long time to write it down. Second, it is easy to make mistakes in keeping track of all those zeros. Besides, all those numbers at the beginning are not really useful when all you want to do is make an estimate. You can learn a more convenient way of handling such numbers. You know that

$\quad\quad\quad 100 = 10 \times 10$
$\quad\quad 1,000 = 10 \times 10 \times 10$
$\quad 10,000 = 10 \times 10 \times 10 \times 10$
$\,100,000 = 10 \times 10 \times 10 \times 10 \times 10$
$1,000,000 = 10 \times 10 \times 10 \times 10 \times 10 \times 10$

And you can see that the number of zeros in the numbers on the left-hand side of the equals sign is the same as the number of times the factor 10 appears on the right-hand side. Mathematicians have invented a convenient way of writing such numbers. It is called the *exponential notation,* and it goes like this:

$\quad\quad\quad 100 = 10 \times 10 = 10^2$
$\quad\quad 1,000 = 10 \times 10 \times 10 = 10^3$
$\quad 10,000 = 10 \times 10 \times 10 \times 10 = 10^4$
$\,100,000 = 10 \times 10 \times 10 \times 10 \times 10 = 10^5$
$1,000,000 = 10 \times 10 \times 10 \times 10 \times 10 \times 10 = 10^6$

The rule is that the smaller number at the upper right-hand corner of the 10 means that 10 is to be taken that many times (in multiplication). This "superscript" number is called the *exponent*. In using it, you merely have to *count* the number of zeros and write the count as the exponent of 10. One million is 1,000,000 and has 6 zeros; it may be written as 10^6.

It's also easy to multiply two such numbers. For example, instead of writing out

$100 \times 10,000 = (10 \times 10) \times (10 \times 10 \times 10 \times 10)$
$= 10 \times 10 \times 10 \times 10 \times 10 \times 10$
$= 10^6$

you may write directly

$10^2 \times 10^4 = 10^6$

You get the exponent on the right simply by *adding* the two exponents on the left. Similarly, you can show by writing it out that

$10^3 \times 10^2 \times 10^4 = 10^9$

One little trick is needed here. The number 10 may also be written as 10^1, or 10 taken only once.

$10^3 \times 10 \times 10^2 = 10^3 \times 10^1 \times 10^2 = 10^6$

What do you do with a number such as 3,000,000?

$3,000,000 = 3 \times 1,000,000 = 3 \times 10^6$

You may always write *any* large number as some small number, such as 3, multiplied by a *simple* large number, such as $1,000,000 = 10^6$. These large numbers, which have only a "1" with many zeros, are also called *powers of ten*. For example, 10^6 is called the *sixth power of ten*. Finally, remember that the same way of writing can apply to numbers with decimals. For example,

$3.2 \times 100 = 320$
$3.28 \times 100 = 328$
$3.284 \times 100 = 328.4$

To multiply by 100, the rule is that you have to "move the decimal point" two places to the right. To multiply by $1,000 = 10^3$, you move it three places. To multiply by $1,000,000 = 10^6$, you move it six places. So if you have a number such as 3,280,000, you may also write it as

$3,280,000 = 3.28 \times 1,000,000 = 3.28 \times 10^6$

Now you're ready for the large number you calculated for the volume of the earth. It may be written more conveniently as

Volume of earth $= 1.047376 \times 10^{21}$

Here, the decimal point was placed after the 1, and then the power of ten was figured by counting the number of places there are from the decimal point to the end of the number. You also *could* have written it as 10.47376×10^{20}, or 104.7376×10^{19}.

But you are, after all, only making an estimate, and the number 1.047376 is more complicated than you need. Besides, it's hard to remember. Suppose someone advertises a television set for $99.95. Most people think, "Well, that's practically the same thing as $100." After all, the difference is only 5 cents, which is not *significant* if you're spending 100 dollars. The scientist, too, will often keep what he calls only the "significant" figures.

For example, you might decide that the last four figures in 1.047376 are not significant for your purposes and you can just as well call it 1.04. Actually, it would be a little better to call it 1.05 because the *next* number is 7, and 47 is closer to 50 than it is to 40. But for this approximate calculation 1.04 is just about as good, particularly because you got that number in the first place from an approximate number for the diameter of the earth. When you want to emphasize that you are making a *rough* calcula-

A Galaxy · A Star · A Planet · The Crust

The Universe

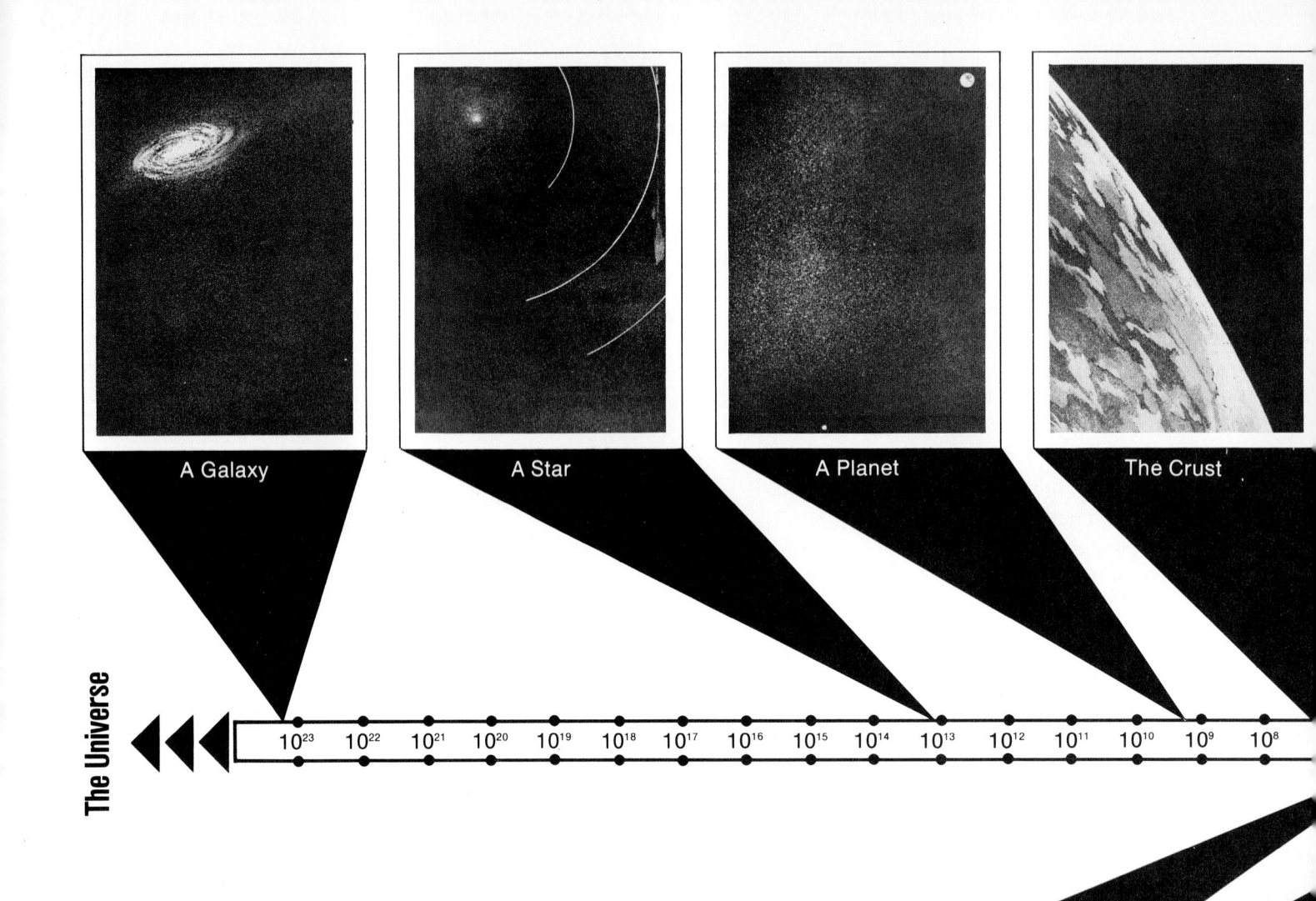

10^{23} 10^{22} 10^{21} 10^{20} 10^{19} 10^{18} 10^{17} 10^{16} 10^{15} 10^{14} 10^{13} 10^{12} 10^{11} 10^{10} 10^{9} 10^{8}

A Community · A Population

Figure 10. Exponential notation can be used to give a panoramic view of the scale of nature. The objects that are represented here are orders of magnitude apart in size, and you might think they do not lend themselves to comparisons, one against the other. But when they are organized on a *scale* from the largest things we know about—the universe, the galaxies, the stars—to the other extreme boundaries of our knowledge, the elementary particles, then we can begin to explore the significance of worlds far larger and far smaller than the world of human experience.

An Organism | A Tissue | Macromolecules | An Atom

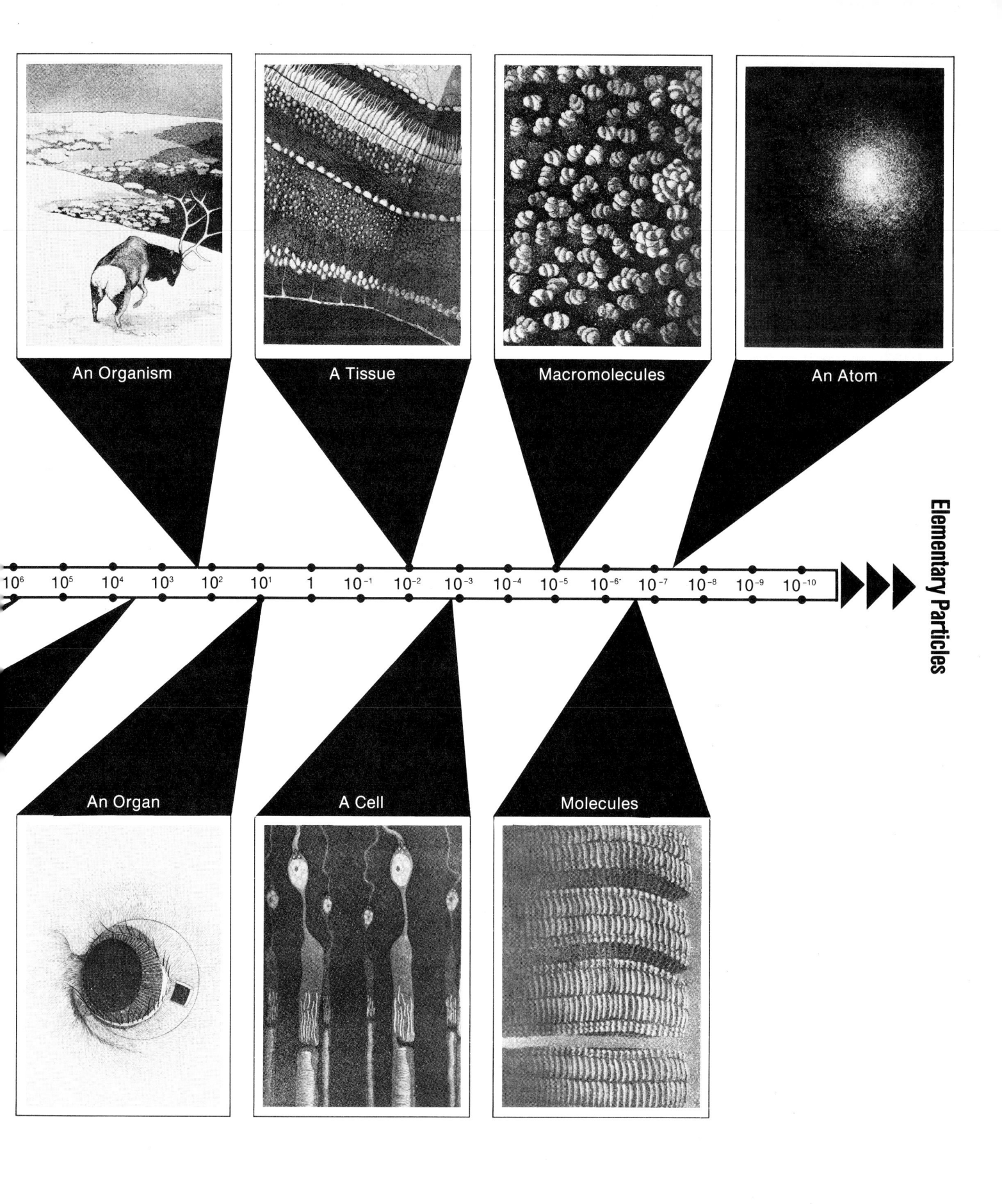

10^6 10^5 10^4 10^3 10^2 10^1 1 10^{-1} 10^{-2} 10^{-3} 10^{-4} 10^{-5} 10^{-6} 10^{-7} 10^{-8} 10^{-9} 10^{-10}

Elementary Particles

An Organ | A Cell | Molecules

tion, you can replace the "equals sign" by the sign ≈, which is read "approximately equal to." So you may write that

Volume of earth ≈ 1.04×10^{21} cubic meters

But what you are after is the mass of the earth. Because each cubic meter is estimated to have a mass of 3,000 kilograms, you multiply to get

Mass of earth ≈ $3,000 \times 1.04 \times 10^{21}$ kilograms

Or, remembering that 3,000 is the same as 3×10^3,

Mass of earth ≈ $3 \times 10^3 \times 1.04 \times 10^{21}$ kilograms

Now what is this number? First, you know that things can be switched around:

$3 \times 10^3 \times 1.04 \times 10^{21} = 3 \times 1.04 \times 10^3 \times 10^{21}$

Next, you can multiply the two numbers on the left, $3 \times 1.04 = 3.12$. Then you can multiply the two powers of ten, $10^3 \times 10^{21} = 10^{24}$. So you get, finally,

Mass of earth ≈ 3.12×10^{24} kilograms ≈ 3,120,000,000,000,000,000,000,000 kilograms

This estimate of the mass of the earth is not bad. Actual measurements of the earth's mass show the mass is actually about twice as large as you calculated—the measured mass of the earth is 5.98×10^{24} kilograms. How can you think about such a number? First, you have to think of something that weighs about 6 kilograms (that's about 12 pounds). A cat, for instance, weighs about 6 kilograms. The mass of the earth is heavier by 10 multiplied together 24 times—or a million, billion, billion times heavier.

The same kind of scientific notation is also used for small numbers. First remember the definition

$0.1 = \dfrac{1}{10} = 1/10$

$0.01 = \dfrac{1}{100} = 1/100$

(Sometimes "one-tenth" is written .1 and sometimes 0.1. It's always useful to put a zero before the decimal point in case the point is not seen clearly.) The mathematical rule now is to write those numbers as

$0.1 = \dfrac{1}{10} = 1/10 = 10^{-1}$

$0.01 = \dfrac{1}{100} = 1/100 = 10^{-2}$

According to that rule, when 10 appears in the denominator (below the line) with an exponent it can be moved to the numerator (above the line) if the *sign* of the exponent is changed: plus to minus, or minus to plus. This also means that

$\dfrac{1}{0.1} = \dfrac{1}{10^{-1}} = 10^1$

$\dfrac{1}{0.01} = \dfrac{1}{10^{-2}} = 10^2$

The nice thing about the system is that the multiplication rule found earlier still works:

$$10{,}000 \times \frac{1}{100} = \frac{10{,}000}{100} = 100$$

becomes

$10^4 \times 10^{-2} = 10^2$

You merely take the *algebraic* sum of the exponents (keeping track of minus signs). Only one more trick is needed. If you want $1/10 \times 10$, you would say it's the same as $10^{-1} \times 10^1 = 10^0$. What is 10^0? It must be just 1. Ten taken *no times at all* is simply 1. So things work like this:

$$10 \times \frac{1}{10} = 10^1 \times 10^{-1} = 10^0 = 1$$

$$100 \times \frac{1}{100} = 10^2 \times 10^{-2} = 10^0 = 1$$

$$1{,}000 \times \frac{1}{1{,}000} = 10^3 \times 10^{-3} = 10^0 = 1$$

$$1{,}000 \times \frac{1}{10} \times \frac{1}{100} = 10^3 \times 10^{-1} \times 10^{-2} = 10^0 = 1$$

Consider just one more example of how you might use this scientific notation. How long would it take a photon of light to travel a distance equal to the diameter of the earth? The time it takes to travel is the distance divided by the speed:

$$\text{Time} = \frac{\text{distance}}{\text{speed}}$$

The diameter of the earth is 12.8×10^6 meters. And the speed of a photon is 300,000,000 meters per second, or 3×10^8 meters per second. So the time would be approximately

$$\text{Time} = \frac{13 \times 10^6}{3 \times 10^8} = \frac{13}{3} \times \frac{10^6}{10^8} = 4.3 \times 10^{-2} \text{ second}$$

or

$\text{Time} \approx 4 \times 10^{-2} \text{ second} \approx 4/100 \text{ second}$

The time is less than the blink of an eye!

3 | Measurements and Numbers

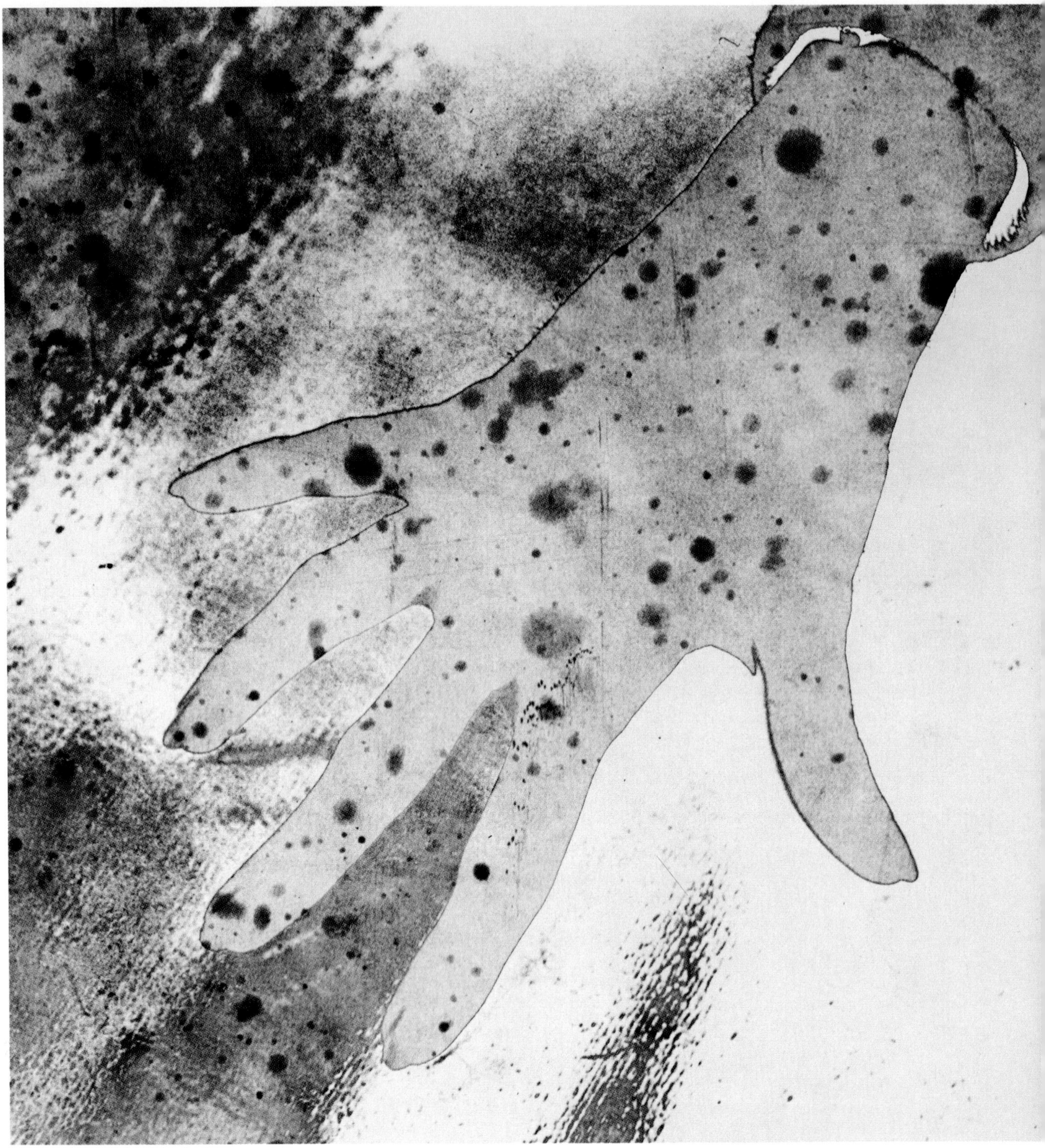

The idea that matter is composed of atoms is as much a part of the intellectual outlook of modern man as the Copernican model of the solar system. Nevertheless, neither of these ideas seems to be in accord with direct experience; a rock or a piece of glass seems continuous and solid to our touch, and the sun seems to circle the earth in its daily path from east to west.

The point is that a single fact or observation often can be explained a number of different ways. When only a few facts are known, there may be no clear reason to prefer one kind of explanation over another. We can account for the apparent solidity of a rock by assuming that matter is fundamentally continuous or by assuming that it is made of atoms so small and so close together that we are not aware of the discontinuities. Many theories can give a plausible explanation of this one fact, but as the variety and accuracy of observations increase, some theories become strained and less believable.

In this chapter you will be tracing the intellectual history of atomism to see why it finally came to be so widely accepted. You will find that two processes were at work: new observations and experiments helped bolster the atomic hypothesis, but at the same time the new facts often showed that the prevailing conception of atoms and how they behaved needed to be changed. When only a few qualitative observations about matter needed explanation, the model of the atom was simple and relatively featureless. Later, as a wealth of measurements accumulated, the model that explained things best was one with an intricate inner structure. However, you will see that even though the models became more complex, the principles involved became simpler and more universal. Using the modern picture of the atom, you do not need one theory to explain the behavior of hydrogen and another to explain the behavior of gold—the same principles apply to all atoms. This is what people generally mean when they say that nature is "simple." The study of nature would not be worthwhile if there were no general principles. All we would have would be a catalog describing every experiment that had ever been done, and there would be no way of telling what might happen in a new situation that had never been tried before.

EARLY PICTURES OF MATTER

Water freezes to a hard solid; a log burns down to a small pile of ashes; salt disappears when stirred in water and reappears when the water is boiled away. Matter can undergo a bewildering variety of change, and the early theories of matter were primarily concerned with accounting for such changes in a qualitative and descriptive way.

About 400 B.C. the Greek philosopher Democritus taught that matter was empty space populated by a vast number of tiny indivisible particles called *atoms*. He wrote that our senses give us imperfect knowledge of the world, and that atoms and emptiness were the only reality. Because only a few sentences from Democritus' writings have survived, it is hard to tell to what extent he applied his idea of atoms to explaining nature. His views were enlarged by Epicurus, who suggested that substances differed in their properties because every substance was made from atoms of a particular shape and size. For example, water was fluid

4
The Atom Hunters

Much of the story of human existence is deeply embedded in our continuing search for explanations of the nature of the physical world.

Figure 2. Epicurus (left) attempted to provide intelligent, practical theories for his observations of the world. Unfortunately, in his era there were almost as many theories as observations.

Figure 3. What intuitions did Democritus (middle) have about the universe? How could he know about the nature of atoms in empty space so early in the history of science?

Figure 4. Aristotle (right) was a powerful, dynamic thinker. His philosophy of the world seemed so self-contained and appealing that it took the unrelenting progress of many centuries of scientific thought and observation to shake its foundations.

because its atoms were round and slippery, and iron was a hard metal because its atoms were jagged and could hook together.

Epicurus' model of the atom really didn't explain very much. If every material has its own kind of atom with its own special properties, we have merely traded one set of facts for another. To explain that blue glass is different from red glass because its atoms are blue instead of red is not much of an accomplishment. It was hard to predict anything with Epicurus' atomic model because there were no general principles relating one kind of atom to another; the theory was overelaborate.

Epicurus' atomic theory was meant to be a description of ultimate reality, but it had so little predictive power in explaining matter and change that it remained a philosophical speculation. Atomism had few followers and played no important role in European scientific thought until it was revived in a new form in the seventeenth century. From ancient times through the Middle Ages the ideas of Aristotle dominated science, and Aristotle did not believe in atoms.

Aristotle held that everything was made of four basic elements: earth, air, fire, and water. He was not the first to advance this idea, but he built it into a coherent picture of chemical and physical change. According to Aristotle's ideas, substances had different properties because they contained different proportions of the four basic elements. For instance, because steam was hot and wet it was considered to be a mixture of fire and water. One substance changed to another because the relative proportions of the basic elements had changed. For example, adding fire to water produced steam, which combined the qualities of heat and moisture.

Aristotle was not an atomist. If you believe in atoms you have to believe there is empty space between the atoms. But Aristotle didn't believe that empty space could exist. He had no vacuum pumps, and every corner of the world around him seemed crammed with matter. To Aristotle, matter was perfectly continuous with no intervening gaps; matter was a fine mixture of the four elements.

Aristotle's idea that all matter was made of only a few different basic materials represented a considerable economy of thought over the vast variety of Epicurus' atoms and was a significant generalization from observation. His picture of chemical change offered a useful guide to the alchemists, and it is not surprising that his views prevailed for such a long time.

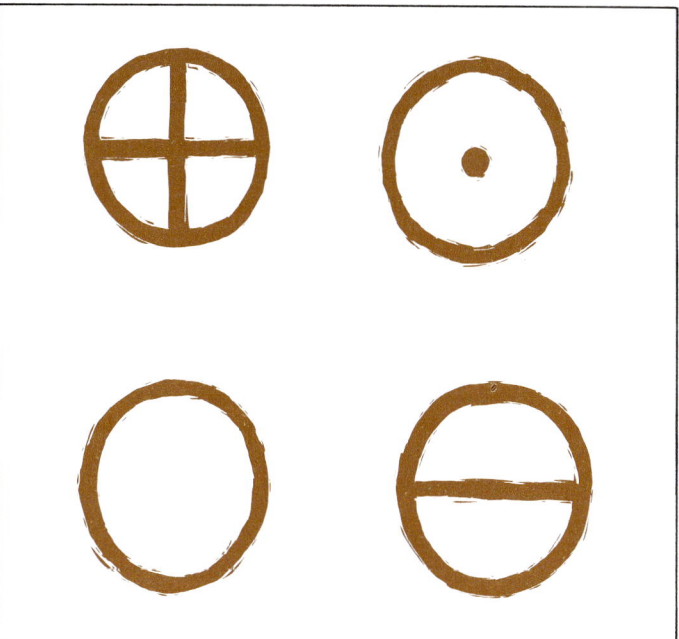

Most of Aristotle's writings were unknown to early medieval Europe because they were lost in the political and economic upheavals that accompanied the breakup of the Roman Empire. Ancient learning had been preserved by the Arabs, however, and while Europe was recovering its energies the Arab world enjoyed a period of brilliant accomplishments in chemistry, physics, and mathematics. Finally, in the twelfth century, Arabic translations of Greek writings filtered back to Europe, but they were accompanied by the Arabs' own contributions.

The Europeans were especially influenced by Arab experimentation in chemistry, which was a mixture of observable fact and magical doctrine known as *alchemy*. Alchemy promised great powers—including the power to change lead into gold. According to Aristotle's teachings, it was not unreasonable to expect that lead could be turned into gold; all that had to be done was to find a way of treating lead to take away some of its earth nature and add a little more fire. Anything could be changed into anything else merely by readjusting the relative proportions of the four elements.

Figure 5. (Left) Epicurus had some interesting ideas about atoms. It was obvious that there were many different kinds, but he could visualize these differences only within the framework of the physical world that he could see directly.

Figure 6. (Right) Each of the four Aristotelian elements had its own sign. These symbols were still in use well into medieval times.

THE CHEMICAL ELEMENTS AND ATOMIC THEORY

For all their mysticism and magic, the alchemists built up considerable knowledge concerning chemical reactions. By the middle of the seventeenth century it was clear to Robert Boyle that the Aristotelian concept of four basic elements was insufficient to explain the body of chemical knowledge. For one thing, no alchemist had ever been able to change lead to gold. In addition, it was apparent that certain substances such as sulfur and mercury played a special role in chemistry. Although these substances took part in chemical reactions to form new compounds, they themselves had never been resolved into other distinct substances by

Figure 7. Robert Boyle was not one to accept without question the concepts of his time; he rattled the status quo with his attack on the Aristotelian belief in four basic elements. With the use of simple but precise equipment (right), he helped reinstate atomism.

Figure 8. The meticulous construction of Antoine Lavoisier's experiment to show that air has weight.

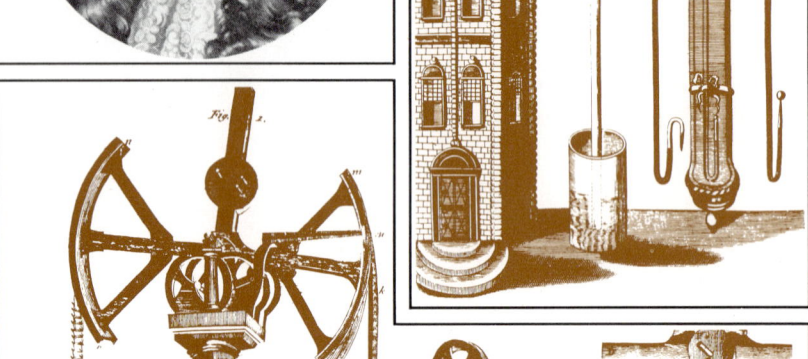

Figure 9. Lavoisier in his laboratory carefully measured the respiration of men in a state of rest; his questions about the nature of the world extended to his fellow man. But genius was not necessarily rewarded in those days, and Lavoisier was one of the few chemists to be guillotined during the French Revolution.

any of the chemical procedures available. Boyle called these fundamental substances *elements,* and suggested that elements could combine to form *chemical compounds.*

At first sight, Boyle's definition of a chemical element seems a little vague. How is it possible to tell whether a substance is broken down as far as possible? For instance, the later chemists Humphry Davy and Joseph Gay-Lussac couldn't agree whether chlorine was an element or not. Furthermore, there was no guarantee that something looking very much like an element today might not be resolved further by a new technique tomorrow. Nevertheless, Boyle's ideas were very much in the modern spirit. Instead of starting from principles that had no connection with experimental evidence, as Aristotle and the scholastics of the Middle Ages had done, Boyle tried to generalize from experiment. Instead of speculating on how many elements there might be, he looked to chemical experiment to find out how many there actually were. Boyle's work prompted a vigorous hunt for elements.

Atomism returned to respectability with Boyle. The invention of the barometer and Boyle's experiments with simple vacuum pumps had shown that Aristotle was wrong in saying that vacuum could not exist. Once the philosophical objection to emptiness had been removed, Boyle was free to infer that matter was made of atoms. It remained an inference, however, for experimental proof was to come much later.

A hundred years after Boyle, Antoine Lavoisier introduced a powerful tool for the hunt for elements: the systematic use of the chemist's balance. Lavoisier's fundamental insight was that matter was not spontaneously created in chemical change. If one substance *gains* weight in chemical change, another substance must *lose* the same amount. Because an element cannot be broken into simpler substances, it must always gain weight by combining with other materials when it engages in chemical change. Lavoisier replaced conjecture about the elements with a quantitative test.

Figure 10. This experiment shows that matter is not destroyed in combustion. (How do you think the candle was lit within this sealed glass chamber?)

Figure 11. John Dalton, an English chemist and physicist whose work enabled the abstract theories of atomism to be tested in the laboratory.

Lavoisier's outstanding accomplishment was his explanation of combustion. The Aristotelians held that burning always resulted in loss of weight; a log, for instance, always burned down to a pile of ashes. Lavoisier took a systematic approach to the problem. He burned simple substances such as sulfur and mercury in sealed containers, and by careful weighing he established that the products of combustion weighed *more* than the original substance. More importantly, he found that the added weight was accounted for by the loss in weight of the air that was originally in the chamber.

In all the combustion experiments, however, not more than about 20 percent of the air was used up. The reason, which John Priestley in England helped supply, was that air was a *mixture* of elements, and only one of the elements, named oxygen, was involved in combustion. Lavoisier's method had led to the discovery of oxygen and to a way of understanding chemical change. Burning and rusting could now be understood as chemical reactions between oxygen and other substances. This way of looking at chemical change originated with Lavoisier. For instance, in 1783 Henry Cavendish burned hydrogen in oxygen and noticed that water was formed, but he could not explain why. The reason was clear to Lavoisier—water was not an element, as Aristotle had said, but was a compound formed by the reaction of hydrogen and oxygen.

THE CHEMICAL ATOM

Lavoisier laid the foundations of modern chemistry through his explanation that chemical change was due to chemical reactions between elements and compounds. He showed the importance of weighing and measurement, and by the beginning of the nineteenth century some 40 elements had been identified, about half the number known today. Nevertheless, atomism played no essential role in any of this activity. Atomism remained a hypothesis for which there was no clear experimental proof.

Combining Weight and Atomic Weight

Lavoisier's weighing techniques helped reveal the rules that govern the way elements combine, and chemists summarized these rules in the concept of *"combining weights."* Lavoisier had shown in his experiments on water that a given weight of hydrogen always combines with about 8 times its own weight of oxygen to form water. Taking hydrogen to be unity by convention, the combining weight of oxygen was set at 8. Correspondingly, the combining weight of any element was the weight that combined with, or displaced from some other combination, a unit weight of hydrogen.

John Dalton in 1808 realized the significance of combining weight for atomism. He argued that the weights measured by the chemists with their chemical balances mirrored the weights of the atoms. According to his viewpoint, oxygen had a combining weight of 8 because the atom of oxygen was 8 times heavier than the atom of hydrogen. Furthermore, two elements A and B could combine to give only the following compounds:

Compound 1: 1 atom of A + 1 atom of B
Compound 2: 1 atom of A + 2 atoms of B
Compound 3: 2 atoms of A + 1 atom of B

and so on. The relative weights of A and B in these compounds can be inferred from their composition, if the relative weights of the A and B atoms are known. Suppose, for example, that the atom of A is twice as

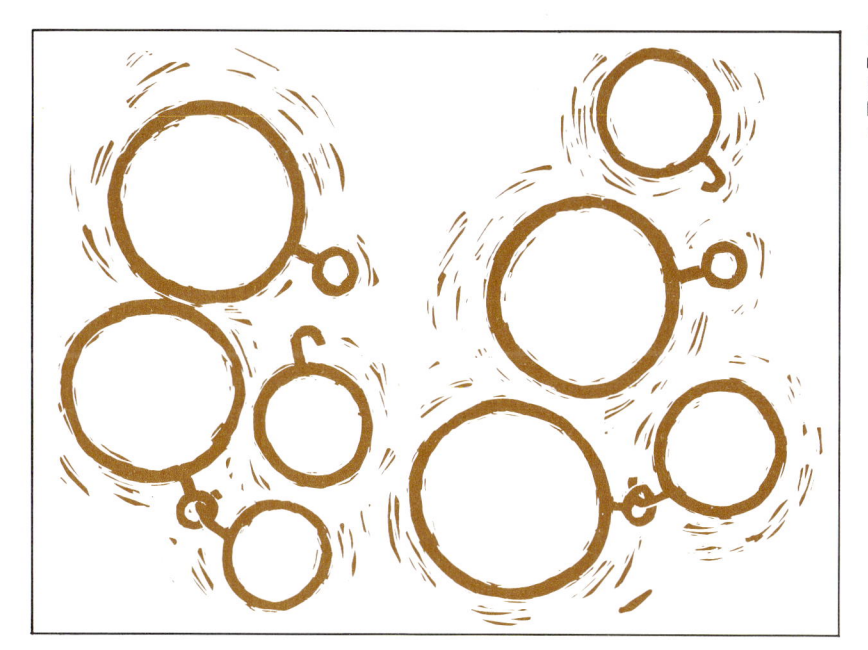

Figure 12. This illustration shows Dalton's model of the atom. It had hooks and eyes for chemical combination and was surrounded by a shell of heat, which might be given off when elements combined.

Atomic Weight of Nitrogen (N) = 14

Atomic Weight of Oxygen (O) = 16

Examples:
1 Atom N + 1 Atom O → 1 Molecule NO (Nitric Oxide)
(14 grams N react completely with 16 grams O to form 30 grams NO)

1 Atom N + 2 Atoms O → 1 Molecule NO_2 (Nitrogen Dioxide)
(14 pounds N react completely with 32 pounds O to form 46 pounds NO_2)

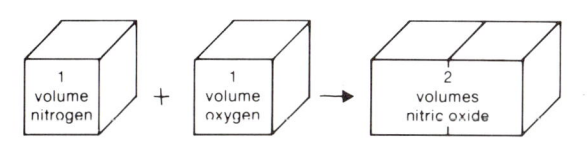

Figure 13. Results from chemical analyses enabled Dalton to deduce the relative weights of atoms of the 30 or so elements that were known at the time. The hydrogen atom proved to be the lightest, so he arbitrarily set its weight equal to 1.

heavy as the atom of *B*. Then in forming compound 1, one weight of *B* would combine with two weights of *A*; in forming compound 2, one weight of *B* would combine with one weight of *A*.

Dalton brought atomism into the laboratory and made it an important working tool, and he is considered to be the father of modern atomism. However, there was a gap in his reasoning. Water was the only compound of hydrogen and oxygen known, and he assumed, on grounds of simplicity, that it consisted of one atom of hydrogen and one atom of oxygen. Simplicity is a useful guide in science, but it is not a standard of proof.

The problem became more acute when careful studies were made of the reaction of nitrogen and oxygen to form the colorless gas nitric oxide. Dalton assumed—correctly, as it later turned out—that a nitric oxide molecule consisted of one atom of nitrogen and one atom of oxygen.

1 atom nitrogen + 1 atom oxygen ⟶ 1 molecule nitric oxide

(N + O ⟶ NO)

But when nitrogen and oxygen react completely to form nitric oxide, the *volumes* of gases involved are

1 volume nitrogen + 1 volume oxygen → 2 volumes nitric oxide

4 | The Atom Hunters

Because the nitrogen and oxygen atoms paired up to form nitric oxide, it was difficult to see how two volumes of gas could be produced. In fact, in 1811 Amadeo Avogadro conjectured that *equal volumes of any gases under the same conditions of temperature and pressure were physically very much alike, and contained the same number of independent molecules.* Avogadro's hypothesis only made the problem worse, and Dalton refused to accept it. Avogadro then added a new idea: that some gaseous elements might be made of molecules consisting of *several* atoms. Now everything worked fine:

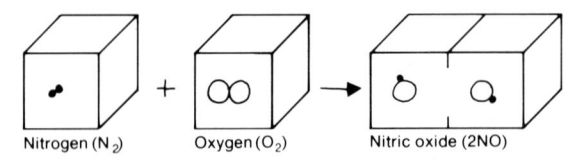

It was known that one volume of oxygen combined completely with two volumes of hydrogen to make two volumes of water vapor. If you believe Avogadro's ideas, here is how you must picture the situation:

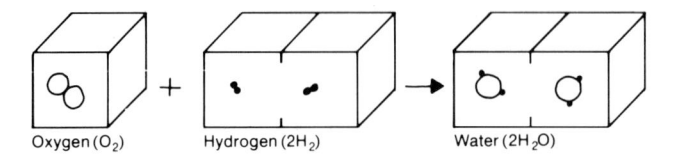

Figure 14. The Italian physicist Amadeo Avogadro perceived that equal volumes of different gases under identical conditions of pressure and temperature contain the *same* number of molecules.

A water molecule must consist of two atoms of hydrogen and one atom of oxygen, H_2O. Avogadro was right, but his hypotheses were not fully accepted until there was substantial independent proof of their correctness, and that would be some time in coming.

If you are confident that one oxygen atom combines with two hydrogen atoms to form water, an oxygen atom must be 16 times heavier than a hydrogen atom because the combining weight of oxygen in a water molecule is 8. By extending the measurement of combining weights and gas volumes to a wide variety of compounds, chemists were able to determine the relative atomic weights of the elements.

In chemistry, only the relative weights of the atoms are important; the actual weight of an atom never enters into it. The modern scale of atomic weights is set up relative to carbon, which has been assigned atomic weight 12. Carbon was chosen as the standard because it forms an incredibly large number of chemical compounds; accordingly, the determination of relative weights of compounds is made easier.

Atomism was not readily accepted in the nineteenth century. Although atoms were useful in explaining a few of the principles of chemistry, the average chemist could go about his work with very little thought about atoms. The theories of Dalton and Avogadro were too limited in scope to help a chemist who was trying to understand the details of a chemical reaction. The Daltonian atom had no inner structure that might explain why hydrogen and chlorine reacted violently or why hydrogen and sodium hardly reacted at all. However, you may leave the inner structure of the atom for later chapters and concentrate here on simple properties of atoms.

It is not worthwhile to try to discuss modern atomism in strict historical order. Progress in science has been so rapid in the last hundred years, and lines of inquiry have overlapped so much, that it is easier to focus on single topics such as the number and the size of atoms.

Most physical scientists don't spend much time studying the history of science. Philosophers still find the writings of Plato, Aristotle, and

Kant useful, but beginners in physics don't start with the works of Galileo or Isaac Newton. Nevertheless, Newton's thoughts continue to be important in physics. The point is that science progresses, and physicists after Newton have significantly added to and clarified his ideas. Unless a scientist has a special interest in history, he seldom has reason to look further than modern books and recent research papers; these are built upon all the work that has gone before. In an active field of research, papers more than 10 or 15 years old generally retain little interest because of advances in theory and experimental technique. There is the danger that good ideas can be lost for a time. Physical science is closely tied to experiment, however, and an idea that is helpful in explaining observation ultimately will be reinvented.

The Number of Atoms and Avogadro's Number

How many atoms are there in a glass of water? The number must be very large, because the tiniest drop of water still looks like water—you can't see the individual atoms. The smallest drop of water you can see under a microscope is about 10^{-4} centimeter across, and its volume is roughly 10^{-12} cubic centimeter. You may conclude that 10^{-12} cubic centimeter of water contains many atoms. A glass of water has a volume of about 200 cubic centimeters, which is enough water to make up 2×10^{14} of the tiniest drops you can see with a microscope. Because each drop has many atoms, you know that there must be considerably more than 2×10^{14} atoms in a glass of water. This value represents a *lower limit*. The number of atoms cannot be less than this, according to the observations, but it could be much greater. Nevertheless, this crude estimate shows that even a small chunk of matter must contain an immense number of atoms.

Figure 15. One of the greatest scientists of all time was Isaac Newton. He conceptualized a framework in which to view the world and expanded that framework with precise analytic techniques.

To make a better estimate of the number of atoms, go back to the idea of atomic weight. The chemist found, for example, that the relative atomic weight of carbon is 12 and the relative atomic weight of sulfur is 32. Suppose you take 12 grams of carbon and 32 grams of sulfur. Because the weights of the individual atoms are in the same ratio (12 to 32), 12 grams of carbon and 32 grams of sulfur contain the *same* number of atoms. Similarly, 12 pounds of carbon and 32 pounds of sulfur also contain the same number of atoms.

The relative atomic weight of carbon is 12, and the number of atoms in 12 grams of carbon is called *Avogadro's number.* The amount of substance containing Avogadro's number of atoms is called *one mole.* One mole of carbon weighs 12 grams, and one mole of sulfur weighs 32 grams.

The mole is only a shorthand expression for a certain number of atoms, and it can be applied to compounds as well as to elements. For example, the molecular weight of water, H_2O, is $1 + 1 + 16 = 18$, so 18 grams of water make one mole and contain Avogadro's number of water molecules.

In its ordinary gaseous state, hydrogen consists of molecules made up of two atoms. Therefore, 2 grams of hydrogen molecules make one mole of hydrogen molecules, but if you wanted a mole of hydrogen atoms you would only need 1 gram of hydrogen gas. It doesn't matter that hydrogen atoms can't easily exist alone—it's just a matter of counting the particles you are interested in. A mole of sand grains is just Avogadro's number of sand grains.

Avogadro's number is very large; there are about 6×10^{23} atoms in a mole of atoms. You can see now why the atomic nature of matter is not apparent to our senses. Even a tiny piece of matter has an immense

Measuring a Film of Oil on Water

When oil is spilled into water, it floats on top and spreads out over a large area. How large an area can be covered by a liter of oil, assuming that the oil film is uniformly one molecule thick?

<u>Step 1.</u> A typical oil molecule consists of a long backbone of carbon atoms, with hydrogen atoms branching off to the sides. The basic building block of an oil molecule is one carbon atom and two hydrogen atoms:

How large an area does one of these building blocks cover? You can make a good estimate by assigning each atom a diameter of about 10^{-8} centimeter and assuming an approximate 10^{-8} centimeter separation between neighboring atoms. So the area covered by each basic unit is approximately:

$$\text{Area} = (5 \times 10^{-8} \text{ centimeter}) \times (2 \times 10^{-8} \text{ centimeter})$$
$$= 10^{-15} \text{ square centimeter}$$

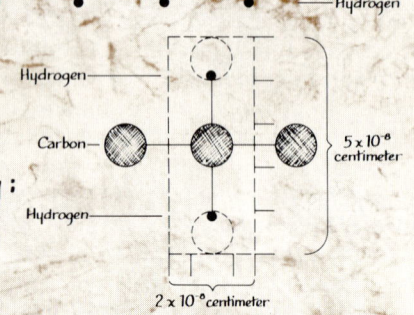

<u>Step 2.</u> How many basic units are there in one liter of oil? One liter is 1,000 cubic centimeters, and 1,000 cubic centimeters of water weigh approximately 1,000 grams. Oil, which is less dense than water, has a specific gravity of 0.9. So one liter of lubricating oil weighs 0.9 as much as one liter of water:

$$\text{Weight of one liter of oil} = 0.9 \times 1,000 \text{ grams}$$
$$= 900 \text{ grams}$$

One carbon atom has a relative atomic weight of 12, and one hydrogen atom has a relative atomic weight of 1. So

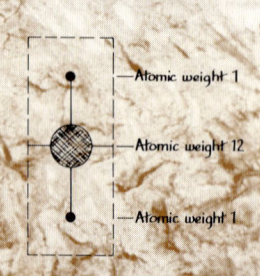

$$\text{Relative weight of oil molecule} = 12 + 1 + 1$$
$$= 14$$

This means that 14 grams of oil contain Avogadro's number of basic units. Taking Avogadro's number to be 6×10^{23}, the total number of basic units in 900 grams of oil is

$$\frac{900 \text{ grams}}{14 \text{ grams}} \times (6 \times 10^{23}) = 3.9 \times 10^{25} \text{ basic units}$$

<u>Step 3.</u> Each unit covers 10^{-15} square centimeter, according to the estimate from Step 1. Therefore, the 3.9×10^{25} units in one liter of oil cover a total area of

$$(3.9 \times 10^{25} \text{ units}) \times (10^{-15} \text{ square centimeter/unit}) = 3.9 \times 10^{10} \text{ square centimeters}$$

This area is equivalent to 3.9×10^{6} square meters, or about a square mile.

number of particles in it, and individual atoms must be very small. The glass of water discussed earlier, 200 cubic centimeters in volume, has about 7×10^{24} water molecules. This is much larger than the crude lower limit of 2×10^{14} molecules based on simple microscopic observation, but both numbers are too big to comprehend and even the rough estimate shows that atoms are numerous indeed.

There are many ways to measure Avogadro's number. It can even be estimated from the fact that the sky is blue, because you saw in Chapter 2 that the blueness of the sky depends on how many atoms there are in the air.

A direct way of measuring Avogadro's number would be to count the number of atoms in a known weight of material, but atoms are so numerous that the method is not feasible. The smallest drop of water that can be handled easily is about 10^{-6} gram, but even a drop this small has 3×10^{16} molecules in it. Although it is possible to count individual molecules under the proper circumstances, it is difficult to count more than 10^8 per second. Because a year has about 3×10^7 seconds, it would take 10 years of steady high-speed counting to count 3×10^{16} molecules.

A more practical way to find the number of atoms in a piece of matter is to use the fact that the atoms in a crystal are spaced in a remarkably regular way. Crystals can therefore act much like the diffraction gratings discussed in Chapter 2, but they work best with x-rays because the wavelength of x-rays is comparable to the spacing of the atoms. By measuring the diffraction pattern the spacing of the atoms can be found, and then it is easy to calculate the number of atoms in a crystal of known size.

Knowing Avogadro's number enables you to determine how much an individual atom weighs. For example, because 6×10^{23} hydrogen atoms weigh 1 gram, a single hydrogen atom must weigh 1.7×10^{-24} gram; viruses, the smallest living things, are at least 10^7 times heavier. Life is complicated — it seems to take at least 10^6 or 10^7 atoms to form something "living," but even then a virus needs parts from a host cell to reproduce itself.

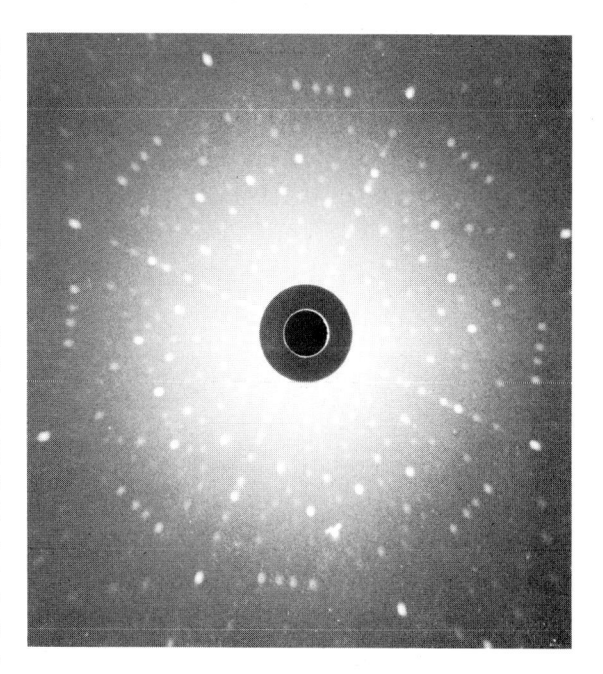

Figure 16. The internal ordering of crystals gives rise to these symmetric x-ray diffraction patterns. From such patterns, the spacing between atoms can be determined.

THE SIZE OF ATOMS

If you want to know the size of a room, you lay out a measuring tape and read off the number of feet and inches from one wall to the other. But rulers won't do for finding the size of an atom, and less direct methods have to be used.

According to the estimate in the above section, 200 cubic centimeters of water contain about 7×10^{24} molecules. Each molecule therefore has 3×10^{-23} cubic centimeter to itself, and that is the volume of a cube 3×10^{-8} centimeter on a side. This represents an *upper limit* on the size of a water molecule, for it neglects the empty spaces between the molecules. Unlike gases, however, water doesn't compress or expand easily, so the spaces between the molecules are not very big.

Look at a pure element such as silver. The atomic weight of silver is 108, so 108 grams of silver contain 6×10^{23} silver atoms. Now 108 grams of silver has a volume of 10 cubic centimeters, so each silver atom occupies 1.7×10^{-23} cubic centimeter. This works out to a size of 2.6×10^{-8} centimeter. All atomic sizes found this way turn out to be about 10^{-8} centimeter or so.

What does size really mean? In the everyday world you gauge the size of doors or chairs by looking at them or by touching them. A door seems to have well-defined edges and a definite size when you look at it. But this is true because the wavelength of light is short, about 5×10^{-5} centimeter. Light rays passing more than a few wavelengths from the

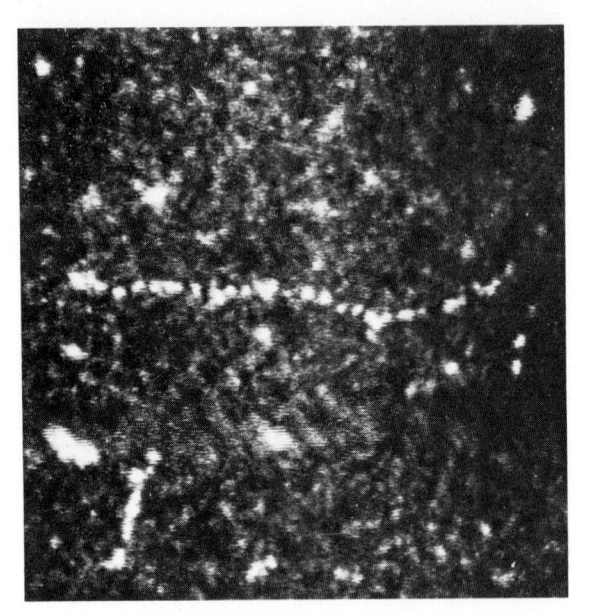

Figure 17. This is a thorium-BTCA chain that was sprayed onto a 25-angstrom-thick evaporated carbon film. The small spots along the chain represent single atoms of thorium. This remarkable photograph was taken with a high-resolution scanning electron microscope in the laboratory of Dr. A. V. Crewe.

edge of the door do not interact with it, and on the human scale the slight blurring of the light near the edge is not noticeable.

The edge of the door seems solid and distinct when you touch it, but this merely reflects the behavior of the forces between atoms. The atoms in the door do not exert strong forces on the atoms in your hand until the separation is about 10^{-8} centimeter or so. The distance is so small and the force increases so rapidly in strength that you have no trouble telling where the edge is.

On the atomic scale, however, the everyday concept of size isn't too useful. You have to be much more careful about what things really mean. Estimation of atomic volumes gives one set of sizes but other approaches give different values. One way of measuring an atom's size is to throw things at it. A boy playing marbles won't hit another marble unless his shot rolls close enough. An atom flying by another atom won't hit something and be knocked aside unless it gets near enough. Sizes found this way are really a measure of the size of the region over which an atom can exert significant pushes or pulls on another particle. The size of an atom "seen" by an electron is different from the size "seen" by another atom, because electrons and atoms do not push or pull each other the same way atoms push or pull other atoms.

SEEING ATOMS

Is there any way to see an individual atom? According to the usual definition of seeing, the answer is no. An object can't be seen unless it is big enough to disturb a light wave noticeably. By analogy, a small rowboat on the ocean barely affects the pattern of water waves but an ocean liner or an island changes them significantly. Because atoms are about 5,000 times smaller than the wavelength of visible light, ordinary optical microscopes can never allow us to see an atom. Nevertheless, a number of devices and experiments can make atoms seem so immediate that it is almost as good as seeing them.

One method is to use a microscope that does not depend on visible light. It is not difficult to produce x-rays that have wavelengths smaller than the size of an atom, but x-rays can't be focused easily—they pass right through glass lenses without being bent. Electrons seem to be the best choice for a high-magnification microscope. Electrons are much smaller than atoms, and electron beams can be focused by carefully designed magnets.

It is difficult to bring objects into sharp focus, however. Until recently no electron microscope was capable of showing individual atoms. An improved type of electron microscope constructed at the University of Chicago is capable of focusing an electron beam down to a spot 5×10^{-8} centimeter in diameter, comparable to the size of a large atom. In operation, the tiny spot of electrons is swept back and forth across a small area of the sample to be studied. At each position, the relative number of electrons bouncing back from the sample is recorded, and a kind of television picture of the area scanned is built up. Heavy atoms are particularly effective in bouncing back electrons, and the microscope has been able to show individual uranium and thorium atoms lying on a carbon plate.

There is another way to "see" atoms. As you recall from Chapter 1, the molecules of gases and liquids are in constant motion. They continually jostle one another, like people moving aimlessly in a crowd. The motion can be made visible by mixing specks of matter, such as some carbon particles from a drop of India ink, with water. The specks act like oversize, but visible, molecules, and their random motion can be followed under a microscope as they are pushed from place to place by the water molecules.

The effect, known as Brownian motion, was first noticed by Robert Brown in 1827 while he was looking at pollen grains in liquid. Brown was unfamiliar with the idea that atoms in liquids and gases are in motion, and he could not explain why his pollen grains never quieted down. Albert Einstein devised a mathematical theory for Brownian motion in 1905, the same year in which he invented the special theory of relativity. Because molecular motions are random, a pollen grain in water will be bumped unequally, with more pushes on one side than another. To calculate the zigzag motion of a pollen grain in detail you would have to (and could not) know the motions of all the billions of water molecules, but Einstein was able to use simple atomism to find the average distance the grain would move in a given time. Einstein showed that the smaller the particle, the farther it would go. When you float in a swimming pool, your body is being jiggled by the water molecules, but you are so big compared to the molecules that the motion is too small to be noticed. Careful experiments verified Einstein's calculation and gave additional support to the atomic theory.

The discovery of radioactive decay presented clear evidence that matter was made of particles. Some kinds of atomic nuclei are unstable and change spontaneously from one form to another. As they do so, they can emit particles of three different kinds, depending on the type of decay. The identity of the particles was not recognized at first, and they were arbitrarily named "alpha," "beta," and "gamma" rays. Soon, however, it was found that the alpha rays were nuclei of helium, the beta rays were electrons, and the gamma rays were photons. The important point is that each particle was ejected with such energy that a single particle could be detected easily. It takes energy to do things, and if a particle has a great deal of energy it can be made to induce physical or chemical changes to show its presence.

Here is a simple experiment you can do to see how particles are detected. At night, after your eyes have become adapted to the dark, hold the luminous dial of a watch directly in front of one of your eyes. Although you will not be able to focus on the dial, you should see blurred pulsations of light. What is happening is that radioactive material in the dial paint is emitting particles, and as each particle collides with molecules of the special paint, it causes the molecules to emit enough visible light to cause a momentary glow. Each flash of light corresponds to a single particle hitting the paint. Are you "seeing" atomic particles? Enough so to make the idea that matter is made of particles seem beyond further doubt.

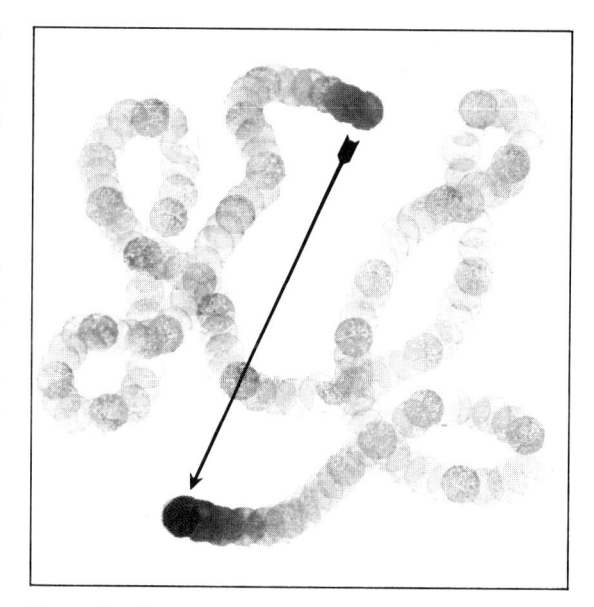

Figure 18. The motion of a pollen grain in water is random, but the *average* distance it travels in a given time can be measured and calculated.

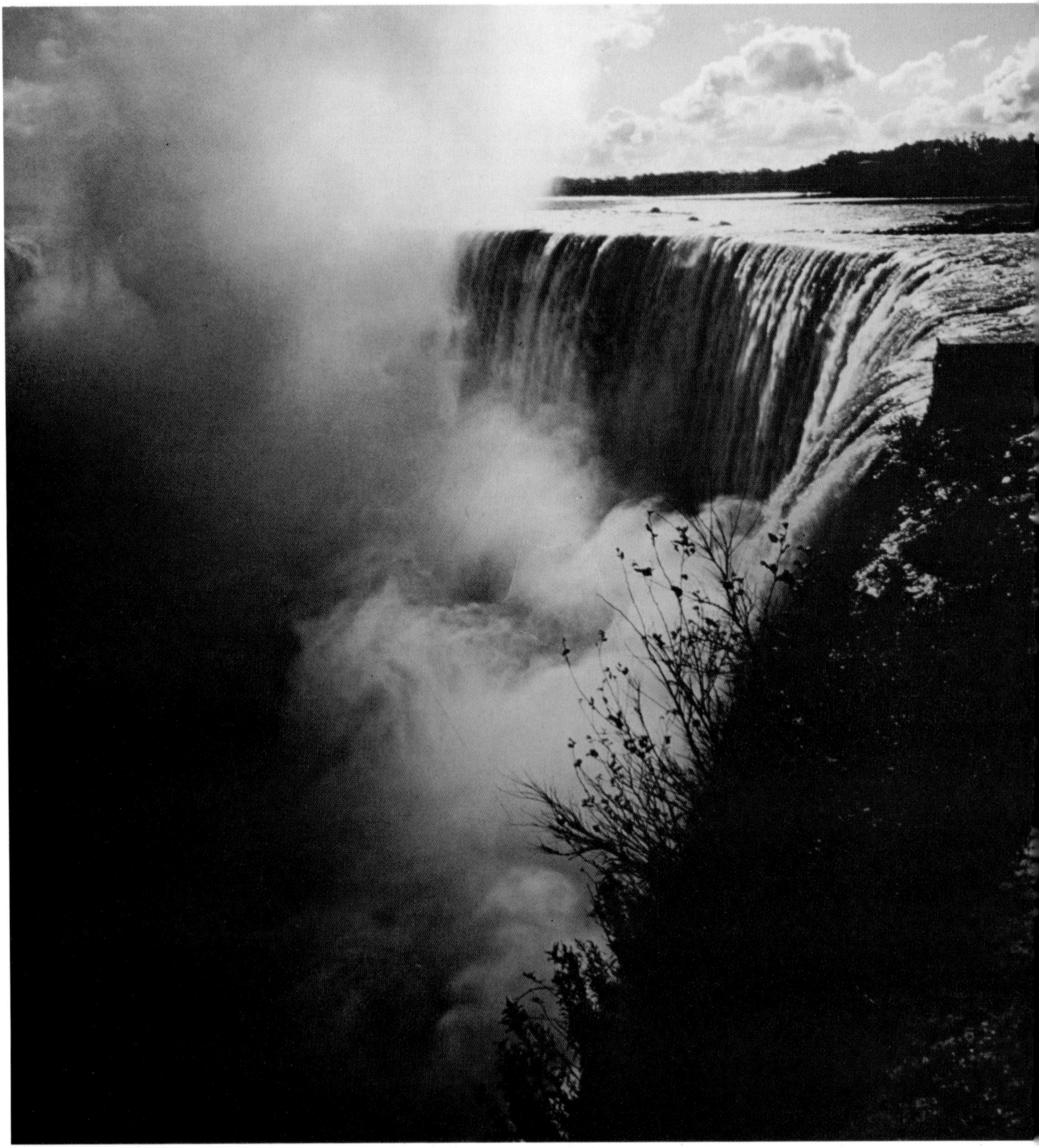

A n ordinary chunk of matter gives no outward sign that it is made of atoms. As you lie on a summer hillside listening to the roll of distant thunder, you are not directly aware of the vast numbers of atoms around you. Yet the strength of the rock supporting you, the feel of the wind, the sound from the thunder—all these things are manifestations of atoms and their forces and motions. That is why the modern atomic picture is so useful and important. It not only describes the ultimate structure of matter, it explains the properties of matter and why matter acts the way it does.

In Chapter 1 you had an orientation to the atomic description of matter, and you read how the atomic picture explains why matter can be solid, liquid, or gas. You read how the hundred different kinds of atoms can combine with one another to form the immense variety of substances in our world, ranging from water to the complex molecules of living organisms.

This chapter is devoted to sharpening your picture of atoms by looking at how atoms engage in physical and chemical change. First of all, look at some of the basic physical properties of matter—density, temperature, and pressure.

DENSITY AND THE SPACING OF ATOMS

Most of us learn through experience to judge the weight of an object by its size. This works well enough most of the time, and it is a surprise when experience breaks down. A suitcase full of books weighs a good deal more than one containing only clothes, and you are surprised to find it weighs so much when you first pick it up. Similarly, a massive gold ring seems too heavy for its size, and an aluminum ring seems too light.

Density is the amount of matter per unit volume of a given substance. In the metric system, the unit of density is the *kilogram per cubic meter.* Another commonly used unit of density is the *gram per cubic centimeter.* Density combines the ideas of size and weight; when you say that something is "heavy," you usually mean that it is unexpectedly dense—that there is a great deal of matter in a small space. A kilogram of feathers and a kilogram of lead both contain the same weight of matter, but the lead occupies a much smaller volume.

Table 1 gives the densities and specific gravities of various materials. The *specific gravity* of a substance is the ratio of its density to the density of water. The specific gravity of aluminum is 2.7, for example. Specific gravity is a ratio and has no units. However, because the density of water is nearly 1.0 gram per cubic centimeter, the specific gravity of a substance is numerically nearly equal to its density in grams per cubic centimeter. The density of water varies somewhat with temperature and pressure, and the precise definition of specific gravity states that the density of water at 4 degrees Centigrade and atmospheric pressure is to be used as the standard.

Isaac Newton realized that atomism could easily account for the difference in densities between different substances. Although each atom has a certain weight, density is related to volume as well as weight, so it must depend on how closely the atoms are spaced. A molecule of air, for example, weighs about as much as an atom of aluminum, but

5
Stuff and Substance

How do materials of the world change forms as they do? Water vapor breaks loose from the falls and winds may carry it poleward, where it may become locked in snow or ice.

Table 1.
Densities of Various Materials

Material	Density (kilograms per cubic meter)	Specific Gravity
Air	1.3	1.3×10^{-3}
Redwood	400	0.4
Water	1,000	1.0
Glass	2,500	2.5
Aluminum	2,700	2.7
Iron	7,900	7.9
Silver	10,500	10.5
Lead	11,400	11.4
Mercury	14,200	14.2
Gold	19,400	19.4
Iridium	22,800	22.8

Figure 2. This glass of a popular gelatin dessert has separated into three layers according to the different *densities* of its components.

aluminum metal is thousands of times denser than air (Table 1) because aluminum atoms are packed together more closely. Only a relatively small change in spacing can cause a large change in density. If the spacing is doubled, the density decreases by a factor of 8. Aluminum is about a thousand times denser than air and has about the same atomic weight, so the atoms of aluminum must be about 10 times closer together than the molecules in air.

Almost all the weight of an atom is concentrated in its nucleus. Because atomic spacing is about the same in all metals, a good rule of thumb is that metals of high atomic weight have higher densities than metals of low atomic weight. Gold, with atomic weight 197, has specific gravity 19.4, and aluminum, with atomic weight 27, has specific gravity 2.7. The spacing of the atoms in gold and in aluminum is nearly the same; in fact, x-ray measurements show the spacing in gold is 2.87×10^{-8} centimeter and, in aluminum, 2.85×10^{-8} centimeter. Atomic spacings in metals range from about 2.5×10^{-8} centimeter to 5.0×10^{-8} centimeter. The range is small because the metal atoms are held together by the interaction of the outer electrons in the electron clouds. Most atoms are about the same size, and to join together in a metal, the atoms must be close enough so that the electron clouds overlap a little between the atoms.

Slight changes in structure can cause large changes in densities, however. An interesting case is lead, atomic weight 207 and specific gravity 11.4, compared to gold, atomic weight 197 and specific gravity 19.4. The atomic spacing in lead is comparatively large, which is reflected by the fact that lead has a low melting point. The atoms of lead are so far apart that the additional motion brought about by an increase in temperature can easily cause the atoms to break apart from one another, and the metal melts.

The densities listed in Table 1 are the densities under ordinary conditions. But density can be changed. When you pump air into a partially inflated tire, the volume changes only a little as you add air molecules, and the density becomes greater than normal. The gas in a diver's air tank can be 100 times denser than in the atmosphere. Similarly, air can be removed from a chamber with a vacuum pump, which reduces the density to low values. The best vacuum pumps can reduce the density to about 10^{-15} times normal; at that density, there are about 30,000 air molecules in a cubic centimeter. In deep space, away from stars, there may be only one or two atoms in a cubic kilometer.

It is easy to squeeze gas molecules together because the molecules are too far apart to exert strong forces on one another. In solids or liquids, the atoms are close together and any attempt to push them still closer is opposed by strong forces between the electrically charged particles. Nevertheless, sufficiently great external forces can push the atoms together. The deep interior of the earth is primarily iron, but the density at the center of the earth is estimated to be 18,000 kilograms per cubic meter, which is much greater than the normal density of iron, 7,900 kilograms per cubic meter. Tremendous gravitational forces push the atoms together. The density at the center of the sun is 160,000 kilograms per cubic meter, yet the sun consists mainly of hydrogen.

What is the ultimate density? Atoms are largely empty space, with most of the weight concentrated in the nucleus. *The nucleus is the densest matter known.* A hydrogen atom nucleus has a radius of 10^{-15} meter and weighs 1.7×10^{-27} kilogram. Its density is therefore 4×10^{17} kilograms per cubic meter. Some burned-out stars consist of closely packed neutrons and have densities close to the ultimate.

Figure 3. Atoms in solids are much closer together than they are in gases. Solids are more than a thousand times denser.

Many of the concepts of physical science are useful only over a certain range of conditions, or in certain contexts, and they lose their meaning if pushed too far. For example, finding the density of a gas represents an *averaging* procedure that is difficult to apply unless large numbers of molecules are involved. A cubic meter of air under ordinary conditions contains about 3×10^{25} molecules. It makes sense to say there are 10^{25} molecules per cubic meter; if you take cubic meters of air at random, the number of molecules in them would not differ significantly from 3×10^{25}. The situation is different if the sampling volume is very small. Suppose you take a volume of 3×10^{-25} cubic meter instead of 1 cubic meter. The expected number of molecules in the tiny volume is three, but if you take random samples of this size your volumes might contain two, four, or sometimes six molecules—or even no molecules at all. Gas molecules are in rapid motion and by chance the number of molecules in a sample may differ considerably from the average.

Fortunately the effect becomes less significant as the sampling volume is made larger, and density is well defined when many molecules are involved. Consider a room containing a single gas atom that is moving about randomly. On the average, it will be in the left half of the room half the time. If there are two molecules, both molecules will be in the left half of the room 25 percent of the time. An ordinary room contains about 10^{27} molecules. It is extremely unlikely that all the molecules will by chance find themselves in the same half of the room; it is the same likelihood as flipping a coin 10^{27} times and having it turn up "heads" each time.

TEMPERATURE AND THE MOTION OF ATOMS

If you have ever climbed to the top of a mountain or visited a town located at a high altitude, you know that you have to breathe more deeply and more rapidly than usual. The air at high altitudes is "thin"—its density is less than it is at sea level. At the top of Mount Everest, about 8 kilometers above sea level, the air is almost too thin to support life and climbers rely on bottled oxygen during periods of great exertion.

But how can air molecules form a layer of atmosphere at all? Why don't they fall to the ground, like everything else? If you toss some sand grains into the air they simply fall back down and show no inclination to hover above the earth.

To understand the atmosphere, you have to understand gases. The behavior of gases was worked out in the last half of the nineteenth cen-

Figure 4. Rudolf Clausius, Ludwig Boltzmann, and James Maxwell—three men who worked out our ideas concerning the behavior of gases.

tury by Rudolf Clausius and Ludwig Boltzmann in Germany and James Clerk Maxwell in England. They used the atomic picture of matter at a time when the existence and properties of atoms still remained a matter of conjecture. The success of their ideas, called the *kinetic theory of gases,* was a powerful argument in favor of atomism.

The kinetic theory of gases introduced a new way of thinking about the physical world. Prior to the work of Maxwell and Boltzmann, the emphasis had been on the detailed and exact calculation of motion using Newton's physical principles and mathematical methods. For certain problems, such as finding the motion of the earth and the moon, the method had been wonderfully successful. With a little care, it was possible to predict eclipses a hundred years ahead of time to an accuracy of a few seconds. Unfortunately, the Newtonian method works best when only a few bodies are involved. In fact, the motion of the earth, moon, and sun, a three-body system, has never been found exactly although it can be calculated approximately to great accuracy. Applying the Newtonian method to five or ten particles was hopelessly difficult; finding the motion of 10^{23} gas molecules was impossible.

Maxwell and Boltzmann realized the impossibility of following the motion of each gas molecule and concentrated instead on the average properties of a gas. They developed a *statistical* theory. You see statistics in the newspaper every day. When you read that the current birth rate is 2.5 births per 1,000 population per year, you realize it represents an average for the whole country. For any particular group of 1,000, the number of births might be 5 or 1, but when millions of people are involved the average represents an accurate value. Because the number of gas molecules is so great, you can similarly expect averages to be quite accurate in the kinetic theory of gases.

Maxwell and Boltzmann pictured the gas molecules to be in constant motion, colliding with one another and with the walls of a container, bouncing off one another and moving randomly. The assumption of random motion was important and it allowed them to apply the theory of probability to the problem. Randomness and probability are related concepts. If you throw perfectly made dice randomly, each of the six numbers will come up one-sixth of the time on the average. If the dice are loaded, however, the same number may come up all the time. Probability can't be applied unless there is a random nature to the events.

Gases under ordinary conditions are easy to compress or expand, which shows that the molecules are so far apart the forces between them are negligible. Therefore, the motion of one molecule has little effect on the other molecules, except during collisions, and it is valid to as-

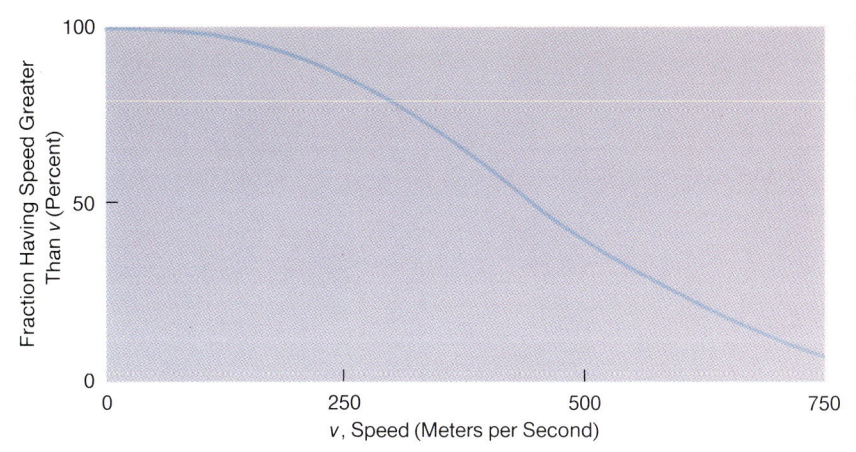

Figure 5. At room temperature, 87 percent of the nitrogen molecules in air have speeds greater than 250 meters per second, but only 10 percent have speeds greater than 750 meters per second.

sume the motions are independent and random. In solids and liquids, on the other hand, the atoms are close together and influence one another strongly. The motion of an atom in a solid or a liquid depends on the motion of the other atoms, so it is not completely random.

If the molecules of a gas are moving randomly, all their speeds cannot be the same. If two molecules having the same speeds make a glancing collision, one will be moving faster and one will be moving slower after the collision. Gas molecules make many collisions of every kind each second. Even if their speeds were all the same to begin with, the collisions would make the speeds change.

If you *were* able to follow the motion of a single gas molecule, you would see its speed change drastically from collision to collision. A molecule moving slowly might collide with a fast molecule and be speeded up in the process; and a fast molecule might be slowed down by colliding with a slow molecule. But the number of molecules is enormous and their motions are independent and random. As one molecule speeds up, then, another molecule somewhere else slows down; *the proportion of fast molecules stays the same.*

Under normal conditions, 50 percent of the nitrogen molecules in air have speeds greater than 450 meters per second. You don't know *which* molecules have high speeds because that value is a statistical result. It means that if you measured the speeds of a great many molecules, half of them would have speeds greater than 450 meters per second. Only a small fraction of the molecules at any time are traveling very slow or very fast. About 5 percent have speeds less than 150 meters per second, and 5 percent have speeds greater than 900 meters per second.

If a small puff of gas is introduced into a vacuum chamber, the density becomes so low that the molecules no longer collide with one another to any extent, and they must therefore retain the speeds they had when they entered the chamber. By timing how long it takes molecules to go from one end of the chamber to the other, the speed distribution can be found.

A fundamental scale of temperature can be based on the properties of gases. The two temperature scales you are most familiar with, the "Fahrenheit" and the "Centigrade," were designed arbitrarily. The freezing point and boiling point of water fall at 32 and 212 degrees on the Fahrenheit scale, and at 0 and 100 degrees on the Centigrade scale. Although water is certainly an important substance, there is nothing particularly fundamental about its freezing and boiling points. In gases, however, a remarkable thing happens as they are cooled: *the average speed of the molecules decreases.* The reason is that heat is a form of

Figure 6. (Below) To calibrate a thermometer, two points must be specified and then the interval between them must be divided in some way. For the Fahrenheit scale, the two points are the temperature of a mixture of ice and salt taken as 0 degree; the normal temperature of the human body is taken as 100 degrees. The Centigrade scale uses the freezing point of water as 0 degree and its boiling point as 100 degrees. Because the Centigrade scale is subject to less variation (what, after all, is the "normal" human body temperature?), the Fahrenheit scale was eventually standardized to the same points. Now 32 degrees is the freezing point and 212 degrees the boiling point on the Fahrenheit scale. Although the Kelvin scale is graduated in the same *increments* as the Centigrade scale, it takes the boiling point of water to be 373 degrees Kelvin.

In this figure, a reading of 122 degrees Fahrenheit is marked off. Using the conversion factors in Appendix 1, can you determine what the equivalent reading is on the other two scales?

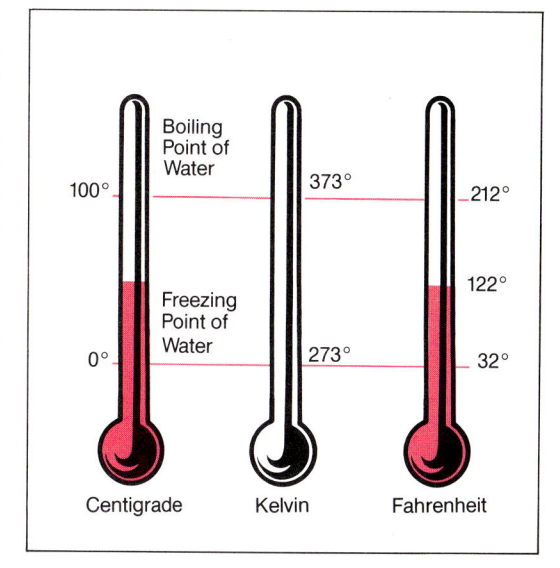

5 | Stuff and Substance

Figure 7. The atoms in a hot gas move faster on the average than do atoms in a cooler gas. Here the hot gas atoms on the left sweep out greater distances per unit time (each shaded sequence) than the cold gas atoms.

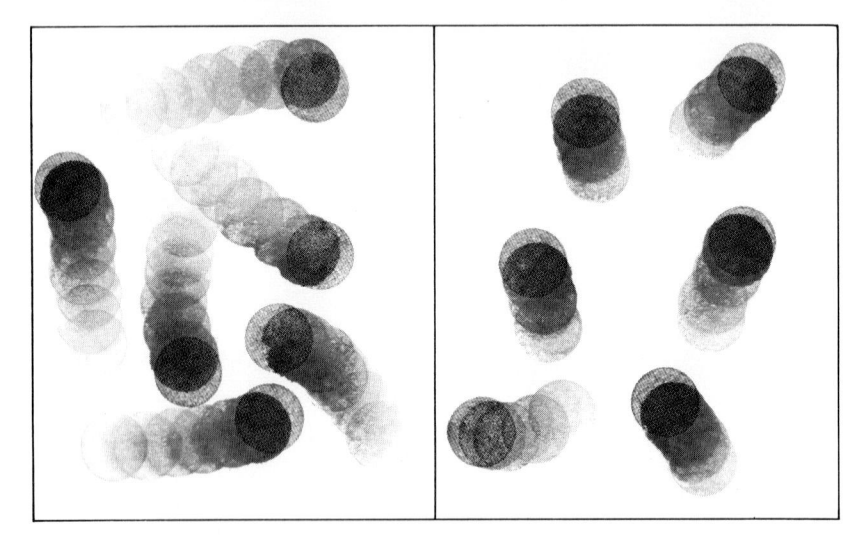

energy, and on the atomic scale, heat energy is translated into energy of motion of the atoms. When a gas is cooled, energy is taken from it and its molecules move more slowly on the average.

Extrapolating the properties of gases leads us to conclude that all motion would cease at a temperature of about −273 degrees Centigrade. That temperature has been made the zero point of the absolute, or *Kelvin,* temperature scale. The Kelvin scale is like the Centigrade scale except for the displacement of the zero point. In other words, 0 degree Centigrade corresponds to 273 degrees Kelvin; and room temperature, 20 degrees Centigrade, corresponds to 273 + 20 = 293 degrees Kelvin. Using the Kelvin scale, the median speed of a gas molecule, in meters per second, can be expressed as

$$\text{Median speed} = 140 \sqrt{\frac{\text{temperature}}{\text{molecular weight of molecule}}}$$

For example, the molecular weight of nitrogen molecules is 28, room temperature is 293 degrees Kelvin, so the median speed is 450 meters per second. The factor 140 comes from the conversion of actual weight to molecular weight and the conversion of heat energy to energy of motion.

The temperatures of solids and liquids can be measured by putting them in contact with gases and inferring the temperature from the properties of the gas. Heat is motion in solids and liquids, too. The atoms in solids are bound in place but they have energy of motion because of their vibrations. The air around a hot stove becomes heated because the speed of the gas molecules increases as they hit the vibrating atoms. The same thing happens to the molecules of the atmosphere. The gas molecules cannot remain on the ground because they are struck by the vibrating atoms and fly upward. In contrast, a sand grain is far too heavy to be boosted up by the collisions. If gas molecules were 10 times heavier than they actually are, the depth of the earth's atmosphere would be only half of what it actually is.

The average speed of a molecule depends on its molecular weight as well as on the temperature. Light molecules such as hydrogen move more rapidly. Near the top of the atmosphere, where molecules can fly long distances without colliding, a hydrogen molecule can move so rapidly that it has a good chance of leaving the atmosphere entirely. That is why the earth's atmosphere has so little hydrogen and helium even though they are the most abundant elements in the universe. The force of gravity

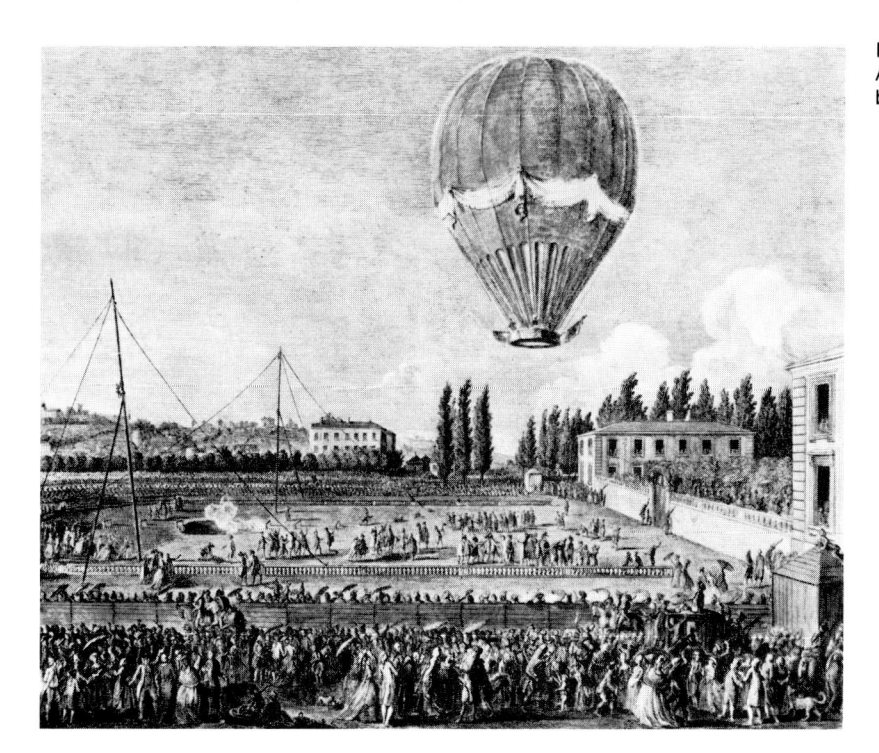

Figure 8. Atoms exert forces by colliding with objects. A balloon is held up because collisions with atoms from below outnumber collisions with atoms from above.

on the moon is weaker than on the earth, which accounts for the moon's lack of an atmosphere: over the ages all the gas molecules have escaped.

PRESSURE AND THE COLLISIONS OF ATOMS

Molecules constantly collide with one another. From the size of molecules and their spacing, it can be estimated that the molecules in air travel 10^{-7} or 10^{-8} meter between collisions. An average air molecule is traveling 450 meters per second, so it must make about 10^{10} collisions each second. That is certainly more than enough to ensure random motion.

The number of collisions depends on the temperature and on the density. In a hot gas, the molecules move rapidly and take only a short time to go from collision to collision. In a dense gas, there are more molecules to collide with.

Without collisions, a molecule in the earth's atmosphere would fall toward the ground. Instead, it is on the average held up by collisions from below. But collisions from above tend to push the molecule toward the ground. The atmosphere is stable only if the collisions from below slightly outnumber the collisions from above—enough to balance the effect of gravity. Neglecting variations in temperature, *the density of the atmosphere must decrease with altitude because collisions are fewer in a less dense gas.* That is the origin of the upward force on a balloon—the number of collisions from below exceeds the number of collisions from above.

Air molecules collide not only with one another but with the ground, with walls, and with you. Every time a molecule bounces off you, it pushes you a little. When you are hit by a basketball or a baseball, you are pushed hard and feel it, but usually you aren't aware of the constant drumming of the air molecules because there is air inside your body drumming away just as hard. When you go up in a fast elevator or an airplane, however, your ears "pop" because the slight decrease in air density with altitude makes the collisions outside your eardrums less effective than the col-

5 | Stuff and Substance

lision inside until the densities have a chance to equalize. Similarly, a balloon or a tire is kept rigid by the increased gas density and increased number of collisions on its inside walls.

The push that a gas exerts on a unit area is called its *pressure*. At the earth's surface, the pressure of the atmosphere is about 10^5 newtons per square meter. The concept of pressure can be applied to liquids and solids as well. You can feel the increase in pressure when you dive under water. Minerals deep in the earth are under terrific pressure from the surrounding rock.

CHANGE OF STATE

One of the most common physical changes in the world is the change of water to the different states we call solid, liquid, and gas. A glass of water left alone for a few days dries up. A pot of water boils dry. Ice cubes melt.

Take evaporation. The molecules in liquid water are held loosely together by interatomic forces. Once in a while, a molecule near the surface gains an unusually high speed in a collision and can escape the water completely, beyond the pull of the molecules remaining in the liquid. Molecules in warm water move faster on the average than in cold water. There is a greater proportion of fast molecules in warm water, and because fast molecules are the only ones able to escape, warm water evaporates more rapidly than cold water. The number of fast molecules available increases rapidly with temperature. At 75 degrees Centigrade, the rate of evaporation of water is only 40 percent of what it is at 100 degrees Centigrade. That is why waiting for a tea kettle requires so much patience; the steam starts rising vigorously only when the water is near the boiling point.

Evaporation is the loss of molecules with energies above average, so the temperature of the liquid must decrease. A drop of alcohol on the skin feels cool as it rapidly evaporates. To keep an evaporating liquid at the same temperature, heat energy must be added to make up for the energy carried away by the fast molecules. Under these conditions, the proportion of fast molecules in the liquid is constantly replenished by the collision mechanism.

Change of physical state involves a change in the disposition of the interatomic forces and is accompanied by a change in heat energy. For example, water at 0 degree Centigrade can exist either as solid ice or as liquid water. To change ice at 0 degree Centigrade to water at 0 degree Centigrade requires the addition of enough heat energy to break the water molecules out of the tightly bound crystalline solid form. The added heat energy, instead of changing the temperature, simply goes into breaking the bonds. Conversely, to change liquid water at 0 degree Centigrade into ice at 0 degree Centigrade, heat energy must be removed.

The most common changes of state are between solid and liquid or between liquid and gas. Other changes are possible as well. Some solids can change from one crystalline form to another with an accompanying release or absorption of heat energy as the atoms rearrange themselves in a new configuration. Under normal conditions, ice exists in only one crystalline form, but at higher pressures there are several other distinct kinds of ice.

CARRYING ENERGY

When you place a silver or aluminum spoon in a cup of hot tea or coffee, the handle of the spoon soon becomes warm to the touch. Matter can carry heat energy from one point to another, and the fundamental reason

Figure 9. The dense mist surrounding a burst of hot water from a steam geyser shows water changing state from liquid to gas.

it can do so is that atoms interact with one another. In metals the mechanism of heat transport is indirect. The atoms of a metal are bound in place but they can vibrate back and forth. About one or two electrons per atom are not held tightly to a particular atom in a metal, however, and they are able to wander long distances through the metal. As they move, the electrons collide with atoms, and a transfer of energy can occur in these collisions. For example, at the hot end of the spoon the atoms are vibrating vigorously and electrons tend to gain energy in collisions with atoms. The energetic electrons wander into the colder parts of the spoon, where the atoms are not vibrating strongly. There, the energetic electrons tend to give energy to the atoms in the cooler parts of the spoon during collisions.

The wandering unbound electrons in a metal play another important role. They enable the metal to conduct electricity, as you will read in a later chapter. Materials such as glass and plastic do not conduct electricity well because they have very few unbound electrons; accordingly, they also do not conduct heat well. The handle of a plastic spoon stays cool. There is always some degree of heat conduction, however, because the vibrating atoms in a solid are close enough together to interact directly with one another to some extent. But electrons carry heat energy far more efficiently. Silver, one of the best metallic conductors, conducts heat about 250 times better than glass.

Liquids and gases conduct heat by collisions between atoms. Because the process of conduction tends to be slow, in many practical applications it is better to move the hot liquid or gas itself. In some home heating systems, for example, water or air is heated in a furnace and then pumped through conduits to the rooms.

One important kind of energy transfer in matter is the carrying of sound waves. To start a sound pulse going in air, all you need to do is make the density of part of the air suddenly different from its normal value. Clapping your hands is one way to do this. Molecules in a region of high density tend to move toward lower density because the greater frequency of collisions in the higher density region tends to push them more toward lower density. A compact region of high density, such as that formed in air when you clap your hands, tends to move outward in all directions. The speed at which it moves has nothing to do with how it is produced. The speed depends on *collisions in the gas* and the *speed of the molecules*. The speed of sound in air is 330 meters per second, comparable to the average molecular speed of 450 meters per second.

Sound travels more rapidly in liquids and solids than in gases because the atoms in solids and liquids are closer together than in gases. As a result, the motion can be transferred more rapidly. The speed of sound in aluminum, for example, is as fast as it is in air.

To produce a steady sound, the changes of density must be produced in a regular way. Musical instruments, loudspeakers, and your voice box all have some sort of vibrating membrane or string or air column to change the density of nearby air in a regular and rapid way. The degree of motion required is not great. Even in a moderately loud sound, such as that produced by a piano, the density doesn't change by more than one-hundredth of a percent or so from normal.

CHEMICAL CHANGE

Maxwell and Boltzmann's kinetic theory needed only a simple model of the atom. Their atoms were only called upon to move and collide. However, observation of chemical change and long experience in chemistry showed that the atoms of different elements have their own characteristic chemical behavior. Atoms cannot be simple hard spheres but must have a complex structure. A far-reaching theory of the atom was not developed until the beginning of the twentieth century. Long before that, chemists found a way of thinking about the elements that at least helped to organize the facts of chemistry, if not to explain them.

Valence

Valence is a measure of the capacity of an atom to combine with other atoms into molecules; it is the number of "hooks" an atom has available. An element has a valence of 1 if one atomic weight of the element combines with or displaces one atomic weight of hydrogen. If one atomic weight of an element combines with two atomic weights of hydrogen, the valence is 2, and so on. In water, symbolized as H_2O, each oxygen atom hooks onto two hydrogen atoms, so the valence of oxygen is 2. Similarly, four atoms of hydrogen combine with one atom of carbon in methane, CH_4, and the valence of carbon is 4.

The concept of valence is useful because the valence of an element tends to be the same in a wide variety of chemical compounds. Because the valence of oxygen is 2 and the valence of carbon is 4, we expect one atom of carbon to combine with two atoms of oxygen; the result is the well-known compound "dry ice," carbon dioxide (CO_2).

A great deal of chemistry can be summed up with the help of valence, but valence is not a hard-and-fast property of elements. For instance, hydrogen and oxygen also form hydrogen peroxide, H_2O_2. In this

Calculating Pressure Changes of a Gas

What happens to the pressure of a gas when its volume and temperature are changed?

Step 1. Consider the mixture of air and gasoline entering one of the cylinders of an automobile engine. As the piston moves up, the volume of the air is decreased, typically by a factor of 8. Then, when the volume is smallest, the air-gasoline mixture is ignited. The rapid burning raises the temperature of the mixture to about 1,800 degrees Kelvin.

Step 2. Although the temperature of the gas increases somewhat during the compression phase, assume for the moment that its temperature remains constant at 300 degrees Kelvin. As the volume of an ideal gas at constant temperature decreases, its pressure increases proportionately. If the volume is decreased by a factor of 8, the pressure increases by a factor of 8. So if the air-gasoline mixture is at atmospheric pressure before compression, the pressure after compression is 8 atmospheres. Note that work must be done to push the piston against the compressed air-gasoline mixture. This work is the main source of "engine braking."

Step 3. When the air-gasoline mixture at 8 atmospheres pressure is ignited, its temperature rises quickly from 300 degrees Kelvin to 1,800 degrees Kelvin. The combustion is so rapid that it is effectively complete before the piston can move much farther, and you may take the volume of the gas to be constant during combustion. If the volume of an ideal gas is constant, the pressure is directly proportional to the temperature:

$$\text{Final pressure} = 8 \text{ atmospheres} \times \frac{1,800 \text{ degrees Kelvin}}{300 \text{ degrees Kelvin}}$$

$$= 48 \text{ atmospheres}$$

The strong forces produced by the heated gas drive the piston down, providing the motive power of the automobile.

Figure 10. Just like their human observers, atoms can be organized into groups on the basis of their unique characteristic behaviors.

compound, oxygen acts as if it has valence 1. Nevertheless, hydrogen peroxide is not a very *stable* compound. One of the oxygen atoms is readily detached, leaving H_2O; oxygen seems to "prefer" a valence of 2.

Valence can be either a positive or a negative number. Negative valence means that the element has a tendency to combine with hydrogen. Oxygen therefore has valence −2, and fluorine, which forms hydrofluoric acid, HF, has valence −1. Positive valence means the element has a tendency to displace hydrogen. Hydrogen has valence +1, and sodium also has valence +1 because it easily forms compounds such as sodium fluoride, NaF, which is analogous to hydrofluoric acid, HF, with the hydrogen replaced by sodium.

Some elements form no compounds at all, and have zero valence. The first to be discovered was argon, which constitutes nearly 1 percent of the earth's atmosphere. It was found in 1894 in experiments on the liquefaction of air at low temperatures. Other elements usually exhibiting zero valence are helium, neon, krypton, xenon, and radon. These elements, together with argon, are called the *noble gases*. However, since 1962 several compounds of xenon with oxygen and fluorine have been produced.

The Periodic Table

It is curious that certain age groups of elements have nearly identical chemical and physical properties. For example, fluorine, chlorine, bromine, and iodine are all gaseous or vaporize readily, and all combine in much the same way with metals to form "salts," which are in general easily dissolved in water. These elements are called the *halogens,* the salt-formers. Similarly, lithium, sodium, potassium, rubidium, and cesium —the *alkali metals*—are all soft metals that combine avidly with oxygen. They all react violently with water, releasing hydrogen and forming power-

Group	I	II											III	IV	V	VI	VII	0
Period																		
1	H 1																	He 2
2	Li 3	Be 4											B 5	C 6	N 7	O 8	F 9	Ne 10
3	Na 11	Mg 12			Transition Elements								Al 13	Si 14	P 15	S 16	Cl 17	Ar 18
4	K 19	Ca 20	Sc 21	Ti 22	V 23	Cr 24	Mn 25	Fe 26	Co 27	Ni 28	Cu 29	Zn 30	Ga 31	Ge 32	As 33	Se 34	Br 35	Kr 36
5	Rb 37	Sr 38	Y 39	Zr 40	Nb 41	Mo 42	Tc 43	Ru 44	Rh 45	Pd 46	Ag 47	Cd 48	In 49	Sn 50	Sb 51	Te 52	I 53	Xe 54
6	Cs 55	Ba 56	* 57-71	Hf 72	Ta 73	W 74	Re 75	Os 76	Ir 77	Pt 78	Au 79	Hg 80	Tl 81	Pb 82	Bi 83	Po 84	At 85	Rn 86
7	Fr 87	Ra 88	‡ 89-															

*	La 57	Ce 58	Pr 59	Nd 60	Pm 61	Sm 62	Eu 63	Gd 64	Tb 65	Dy 66	Ho 67	Er 68	Tm 69	Yb 70	Lu 71
‡	Ac 89	Th 90	Pa 91	U 92	Np 93	Pu 94	Am 95	Cm 96	Bk 97	Cf 98	Es 99	Fm 100	Md 101	(?) 102	

ful alkaline hydroxides that can react with grease or oil to form soaplike compounds.

These chemical groupings have little to do with atomic weight. The noble gases, for example, range from helium with atomic weight 4 to radon with atomic weight 222. Chemical properties are evidently not determined by atomic weight.

By 1865 some 65 elements had been identified, and valences and atomic weights had been assigned to most of them. John Newlands noticed, when he listed those elements in order of their atomic weights, that an element with similar properties tended to appear periodically in the list. He summarized that periodicity in a "Law of Octaves," by analogy with the notes of the diatonic musical scale. To many of his colleagues his idea seemed capricious: it drew from the chemist Carey Foster the scornful question of whether Newlands had thought of listing the elements in alphabetical order instead.

The climactic achievement of the search for periodic order among the elements came 4 years later at the hands of Lothar Meyer and Dmitri Mendeleev, who were working independently in Germany and Russia. Meyer's correlations emphasized the physical properties of the elements; Mendeleev's, their chemical properties. Mendeleev's scheme of listing, with modifications and additions dictated by later work, is now called "the periodic table of the elements." A detailed version of the periodic table is presented at the back of this book, but the outline form in this chapter (Figure 11) is sufficient for now. When the elements are listed in a horizontal row in order of increasing atomic weight, it's clear that the valence goes through cyclic variations. If a new row is started every time a valence cycle begins again, the vertical columns in the table form the *groups* of similar elements. The horizontal rows are called *periods*.

For example, you can see from the periodic table that the noble gases are all in the same column. Similarly, the halogens have their own

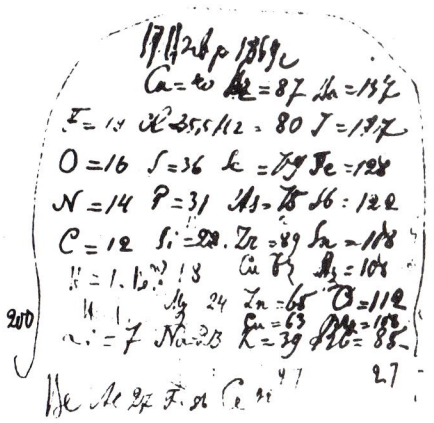

Figure 11. (Above, left) The cyclic variations in valence give rise to a high degree of periodicity among the elements, and that is why the elements can be organized into groups. Dmitri Mendeleev worked out the first systematic periodic table. The foldout in the back of this book illustrates the degree of sophistication that the periodic table of the elements has since achieved.

Figure 12. (Above) Dmitri Mendeleev, the first man to organize the findings from years of chemical analyses of the properties of the known elements. (Below) A page from Mendeleev's notes.

5 | Stuff and Substance

Figure 13. The similar chemical and physical properties of elements in the same group enable us to predict the properties of unknown elements.

SILICON
Steel Gray Metal
Valence 4
Density 2,300 kg/m³
Forms SiO$_2$
Melting Point 1700°C

TIN
Silvery White Metal
Valence 2, 4
Density 7,300 kg/m³
Forms SnO$_2$
Melting Point 1130°C

column, as do the alkali metals. The elements in the same column or group often have striking similarities. Copper, silver, and gold are all in the same group; all are shiny, ductile metals that are easily worked. All three are among the earliest metals used by man because they resist chemical combination and can be found in pure form in nature.

Going across the first long period from lithium, atomic number 3, to neon, atomic number 10, here is how the valences change:

Lithium	+1
Beryllium	+2
Boron	+3
Carbon	+4, −4
Nitrogen	−3
Oxygen	−2
Fluorine	−1
Neon	0

The regular change of valence is striking.

Mendeleev provided some dramatic illustrations of the power of this tool. Noticing gaps in the listed sequence of elements, he confidently predicted that new elements would be discovered to fill those gaps, and he even predicted what their properties would be. For example, a gap between zinc (atomic weight 65, valence 2) and arsenic (atomic weight 75, valence 5) suggested that two elements were missing, one of valence 3 and atomic weight 68 or 69, and one of valence 4 and atomic weight 72 or 73. Mendeleev called the second missing element eka-silicon (Es) because it would fall directly beneath silicon in his table. Mendeleev predicted that eka-silicon would turn out to be a gray metal having valence 4 and a specific gravity of about 5.5, and that when heated in air it would react with oxygen to form a dioxide, EsO$_2$, which would have a high melting point.

To understand Mendeleev's reasoning, look at the elements silicon (Si) and tin (Sn) which fall above and below "eka-silicon" in the same column of the periodic table (Figure 11). Silicon is a steel-gray metal of valence 4 and a specific gravity of 2.3. It forms a dioxide, SiO$_2$, which has a melting point of 1,700 degrees Centigrade. Tin, below eka-silicon, is a silvery white metal of valence 4 and a specific gravity of 7.3; it forms a dioxide, SnO_2, which has a melting point of 1,130 degrees Centigrade. In 1886 Clemens Winkler discovered the element now called germanium (Ge). It is a gray-white metal with atomic weight 72.6, valence 4, and a specific gravity of 5.3. And it forms a dioxide, GeO_2, which has a melting point of 1,090 degrees Centigrade.

The design of the periodic table is not straightforward, however. The table starts out well enough, but problems arise toward the end of the fourth period. The elements iron, cobalt, and nickel follow in order when the elements are listed by atomic weight, and they would be expected to fall into different groups of the periodic table. But these three elements are similar chemically and physically—they often occur together in the same ores and they are the only elements that are strongly magnetic. A similar difficulty arises in the sixth period: the element lanthanum, atomic weight 139, is followed in atomic weight by 14 elements that are chemically so identical they were not separated from one another until the 1940s.

You will see later in this book how these problems were resolved, but at this point a few comments on the present status of the periodic table are in order. You have read many conjectures about the physical world that were later superseded when more complete knowledge was available. Is the periodic table only an interesting historic relic as well? Certainly not. First of all, the periodic table obviously summarizes a considerable amount of chemistry, and it is one of the tools of the trade for every working chemist. Second, the periodic table also says a great deal about physical properties such as density and melting point.

The principles of atomic physics are well understood today and can be used to give a good account of the main features of the periodic table. But atoms have such a complicated structure that the theory can be fully applied in only the simplest cases. Accurate calculations of most chemical and physical properties are not possible. That is the reason the periodic table retains its importance: it expresses relationships too subtle for present theory to encompass. For example, the periodic table has been of great help in the search for new superconducting alloys and for compounds useful in solid-state electronics.

Figure 14. Spectra are the characteristic signatures of the elements. Sodium is represented by the unique double yellow lines at 5889.95 and 5895.92 angstroms. Mercury is represented by two yellow lines, a green line, and several blue lines; and neon "signs in" with many weak red lines and three strong yellow ones.

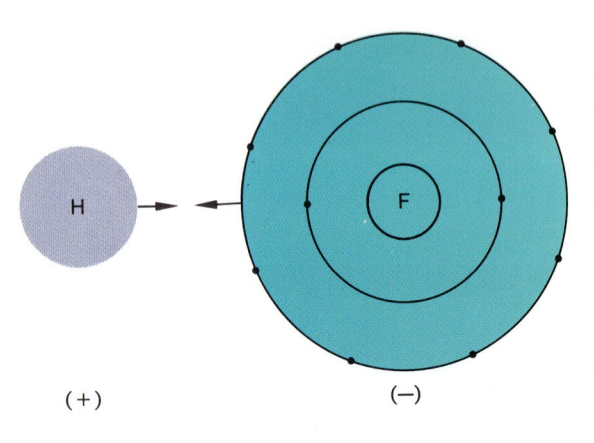

Figure 15. Hydrogen and fluorine are held together by electric attraction because the electron from the hydrogen atom spends most of its time filling the outer shell of fluorine.

ATOMIC SPECTRA

How far a science has progressed can be judged by the number of phenomena it has yet to explain. In our own day, the unresolved questions include the relation of chemistry to life, the nature of the elementary particles of matter, and the energy sources of pulsars, but the physical principles underlying most everyday phenomena are understood.

A hundred years ago, however, it was not necessary to turn to far-off space to be confronted with mysteries. Although Mendeleev's periodic table had demonstrated there was order in the properties of the elements, the structure of the atom remained completely unknown. There was no coherent picture of the atom capable of accounting for the periodic table or for any of the facts of chemistry and chemical binding.

A technique developed in the middle of the nineteenth century ultimately helped unravel the mystery of the atom, but at first it seemed to compound the difficulties. Gustav Kirchhoff and Robert Bunsen put chemicals in a hot, nonluminous flame and looked at the light that was given off. You have seen the bluish-green light from a mercury vapor street lamp and the red light from a neon sign. Similarly, sodium gives a strong yellow color to a flame and potassium gives a purple tinge. However, Kirchhoff and Bunsen did not stop with such simple observations. They set up a glass prism and decomposed the light into its component colors.

When passed through a prism, a beam of light consisting of a pure color is bent by an amount depending on the wavelength of the light. Kirchhoff and Bunsen found (as others had earlier) that the light from their colored flames consisted of only certain wavelengths. However, they recognized that each element has its own distinct pattern of wavelengths, called its *spectrum.* Most chemical compounds are decomposed by hot flames and give off the spectra of their constituent elements.

In short order Kirchhoff and Bunsen discovered two new elements, rubidium and cesium, while analyzing water samples. And they showed that lithium, once thought to be rare, could be found in a wide variety of substances including cigar ashes. When light from the sun and the stars was analyzed, the same spectra that had been observed on earth were seen—atoms were the same everywhere. In fact, one element, helium, was discovered in the sun before it was found on earth.

Soon many spectra were measured and accurate tables of the wavelengths were painstakingly constructed. The atomic spectra were the *first indication* that atoms had a complex inner structure.

Aside from the main problem of what kind of structure could account for the spectra, many other questions were raised. Even the smallest sample of an element contains billions and billions of atoms, yet in a hot flame each of the atoms independently gives off the same spectrum. The atoms of a given element must be identical to a degree that classical mechanics cannot explain. For example, the solar system seems comparatively stable, but every time a comet passes by, its gravitational attraction slightly changes the orbits of the planets. Yet in a flame atoms undergo millions of violent collisions each second—and they remain the same. One sodium atom is just like any other sodium atom: there is no way to tell one from another.

THE STRUCTURE OF THE ATOM

Newtonian mechanics had no way of explaining such incredible stability or any of the facts of atomic spectra. A wide variety of experiments and theories in the early part of the twentieth century showed that new principles were needed to understand the behavior of matter on the atomic scale. Leaving a more complete story to later chapters, you may use the following summary of a modern picture of atomic structure to gain understanding of the physical basis of chemical change and the periodic table:

1. *An atom consists of a tiny dense core—the nucleus—surrounded by electrons.*

2. *The number of electrons in a normal atom is equal to the atomic number of the element. The electrons are bound to the nucleus by electric forces.*

3. *The electrons are arranged in groupings called shells. The shells have reasonably well-defined radii.*

4. *There is a limit to the number of electrons that can be grouped into a given shell. The innermost shell, called the K shell, can contain only two electrons at most. The next shell, called the L shell, can hold no more than eight electrons, and so on.*

5. *The electrons in a given atom tend to be arranged with the shells of smallest radii filled to capacity, insofar as this is possible.*

To see how these principles work, take the lightest elements, hydrogen through neon:

Hydrogen: One electron, in the *K* shell.

Helium: Two electrons, both in the *K* shell, filling it to capacity.

Lithium: Three electrons, two filling the *K* shell and one in the *L* shell.

Beryllium Through Neon: Each of these atoms has two electrons in the *K* shell, and the remainder in the *L* shell. At neon, with ten electrons, the *L* shell becomes filled.

The chemical properties of an atom depend largely on the electrons in its outermost shell because when two atoms come together, the outer shells influence each other most strongly. You can see that the number of electrons in the outer shells of the light elements bears a strong relationship to the *valence* the element exhibits in chemical reactions. Helium and neon, which have completely filled shells, have valence 0. Hydrogen and lithium, both with one outermost electron, have valence +1. Oxygen, with six outer electrons, has valence −2, corresponding to the number of unoccupied spaces in the *L* shell. Fluorine, with seven electrons in the *L* shell, has valence −1.

When atoms enter into chemical combination, they exhibit a strong tendency to try to fill the unoccupied spaces in their outermost shell. For example, in the formation of hydrofluoric acid (HF), the single electron of the hydrogen atom is attracted to fill the one empty space in the shell of seven in fluorine. How does this process result in the formation of a stable molecule? Atoms are normally electrically neutral; the positive electric charge on the nucleus is equal in strength to the total negative electric charge of the electrons. In hydrofluoric acid, the electron from hydrogen spends most of its time around the fluorine atom, filling its *L* shell. The fluorine is therefore more negative than normal. Positive and negative charges attract, so the hydrogen atom is attracted electrically because it is more positive than normal, having lost its electron.

This, then, is the physical basis for Dalton's "hooks and eyes." The chemist's model of the atom and the physicist's model of the atom are united into a coherent picture of great explanatory power. Later chapters elaborate the simple picture presented here.

5 | Stuff and Substance

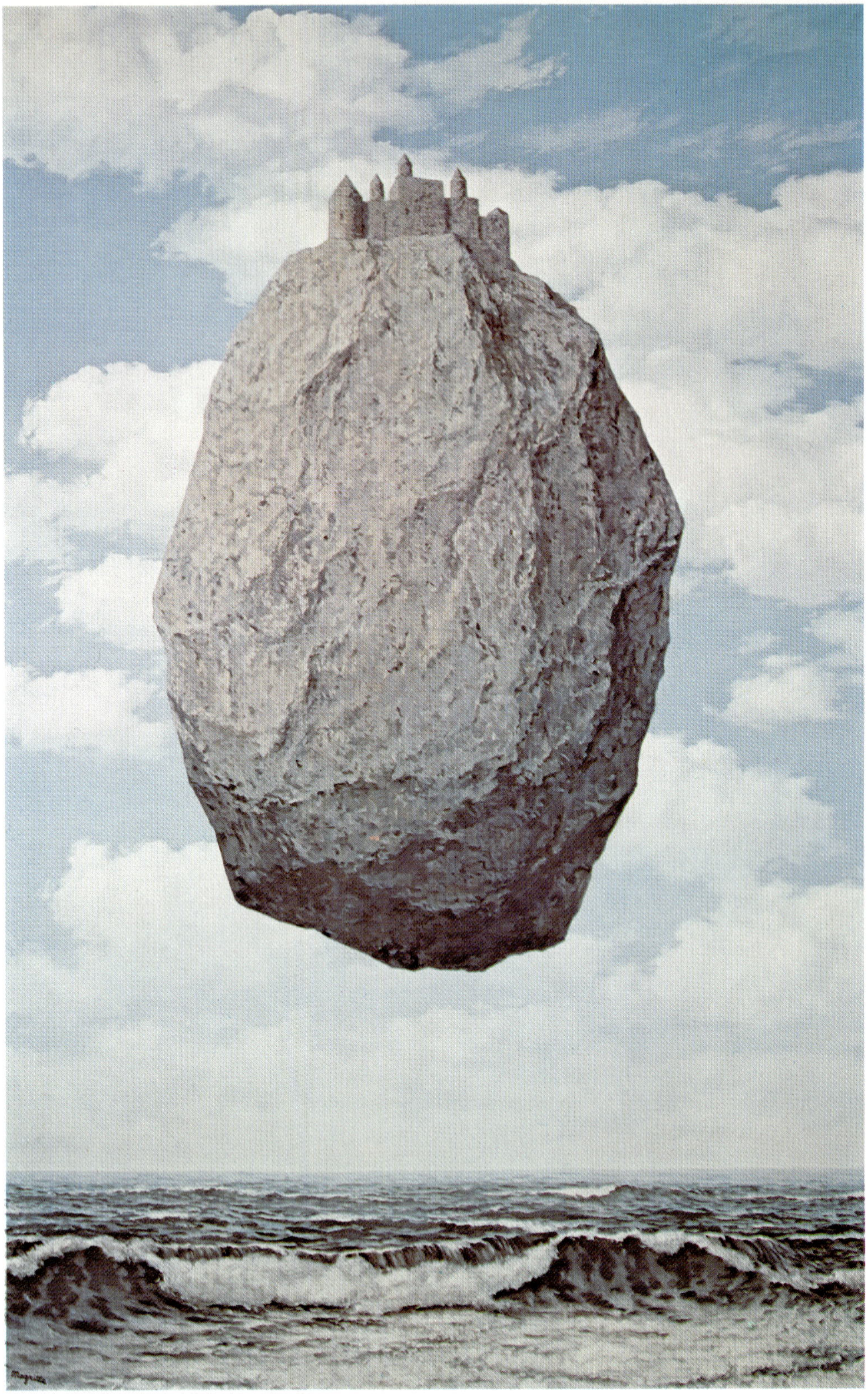

Castle in the Pyrenees
René Magritte (1898–1967)

Is it possible that the world around us exists only in our imagination? Even though we know little enough of the workings of our consciousness, we would be hypocrites to pronounce that matter does not exist; our faith in its existence as well as in its generally predictable behavior governs everything we do. *Castle in the Pyrenees* plays with what we know of matter and the forces that act upon it. We suspect intuitively, for example, that gravity should bring down that massive mountain—and yet there it is, hovering outrageously in the sky.

The impact of Magritte's painting is ensured precisely because of our awareness of *scientific laws*, whether we derive that awareness from years of deliberate study or simply by nurturing it quite unconsciously as a result of day to day observations. The scientific laws that span the entire universe are integral with the human experience. The more we know of them, the greater is the possibility we have of enriching the world of our imagination. Similarly, the more we free our imagination from mundane patterns of thought and observation, the greater is the possibility for discovering just how intricately these laws permeate the very substrate of our material world.

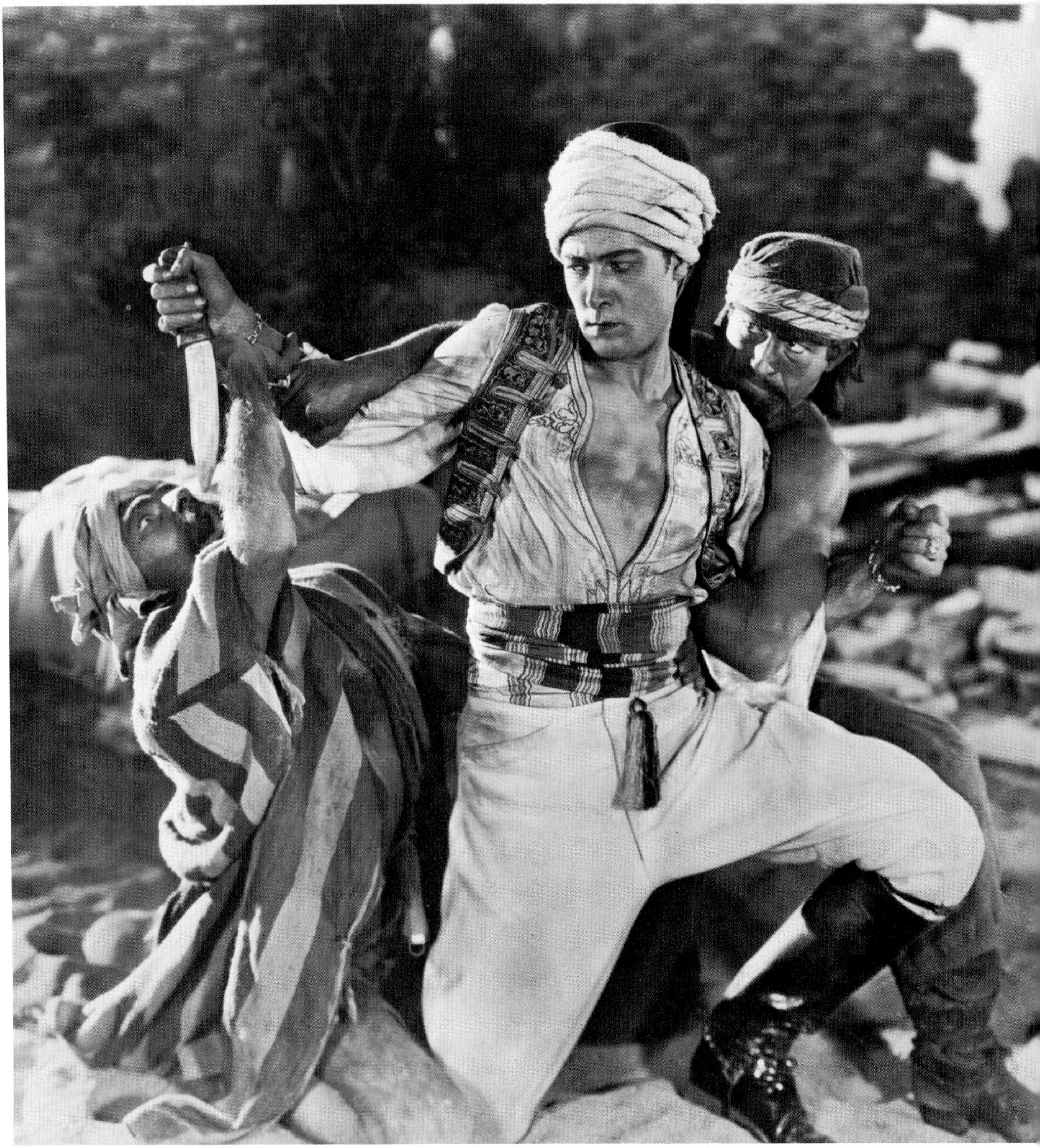

6
Pushes and Pulls

Forces are the pushes and pulls that set things in motion, stop their motion, or hold them in place.

Why do things move? Why do they stand still? All motion and nonmotion begin with the forces that are at work in the world. Forces are the pushes and pulls that set things into motion, or stop their motion, or hold them in place.

If someone wants you to move out of his way he may give you a push, but if he wants to help you climb a mountain he may give a pull on your hand. To move a book across a table you push against one side of the book until it moves. To carry a book into another room you first lift it— you put an upward force on it in order to overcome the downward force of gravity—then you put a small sideways force on it to get it going in the right direction. To throw a ball you push on it with your hand; to catch it you stop its motion by a force from your hand.

When you are standing still you feel the downward tug of gravity and the upward push of the floor. It is the combination of forces that keeps you in place. Similar forces keep the walls of buildings in place and keep all the parts of a bridge in place even as heavy cars travel over it.

People generally have an awareness of many kinds of forces. When you watch a bat slamming against a baseball, you are aware of the force of one solid object against another object—often called the *contact* force. You have heard about the gravitational force that pulls all things toward the earth. It doesn't need contact: it can exert its pull across empty space. You know there are forces of pressure—you feel the air pressing against your eardrums when you dive into water; you feel it pulling against your ears when you drive up a mountain. There are forces of friction between your feet and the earth. These forces not only work to keep you from sliding down a slope, they also are at work even when you are walking or standing still. The same kind of friction forces bring a car to rest when you apply the brakes. Friction forces in water stop a boat when you turn off the motor, or when you stop rowing. There are forces inside a piece of wood (or metal, or concrete) that work to maintain the object's shape or bring it back to a natural position if it has been bent. There are even "sticky" forces you come across when you use adhesive tape and glue.

Clearly many different forces are encountered every day. Yet all of them arise from just *three fundamental forces in nature:* the gravitational, the electric, and the nuclear force.

The gravitational force pulls us and all things toward the center of the earth. It pulls on rocks and water, helping to maintain the surface features of our planet and to maintain its roundness. It pulls the materials of mountains and hills downward as they are eroded away; it draws the rivers toward the seas. It tugs even on the air, cloaking the earth with the atmosphere that is essential to life. And through its ever present force, it controls the way we live and move.

The force of gravity spans the time and space of the universe. Without it, our sun and its planets would not have been created, nor would the planets maintain their orbits around the sun today. This force controls the course of our sun and its planets in their motion through our Galaxy; it controls the course of our Galaxy as it swings slowly around our sister galaxies. And it acts across the most distant reaches of space, working to hold the entire universe together—although we don't know yet whether it will succeed or not.

Almost all other forces encountered every day are electric in nature.

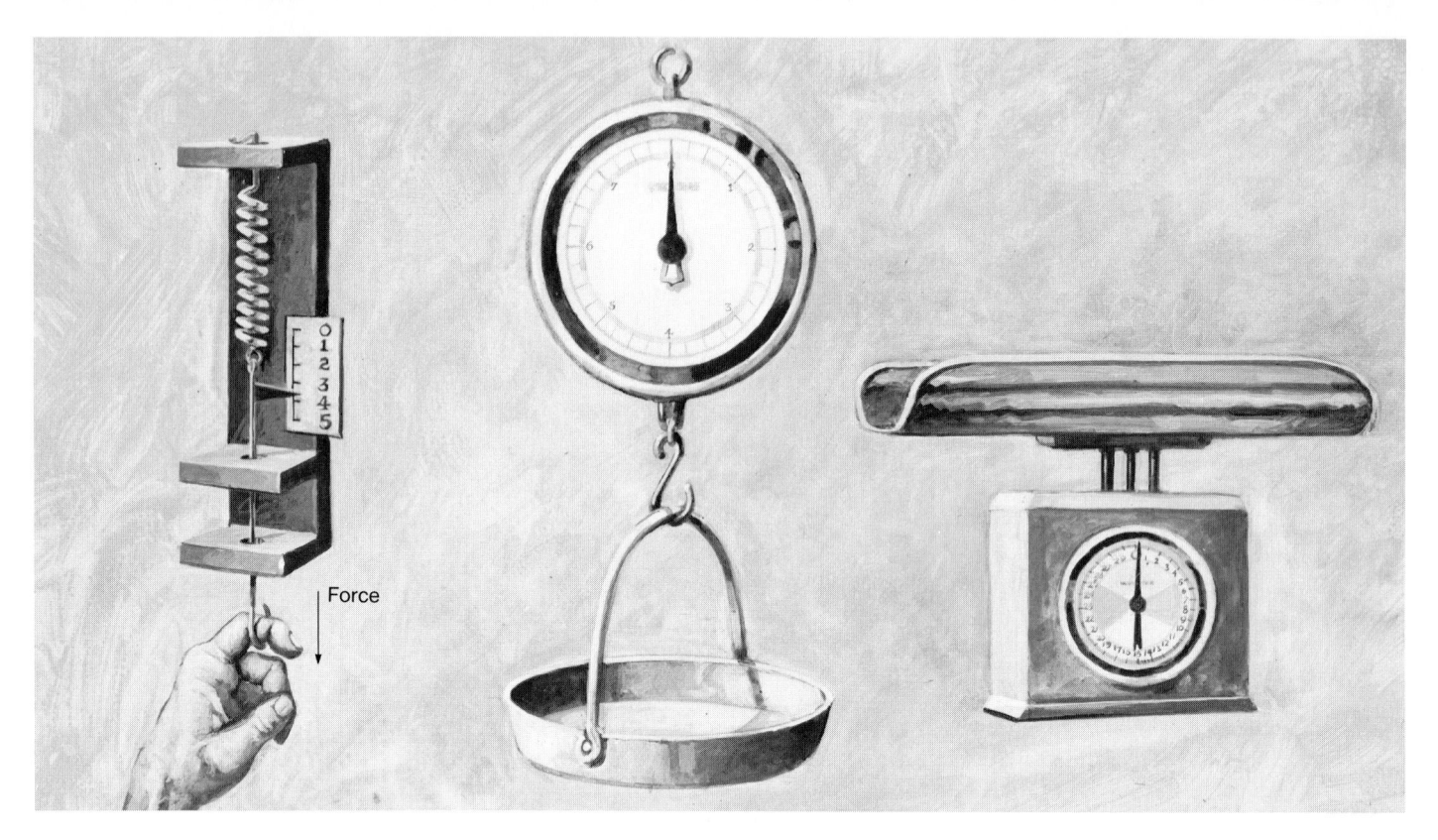

Figure 2. (a) This sketch illustrates the principle of the spring scale: pull twice as hard, and the stretch is twice as great. (b) Examples of spring scales.

The *electric force* holds the parts of an atom together and also holds each atom in place next to other atoms, giving solid matter its shape and strength. Because of this force, the atoms of a liquid are kept a certain distance apart even as they roll and slide over one another, and the atoms of a gas are kept flying. Through the electric force, light is generated and the sun's energy is carried to the earth. Not only is it the force underlying friction and adhesion, it is also the force underlying all the electric machines of our time—the motors, stoves, lamps, radios, phonographs, and television sets. The electric force even governs all the chemical machinery of our bodies—the pumping of our heart, the motion of our arms and legs, our seeing and touching, and the processes of our brain.

Inside the atomic nucleus is a force that is not usually encountered in our daily experience. The *nuclear force* holds the parts of the nucleus together. It is the source of radioactivity and it is the source of nuclear energy, which man may harness for power plant or bomb. The nuclear force is responsible for the creation of all the forms of matter—the chemical elements—that we know.

MEASURING FORCES

The gravitational force, the electric force, and the nuclear force govern all that happens in the world. How do they work? What effects do they have? To be able to talk about such things we need to be able to think with numbers about the "strength" of a force, and to measure it against the strength of another force.

One way to measure a force is to use a *spring scale,* which is the same kind of scale used in markets to weigh fruits and vegetables. The spring scale is based on a simple idea: the more you pull on a spring, the more you stretch it. Pull twice as hard and the stretch is twice as great;

in other words, the stretch is *proportional* to the force applied. To make a spring scale for measuring force, then, all you need is a stretchable spring and a pointer to indicate the amount of stretch. You could build a spring scale as shown in Figure 2, but for practical purposes they usually are built in one of the ways shown in Figure 2.

How would you know your spring scale is accurate—that, say, twice as much force will indeed pull the pointer twice as much? Your scale can be *calibrated*—its graduation can be checked out. First you have to find a way to put a certain definite force on the spring. For example, you can take some object (usually a piece of metal) to be the *standard object*. You can then take the pull of gravity on it to be your standard force, or your *unit of force*. You might call it "the pound." When you hang the standard object your spring scale should pull the pointer to the "1-pound" mark. Next, remove the standard object from the spring scale and hang another, identical object from it. A truly identical object will have the same gravitational pull on it and will also pull the pointer to the "1-pound" mark. Now you can place both objects together on the scale. If the scale is accurate the spring will be pulled to the "2-pound" mark. By following the same procedure you can check the marks for 3 pounds, 4 pounds, 5 pounds, and so on.

But wait a moment. How do you *know* that two 1-pound objects together pull with a force of 2 pounds? You might say, "That's just common sense." It is more than that; it is really a *definition* of a measure of force. A 2-pound force is defined to be what you get when two 1-pound forces act together, or any force of equal pull. And this definition is part of a more general rule:

■ **Addition of Forces.** If two forces act together and in the same direction, the combined effect is a force that is the sum of the two original forces.

Calibrating a scale is one thing, but finding a good, reliable unit of force is more difficult. A common unit of force in the United States is the *pound*. Originally a pound was defined as the gravitational pull on 1 pint of water, but people later discovered it's not easy to be sure just how much a *pint* is. Besides, the gravitational force on any object depends on where in the world you happen to be when you make the measurement! The gravitational pull is greater at sea level than it is in the mountains, and it is not exactly the same in New York and San Francisco. So now the pound is defined as the gravitational pull on a certain standard block of platinum (kept in the Standards Office in London) when that object is measured in a certain standard place.

In the metric system of units, the unit of force is the *newton*. It is approximately equal to one-fifth of a pound. Five medium-size apples weigh about a pound, so a newton is about the weight of an apple. Because the precise definition of the newton is expressed in terms of the *motion* such a force produces on a standard object, it will be discussed in the next chapter.

FORCE IS A VECTOR

When you push on something you not only push a certain amount, you also push in a certain direction. To describe a force, then, you have to know several things about it:

Where and on what is the force acting?

How large is the force?

In what direction is it acting?

Figure 3. (Above) The girls' feet push down *on* the rock. (Middle) The man's hands push upward *on* the ashtray. (Below) The man behind pulls *on* the other man's jacket. An arrow is used to represent a force. The tail of the arrow, shown by a dot, is placed on the object *receiving* the force. The direction in which the arrow points shows the *direction* of the force. The length of the arrow is proportional to the strength, or *magnitude,* of the force. In these photographs, only a few of the forces are shown.

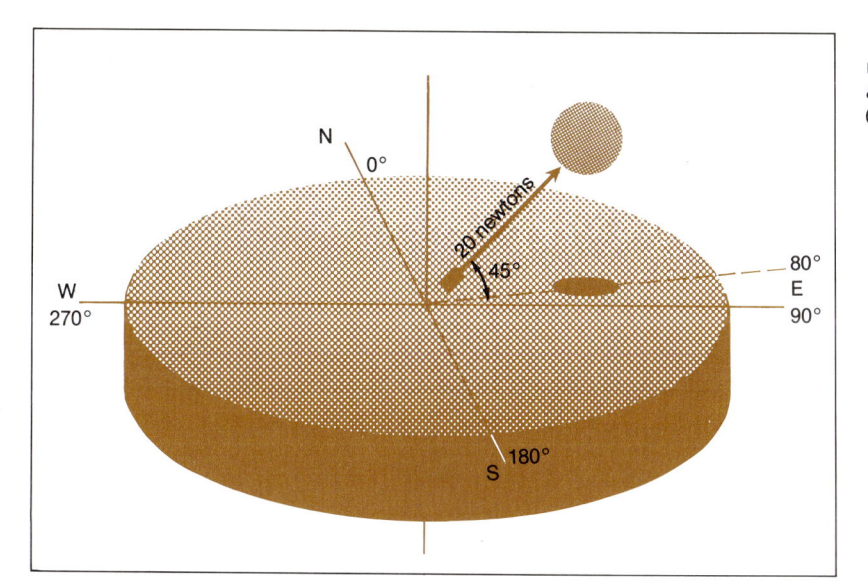

Figure 4. (Left) Force is a vector and requires three numbers to define it. Here those three numbers are the *angle* up from the horizontal (45°), the *magnitude* (20 newtons), and the *direction* (80° northeast).

The force may be upward, downward, or at some angle. It may be pushing toward the east or the south, or it may be pulling downward at 30 degrees toward the northwest.

Forces are commonly represented by an arrow. The direction in which the arrow points shows the direction of the force. The tail of the arrow is placed on the object that is *receiving* the force, and the length of the arrow is drawn to represent the *strength* of the force—for example, how many pounds or newtons are involved. Of course a pound or a newton is quite different from a length, so the length of arrow used to represent exactly 1 pound or 1 newton is quite arbitrary. You simply choose some convenient length to represent your unit of force. Some scale is selected that is convenient. For example, an arrow 1 inch long might stand for a force of 10 pounds; or an arrow one-tenth of an inch long might stand for a force of 100 newtons. To indicate the scale, the amount —or magnitude—of the force is usually marked next to the arrow.

A force is an example of a physical quantity called a *vector.* Other examples that will be described later have to do with motion and velocity. All such vector quantities have both a strength or size—usually called the magnitude—and a direction. And to *completely* describe such a vector you have to give three numbers: one to indicate the magnitude, and *two more* to tell the direction of the force. For example, a particular force might have a magnitude of 6 newtons; it might be pushing upward at a 30-degree angle, with respect to the horizontal plane; and it might have a direction that is 35 degrees northeast. There are other ways of describing a force, but each complete description always requires three numbers.

Those quantities in science that require only one number to tell the whole story, such as temperature, are often called *scalars* to distinguish them from vectors. Other examples of scalars are volume, brightness, and density.

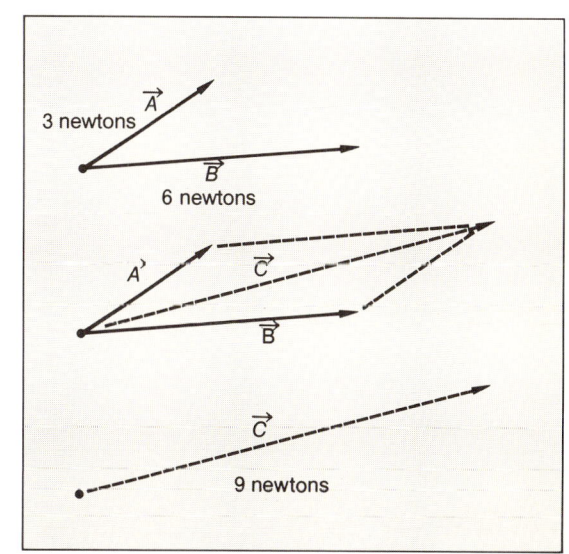

COMBINING FORCES

Earlier you read that if two forces pull in the same direction, they are said to produce a force equal to the sum of the two. For instance, a 6-newton force and a 3-newton force working together make a 9-newton force. But what happens when two forces don't exactly work together—

Figure 5. What is the effect of two forces, $\vec{A}$ and $\vec{B}$, acting together? To follow the rule of vector addition, first make a parallelogram (shown by the dashed lines), then draw a diagonal from the tails of the arrows to the opposite angle. The resultant force $\vec{C}$ is the same as $\vec{A}$ and $\vec{B}$ acting together.

6 | Pushes and Pulls

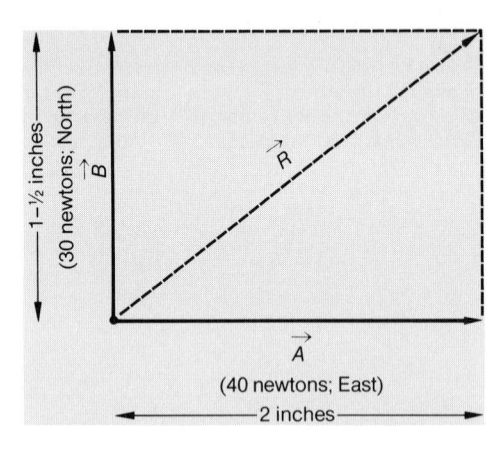

Figure 6. (Above) In this diagram, the scale is ½ inch = 10 newtons. First check that $\vec{A}$ and $\vec{B}$ are drawn correctly. Do you find that $\vec{R}$ is 50 newtons? In this special case, the two forces $\vec{A}$ and $\vec{B}$ are at *right angles*, so you can calculate the magnitude of $\vec{R}$ from Pythagorean's theorem for a right triangle. The "length" of $\vec{R}$ is obtained from $30^2 + 40^2 = 50^2$.

Figure 7. Examples of vector addition are shown to the right.

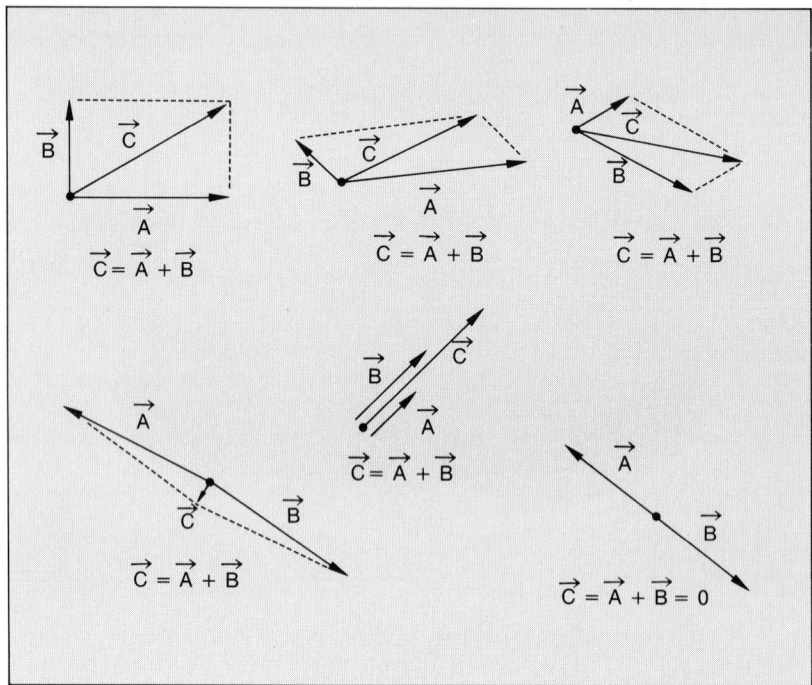

that is, when each force pulls in a different direction? Experiments can be done to show that two forces acting on the same thing are equivalent to one new force, and that the magnitude and direction of the new force are determined by the *rule of vector addition*.

The rule works this way. First you draw two arrows to represent the two original forces. Second, you complete the "parallelogram" (Figure 5) that is started by these two forces. And finally you draw an arrow across the diagonal of the parallelogram. This diagonal arrow represents the combined effect of the two original forces. It is the *resultant vector* of the two original vectors. It's convenient to think that the resultant of two vectors is really just the result of "adding together" the two vectors, and a simple way of representing this idea has been developed. First, a vector such as a force is depicted as a letter with an arrow over it. For example, $\vec{A}$ stands for the vector force $\vec{A}$ including both its strength *and* its direction. And the same for $\vec{B}$. The resultant vector $\vec{R}$ is then written as the sum of $\vec{A}$ and $\vec{B}$, or $\vec{R} = \vec{A} + \vec{B}$.

But this equation is not merely simple addition of the magnitudes of the force. The equation represents the *process* of using the rule of vector addition. So if $\vec{A}$ is 40 newtons east and $\vec{B}$ is 30 newtons north, $\vec{R}$ is 50 newtons northeast (and not 70 newtons). See Figure 6.

Of course if two forces point in the same direction they simply are added (see Figure 7). If they point in opposite directions, however, their magnitudes are subtracted. And if two forces are exactly equal in magnitude but opposite in direction, then the vector rule of addition tells us that the resultant disappears. We say the vector sum is a "zero vector." The two forces cancel each other; when they are acting together on the same object, it is as if there is no force at all.

TAKING FORCES APART

If someone gives you 5 dollars and someone else gives you 2 dollars, you can just pocket the money and remember that you have 7 dollars; you don't have to worry about where each dollar came from. You may

Finding the Vector Sum of Several Forces

How do you find the vector sum when there are more than two forces acting on the same thing?

Step 1. Pick any two forces and combine them to find out their resultant, and replace them by the resultant force:

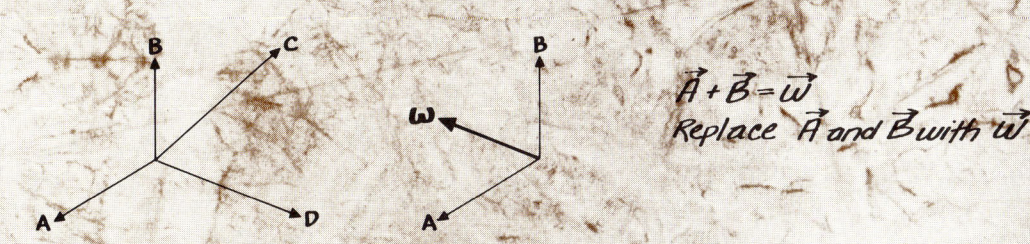

$\vec{A} + \vec{B} = \vec{W}$
Replace $\vec{A}$ and $\vec{B}$ with $\vec{W}$

Step 2. Now it's as though you have one force less than you did when you started out. You do the same thing again, treating the resultant as if it were like any of the other forces:

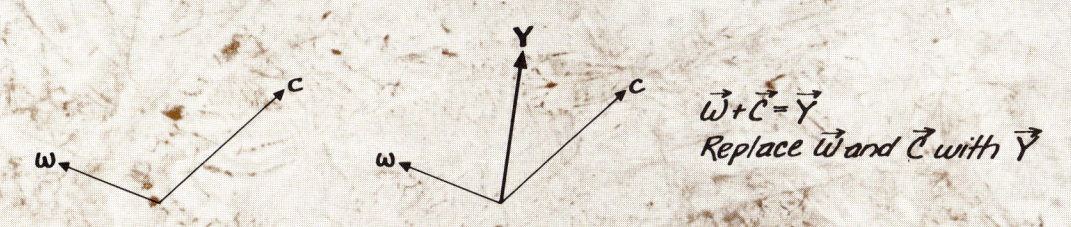

$\vec{W} + \vec{C} = \vec{Y}$
Replace $\vec{W}$ and $\vec{C}$ with $\vec{Y}$

Step 3. You continue doing this for as many forces as are present:

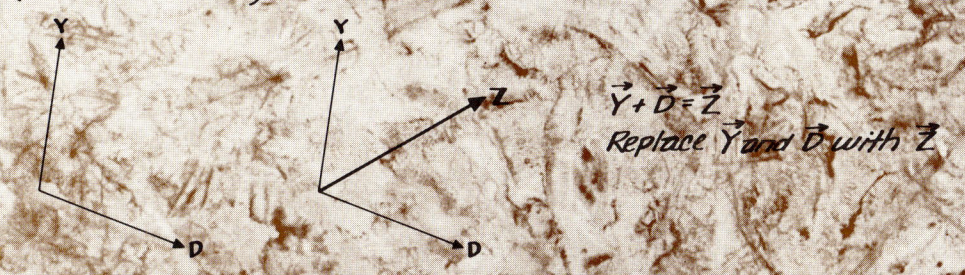

$\vec{Y} + \vec{D} = \vec{Z}$
Replace $\vec{Y}$ and $\vec{D}$ with $\vec{Z}$

Step 4. Now you have only one force left — the grand resultant, or total vector sum of all the forces.

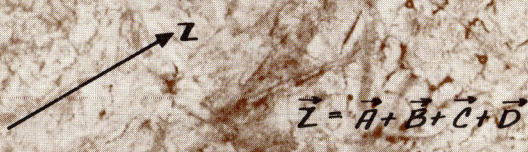

$\vec{Z} = \vec{A} + \vec{B} + \vec{C} + \vec{D}$

then spend, say, 1 dollar in one place and 6 dollars in another place. You can "take apart" the total amount of money any way you like to make two amounts that add up to the total.

A similar thing can be done with forces. Just as you can combine two forces to make a resultant, so also can you take apart any force into two pieces that are equivalent to the original force. The only requirement is that you must be able to recombine the two parts (as vectors) so that they exactly make up the force you started with. When you do that, we say that the force is broken into *components.*

Why would you ever want to take apart a force? Sometimes the parts of a force that act in different directions may have different effects, and perhaps the different effects should be thought about separately. For example, in Figure 8 a man is clinging to a slanted floor. The gravity force is pulling straight downward on him, but you sense from the picture that he may at any moment start to slide down. Why? Imagine that the downward gravity force is broken into two components—one component $\vec{P}$ pointing directly toward the floor and another component $\vec{S}$ pointing parallel to the sloping floor. The force $\vec{P}$, directed toward the floor, presses the man against the floor boards. But the component force $\vec{S}$ points in the direction that he can move. That is the force that will cause him to slide downward.

FORCES ALWAYS COME IN PAIRS

You know already that whenever there are forces at work, there must always be two different things working. If a force is acting *on* something, there is always *something else causing the force.* If you feel a force, you know that something else is pushing or pulling you. But take a closer look at what happens. Suppose that Gerald is pushing on Camille with a force of 20 pounds (Figure 9). If you think of the force *on* Camille you would draw the force *on* her. At the same time, however, Gerald is saying that Camille is pushing against *him.* So you could draw another diagram for him, as in Figure 9(b). And the force on Gerald is also 20 pounds. This example illustrates an important law of nature:

■ **Reaction Equals Action.** Whenever there is a force acting on something, that thing always pushes back just as hard.

The "pushing-back" force is called the *reaction* force, and it is always precisely equal in magnitude, but opposite in direction, to the original force.

Consider what happens when you push against someone's hands. If the person doesn't cooperate by pushing back, you can't really push at all. (Or if you try you may land on your face.) Suppose someone knocked you down with a hard shove from his shoulder. Do you feel that you gave him an equal shove back? The law says that you did. Why didn't he fall down, too? Because he was ready for it, and had his feet placed differently on the floor.

But where is the "equal" and "opposite" force when you push against, say, a wall? Does the wall really push back? It does. The wall pushes on you just as hard as you push on it (see Figure 10); if it did not, you would go right through it.

How does the wall push back? You must first recognize that the wall is not perfectly solid; it can be likened to a piece of hard rubber. When you start pushing, it yields slightly but then the parts of the wall are under tension so it starts pushing back. The more you push the more

Figure 8. Components of forces: The gravitational force $\vec{F}$ is equivalent to the sum of the two forces $\vec{P}$ and $\vec{S}$. Here, $\vec{S}$ is chosen to be parallel to the floor and $\vec{P}$ perpendicular to the floor; together they represent the *components* of force $\vec{F}$. The component $\vec{S}$ tries to cause the man to *slide* along the floor. The component $\vec{P}$ does not cause motion but *presses* him toward the floor.

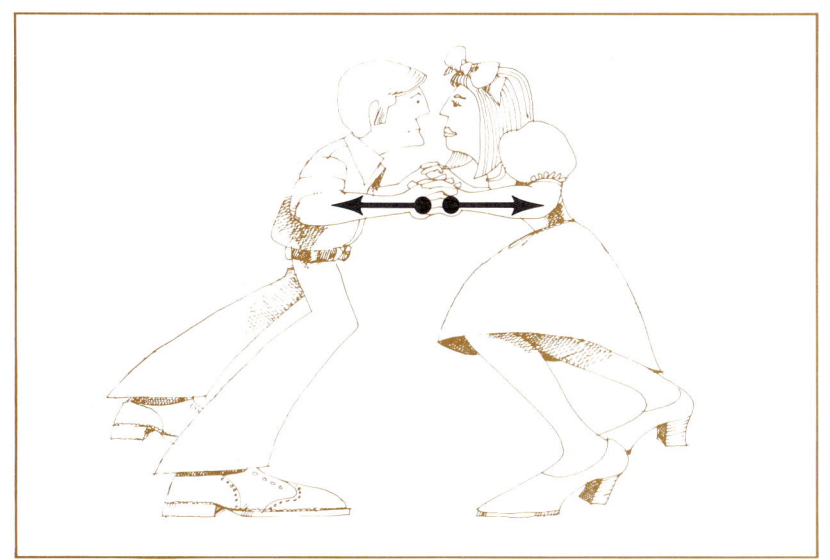

Figure 9. The force Gerald exerts on Camille is equal to the force Camille exerts on Gerald. Gerald's force on Camille is *her* reaction force, and Camille's force on Gerald is *his* reaction force.

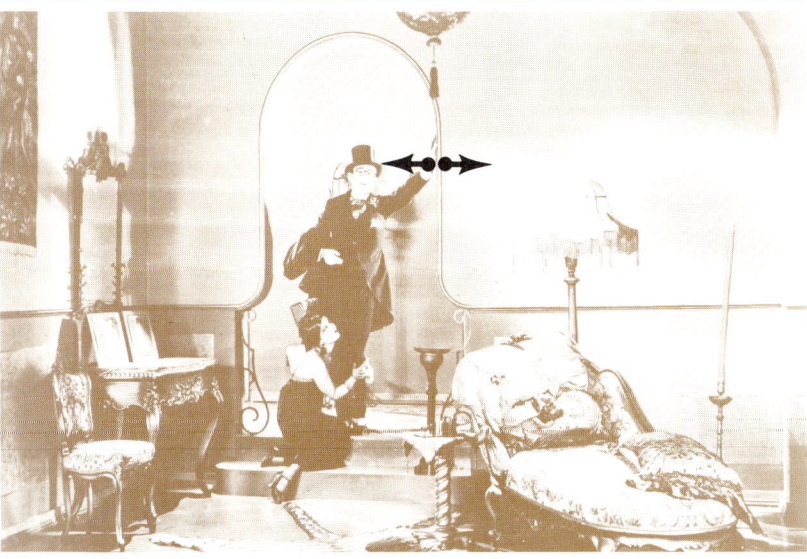

Figure 10. The wall holds up the man by exerting a reaction force equal and opposite to the force that the man is applying to the wall.

6 | Pushes and Pulls

97

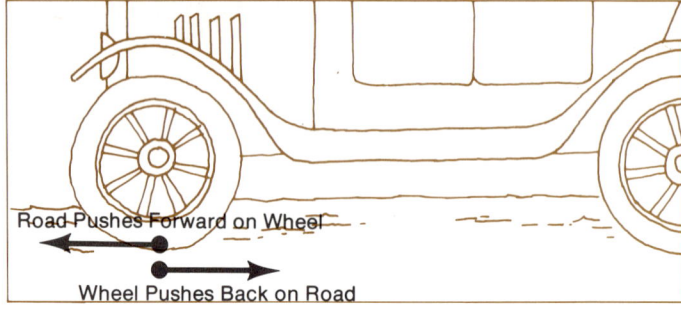

Figure 11. (Above) Forces come in pairs. When the hero pulls on the villain with a force $\vec{F}$, the villain pulls back on the hero with an equal and opposite force $-\vec{F}$.

Figure 12. (Above, right) The force that pushes an automobile forward is the reaction force of the wheel pushing backward on the road.

Figure 13. (Below) The upward forces exerted by the chair ($\vec{X}$) and the floor ($\vec{C} = 20$) and the downward forces exerted by the mother and the cat ($\vec{A} + \vec{B} = 110 + 12 = 122$) must add to zero, because the mother is at rest. Therefore the upward force from the chair must be $\vec{X} = 122 - 20 = 102$ pounds.

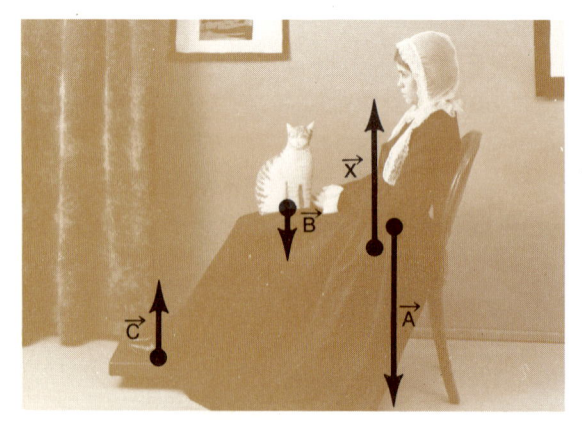

it yields—but the tension increases and the wall pushes back more and more. On a microscopic scale, it can be likened to what happens when you pull on a rubber band to stretch it. You can feel it pulling back on you. And if you let go, its force causes it to snap back to its natural length. When you stop pushing on a wall, it pushes you back until it has snapped back to its normal position.

For every force, the law of reaction is always at work. In Figure 11, the force that the hero is pulling on his rope is equal (but opposite) to the force that the villain is pulling on him. Or consider more common experiences. When you knock on a door, the door is also "knocking" with the same force on your knuckles. (If you don't believe it, knock harder and see what happens to your hand.) As another example of how forces come in pairs, think of the force that drives a car forward. When it is at rest the car pushes downward on the pavement—through the four tires—with its weight; and the pavement pushes upward on the car by the same amount (Figure 12). When the car is running, or accelerating, the motor of the car drives the wheels around, causing the tires to push backward on the pavement. This *backward* push is matched by an equal *forward* push by the pavement *on* the tires. It is this "reaction" force that drives the car forward.

THE LAWS OF STATICS

You are now ready to look at another of the basic laws of nature regarding forces. It is this:

■ **The First Law of Statics.** When anything is sitting quietly without motion the total force on it must be zero.

Most things you see spend a large part of their time sitting still. People, houses, cars, furniture, trees, rocks, animals, nearly everything. The law

Physical Science Today

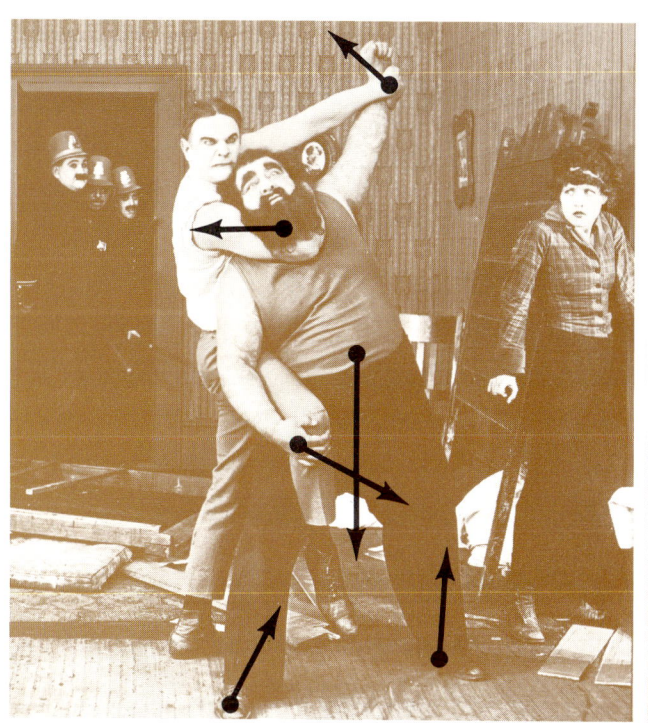

says that when something is *at rest,* the total force acting on it is zero. The "total" force means the vector sum of all the individual forces acting on the object.

Suppose you are sitting quietly in a chair with a cat in your lap. What are all of the forces acting *on you?* The gravitational force pulling downward on you is equal to your weight—say, 110 pounds. The force of the cat pushes downward on your lap—say, 12 pounds. The upward force from the chair pushes upward on you. And if your feet are on the floor, the floor pushes upward on your feet with a force—perhaps 20 pounds' worth. There are other forces, too, such as the force of the light photons that land on you. But that force is very small and you can forget it for now. There is also the force of air pressure acting all over your body. But this force presses nearly as much upward as downward and its total effect on you is not very large. So you can forget it, too.

The important forces are shown in Figure 13. According to the first law of statics, all these forces must add to zero. Said another way, they must work to cancel each other. From that law, you can figure out that the upward force on you from the chair must be 102 pounds. (For a somewhat more complicated situation, think of the forces acting on the bearded man in Figure 14.)

But the first law of statics doesn't tell the whole story. Even if all of the forces on an object add to zero, it is possible that the object can still move. It won't *go* anywhere, but it could twist around. If something is really at rest—not going anywhere *and* not twisting—one more thing can be said. It is called the second law of statics:

■ **The Second Law of Statics.** When something is at rest, the twisting effects of the forces on it balance to zero.

The definition of the "twisting effect" of a force, called the *torque,* is a little too complicated to go into now. But the engineer must carefully

Figure 14. (Above, left) The bearded man is at rest. If you put all the tails of all the arrows together and follow the process of vector addition, you will see that the vector sum of all the forces on him is zero. All of the important forces are shown.

Figure 15. (Above) The laws of statics:
I. When an object is at rest, the vector sum of all the forces is zero. (The upward forces balance the downward forces.)
II. The total twisting effect, or *torque,* of the force on the object at rest is zero.

6 | Pushes and Pulls

Figure 16. The gravitational force *between* two people. (Other forces, such as the gravitational pull of the earth, are not shown in this illustration.)

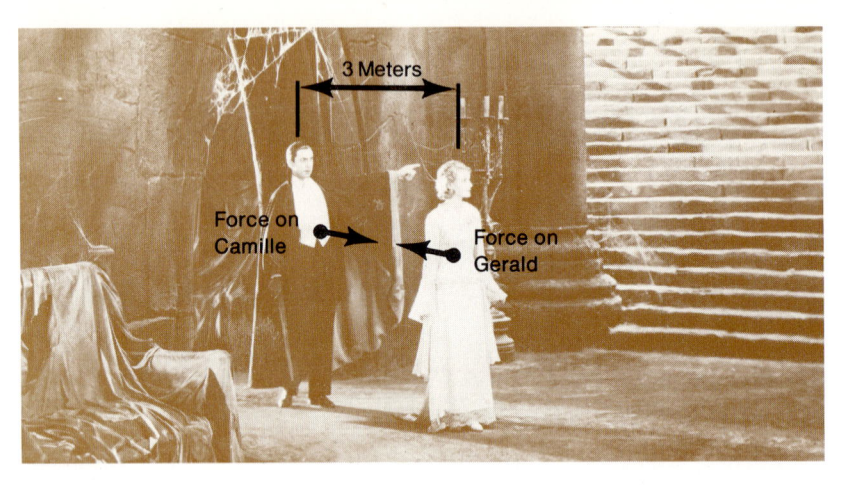

use both laws of statics when he is designing a structure he wants to stay put.

GRAVITY

Everything we know is pulled toward the ground by what is sometimes called "the force of gravity." It is perhaps more precise to say it is pulled by the gravitational force from the earth. It was Isaac Newton who first realized this force is not unique to the earth but acts throughout the solar system, controlling the shape and motion of the sun, the moon, and the planets. And we now know that it is the main force controlling the evolution of the galaxies and of the universe itself.

Newton not only realized the importance and universality of the gravitational force, he was able to describe how it works in precise, quantitative terms which he set forth as a grand law of nature. He called it The Universal Law of Gravitation:

■ **The Law of Gravitation.** Every two objects are pulled toward each other with a force on each that increases in proportion to the mass of each object and decreases in proportion to the square of the distance between them.

What does all that mean? First, think of two objects—perhaps Gerald and Camille as depicted in Figure 16. Call Gerald *"A"* and Camille *"B."* The law says that there is a gravitational force on each person due to the presence of the other; that the force on each acts toward the other; and that although the forces are opposite in direction their magnitudes are the same.

Next, the law says that every object has a quantity associated with it that is called its *mass.* The mass is a measure of the "quantity of matter" in an object, as measured by its ability to attract other objects. The easiest way to find out the mass of an object is to measure directly its gravitational attraction toward another object, or to compare its gravitational attraction to the attraction of some other object of known mass. The second way is the one most commonly used for objects on the earth—and is used, in fact, to establish a basic unit of mass.

The chosen unit of mass in the metric system is now usually taken as the *kilogram.* (The name may sound like a strange choice; it means "one thousand grams." The "gram" once was the standard, but then a unit 1,000 times bigger was found to be more convenient.) The standard kilogram is the mass of a special block of metal kept in a laboratory in France. Copies can be made of equal mass as it can be determined by comparing the gravitational pull on the two with a balance. The mass of any object of reasonable size can be determined by "weighing" it on some kind of a balance or scale, which compares its weight to the weight of one or more "1 kilogram" weights—or to some other smaller or larger objects that have also been calibrated with respect to a standard 1 kilogram mass. So you can find the masses of Gerald and Camille in Figure 16 by weighing them against some standard masses.

Finally, the law of gravitation says the magnitude of the gravitational force acting on the man and on the woman (due to the other person's presence only) depends on the distance from one person to the other; and the dependence is such that the force *decreases in proportion to the square of that distance.* For now, call the distance $\overline{AB}$ and say it is 3 meters. If Gerald and Camille move farther apart to 6 meters, doubling the distance, then the forces of attraction would be decreased by the amount you get if you divide by $2^2 = 4$. The force is only one-fourth as great. This kind of variation with distance is called an *inverse square law*. The way the force changes with distance is shown in Figure 17.

All of the law of gravitation can be written compactly, as well as more completely, in a simple formula:

Force of attraction = special number $\times \dfrac{\text{mass } A \times \text{mass } B}{(\text{distance } \overline{AB})^2}$ (*equation 1*)

In other words, to find the gravitational force acting on each person, A and B, you multiply mass A times mass B, divide by the square of the distance, and multiply by a special number. Then you have the force.

What is that special number? It is a property of nature and must be found by a measurement of the force, using objects of measured masses placed a measured distance apart. The number that is found will, of course, depend on what units are chosen for measuring masses, distances, and forces. In the metric system, the special number is found to be 6.7×10^{-11}. The kind of experiment used to find this number is described in Figure 18.

This special number is called the *gravitational constant*. Usually some convenient letter such as "G" is chosen to stand for such a number so that you don't have to write out the number until you are ready to calculate with it. And the equation that expresses the law of gravitation is often written this way:

Force of attraction = $G \times \dfrac{\text{mass } A \times \text{mass } B}{(\text{distance } \overline{AB})^2}$ (*equation 2*)

where G stands for 6.7×10^{-11} provided you are using the units "meters," "kilograms," and "newtons."

Now you can use this formula to calculate the gravitational force between Gerald (A) and Camille (B). You might use the following numerical values: mass A = 65 kilograms, mass B = 45 kilograms, and distance $\overline{AB}$ = 3 meters. Putting these numbers into equation (2), you get:

Force = $6.7 \times 10^{-11} \times \dfrac{65 \times 45}{9}$

$= 6.7 \times 10^{-11} \times 325$

$= 2{,}177.5 \times 10^{-11}$

$= 2.2 \times 10^{-8}$

The force is about 2.2×10^{-8} newton. That is a very small force, but not zero force. It is something like the weight of a flea.

The law of gravitation is not entirely correct or precise as it has

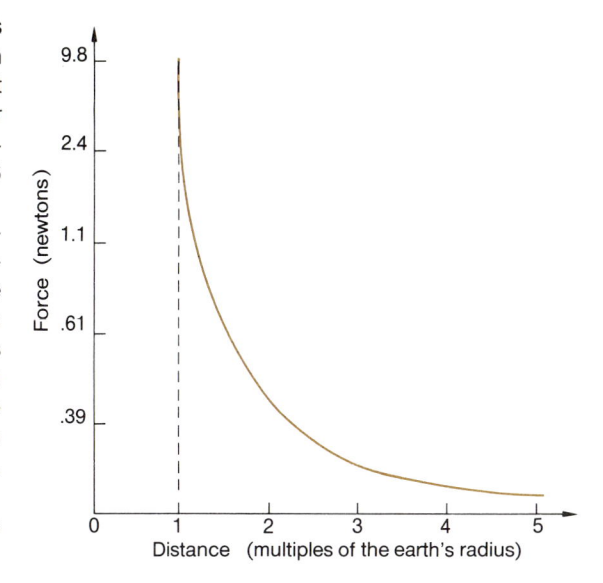

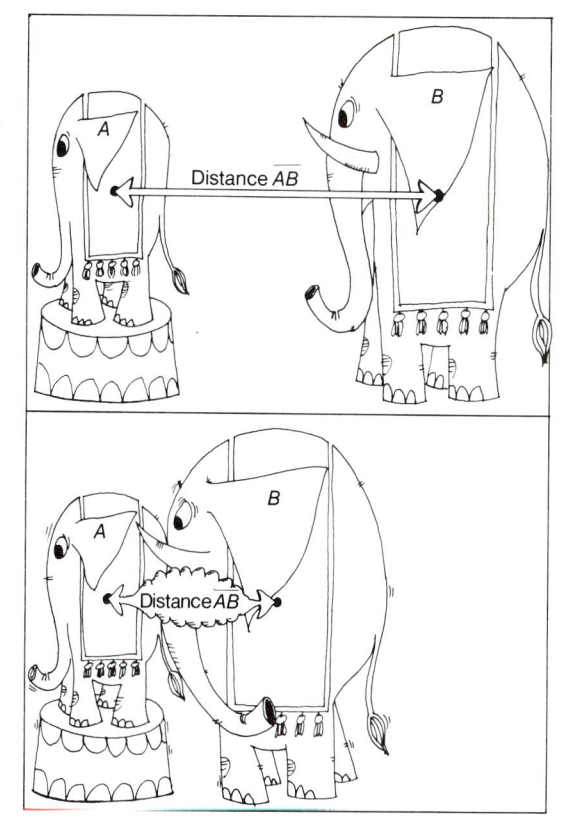

Figure 17. (Above) The magnitude of the gravitational force between two objects is inversely proportional to the square of the distance between them. (Below) The gravitational force between elephants A and B is given by the expression $\vec{F} = G \times \text{mass } A \times \text{mass } B / (\text{distance } \overline{AB})^2$. The force between the elephants is greater when they are closer together because the square of the distance between them is smaller.

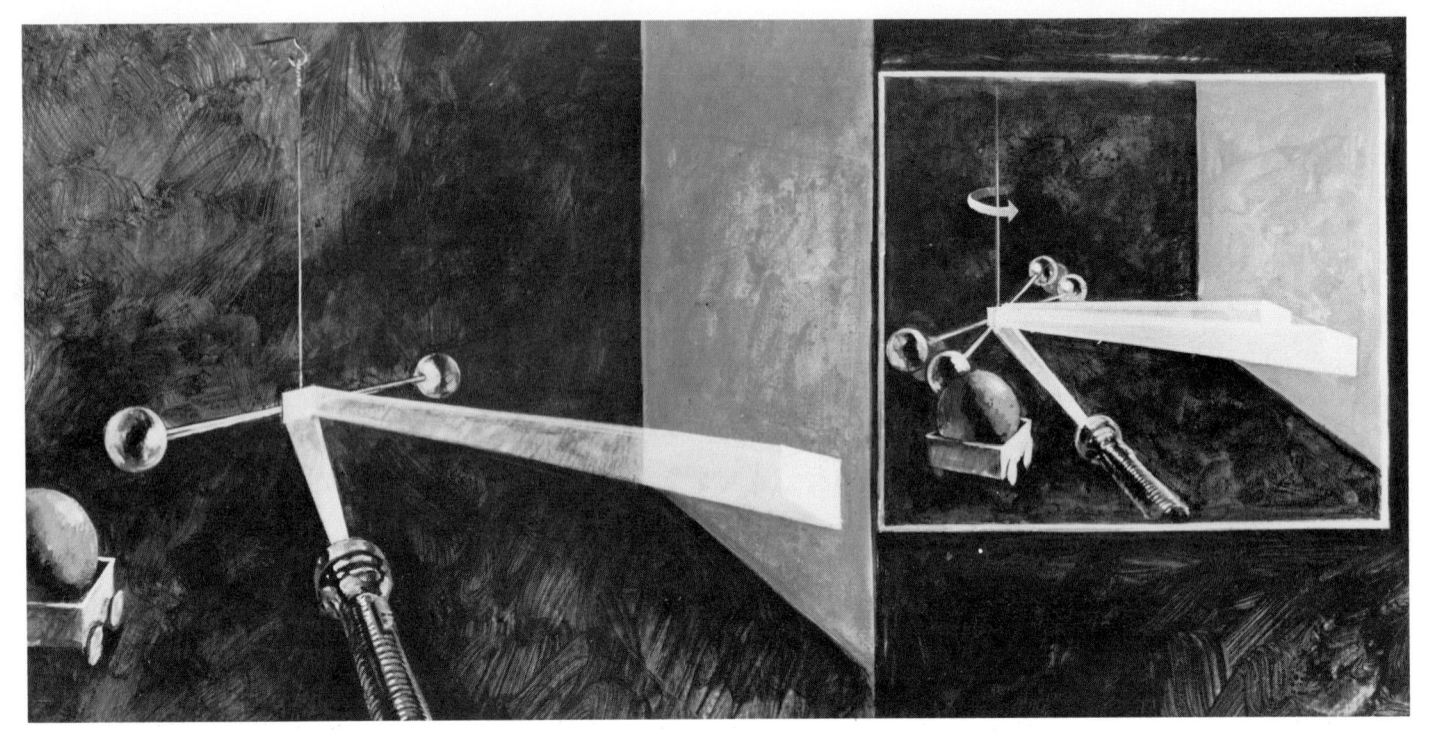

Figure 18. Measuring the gravitational constant G:

1. Two small, heavy balls made of gold are fastened to the ends of a thin rod that is suspended from its center on a long, thin wire.

2. A tiny mirror is glued to the rod. A small beam of light is pointed at the mirror, which reflects the light to make a spot of light on a wall a distance away. A slight twisting motion of the rod will cause the spot to move.

3. A large lead ball is placed near one of the small gold balls. The force of attraction pulls the smaller ball closer, twisting the rod and changing the position of the spot of light by a small distance. The large ball is moved to the other side of the smaller ball and the light spot moves a little bit in the other direction.

4. We can calculate how much twisting there is in the long, thin wire from how far the light spot moves.

5. By a direct measurement (or by calculation), we can determine how much force between the balls would cause such a twist.

6. Once we know the force, the masses of the large and small balls, and the distance between them, we can calculate the gravitational constant G

Note: To the scientist, a "constant" is a *number* that is always the same; it does not change as other things change.

been described. You can see the difficulty by referring back to Figure 16, which was based on the idea that the force depends on the distance $\overline{AB}$, the distance between the people. But what does that mean? Between their middles? But where is the "middle" of a person? Between one person's head and the other person's feet?

The law of gravity will now be stated in a more correct way. It has three parts:

1. *Every small bit of matter attracts every other small bit of matter with a gravity force proportional to the product of the masses of the two bits and inversely proportional to the square of the distance between them:*

$$\text{Force} = G \times \frac{\text{mass of bit } A \times \text{mass of bit } B}{(\text{distance } \overline{AB})^2}$$

2. *The gravity force between any two bits of matter is not changed by the presence of other matter between or nearby.*

3. *The total gravity force on any object from another object is the sum of all the forces on each of the bits of matter in the first object from all of the bits of matter in the second object.*

Part 2 means, for example, that if a dog runs between Gerald and Camille in Figure 16, the forces *between* them do not change, although each person will feel an additional force of attraction toward the dog. Even if the two people were standing on opposite sides of a concrete wall the gravitational force of each on the other would not change.

Part 3 means that to find precisely the force on, say, Camille, you must first find the force on her nose, then her ears, then each part of her head, then each hand, and so on. In the end you add up all the forces to find the total "force on *B*." But to find the force on her nose, you have to calculate the force *from* Gerald's nose, *from* his ears, and so on. It's a complicated process that would require the use of a computer! But it can be done. The approximate formula given earlier is much simpler and it gives a result that is good enough for most purposes.

Fortunately, the more precise law gives a simple result if the shapes of the objects are simple. If you have two spheres that are uniformly filled with matter, then equations (1) and (2) are precisely correct as long as you understand "the distance" means the distance between the *centers* of the two spheres (see Figure 17).

This result can be obtained from the precise statement of the law of gravity only by using the methods of calculus. (In fact, calculus was *invented* by Newton just to solve this problem.) It is an important result, for it allows us to understand easily the nature of the gravitational pull of the earth. The earth is fairly close to being a sphere. Each bit of matter above the surface of the earth is attracted toward each of the innumerable bits that go to make up the earth (see Figure 19). But the total effect is a pull toward the *center* of the earth as if the whole mass of the earth were concentrated there. So the gravitational pull of the earth is always approximately toward its center—approximately because the earth is not really a sphere and it is *not* filled uniformly with matter.

What is the strength of the gravitational pull of the earth for a small object near its surface? For any object near the surface, the distance to the center of the earth is always nearly the same as the radius of the earth. So you may rewrite equation (1) in the following way:

Gravity force on object at earth's surface $= \left[\dfrac{g \times \text{earth's mass}}{(\text{earth's radius})^2} \right] \times \text{object's mass}$ (*equation 3*)

All of the quantities in brackets are fixed numbers. Together they make a new number that is nearly the same everywhere on the earth. The number may be written as g, so the equation may be written in a simpler way:

Gravity force on object at earth's surface $= g \times \text{object's mass}$ (*equation 4*)

Figure 19. The total force on a bit of matter above the surface of the earth (the peanut in the sky) is the sum of all the small forces toward each bit of matter in the earth, represented here by a few elephants. The total force is equal to a pull *toward* the center of the earth, as if all the mass of the earth were concentrated there.

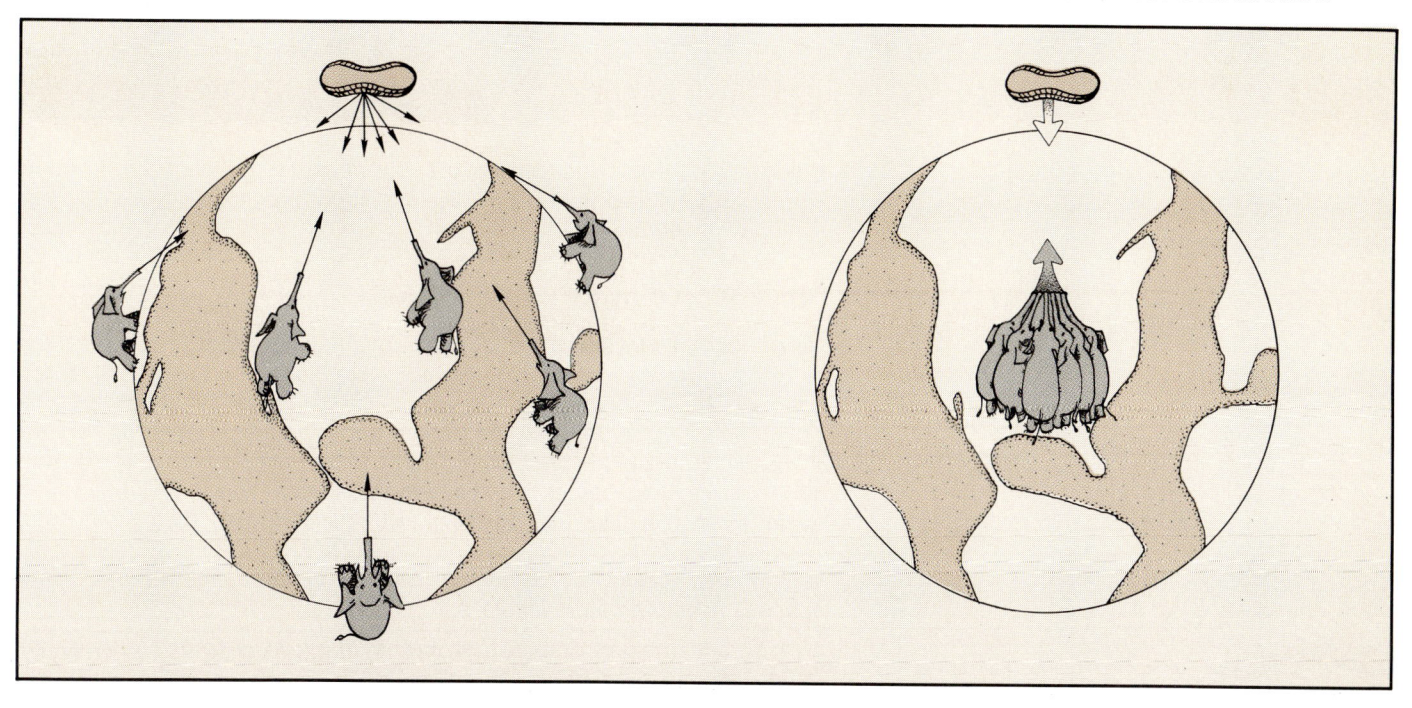

Figure 20. Frank is being pulled by the earth's gravity field. If the chains were not there, he would fall straight back.

If you figure it out, the number g is 9.8 newtons per kilogram. In other words, for an object that has a mass of 1 kilogram, the earth's pull (at sea level) is 9.8 newtons.

THE IDEA OF A FIELD OF FORCE

Scientists have invented a useful way of thinking about the forces such as gravity that appear on an object because it is in a certain place. So far, you have read that the pull of gravity on a person exists because of the gravitational force between the person and the earth. The earth pulls on the person, the person pulls on the earth, and the two pulls are equal. That situation may also be described in a quite different way that still gives the same result. The force on a person may be described with two statements:

1. *The earth causes something to happen in the space around it; the space is filled with a gravity field.*

2. *A person (or any object) that sits in this gravity field is pushed (or pulled) toward the earth.*

At first this may seem to be a strange way of thinking, but it has some great advantages. In fact, after thinking this way for a while the gravity *field* seems to become just as real as the earth or a person. One advantage is that you can take into account the gravitational effect of the earth on *all* objects by figuring out once and for all what the earth does to the space near it by describing the gravity field it produces. Then, when you want to know what the gravitational pull of the earth is on another person, a car, or anything at all, you can use the gravity field to find out.

Look back, for a moment, at equations (3) and (4). These two equations each say exactly the same thing but they can be thought about in different ways. Equation (3) represents the force between two objects. But equation (4) can be read to mean that the earth has a gravity field at its surface. The strength of the field is g, and the field direction is toward the center of the earth. Any object is pushed by this field with a force equal to the product of the field strength and the mass of the object:

Force = field strength × mass of object

The field strength (the same as the number g) can be calculated from the law of gravity. It is the same quantity that was put into the brackets in equation (3).

When you climb a mountain or fly up in an airplane, the gravity force on you diminishes because you are going farther from the center of the earth. You can account for this effect by saying the *strength* of the gravity field g decreases as you go higher in altitude. Then the field strength g is a number, and the *value* of that number depends on where you ask about it. Every point in space near the earth has associated with it some particular value for g. Then, once you know the value of g at every point—on Mount Everest, at 20,000 feet in an airplane, or at the moon—you can say any object found at that point is pulled toward the earth with a force proportional to g—the force is simply g multiplied by the mass of the object. Once you know about the field you can almost forget about the earth that caused it; you can think of the *field* as pulling on anything placed in it (see Figure 21).

THE ELECTRIC FORCE

Turn now to the second of the three basic forces—the electric force. Atoms are held together by the electric force between the central nucleus

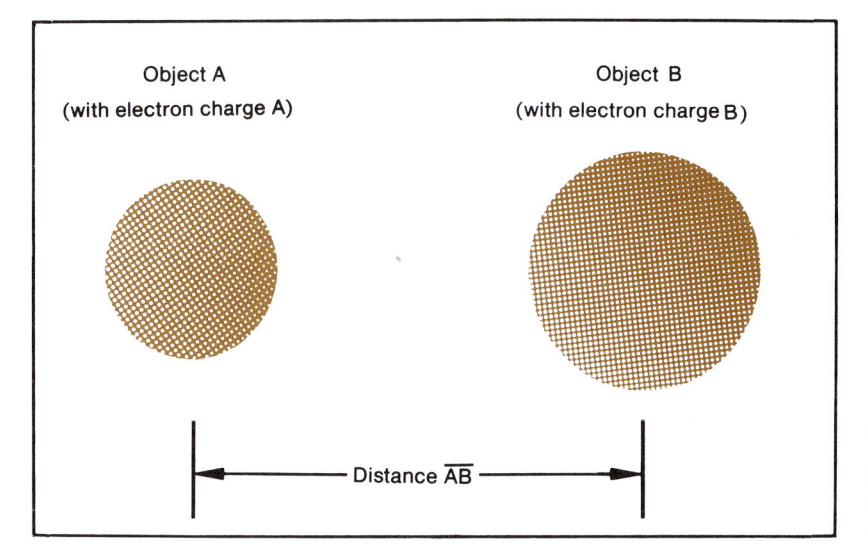

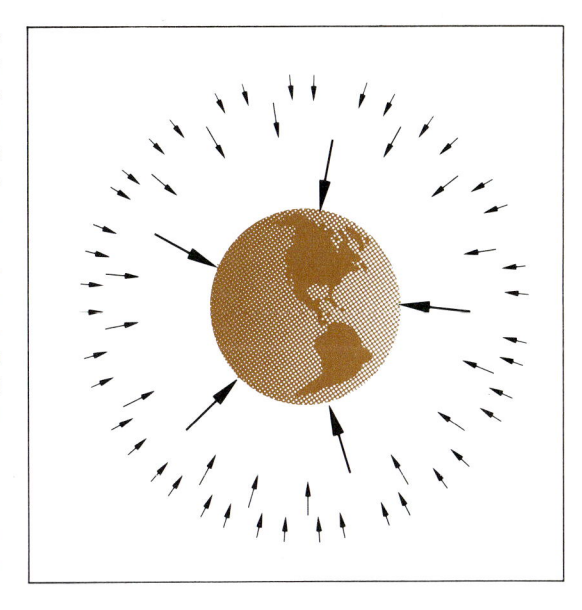

Figure 21. The earth produces a gravity field in the space around it. It is the gravity *field* that pulls objects toward the earth.

Figure 22. The electric force (in meter-kilogram-second units) between objects A and B is equal to 9×10^9 times the product of the charges on A and B, which may have nothing to do with their sizes. This product is divided by the square of the distance $\overline{AB}$.

6 | Pushes and Pulls

Figure 23. Electrified object A pulls on object C with twice the force that it pulls on object B. It does this because object C has twice as much charge as object B.

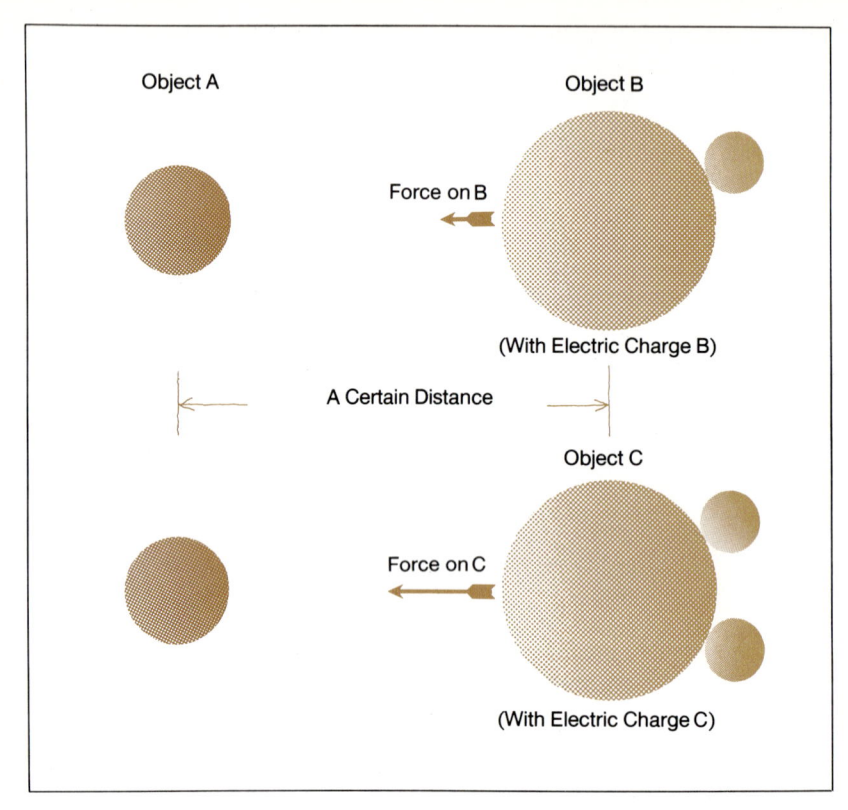

and the surrounding electrons. How does this force work? The nucleus has a certain property, called its *electric charge,* and each electron has a similar property, called *its* electric charge. These electric properties of the nucleus and the electrons cause them to be strongly attracted toward each other. The attraction is called the electric force.

The strength of the force acting on each of two electrified objects is in many ways similar to the gravitational force. It is, first of all, proportional to a certain property of each object. This time it is an *electric property* of each object that is called its *electric charge.* Second, the force is inversely proportional to the square of the distance between the two objects. The electric force has the same "inverse square" dependence on the distance as the gravity force. You can write an equation for the electric force in the following way (see Figure 22):

$$\text{Electric force on } A \text{ or } B = \begin{pmatrix}\text{special}\\ \text{number}\end{pmatrix} \times \frac{\text{charge in } A \times \text{charge in } B}{(\text{distance } \overline{AB})^2} \qquad (equation\ 5)$$

Notice how similar in form this equation is to equation (1) for the gravitational force.

How can you know what the electric charge is in something? In principle, you can use the same sort of procedure that was used to determine how much mass there is in an object. You can compare the *electric* pull on the object in a standard place to the electric pull on a special electrified object that is defined as having a standard amount of electric charge.

Using Figure 23, begin with some electrified object *A* (it can be any electrified object). A certain distance away, put a second object *B* that has some definite amount of electric charge, obtained in a definite and repeatable way. For example, object *B* might be a small metal ball on an insulating stick that you can "charge up" in a standard way, perhaps by touching it to a standard battery in a prescribed arrangement. You may define the electric charge carried by object B to be your *unit* of electric charge. Now place object B at a certain distance from object A and measure the push or pull on it. Next, take object B away and put any object — call it object C — at the same place and measure the force on it. If the force is twice as big, then you say that object C has twice the amount of electric charge as object B. If the force is 3 or 4 times as big, the electric charge is 3 or 4 times as big.

The electric charge carried by object C is *proportional* to the force on it in the standard place. And by comparing the force on object C to the force on the standard *unit charge* object B, you know the *amount* of the electric charge on object C. The amount of charge on object C is the ratio of the force on it to the force on the standard object B. Once you have this way of determining the amount of charge, you can check out the rest of equation (5) by measuring the force for various charged objects and distances.

In the metric system the unit of charge is called the *coulomb*. A coulomb is roughly equal to the amount of electric charge that passes through a 100-watt light bulb in 1 second. For practical reasons, its precise definition is given in terms of the force between wires when electric charges are moving in them.

Once the unit of electric charge is chosen, you can do the experiment to determine the special number in equation (5). This number is called "Coulomb's constant" and is written C for short. In the metric system of units, C has the value 9.0×10^9 (in newtons, meters, and coulombs). The basic law of electricity may then be written completely and compactly:

■ Fundamental Law of Electricity

$$\text{Electric force} = C \times \frac{\text{charge in } A \times \text{charge in } B}{(\text{distance } \overline{AB})^2} \qquad \text{(equation 6)}$$

$$C = 9 \times 10^9 \text{ newtons, meters, coulombs}$$

This law is often called Coulomb's law in recognition of the man who first discovered that electric forces work this way.

The electric force is different from the gravitational force in one very important way: the electric force may be either *attractive* (pulling the two objects together) or *repulsive* (pushing the two objects apart). Now it has been observed that if some object A *pulls* both object B and object C, then objects B and C will *push* each other apart. Objects B and C are said to have "like" charges, and "like" charges repel each other. If, on the other hand, object A pulls object B but pushes object C, then objects B and C are said to have "unlike" charges, and they will be *attracted* toward each other. Like charges repel; unlike charges attract each other.

For these two rules about charges to be always correct, there must be only two *kinds* of charges: "one" kind and "the other" kind. It is convenient to call one kind of charge *positive* and the other kind *negative*. Usually "positive" is chosen to be the kind of charge found on the atomic nucleus, and "negative" refers to the kind of charge found on the surrounding electrons.

It is convenient to call charges "positive" and "negative" for the following reason. When equal positive and negative charges are put together in the same object, they are equivalent to no charge at all. Their effects cancel out because the positive charge pulls (or pushes) as much on some other electrified object as the negative charge pushes (or pulls). So if we have both positive *and* negative charges in an object (as we do in every object in normal life), the total charge is the "algebraic sum" of all the charges—taking into account the positiveness and negativeness. For example, suppose you start with a *positive* charge of 6 coulombs in an object and you do something that puts into the object an additional *negative* charge of 4 coulombs. The actual charge that is now in the object is found by writing:

First charge   = +6 coulombs

Second charge = −4 coulombs

Total charge   = first charge + second charge

= +6 coulombs −4 coulombs

= +2 coulombs

In an orange, all of the nuclear charges add up to a positive charge of $+10^6$ coulombs. Why don't two oranges then attract each other with a tremendous electric force? Because the electric charges of all the electrons in an orange add up to a negative charge of -10^6. The total charge (or *net* charge) in an orange is zero. That is why there is not usually any electric force between two oranges.

But the electric charge in an orange is not always exactly zero. It is possible to electrify an orange either by adding extra electrons or by taking away some of the electrons that are there. In fact, that is exactly what has happened whenever you see electric forces between ordinary objects —when, for instance, you run a comb through your hair and it is attracted to the comb. When the comb passes through your hair, some of the atomic electrons of the comb are rubbed off onto your hair. There is no longer a perfect balance between the positive and negative charges on the comb (or on your hair) and the electric force is readily apparent.

The electric force appears in other places, too. When you stick a piece of adhesive tape on a sheet of paper, what is the force holding it there? It is the combination of billions of tiny electric forces between atoms of the tape and the atoms of the paper. The sticky side of the tape has a special kind of chemical substance that easily lets go of some of its atomic electrons. They jump over to the paper. The surface of the paper becomes negatively charged and the surface of the tape becomes positively charged, and the attraction holds the two together.

The electric force is also responsible for friction. If you slide your feet along the floor, the atoms of the floor and the atoms of the bottom of your shoes are pushed out of shape, and there is a resulting electric force of one on the other. In fact, *all* contact forces are really electric forces among atoms. When you shake hands with someone, your hand does not "go through" the other person's hand because of the strong repulsion force between the positive electric charges of the nuclei of the atoms of the two hands.

Finally, one other important force is electric at its base. When you use your muscles to move your body, you are really just changing the arrangements of atoms so that electric forces pull some atoms to new positions. When you turn a page of this book it's all done with electricity!

As was the case for the gravitational force, however, the law expressed by equation (6) cannot be the whole story. If the electrified objects are large, what is meant by "the distance from A to B"? A more correct way to deal with the electric force is the same way that is used to deal with the gravity force; and the precise statements about it may also be written out in more detail as was done for the gravitational force. However, electric forces are more complicated than gravity forces, because they depend on how the objects are moving as well as on where they are located. The law given in equation (6) is true only when both of the charged objects are sitting still. It is therefore called the law of electrostatics ("static" means "still"). You will be reading a complete description of how electric forces work later on in the book.

You may have wondered why the magnetic force was not included in the list of basic forces. That is because magnetism is just one part of the electric force between charged objects. It may be the main part when electric charges are moving. A "magnet" behaves the way it does because there are electric charges forever moving inside it.

NUCLEAR FORCE

The nature of the nuclear force is still one of the unsolved mysteries of science. We know some things about it, but we do not understand completely how it works. The atomic nucleus is a very small tight ball that contains two kinds of still tinier objects: protons and neutrons. The protons and neutrons seem to be identical in most respects except for the fact that the proton has a positive electric charge and the neutron has no electric charge (it is electrically "neutral"). There are, of course, electric forces among the protons of the nucleus. These forces are repulsive and would like to make the nucleus fly apart. But the neutrons and protons are stuck together by the strong nuclear force.

The nuclear force is quite different from the other forces you've looked at so far. It is very strong; also, it acts only across *very* short distances—only about 10^{-15} meter, roughly the size of the nucleus. It is, of course, attractive, because it holds the nucleus together. And the attraction is the same between all combinations of neutrons and protons.

That's all that you have to think about, for now, except for one interesting fact. A part of the nuclear force is very weak—it is called the *weak interaction*. And does strange things. It can change a neutron into a proton, and vice versa. It is this process that is responsible for most kinds of radioactivity. It is also the only force in nature that is different when things move in a right-handed circle and when they move in an exactly similar left-handed circle. Until this effect was discovered in 1957, it had been thought that the world was basically symmetric with respect to right- and left-handedness. But the nuclear weak interaction force has taught us that in one respect, at least, a right-handed universe and a left-handed universe would be somewhat different.

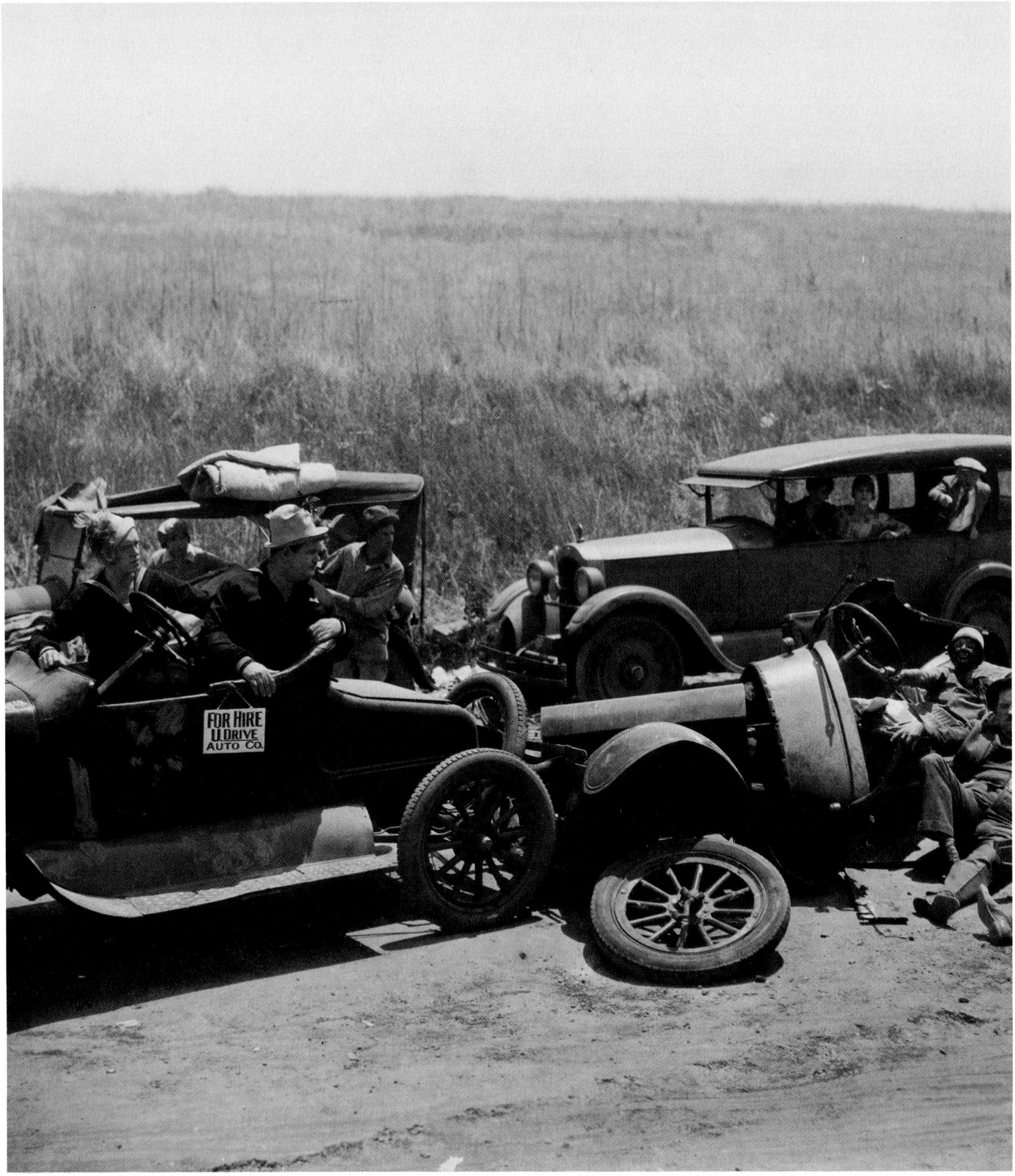
FOR HIRE
U. DRIVE
AUTO CO.

L ife is interesting and exciting to us because the world is always changing. Days are followed by nights and new days; seasons come and go. Each day, each year brings new sights of things we have not yet encountered. Old friends move with and around us, and strangers move into and out of contact in sometimes unexpected ways.

Changes come about because there is movement of matter. In all instances of change, some atoms are moving toward or away from other atoms. Such movement causes one part of our body to move with respect to another part; one car to move with respect to another car, or to a truck, or to the earth beneath it; the earth to move with respect to the sun; and the sun to move with respect to our Galaxy. Whether the change is on an everyday scale or part of the immense scale of the universe, the process of change itself is the motion of the parts.

DESCRIBING MOTION

When something moves, you are conscious of its motion because you notice that its position changes from one moment to another. You know that a car is moving rather than standing still when it is nearby one moment and farther away the next, or when it first appears on the horizon and then approaches the place where you are standing. You are conscious of the change in its position *relative to your own position.* To describe the motion of anything at all, you have to start with its position in relation to other objects, then you will be able to talk about the change of the position as time goes on.

Suppose you want to describe the position of a moving object as it arches through the air—for example, the position of the rose in Figure 2. At some instant in time it is at a certain place in the air—but how can you determine exactly where that place is? You can begin by describing its position *in relation to other objects*—in relation to the balcony, or to the woman, or perhaps in relation to the man who threw the rose. The choice itself is arbitrary.

Say you choose a particular spot on the ground at the feet of the person who threw the rose. That spot would be your *reference point* for the description of the position of the rose. You may label it "O" to stand for the originating point, or "Origin," for all of your measurements. The position of the rose with respect to this point may then be described in terms of the arrow extending from this point to the rose, as shown in Figure 2. This arrow is the *vector position* of the rose. It's conventional to represent vector position by the symbol $\vec{S}$. Why is the letter S used? Actually the letter P might have been better (for "Position"), but scientists use P to represent another quantity. Nevertheless, you can think of the letter S as standing for "Space position" because it does, after all, represent position in space.

Like forces, positions are vectors. And to describe the position vector $\vec{S}$ completely, you must specify certain quantities. One is the *magnitude* of the vector position and another is the *direction* in which it points. The magnitude is specified with one number—in this case, it is merely the straight-line *distance* from the reference point O to the rose. The direction is specified with two numbers. One might be the angle up

7
Motion

The total momentum in the universe doesn't change; it just moves around a lot.

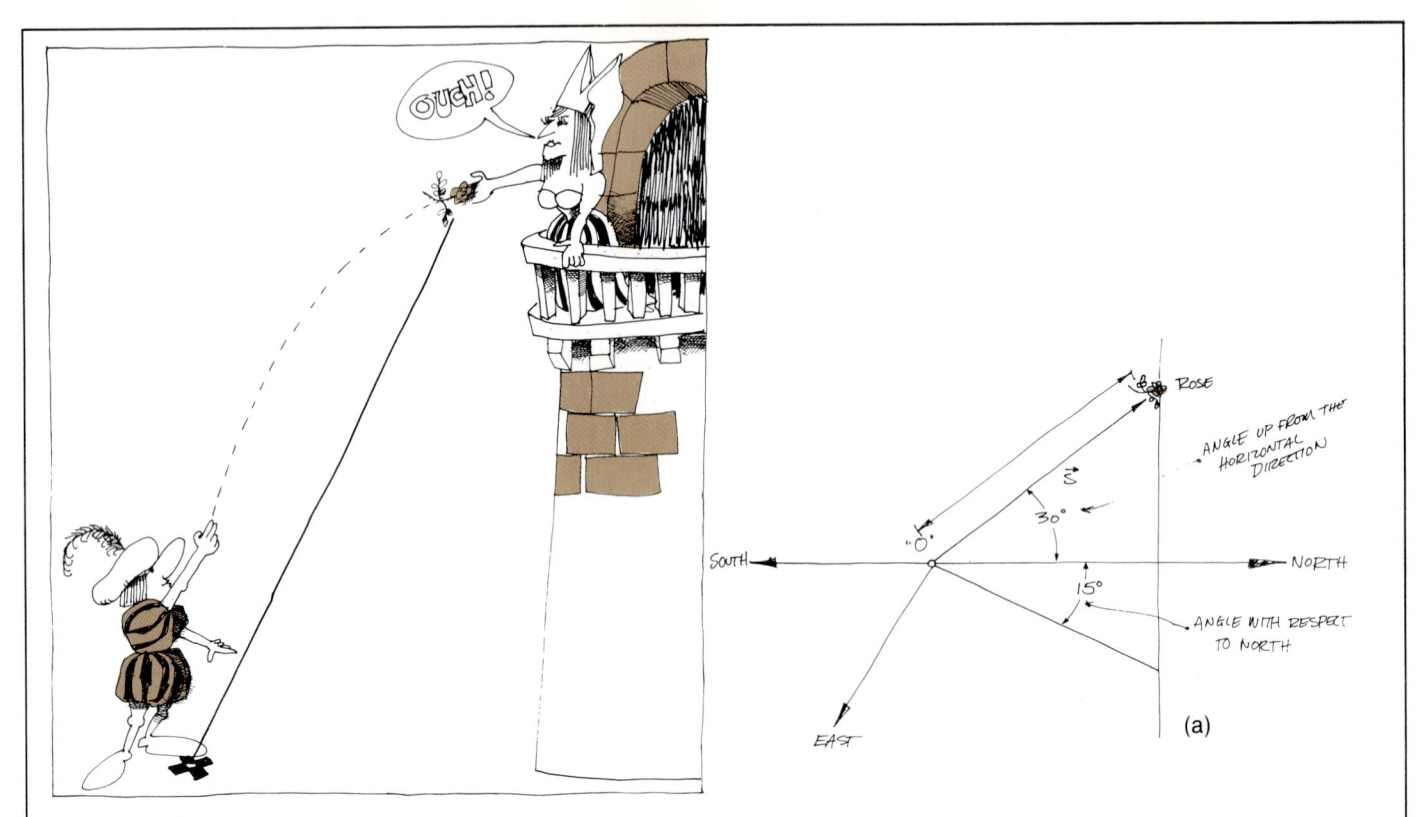

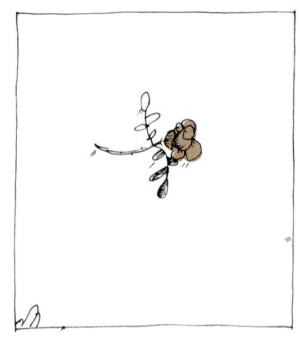

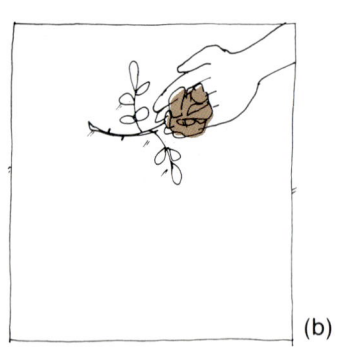

TIME	MAGNITUDE OF $\vec{S}$	ANGLE WITH RESPECT TO HORIZONTAL	COMPASS DIRECTION
0⁺ SEC.	0 METERS	60° 0'	NORTH
25 SEC.	2.45 METERS	56° 30'	NORTH
55 SEC.	3.97 METERS	51° 10'	NORTH
75 SEC.	5.27 METERS	44° 35'	NORTH

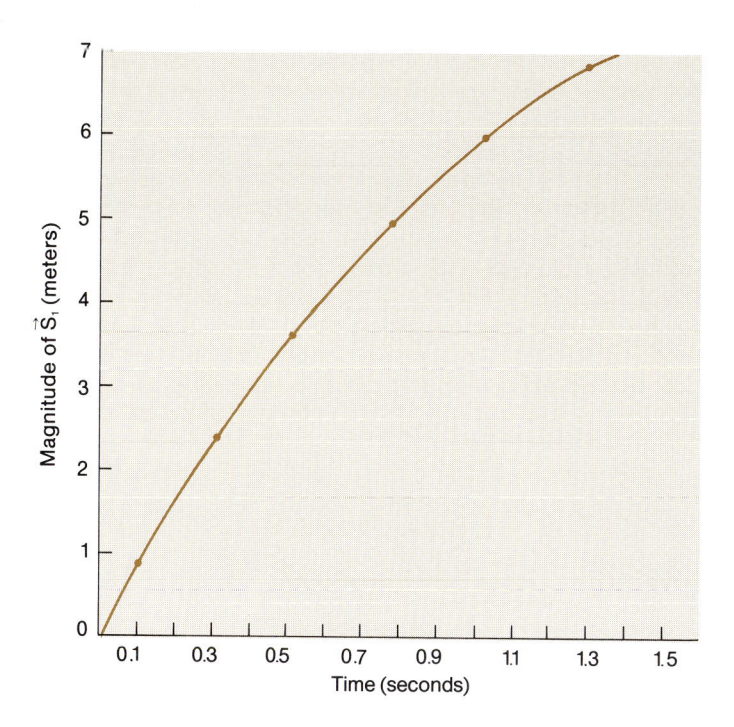

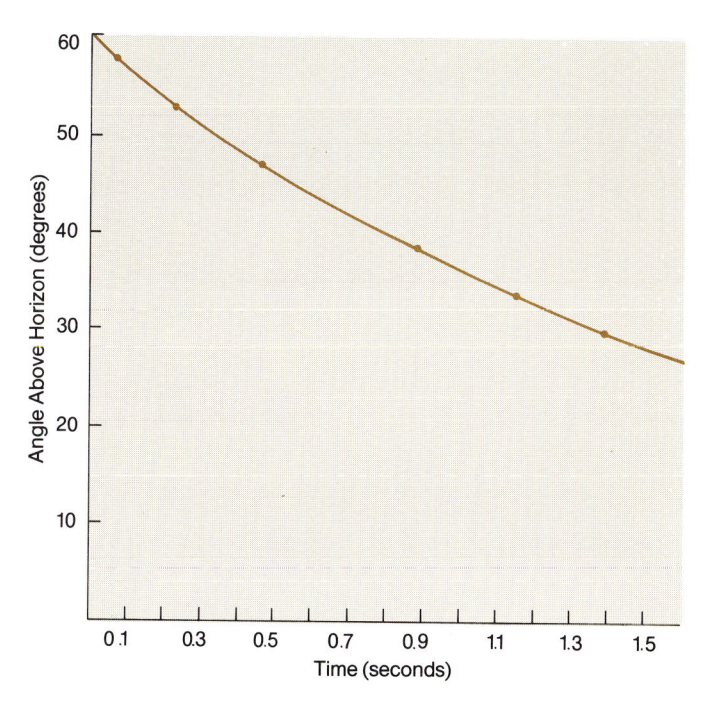

from the horizon, and the other might be the compass reading—say, the number of degrees of angle away from the direction called "north" (see Figure 2).

Once you have decided on a way of measuring the position of the rose, you will then be able to describe its motion by saying what its position is at various *times*. To specify the time you simply need some kind of a clock—perhaps a wristwatch—to mark off seconds, minutes, and so on. Then, at certain moments indicated by the clock, you may note the position of the rose.

Of course, it might be a little difficult to measure and write down the positions of the rose while the rose is moving through space. You could, however, record its positions with a motion picture camera, which would give you a sequence of pictures in rapid succession. Once you have the pictures you can look at them and measure off times, distances, and angles to describe the motion. If you put these measurements together in tabular form, as in Figure 2, you have a quantitative description of the motion. The description is not complete, of course, because there are instants of time that have not been recorded in the motion picture frames. But you can imagine what happened in between the frames because intuition tells you the rose moved smoothly from one position to the next.

It's even possible to give a more complete description of the motion by drawing a graph of the position for various instants of time. For example, the heavy, round points in the three graphs of Figure 3 depict the same information conveyed in Figure 2. And the fine lines that have been drawn through these points "fill in" the rest of the motion. By reading the graphs, the position of the rose at any time can be found from the three numbers that tell the position at any one instant of time.

CHANGES IN POSITION

Consider some other object—say, a golf ball. If the golf ball is at some position $\vec{S}_1$ at one time, let's say at the instant t_1, and is at a different position $\vec{S}_2$ at a later instant t_2, then there has been a certain change of position between the instants t_1 and t_2. Such a *change in position* will be

Figure 2. (Far left, above) The position of the rose is described by the vector position $\vec{S}$ with respect to some chosen reference point, the origin O. In (a), to describe the vector position $\vec{S}$, we give the *magnitude* of $\vec{S}$, the *distance* from O to the rose, and the two *angles* that describe the direction of the arrow with respect to some other agreed-upon directions—such as the horizontal direction, and the direction "north." In (b), a sequence of drawings depicts the flight of the rose through the air. The numbers in the table at the bottom of the page correspond to the position of the soaring rose at different instants in time.

Figure 3. (Above) Because experience tells you that the flight of the rose is smooth and that the calculations match the real situation, you can draw a straight line through the points calculated on a graph to show the motion of the rose *as a function of* time.

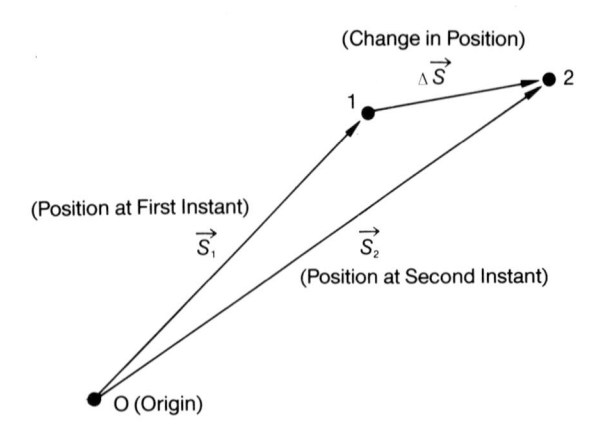

Figure 4. The change in the rose's position in space between two intervals of time can be calculated if you know where the rose is at each instant in time with respect to origin O. Here, $\vec{S_2} = \vec{S_1} + \Delta\vec{S}$ in vector notation.

represented by $\Delta\vec{S}$, a symbol which is to be read as "delta S." The symbol Δ is the Greek letter "delta," from which our letter D was derived, and it is now most commonly used to designate the change (or "Difference") in any quantity.

Now the change in position is also a vector quantity because its magnitude and its direction must be specified in order to describe it. Figure 4 shows the position of the golf ball at two different instants in time. The change in position $\Delta\vec{S}$ is the arrow from the first position to the second position.

Recalling the rule for vector addition described in the preceding chapter, you will see that the three vectors in Figure 4 are related in a logical way. The second position is the *vector sum* of the first position and the change in position. You may even write this relationship as a vector equation:

Position at second instant = position at first instant + change in position

or

$$\vec{S_2} = \vec{S_1} + \Delta\vec{S}$$

The equation says just what you might think: the second position can be reached by starting *from* the first position and adding to it the change in position (see Figure 5).

CHANGE IN TIME

As the golf ball travels from one position to a different position, a certain amount of time passes. Say your wristwatch indicates that the time when the golf ball is at the first position is 11 seconds and the time when the golf ball reaches the second position is 13 seconds. It's useful to think of the change in time—also called a *time interval*—from the first instant to the second instant, and to use a shorthand way of writing it as Δt. This change in time is merely the amount of time you add to time 1 to arrive at time 2. Written in equation form,

Time at second instant = time at first instant + change in time

or

$$t_2 = t_1 + \Delta t$$

Both are equivalent ways of saying the same thing.

SPEED AND VELOCITY

If you are driving somewhere in your car and you happen to notice the speedometer reads 30 miles per hour, you may think: "If I continue at this speed for 1 hour I will travel 30 miles." Similarly, if you kept driving for only 1/2 hour, you would travel 15 miles. Speed is, in fact, *defined* as a *number* that you multiply by a change in time to get the distance traveled. You may write this relationship in the following way:

Distance traveled = speed × change in time   (*equation 1*)

Figure 5. In order to measure $\Delta\vec{S} = \vec{S_2} - \vec{S_1}$, follow the rules of vector addition. You can do it simply by filling in the parallelogram with dotted lines.

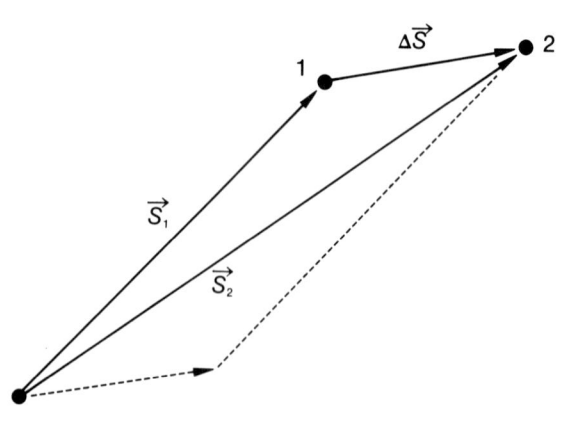

The distance traveled is related to the change in position that occurred during the change of time being considered. But the change in position is something *more* than the distance traveled, because the change in position must also include the *direction* of travel. Change in position is a vector quantity, whereas the distance traveled is simply a length—it is, in fact, simply the magnitude of the change in position. So

Figure 6. *Magnitude:* How hard he bends the spoon determines how far the beans travel. *Direction:* Which way he aims the spoon determines who he hits.

the change in position is a vector where length—or magnitude—is the distance traveled.

When you are traveling at some speed, the place where you eventually will end up will depend on the *direction* of travel. Scientists find it convenient to consider both the direction and the speed together in one idea, called the *velocity*. Velocity is a vector quantity—like an arrow—where direction is the direction of travel and where magnitude is the speed of travel. More precisely,

The velocity of an object is a vector that can be multiplied by a change in time to get the change in position.

Written in equation form,

Change in position = velocity × change in time

or

$$\vec{\Delta S} = \vec{v} \times \Delta t \qquad \text{(equation 2)}$$

Before this definition can really make sense, you have to know what it is supposed to mean when you multiply a *vector quantity* (velocity) by a *number* (change in time). The product is defined to be a new vector having the same direction as the velocity but a magnitude that is equal to the magnitude of the velocity multiplied by the change in time.

Suppose the brass pot being thrown in Figure 7 has a velocity in a certain direction and a magnitude of 1 meter per second. Suppose you watch it for 2 seconds and its velocity doesn't change. Where will it end up? Equation (2) says the change in position will be in the *direction* of the velocity and will be of the magnitude 1 meter per second × 2 seconds = 2 meters. The brass pot will travel a distance of 2 meters and will end up on the other side of the door.

(a) Velocity
1 meter/second
(Magnitude of Velocity)

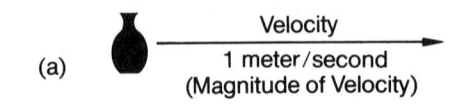

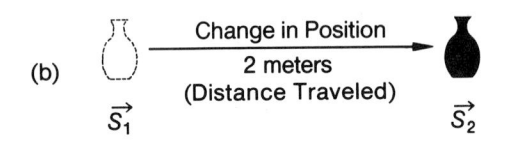

(b) Change in Position
2 meters
(Distance Traveled)
$\vec{S_1}$ $\vec{S_2}$

Figure 7. The distance the brass pot travels can be calculated by seeing how long it takes to travel from $\vec{S_1}$ to $\vec{S_2}$ at a constant velocity of 1 meter per second.

The unit chosen for measuring speed or the magnitude of velocity is usually a derived unit. It is obtained from a unit of distance divided by a unit of time. In the metric system, it would be 1 meter divided by 1 second, which may be written in various ways:

$$\text{Unit of speed} = \frac{1 \text{ meter}}{1 \text{ second}} = \frac{\text{meter}}{\text{second}} = \text{meter/second} = \text{meter per second}$$

They all mean the same thing. In other words, if you multiply a speed in meters/second by a time in seconds, you get the answer for a distance in meters. Suppose you are measuring a velocity and the speed is 20 meters/second and the time, 2 seconds. You would write

Distance = speed × time

$$= 20 \, \frac{\text{meters}}{\text{second}} \times 2 \text{ seconds}$$

$$= 20 \times 2 \, \frac{\text{meters}}{\cancel{\text{second}}} \times \cancel{\text{second}}$$

$$= 40 \text{ meters}$$

By writing out the *units* in the arithmetic you may cancel the "second" in the units and get "meters." But the cancellation works only if you use meters per *second* for speed and *seconds* for time. It wouldn't work if you tried multiplying 20 meters per *hour* with 2 *seconds*. Can you figure out how many meters per second are equivalent to the speed 60 miles per hour?

UNIFORM MOTION

Uniform motion means motion with a constant velocity. If the velocity is constant the *speed* does not change; neither does the *direction* of the motion. A car traveling 60 miles per hour on a perfectly straight and smooth road is an example. Other examples are a boat or an airplane going at a fixed speed in a fixed direction.

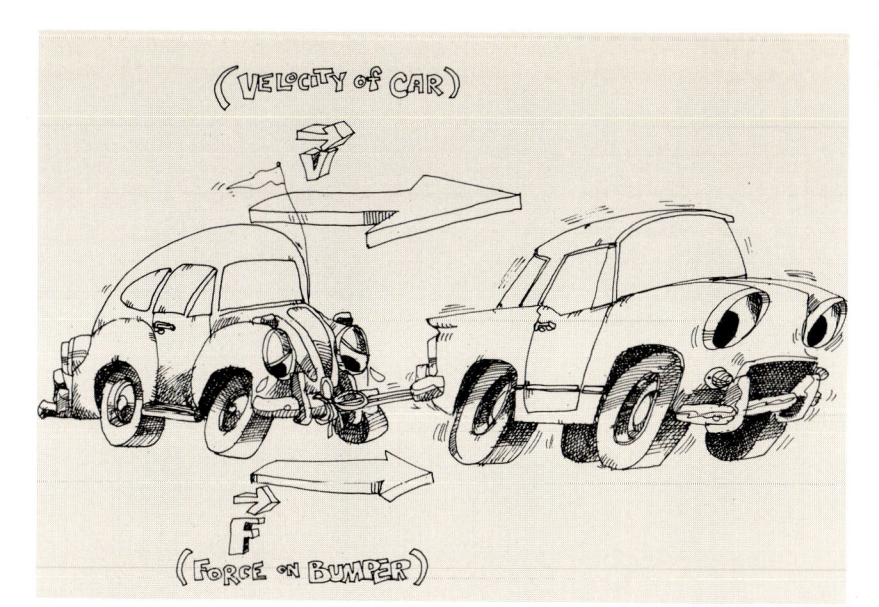

Figure 8. In towing an automobile at a constant velocity, the pull on the second automobile by the rope is a constant force $\vec{F}$.

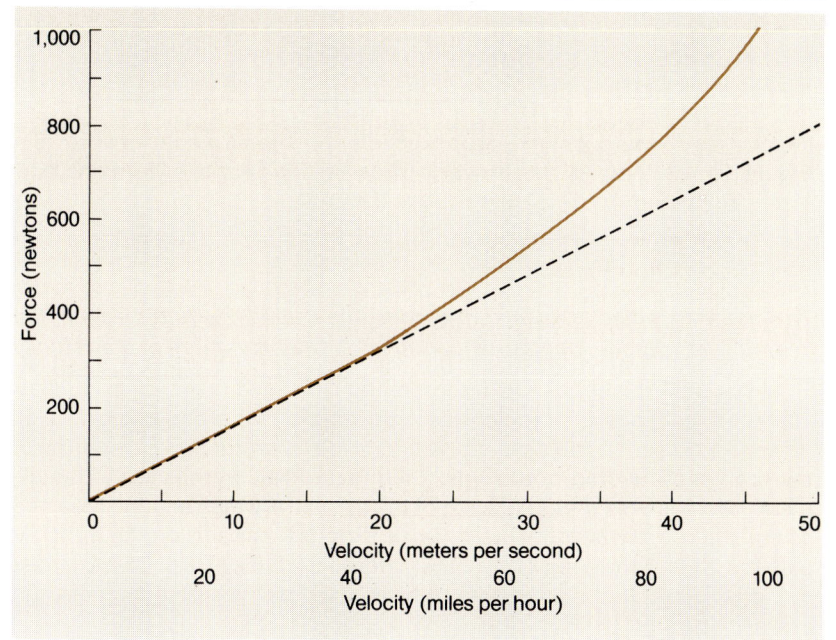

Figure 9. As an automobile goes faster, more force is needed to overcome friction in the engine and, at high speeds, air friction. This graph shows the force required to keep a particular automobile going at a fixed velocity.

Imagine towing a car along a straight road at a fixed speed (see Figure 8). To keep the second car moving, there must be a pull on the rope. The rope *pulls* on the towed car with a force $\vec{F}$. Experience tells you that once the car is moving, the force is some constant force. Both the magnitude and direction of the force are unchanging.

Or consider what happens when you are driving along a straight and level highway at a constant speed of 60 miles per hour. You know that your foot must keep the accelerator pressed down to a fixed position. The engine runs at a constant power; the wheels push against the road surface with a constant force; and the reaction from the road pushes the car ahead with a constant force. The force required to keep a typical car moving at a constant speed is shown in Figure 9.

As another example, consider a small airplane flying at a constant speed in a straight line. The propeller pushes backward on the air and the

7 | Motion

Figure 10. The air driven back by the propeller exerts a reaction force on the airplane, which helps to move the airplane forward.

air pushes forward on the propeller, driving the airplane forward. See Figure 10. As long as the forward force on the propeller remains constant, the speed of the airplane stays the same.

From these three examples, you can make the following generalization about uniform motion:

To keep any object moving at a constant velocity requires a constant driving force whose magnitude depends on the magnitude of the velocity.

But what determines the driving force that is needed to maintain constant speed? It depends on the various drag forces that are working to slow down the object. In the case of the car, there is the friction of the tires on the road and the drag of the wind on the car itself. In the case of the airplane, there is only the drag force of the air on the wings and the fuselage. These drag forces are pulling *backward* on the car and on the airplane, and the forward driving force must overcome their effect. There is also the *downward* force of gravity, which must be balanced by an equal *upward* lift force on the wings if the airplane is to stay in the air.

As far back as the 1600s, Galileo was already thinking about uniform motion. He proposed that the constant driving force required to keep an object in uniform motion is exactly equal to the total drag force trying to slow it down, so that the *total force* on the object is zero. This idea, often called "the first law of motion," may be written in this way:

■ **First Law of Motion.** As long as the total force on an object is zero, the object will remain in a state of uniform motion.

What happens when an airplane flies at constant velocity may therefore be interpreted to mean that the forward driving force from the propeller is exactly equal in magnitude but opposite in direction to the drag force of the air on the parts of the airplane (see Figure 11).

But what if the airplane is sitting still on the ground? Even then it has a "constant velocity"—the velocity stays at the special value *zero.* So the total force on the airplane is again zero, which is just another way

Figure 11. The total horizontal force on the airplane is the sum of the forward driving force $\vec{F_1}$ and the backward drag force $\vec{F_2}$. When the airplane is flying at a constant velocity, they add to zero. And when the airplane is flying at a constant altitude, the vertical forces (gravity and lift) are also zero.

of stating the first law of statics described in the preceding chapter. The first law of statics is merely a special example of the first law of motion.

To complete the story of uniform motion, consider briefly how the drag forces—often called *frictional* forces—arise. The drag forces result when some object contacts other objects—when, for example, the wings of an airplane rub against the atoms of the air. If there were no other objects around there would be no *interaction* and there would be no drag force. And if that were the case, a driving force would not be needed to keep the airplane moving. A satellite moving around the earth operates on this principle. If the orbit of the satellite is high enough, the atmosphere will be so extremely thin that, for all practical purposes, a drag force will no longer exist. The satellite will move forward at an unchanging speed without the need for a driving force.

But what has just been said about constant velocity is only approximately true. Even without drag, the velocity of a satellite cannot really be a constant velocity because the *direction* of the velocity changes slowly as the capsule swings around the earth. How this change comes about will be considered later in this chapter.

MOMENTUM

Now that you have considered the special circumstance of uniform motion, you can turn to more complicated situations in which the state of motion can be changing. The idea of changing motion begins with the idea of "momentum."

Momentum is a kind of "quantity of motion" that can be associated with every moving object. The velocity $\vec{v}$ is one kind of quantity connected with motion. But it doesn't tell the whole story because it doesn't say anything about the *object* that is moving. You know intuitively that if a large object (a car) and a small object (a person) are each moving at the same speed—say, 5 miles per hour—the larger object has "more motion." At least it has more effect if it hits something (Figure 12). The quantity of motion called momentum takes this relation into account.

Figure 12. The larger the object, the more effect it has if it hits something.

The momentum of an object is defined to be the product of the mass of an object and its velocity:

Momentum of object = mass of object × velocity of object

By using $\vec{P}$ to stand for momentum, m for mass, and $\vec{v}$ for velocity, you can write the above equation in a simple way:

$$\vec{P} = m \times \vec{v} \qquad \text{(equation 3)}$$

Here, the momentum of an object is a vector quantity, also. What happens when the scalar quantity *mass* is multiplied by the vector quantity *velocity*? The result is a new vector, momentum, where the direction is the same as the velocity but where the magnitude is equal to the mass multiplied by the magnitude of the velocity. In Figure 13, the airplane has a mass of 2,000 kilograms (about 4,000 pounds) and a speed, or velocity magnitude, of 90 meters per second (about 200 miles per hour). The momentum of the airplane is then found from:

Momentum = mass × velocity

= 2,000 kilograms × 90 meters per second

= 180,000 kilograms × meters per second

Notice that the unit measure of momentum is a derived unit, obtained by multiplying the unit of mass (kilogram) by the unit of speed (meter per second):

Unit of momentum = kilogram × meter per second

$$= \text{kilogram} \times \frac{\text{meter}}{\text{second}}$$

$$= \frac{\text{kilogram} \times \text{meter}}{\text{second}}$$

The unit has been written in several equivalent ways.

Why is momentum a useful quantity of motion? You can use it to describe one of the most profound laws of nature. Every measurement

Figure 13. The momentum of an object is the product of its mass and its velocity.

7 | Motion

Figure 14. Because the cowboy's momentum is conserved, he continues across the gap. The momentum he had before he left carries him across.

that has ever been made—from measurements of the motion of the smallest subatomic particles to measurements of the motion of the largest galaxies—shows that nature always follows a certain law. The "law of momentum," as it is called, can be described in this way:

■ **The Law of Momentum.** When moving things interact with one another, no matter what happens the sum of all of their momenta never changes.

As a simple example of this statement, think of a marble rolling along the floor and bouncing off a stationary marble (Figure 15). The two marbles have collided. Even though we don't know any of the details about what is happening to the marbles during the collision, the law of momentum tells us that *after* the collision, the sum of the momenta of the two marbles is the same as it was *before* the collision.

Remember that momentum is a vector quantity. When momenta are added, then, the "sum" means the result taken according to the rule of vector addition given in the preceding chapter.

In the collision shown in Figure 15, one of the marbles was sitting still before the collision, and after the collision the two marbles moved off in a symmetric way with equal speeds. That is one special kind of collision; several examples of how the law of momentum works for other collisions between two equal marbles are shown in Figure 16.

If two colliding marbles have different masses, the collision occurs in a different way but the law of momentum still works. Two examples are given in Figure 19.

Think about what can happen if a small steel ball has a hard collision with a glass marble. It's possible that the glass marble may break into many pieces. In this case, the total momentum *after* the collision is the sum of the momentum of the steel ball and the momenta of all the broken pieces of the glass marble (see Figure 17).

After any of the collisions just described, the marbles eventually will be slowed down and will come to rest. In each case the momentum will become zero. What happened to the momentum? The law of momen-

Figure 15. (Far right, above) A red marble collides with a blue marble of equal mass. The angle formed by the rebounding marbles is 90 degrees for collisions of equal mass. The vector sum of the momenta is the same before and after the collision.

Figure 16. (Far right, below) These drawings illustrate the collision between two identical marbles. In (a), with one marble initially at rest, the two marbles come out of the collision at a relative angle of 90 degrees. In (b), both marbles rebound at an angle that is the same as the angle of incidence. In (c), if the marbles collide head-on, then they rebound back down their original paths.

BEFORE

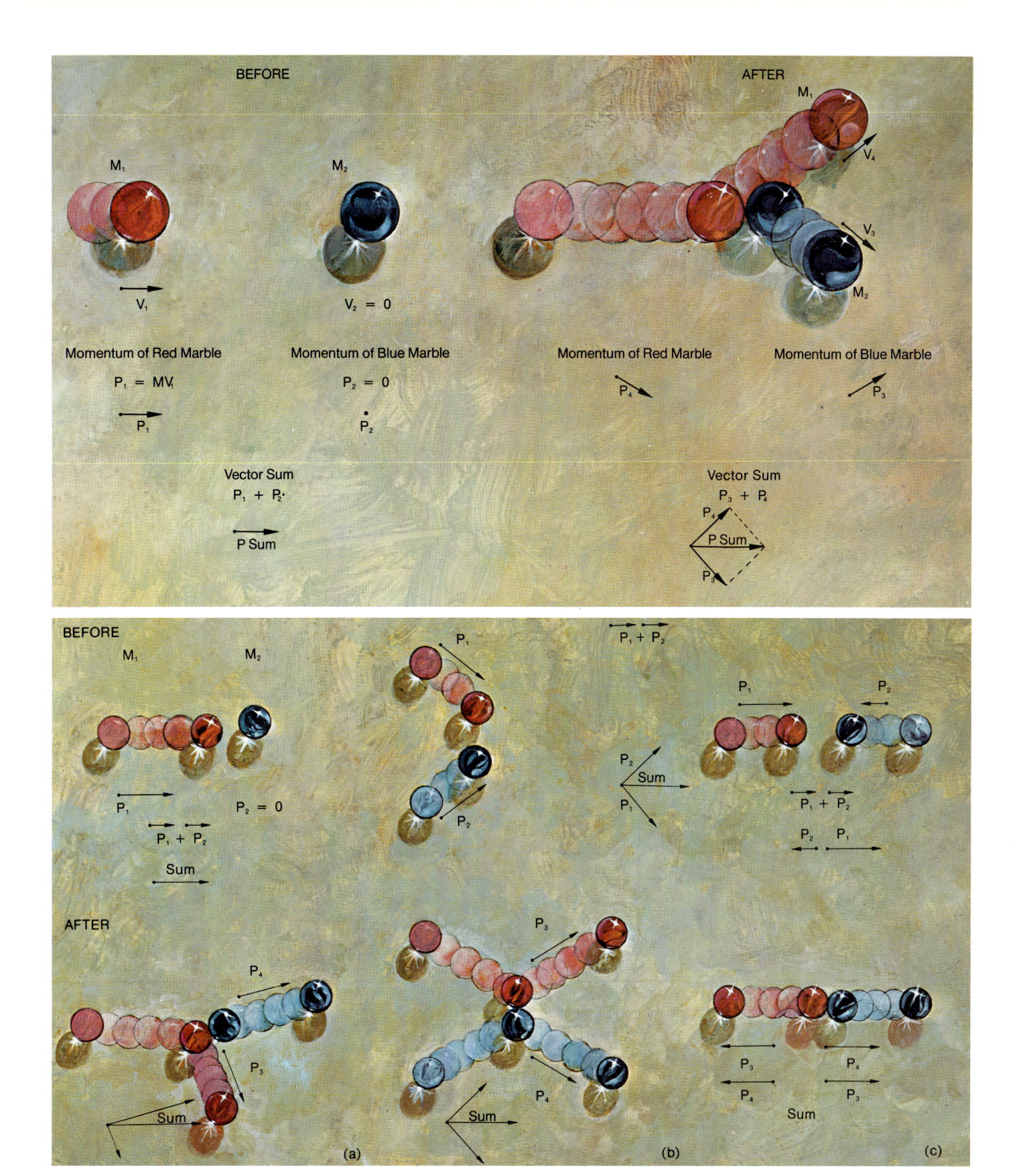

M_1 M_2

$\xrightarrow{V_1}$ $V_2 = 0$

Momentum of Red Marble
$P_1 = MV_1$
$\xrightarrow{P_1}$

Momentum of Blue Marble
$P_2 = 0$
$\cdot P_2$

Vector Sum
$P_1 + P_2$
$\xrightarrow{P\ Sum}$

AFTER

M_1

$\searrow V_4$

$\searrow V_3$

M_2

Momentum of Red Marble
$\searrow P_4$

Momentum of Blue Marble
$\nearrow P_3$

Vector Sum
$P_3 + P_4$

P_4
$\longrightarrow$ P Sum
P_3

BEFORE

M_1 M_2

$\xrightarrow{P_1}$ $P_2 = 0$

$\xrightarrow{P_1 + P_2}$
Sum

$\nwarrow P_1$

$\swarrow P_2$

$P_1 + P_2$

$P_2 \nwarrow$ $\nearrow P_1$
Sum

$\xrightarrow{P_1}$ $\xleftarrow{P_2}$

$\xrightarrow{P_1 + P_2}$

$\xleftarrow{P_2} \xrightarrow{P_1}$

AFTER

$\nearrow P_4$

$\searrow P_3$

Sum

$\nearrow P_3$

$\searrow P_4$

Sum

P_3

P_4

$\xleftarrow{P_3}$ $\xrightarrow{P_4}$

$\xleftarrow{P_4}$ $\xrightarrow{P_3}$

Sum

(a) (b) (c)

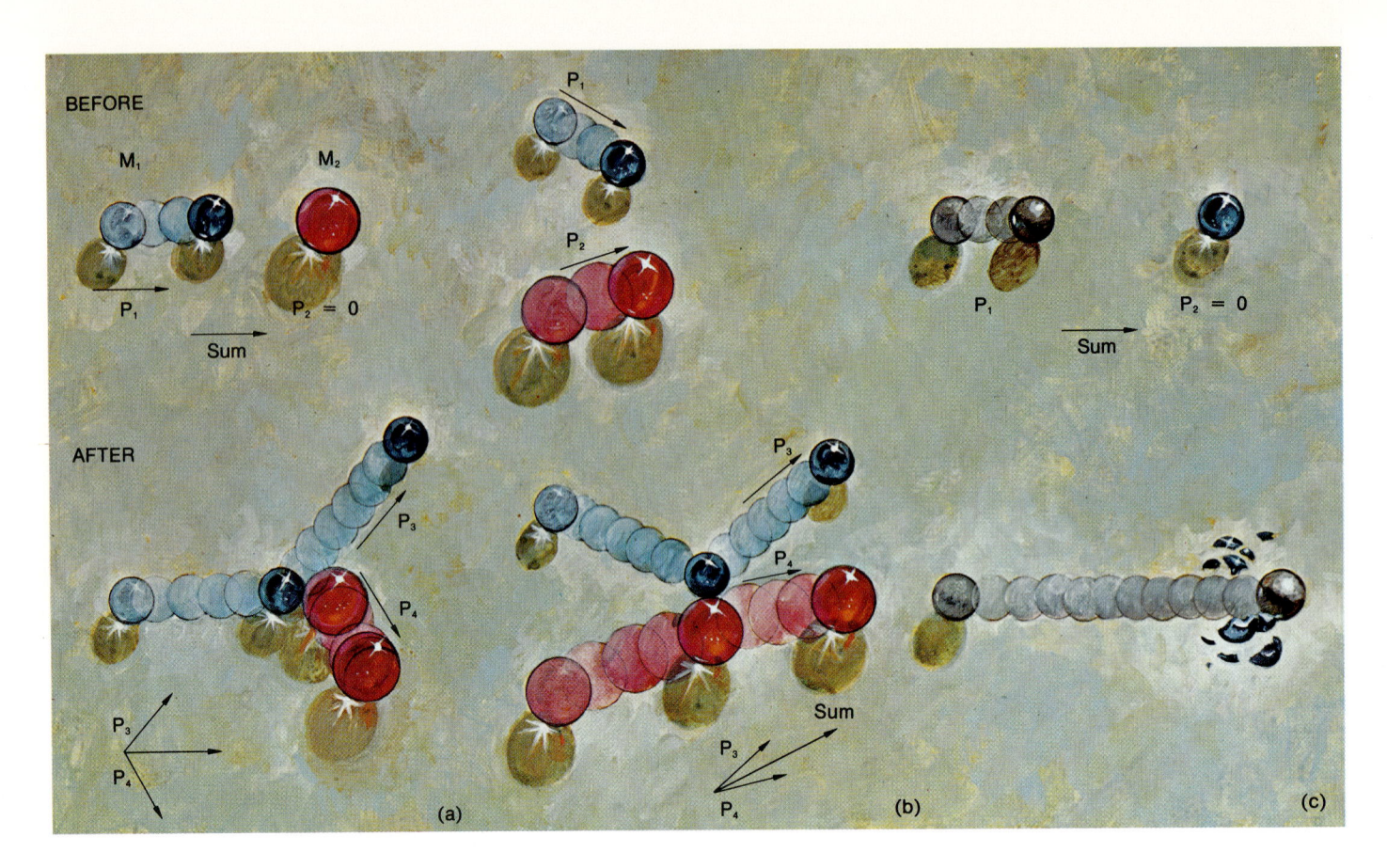

Figure 17. What happens when a large marble and a small marble collide? The momentum of the small marble is altered more radically relative to the change in momentum of the large marble. In (a), the small marble is deflected at an angle greater than 90 degrees; in (b), the angle of incidence and reflection of the small marble is greater than that of the large marble. What happens when a steel ball collides with a glass marble? In (c), the steel ball bursts through its target. The vertical components of momentum of all the fragments of the glass must add to zero.

tum says that it must go somewhere—it must end up in some other object. In fact, it goes into whatever object happens to be slowing down the speed of the marbles. If the marbles are rolling along the floor, the interaction of the marbles with the floor slows down the marbles through the forces of friction. *But in the process, the momentum of the floor must change.* In the end, all of the momentum that was in the marbles is given to the floor. It is only because the *mass* of the floor is so large that you don't notice the momentum it has acquired. It needs to have only a tiny velocity in order to have a momentum that equals the initial momentum of the marbles.

But surely the floor doesn't keep moving. Where does its momentum go? The floor is fastened to the earth. The momentum goes into the earth and the earth's motion is changed slightly—much too slightly ever to feel, of course, because the mass of the earth is huge compared with the mass of a marble. But change it does.

Consider the scene in Figure 18. The irritated man picks up a plate and throws it across the room, where it lands on the floor. He flings another plate into the air and it soars into the window pane; the glass shatters and the pieces fall to the floor. Becoming increasingly enraged, he lifts up a chair and throws it across the table; it bounces a few times and comes to a stop on the floor. Can you imagine how the momentum has moved around in this scene?

The word "conserve" means "to keep from being lost or wasted," or "to save." Whatever happens, the total momentum in the universe doesn't change. The momentum in a part of the universe is conserved until there is some interaction with other parts that allows some momentum to be shifted around. So the law of momentum says that nature always

Figure 18. You might think that considerable momentum is being transferred to the earth, but the earth's momentum already is so enormous, due to its large mass, that you would hardly notice the effects of all this destruction.

conserves its momentum. For this reason, it is often called "the law of the conservation of momentum."

FORCE AND MOMENTUM

Although the sum of all momenta never changes, the momentum of any particular object can change if it interacts with some other object; that is what happened in the marble collisions described earlier. During a collision, one marble interacts with another. During the interaction there are *forces* between the two marbles. It is the force of one marble on the other that changes their two momenta.

The first law of motion says that when there is *no force* acting on an object—when the net force is zero—the object continues to move uniformly; in other words, it moves with a constant velocity. And if the velocity is unchanging, the momenta, too, are unchanging. Only when a force acts does momentum change. There is a simple rule that relates the change in momentum to the force that is acting. This rule was discovered by Isaac Newton, and it is known as the second law of motion. It can be said in this way:

■ **The Second Law of Motion.** The change in the momentum of an object in any small interval of time is proportional to the total force acting on the object, multiplied by the amount of time the force acts.

Consider a total force acting on an object—say a baseball—and think about what happens to it during a small interval between two instants of time. At the first instant of time, the ball has some momentum. After the

The second law of motion says that:

$$\Delta \vec{P} = \vec{F} \times \Delta t$$

Start with $\vec{F}$:

and multiply by Δt:

Then: $\Delta \vec{P} = \vec{F} \times \Delta t$

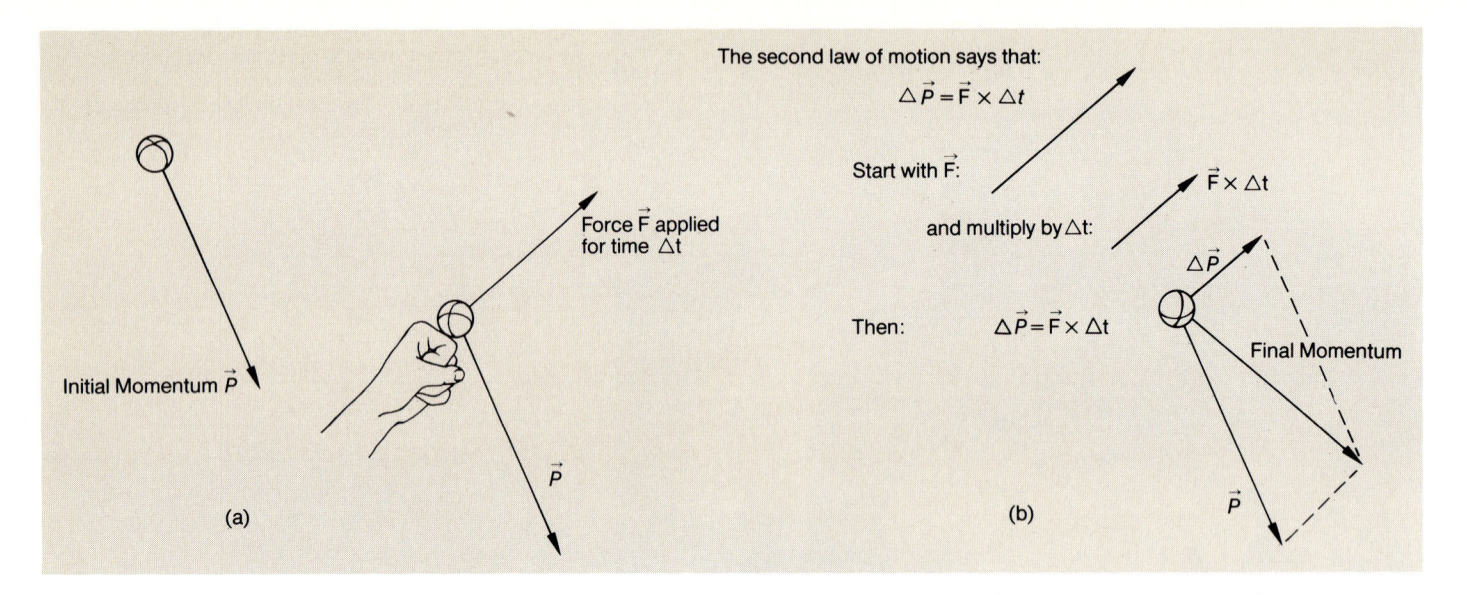

(a)   (b)

Figure 19. (a) The momentum of the ball can be changed by applying a force $\vec{F}$ for a time Δt. (b) The change in momentum is $\Delta \vec{P} = \vec{F} \times \Delta t$, and the final momentum is the vector sum of $\vec{P}$ and $\Delta \vec{P}$.

force has been acting through the time interval, the ball will, at the second instant of time, have a different momentum that is related to the *change in momentum*. In equation form,

Momentum at second instant = momentum at first instant + change in momentum

or

$$\vec{P}_2 = \vec{P}_1 + \Delta \vec{P}$$

which is a vector sum. The second law of motion says that the change in momentum is proportional to the force multiplied by the change in time. You may write this as

Change in momentum $\propto$ force $\times$ change in time

or

$$\Delta \vec{P} \propto \vec{F} \times \Delta t$$

where the special sign $\propto$ merely stands for the words, "is proportional to."

Now when someone says that one quantity A *is proportional to* another quantity B, they mean that if A doubles, so does B. Or if A triples, so does B. And so on. Mathematically, B is always equal to A multiplied by some definite, special number. So if you have the relationship $A \propto B$, you could also write an equation

A = (special number) $\times$ B

You could then rewrite equation (3) as

$$\Delta \vec{P} = \text{(special number)} \times \vec{F} \times \Delta t$$

The letter *k* can be used to stand for that special number. Now you may write the second law of motion as:

$$\Delta \vec{P} = k \times \vec{F} \times \Delta t$$

The second law of motion was discovered by observing how the momenta of various objects changed when forces were acting on them.

The same measurements will give the special number *k*. The particular number you get for *k* depends, however, on what *units* you choose for measuring force, the time, and the change in momentum.

The metric system has been worked out so that calculations can be made as simply as possible. In particular, the unit of force in the metric system, the newton, has been chosen so that *k* turns out to have a numerical value exactly equal to 1. In other words, if you measure the force in newtons, the time in seconds, and the momentum in kilograms multiplied by meters per second, then *k* = 1. So in the metric system, we have the simpler form to represent the second law of motion:

$$\Delta \vec{P} = \vec{F} \times \Delta t$$

In other systems, *k* will have a different value. For instance, if you measure speed in "feet per second," and use "pounds" both to measure masses and to measure forces (as some people do), then *k* has the value 32.

The form we have given for the second law of motion is not the one you will find in most books. Most often the law is written as

Force = mass × acceleration

or, in shortened form,

F = *ma*

This equation means that when an object moves, the force on it is equal to the product of its mass and the acceleration it experiences.

This form of the second law of motion is less exact than the equation given above, but the two are equivalent for objects that are not moving too fast. The *acceleration* of an object is the change in its velocity Δv divided by time interval in which the change occurs; that is,

$$a = \frac{\Delta v}{\Delta t}$$

If you put this equivalence to *a* in *F* = *ma* you get

$$F = m \times \frac{\Delta v}{\Delta t} = \frac{m \Delta v}{\Delta t}$$

Now if an object's mass doesn't change, the quantity $m \Delta v$ is just equal to the change in its momentum, $m \Delta v = \Delta \vec{P}$. So the above equation is equivalent to the equation $\Delta \vec{P} = \vec{F} \times \Delta t$.

MOTION WITH A CONSTANT FORCE

When a constant force such as gravity acts on an object, the motion it causes is rather simple. Suppose you hold a tomato in your hand and drop it out of a third-story window. When you first let go, the tomato is not moving—it has no momentum. At that instant the momentum is zero. Some time later, after the tomato has fallen part way to the ground, you could identify its momentum at a second instant in time. Because the original momentum was zero, the *change* in momentum is simply equal to the momentum at the second instant in time. And the change in momentum is simply the gravitational force on the tomato multiplied by the amount of time it has been falling.

Consider a particular example. Suppose the tomato weighs about 1/5 of a pound, so that the gravitational force on it is just 1 newton. After the tomato has been dropping for a time interval of, say, 1 second, its change in momentum will be found from

Momentum at second instant = change in momentum = force × change in time

$$\vec{P_2} = \Delta\vec{P} = \vec{F} \times \Delta t$$
$$= 1 \text{ newton} \times 1 \text{ second}$$
$$= 1 \text{ newton} \times \text{second}$$

Now remember that the newton was *defined* so that 1 newton × second is the same thing as 1 kilogram × meter per second. So

$$\vec{P_2} = 1 \text{ kilogram} \times \text{meter per second}$$

Now if the gravitational force on the tomato is 1 newton, you can use what you learned about gravitational forces in the preceding chapter to determine that its mass is about 1/10 of a kilogram. Remember that momentum is mass times velocity:

$$\vec{P_2} = m \times \vec{v_2}$$

If the momentum is 1 kilogram × meter per second and the mass is 1/10 of a kilogram, the velocity is 10 meters per second. Or you may rewrite the equation as

$$\vec{v_2} = \frac{\vec{P_2}}{m}$$

Then, using the known values of $\vec{P_2}$ and m, you would have

$$\vec{v_2} = \frac{1 \text{ kilogram} \times \text{meter per second}}{1/10 \text{ kilogram}}$$

If you had watched the tomato fall for 2 seconds, the change in time would have been 2 seconds and

$$\vec{P_2} = \Delta\vec{P} = 1 \text{ newton} \times 2 \text{ seconds}$$
$$= 2 \text{ newtons} \times \text{seconds}$$

or

$$\vec{P_2} = 2 \text{ kilograms} \times \text{meter per second}$$

For this case,

$$\vec{v_2} = \frac{\vec{P_2}}{m} = \frac{2}{1/10} = 20 \text{ meters per second}$$

The second law of motion says that when the tomato falls with the constant force of gravity, it picks up a velocity of 10 meters per second after the first second and a velocity of 20 meters per second after the next second. If it continues to fall, it will have a velocity of 30 meters per second after the third second, and so on.

What happens when you *throw* a tomato out of a window is shown in Figure 20.

The motion described for a falling tomato *is the same for any object falling freely (without any drag from the air) in the gravity field at the sur-*

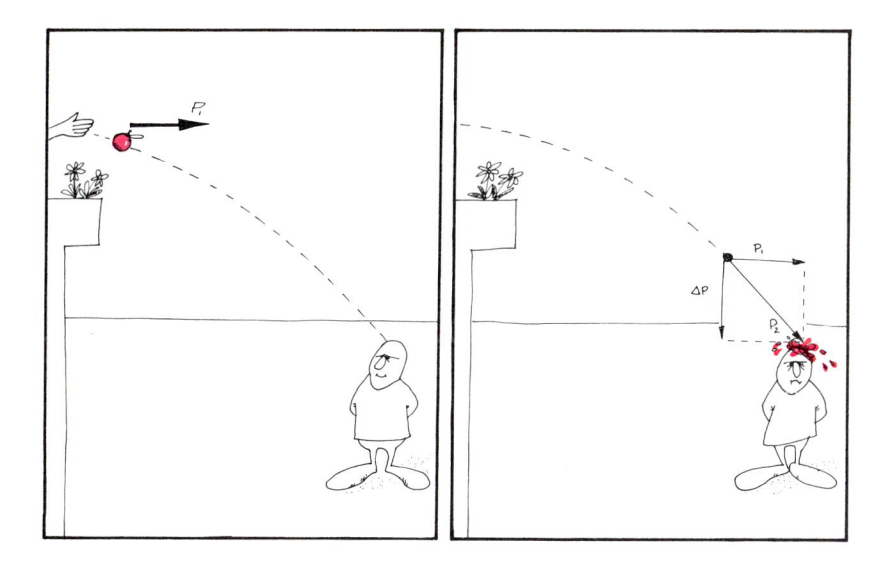

Figure 20. A tomato is thrown out of a window. In (a) you see the path taken by the tomato; in (b) you see the tomato at t_2 after a time Δt. Because the gravitational force $\vec{F}$ is downward, $\Delta \vec{P}$ is also downward.

Figure 21. As long as the drag force of the air is negligible, all objects at the earth's surface fall in such a way that their downward velocity increases by *10 meters per second* for each second they are falling.

face of the earth. Suppose the object is 5 times as heavy as the apple. The force will be 5 times as great, so the momentum gained in 1 second will also be 5 times larger. But because the *mass* is also 5 times larger, that momentum corresponds exactly to the same velocity the tomato had: 10 meters per second. As long as the drag force of the air is not an important effect, all objects at the earth's surface fall in such a way that their downward velocity increases by 10 meters per second for each second of fall.

If any other kind of *constant* force acts on any object, the motion is similar. For each second the force acts, the object will change its momentum (in the direction of the force) by the same amount.

CHANGING FORCES

All motion can be related to the force that produces it. The most complicated motions result when the resultant force on an object is *not* a constant force. There are forces where magnitudes or directions change in a specified way with time—for example, when you press your foot on the accelerator of a car. Or there are forces where magnitude and direction will depend on the position of the object. An example is the gravitational force of the earth, where magnitude varies with the distance from the center of the earth and where direction points back toward the center of the earth. For a moving satellite, both of these quantities depend at each instant on the satellite's position in space. In the most complicated of all situations, the force may depend on *both* the position of the object in space and on the time it arrives there.

In all of these situations, however, the path taken by the object is still governed by the second law of motion. It is *always* true that in any small time interval, the momentum changes according to

$$\Delta \vec{P} = \vec{F} \times \Delta t$$

But what do you use for $\vec{F}$ if the force is changing? What you have to do is think of the motion in many small segments. Suppose an object is moving along some curved path as shown in Figure 22(a). You can, on this curve, mark the position of the object at a whole sequence of instants t_1, t_2, t_3, t_4, t_5, and so on. It is convenient to take the interval between each successive instant as being always the same. But most importantly,

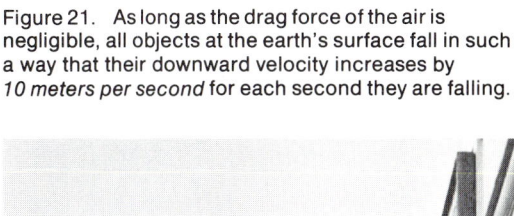

7 | Motion

Figure 22. When an object moves with a constant speed in a circular path, the *magnitude* of the force causing the motion does not change but its *direction* is always changing so that it points toward the center of the circle. In (a), as the smoothly varying force changes, so does the momentum giving a smooth *trajectory* of the motion of the object. (b) If you magnify a piece of the trajectory enough, you will find a small enough Δt such that the force is practically the same value inside the whole interval. (c) In this interval, the change of momentum of the object can be calculated from the formula $\Delta P = \vec{F} \times \Delta t$.

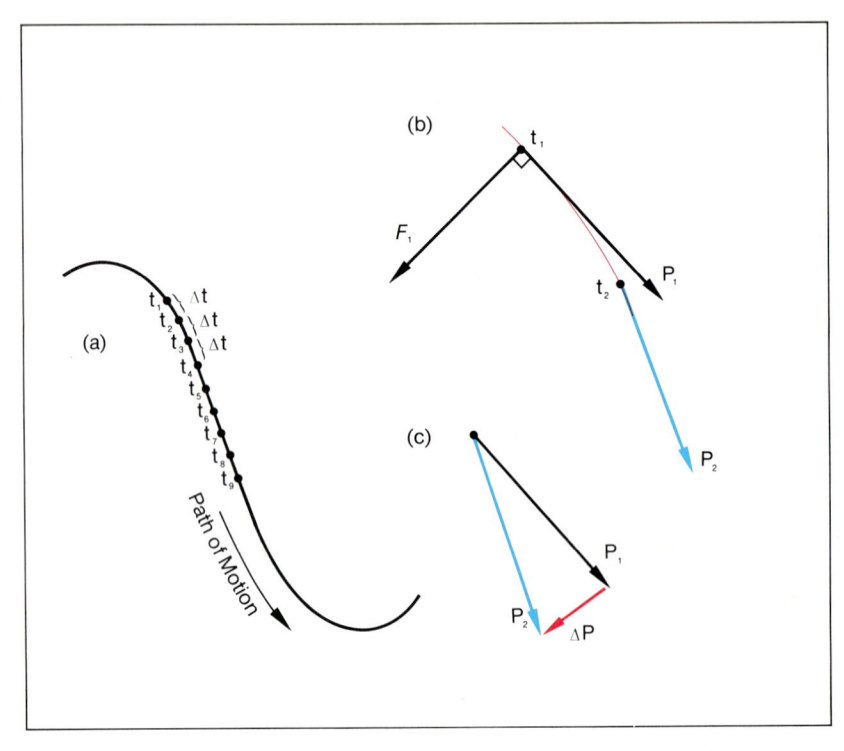

the *size* of the time interval must be small enough so that the force on the objects does not change by a significant amount during the change in time.

Now because the momentum of an object is in the same direction as the velocity, the direction of the momentum at each instant must be *tangent* to the curve of motion (see Figure 22b). As the object moves from one instant to the next, the momentum changes slightly, as shown in Figure 22(c). Because the change in momentum is equal to the product of the force multiplied by the change in time, the force acting on the object must be in the same direction as the change in momentum.

Suppose you knew what the force on the object was going to be at every instant. Could you predict what the motion would be? You could if you use the second law of motion over and over again. Suppose you start to follow the motion at t_1, when the force is $\vec{F}_1$. Then a short time later at $t_2 = t_1 + \Delta t$, the force is changed slightly to $\vec{F}_2$. A short time after that, at $t_3 = t_2 + \Delta t$, the force will have changed to $\vec{F}_3$, and so on. You can follow the motion by working out a table along the lines of Table 1. Figuring out such a table can be a lot of work, particularly when you remember that each sum means a *vector* addition; you would have to keep track of the angles of all of the different changes in momenta.

In practice, such calculations are done with an electronic computer. The computer is given all of the information about the force and is given instructions on how to do each step in Table 1. The computer never gets bored and calculates rapidly. In fact, not only can it calculate a table such as Table 1, in a small fraction of a second it can draw a picture of the motion on a television screen. When the Apollo spacecrafts travel from the earth to the moon, electronic computers are incessantly calculating their motion in space, using the second law of motion.

There is a further complication that must be taken into account. The force on a spacecraft depends on the *position* of the spacecraft as it travels through space. To find out what the force is, you need to know *where* the object is. Suppose the object starts out at a certain place—say at position $\vec{S}_1$—at time t_1. How do you know where it will be a short time

Table 1.
Working Out the Motion When the Force Depends on Time

Time	Momentum at Start of Interval Δt	Force	Change of Momentum	Momentum at end of Interval Δt
t_1	$\vec{P}_1$	$\vec{F}_1$	$\vec{F}_1 \Delta t$	$\vec{P}_2 = \vec{P}_1 + \vec{F}_1 \Delta t$
t_2	$\vec{P}_2$	$\vec{F}_2$	$\vec{F}_2 \Delta t$	$\vec{P}_3 = \vec{P}_2 + \vec{F}_2 \Delta t$
t_3	$\vec{P}_3$	$\vec{F}_3$	$\vec{F}_3 \Delta t$	$\vec{P}_4 = \vec{P}_3 + \vec{F}_3 \Delta t$
t_4	$\vec{P}_4$	$\vec{F}_4$	$\vec{F}_4 \Delta t$	$\vec{P}_5 = \vec{P}_4 + \vec{F}_4 \Delta t$
t_5	$\vec{P}_5$	$\vec{F}_5$	$\vec{F}_5 \Delta t$	$\vec{P}_6 = \vec{P}_5 + \vec{F}_5 \Delta t$
t_6	$\vec{P}_6$	$\vec{F}_6$	$\vec{F}_6 \Delta t$	$\vec{P}_7 = \vec{P}_6 + \vec{F}_6 \Delta t$
t_7	$\vec{P}_7$	$\vec{F}_7$	$\vec{F}_7 \Delta t$	$\vec{P}_8 = \vec{P}_7 + \vec{F}_7 \Delta t$
t_8	$\vec{P}_8$	$\vec{F}_8$	$\vec{F}_8 \Delta t$	$\vec{P}_9 = \vec{P}_8 + \vec{F}_8 \Delta t$
t_9	$\vec{P}_9$	$\vec{F}_9$	$\vec{F}_9 \Delta t$	$\vec{P}_{10} = \vec{P}_9 + \vec{F}_9 \Delta t$
t_{10}	$\vec{P}_{10}$	$\vec{F}_{10}$	$\vec{F}_{10} \Delta t$	$\vec{P}_{11} = \vec{P}_{10} + \vec{F}_{10} \Delta t$

later, at t_2? If you know its momentum at the start, you also know its velocity from

$$\vec{v} = \frac{\vec{P}}{m}$$

Knowing its velocity, you can figure out its change in position during the change in time from

$$\Delta \vec{S} = \vec{v} \times \Delta t$$

Once you know $\Delta \vec{S}$ you can get the new position at t_2 from

$$\vec{S}_2 = \vec{S}_1 + \Delta \vec{S}$$

And by using the second law of motion, you can figure out the change in momentum during the change in time so that you also know what the momentum is at t_2. Because you now know the position of the object at t_2 you should be able to know the new force $\vec{F}_2$. You can repeat the whole process for the time interval from t_2 to t_3. The process just described is summarized in the problem on the following page.

CIRCULAR MOTION

When an object moves in a circular path—as when a satellite moves in a circular orbit around the earth—the motion is caused by a special kind of varying force. The magnitude of that force does not change but the direction of the force is always changing so that it points toward the center of the circle.

Consider an object moving around a circular path with a constant speed, as shown in Figure 23. At any instant t_1, it has a momentum $\vec{P}_1$ tangential to the circle because the momentum is in the same direction as the motion. A short time later, at time t_2, the object has moved a short distance around the circle. Suppose the force on the object points toward the center of the circle. Then the change in momentum will be a small

Working out the Motion When the Force is Different for Each Position

<u>At time t_1,</u>

    Position is $\vec{S}_1$

    Momentum is $\vec{P}_1$

    Force is $\vec{F}_1$

    Mass is M

<u>During the time interval Δt,</u>

    The velocity is $\vec{V}_1 = \dfrac{\vec{P}_1}{M}$

    So the object moves $\Delta \vec{S} = \vec{V}\Delta t$

    And gets to the new position $\vec{S}_2 = \vec{S}_1 + \Delta \vec{S}$

    The momentum changes by $\Delta \vec{P} = \vec{F}_1 \Delta t$

    So the new momentum is $\vec{P}_2 = \vec{P}_1 + \Delta \vec{P}$

<u>So at time t_2,</u>

    Position is $\vec{S}_2 = \vec{S}_1 + \Delta \vec{S}$

    Momentum is $\vec{P}_2 = \vec{P}_1 + \Delta \vec{P}$

    The force is $\vec{F}_2$

Repeat process for next time interval Δt from t_2 to t_3

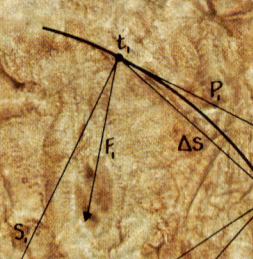

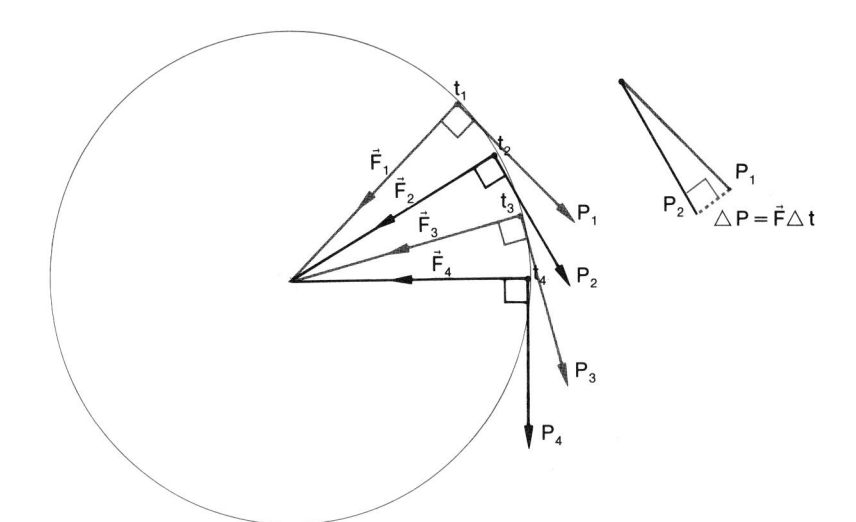

Figure 23. Circular motion of an object about the center of a circle.

vector $\vec{\Delta P}$ that also points toward the center; therefore, it is at right angles to $\vec{P}_1$ (see Figure 23). When that happens, the new momentum $\vec{P}_2$ has the same *magnitude* as $\vec{P}_1$. Only the direction has changed. Because there was no change in the magnitude of the momentum, the speed has not changed. If the force is just the right size, the size of $\vec{\Delta P}$ will be just enough to make the new momentum $\vec{P}_2$ tangential to the circle when the time t_2 is reached. If the same thing happens from t_2 to t_3, and from t_3 to t_4, the object will move in a circle without end.

How big must the force be? It can be worked out from the geometry of Figure 23 and the second law of motion. If the radius of the circular path is R, if it makes one revolution in a time T, and if the mass of the object is M, you would find out that the magnitude of $\vec{F}$ must be determined by

Magnitude of $\vec{F} = 4\pi^2 \times \dfrac{M \times R}{T^2}$

When a satellite is put into a given circle around the earth, its speed must be just right so that this formula is correct when the force $\vec{F}$ is the gravitational pull of the earth.

7 | Motion

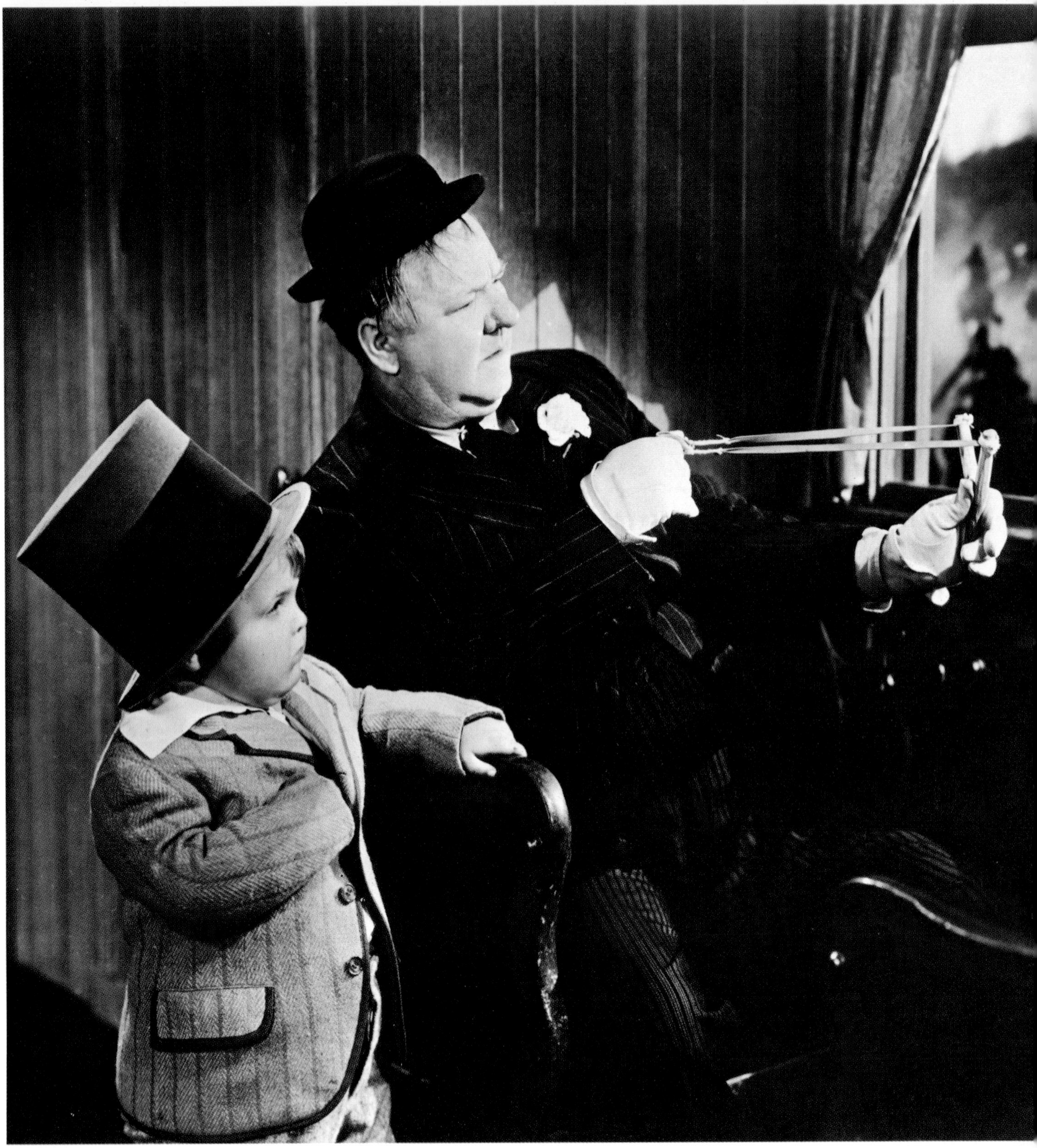

In the previous chapters you investigated the relationship between forces and changes in motion. Things were pushed and pulled and their directions changed; they were made to move faster and to slow down. Other things were analyzed to show that it is possible to predict what will happen when forces are applied to them.

So far the discussion has taken a "causal," or predictive, view of nature: "Tell me the forces and I will tell you what happens" and, alternatively, "Tell me how things move and I can tell you what the forces are." In physics the causal approach comes as close as can be expected to answering the "why" of things—why the earth travels in an ellipse around the sun, why an arrow travels in an arc through space, why a spacecraft follows the path that it does to the moon, or to Mars.

But some things are too complicated for the detailed cause-and-effect analysis of force and motion. Although that kind of analysis works well enough in the study of the motion of oil drops, or of cars, or even of galaxies, in many other cases it just does *not* work. Some objects are so small, and move so fast, that it is no longer possible to make the necessary measurements that will tell us what their motion is like. The sheer numbers of molecules or atoms that comprise a macroscopic object cannot even be counted in detail.

To make sense out of what happens in a gas where 10^{20} molecules are bouncing about, we have to give up all hope of predicting accurately what any individual molecule will do. Instead we must rely on fairly general before-and-after pictures that describe an entire system of particles. This means, of course, that we have to find some way of characterizing the state of the entire system all at once. There exists a special physical quantity that becomes useful in the description of nature. That quantity is called *energy*.

The notion of energy has been around for a long time. Today it has become one of the most fundamental concepts in all of science. It has weathered several storms of opinions and many profound revisions of physical laws. The concept of energy has even survived Albert Einstein's theory of relativity and the subsequent revolution of quantum mechanics. But energy has another, perhaps more familiar, meaning for us. Energy is the basis of life. Plants convert energy from the sun into the chemical energy of food, and food chains tie together the complicated ecological systems we are now trying to understand. Energy powers our machines, heats our homes, and moves us about the earth, as well as away from it.

Energy is also a commodity that we buy. You've heard of terms used to describe it. Electric companies charge us for kilowatt-hours of electric energy. Hot fudge sundaes, martinis, and cottage cheese are rated in Calories; gas and oil companies, if you look at their bills, sell BTU's of energy. All such terms are different but equivalent ways of measuring energy.

The evolution of the universe may be sketched out by means of the concept of energy. That picture would include a discussion of the conversion of energy into its various forms, and of the natural laws that encourage or hinder such conversions. The evolution of man, too, may be traced in terms of the consumption of energy. In the prehistoric past, man consumed little more than the 2,000 Calories or so each day necessary to keep

8
Energy

The slingshot has been pulled backward with a force $\vec{F}$. The force multiplied by the distance is the work done. There is potential energy equal to the work stored in the rubber band.

Figure 2. (Below) Man's use of energy is increasing at a phenomenal rate.
From "The Flow of Energy in Industrial Society," by Earl Cook. Copyright 1971, Scientific American, Inc. All rights reserved.

Figure 3. (Right) In an *elastic collision* between two steel balls of mass 1 and mass 2, both momentum and energy are conserved. The sum of the kinetic energies *before* the collision is equal to the sum of the kinetic energies *after* the collision.

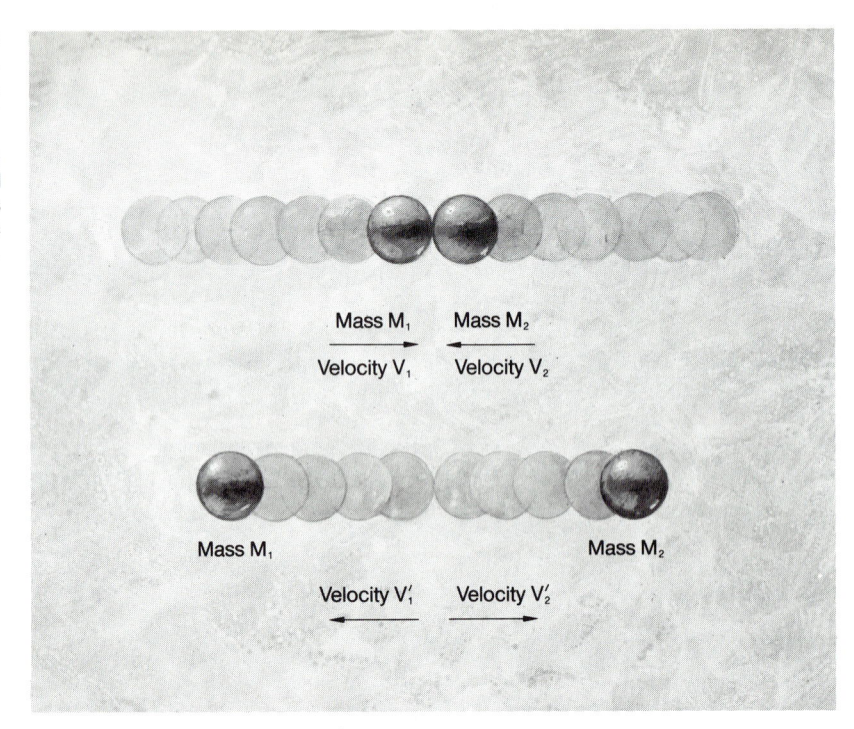

alive. Much later he left the grasslands and embarked on an increasingly complex journey fueled by ever increasing sources of energy.

The growth in energy consumption (Figure 2) was paralleled by and, to a great extent, was caused by the cumulative discovery and invention of various useful forms of energy. Hunters used fire for warmth and for cooking food; early agriculturalists bred and domesticated animals, which also had to be fed; advanced agriculturalists employed coal for heating, and some water power; industrial man had the steam engine, and so on. Technological man has electricity, internal combustion engines, heating, and air conditioning for his own comfort and myriad energy-dependent industrial processes. Each person in the United States now has at his disposal an average of 900×10^6 joules of energy each day, or approximately 200,000 Calories per person. Compare this with the 2,000 Calories for prehistoric man.

Because energy comes in so many different forms, different names are used to distinguish one kind from another. "Mechanical energy" is related to the motion of objects. "Electric energy" is associated with electric charges; it is the form of energy transmitted through wires to your home. There is also something called "gravitational energy." You encounter "chemical energy" whenever you start a fire; you bask in "thermal energy" when you sunbathe on the beach. You have read about "nuclear energy" in a variety of contexts. And there are many other forms as well, because one form of energy can be converted into another.

This chapter makes connections between the various forms of energy; subsequent chapters describe some applications of the energy concept. There are ways of comparing and even relating such seemingly different forms as the gravitational energy of a collapsing star and the energy attributed to your breakfast cereal in television commercials. And there are laws that restrict the utilization of energy not only by man, but even by the entire universe.

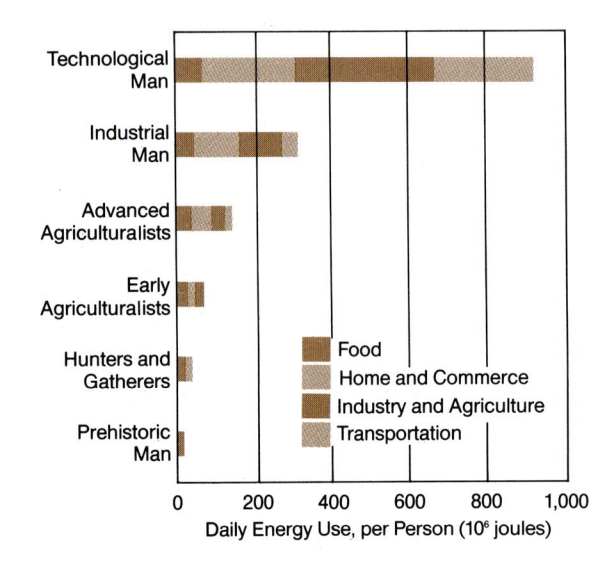

Figure 4. A woman bouncing on a trampoline is an example of the interplay between kinetic energy and the potential energy of the springs in the trampoline and of gravity.

MECHANICAL ENERGY—KINETIC AND POTENTIAL

In the preceding chapter you read that the momentum of an object is defined as the product of its mass and its velocity. Christian Huygens in the sixteenth century was able to show that in the "elastic" collision between two hard steel spheres (see Figure 3), the total momentum is conserved, and moreover that another quantity characterizing such a system is similarly conserved. He found that one-half the mass of each ball times the square of its velocity, when summed over the two steel balls, has the same total value *after* the collision as it had before. He called this new quantity *vis viva,* or "living force." The quantity is now identified as the *kinetic energy,* or the energy of motion, and it is expressed as:

Kinetic energy $= \frac{1}{2}$ mass $\times$ (velocity)^2 *(equation 1)*

or

Kinetic energy $= \frac{1}{2} \times m \times v^2$

This principle of energy conservation applies regardless of the number of objects involved in the collision. However, kinetic energy is not the only form of energy, and there exist situations in which it can be converted into other forms of energy. In such cases, the kinetic energy alone is *not* conserved—it does *not* remain constant.

For example, think about what happens when someone is bouncing on a trampoline. In Figure 4(a) a woman is at the peak of her first bounce; in the next frame she lands on the trampoline. She then drives the canvas down under her weight (Figure 4c), rebounds from it (Figure 4d) and finally soars up again (Figure 4e). As the woman falls, she has an amount of kinetic energy that is given (as always) by one-half the product of her mass times her velocity squared. But look at the drawings. In the first, where the woman is at the top of her bounce, her velocity is zero so the kinetic energy must be zero, also. It has its greatest value in Figure 4(b), just before she reaches the trampoline, because after she hits the canvas the springs will begin to slow her down and eventually bring her to a complete halt. At this point the kinetic energy will again become zero. But then the stretched springs of the trampoline snap her back into the air, as you see in the subsequent frames; again her kinetic energy decreases as she rises, and again it becomes zero as she reaches the top of her flight.

8 | Energy

Thus it becomes quite clear that (at least on a trampoline) kinetic energy of motion is *not* conserved. However, rather than considering energy as something that just appears and disappears, you can choose instead to see in this cycle evidence for another form of energy, called *potential energy.* When the woman is at the top of her bounce, even though she has lost all her kinetic energy, she has gained something in exchange—altitude. This increase in altitude is like money in the bank; any time she comes back down she recovers her lost kinetic energy. In other words, even though she is no longer in motion at the top of her flight, the very fact that she is higher than she was formerly gives her the *potential* of regaining the speed that initially sent her there. It is this potential energy which is converted back into kinetic energy as she falls down again. You can keep her energy account in order by assigning an appropriate amount of potential energy to her height above the trampoline; the expression that will make it correct, and keep her *total energy* constant, is

Potential energy = mass × g × height

where g is the gravitational field described in the preceding chapter (9.8 newtons per kilogram), and the height in question is the distance between the woman and the ground. Note that her mass is included as a factor in this expression, just as it was for kinetic energy.

At the top of her bounce she has a certain potential energy; as she begins to fall toward the trampoline her height decreases and so her potential energy decreases, also. At the same time, however, she is beginning to build up kinetic energy. *The total energy—kinetic plus potential—remains the same.* The conversion of the two forms of energy is shown in Figure 6.

However, another problem appears after she strikes the trampoline. At this point she begins to slow down, thereby losing kinetic energy as the springs tighten. But now you can't explain the loss in terms of a gain in altitude. Where does the missing kinetic energy go this time?

The answer is to be found by looking at the trampoline springs. They are stretched; they have the potential of *giving back* kinetic energy as they snap back. When the woman lands on the canvas, the trampoline begins to slow her down. Now her lost kinetic energy is taken up by the trampoline and stored as *elastic potential energy* in the stretched springs. When she finally stops at the bottom of her fall, the potential energy stored in the springs has reached its maximum value. It is this potential energy of the springs that is converted back into kinetic energy of the woman as the trampoline resumes its flat shape and throws her back up into the air. *But the total energy—kinetic, plus gravitational potential, plus elastic potential—remains the same* throughout the entire performance. Thus, although each individual form of energy does not itself remain constant, the total energy is conserved.

FORCE AND WORK

The preceding examples dealt with *changes* in energy—changes from some form of potential energy to kinetic energy and back again. Now whenever there is a change of energy, that change is necessarily caused by something. It is the force of gravity that pulls the woman down toward the trampoline and causes her kinetic energy to increase. When she moves upward *against* this gravitational force, her kinetic energy is turned back into gravitational potential energy. Thus potential energy may be stored when motion is resisted by a force. By the same token, there is a

Figure 5. Potential energy is the energy of position.

Figure 6. Kinetic and potential energy. The woman leaps into the air. The kinetic energy of *motion* changes to the potential energy of *position* in the earth's gravity. At the top of her flight the kinetic energy is zero and the potential energy is at a maximum. Then, as she falls, the potential energy decreases because it is converted into kinetic energy, which has *its* maximum as she lands on the trampoline.

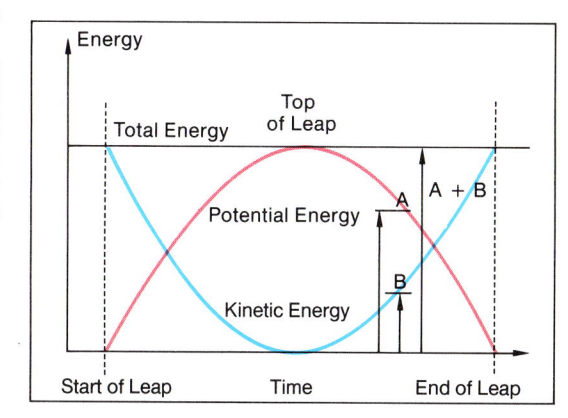

8 | Energy

Figure 7. Work = force × distance raised.

resisting force associated with the storing of elastic potential energy in the springs of the trampoline. The magnitude of this force depends on how much the springs have been stretched — in other words, on the *position* of the jumper who is stretching these springs. Actually, this elastic force is an electric force. The atoms of the spring are pulled apart or pushed together as the spring is stretched or compressed. Charged particles — electrons and protons in the atoms and molecules of the material — resist this distortion, exerting electric forces on one another and eventually on the jumper.

The resulting net force plays the same role in converting kinetic into elastic potential energy (and back) that the gravitational force plays in conversions between kinetic and gravitational potential energy. Elastic potential energy is stored whenever the spring *resists* the motion, just as gravitational potential energy is stored whenever the gravitational field resists. In other words, potential energy may be thought of as a reserve form of energy that is due to a force whose magnitude depends on the position of the object.

Another way of expressing this production of energy, or the transfer of energy from one form to another, is in terms of the *work* required to produce the same effect. The kinetic energy a moving object acquires is equal to the work required to produce that motion. The potential energy in a given shape (of a spring) or position (of an object) is the work required to bring things to that shape or position.

Work done on an object is *defined* as the force applied to that object, multiplied by the distance the object moves in the direction of the force:

Work = force × distance in the direction of the force .

Figure 4(a) showed a woman suspended momentarily above the trampoline. How did she get there? In terms of what happens subsequently, it really doesn't matter. She managed somehow to bounce up that high. If instead she had been lifted with a pulley and rope to the same height, the entire process of falling and bouncing that follows thereafter would still occur in precisely the same way. In other words, her gravitational potential energy is the work it would take to move her there against the force of gravity—regardless of how this was accomplished or what path she followed—regardless of whether she was given one fast shove by the springs or lifted gingerly into position by her acrobatic partner. Because it all comes to the same thing in the end, you might just as well compute this work in the easiest possible way—by assuming that she was lifted very slowly by a force just barely great enough.

The force required to lift the woman is her weight, which, you recall, is her mass multiplied by the gravitational field g. This force operates over the entire distance she is raised. The work in that case would then be:

Work = force × distance raised = mass × g × height (equation 2)

This work is precisely equal to the change in her potential energy. In this example, it causes an increase in gravitational potential energy because the motion is resisted by the force of gravity. As the mass of the woman moves away from the center of the earth, the force of gravity resists, and gravitational potential energy is stored in the system. Once stored, of course, it can always be recovered. The system converts the stored potential energy into an equivalent amount of kinetic energy when the two masses (the woman and the earth) come back together. In other words, whenever the jumper gets closer to the earth, she speeds up, regaining the kinetic energy that had been stored as gravitational potential energy.

This example is a relatively simple one. The force is a constant that doesn't change with position, so a simple multiplication of force by distance suffices. Moreover, the force and the direction of motion are parallel to each other, which also simplifies matters. But this definition of work can be extended to include more complicated situations as well.

Work is done only by that *component* of the force which is parallel to the direction of the motion. If the force is *perpendicular* to the direction of motion, *no work* is done, because in that case the force has no parallel component. For example, consider an ice skater gliding across the smooth, frictionless ice of a frozen pond. There is a force—the skater's weight pressing down through his skates onto the ice. The skater moves across the ice at right angles to this force. But there exists no force in the direction of his motion; hence, no work is done. The velocity remains constant, the kinetic energy remains constant, the potential energy remains constant. It is only when the ice is *not* so smooth—when it pushes against the skater, *parallel* to his direction of motion—that he speeds up or slows down and changes his kinetic energy. This effect of the angle between the direction of the force and the direction of the velocity is taken into account when we compute the work by multiplying only that component of the force in the direction of the motion by the distance traveled (see Figure 8).

Another example: A spacecraft in a circular orbit around the earth maintains the same potential energy because its altitude never changes. But there is a force operating, the force of gravity. However, because gravity always acts toward the center of the earth, this force always remains perpendicular to the direction of horizontal circular motion of the spacecraft; hence, no work is done. See Figure 9. It is only when the capsule begins to fall toward the earth that work is done. But in this case, the gravitational force is *doing* the work, instead of resisting it. Therefore, rather than *increasing,* the potential energy *decreases.*

If the force on an object is changing, or if the object is moving along a curved path, the work done must be evaluated along each small segment of its path. Looking at Figure 10, the work done between the time t_1 and

Figure 8. Work is the parallel component of the force ($\vec{F}$) in the direction of motion times the distance traveled.

Figure 9. (Above) The force of gravity does no work on spacecraft in circular orbits, so there is no change in energy.

Figure 10. (Above right) The component of $\vec{F}$ in the direction of $\vec{S}$ is the perpendicular projection of the length of $\vec{F}$ along the line of $\vec{\Delta S}$.

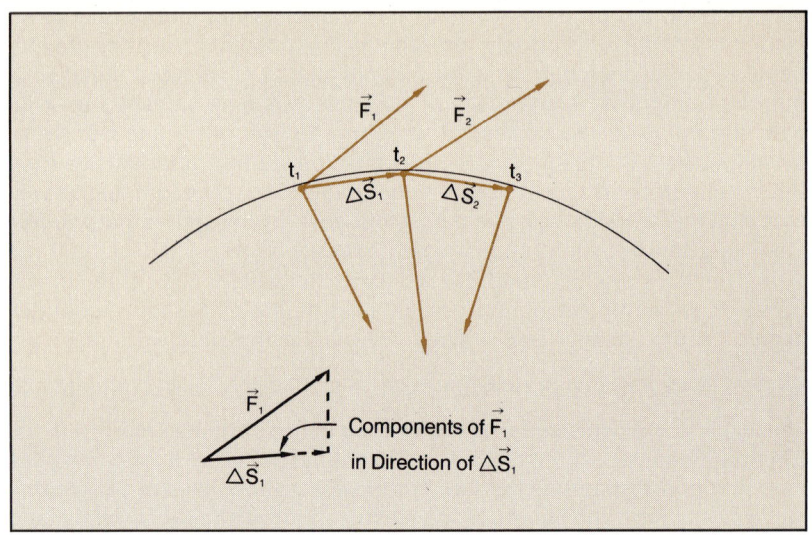

the time t_2 is the *component* of the force $\vec{F}_1$ in the direction of $\vec{\Delta S}_1$, the *change* of position from t_1 to t_2. Then

ΔW = (component of $\vec{F}$ along $\vec{\Delta S}_1$) × (magnitude of $\vec{\Delta S}_1$)

Notice that work is a *scalar* quantity: it is equal to a *component* of $\vec{F}$ multiplied by a distance. This distance is the *magnitude* of $\vec{\Delta S}_1$, the change in position.

To find the total work done, you need to add up all of the pieces ΔW obtained for each piece of the path. And because the total work W is a scalar, you use simple addition (no need of a vector sum here).

Having defined work in terms of force and distance, it is now possible to say something about the units used to measure it. A unit often used is the "foot-pound," which is the work done by a 1-pound force operating over a distance of 1 foot. In climbing a 10-foot staircase, a 175-pound man does 1,750 foot-pounds of work.

The metric unit for work is the newton-meter, which is the work done by 1 newton of force operating over a distance of 1 meter. For the sake of convenience, the metric unit of work has also been given another name: the *joule*.

1 joule = 1 newton-meter

Because a quantity of energy is defined as equal to the work required to produce that amount of energy, the unit of work is also the unit of energy. The unit of energy is 1 joule.

Another quantity related to energy is *power,* which is defined as the amount of work done in 1 unit of time. In general, work and power are related by:

Power = $\dfrac{\text{amount of work}}{\text{time}}$

Because work done is also equal to the change in energy, power is also a measure of the flow of energy. It is, in fact, the amount of energy that flows during 1 unit of time.

An automobile traveling 60 miles an hour has a kinetic energy of 1/2 mass × (velocity)^2, so a definite amount of work must have previously been expended in accelerating the car from 0 to 60 miles per hour. The amount of *work* is independent of how long it took to attain this speed. Because power is work per unit time, however, the power *is* affected by the time spent. The shorter the time taken to expend a given amount of work, the greater the power required.

In the metric system the unit of power is called the *watt.* It is the power flowing when work amounting to 1 joule is done in 1 second, or when the energy is converted from one form to another at a rate of 1 joule per second. A more familiar unit of power is the *horsepower:*

1 horsepower = 746 joules per second = 746 watts

The term "horsepower" actually derives from the power an average horse can deliver. The pertinent measurements were done by James Watt (of steam engine fame) on the patient (and no doubt tired) horses that were used to turn huge water-lifting devices in English mines.

A *kilowatt* is 1,000 watts. Most of our electric appliances are rated in watts to indicate the rate at which they expend electric energy while they are in service. Thus a light bulb might be rated at 100 watts, a toaster at 1,400 watts. Even though electric devices are rated by their power, the amount we pay is determined not by the power but by the amount of energy used. If you check your electric bill you will find that what you are paying for is the kilowatt-hour. One kilowatt-hour is the energy equivalent to 1 kilowatt (1,000 watts) flowing for 1 hour (3,600 seconds). Because a watt is 1 joule per second, 1 kilowatt-hour is the same thing as 3,600,000 joules.

HEAT ENERGY

If the woman on the trampoline stops using her muscles she will bounce only a few times and end up, before very long, just sitting still on the trampoline. At this point, her potential energy is zero; her kinetic energy is zero. There is a little potential energy stored in the springs as the result of her weight, but this cannot throw her into the air. Where did all her previous energy of motion go? It went into *heat* that appears in the canvas and springs of the trampoline. Her former mechanical energy has been converted into *heat energy.* This means that the temperature of the material of the trampoline has been increased. Heat energy is the energy of the rapid, microscopic motion of the atoms. The energy of the visible motion of the woman has been changed into the energy of the invisible motion of the atoms, which we call "heat." Rub your hands together rapidly and notice that your skin immediately gets hot. The work of the motion of your hands increases the motion of the atoms in your skin, and as a result your skin temperature increases.

The hotter an object is, the more heat energy it has. If you have an object and you want to raise its temperature, you have to put some heat energy into it. One way is to put it next to a hot object—for example, the burner on a stove. Then the fast-moving vibrating atoms of the hotter object will bump against the slower-moving atoms of the colder object and cause them to move more rapidly. In the process, some of the energy of atomic motion is transferred over to the cold object, which gets more energy and gets warmer.

For any object, a certain amount of heat energy is needed to raise the temperature 1 degree; twice that amount is needed to raise the temperature 2 degrees; and so on. The heat energy required *is proportional to* the temperature change. The amount of heat energy required for each degree depends on the kind of material—wood, glass, air, iron, water, and so on—and on how much material there is in the object. The heat energy required to raise 1 kilogram of the material 1 degree Centigrade (or 1 degree Kelvin, which is the same thing) is called its *specific heat*

Table 1.
Specific Heat Capacity of Some Common Materials

Material	Specific Heat Capacity (joules per kilogram per degree Centigrade)
Silver	23
Lead	125
Iron	480
Aluminum	900
Wood	1,680
Water	4,180

capacity. If you have 2 kilograms, it takes twice as much energy, so the specific heat capacity is the heat required per kilogram and per degree of temperature rise.

For example, it requires 4,180 joules to raise the temperature of 1 kilogram of water 1 degree Centigrade. So the specific heat capacity of water is 4,180 joules per kilogram per degree Centigrade.

Before people knew that heat was energy, they measured heat in Calories. The habit of using the Calorie unit became so strong that it is still used today in many practical activities, such as engineering. It is also used in nutrition studies as a measure of how much heat energy can be obtained from various foods. You probably know, for instance, that a slice of chocolate cake is worth 200 Calories. A Calorie (more correctly, a "kilogram-calorie") was defined as the amount of heat needed to raise the temperature of 1 kilogram of water by 1 degree Centigrade. Evidently, a Calorie of energy is about equal to 4,180 joules.

If you know the mass and the specific heat capacity of an object, you know the heat energy required to change its temperature by a given amount. The relationship may be written as:

Heat energy required = mass × specific heat capacity × change in temperature

which follows simply from the definition of specific heat capacity. Most materials take less energy to heat up than water does. Wood takes somewhat less; iron and lead take much less. The specific heat capacities of several common materials are listed in Table 1.

How do we know how much heat energy goes into an object when its temperature rises? It is simply the amount of mechanical energy (for example, kinetic energy) it takes to heat up the object—the amount of work done. In other words, the heat energy given to the object is necessarily equal to the mechanical energy that was expended in heating it. That is how the specific heat capacities in Table 1 can be determined.

HEAT ENERGY TO MECHANICAL ENERGY

When heat energy goes into an object, it takes the form of mechanical energy (kinetic and potential energy) of the atoms and molecules of the material. But the motions are microscopic and invisible. (You can *feel* them, however, if you put your hand on a hot stove.) It is even possible to get the heat energy out again into a more "useful" form, such as the kinetic energy of a moving, large-scale object. Many kinds of *engines* change heat energy to large-scale mechanical energy—the gasoline engine of an automobile and a steam engine are examples.

Such engines generally work by using the forces created by a hot gas. You read earlier that, in a gas, the molecules bouncing off the walls of the container exert a push against the wall. This push is called the gas *pressure*. Suppose the gas is contained in a cylindrical tube closed at the bottom and with a moveable wall called a *piston* at the top (see Figure 11). The gas pushes in all directions, but in particular it pushes upward on the piston. If this upward push is great enough to overcome any resisting forces that may be present, the piston will move upward.

It is the kinetic energy of the gas molecules in the cylinder that causes the upward push. Hence, the *heat energy* in the gas is converted into the *kinetic energy* of motion of the piston. The piston can, in turn, move a car by putting its energy of motion into the kinetic energy of motion of the car.

When the gas moves the piston, does the gas lose energy? Indeed it does. Each molecule that bounces off the moving piston comes back with less energy. The *average* kinetic energy of the molecules of the gas is decreased. The temperature of a gas depends on the average energy of its molecules. As the piston is moved, then, the temperature of the gas decreases. Measurements show that the decrease of the heat energy in gas is exactly equal to the amount of energy given to the piston.

Heat energy can, in principle, be taken out of any object this way. You can use the hot object to heat the gas—giving its heat energy to the gas. Then the gas will push the piston, changing its heat energy to mechanical energy.

The recognition in the early 1800s that mechanical energy could go into heat and come back again led to the first statement of what has become a fundamental scientific principle, called the first law of thermodynamics:

■ **Conservation of Energy—The First Law of Thermodynamics.** Energy can neither be created nor destroyed.

This law tells you that associated with every object is a physical quantity called its energy. If the object is moving, a part of this energy is kinetic energy. If the object is a spring, or sits in the earth's gravitational field, there may also be potential energy. And finally, some of the energy will appear in the form of heat energy of the object due to its temperature.

The first law of thermodynamics says that the energy can move around—from one object to another, or from one kind of energy to another—but that the total amount of energy *always* stays the same. It is never "made from scratch." It never disappears—although it may go somewhere else.

The law says, for example, that you can never build a "perpetual motion machine." Such a machine is an imaginary engine that runs forever with no energy coming in. But we know that if an engine is running, it is necessarily giving up energy to something. Even if it's not doing anything useful, it is at least creating heat due to the friction of its moving parts. This law says that no machine can give out energy forever unless it is taking it in somewhere. So "perpetual motion machines" are impossible.

CHEMICAL ENERGY

You may be thinking, "But where does the energy come from that drives the engine of my car? The gasoline isn't hot." That's true, but the gasoline has a kind of *potential* energy in the arrangement of its molecules. When those molecules are combined in the right way with molecules of the air, they become rearranged into different chemicals in the process called *burning*. In that process the atoms "snap" and are set into rapid motion. The potential energy of the molecules is changed into kinetic energy of the moving atoms. The gas gets hot. So chemical energy can go into heat energy in the engine of the car. Then some of the heat energy goes into the energy of motion of the car.

If you keep careful records, you discover that the total energy never varies. It just changes form: chemical to heat to motion.

OTHER FORMS OF ENERGY

When you heat water on an electric stove, where does the heat energy come from? It comes to your house through power wires in the form of *electric energy*. This energy is again a form of potential energy given to the electrons in the wires at the power station. In the heating element of

Figure 11. The heat energy in a piston is proportional to both the pressure and the column of the amount of gas enclosed. In this diagram, the gas intake is regulated by a valve. The arrows signify the pressure caused by the hot gas inside the cylinder.

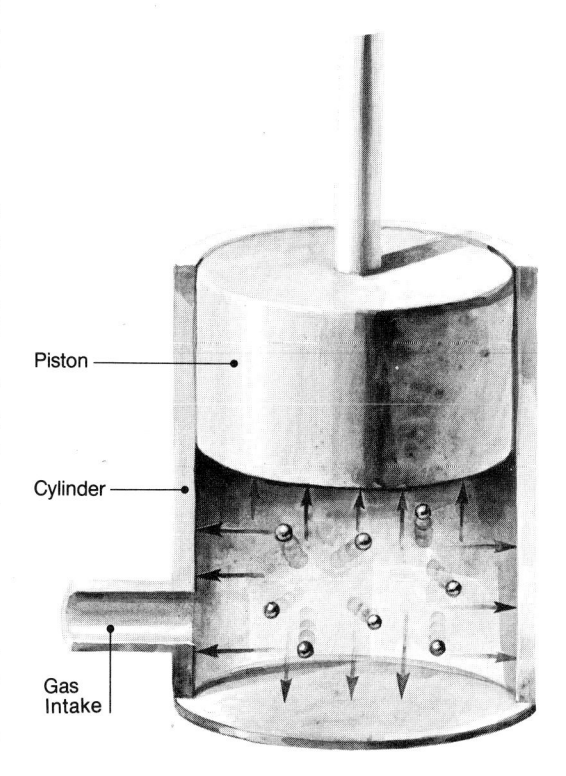

Figure 12. Energy moves downhill, usually in ways more subtle than this.

the stove, this potential energy is changed into kinetic energy of the atoms—heat again.

Probably the electric energy was "manufactured" at a power station that burns natural gas. It was obtained by changing the chemical energy of the gas into heat energy; the heat energy was changed into mechanical energy in an engine; the mechanical energy was changed into electric energy in a generator; and you changed the electric energy back into heat when you turned on the stove. And if you keep careful track you will find that all along the line the energy that is moving around always shows up somewhere. The total amount never changes.

And the story is the same for other forms of energy. Nuclear power is the release of energy that is stored up inside the nuclei of certain kinds of atoms, such as uranium. At a nuclear power plant the energy is unlocked and converted into other forms—heat and electric energy.

ENERGY RUNS DOWNHILL

If energy is neither created nor destroyed, why do we always need *more* energy? Because some kinds of energy are "better" (in the sense of being more useful) than other kinds. There are high-grade forms of energy and low-grade forms of energy. And whenever we "use" energy to do something useful for ourselves, we degrade the energy we use to a less useful form.

The potential energy of gravity is an example of the highest grade of energy. Heat energy is a low-grade energy. And the lower the temperature of an object, the lower is the grade of energy contained in it.

And the universe runs because energy is always flowing "downhill" —that is, it is always moving, in part, toward a lower grade of energy. Mechanical kinetic energy eventually disappears because it is changed into heat energy. Hot objects cool off, passing their energy into cooler objects. Both are examples of energy running downhill.

There is another law about energy, which is called the second law of thermodynamics. That just means the "second law of heat energy." It says that heat energy is the lowest grade of energy and can therefore never be completely converted into mechanical energy. Whenever energy is changed in form, some energy always ends up as heat. And the total effect is invariably a lowering of the average grade of the energy involved.

- **The Second Law of Thermodynamics.** The negentropy of the universe is always decreasing. Energy runs downhill from higher to lower grades.

When the universe was created, all its energy was in a high-grade form. All matter was spread out so that it had the highest amount of gravitational potential energy. It was in the form of hydrogen atoms, which have the highest form of nuclear potential energy. And the universe has been running down ever since.

The hydrogen atoms got pulled together to make the stars and their energy went into the heat of the stars—including our sun. At its extremely high temperature our sun radiates energy of medium-high grade. It comes to the earth and, through photosynthesis, changes its form into chemical energy. We eat the food that is eventually formed and use the energy for living. But in the process, we change the energy to heat that we give to the atmosphere. The atmosphere radiates its energy as low-grade heat energy into the cold reaches of distant space. What we call life is a brief game played with energy—a temporary holding action in the course of its natural journey down to the lowest form.

Ultimately all of the energy of the universe will have run downhill to its lowest form—heat energy at some low temperature. All activity, including life, will have stopped—except in the gentle motion of the vibrating atoms at the cold temperatures of dark space.

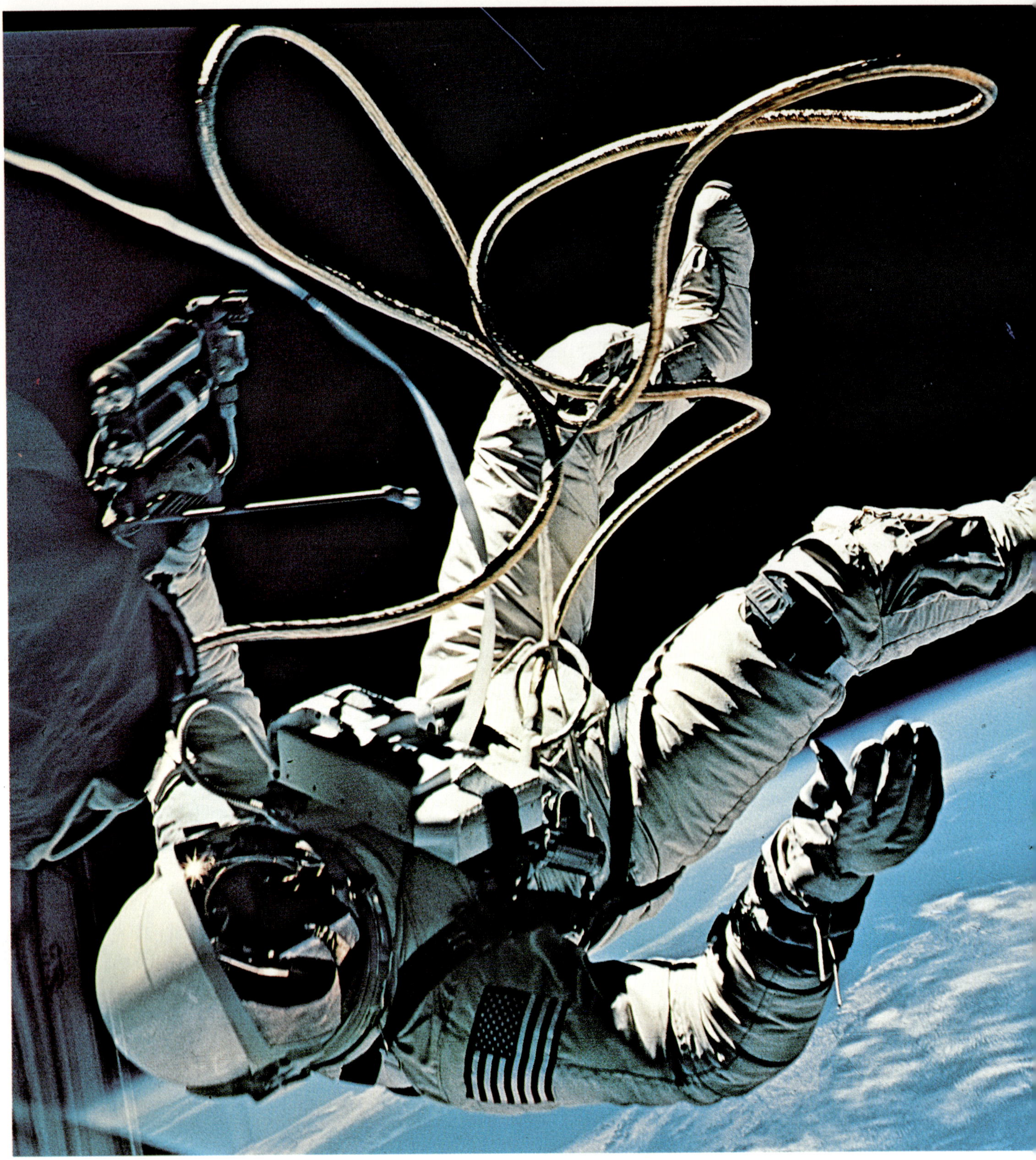

9
Relativity

If the astronaut moves at the same velocity as the earth, how would he know he was moving at all?

When you are moving along smoothly in a car you probably have very little sense of your body being in motion. Usually the effects of motion become apparent only when you press your foot on the accelerator and the car speeds forward. This effect is expressed in one of Newton's laws. If there is no acceleration there cannot be any net force acting on you or the car. In other words, events inside the car take place just as if the car were at rest. In fact, no experiment you could ever do *within* the car would give you the slightest inkling of whether it was at rest or in steady, straight-line motion with constant speed. Motion in a straight line at a steady speed is called *uniform* motion. Newton's laws say that uniform motion is indistinguishable from a state of rest because no net force acts in either case.

So how can you decide what is moving and what is not? It's a matter of convenience, whichever attitude you adopt. And the undefinable aspect of this situation is a "principle of relativity," because what it means is that *"absolute" uniform motion is a meaningless concept.* As long as the car moves uniformly—with no forces acting on it—you will never be able to find out its speed.

At first glance this principle may seem to be pure nonsense, whatever Newton's law seems to say. You can look out the car windows, for instance, and *see* that you are moving so you *can* measure your velocity. Of course, what you are measuring when you do this is your velocity relative to the ground, which you assume is stationary. But the principle of relativity says that, as far as the laws of physics are concerned, you can just as easily describe this measurement by saying your car is stationary and the *ground* is moving uniformly.

Suppose there *were* such a thing as "absolute" speed and there *were* some way to tell when you were absolutely at rest. There would have to be some physical observation you could make that would single out zero absolute speed. But not one of the principles of physics implies there is a way to make such an observation. Every known physical effect looks as reasonable to an observer moving uniformly as it does to an observer thought to be at rest. There are no differences to single out a special zero speed.

To understand this point more clearly, consider a "thought experiment." A thought experiment is a description of an idealized situation designed to illustrate a physical principle. It is not meant to be actually performed, but it could be done if necessary. Suppose there are two parallel railroad tracks with a flat car on each track (Figure 2). An observer (someone with a pencil and notebook) is seated on each flat car. It is night, and it is totally dark. Now imagine that the first observer suddenly turns on a flashlight and lets it fall at the same instant. Suppose the second observer sees the dim light falling straight down. What can he conclude about the motion of the flat cars?

One possibility is that *both* flat cars are at rest and the first observer merely let the flashlight slip from his hand. But another equally valid possibility is that both flat cars are *moving along at the same steady speed.* A third possibility is that the first observer threw the flashlight forward just after he turned it on, and that the flat car supporting the second observer happens to be moving with the same horizontal speed as the flashlight. As far as their observations are concerned, there is no way to single

Figure 2. The two observers cannot determine whether either one is at rest on the basis of any kind of experiment.

out one of these possibilities as being in any way "special." None of them appears to disobey the principles of physics and there is no reason to choose one above another. The relativity principle, then, is that no experiment conducted in a uniformly moving system can distinguish this system from a system that is taken to be at rest.

Perhaps the thought experiment with the flat cars seems too artificial to you. If so, try answering this question: How would you determine whether the sun was in uniform motion through space or whether it was "at rest"? Your first thought might be to start observing the stars to see evidence of the sun's motion. But that would only indicate motion relative to other stars, not absolute motion. There do not seem to be any mileposts in space that would let you gauge absolute speed.

Despite the lack of any means of detecting absolute motion, it might be maintained that absolute rest "exists" but cannot be measured—which is a weak philosophical position. It is true that important things such as love and beauty exist and cannot be measured, but no one claims that love and beauty are concepts of physical science. Velocity *is* a physical concept, however, and if there is no way to measure absolute velocity, there seems to be no reason to try building it into the structure of science.

Newton was aware of the relativity principle. He was, in fact, disturbed by the idea that rest and motion, at least uniform motion, could not be distinguished within his physics. To resolve what he felt to be a problem this idea created, he went outside physics to theology. Newton held fundamentalist religious views and in fact spent much of his time searching the Bible for clues about the evolution of the universe. It was natural, to him, to introduce God as the absolute reference frame, the final arbiter to decide the states of absolute motion of physical systems.

Most of the followers of Newton in the seventeenth and eighteenth centuries believed in absolute space and identified it with the sphere of fixed stars. It was a way of sweeping the problem under the rug. On close examination the stars are by no means fixed—they move relative to one another, and great clusters of them are sailing out into space at high speed relative to the earth. But the stars are so far away they do not seem to be moving much from year to year, and the fundamental problem of absolute motion could be ignored.

Newton died in 1727 and was buried, as an Englishman of universally recognized greatness, in Westminster Abbey. In fact, so overwhelming was Newton's influence on science that during the next century it was difficult for scientists to examine his work critically, and the theological and metaphysical assumptions forming the substrate of parts of it were consciously or unconsciously ignored. Among these assumptions were some that seemed too self-evident to be worth thinking about explicitly. It was taken for granted that two observers moving uniformly at different speeds would agree on measurements of length, time, and mass. One consequence of these assumptions is that speeds should add directly. If you are walking down the aisle of a bus toward the front at 3 miles an hour, and if the bus is moving at 20 miles an hour, your speed relative to the street is 23 miles an hour.

It is worth examining how a Newtonian physicist would express the connection between two moving systems because these relationships were drastically altered by the work of Albert Einstein. Figure 3 shows the coordinates x, y of an observer at rest and the coordinates x', y' of an observer moving uniformly down the x axis at velocity v. The coordinates are connected by the relations

$x' = x - vt$

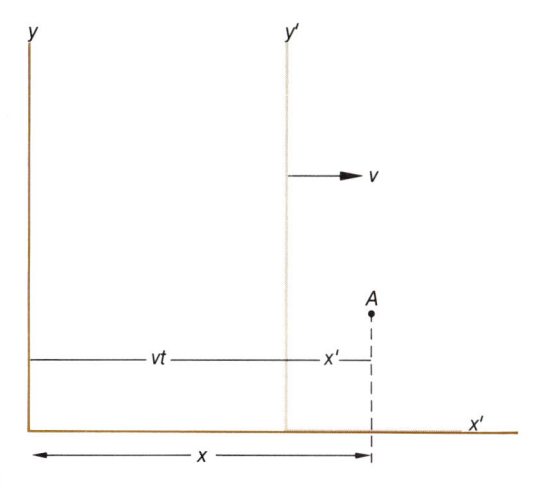

$y' = y$

Figure 3. Coordinates are shown for an observer at rest. Particle A has coordinates x' and y' as read in the moving coordinate system, and coordinates x and y as read in the stationary system.

Here x' represents, for example, the coordinates of a particle as measured by the moving observer, and x represents the coordinate of the same particle as measured by the observer at rest. The additional assumption that time is the same for both observers, $t' = t$, is taken for granted. These relationships are usually called the *Galilean transformations*, after Galileo. Notice, incidentally, that the times sign is not included between the letters v and t in the first equation. Times signs have been left out of the equations in this chapter as a way of illustrating still another shorthand way that scientists commonly use to write equations, but it's understood that the times sign is there.

The striking new discoveries not at all anticipated in Newton's physics began in the early nineteenth century. These discoveries had to do with electricity and magnetism. As the British physicist James Clerk Maxwell showed, they included the theory of the propagation of light — a form of electromagnetic radiation. Newton had made significant discoveries about light, notably that the "white light" from the sun could be broken up into a spectrum of colors by a prism. He held an atomic view of both matter and light and speculated that light propagated in beams of *particles*.

By the beginning of the nineteenth century, however, it had become clear that this notion could only be part of the story. Several physicists had shown by experiment that light beams can interfere with one another in a manner somewhat analogous to the way patterns of water waves interfere with one another. This interference is a characteristic feature of wave motion. What Maxwell did was to develop a mathematical theory that described the propagation of electromagnetic waves. An important result of his theory was that, in vacuum, electromagnetic waves should propagate with a speed of 3×10^8 meters per second (about 186,000 miles per second), a speed that turns out to be the same as the measured speed of light. Maxwell considered this correlation a compelling argument for identifying light as a form of electromagnetic radiation.

We now take the vacuum propagation of light almost for granted. After all, "outer space" is a nearly perfect vacuum and starlight propagates freely through it. Aside from this relatively recent understanding, however, all the wave motion we perceive usually takes place by the periodic motion of some *medium.* Sound waves propagate by the vibra-

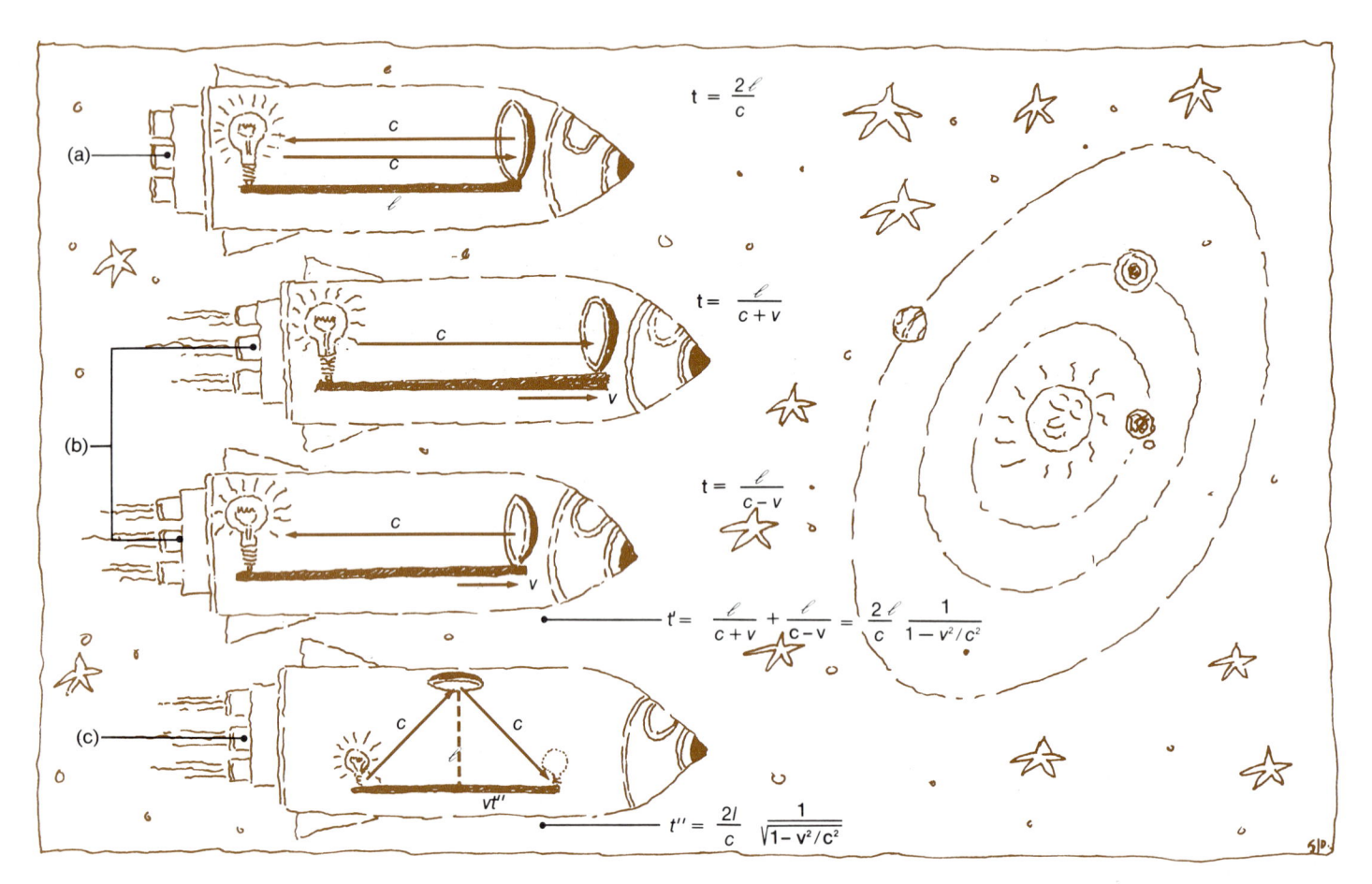

Figure 4. In (a), the apparatus is stationary with respect to the ether because the spaceship isn't moving. In (b), and (c), it is moving with speed v, which is the speed of the spaceship through the ether. The letter c is always used to designate the velocity of light.

$$t = \frac{2\ell}{c}$$

$$t = \frac{\ell}{c+v}$$

$$t = \frac{\ell}{c-v}$$

$$t' = \frac{\ell}{c+v} + \frac{\ell}{c-v} = \frac{2\ell}{c} \frac{1}{1-v^2/c^2}$$

$$t'' = \frac{2l}{c} \frac{1}{\sqrt{1-v^2/c^2}}$$

tions of the air; water waves by the vibrations of the sea. So it was quite natural for the nineteenth-century physicists to ask what medium was vibrating during the propagation of light waves. They responded to their question by inventing a medium, which they called the "ether." It was a most peculiar entity. On the one hand, the ether had to be transparent and frictionless because the planets moved through it without any noticeable effect. On the other hand, the ether had to be extremely dense and rigid to allow light vibrations to propagate a million times faster than the speed of sound in air. Einstein and his successors were convinced it served no real purpose in the physics of Maxwell's theory. Maxwell's equations merely could be accepted as the true laws of electromagnetic wave propagation and everyone could forget all about the ether.

But this realization came several years later. Meanwhile, most of the physicists of the period were convinced the ether had some sort of existence. In particular, it was natural to imagine that the ether was absolutely at rest in the universe and that *it,* whatever *it* was, provided the absolute reference frame for defining all velocities.

THE MICHELSON-MORLEY EXPERIMENTS

Physics is an experimental science, and near the end of the nineteenth century the physicist Albert Michelson—the first American scientist to win a Nobel Prize—and a young chemist named Edward Morley began a series of experiments designed to measure the velocity of the earth as it moved through the allegedly stationary ether. To see what Michelson and

Figure 5. (Left) Albert Michelson and (right) Edward Morley, two men who wondered what the ether was.

Morley wanted to do, consider a simplified form of their apparatus: a flashing lamp and a mirror a distance ℓ away. When the lamp flashes, a pulse of light travels to the mirror and bounces back to the lamp. Figure 4(a) illustrates what happens if the apparatus is stationary with respect to the ether. The light pulse travels with speed c a total distance 2ℓ and returns to the lamp after a time interval $t = 2\ell/c$. (Physicists always use the letter c to denote the speed of light in vacuum.)

According to the ether theory, the speed of a light pulse is c relative to the ether. If the apparatus is moving with respect to the ether, it can partly catch up with the light pulse. Figures 4(b) and 4(c) show the situation when the apparatus is oriented along the direction of motion. The time for the round trip in this case is called t'.

When the apparatus is at right angles to the direction of motion (Figure 4d), the light has to follow the triangular path shown in order to return to the lamp. The time required to go from the lamp to the mirror and back to the lamp is called t''.

The crucial conclusion is that when the earth's velocity is not zero with respect to the presumed ether, t' and t'' are necessarily different and, in fact, t' is larger than t''. Moreover, the formulas cease to make sense if the velocity of the earth through the ether were to exceed the velocity of light. Nothing in the classical mechanics of Newton prevents such a speed from being the case: according to Newton's law, if even the smallest sort of force is applied indefinitely to an object, that object can be accelerated to any speed — even speeds *greater* than the speed of light. In any event, these considerations became the basis of the Michelson-Morley experiment, which is diagrammed in Figure 6.

The essence of the experiment is to launch two steady light beams at right angles along distances that are as nearly equal as can be made in practice. The light beams travel up their respective paths, bounce off mirrors, and return to their source. Now if this apparatus is stationed on the earth and is moving through the ether, then the two light beams can't return to the central point simultaneously. This effect can be measured with great accuracy because, as Figure 7 shows, the two light beams would interfere with each other, thereby producing characteristic interference fringes of the type discussed in other chapters. The two arms cannot be made exactly equal in length. Michelson rotated the apparatus

9 | Relativity

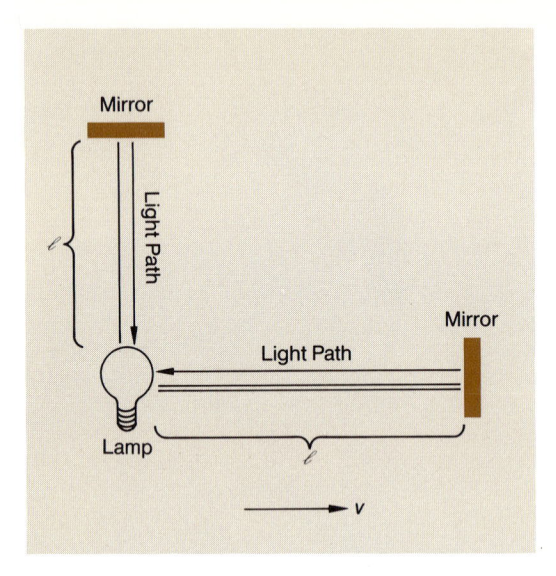

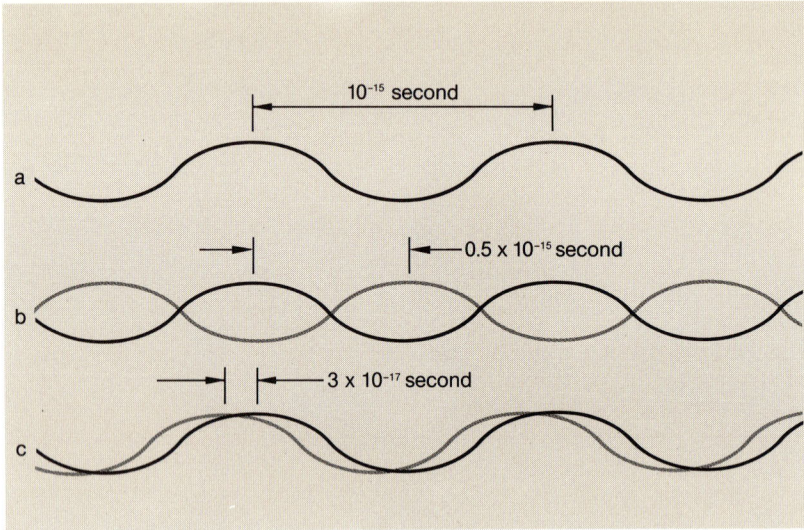

Figure 6. (Above) The Michelson-Morley experiment attempted to find the effects of the ether on light propagation by comparing the time for light rays to move in the *same* direction as a moving source with the time for light rays to move *perpendicular* to the direction of a moving source. In the experiment, the whole apparatus is attached to the earth, which moves with a velocity v in the ether.

Figure 7. (Right) In (a), the time between vibrations in light is about 10^{-15} second. In (b), complete interference occurs for a time difference of 0.5×10^{-15} second. In (c), partial interference is visible even when the time difference is 3×10^{-17} second.

between measurements to make first one arm and then the other point along the direction of motion, then he looked for a change in the interference pattern.

How do you go about determining how large an effect might be measured? Michelson thought the sun was at rest in the ether, so he assumed the speed of the earth in the ether was the speed of the earth itself in its orbit around the sun—about 3×10^4 meters per second. The arms of Michelson's apparatus were about 1 meter long. If you use $\ell = 1$ meter and $c = 3 \times 10^8$ meters per second, you can calculate the difference between t' and t'' from the results in Figure 4. It comes out to be 3×10^{-17} second. This time interval is extremely short, but Michelson was a superb experimenter and his techniques would have permitted him to measure interference fringes caused by even shorter times. The reason his apparatus was so sensitive is that light makes about 10^{15} vibrations per second, so one vibration takes only 10^{-15} second. As Figure 7 shows, complete interference occurs when the two beams merge after a time interval of 0.5×10^{-15} second, so an interval of 3×10^{-17} second produces enough interference to be easily detected.

However, when the experiment was performed Michelson and Morley found *nothing!* There was no effect. Michelson and Morley repeated their experiment several times and it has been repeated a number of times in recent years with modern techniques. And neither they nor any other observers have been able to discover the slightest effect of the earth's motion with respect to a hypothetical ether.

FITZGERALD AND THE CONTRACTION OF MOTION

The physicists of the late nineteenth century found Michelson and Morley's result utterly baffling and Michelson himself was disappointed over the failure of his experiment to measure the speed of the earth through the ether. The earth certainly could not be at rest with respect to the ether because as the earth goes around the sun it moves in many different directions. Repeating the experiment at different times of the year always gave the same zero result. In 1895 the Irish physicist George Fitzgerald proposed a remarkable explanation. Go back, for a moment, to Figure 4. Michelson assumed the distance ℓ between the lamp and mirror in each arm stayed the same no matter how the apparatus was oriented. Fitzgerald, however, noticed that Michelson's zero result could be accounted for if the arm pointing in the direction of the motion got *shorter*. Instead of length ℓ, Fitzgerald assumed that the length of the arm became

$$\ell' = \ell \sqrt{1 - v^2/c^2}$$

It's easy to see that this makes t' and t'' equal:

$$t'_F = \frac{2\ell}{c} \frac{1}{1 - v^2/c^2} = \frac{2\ell}{c} \frac{\sqrt{1 - v^2/c^2}}{1 - v^2/c^2} = \frac{2\ell}{c} \frac{1}{\sqrt{1 - v^2/c^2}}$$

where t'_F represents t' calculated with the Fitzgerald contracted length ℓ'. With this equation the Michelson-Morley experiment was accounted for (Figure 8).

In Fitzgerald's thinking, this assumption more or less appears to have been designed to explain one particular experiment. You might be tempted to dismiss such a notion altogether, knowing from common experience that a ruler, for example, doesn't appear to shrink when you move it. But the crucial quantity in the shrinking formula is $1 - v^2/c^2$. Inasmuch as the velocity of light is so large, the departure from ℓ is miniscule for all the motions familiar to us in daily life. In fact, if the earth were to be moving with respect to the presumed ether with the same velocity that it moves with respect to the sun, a 1-meter ruler would be contracted by only about a 10^{-8} meter. This contraction is so small that it could only be measured in an optical interference experiment of the type that Michelson was performing.

THE LORENTZ TRANSFORMATIONS

In the same year, 1895, the Dutch physicist Hendrick Lorentz had also independently hit on the contraction hypothesis and he made a significant attempt to explain it. His idea was that the forces holding matter together are principally electric. From Chapter 1 you have an idea that the chemical bonding of different atoms is caused by the charged cloud of electrons around an atomic nucleus. Lorentz didn't know of the nucleus because it hadn't been discovered yet. But he did recognize the basically electric structure of atomic forces. He also discovered the force law by which electrically charged particles interact with one another. Part of this force is the simple electric force (see Coulomb's law, Chapter 6), which in its mathematical form closely resembles Newton's law of gravitation. The other part is a new force, arising from the magnetic fields produced by charged particles in motion.

It's not important, at this point, to give the precise mathematical characterization of these force laws. It's enough to remark that Lorentz was able to show that a stick would undergo the Lorentz-Fitzgerald contraction if the atomic forces holding the stick together obeyed his force law. In the course of his work he had to consider two sets of equations. One set was expressed in terms of the coordinates x, y, z, and t (space and time) of a particle at rest with respect to an observer. A second set involved coordinates x', y', and z', and t', which an observer at rest would attribute to a particle in uniform motion. In all further considerations, this motion is assumed to be along the x-axis. The y and z coordinates won't be affected, so, for simplicity, they simply will be left out of the discussion.

Earlier you were led to consider the connection between these two sets of coordinates, which a classical Newtonian physicist would have given as $x' = x - vt$ and $t' = t$. This is the set of transformation equations

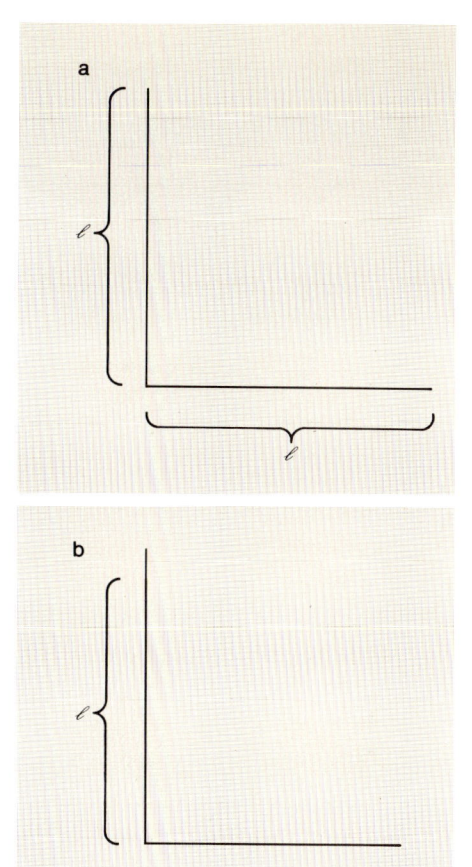

Figure 8. The contraction of motion. In (a), the apparatus is at rest. In (b), it is in motion with velocity v. Here, $\ell' = \ell \sqrt{1 - v^2/c^2}$.

Figure 9. Hendrick Lorentz (1853–1926) built a sound theory of spatial contraction and time dilation in moving reference frames.

Figure 10. Albert Einstein isn't testing the laws of motion; he's just enjoying himself. This photograph was taken at the home of friends near Los Angeles in the spring of 1933.

that expresses the relativity principle in Newtonian physics. But Lorentz found these transformations were not appropriate for his considerations, which involved electricity and magnetism. He determined that his equations maintained their form in the two coordinate systems only if he introduced the following set of transformation equations:

$$x' = \frac{x - vt}{\sqrt{1 - v^2/c^2}} \quad \text{and} \quad t' = \frac{t - v^2/c^2}{\sqrt{1 - v^2/c^2}}$$

These equations disturbed Lorentz because time in the rest frame was no longer identical with time that an observer in the rest frame would attribute to a clock attached to a moving system. Lorentz didn't attempt a systematic analysis of this curious aspect of his transformations. He merely regarded his equations as a neat mathematical trick for simplifying the analysis of the problem of the electromagnetic forces when they are studied in various frames of reference. His work took on a totally different aspect with Einstein's formulation of the special theory of relativity in 1905.

THE SPECIAL THEORY OF RELATIVITY

Albert Einstein was born in 1879 in Bavaria. As far as he was able to determine, no one in his family history had shown any particular mathematical or scientific aptitude, although his father Hermann was involved (for the most part unsuccessfully) in various business enterprises making use of electrochemistry. The young Einstein hardly appeared precocious; he didn't learn to talk until the age of 3. During his early education he didn't especially distinguish himself except in mathematics, probably because he hated the rote character of the teaching.

When Einstein was 15, his father's business in Munich failed and the family moved to Italy; Einstein was left behind to finish his schooling. He soon managed to drop out and to follow his parents to Italy where, like many more recent "dropouts," he spent an exceedingly happy period of several months wandering about on foot. His idyll came to an end when his father could no longer support him. With the help of some wealthy relatives, his father managed to send Einstein to Zurich, where he was meant to enter the Swiss Federal Polytechnic School and begin a course of study that would have trained him to become an electrical engineer. Einstein failed the entrance examination at the Polytechnic and was sent to a progressive high school in nearby Aarau to continue his work. A year later he entered the Polytechnic and was graduated in 1901.

During this period he attended few lectures and spent most of his time in self-study, both in the laboratory and with textbooks. In fact, the Polytechnic offered no formal course on the Maxwell equations of electricity and magnetism, so Einstein taught them to himself. Because he made almost no impression on his professors, when he was graduated he could get no real scientific job. Finally he drifted into the Swiss Federal Patent Office in Bern and became a technical examiner for patent applications. There he remained until 1909. During that time, when he was almost totally isolated from the European scientific community, he invented the theory of relativity and made a series of contributions to other branches of physics. His work became the foundation for much of twentieth-century science.

This brief account of Einstein's early life might make somewhat more comprehensible the unique character of his early work on relativity. Possibly one of the reasons he was able to look at physics problems in such a completely novel way was that, during his formative scientific years, he was isolated from people who might have tried to convince him to bring his ideas into a more conventional framework. By the time Einstein did become involved with the rest of the scientific community, his habit of independent thought was so thoroughly formed there wasn't much chance of such contact impeding his scientific creativity. And late in his life, when he turned against the modern formulation of the quantum theory, most physicists came to feel that he had isolated himself too completely from the new trends in physics.

In any event, when he was 16 years old Einstein began thinking about what eventually developed into the special theory of relativity. Although Michelson and Morley had performed their experiment a few years earlier, the result of that experiment was not the basis of Einstein's thinking. Indeed, various statements he made during his lifetime imply it is quite likely he had never even heard of the Michelson-Morley experiment prior to the 1905 publication of his first paper on relativity theory! There is no mention of the experiment even in this paper.

Einstein had an *intuitive conviction* that the principle of relativity was true—that no experiment could detect effects of uniform motion. And he was also convinced this principle held not only for the motion of particles considered by Newton but for the then-recently discovered phenomena of electromagnetic radiation. In Einstein's view, *there is nothing to explain* as a result of the Michelson-Morley experiment. The principle of relativity states that no experiment can detect the speed of uniform motion when you are stationed on a uniformly moving system, which is just what Michelson and Morley found.

But the young Einstein was troubled. If relativity were applicable to electromagnetic phenomena, then Newton's laws of motion could not be strictly valid for velocities comparable to the velocity of light. At the age of 16 he constructed some paradoxes—some "thought experiments" —that made this problem clear to him. All of these thought experiments are based on the assumption that it is possible for an observer to move with the speed of light—something that is completely consistent with Newtonian physics.

A simple example of one of Einstein's paradoxes is illustrated in Figure 11, which depicts a light wave. A stationary observer would see

Figure 11. One of Einstein's paradoxes. The stationary observer will see the wave move past him, but the moving observer can move along with the wave and not see any oscillation.

EINSTEIN
IN RELATIVITY
AS WE BEGIN OUR TALE, WE FIND ALBERT FIDDLING HIS WAY TO RELATIVITY
AM I MOVING TOWARD THE ISLAND...
OR IS IT MOVING TOWARD ME?
AS YOU CAN SEE - WE LIVE IN A WORLD SHROUDED IN ETHER. IT'S THE MEDIUM FOR ALL OF OUR MOTION.
CAN'T HE SEE THE ETHER?
HE'S NUTS!
HE'S BLIND
HE'S A FIDDLER
HERESY
WAIT. LISTEN. THE MICHELSON-MORLEY EXPERIMENT PROVES THAT THERE IS NO ETHER.
HE'S RIGHT.
WE'RE FREE
IT'S GONE
BECAUSE YOU ARE ALL-KNOWING, PERHAPS YOU WILL HONOR US BY RELATING THE RELATIVITY GAMES TO US.

HERE AT THE 30 YEAR LIGHT RACE, THE RUNNERS ARE OFF AT NEARLY THE SPEED OF LIGHT. THIS MAY SOUND STRANGE, BUT BY THE END OF THIS GRUELLING EVENT THEY WILL HARDLY HAVE AGED. THE SPECTATORS ON THE OTHER HAND...
"THIS YEAR'S MASS CONTRACTION POLE VAULT EVENT IS MARKED BY THE INTRODUCTION OF THE FITZGERALD TELESCOPING POLE, WHICH SHOULD LAUNCH THE COMPETITORS TO RELATIVELY NEW HEIGHTS."
TODAY AT THE MASS TO ENERGY CONVERSION EVENT WE ARE SEEING GRAPHIC PROOF THAT MASS DOES IN FACT CONVERT TO ENERGY.
YOUR MOTHER SAYS SHE'S PROUD OF YOU BUT YOU LOOK SKINNY.
E=MC²

the wave pass by with a regular frequency and he would record a series of optical vibrations on a suitable measuring instrument. But now imagine there is another observer moving at the velocity of light with the wave. The second observer could attach himself to a maximum or minimum, or some intermediate point on the wave, and he would record no vibrations. From such an experiment he could conclude that he was moving uniformly with the speed of light, *in violation of the relativity principle.* This paradox is unavoidable as long as Newtonian mechanics applies at the speed of light.

It took Einstein nearly 10 years before he arrived at his formulation of the relativity theory. In 1905 he published his formulation in an extraordinary paper entitled "On the Electrodynamics of Moving Bodies." Most of this paper doesn't involve mathematics any more sophisticated than high school algebra. The complexity lies in the subtle logic of the argumentation, which leads relentlessly from the two basic and seemingly straightforward hypotheses of the theory to the almost incredibly radical conclusions that have completely modified our perception of space and time. The first of these hypotheses is the principle of relativity itself:

■ **The Principle of Relativity.** No experiment conducted in a uniformly moving system can distinguish this system from a system that is taken to be at rest. The only physically meaningful velocities are the relative velocities of one system with respect to another. Absolute velocities are physically meaningless.

The second hypothesis is a statement about light that was already contained in Maxwell's equations, but its precise significance had not been emphasized prior to the work of Einstein. This hypothesis is called the principle of the constancy of the speed of light. It states that:

■ **The Principle of the Constancy of the Speed of Light.** All uniformly moving observers in a vacuum, if they measure the speed of light with respect to their systems, will find exactly the same answer; namely, $c = 3 \times 10^8$ meters per second.

A few comments about the second hypothesis are in order. First of all, in this discussion the hypothesis is limited to observers in a vacuum because the speed of light in a material medium is slower than the speed

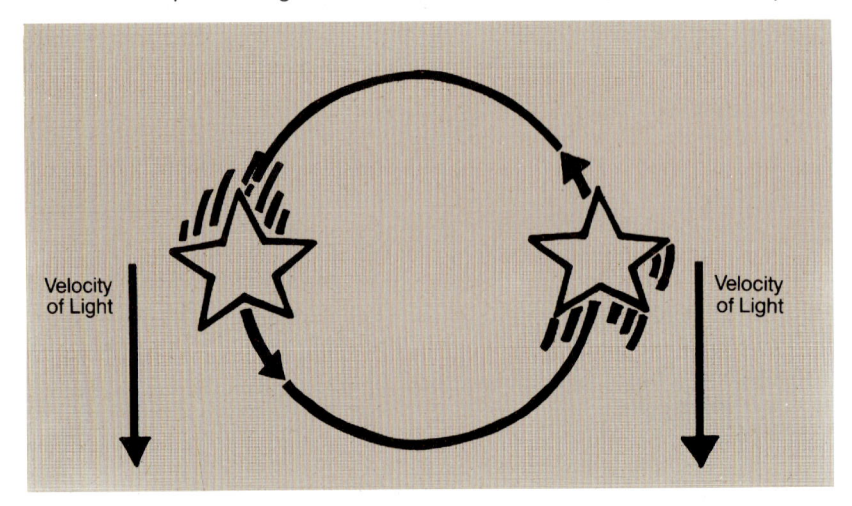

Figure 12. In a binary star system, the velocity of light that reaches an observer is the same whether either star is moving toward or away from the observer.

of light in a vacuum. Considering light propagation only in a vacuum doesn't change anything essential to the ideas. Second, this assumption is directly accessible to experimental test. The most graphic example of such tests is given by "double stars." These astronomical phenomena are pairs of stars that circle rather closely around their common center of mass, as depicted in Figure 12. If the velocity of light depended on the velocity of the source, an observer on earth would see multiple images from the moving stars because the light would arrive at different times from various points on the orbit. However, such images have not been observed. And the conclusion is that the velocity of light from the star is the same whether it's moving toward or away from the observer.

In itself, the second hypothesis is already inconsistent with Newtonian physics. Suppose a man on a space platform turns on a light. The light travels out from the platform at speed c. Suppose, at the same time, a spacecraft goes by at speed $3/4\,c$. According to Newton, the men on the spacecraft would see the light traveling at speed $c - 3/4\,c = 1/4\,c$. But according to Einstein, the men on the spacecraft would *also* see the light move at speed c.

The "intuitively obvious" way of adding velocities directly is accurate only if the magnitudes are small compared to the speed of light. Einstein discovered that the true addition formula for velocities, which follows from his two assumptions, is

$$w = \frac{u + c}{1 + uv/c^2}$$

where w is the sum of two velocities u and v. If one of these velocities has a magnitude equal to the speed of light c, the interesting result is that

$$\frac{u + c}{1 + uc/c^2} = c\left(\frac{u + c}{u + c}\right) = c$$

This equation is another, and more precise way, of saying that the speed of light is the same to all observers, regardless of their relative speeds.

The first things Einstein was able to derive from his two hypotheses were the Lorentz transformation equations themselves. In particular, he was able to show the time transformation is not some sort of obscure mathematical trick but rather is an important consequence of how we actually measure time.

Einstein's principal insight was that the measurement of the time of an event is actually a measurement of the simultaneous occurrence of two different events: some happening—say, the arrival of a train at a certain place—and the position of the hands of a "clock." (For this discussion a "clock" can be any phenomenon that repeats itself periodically, and the "hands" are a figurative expression for some way of labeling these periodic events. How the clock is actually constructed isn't important here.) When these two events are located at the same point in space, this intuitively accepted procedure poses no difficulties.

But consider what would happen if you were to synchronize two clocks by comparing them when they are together on earth and if you were then to put one of the clocks on a rocket to the moon. How would you know the two clocks remain synchronized? The simplest thing to do would be to have someone traveling along with the clock to the moon; he could radio back the time from his rocket as he speeds through space. Now radio waves move with the speed of light, so there will be an inevitable delay while the signal from the rocket propagates to the earth. Because the "law" of the radio wave (or of light propagation) is known, the

Figure 13. In (a), two lightning bolts strike at time zero. Because the speed of light is the same in all reference frames, light will come to ⊗ simultaneously at the origin O. In (b), light will not arrive simultaneously at the origin ⊗ because the two rays must travel different distances.

Figure 14. (a) The "clock" at rest in the bottom mirror records twice the time necessary for light to travel the vertical distance. (b) The moving clock records twice the time necessary for light to travel the diagonal distance.

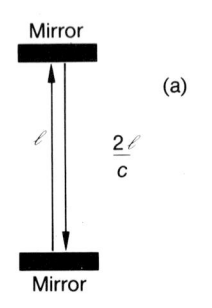

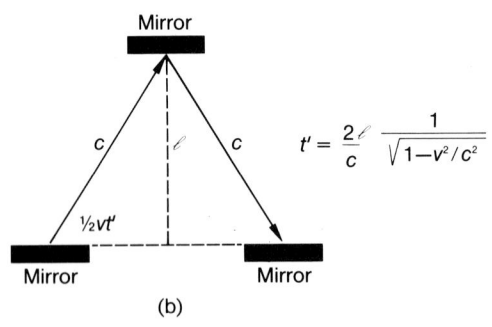

$$t' = \frac{2\ell}{c} \frac{1}{\sqrt{1-v^2/c^2}}$$

delay can be taken into account and you can determine whether or not the two clocks have remained synchronized.

What this example illustrates is that signals must be used to compare clocks at distantly located points. For this discussion, light or radio signals are chosen because they follow simple laws. Einstein proposed a general method for determining whether two events taking place in a given frame of reference were, or were not, simultaneous. In your space-traveling clock example, you can measure the distance between the two events and send an observer to the midpoint between them. When the events take place, an observer stationed at each event can send a light signal to the central observer. If the signals meet at the midpoint simultaneously, you know the two events were simultaneous.

Einstein now asked the crucial question of whether a *moving* observer would say the same two events occurred simultaneously (see Figure 13). The answer clearly is that he would not: the light coming from one event would arrive at the moving observer's origin before the light coming from the second event. To state that the two events are simultaneous, then, is not enough. The statement must be qualified by asking in which frame of reference the events are simultaneous. But all time measurements are measurements of simultaneous events; therefore, *time cannot be given an absolute quantity but must be specified with respect to a given frame of reference.*

Look at the problem in another way by imagining a simple version of a clock. Take two mirrors and separate them by a distance ℓ, as depicted in Figure 14. Next, send a light beam bouncing back and forth between the two mirrors. If the apparatus is at rest there will be a complete passage of the light beam every $t = 2\ell/c$ seconds. This is the "beat" of the clock in the rest frame. If the contraption is set in motion, as indicated in Figure 14(b), then the light will appear to follow the triangular path as shown and each "beat" will now be

$$t' = \frac{2\ell}{c} \frac{1}{\sqrt{1-v^2/c^2}}$$

According to this argument, light takes a longer time to go between the mirrors when the clock is moving. The beats of the moving clock come less often, and the moving clock runs slower than when it was at rest. Although the argument refers to a special sort of clock, it can be made completely general. And the theory predicts that *any* moving clock will appear to be "slow" when compared to a clock at rest.

This remarkable prediction of the relativity theory can be, and has been, checked innumerable times in the following way. Many of the celebrated subatomic particles of modern physics are known to be *unstable*. Such particles can be produced copiously in the large particle accelerators. But after they are produced they spontaneously decay into particles that, for one reason or another, are really stable. The decay is characterized by a "half-life," which is the time it takes for half of a given sample of particles to decay. The process is also a sort of clock in that the half-life can be thought of as a "beat."

If Einstein's reasoning is correct, the half-life of a particle can be expected to be longer when the particle is in motion than when it is at rest. It turns out that the half-life of a moving particle *is* longer, and an increase in lifetime by a factor of 10 or more has frequently been observed for fast particles. Because of this effect, when a physicist talks about the half-life or lifetime of a particle he is probably referring to these quantities as measured in a frame of reference in which the sample is at rest.

The Lorentz-Fitzgerald length contraction also emerges as a consequence of Einstein's general arguments. The important point here is that nothing in his argument refers to a particular theory of matter. Any theory of matter that is consistent with the postulates of the relativity theory will yield a Lorentz-Fitzgerald contraction. Lorentz's electromagnetic model of matter is such a theory and that's why he got the answer that he did.

Recall that the Lorentz-Fitzgerald length contraction is given by

$$\ell' = \frac{\ell}{\sqrt{1-v^2/c^2}}$$

This is mathematically similar to the clock rate formula

$$t' = \frac{t}{\sqrt{1-v^2/c^2}}$$

Many of the fundamental results of relativity depend on the factor $\sqrt{1-v^2/c^2}$. The results of relativity seem contrary to ordinary experience because the factor $\sqrt{1-v^2/c^2}$ is close to 1 unless v is comparable to the speed of light. The speeds you encounter in everyday life are much less than c. One of the fastest man-made objects, a space rocket, has a top speed of only 400 miles a minute or so. Even for this $\sqrt{1-v^2/c^2}$ differs from 1 only by 10^{-10}, so relativistic effects are not obvious in ordinary life. Only atomic particles go fast enough to make relativistic effects really important.

Einstein also applied his theory to Newton's laws of motion and found they had to be modified. One of the important modifications was that mass—considered to be independent of motion in Newtonian physics—had to be replaced by relativistic mass:

$$\frac{m_0}{\sqrt{1 - v^2/c^2}}$$

where m_0 is the mass the object has when it is measured at rest. If m_0 is not zero, then clearly this inertial mass becomes larger and larger as the light velocity c is approached. In high-energy accelerators, electrons have been speeded up so close to the speed of light that their relativistic mass has been *40,000 times greater* than their mass at rest. Thus the particle becomes more and more difficult to accelerate with a given force and, indeed, at the light velocity itself *no force* can accelerate the particle. A particle with m_0 greater than zero can never reach the speed of light. But there is still the possibility that particles may exist with m_0 equal to zero. In fact, such particles do exist and, as you might imagine, light that clearly moves with the speed of light is composed of such particles—the *photons* described in Chapter 2. They always move with the speed of light and can never be slowed down.

For better or worse, the best-known consequence of Einstein's relativity theory is the equation $E = mc^2$. This equation, more often quoted than understood, does not appear in Einstein's original 1905 paper on relativity. He published it shortly thereafter in a remarkable 2-page paper. It may be that never in the history of science has such a portentous scientific discovery been stated more economically. Proceeding in his usual fashion, Einstein derived this equation by considering a simple thought experiment that illustrated the essential physics and was, when properly considered, the fully general situation. In essence, his example involved an atom that could decay spontaneously into two photons. (Einstein didn't phrase his example quite this way but this is the nub of it.)

There are, as we know, particles in physics that decay in exactly this way. He assumed that in such a decay process both energy and momentum were conserved. In pre-relativistic physics, the only energy associated with such a particle was its *kinetic energy*, which is given in terms of the mass and velocity of the particle by the expression $E = 1/2\, mv^2$. In such physics, conservation of energy meant simply the equation $E_1 = E_2 + E_3$, where the E's refer to the kinetic energies: E_1 for the original decaying particle, and E_2 and E_3 for the newly created and emerging particles. A frame of reference can always be chosen in which the particle that decays is at rest, so that E_1 equals zero. However, E_2 and E_3 are both greater than zero unless both particles emerge from the decay at rest, which does not happen in the experimental situation. So this decay process cannot conserve energy if the kinetic energies as considered by the pre-Einsteinian physicists are the whole story.

At the time of Einstein's first papers, the whole subject of radioactivity was so new and so mysterious that this contradiction seems to have escaped most people's attention. In classical pre-Einsteinian physics, the momentum of a particle of mass m is given by the equation $P = mv$. In relativity theory, the relativistic momentum is defined as

$$P = \frac{m_0 v}{\sqrt{1 - v^2/c^2}}$$

For velocities small compared to the velocity of light, this formula goes over to the Newtonian form. Einstein showed that, in the relativity theory, the energy of a particle can be written in two equivalent forms:

$$E = \frac{m_0 c^2}{\sqrt{1 - v^2/c^2}} = \sqrt{m_0^2 c^4 + p^2 c^2}$$

Establishing the equivalence requires the use of the definition of the relativistic momentum and a little elementary algebra. One important consequence of this formula is that, in relativity, a particle at rest has an energy given by $E_0 = m_0 c^2$. It is this energy that is available and given up when a particle decays. In Einstein's original example, all the energy is given up and shared equally between the two photons (see Figure 15).

The applications of Einstein's discovery of the existence of a "rest energy" associated with every object in the universe that possesses mass could fill a book. It is beyond the subject of this chapter to try describing them, except to make the remark that this energy is enormous because c, the velocity of light, is so large. This is the real "secret" of nuclear power, where a small amount of rest energy is converted into enough energy to light a city.

Figure 15. A particle with rest mass m and energy $m_0 c^2$ can decay into two photons, each having an energy $m_0 c^2 / 2$. The total momentum of the system still remains zero.

THE GENERAL THEORY OF RELATIVITY

This chapter will conclude with a brief and necessarily superficial account of Einstein's greatest discovery—the general theory of relativity. It is, from the viewpoint of symmetry, the most beautiful theory ever invented in physics. The theory of relativity so far considered is called "special relativity" because it restricts itself to the consideration of special frames of reference that move with uniform, nonaccelerated motion.

In the real physical universe there are no such frames of reference. There are parts of the universe, far away from matter, that are very nearly free of forces, and there are situations even on earth where the forces acting on an object become very small. But in precise terms, the effects of gravity, for instance, penetrate all corners of the universe and we are affected however weakly by the gravitational forces of the most distant stars. Einstein was, of course, well aware of this interaction, and almost as soon as he had finished the special theory he began work on a theory of relativity that would also include general motions. It took him nearly 10 years of unceasing work before he found the final formulation of such a theory. And this work, surprisingly enough, led him to a completely novel view of gravitation that has now replaced Newton's "universal" law of gravitation.

Einstein's considerations began with the contemplation of an equality in physics that, although well-known, had never been seriously explored. From earlier discussions, you know that the term "inertial mass" has been used when referring to the quantity m occurring in Newton's law $F = ma$ or its relativistic generalization. However, in Newton's law of universal gravitation, in which the force is given by $F = mMG/R^2$, the mass m appears once again. In Newtonian physics there was no particular reason why the mass m (the inertial mass) appearing in $F = ma$ should be identical in value to the mass m appearing in the gravitational equation $F = mMG/R^2$. This mass is usually called the "gravitational mass." It turns out, on the basis of experiment, that these two masses are identical to at least one part in 10^{11}. This equality is not, and cannot, be accounted for in Newtonian physics, where it was taken merely as a fortunate empirical coincidence. This point of view Einstein could not accept. He felt that such a striking observation should not be put down as mere coincidence but should emerge as part of a more fundamental theory of gravitation.

In 1911 Einstein published a short paper in which he gave the first steps toward the construction of such a theory. His line of reasoning went roughly as follows. On the surface of the earth, the force of gravity on any object of gravitational mass is given, according to Newton's law, by the equation $F = mMG/r^2$, where M is the mass of the earth and r its radius.

If inertial mass and gravitational mass are identical, it follows from the equation $ma = mMG/r^2$ that every object on the surface of the earth falls toward the center of the earth with a gravitational acceleration $a = MG/r^2$. (The effects of air resistance and the fact that the earth is not exactly a sphere are ignored here.) This remarkable insight may be viewed in a somewhat different light by using an example that Einstein gives in his 1911 paper.

Imagine an elevator attached to a cable. Suppose you are inside the elevator and some mechanism begins to pull the elevator upward with a uniform acceleration a, the gravitational acceleration. If you really are enclosed in the elevator all you will notice is that every object, yourself included, is accelerated toward the floor of the elevator with the acceleration a. You will feel your feet being pressed to the elevator floor. Therefore, you will have no way of deciding whether you are in such a uniformly accelerating frame of reference or whether you are, in fact, stationary and that someone has moved a large gravitational mass in the neighborhood of your elevator, which has produced a downward gravitational force.

Clearly this is a kind of generalized relativity principle—in other words, a coordinate system in uniform acceleration cannot be distinguished from a stationary coordinate system acted on by a uniform gravitational force. Such an equivalence is only possible if the gravitational and inertial masses are exactly equal; otherwise each object in a uniform gravitational force would fall with a different acceleration. Einstein turned things around and said this relativity principle was really the *explanation* of the equality of inertial and gravitational mass.

His explanation would be somewhat hollow if that were all there was to it. But Einstein was able to draw two further conclusions, which have been tested experimentally and found to be true. The first of these conclusions is that light rays must necessarily be bent by the force of gravity. Einstein's argument can be illustrated again by the elevator (see Figure 16).

The light enters a window at one side of the elevator. If the elevator were stationary, the light would move in the straight line shown and hit the other side at a point above the floor, the same height as the height at which it entered the elevator. On the other hand, if the elevator is being accelerated upward, the light will hit at a lower point and, to the occupants of the elevator, it will appear as if the light has been bent. Because these occupants can't distinguish this situation from one in which they are under the action of a uniform gravitational force, they will conclude gravitation bends light rays.

In his 1911 paper, Einstein suggested this notion could be tested during a total eclipse of the sun, when the light rays from certain stars become visible as they pass by close to the sun. These stars should appear slightly "displaced" in comparison with their positions when the sun is not in their neighborhood (see Figure 17).

Einstein gave a value of this displacement in his 1911 paper, which he modified in 1916 when he published his fully generalized theory of relativity. This theory takes account not only of uniformly accelerated frames of reference and uniform gravitational forces but of *all* frames of reference and *all* gravitational forces. It is his 1916 prediction that has been confirmed by astronomical experiments of ever increasing accuracy.

The second effect discussed by Einstein is called the gravitational "red shift." You are familiar with the fact that, when an object that's emitting a sound is moving past you, the pitch of the sound is altered. If the object is moving toward you the pitch is raised; as the object moves away the pitch is lowered. The same effect also occurs for light, despite the fact

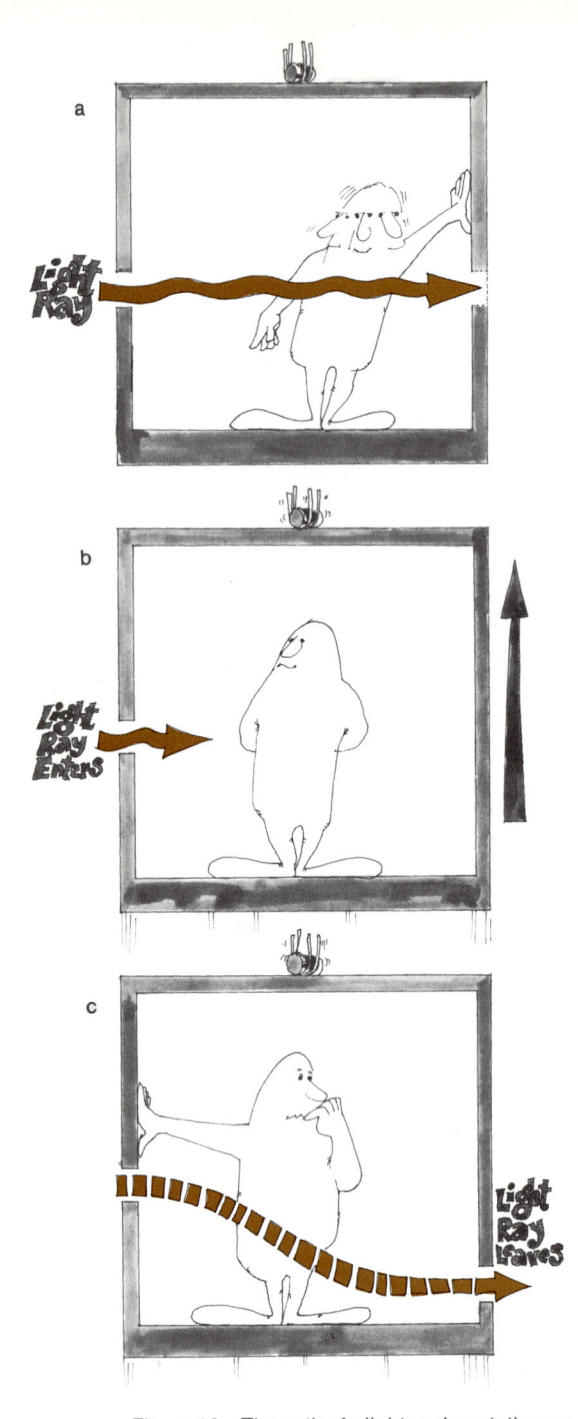

Figure 16. The path of a light ray in a stationary elevator is shown in (a). In (b), a light ray enters an elevator that is accelerating upward, and (c) depicts the trajectory of the light ray as seen by the occupant of the elevator.

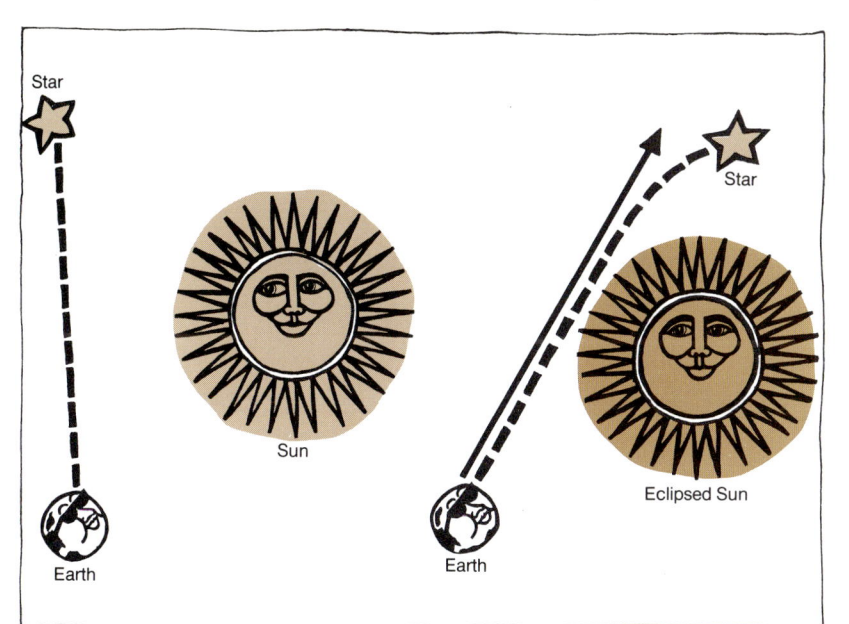

Figure 17. During a total eclipse of the sun, an observer on earth will see the star along the solid line. The dashed line in (b) shows how the trajectory is bent. The observer will see the star displaced *away* from the sun.

that the speed of light is the same for all observers in uniform motion. When an object emitting light moves away from an observer, the color of the light is shifted toward the red. Einstein argued this effect would occur for an object undergoing a uniform acceleration away from an observer and, by his generalized relativity principle, also for an object that is stationary but in the presence of a gravitational mass. In fact, because the periodic motion of the light waves constitutes a sort of clock, Einstein reasoned that any clock placed in the presence of gravitating masses should be slow compared to a clock in gravitationally free space. This prediction of his theory has also been confirmed both in experiments on earth and, in the case of the red shift, in data involving the spectrum of light coming from very dense and massive stars.

Because gravitation changes both the straight line orbits of light rays and the motion of clocks, we can say the presence of gravitating matter changes the "geometry" of space and time because, physically, this geometry is determined by the behavior of clocks and light rays. Einstein derived the most general equations relating this geometry to gravitation and indeed, from his point of view, the laws of gravitation become the laws of the geometry of space and time.

No doubt these remarks seem somewhat mysterious, and perhaps that's the way it should be. No passing review can make the more subtle and abstract points of modern physics completely clear—even when they are well-understood by the physicist himself. Such understanding takes many years of serious study. Nevertheless, the more we understand these theories, the more imposing does the prospect appear of what must lie outside our present boundaries of knowledge and of what discoveries are still to be made.

9 | Relativity

Wishing Won't Make It So

Long before this book was written there was a popular song called "Wishing Will Make It So." The idea was that you can make things happen by wishing for them hard enough, and anyway the wish itself is as good as the real thing (perhaps better). Conversely, one might wonder if the universe exists only because someone wished for it.

This philosophy is almost precisely the antithesis of what normally constitutes a scientific approach to problems—although scientists wish for things as much as anyone else. In fact, the scientist has one special wish of his own in addition to all the others; namely, that his hypotheses be supported by subsequent observations. Occasionally this may introduce a bit of strain on his powers of observation. But it helps to remember that nature universally abhors lies and punishes liars and moreover retains a large contingent of agents who are always trying to ferret them out.

However, the real motivation of scientific pursuit is that one of the great joys of existence for each individual—although admittedly not the only one—is his awareness of the immense world outside himself. This knowledge represents an important point of departure in the life of man from that of the animals. It means knowing about other creatures than oneself, other places than this one, and other times than now. Such knowledge makes it possible for (some) human beings to extrapolate beyond their own existence, to care what happens before and after their lifetime, and to impart to the chaotic world confronting them a sense of purpose.

Although a search for values is certainly not the primary occupation of a scientist, there can be no science at all without at least one basic value: an unwavering respect for objective truth. And because there can be no lasting moral structure without the same principle, this is perhaps the real point of contact between science and humanity. The scientist who misrepresents his findings because it is for some reason convenient to do so is no longer practicing science. And the politician who misleads the public as a matter of expediency is committing an immoral act. If scientists lied about their science there would be no scientific achievements today. And if politicians continue to lie about their politics, there may be no human race tomorrow. Thus the same principle of integrity is essential to both scientific progress and human survival.

Unfortunately, even when people want to answer questions truthfully it is not always easy to do so. Moreover, scientists and mathematicians have found that proper formulation of a question actually may be even more difficult than the search for a truthful answer. In other words, the way a question is asked often predisposes the reply. And both may add up to nothing at all. The difficulty is that all of us are susceptible to the same universal pitfall—the belief that wishing will make it so.

Most of this book consists of answers. This essay examines rather *the nature of the questions.* In other words, how do we decide what it is possible to know? And if wishing won't make things so, then what will?

There is really nothing wrong with wishing, even in science. In fact, if you could somehow reconstruct the various processes of discovery, I suspect you would find a wish at the root of each one. What makes a procedure fall within the province of science is that one is fully prepared to have his wish not come true and moreover has the means for recognizing when this has happened.

If I suspect that I have the ability to move distant objects merely by thinking very hard about it, I shall probably carry out a large number of tests to explore this conjecture. After several years of unsuccessful experimentation one of the objects I have carefully balanced on a table finally topples over. Now I have two courses of action open to me; let us call them the optimistic and the suspicious. If I am sufficiently optimistic, I call in the representatives of the press, show them the object lying on its side, and tell them about all the difficult, unsuccessful years through which I persevered until the day it finally happened. Of course there may be people who will call me a crank, but hopefully not everyone has a closed mind. Eventually someone actually manages to duplicate my success. We form a society of people-who-can-sometimes-topple-things-over-just-thinking-about-it; the society grows, and with a little luck an eccentric millionaire or perhaps a particularly optimistic representative of the defense establishment provides the necessary research funds to produce many successes.

On the other hand, if I am inclined to be suspicious I will instead try to think of all possible explanations within the scope of present knowledge. If I am nevertheless still unable to explain my astonishing but impressive result, I call in my most cynical friend. He says, "Try it once again. But this time close the window."

Scientists are indeed dreamers, but they are suspicious dreamers. Three important tests must be applied to a new theory even after it has once been verified. The first test is *repeatability.* It must be confirmed many times, especially when the experiment is carried out by people who may not believe in the theory. If the claim is made

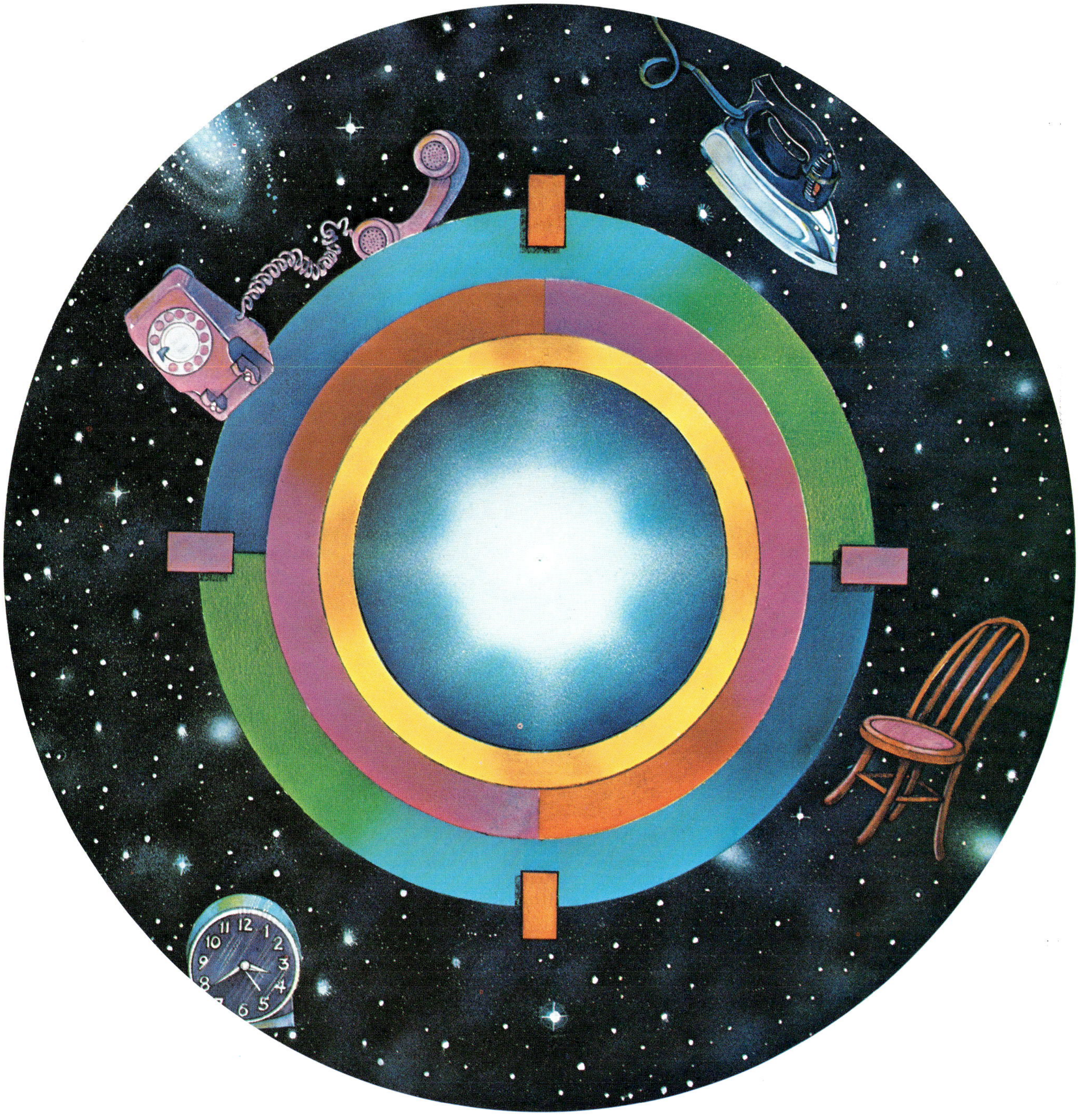

by Americans, it had better be sustained by Russians, and vice versa.

The second test is *simplicity.* What do we mean by a *simple* theory? It is one that depends on very few assumptions, or *postulates.* For example, suppose someone tells you he has discovered that earthquakes are caused by people with green eyes.

"The pigment," he says. "It sends out waves. I was in an earthquake once, and there was this shifty-looking character with green eyes hanging around."

"I saw an earthquake in India," you say. "And there was nobody with green eyes within a hundred miles."

"Never works in India," he says. "There it's another thing, because it's so hot."

"But there was one in a remote section of the Andes, far from all human habitation," you point out.

"The altitude," he says. "The waves travel really far at that altitude."

"My wife has green eyes," you say. "And she's never been near an earthquake."

"Only men," he says. "Women can't do it. Not enough testosterone."

Before long you come to the conclusion that there ought to be a better explanation for earthquakes. Why don't you like the green-eyed theory? It may be "right," in that it actually can be adjusted to fit all the data on past earthquakes. But it is not a simple theory; there are too many postulates, probably one for each event. A good theory should introduce not much more than one or two new postulates; they should be capable of being stated concisely, and preferably be not hard to understand even though they may be quite difficult to believe.

Thus before Einstein it was always "obvious" that light waves approach a stationary observer at *one* speed, whereas they approach a moving observer (who is rushing forward to meet the waves) at *another* (greater) speed. Einstein's postulate was simply that this was not so. Light has the *same* speed for someone moving at a constant velocity as for someone who is not moving at all, and the laws of nature must be the same for all such observers. The entire basis for the theory of special relativity was just so simple to state, although not at all easy to believe, and it agreed with the experimental evidence.

Hence the theory of relativity is a simple theory, despite the fact that it has led the human race to some startling conclusions.

Finally, a good theory, in addition to "predicting" results that were obtained in the past, must have the ability to predict future events as well. It has undoubtedly occurred to you by now that before the scientific community becomes willing to give serious credence to the green-eyed earthquake theory, our theorist is probably going to have to produce an earthquake on demand. Almost anybody with a little practice can learn to predict the past; the future is something else again.

We do of course run the risk in exposing a fledgling theory to such a hostile environment that it may be stifled before it has had a chance to take proper root. Unfortunately there is little choice; the risk must be taken. There has to be some way to eliminate irresponsible conjectures from the evaluation process. Otherwise the community will be inundated with a flood of pseudo-scientific "theories" and self-fulfilling prophecies, with no possible means for isolating the theories that have actual merit and giving them the critical attention they require.

The usual procedure is for a scientist to submit his theory in the form of a paper to a specialized publication in his field. The editor of a responsible research journal customarily forwards such a paper to a (preferably suspicious) referee who is an expert in the field and who proceeds to look for something wrong. If the referee rejects the paper, giving his reasons, the author can protest the decision, whereupon a correspondence may take place involving a succession of referees. Eventually something is worked out, or the author has the option of trying to publish his work in another journal with somewhat less extensive circulation. The system is by no means perfect, but it is at least viable. It is my opinion that if green-eyed people do in fact cause earthquakes we shall probably discover this calamity before long in any case.

The problem of formulating a question or carrying out an experiment in such a way as to predispose the answer is not the only difficulty in scientific investigations. There is an even more insidious danger. It is possible for a line of questioning to have *no answer at all*—right or wrong, good or bad. Here, too, we must provide tests to prevent ourselves from becoming tangled in webs of our own creation. Consider how some of these traps are laid.

Every field has a primary goal. In medicine it is the preservation of life. In science it is knowledge. In military affairs it is security. The concern with security is understandable, although perhaps a bit paradoxical in the context of that particular field.

In the course of one of this nation's wars I found myself in charge of a supply room with a rack containing a large assortment of rifles, shotguns, and various other dangerous weapons. A visiting inspecting officer who was concerning himself with the matter of the security of all this equipment had his own objective tests.

"Those rifles," he said. "Do you always keep them locked up?"

"Yes sir," I said.

"And where do you keep the key?" he asked me.

"In my desk drawer," I said.

"Then tell me," he said, his eyes lighting up. "Do you keep the drawer locked?"

"Yes sir," I said enthusiastically.

"And where do you keep the key to the drawer?"

"It's hanging on this hook, sir."

"Let's get that key locked up right away," he ordered.

"Yes, sir!" I said.

I wish I could assure the reader that scientists (unlike inspecting officers and other mortals) are incapable of falling into logical traps of their own setting. Unfortunately this is not the case. Just as explorers are the first to encounter physical dangers, it is the scientists who usually make the first mistakes. What is often called "scientific method" is really just a procedure evolved to avoid repetition of past mistakes.

Some of these procedures have become so much a part of our lives that we take them for granted. Before Galileo, scientists made observations, but it evidently never occurred to anyone to set up controlled experiments to test the conclusions they drew from their observations. Thus Aristotle *assumed* that an object has velocity only when acted upon by a force. But it was Galileo who demonstrated *experimentally* that this is not so—rather it is the *change* in velocity that is the result of force. It was Galileo also who taught us that simplicity, not complexity, is the test of a good theory. And it was Galileo who had to contend with the problem of knowledge as a threat to the established order.

But the discovery of what can happen as the result of a badly worded question arose much later. It was Newton who made the "mistake," and Einstein who taught us the lesson. Newton had concluded from his laws of mechanics that there was no mechanical means of measuring "absolute" motion at constant velocity. In other words, if you are riding in a train moving at fixed speed along a perfectly straight and smooth track, and all the window shades are drawn, then there are no possible experiments that will tell you whether the train is in motion or at rest and what its absolute speed is. Of course if you want to cheat and pull up the shades you will indeed see houses and trees going by. But for all anyone knows, perhaps it is the *houses and trees* that are really in motion, and not the train at all. Even though there is clearly *relative* motion between the earth and the train, one would be hard put to establish which one has the *absolute* motion; that is, which one is "really" moving.

Newton, however, refused to believe that the earth is no better than the train. (And perhaps the reader sympathizes.) Even if *you* can't tell which one is moving, *somebody* ought to be able to, was his reasoning. Despite the impossibility of anyone actually measuring such absolute motion and thereby establishing once and for all whether it is the train that is really moving or the trees and earth, down in the pit of one's stomach one feels he knows very well what the answer is. And just because people with their meter sticks and inexpensive clocks sometimes make mistakes is no reason to claim there's no such thing as right and wrong. This was the reasoning of all mankind before Einstein.

The only disturbing thing in Newton's day about this idea of an absolute frame of reference was that there was apparently no *experimental* means of establishing which object was in motion and which was at rest. But in the nineteenth century, when light was discovered to consist of electromagnetic waves traveling at a theoretically predictable speed *in vacuo*, it was only natural to assume that this speed established an absolute frame of reference with respect to which the light waves were traveling. Because one speed was known to be correct (namely, the speed of light), it followed that there had to be an absolute reference frame (called the "ether") in which this speed could be measured correctly. It seemed as if it was finally possible to settle arguments about what was in motion, and what was not.

If someone riding the roof of the train measured the speed of light and found it to have a *different* speed from the theoretically predicted one, he obviously would have to be experiencing an "ether wind"; it would therefore follow that his train was moving. If, instead, someone sitting on the branch of a tree were to feel this ether wind, then it would have to be the tree and the earth moving, not the train. And when experimenters such as Michelson and Morley established that neither trains nor trees encounter any ether wind—that light has the same predicted speed regardless of whether the observer is "at rest" or "in motion"—everyone tried without success to explain away (or wish away) the evidence. Everyone except Einstein, that is. Michelson himself believed the experiment had failed in some inexplicable way even though he had established without doubt the correctness of his findings.

Einstein simply said that it was pointless to ask questions involving the ether and its "absolute" frame of reference. Because there is no way even *in principle* to detect absolute motion, what meaning can there be in asking whether an object has such motion? The theory of relativity bypassed all the metaphysical questions and addressed itself instead to those questions that could be formulated in terms of conceptual if not actual experiments.

This philosophical approach to questions and answers is called *operationalism* because it defines everything in terms of operations. One need not actually have the capability of performing the experiment in order to *ask* the question; but at least one must be able to *think* about how it might be done if the means were available. Thus distance may be defined as something to be measured by a meter stick. Time is measured by a clock. Velocity is obtained by dividing distance by time. It does not necessarily follow that your meter stick and my meter stick, your clock and my clock, must always agree, or that one has a claim to the "right" answer and the other is "wrong." And if one cannot even in principle conceive of a test for the presence or the absence of an ether, he would probably do well not to spend his time asking whether it actually exists. The answer to such a question is neither yes, nor no, nor even maybe; it is none of these, since we do not understand the question.

Have I perhaps in another age lived another life, on another planet, with another body, and altogether different memories of totally indescribable experiences? Yes or no? Without the operational requirement I may be inclined to say, "Perhaps. Who knows?" And my friends may nod their heads, and say, "What an interesting thought." But if one of them is a hard-headed operationalist he will instead ask, "Suppose it is as you say; how could you ever know? And if it is *not* true, then how could you ever know that?"

Problems normally considered to be the province of science are usually formulated in operational terms because then one can do something, or at least think about doing something, to find an answer. There is of course no law against thinking about questions that do not meet the operational test. But then one proceeds at his own risk. Before rushing off to find an answer, one should at least try to understand the question. If that is not possible, at least one knows where things stand.

Meanwhile, recall that we left our inspecting officer in the supply room without his security. What could he have been trying to tell me? Was it that keys should always be placed in drawers, which are promptly locked with other keys, which are then placed in other drawers, and so on ad infinitum? Or that security means a person has to have a minimum of at least 45 keys before obtaining a dangerous weapon? Or shall we rather say that security will be attained only when no one can ever get his hands on such a weapon, no matter how many keys he has? Perhaps this is the only meaningful definition. I submit this answer to the people who hold the keys to the supply rooms of the world.

Adolph Baker

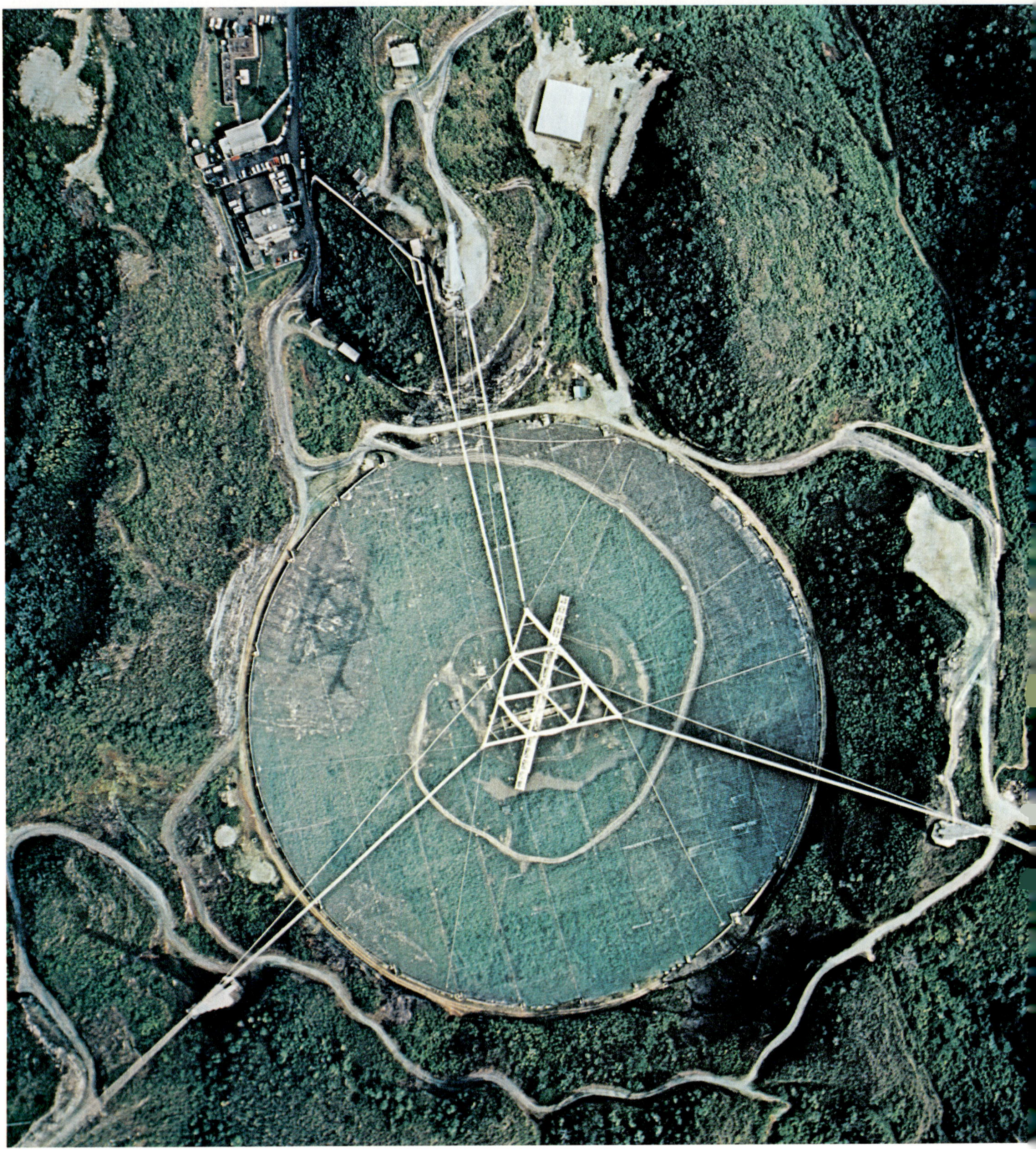

The study of the universe is, in a fundamental sense, an observational science. It is true that in *all* areas of science, attempts to understand the physical world by pure thought alone usually lead nowhere. But there is in the laws of science a constant interplay between the way things "ought to be" and the way they actually are, as revealed by experiment. Moreover, the simpler the theories and descriptions are, the more rewarding they often prove to be in the search for fundamental principles.

In contrast, any simple description, interpretation, or theory of the physical universe can be immediately judged to be incorrect. In the study of the universe, we must always be guided primarily by what we observe. From understanding the basic laws of science alone, no one would ever have predicted the existence of a star surrounded by nine planets of significant size, all of which are characteristically different. Who would have predicted that one would be a miniature star, emitting more radiation than it absorbs from the central star of the system; that another would be searing hot and dry beneath an incredibly dense atmosphere; that another would be surrounded by a beautiful set of thin, wide rings; and that still another would have an atmosphere containing 20 percent oxygen, an extremely unstable substance that *should* react rapidly with most planetary matter, rather than be present in the gaseous form? And who would have predicted that one of those planets—the earth—would be inhabited by a wide variety of fantastically complex organisms, some of which read and write books on the nature of the physical universe?

When faced with complexities of this kind, we must first of all be committed to observing as carefully as possible what there is to observe; and then we must try to fit the observations into a rational interpretation of the universe. It is in the interpretation that we must be strongly guided by the fundamental laws of physics and chemistry. Without the use of those principles our observations would yield nothing.

Despite the difficulties accompanying these observations and the interpretation of them, the scientist strives to go further. Not content with the difficult task of understanding the present state of the universe, he tries to understand the processes and sequence of events leading up to the present state of the universe and the processes and events that will carry it into the future. Here he treads on even more uncertain ground.

For example, our solar system is generally thought to have formed from a cloud of gas, dust, and solid bodies swirling in turbulent eddies, perturbing the motions of one another by their gravitational fields. Those bodies collided with one another, producing explosive bursts of light and heat. The dusty fragments from the explosions obscured the sun, at times producing periods of extreme cold. Magnetic and electric fields intertwined the entire complex medium, accelerating and guiding electrically charged particles through it. It is likely that all of this matter was continually agitated by gigantic bolts of lightning that would dwarf any in our earthly experience.

How would you go about describing these awesome complex processes by a set of mathematical equations? Even far simpler problems—for instance, the dynamic evolution of three bodies, moving as a consequence of their mutual gravitational forces—are full of complexities, and despite study by generations of such great scientists as Karl Gauss, the Marquis de Laplace, and Jules Poincare, many aspects of even the more simple motions are still obscure.

10
Exploring Outer Space

The radio telescope at the Arecibo Ionosphere Observatory, Puerto Rico, picks up signals from astronomical phenomena that lie beyond the limits of optical telescopes.

Figure 2. (Above) The German astronomer Johann Kepler (1571–1630) first determined that the sun is not at the center of our solar system. This discovery was based on measurements of the orbit of Mars—measurements that Kepler inherited when he succeeded Tycho Brahe as Court Astronomer in Prague. When he realized the significance of Brahe's extensive empirical data, Kepler abandoned his previous efforts to understand the planetary orbits with Euclidean geometry and turned to the post-Euclidean geometry of Appolonius, who first studied conic sections (elliptical slices of a cone). Brahe's data substantiated Kepler's new and radical theory that the sun forms one focus of the elliptical trajectories of the planets.

Nevertheless, the scientist still tries to understand the difficult problems concerning the origin and evolution of the planets, the solar system, the stars, and the universe itself. And he has made some significant achievements, which will be discussed as examples in the next two chapters. In this chapter the primary concern will be the observations themselves, and the use of observations in forming a coherent picture of the present state of the universe. Many can be made with the unaided eye; they were made by men in ancient times and you can make them today. Other observations require instruments of great complexity and precision.

It would be deceptive to pretend that all the information and interpretations in this chapter follow directly from these observations. In ancient times men saw the same bright and fainter lights we call the sun, moon, planets, and stars, and they had considerable understanding of the regularities of the motions of those lights. But their interpretations of these phenomena differed greatly from the interpretations given here. Quite naturally, they regarded their own position as the center of the universe. From their point of view it was indeed the center, from which they could look out and see everything else. The stars were believed to be imbedded in a great inverted bowl, the "firmament" that rotated about the earth. Within this firmament were other bodies—the sun, the moon, the planets—that had motions of their own in addition to the motions of the firmament. From time to time transient events occurred. Comets with hairy tails swept by and "guest stars," now named novas and supernovas, made their appearance among the familiar stars, only to fade from view after a few months or years.

Today we describe the same observations from an entirely different point of view. It would require too much space to recount all the errors of the past and to trace the evolution of our thinking step by step to our present state of understanding. Consequently, many of the statements made here will seem to go beyond the inferences that follow directly from the observations. To some extent, that is a fair criticism, and the scientist must accept it as a fair request to distinguish between what he sees and what he thinks. It may not always be possible to make this distinction clearly, but the attempt frequently affords useful insights both to the questioner and to the person who is asked the question.

THE SOLAR SYSTEM

We are living on a planet that once a year travels hundreds of millions of kilometers at a rate of more than 50,000 kilometers per hour. It moves through interplanetary space around a central star, the sun. In one sense, the notion that the earth revolves about the sun rather than vice versa is merely a matter of one's vantage point. The choice of the frame of reference is, in the last analysis, arbitrary. From this point of view, there is still considerable legitimacy to the notion that it is the sun that revolves. In many practical problems of determining the orbits of spacecraft it is mathematically more convenient to use equations of motion based on an earth-centered, or *geocentric,* frame of reference rather than the sun-centered, or *heliocentric,* frame of reference. And in terms of the mathematics involved, converting from the one frame of reference to the other is a simple matter.

Nevertheless, the sun *is* the center of our solar system. From its great reserves of thermonuclear energy we derive the light and heat necessary for life. The planets would not be visible if they were not illuminated by the sun. Its great mass determines, for the most part, the motions of our earth and the other planets. If one of the planets for some reason failed to exist, we would still have a solar system; without the sun there would be none.

An astronomer looks upon the sun as a rather ordinary star. In his own peculiar but useful jargon he would classify it as a "G2V" star—fairly small and cool, with a surface temperature of about 6,000 degrees

Figure 3. Both the Ptolemaic and the Copernican system account for the "geocentric" motions of the planets. The Ptolemaic system (far left) is actually useful for describing such phenomena as the orbit of the moon about the earth; the Copernican system (above left) is more useful for showing the ordering of the solar system. (Above) Awareness of the heavens was not a uniquely European occurrence, as history books might imply. This Arkansas-Sioux Indian buffalo skin depicts the central importance of the moon and the sun in this culture.

10 | Exploring Outer Space

Centigrade, a central temperature of about 10 million degrees Centigrade, and a mass about 300,000 times that of the earth (about 2×10^{33} grams). Its enormous mass creates an internal pressure hundreds of millions of times greater than the pressure of the earth's atmosphere at sea level. And yet its average density is surprisingly ordinary. At about 1.4 grams per cubic centimeter, its density is about halfway between the density of gelatin and sugar.

The sun's chemical composition, too, is ordinary. It consists almost entirely of the light gases hydrogen and helium; only about 2 percent of its mass is in the form of the relatively light elements carbon, nitrogen, and oxygen, with smaller quantities of the other elements. In this regard it is again a typical star, one of hundreds of millions of such stars making up the Milky Way—the diffuse band of light visible on clear nights.

From our provincial viewpoint, it's probably a good thing that the sun is such an ordinary star. The mass of the sun is small enough so that its nuclear fuel is being used up in a relatively frugal manner; in fact, the sun will be about 10 billion years old before it becomes unstable. That timespan has been enough to permit the evolution of human life during the 4.6 billion years the earth and solar system have existed, and it still leaves adequate reserves for the future. Other larger and consequently more profligate stars run through their ration of thermonuclear fuel in a few million years—barely a moment in time in comparison with the age of the earth, sun, and solar system.

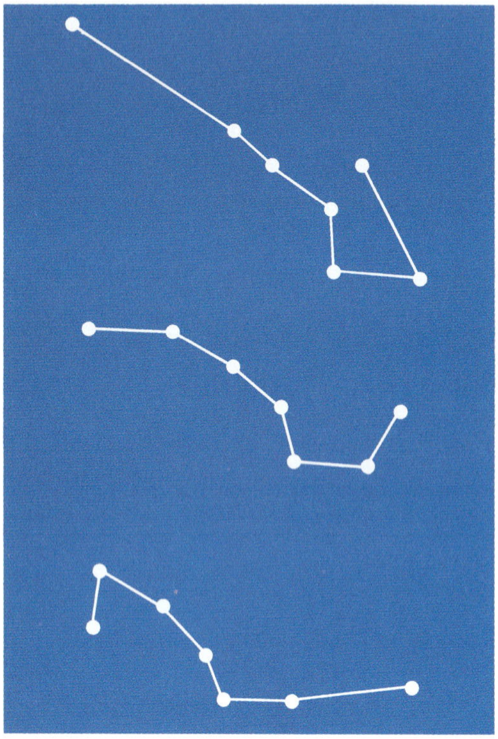

MOTIONS THROUGH THE ZODIAC

All the stars are moving at incredible velocities relative to one another— about 20 kilometers per second. They are so far away from our solar system, however, that for thousands of years people thought they were "fixed" in their positions with respect to one another. Together they formed the configurations of stars that came to be known as the *constellations*. Many of the constellations were named by the inhabitants of ancient Greece although some, such as Microscopium (the Microscope) and Telescopium (the Telescope), obviously were named in more modern times. The names of the constellations have had the somewhat odd consequence of introducing such creatures as centaurs, unicorns, flying horses, and the Princess of Ethiopia (Andromeda) into the terminology of a complex modern science.

People in ancient times used such names to make the unfamiliar seem less remote. And sometimes it's still reassuring to observe familiar signs in the night sky. People far from home often look for Lyra and Vega or Orion and Betelgeuse when they find themselves in somewhat alien surroundings, such as the Karelian Arctic or the *altiplano* of Peru, not far from the dwellings of justifiably xenophobic local inhabitants.

Even today, the constellations and the relatively "fixed" stars comprising them serve as a useful frame of reference when we wish to describe the position of the planets. All the planets lie very nearly in a flat disk, defining a plane in space (Figure 5). If we stand on a particular planet (most conveniently the earth) and look out at the other planets, they will not appear just anywhere in the sky. They will appear only in a few of the constellations, known as the constellations of the *Zodiac*.

Viewed from the earth, the sun, moon, and each of the planets will have in its background the stars of one of the constellations of the Zodiac; to us, a planet looks like an "intruder" into that constellation. In fact, the planets look like stars that don't belong to the constellation. Now the earth and the planets move around the sun, and the moon revolves around the earth. Consequently, as we observe those bodies from the earth they

Figure 4. In geologically "short" times, the "proper motions" of stars will cause constellations to change their shape. The star group known as Ursa Major did not look like a dipper in the past (above), nor will it look like one a thousand years from now. The temporal nature of the universe is somewhat disconcerting:
The stars of the Great Bear drift apart
The Horse and Rider together northeastward
Alpha and Omega asunder
The others diversely
There are rocks
On the earth more durable
Than the configurations of heaven
Species now mobile and sanguine
Shall see the stars in new clusters . . .
From Kenneth Rexroth, Collected Shorter Poems. Copyright 1940 by Kenneth Rexroth. Reprinted by permission of New Directions Publishing Company.

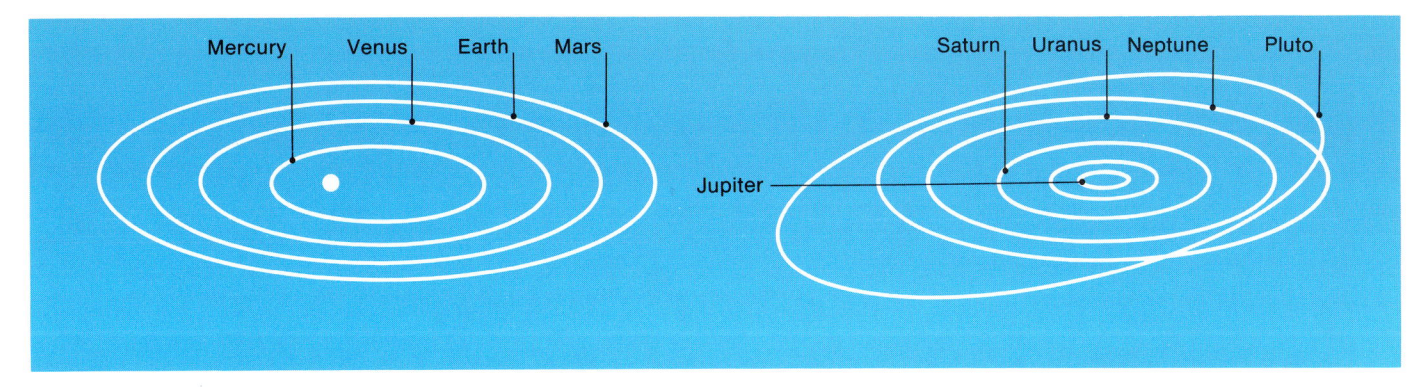

Figure 5. Sketches of the planetary orbits about the sun. The orbits swept out by the planets listed in the drawing to the left actually are *contained within the center of the innermost orbit* (Jupiter) in the drawing to the right.

appear to move through the various signs of the Zodiac because our "line of sight" changes. Using a little mathematics and the laws of motion and gravitation, it's easy to calculate the distance of the various planets from the sun, using the distance between the earth and the sun as the unit of measurement.

The apparent motion of the moon is the easiest to observe. Because it is so close to the earth in comparison with the sun and planets, it appears to move rapidly through the Zodiac. You can watch the change of its position relative to the stars in a single evening.

Once the motion of the moon through the Zodiac is observed and understood, it's easy to go on to other celestial motions. As shown in Figure 6, the sun also moves through the Zodiac. Obviously it's difficult to observe the sun's position in the Zodiac because you can't see stars in the daytime. But there is an indirect way of determining which constellation the sun happens to be moving through. As Figure 6 shows, when you look south at midnight you are looking in the direction opposite the direction of the sun. Because it takes the earth 12 months to revolve around the sun, the constellation of the Zodiac determined by this observation is where the sun was 6 months ago, and by simply counting forward 6 months you can determine the present position of the sun in the Zodiac.

(Interestingly enough, there is a discrepancy between the "signs" of the Zodiac of astrology and the present location of the sun in the Zodiac during a particular month. This is caused by a phenomenon known as the "precession of the equinoxes," which is a result of motion of the earth's axis similar to the motion of a top when its axis of rotation is not perpendicular to the surface on which it is spinning. The *vernal equinox* is the first day of spring: the day on which the sun rises directly in the east and sets directly in the west and the day for which there are exactly 12 hours of daylight and 12 hours of night. Two thousand years ago the sun entered the constellation of Aries on that day, and the astrological "signs" are still based on this. Actually the position of the sun among the constellations drifts a small amount each year. For a long time it has been in Pisces on the first day of spring. As a result of the continuing precession of the equinoxes, it is now on the "edge" of Aquarius.)

The motion of the planets through the Zodiac may also be determined. First, divide the planets into two categories: those closer to the sun than the earth (Mercury and Venus) and those farther from the sun than the earth (Mars, Jupiter, Saturn, Uranus, Neptune, and Pluto). The inner planets Mercury and Venus never are far away in the sky from the sun. In fact, their motion often brings them so close to the sun that we cannot see them. At times, however, Venus is more dazzling than any other star in the western sky for several hours after sunset. It is equally conspicuous in the morning sky, rising in the east several hours be-

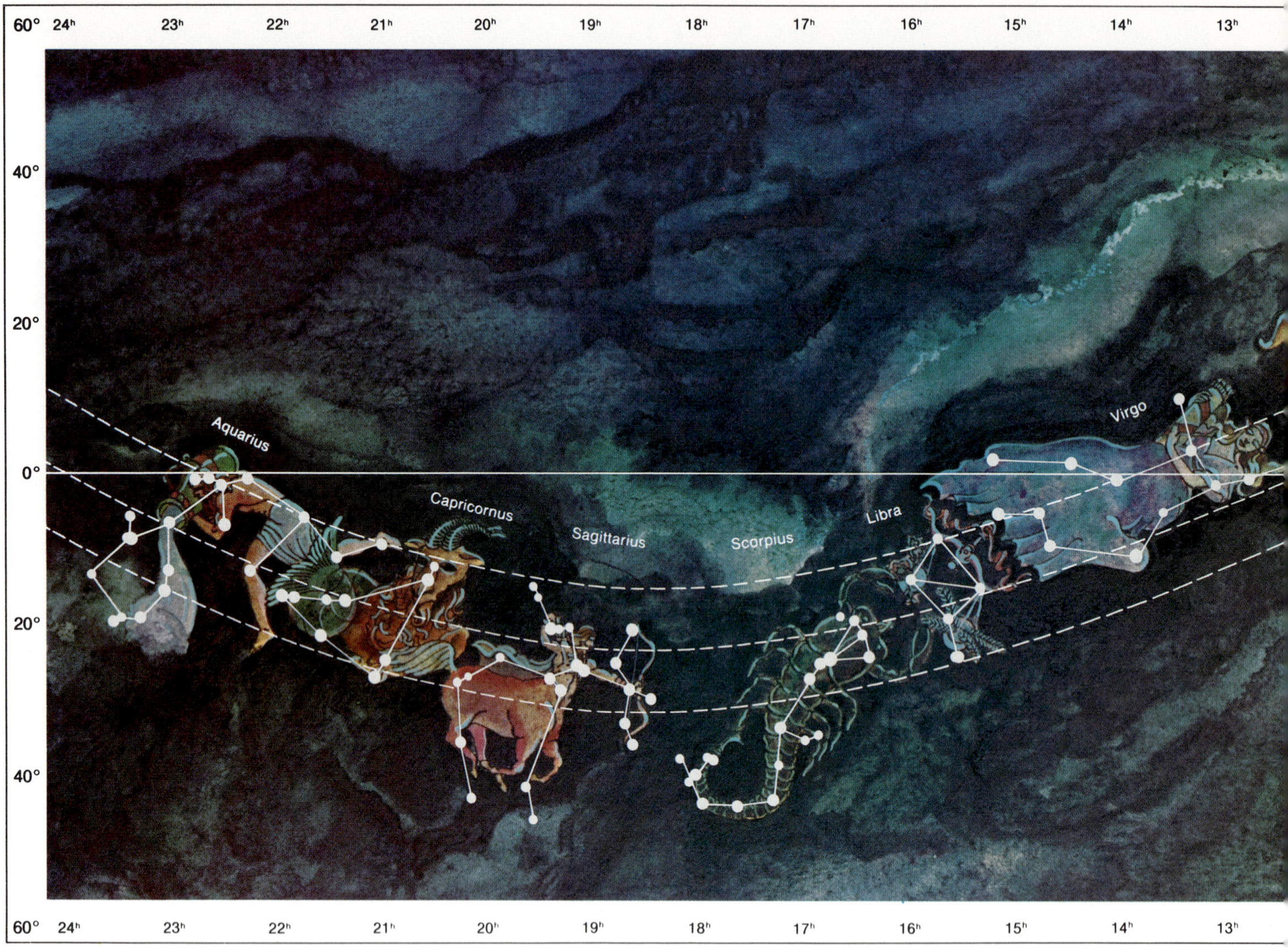

Figure 6. This drawing depicts the motion of the sun through the Zodiac. The straight horizontal line is the *ecliptic*. It is the apparent path of the sun in the sky due to the earth's annual revolution about the sun. The spread in the sky is marked according to the earth's daily rotation as well as the annual apparent motion of the sun.

The sun passes through about one constellation every month. The dates along the ecliptic show its positions at the first day of every month. At the *summer solstice,* the sun reaches 23 degrees 27 minutes north at the equator due to the declination of the earth's axis. At the *winter solstice,* it is 23 degrees 27 minutes south. The sun crosses the equator at the *equinoxes*—the dates when days and nights are of equal length.

With the exception of Pluto, the planets as well as the moon lie in the same galactic plane as the sun (within 7 degrees). Therefore, they would also appear to sweep out similar trajectories across the sky, but they would do so in different time periods.

What are the creatures and beings that "populate" the sky above the earth? The ancient Babylonians, Chinese, and Egyptians identified fixed stars or groups of stars as points of reference, divided the sky into twelve equal parts along the celestial equator, and built up a rich store of mythology about the heavens. Greek and Roman mythology added to these signs of the Zodiac.

You can use the same stars to orient yourself at night. Appendix 2 to this book contains seasonal star charts that can be used as guides through the night sky.

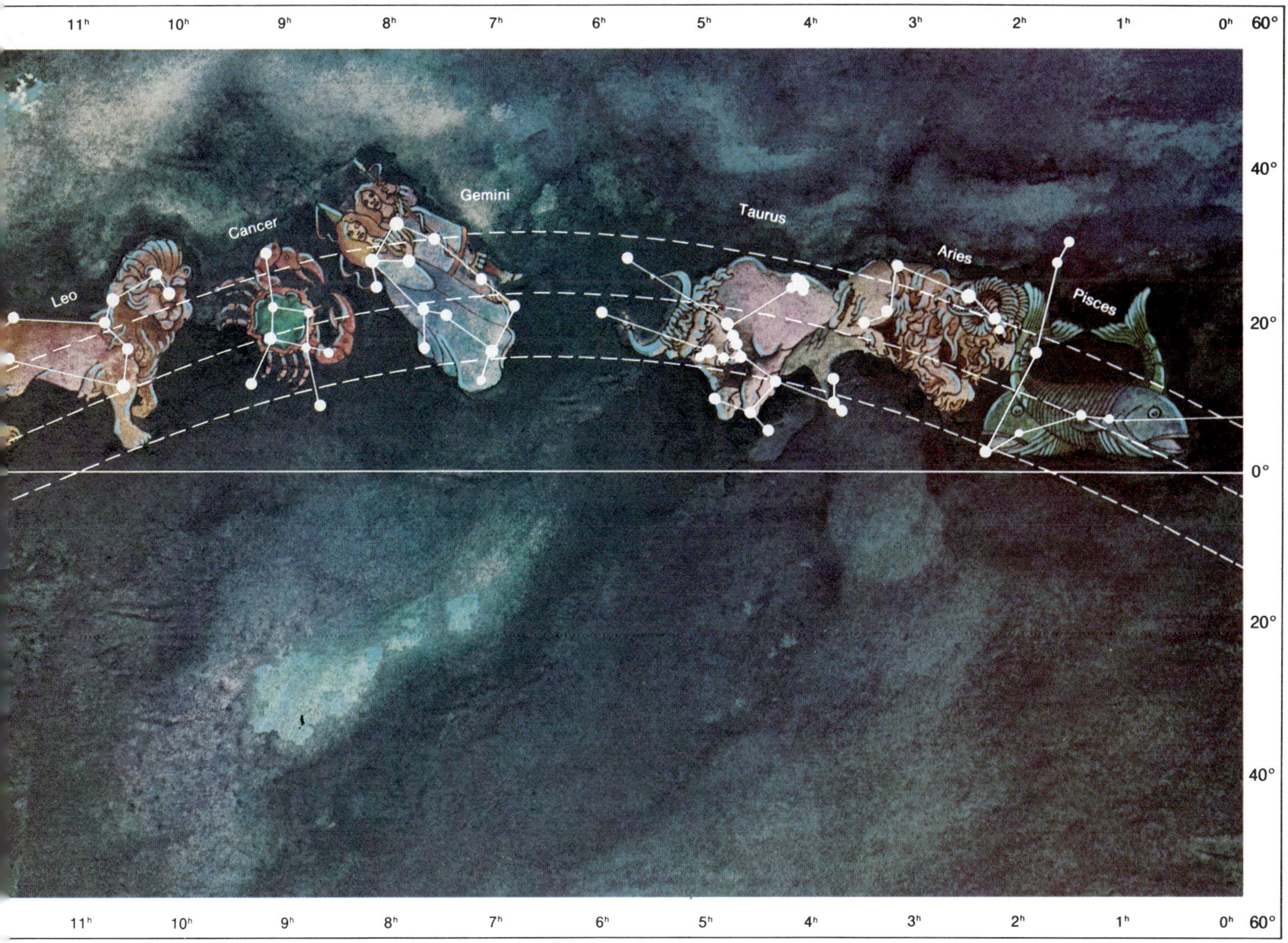

11h
10h
9h
8h
7h
6h
5h
4h
3h
2h
1h
0h
60°
40°
20°
0°
20°
40°
60°
Leo
Cancer
Gemini
Taurus
Aries
Pisces

fore sunrise. In fact, prior to 6 B.C., astronomers thought Venus was two planets: the evening "Phosphorus" and the morning "Hesperus." Why is Venus so exceptionally bright? Not only is it the planet closest to the earth; its proximity to the sun, combined with the high reflectivity of the dense clouds that enshroud it, also contribute to its brilliance.

Mercury is more difficult to observe than Venus because it is never far from the sun. But with a little perseverance you can find it just after sunset during twilight and in the dawn just before sunrise. Unlike Venus, Mercury is not covered with clouds and is not really large enough to compare with the dazzling appearance of Venus. (It is said that the great astronomer Copernicus never saw Mercury, which may offer some incentive to those motivated by a competitive spirit.)

In contrast to the inner planets, the motion of the planets farther away from the sun than the earth is not confined to a direction near that of the sun; at various times these planets may be observed in any direction relative to the sun. Of these planets, Mars, Jupiter, and Saturn appear as bright stars. The more distant planet Uranus can be seen with field glasses or a small telescope. The two more distant outer planets, Neptune and Pluto, require larger telescopes to observe them.

The observation of the planets farther away from the sun provides an interesting opportunity to recognize the consequences of the earth being a "spaceship," moving through interplanetary space in its own orbit. Figure 7 shows how a planet appears to move. The planet travels in the same order already observed in the case of the sun and the moon. But the earth is closer to the sun than that planet, so periodically it will overtake the planet. When the earth is about to overtake it, the planet will appear to reverse its direction of motion through the Zodiac. And this apparent reversed motion will continue beyond the actual moment at which the earth overtakes the planet; thereafter the planet will resume its usual motion through the Zodiac.

This apparent "retrograde" motion can be observed most readily in the case of Mars because its normal direct motion through the Zodiac is so rapid, owing to its nearness to the earth. When it is being overtaken by the earth, Mars appears as a bright red star rising in the east about sunset, and therefore it is easily observable throughout the evening. Despite its distance, Jupiter is also a conspicuous object because of its large size, and when visible it frequently appears as one of the brightest stars in the sky. Saturn is less conspicuous but still appears as one of the brightest stars.

EARTH'S NEIGHBORS

When you look at them without a telescope, the five visible planets merely seem to be bright stars; only their motion among the constellations gives away their planetary nature. When you look at them through even a small telescope, however, you can see how different they are. Unlike stars, which become brighter when viewed through a telescope, planets appear as small, round disks. In fact, if you were to look at them through the largest instruments, stars would still remain points of light even though they are so much larger than the planets. The reason is that the planets are so much closer to the earth. When their image is magnified, their actual nature as spherical bodies becomes evident and it's even possible to see planetary features—clouds, for instance, and surface markings.

As you read earlier, the distance to a planet may be calculated from its rate of motion among the stars. Once you know how far a planet is from the earth and what size its disk appears to be through a telescope, you can measure its diameter. And from the diameter, you can calculate

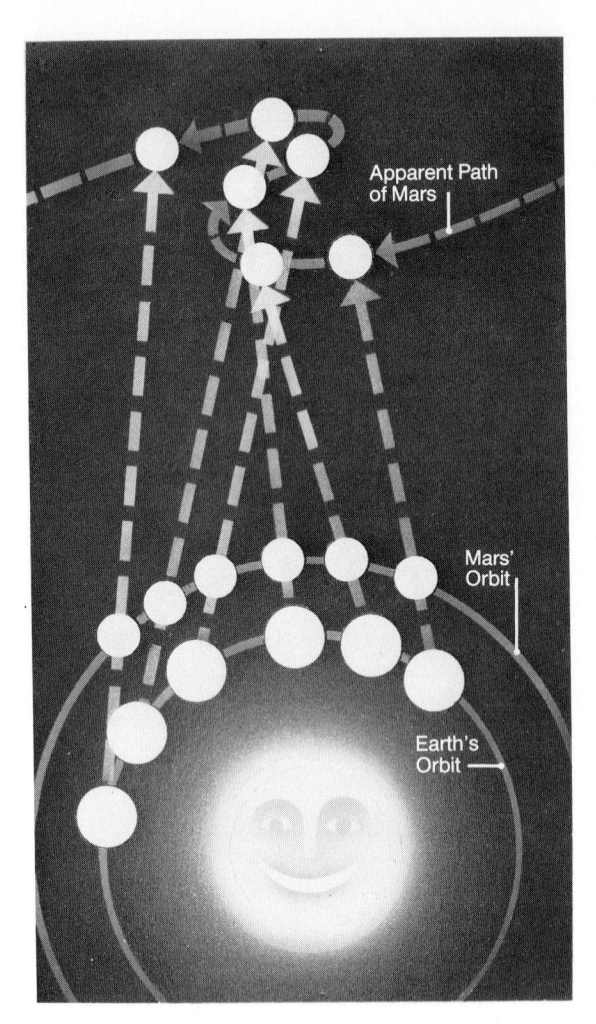

Figure 7. Because the earth's solar year is shorter than that of Mars, the orbit of Mars as seen from the earth appears to be much more complicated than the circular motion you would (if you *could*) observe it from the sun. The dashed lines represent an observer's lines of sight and indicate why the motion of Mars appears to be retrograde.

Calculating Planetary Distance

How do you find the distance from the earth to a planet at a particular time if you know the distance from the sun to that planet and the distance from the sun to the earth? Although modern methods of measurement use sophisticated instruments and calculations, there is a simple way of finding such distances using vectors. Take the distance from the earth to Jupiter as an example.

<u>Step 1.</u> The angle between the sun and Jupiter as seen from the earth will vary, as will the distance between the earth and Jupiter, depending on the day of the year you measure it. Suppose the angle is 80 degrees on a particular day.

<u>Step 2.</u> The mean, or average, distance from the sun to the earth is about 1.5×10^8 kilometers, or 1.0 astronomical unit (1 A.U.). The mean distance from the sun to Jupiter is approximately 7.8×10^8 kilometers, or 5.2 A.U.

To make a scale diagram of these distances, use a 1 inch vector to represent the distance from the sun to the earth and a 5.2 inch vector to represent the distance from the sun to Jupiter.

<u>Step 3.</u> Draw the 1 inch vector. Then use a protractor to measure the 80 degree angle and draw a line at that angle extending from the earth toward Jupiter. To find its length, find the point at which the 5.2 inch vector representing the sun-Jupiter distance crosses the line. Keep one end of your ruler at the point symbolizing the sun and move it in a circle until the 5.2 inch mark crosses the line. The point at which they cross represents Jupiter. Now measure the line on your diagram from the earth to Jupiter and you will have the distance between these two planets in astronomical units. In this problem, the distance will be 5.3 A.U. To convert this to kilometers, simply multiply your result by 1.5×10^8 kilometers, which is the number of kilometers in 1 A.U.

$$(5.3) \times (1.5 \times 10^8 \text{ kilometers}) = 8.0 \times 10^8 \text{ kilometers}$$

<u>Step 4.</u> If you had measured the angle on a different day, you probably would have gotten a different value. Correspondingly, the distance between the earth and Jupiter would also be different. The smallest distance between the earth and Jupiter occurs when the earth is positioned between the sun and Jupiter is aligned with them:

SUN   1.0 A.U.   EARTH   X   JUPITER
                              5.2 A.U.

$$x = 5.2 \text{ A.U.} - 1.0 \text{ A.U.} = 4.2 \text{ A.U.}$$

The greatest distance occurs when the earth is aligned with the sun and Jupiter but is on the opposite side of the sun from Jupiter:

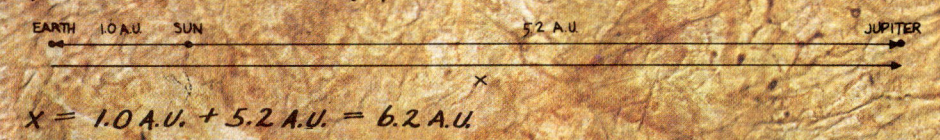

EARTH   1.0 A.U.   SUN   5.2 A.U.   JUPITER
                              X

$$x = 1.0 \text{ A.U.} + 5.2 \text{ A.U.} = 6.2 \text{ A.U.}$$

Figure 8. The size relationship between the celestial bodies of the solar system, drawn to scale. The distances between planets are not drawn to scale (refer to Figure 5). Our *sun,* depicted at the bottom of the page, lies 30,000 light years from the center of our galaxy and completes one journey around the center every 225 million years. A *solar prominence* flares up from the sun. Some prominences have extended to lengths greater than 160,000 kilometers. *Mercury,* the planet closest to the sun, is only 4,800 kilometers in diameter. It is so close to the sun that its entire atmosphere has been burned off. *Venus* is similar to the earth in size and mass but its atmosphere, composed primarily of carbon dioxide, is hot and dense. Venus has no moon. The *earth-moon system* may be thought of as a double planet, for the moon is a "mere" 384,000 kilometers from the earth. When viewed from the earth, *Mars* appears as a red planet. It has two small moons, and its atmosphere is, for the most part, carbon dioxide. The chunks of matter that form the *asteroid belt* range from dust particles to one lump that is as large as the British Isles. The giant planet *Jupiter* has a diameter of 142,750 kilometers yet it makes one complete revolution about its axis every 9-3/4 hours. Cloud belts of what appears to be liquid ammonia enshroud Jupiter. A family of twelve satellites orbits around this planet. *Saturn,* the second largest planet, measures 112,300 kilometers. Ten satellites as well as unique rings of particles encircle this planet. *Uranus,* which is much denser than Jupiter or Saturn, has five satellites. *Neptune* orbits the sun about every 166 years. It is a gaseous planet with two satellites. Discovered as recently as 1930, the planet *Pluto* is extremely cold and probably has no atmosphere.

its volume. Now if you know the mass of the planet, you can calculate its density, which is an important property because it provides a clue to the planet's chemical composition. Once its chemical composition is known, it's possible to learn whether or not a planet is similar to the earth.

The mass of the planets actually can be measured and the density calculated. The principal gravitational force acting on a planet is that of the sun. All the planets move in elliptical orbits about the sun in accordance with Newton's laws of motion, but careful measurements show that the details of their motions cannot be understood only in terms of the sun's gravitational attraction. Another factor that must be taken into account is the gravitational force that the planets exert on one another — a factor that explains the so-called "perturbations" of their motions. If you know the amount of these perturbations you can calculate the mass of the planet producing the perturbation, because Newton's law of gravity indicates that the gravitational force is proportional to the mass of the body producing the force. Any time a planet perturbs the motion of another body in the solar system — be it another planet, a comet, an asteroid, or a satellite (or "moon") orbiting the planet itself — measurement of the perturbation permits measurement of the planet's mass. It is in this way that the mass, and therefore the density, of the planets has been determined.

What will you find when you compare the sizes and densities of the planets? With the exception of the most distant planet, Pluto, of which little is known, the planets fall into two distinct classes. Those nearest the sun — Mercury, Venus, the earth, and Mars — are comparatively small. The earth is the largest, with a radius of 6,370 kilometers; Mercury is the smallest, with a radius of 2,400 kilometers. The inner planets are also quite dense. In contrast to the mean density of the massive sun, which is only 1.4 grams per cubic centimeter, the mean density of the earth is about 5.5 grams per cubic centimeter. Because their radius is smaller, the internal pressures of Venus and particularly of Mars are less. It turns out that the chemical composition of the earth, Venus, and Mars are similar; the differences in density can be largely explained in terms of their varying degrees of compression.

Mercury doesn't fit into this simple picture, however. Even though Mercury is the smallest of the four inner planets and its internal pressure quite low, its density is about the same as that of the earth. The most probable explanation for this discrepancy is that Mercury contains much more metallic iron than the other three planets. When considered together, the four inner planets are characterized by high density and small diameter. Because of their approximate similarity to the earth they are frequently called the *terrestrial planets.*

The next four planets are entirely different. Jupiter, Saturn, Uranus, and Neptune are all much larger than the earth. The largest, Jupiter, has a radius more than 10 times that of the earth and a volume more than 1,000 times greater. Because of their tremendous size, these planets have much greater mass than the terrestrial planets; the mass of Jupiter, for example, is more than 300 times that of the earth. If an observer could study our solar system from a nearby star, he would conclude that the

Pluto
Neptune
Uranus
Saturn
Jupiter
Asteroids
Mars
Earth
Venus
Sun
Mercury

Figure 9. (Far right) The unmanned satellite Mariner 9 (a) provided the first "close-up" shots of Mars. Dozens of scientists and technicians, working with complex instruments and computers, are involved in gathering the data that eventually become the photographs we see of Mars. The 210-foot-diameter antenna dish at Goldstone Station in California's Mojave Desert (b) received over 580,000 "words" per picture from the satellite. Each word consists of 9-bit data representing varying levels of brightness of the Martian landscape. The data were then organized and filtered to eliminate "noise" (interference) with the aid of a computer (c). Another computer was used to "contrast sketch" or translate the data into a picture, because the human eye does not perceive variations in brightness equally at different levels of illumination. A digital-to-analog converter prepared data that could be shown on a television screen and photographed (d). These individual photographs were pieced together into a composite view of Mars (e). Hundreds are required to record the surface area accessible to the scanners on Mariner 9.

sun was associated with two large planets (Jupiter and Saturn), two planets of moderate size (Uranus and Neptune), and some minor bodies probably of little significance—among them, our earth.

Because these four planets are much larger than the terrestrial planets, they are often designated the *major planets.* However, if you look more closely at the numbers given above, a curious fact emerges. Although the volume of Jupiter is more than 1,000 times that of the earth, its mass is only 318 times greater. In other words, its average density is only about one-fourth the density of the earth, or about 1.34 grams per cubic centimeter, despite the fact that its central pressure is enormously greater than that of the earth. If Jupiter were made of the same materials as the earth, its average density would be much greater than the earth's density, so it cannot be made of rock and iron. Careful studies of the properties of matter at high pressure lead to the conclusion that Jupiter must be made almost entirely of the lightest elements —hydrogen and helium—and therefore must be similar to the sun in composition. Jupiter is a star that "never made it." Although it is immense compared to the earth, it was not large enough for its gravitational energy to reach temperatures that could be sustained by thermonuclear reactions.

Saturn is even less dense than Jupiter at 0.7 gram per cubic centimeter. It, too, must be made primarily of hydrogen and helium. Its lower density is attributable to its smaller mass and smaller internal pressure. Although smaller than Jupiter and Saturn, Uranus and Neptune are more dense than these two giants, with densities of about 2 grams per cubic centimeter. So they can't be made simply of hydrogen and helium; they must consist of a mixture of heavier matter—perhaps water, methane, and ammonia, which contain the somewhat less abundant elements oxygen, carbon, and nitrogen.

The four large planets are sometimes called "gas giants" because they are composed largely of compounds that are gaseous at low pressures. Oddly enough, at the high pressure found in the interior of the giant planets these normally gaseous compounds would be compressed into strange solid or liquid substances that never would be seen on earth. For example, it is likely that the deep interior of Jupiter consists of hydrogen in the form of a liquid metal. Such a substance would be more like mercury than the familiar gaseous hydrogen.

Far beyond the terrestrial and giant planets is the ninth planet, Pluto. It is more like a terrestrial planet in size, but its size and mass are so poorly known that little can be said about its density and composition.

Although it's sometimes convenient to lump the planets together into two categories, it's also important to recognize that each planet is unique—that nature has provided examples of the various forms that planetary bodies can take. Comparing the various planets in order to understand why they are different leads to a more general understanding of the earth. If all the planets were alike, geologists studying the earth would be in the position of linguists living in a world of only one language. Although planets are of great interest in themselves, one of the principal motivations of planetary studies is to obtain deeper insights into our own planet, the earth.

OTHER STARS, OTHER PLANETS

The sun is a rather ordinary star, and as you look at the sky on a clear night you can see thousands of other stars, many of which are similar to the sun. The brightest star in the sky, Sirius, appears bright primarily because it is relatively close to the earth—if 7×10^{13} kilometers seems close. At that distance it would take light 8 years to reach the earth from

(a)

(b)

(c)

(d)

(e)

Figure 10. Barnard's star is shown changing position over a period of 22 years—from August 4, 1894 to May 30, 1916. The arrows in the photographs indicate the shift in position.

Figure 11. The orbits of the two perturbations postulated for Barnard's star are diagrammed with circular orbits (above) and the yearly mean average of time-displacement curves resulting from the two circular orbits are plotted below.

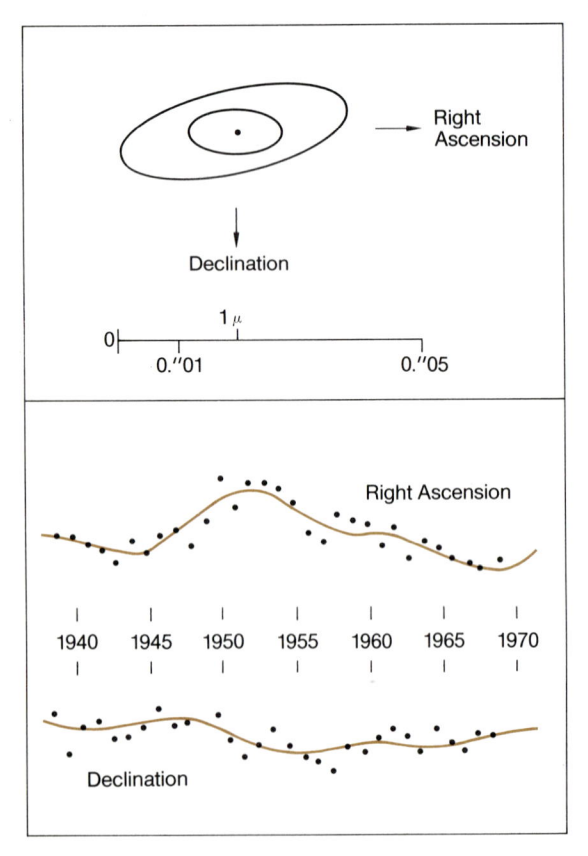

Sirius, in contrast to the 8 minutes it takes light to reach us from the nearest star—the sun. (It is much more convenient to describe the distance of stars in terms of the time it takes for light to travel to the earth from the star. Thus Sirius is said to be 8 *light years* away, whereas the sun is only 8 "light minutes" away.) There are a few stars closer to the earth. For example, the bright star Alpha Centauri in the Southern Cross, visible only in southern latitudes, is "only" 4 light years away.

Whether or not other stars have planetary systems similar to ours is an interesting question. It would be remarkable if the sun were unique in this regard. However, even if the nearest star had a planet as large as Jupiter, our largest telescopes would still be unable to detect it. It has been calculated that even if you were on an orbiting astronomical observatory equipped with a telescope several times larger than those presently used on earth, you would barely see such a planet.

On the other hand, there is indirect evidence that other stars have planets of their own. For many years, Peter van de Kamp has been studying the motion of a faint nearby star—*Barnard's star,* which is at a distance of about 6 light years. To the unaided eye, stars appear to be fixed relative to one another in the form of constellations, but careful telescopic observations permit measurement of their motion with respect to one another. Van de Kamp's measurements of Barnard's star show that it does not move through interstellar space in the straight line that would be expected from Newton's laws of motion for a body that is not influenced by the gravitational forces of other bodies. Instead, Barnard's star moves in a wavy path as if its motion is being perturbed by another body moving in an orbit around it (Figure 11).

Van de Kamp's mathematical analysis of this motion first led him to conclude the hypothetical orbiting body should have a mass about twice that of Jupiter, and unlike Jupiter (as well as the other planets of our solar system), it should be moving in a highly elliptical orbit. Further analysis of his data showed it was equally possible to explain the motion of Barnard's star by assuming it was perturbed by *two* planets of masses very similar to those of Jupiter and Saturn and at the equivalent distance from their parent star (Figure 11). Because the masses of the other planets of our solar system are negligible in comparison with the combined mass of Jupiter and Saturn, the result is exactly what we would expect if Barnard's star possessed a planetary system similar to ours. Therefore, although this interpretation of his results is not unique, it is strongly suggestive that planetary systems are rather common in the universe.

STAR POPULATIONS

Many stars, such as Sirius and Alpha Centauri, appear to be bright because they are comparatively close to the earth. And yet some extremely distant stars are bright because they are extremely luminous. For example, the bright star *Rigel,* defining the left leg of Orion (the Hunter) is at distance of about 1,000 light years; it is bright because its surface temperature is so high—about 12,000 degrees Centigrade—that it is radiating more than 10,000 times more light as the sun. Similarly, the bright star in Orion's right shoulder, *Betelgeuse,* is about 500 light years away. It is a good example of the interesting class of stars called *red giants,* which are cool stars with surface temperatures of about 3,000 degrees Centigrade. They are extremely luminous because of their great size. If the sun were one of the larger stars of this kind, its surface would extend beyond the orbit of Mars.

In addition to the luminous stars and intermediate-size stars such as the sun and Alpha Centauri, there are many fainter stars. In fact, there are almost certainly more faint stars than any other kind. We don't see more of them because their luminosity is so low they are too faint to be seen even with the largest telescopes, unless they are very close to the earth (within about 15 light years). All but three of the twenty-five closest stars are fainter than the sun—one of them is more than 10,000 times fainter.

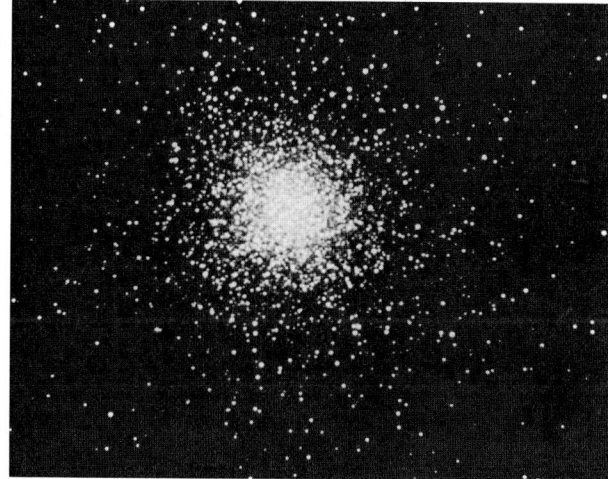

It is not clear nor even clearly defined, for that matter, just how small and faint a star can be and still deserve the title of "star." Theoretical calculations show that if a star is smaller than about one-tenth the mass of the sun, its central temperature will never become high enough to sufficiently ignite the nuclear reactions that provide the source of light and heat in normal stars such as the sun. Such stars will slowly contract under their own weight, and the release of gravitational energy from the contraction will provide enough energy to allow the star to be faintly luminous for periods of time as long as the age of the sun and solar system. In fact, the planet Jupiter, of a mass one-thousandth that of the sun, may be an extreme example of such a star. Recent measurements indicate its radiation of internal energy, which is probably resulting from gravitational contraction, is still several times as great as the quantity of energy it absorbs from the sun.

Many wondrous kinds of objects form the population of the stars. They include double and multiple star systems. In these systems, stars are surrounded by other stars, which, like our planets, are moving in orbits determined by the fundamental laws of motion and the law of gravitation. In fact, there is no fundamental reason why double and multiple stars could not have planets, even though the motion of those planets would be complicated by the fact that they would be moving in orbits determined by the gravitational attraction of more than one central star. There may even be worlds that have several "suns," possibly of greatly different brightness and color, that rise and set at different times.

Stars, too, occur in clusters. The Pleiades in Taurus is one (Figure 12). This attractive group of stars is popularly called the "Seven Sisters"—even though most people can only see six. With binoculars it becomes apparent that many fainter stars are clustered in this group.

There are star clusters that contain many more stars than the Pleiades. These are the *globular clusters* (Figure 12), spherical associations of hundreds of thousands of stars. Considering how many stars are packed into a globular cluster, the diameter of the cluster is not very large—about 100 light years. The average density of the stars comprising

Figure 12. (Left) The Pleiades in Taurus are called the "Seven Sisters" even though only six stars in the cluster are clearly discernible. (Right) The globular star cluster M13 in Hercules, which cannot be seen without a telescope, contains many thousands of stars. The "M" refers to the list made by the eighteenth-century astronomer Charles Messier in his attempt to classify objects that he observed when he was searching the sky for comets.

Figure 13. Innumerable stars and great clouds of dust give rise to the milky appearance of our Galaxy, the Milky Way. This photographic map shows in flat projection what we see as a ring around the earth. The coordinates refer to the latitude and longitude measured from the galactic plane. In this view, Sagitarrius is in the center and our solar system lies near the main plane but 30,000 light years from the center.
Courtesy Lund Observatory, Sweden.

them is about 10 per cubic light year. In the more crowded center of the cluster, the density is even higher—about 100 or more stars per cubic light year. Viewed from a planet within such a cluster, the sky would be a star-gazer's paradise. Unlike the earth, where the nearest star is 4 light years away, there would be thousands of stars close to such a planet. And the space between the brilliant nearby stars would be filled with fainter stars, and nights would never really be dark.

There are also many varieties of pulsating stars, known as *variable stars*. Some pulsate in such a way that their luminosity varies by a factor of about 2 in a periodic way about once a month. The luminosity of others varies irregularly. An example of this group is the star in the constellation Cetus (the Whale) named Mira (the Wonderful). Most of the time this star is no brighter than the distant planet Neptune, and a telescope of moderate size is needed to observe it. But every year or so it brightens up and becomes one of the more conspicuous stars in the sky, about as bright as the North Star, Polaris. Other variable stars, named *novas* under the earlier but mistaken belief that they represent new stars, show even more dramatic and irregular variations in luminosity. Their brightness changes by a factor of thousands in a single day. The extreme case of variability of the luminosity of stars are the *supernovas,* in which the luminosity of the star suddenly increases to hundreds of millions of times its former intensity. Such stars should not really be included in the class of variable stars, however, for they do not represent a continuing pulsation in the life history of a star. Instead, they are the dramatic mode by which many stars end their lives, exploding most of their substance into interstellar space.

GALAXIES

When you look at the sky before it becomes completely dark or in the glow of city lights, you can see only the planets and brightest stars. Even some of the constellations cannot be seen at all because none of their stars are sufficiently bright. In order to see all the sky has to offer, you must get away from the city—into the mountains, the desert, or the countryside, where the number of visible stars will seem uncountable. Away from the city, there is more to be seen than merely more stars.

Passing across the sky from horizon to horizon is a band of light that may appear at first to be a thin cloud. It passes through the constellations of Cygnus (the Northern Cross), Aquila, and Ophiuchus, then crosses the Zodiac near Sagittarius and Scorpius. It moves on through such constellations as Norma and Crux (the Southern Cross, containing Alpha Centauri), and back through Orion, Perseus, and Cassiopeia to Cygnus again. As these constellations rise and set with the rotation of the earth, the band of light rises and sets with them. Therefore, it must be a permanent feature of the sky, not a transient phenomenon of our atmosphere such as a high, thin cloud.

The band of light has been given the somewhat odd name, the "Milky Way." When you look at it with binoculars or a small telescope, you see that much of the diffuse light is caused by an enormous number of faint stars; together the light from each of them blends to form a luminous band. The Milky Way has a rather ragged appearance, and even with binoculars you can see that in some parts of it there appear to be many more stars than in others.

We now know why there is a "Milky Way." Our sun is a member of a large family of stars we call "our Galaxy," and our Galaxy has the shape of a flat disk (Figure 13). The sun is not in the center of this disk but is about halfway toward one edge. Looking out into space from our position on the earth, there are many more stars in the plane of the disk of our Galaxy than in the direction perpendicular to the disk. The stars forming the outlines of the constellations are for the most part near the sun, so they appear to be distributed uniformly in space. But there are few distant stars in directions *perpendicular* to the galactic plane; the light from the distant stars *in* the galactic plane merges to form the Milky Way.

The Milky Way looks ragged not only because our Galaxy contains many stars; it also contains great clouds of dust that obscure the more distant stars. Some of these dust clouds may be stars in the process of formation but not yet hot enough to emit visible light.

The diameter of our Galaxy is estimated to be about 100,000 light years. By comparison, the distance of the nearby stars that form the constellations are typically up to about 1,000 light years from the earth. The mass of our Galaxy is about 4×10^{44} grams; and if all the stars had the same mass as the sun (about 2×10^{33} grams) there would be about $2 \times$

Figure 14. These photographs are examples of some of the configurations that galaxies have taken. This *spiral* galaxy (left) belongs to Ursa Major. The *bar* formation in Pegasus has only two noticeably developed arms, which extend from the ends of the bar (right). The *ellipsoid* galaxy shape is represented by a galaxy in Cassiopeia. The *irregular* shape is the form taken by the Magellanic Clouds (lower right).

10^{11} stars in the galaxy. But the total number of stars in our Galaxy must be considerably more than that because a greater number of stars are smaller, not larger, than the sun.

The large number of dust clouds in the plane of our Galaxy prevents us from seeing most of its stars, so it has been difficult to use observations made with ordinary optical telescopes to learn much about the details of its shape and structure. In fact, the center of our Galaxy has never been seen, so great is the obscuration resulting from the clouds of dust.

Fortunately, observation of the universe is not limited to the use of optical telescopes. The electromagnetic spectrum extends far beyond the narrow range of wavelengths visible to our eyes, and in recent years the use of precise instruments that detect the longer and shorter wavelengths has provided major insights into the structure of the universe. These developments have led to whole new fields of astronomy such as infrared and radio astronomy, which use wavelengths longer than visible light; and x-ray astronomy, which involves the use of shorter wavelengths. In addition to a number of other unique features, radiation at some wavelengths can pass through matter, such as dust clouds, that is opaque to visible light.

In studying the structure of our Galaxy, the astronomer has used a particular spectral line of hydrogen occurring at a wavelength of 21 centimeters. It is an exceptionally long wavelength, compared to the visible radiation that lies in the range of 4×10^{-5} to 7×10^{-5} centimeter. This spectral line falls in the region of electromagnetic radiation that is detectable by special radio receivers that use large dish-shaped antennas known, appropriately, as *radio telescopes* (see Figure 1). With it, astronomers have mapped out the distribution of the most abundant element hydrogen, and thus we know the general galactic structure.

This work indicates that our Galaxy is not uniformly homogeneous in structure, but has a "whirlpool," or spiral, appearance. Much of the matter of our Galaxy is concentrated in the form of "spiral arms." Our sun is located in one of these spiral arms.

As a result of these studies, we have learned that the sun is not simply one among many stars uniformly distributed throughout space but is part of a large family of stars that form, in the expression of Immanuel Kant, an "island universe." Is our "island universe" unique? Are there other galaxies beyond our own?

Although the answer to those questions was unsettled as recently as 1924, it is now clear that our Galaxy is but one of billions of galaxies—some larger, many smaller than our own. Many exhibit the spiral structure of our Galaxy (Figure 14a), whereas others have a barred appearance (Figure 14b). Other galaxies are not even shaped like disks but are in the form of spheres, or ellipsoids (Figure 14c); still others are completely irregular in shape (Figure 14d). Like stars, there are double and multiple galaxies, and clusters of galaxies. The other galaxies are probably distributed uniformly in space, although they *appear* to be concentrated in the directions perpendicular to the plane of our own galaxy because we cannot see as many through the obscuration caused by our own Galaxy.

Typical distances to the nearest galaxies range from a few hundred thousand to a few million light years. Only a few are bright enough to be seen without a telescope. The *Andromeda* galaxy, similar in size and structure to our own, can be seen without a telescope on a clear night. It is an even more interesting object when observed with binoculars. Its distance from us is about 2 million light years, so even though it may be an exaggeration to sing that "on a clear day you can see forever," it is a fact that on a clear night you can see 2×10^{19} kilometers.

The nearest galaxies are two small, irregular galaxies, known as the Magellanic clouds. They are observable only in the Southern Hemisphere of the earth, where they appear as cloudy patches similar to small pieces of the Milky Way. Their distance from the earth is about 200,000 light years.

The most distant objects recognizable as galaxies are between 2 billion and 3 billion light years away. However, there are even more distant objects, known as *quasi-stellar galaxies,* that are still detectable by both optical and radio telescopes because of their great luminosity. Some of these objects may be more than 10^{10} light years away.

It appears that the distances between the planets of our solar systems—distances of the order of 10^8 to 10^9 kilometers—are small in comparison to the distance between neighboring stars, which are on the order of 10^{13} kilometers. And those distances in their turn are small in comparison to the distance between neighboring galaxies (10^{19} kilometers) which in turn is small compared to the size of the observable universe, 10^{23} kilometers. If we built longer and more powerful telescopes, would we be able to see farther and farther? From our past experience of looking out into space, you might guess that the answer to this question should be yes. But as the next chapter will explain, we may have by now seen as far as we possibly can from the earth.

One of the great truly scientists. Galileo recognized the importance of observation to support theory. The church branded Galileo a radical for denying the earth-centered view of the solar system, but there was little Galileo could do but believe in what he saw to be true through his telescope.

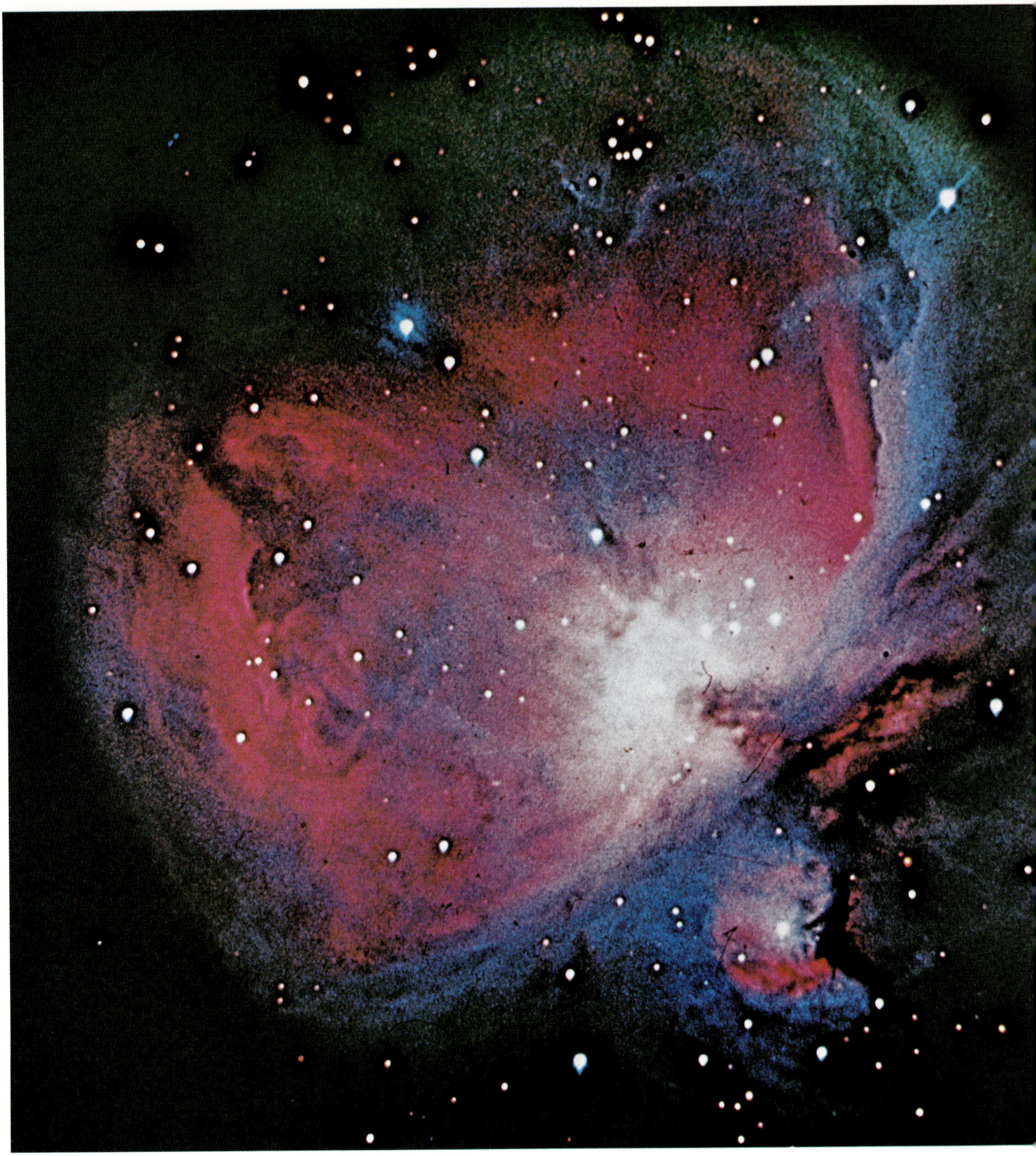

In the preceding chapter you read a description of the universe as it exists today. Any attempt to describe something as vast as the universe is in itself a difficult problem, and anything more than a cursory examination leads quickly into the realm of speculation. Many regions of the universe are nearly inaccessible to observation. Only by indirect techniques is it possible to understand something as proximate as the structure and composition of the earth a mere 30 kilometers beneath its surface—a distance less than half the width of the city of Los Angeles. Even in principle it is impossible to observe the present state of the more distant objects in the universe. The radiation we now receive from the distant optical and quasi-stellar galaxies originated *billions of years ago;* it is a ghostly traveler from events that occurred long before the present epoch in the history of the universe.

Perhaps we should content ourselves with describing the way things are and assume they more or less have always been that way. But there is tantalizing evidence that keeps us from taking the easy way out. Our universe is *not* static; it is evolving. And the remarkable fact is that we have come to understand some of the processes underlying this evolution.

THE EXPANDING UNIVERSE

One line of evidence for a dynamic universe is based on observations of distant galaxies. The chemical composition of a single star can be measured by studying the spectrum of the light it emits, because the chemical elements contained in that star give off spectral "lines" of characteristic wavelengths. In a distant galaxy, the spectral lines from the billions of stars comprising it merge together into a spectrum for the entire galaxy, and the wavelengths of its various spectral lines may also be accurately measured.

In the early 1930s, E. Hubble and M. Humason used the spectra of galaxies to point out an incredible phenomenon that has since dominated our thinking about the origin and evolution of the universe. Their work showed that the more distant the galaxy, the more rapidly it appears to be moving away from us. Their original data are shown in Figure 2. Today their work has been overwhelmingly confirmed: some remote galaxies are receding from us at nearly 150 million meters per second—nearly half the velocity of light!

This observation is based on the Doppler effect, whereby the wavelength of a spectral line radiated from a moving light source appears to a fixed observer to have a different wavelength than when it is radiated from a stationary source. If the source is moving toward the fixed observer, the wavelength is shortened; if the source is moving away from the observer, its wavelength is lengthened. Now light of longer wavelength corresponds to the red end of the visible spectrum. As a consequence of the Doppler effect, then, light sources moving away from us have their spectral lines shifted toward the red. In terms of the moving light source, the waves are emitted at the same rate—the same number in each second. But in terms of the stationary observer, the distance between successive wave crests increases, so the *observed* wavelength is longer (see Figure 3). The general result is that light being sent to us from a receding source will

11
Origin and Evolution of the Universe

The Great Nebula in Orion, as it appears when photographed in red light.

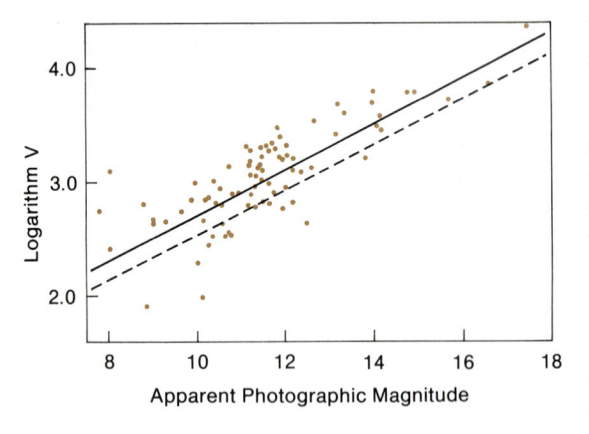

Figure 2. Relation between distance and apparent photographic magnitude of nebulas. For nearly 100 nebulas the formula log $v = 0.2\,m + 0.71$ works well. (v is the velocity causing the red shift and m is the apparent photographic magnitude.) The scale is such that each successive order of magnitude represents a star 2.5 times brighter, and the intensity of brightness on a photographic plate is measured by the size of the image it makes. The apparent faintness of individual nebulas therefore can be taken as a statistical measure of distances.

be shifted toward the red, and exact measurement of the magnitude of the red shift permits calculation of the velocity of recession.

The velocity of recession is tremendous for the most distant galaxies —nearly one-half the velocity of light—and is less for galaxies that are nearby. Something called *Hubble's law* gives the linear relationship between distance and the velocity of recession: it is expressed symbolically as $v = Hr$, where v is the velocity of recession, H is a proportionality constant, and r is the distance between the galaxy and the observer.

Of course, Hubble's law should not be interpreted to mean that all the other galaxies are avoiding us. Rather, all galaxies are moving away from one another as if at one time they had all been at the same place. At that time and place an incredible explosion occurred and all the matter of the visible universe has been moving apart ever since. From the constant of proportionality in Hubble's law, it's possible to calculate that catastrophic explosion (or "big bang," as it is sometimes understated) took place approximately 10^{10} years ago. This means that the universe may be about *twice* as old as the earth and the solar system. There is considerable uncertainty about the exact time, partly because of difficulties in measuring how far away from us the more remote galaxies really are. But the true value for the time of the big bang lies somewhere between about 7 billion and 20 billion years ago.

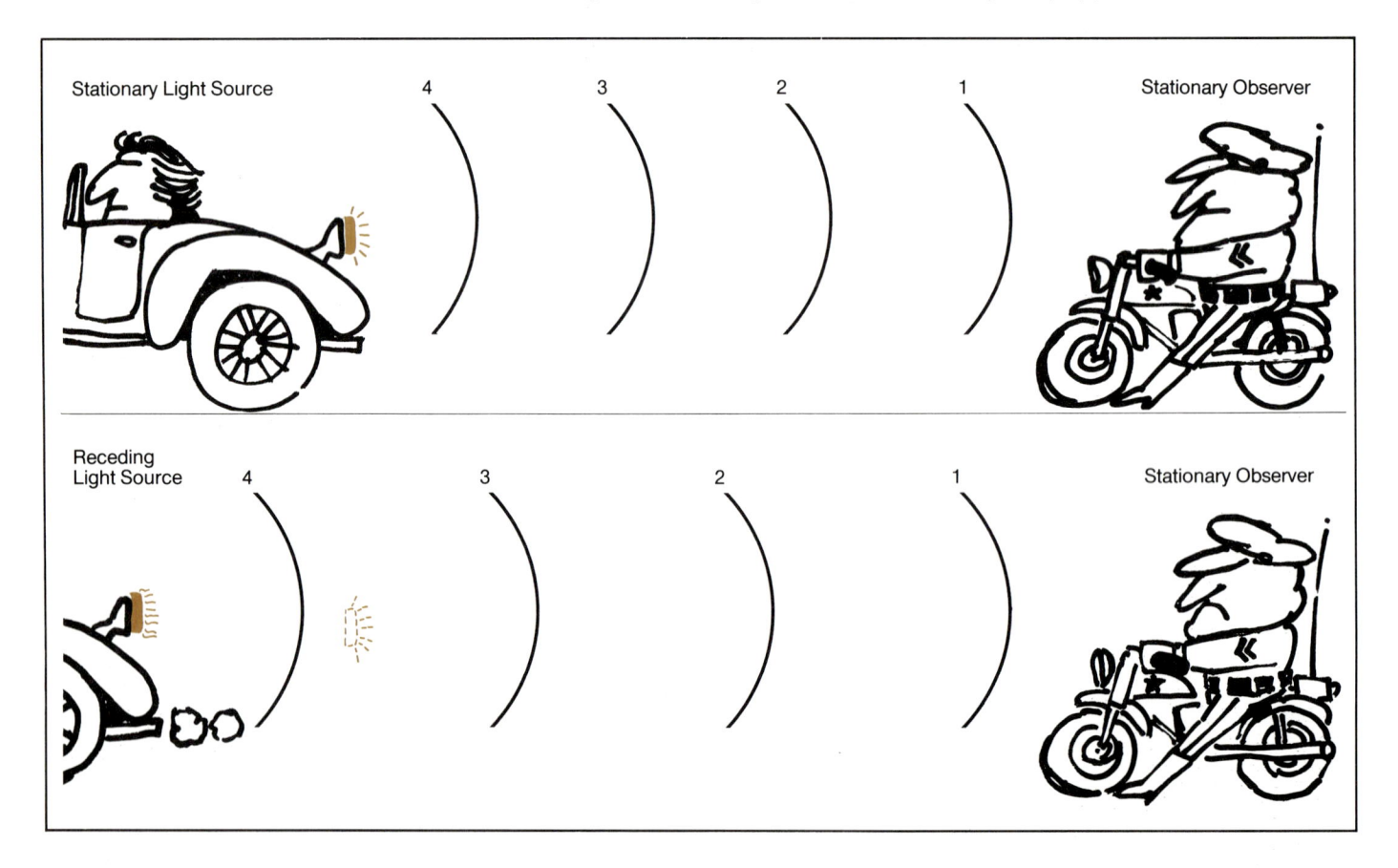

Although the discovery of Hubble's law required many years of painstaking work, not to mention use of the largest telescopes available, there is a simple observation you can make that is related to Hubble's law and in a way provides an observational test for it. All you have to do is go outside at night and see that the night sky appears to be dark between the stars. Far from being trivial, the dark night sky was recognized as something of a problem as long ago as 1826 by Heinrich Olbers and has since been known as Olbers' Paradox.

Imagine the universe to be filled with galaxies, as it apparently is. Next, imagine the earth to be surrounded by spherical shells, as depicted in Figure 4. The total light produced from the galaxies in the first shell may be found by multiplying the volume of the shell by the number of galaxies per unit volume and then multiplying the product by the average light received from each galaxy. Because the volume of the shells increases with the square of the distance, the amount of light produced by each shell also increases with the square of the distance (r^2). At the same time, the amount of light received by the earth falls off with the inverse square of the distance ($1/r^2$). The r^2 cancels out, so the total amount of light that reaches the earth does not depend on the distance. Therefore, the next thin shell should contribute an equal quantity of light, as well as the next, and the next, and each more distant shell. Following this line of reasoning, you would expect the night sky to be infinitely bright. Of course, it isn't.

The red shift of the distant galaxies provides a resolution to this paradox. Distant galaxies do not contribute as much light because their light is shifted out of the visible portion of the electromagnetic spectrum into the infrared; even more-distant galaxies would be known about only through the long, "invisible" wavelengths of the radio spectrum. Eventually, if the velocity of recession were to reach the velocity of light, the most distant galaxies would not be detectable by any means. Furthermore, the energy associated with the photons of the extremely red-shifted distant galaxies is much less than the energy of galaxies known to us through the visible spectrum, and the total energy the earth receives is also finite.

In 1965, radio astronomers provided further confirmation of the big bang theory when they actually succeeded in measuring the extremely red-shifted thermal radiation that accompanied the primeval explosion. The temperature in the region of the fireball must have been extremely high and its radiation associated with very short wavelengths of the electromagnetic spectrum: x-rays and gamma rays. Today, however, we must look far out into space and, therefore, far back into time in order to glimpse its light—light that has required the entire age of the observable universe to reach us. According to Hubble's law, the radiation from such distant sources must be extremely shifted to the red and now is visible only in the radio region of wavelengths. It can be studied only with radio telescopes capable of measuring this long wavelength.

THE EXISTENCE OF RADIOACTIVE ISOTOPES

Another kind of evidence that tells us the universe is a dynamic one is the existence, in nature, of radioactive isotopes of certain elements. The nuclei of the atoms that make up the chemical elements are composed of protons and neutrons bound together by forces attracting them to one another. All nuclei containing the same number of protons are nuclei for atoms of the same chemical element. But the number of *neutrons* may vary, and this variation means that atoms of the same element may have different mass, or different atomic weight. These atoms are known as isotopes.

As they occur in nature, most elements are mixtures of more than one isotope. Tin, for example, consists of a mixture of as many as ten isotopes. Most of these naturally occurring isotopes are *stable*. In other

Figure 3. (Far left, below) The Doppler effect. At a given instant in time, the first wave crest emitted from the fixed light source in the upper sketch will have propagated to position (1) in front of the policeman. At that very same instant in time, the second, third, and fourth wave crests will have propagated to the positions labeled (2), (3), and (4), respectively. In the lower sketch, where the light source is moving *away* from the observer, the wave crest emitted at a given instant in time will propagate to the same position as before—position (1). But in the next instant of time, the moving light source will have *receded* from the policeman. Because the rate at which the waves are emitted doesn't change, the wave itself will propagate over the same distance in a given time. However, because the point of origin is different, the wave crest at position (2) will now be to the *left* of where it was in the upper sketch, and at position (3) it will be still farther left, and so on indefinitely as long as the light source continues to recede. To the fixed observer, the distance between successive wave crests—and the frequency with which wave crests reach him—is increasingly stretched out as the light source moves away from him.

Figure 4. How can you calculate the amount of light arriving at earth from the night sky? The center sphere represents the earth. The small circles represent galaxies. The shell of galaxies nearest the earth appears brighter than those in more distance shells, even though they contain fewer galaxies than are contained in the remaining shells. The faintness in the visible spectrum of the more distant galaxies is attributed to their recession.

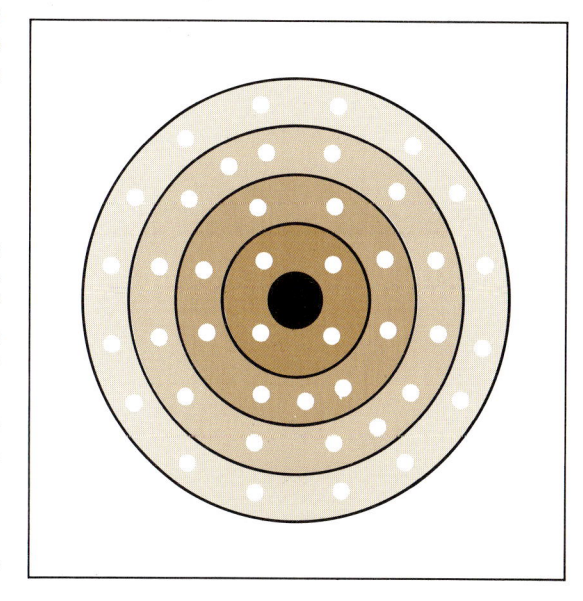

Figure 5. The time it takes for gold 192 to turn into platinum 192 is 4.2 hours. The *half-life* is the amount of time necessary for the "parent" substance to lose one-half of its atoms by the process of conversion into its "daughter." In contrast, the burning of the candle represents depletion as a linear function of time.

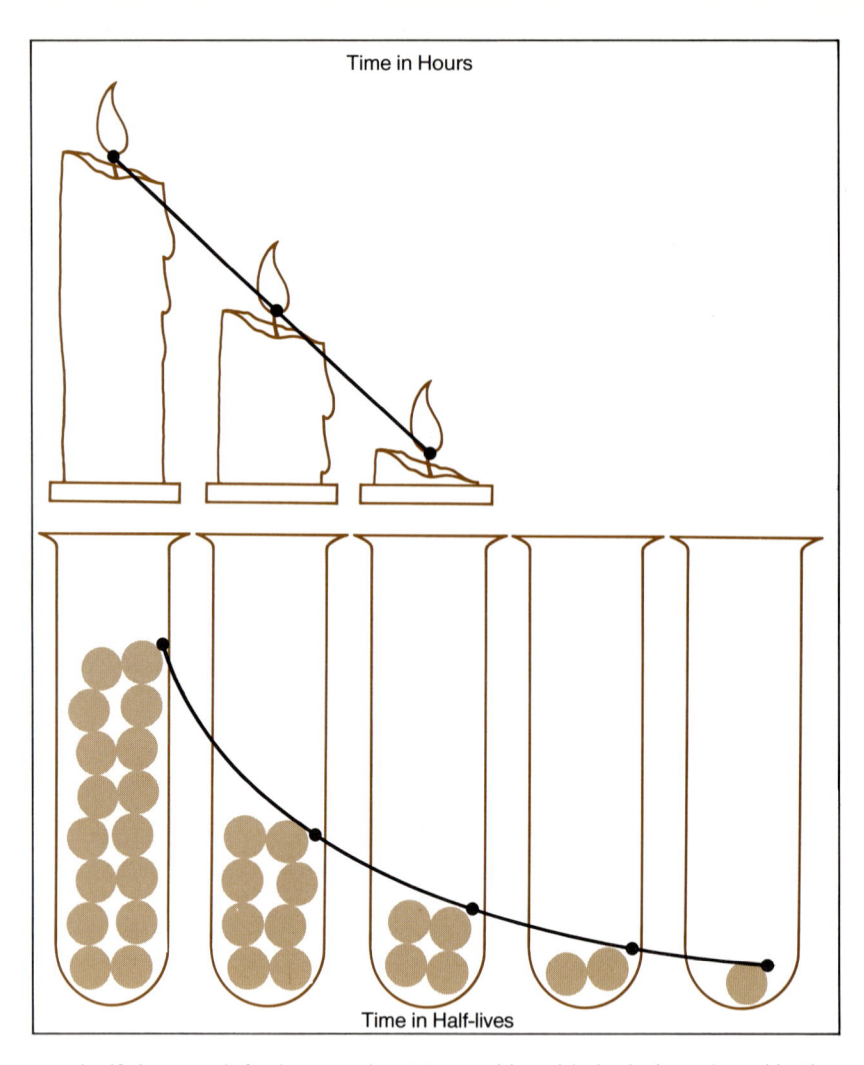

words, if they are left alone and not treated harshly by being placed in the beam of a cyclotron, in the core of a nuclear reactor, or in the center of a very hot star, they will remain unchanged forever.

However, some isotopes exist in nature that are *unstable,* or *radioactive.* In these isotopes, the number of neutrons and protons changes spontaneously—the isotope is transformed into an isotope of a different element. The resulting isotope, in turn, may be radioactive and may "decay" still further into an isotope of still another element. This process of radioactive decay continues until a stable isotope is finally formed.

The natural occurrence of such radioactive isotopes means some natural process must exist to create them, otherwise all the isotopes would have transformed into their final stable decay products.

For most naturally occurring radioactive isotopes, it is essentially impossible to suggest a natural process that produces only the radioactive isotopes of a given element without producing stable isotopes at the same time. Both radioactive *and* stable isotopes and, therefore, the elements must have been produced at some finite time in the past; they did not always exist.

Radioactive isotopes have characteristic *half-lives,* which is the amount of time required for an isotope to be reduced to one-half its original quantity as a consequence of its natural radioactive decay. For example, potassium 40 transforms into stable argon 40 and stable calcium 40 with a half-life of 1.3×10^9 years; uranium 238 transforms into stable lead 206 with a half-life of 4.5×10^9 years, after transforming through a chain of intermediate isotopes which are themselves radioactive. The half-life of these radioactive isotopes is sufficiently long that they could have formed *before* the earth and solar system came into being 4.6 billion years ago. The shortest lived of these naturally occurring, long-lived isotopes is uranium 235, with a half-life of 700 million years. That is less than one-sixth the age of the earth, which means that at the time the earth was formed there must have been more than $2 \times 2 \times 2 \times 2 \times 2 \times 2 = 2^6 = 64$ times as much uranium 235 as there is today.

All of the long-lived radioactive isotopes found on earth today could have been here when the earth was formed, so it isn't necessary to suppose there is any terrestrial process manufacturing them. Neither is it necessary to assume they have been supplied from outer space since the formation of the earth and solar system. (This is fortunate, because it would be very difficult to imagine how that could take place.)

On the other hand, there is evidence that the formation of elements was not confined to the time in the history of the universe prior to the formation of the earth but is still going on. For example, the spectrum of the element technetium has been identified in stars. The longest-lived isotope of this element, technetium 99, has a half-life of only about 1 million years—which indicates that in some places element formation is also occurring in "modern" times.

The existence of radioactive isotopes, together with the stable isotopes that must have accompanied their formation, is further evidence that the universe is dynamic. Any theory of the origin and evolution of the universe must be called upon to explain the changes occurring within it.

At one time there was hope that the elements originated in the primeval fireball of the big bang. At the extremely high temperatures of the fireball, the thermal motions of matter would be great enough to initiate thermonuclear reactions, and perhaps the elements were formed by those reactions. At least that's what people thought.

At the temperature of the fireball, all the elements would be broken down into their constituent particles—protons and neutrons—which would set the stage for the following nuclear reactions:

Proton + neutron ⟶ deuteron

or

Hydrogen 1 + neutron ⟶ hydrogen 2

Here, hydrogen 1 is the hydrogen isotope of atomic weight 1 and hydrogen 2 represents the hydrogen isotope of mass 2 (deuterium, or "heavy hydrogen").

The deuterium could then capture a proton to form the helium isotope, of mass 3:

Hydrogen 2 + hydrogen 1 ⟶ helium 3

which could then capture a neutron to produce the most common isotope of helium:

Helium 3 + neutron ⟶ helium 4

At this point, the process of element formation bogs down. You might think it possible that the build-up of the elements could continue, with protons and neutrons being added to form heavier and heavier elements. The problem is that *no* isotopes of mass 5 are stable for any longer than it takes for them to come apart—about 10^{-16} second. Therefore, a reaction such as

Helium 4 + neutron $\longrightarrow$ helium 5

would immediately be undone by

Helium 5 + neutron $\longrightarrow$ helium 4

In the same way,

Helium 4 + proton $\longrightarrow$ lithium 5

would be followed by:

Lithium 5 $\longrightarrow$ helium 4 + proton

Many more complicated reactions can be conjured up, but they all will fail because there is not even a slightly stable isotope of either mass 5 or mass 8, and all routes lead back to helium 4. Although the big bang was probably important in establishing the primordial ratio of helium to hydrogen (10 to 25 percent helium) and may have played a minor role in the formation of lithium, *it did not contribute to the synthesis of the heavier elements*. After about 1/2 hour, the primordial fireball had expanded to the degree that its density fell to about 10^{-8} gram per cubic centimeter. Nuclear reactions cease to be important at that density.

As the fireball continued to expand and cool, it must have fragmented into vast clouds that were to evolve into galaxies. Our knowledge of this phase of the evolution of the universe is sorely lacking. Only in the past 20 years have we reached any sort of understanding of the origin and evolution of stars, and the origin and evolution of galaxies is still in the rudimentary stage characteristic of our understanding of the origin of stars a short while ago.

The primitive galaxies probably contained great clouds of dust and gas which were to evolve into the first stars, perhaps by processes not too different from the process giving rise to stars today. But these primitive, first-generation stars contained only hydrogen and helium rather than a full complement of the elements found in more "modern" stars.

THE ORIGIN AND EVOLUTION OF STARS

Within our Galaxy there are numerous clouds of dust and gas that obscure the more distant parts of our Galaxy from view. Some of these clouds are large and diffuse; others are much smaller and more dense. When a cloud becomes massive, dense, and cool enough, it will start to contract under its own gravitational attraction. The exact conditions under which this "gravitational instability" will occur are not entirely understood. But the initial size of an unstable cloud is probably about 10 light years—about 1 million times the distance from the earth to the sun; the initial mass is about 10,000 times the mass of the sun; and the temperature is about 100 degrees Kelvin, or 100 degrees above absolute zero. If the temperature of the material is extremely low, the gas molecules and dust grains will have insufficient kinetic energy to overcome the gravitational attraction of the rest of the cloud and will be unable to escape into interstellar space.

As the cloud contracts, collisions between particles release energy that radiates out of the cloud, causing the particles of the cloud to lose energy and become trapped even more within the collapsing mass of

Figure 6. The nebulous matter in Monoceros appears in this photograph, taken in red light.

dust and gas. As the cloud collapses and becomes more dense it becomes unstable, for it fragments into smaller clouds. The process of fragmentation may be repeated several times. Eventually fragments evolve that have the mass of a star, perhaps several times larger or several times smaller than our sun. The continued gravitational contraction of the smallest fragments culminates in a star. In this manner, thousands of stars may be produced in a relatively small region of interstellar space from an original larger cloud of dust and gas.

The notion of gravitational instability is not entirely theoretical. More than once we have identified stars that are being formed in the midst of dust and gas clouds. The best known examples are in the Great Nebula in Orion, visible without a telescope as a fuzzy region in Orion's belt and a breathtaking spectacle when photographed through a large telescope (Figure 1).

In thinking about the time scales associated with the formation of the universe, the galaxies, the earth, the sun, and the solar system, we have become accustomed to thinking of events occurring over vast spans of time—on the order of billions of years. And it's true that the first stages of the collapse of a gas cloud may require millions of years. But once it contracts to about one-tenth of a light year (about 10,000 times the distance from the earth to the sun), the remaining time may be measured in a few thousand years. Why is the last stage so brief? During the gravitational collapse, the material is free-falling. It is not difficult to show that the time required for this free fall is very short.

The rapid free-fall might cease when the density of the collapsing cloud becomes large enough to be opaque to radiation. At that point, the collapse could slow down rapidly because it would no longer be possible for the gravitational energy released by the collapse to be radiated freely into space. The energy would have to be carried up to the surface of the newly formed star by the relatively slow convective motion of the material itself before it could be radiated into space. However, at the high temperature reached in the cloud at this stage, it's likely that the gravitational energy is absorbed by the energy, which separates the hydrogen molecules into atoms and removes electrons from hydrogen atoms to form positive hydrogen ions. So the free fall probably proceeds unchecked. When the free fall ends, after perhaps 1,000 years, a star as massive as the sun would have a radius about 14 times the present radius of the sun; the surface of such a "proto-star" would be within the present orbit of the planet Mercury.

From that time on, the newly formed star would contract much more slowly. However, the rate at which gravitational energy is released is great enough for the star to be quite luminous. It would shine brightly for millions of years, using as its energy source its remaining store of gravitational energy. Even though gravitational energy will permit the star to be highly luminous for long periods of time, it is not the most important source of stellar energy. If gravitational energy had been the only source of solar light and heat, the surface of the earth would have become cold and frozen soon after its formation, and the evolution of life would have been snuffed out at its most rudimentary level.

Fortunately for us, another energy source supplements the gravitational energy source and subsequently supplants it as the principal source of stellar energy production. It arises from *thermonuclear reactions* between the nuclei of atoms.

THE CONVERSION OF HYDROGEN INTO HELIUM

By far the most important thermonuclear reactions in the early history of a star are the ones involving the conversion of hydrogen into helium:

(4) Hydrogen 1 $\longrightarrow$ helium 4

For each four hydrogen nuclei (which are simply protons) transformed into a helium nucleus, about 4×10^{-5} erg of energy is released. That amount may not seem like much, but when you stop and realize that the transformation is occurring throughout the entire central region of a star, the total rate of energy production is enormous: about 2×10^{35} ergs *each second*. That is about 1 million times the energy that would be produced by burning all the known reserves of fossil fuels on the earth.

Large as this total rate of solar energy production may be, the rate of energy production *per unit volume* is extremely low. When you think about thermonuclear reactions, you probably think of hydrogen bombs. In those devices, enormously destructive energies of about 10^{23} ergs are released in a small volume of space within a fraction of a second. In contrast, the rate of energy production *per unit volume* within an ordinary star is somewhere between that of a firefly and a candle, depending on the size of the star. This small rate of energy production maintains temperatures of millions of degrees in the center of a star only because of the great volume in the stellar interior and because of the long time required for this heat and light to be transported to the surface of the star.

The previous expression for the fusion of hydrogen to helium is highly simplified. First of all, four hydrogen nuclei (protons) could not simply combine to form a nucleus of helium 4 because that nucleus consists of two protons and two neutrons: the fusion reaction indicated would also have to involve changing two of the protons into neutrons. Further-

Calculating the Gravitational Collapse of a Star

The gravitational force between a dust and gas cloud in space and a fragment of it may be calculated from the law of gravitation:

$$\text{Gravitational force} = \frac{G\,(\text{mass of cloud} \times \text{mass of fragment})}{\text{distance}^2}$$

<u>Step 1.</u> Remember that G is the universal constant of gravitation, 6.7×10^{-8} erg-centimeter per gram squared. From Newton's second law, this force will be equal to the product of the mass and the acceleration of the small fragment of the cloud:

$$\text{Gravitational force} = \text{mass of fragment} \times \text{acceleration}$$
$$= \frac{G\,(\text{mass of cloud} \times \text{mass of fragment})}{\text{distance}^2}$$

<u>Step 2.</u> Now substitute the following approximate numbers into the last equation: mass of cloud = 10^{37} grams, the approximate mass of 10^4 stars; distance = $10,000 \times 1.5 \times 10^{13}$ centimeters, or 10,000 times the distance from the earth to the sun; and the value of G given above. The acceleration is then:

$$\text{Acceleration} = 3 \times 10^{-5} \text{ centimeter/second}^2$$

That acceleration is certainly small compared to the acceleration of a ball dropped near the surface of the earth (980 centimeters per second squared) or that of a car accelerating to 60 miles per hour in 20 seconds (about 300 centimeters per second squared). But it is still large enough to permit the cloud to collapse rapidly.

<u>Step 3.</u> If you assume that acceleration is constant, you may calculate the time required for the small fragments of the cloud to fall all the way to the center. The distance traveled in a specified time at a constant acceleration will be:

$$\text{Distance} = \tfrac{1}{2}\,\text{acceleration} \times \text{time}^2$$

<u>Step 4.</u> Solving that expression for time, you obtain

$$\text{Time} = \sqrt{\frac{2 \times \text{distance}}{\text{acceleration}}}$$

Substituting the value $10,000 \times 1.5 \times 10^{13}$ centimeters for distance, and the value 3×10^{-5} centimeter squared for the acceleration, you obtain $\sqrt{10^{22}}$ seconds, or about 10^{11} seconds. The number of seconds in a year is about 3.15×10^7 seconds. So 10^{11} seconds will be about 3,000 years. Actually, the acceleration will not be constant but will increase as the cloud contracts and the distance becomes smaller. Therefore, the time required for it to fall all the way to the center will be even less.

Sun 10^{-5} joules/cubic centimeter – second

Hydrogen Bomb 10^{21} joules/cubic centimeter – second

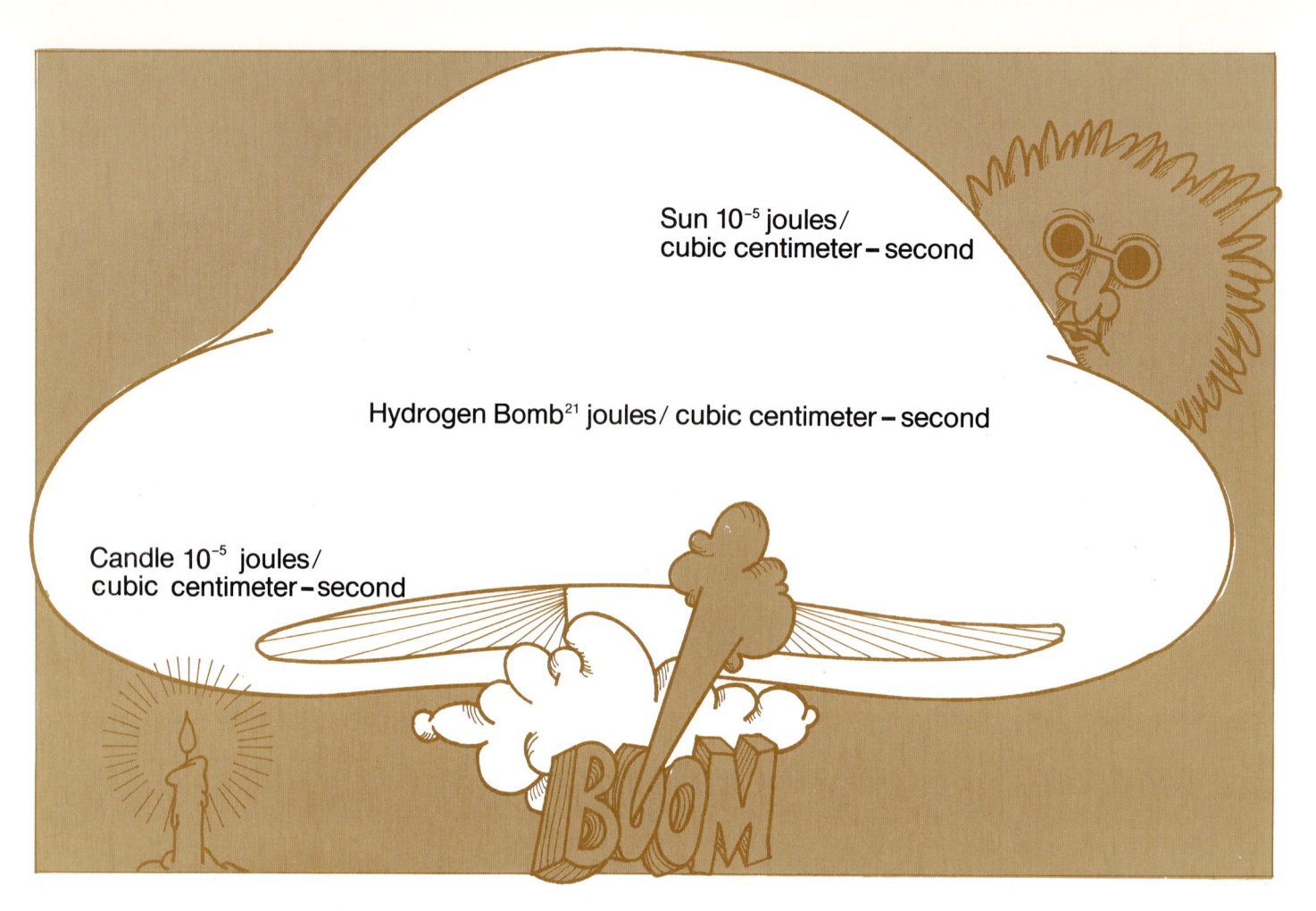

Candle 10^{-5} joules/cubic centimeter – second

Figure 7. The *rate* of energy release of the sun is actually less than that of a candle. A hydrogen bomb releases much more energy per unit volume than either the sun or the candle. The rates in this sketch are expressed as the amount of energy given off per second per unit volume.

more, the probability of four hydrogen nuclei (protons) all colliding at exactly the same time to form a helium nucleus is extremely small. The expression, then, represents a "synopsis" of a more complex process.

The hot interior of a newly formed star is composed primarily of hydrogen nuclei, regardless of whether the star is of the "first-generation" (formed shortly after the primeval big bang) or is a relatively recent star such as our sun. We know this because hydrogen has always been, and probably always will be, the most abundant element in the universe. In the star's interior, hydrogen nuclei, or protons, move about at an average velocity of hundreds of kilometers per second because of their high thermal velocities. When they collide, they usually bounce off one another. In the language of nuclear physics, that collision is referred to as *proton-proton scattering* because no changes in the nature of the nucleus take place. It may be written as:

Hydrogen 1 + hydrogen 1 ⟶ hydrogen 1 + hydrogen 1

However, once two protons do collide, they represent the extremely unstable "compound" nucleus helium 2, which consists of two protons. That compound nucleus can disintegrate in two ways. It can simply fall apart and become two protons again. But in a small proportion of the collisions, particularly those involving protons having greater than average kinetic energy, one of the protons may change into a neutron plus an electron as well as a weightless particle known as a *neutrino.* A neutron and a proton will be bound together in the resulting nucleus. It is the nucleus of deuterium, or "heavy hydrogen." The neutrino will speed off with the velocity of light and will play no further role in the evolution of the star. This reaction is written symbolically:

Hydrogen 1 + hydrogen 1 ⟶
  hydrogen 1 + neutron + electron + neutrino ⟶
    hydrogen 2 + electron + neutrino

This reaction has never been observed in the laboratory because its occurrence is too improbable compared with the more common proton-proton scattering. However, the theory of these nuclear processes is well understood and has been tested, so it's possible to calculate accurately how fast the reaction should occur. This highly improbable reaction is the "bottleneck" in the fusion of hydrogen into helium in the interior of stars, for it determines the rate at which the transformation takes place. And that rate is in good agreement with the observed rate of energy production in stars such as the sun.

This reaction occurs only because the transformation of two protons into a deuterium, or a heavy hydrogen, nucleus involves the release of energy—about 3×10^{-6} erg each time this reaction occurs. This energy release occurs despite the fact that a proton normally cannot transform spontaneously into a neutron; the neutron is about 0.2 percent heavier than a proton and, in accordance with Einstein's relationship $E = mc^2$, the release of energy should favor the opposite reaction. In other words, a neutron should change into a proton, instead of a proton into a neutron. That change is indeed observed for free neutrons, but it happens that the energy associated with the attractive nuclear force binding a neutron and a proton into a deuterium nucleus barely overcomes (by 3×10^{-7} erg) the energy released by the transformation of a free neutron into a proton. If the neutron's mass were only 1 percent greater, the energy of formation of a deuterium nucleus would not overcome the tendency for a neutron to transform spontaneously into a proton, and the fusion of two protons into a deuterium nucleus would not be possible.

If that were the case, no further build-up of the heavier elements would be possible and the universe would be cold, simple, and relatively uninteresting—a lifeless place consisting of cold clouds of hydrogen and possibly very dense stars that would continue to collapse gravitationally because the thermonuclear reactions converting hydrogen to helium could never take place.

Once deuterium nuclei are produced, the remaining nuclear reactions converting hydrogen into helium proceed rapidly. The remaining reactions can follow several possible sequences. Including the first step already discussed, a simple form of the entire sequence is:

Hydrogen 1 + hydrogen 1 ⟶ hydrogen 2 + neutrino + electron

Hydrogen 2 + hydrogen 1 ⟶ helium 3

Two isotopes of helium 3 can then combine:

Helium 3 + helium 3 ⟶ helium 4 + (2) hydrogen 1

Altogether it takes three hydrogen nuclei to produce each of the helium 3 nuclei, and therefore six are required to produce the helium 4 nucleus. But the last reaction released two hydrogen 1 nuclei, so it really required only four hydrogen nuclei to produce the helium nucleus, in accordance with the expression

(4) Hydrogen 1 ⟶ helium 4

The fusion of hydrogen to helium can take place by several other series of reactions, or "routes." In stars that are more massive than the

Hydrogen 1 + Hydrogen 1 → "Unstable" Helium 2 → "Heavy Hydrogen" Hydrogen 2
(Proton)   (Proton)   (Proton, Proton)   (Proton, Neutron)

Hydrogen 2 + Hydrogen 1 → Helium 3
(Proton, Neutron) (Proton)   (Proton, Proton, Neutron)

Helium 3 + Helium 3 → Helium 4 + Two Hydrogen 1
(Proton, Proton, Neutron) (Proton, Proton, Neutron)   (Proton, Proton, Neutron, Neutron)   (Proton) (Proton)

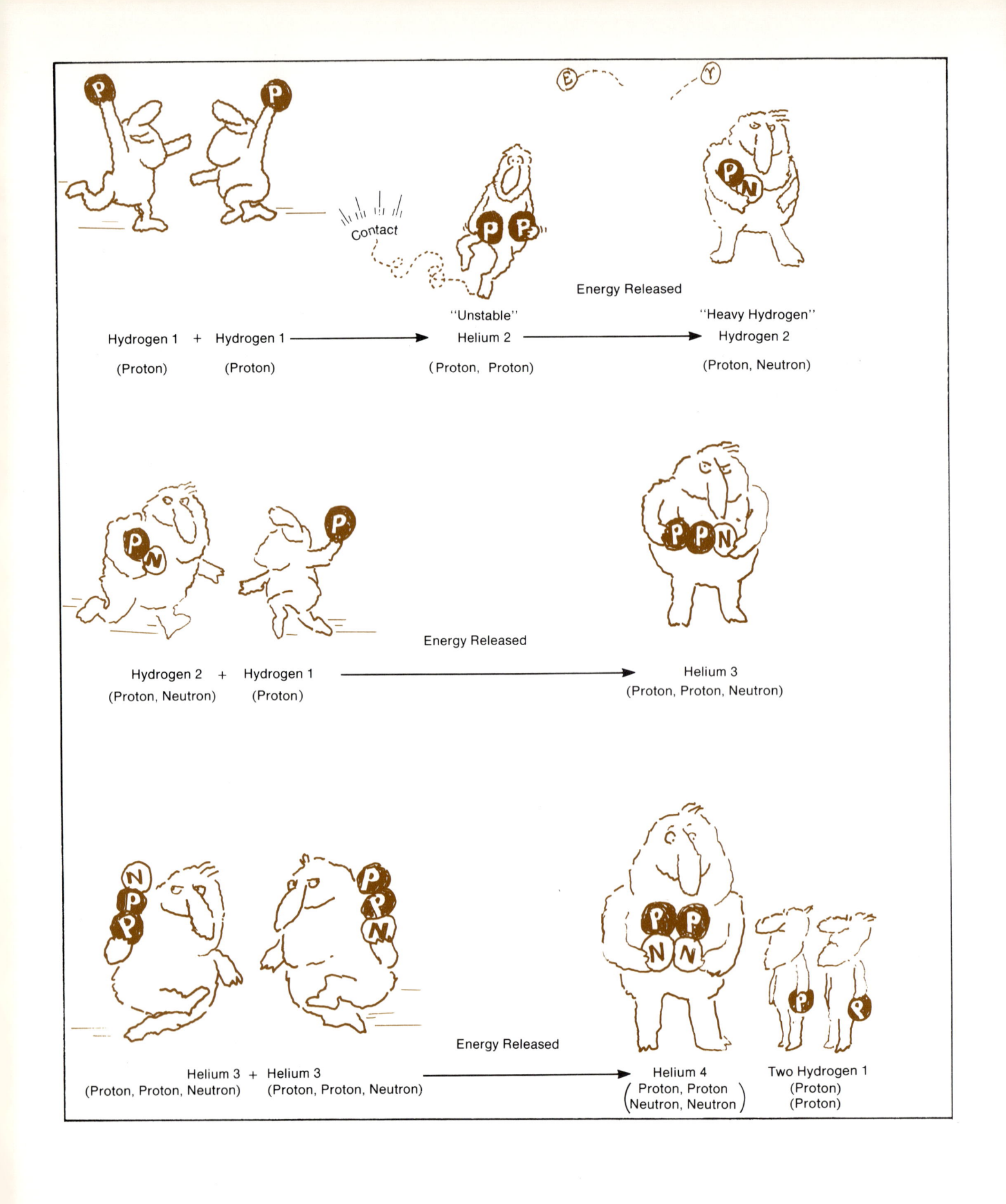

sun and at least "second generation" in that they contain heavier elements such as carbon, nitrogen, and oxygen, the fusion of hydrogen into helium takes place by way of a series of thermonuclear reactions. These three heavier elements participate in the reactions but there is no change in the total quantity of them. In any case, the overall result is the same:

(4) Hydrogen 1 $\longrightarrow$ helium 4 + energy

where the energy is 4×10^{-5} erg for each helium 4 isotope produced.

As the temperature in the interior of the star rises, first from gravitational attraction and later by the fusion of hydrogen into helium, the pressure of the gas in the interior also increases in accordance with the gas laws. Eventually a condition is reached where the thermonuclear reactions are producing enough energy to cause the internal gas pressure to balance the great weight of the outer regions of the star, and gravitational contraction and compression stops.

When that condition is reached, we have one of the most wonderful things in nature—a stable star. Suppose the temperature of the central region of the star were to increase slightly. In accordance with the gas laws, the central pressure would rise a little, the star would expand a little, and the central density would fall slightly. As a consequence, the hydrogen nuclei would collide less frequently with one another and the rate at which thermonuclear energy is produced would fall accordingly. The slowing down would cause the central temperature to fall, offsetting the rise in temperature that initiated the entire sequence of events. On the other hand, if the central temperature were to fall, an opposite sequence of events would occur, the rate at which the thermonuclear processes occur would increase, and the fall in temperature would be corrected. Thus, by entirely natural processes, the center of the star operates as an accurately controlled thermostat.

The variable, or pulsating, stars discussed in the preceding chapter do not exhibit this stability. In some cases, they are in a more advanced stage of stellar evolution, reached only after their hydrogen fuel has been exhausted. But most stars, including our sun, become stable. The change in their rate of energy production is gradual because the quantity of hydrogen fuel in their cores decreases gradually as it is converted into helium.

As long as an adequate supply of hydrogen—the nuclear fuel— remains, the star remains stable. In the case of a star the size of our sun, the fuel is sufficient to maintain stability for about 10^{10} years. For large stars of about fifty solar masses, the central pressures and temperatures are much higher and the fuel is consumed at a much greater rate. Such stars consume their supply of hydrogen in as little as 1 million years.

The sun, like other stars, was formed from a dust cloud that fragmented, producing a large number of smaller clouds and ultimately producing a cluster of stars as neighbors in space. Some of the stars in the cluster were much larger than the sun, many were of similar size; most were smaller. The larger "brothers" of the sun long ago exhausted their supply of hydrogen and passed on to the later, terminal stages of stellar evolution. The smaller members of our sun's family can continue to shine even if our sun becomes a cold, dead cinder of a star.

The conversion of hydrogen into helium in the interior of the sun and other stars is an excellent way of keeping us and the possible inhabitants of the planets of other stars comfortably warm, but it does not contribute very much to the formation of the heavier elements; it merely adds more helium to that produced in the primitive fireball. The formation of heavier elements must await later stages of stellar evolution—after the

Figure 8. (Far left) Simplified version of the formation of nuclei. There are always electrons around to make up the nuclei but they have been left out of these sketches for the sake of simplicity. The top reaction shows the conversion of the two hydrogen 1 atoms into hydrogen 2; the middle reaction shows the conversion of a hydrogen 2 atom and a hydrogen 1 atom into helium 3; and the lower reaction shows the conversion of two helium 3 atoms into a helium 4 atom plus two hydrogen 1 atoms.

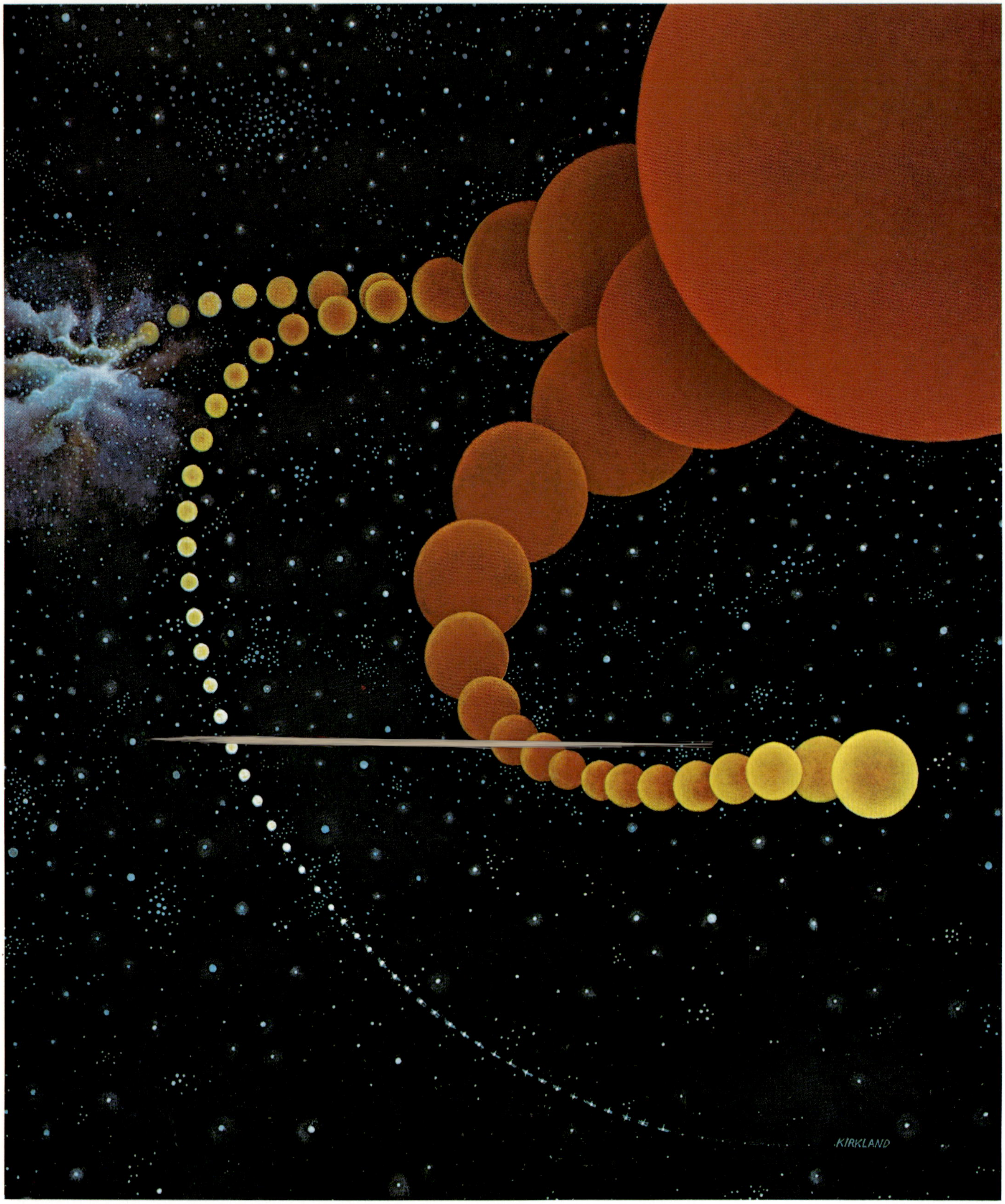
KIRKLAND

supply of hydrogen fuel in the central portion of the star has been consumed.

There is sufficient hydrogen fuel to sustain a star the size of our sun for billions of years. Eventually, however, there will come a time in the life of any star when so much of the hydrogen in its core has been converted into helium that the rate of thermonuclear energy production will be insufficient to maintain the temperature and pressure necessary to support the great weight of the outer portion of the star. Even though most of the star will still consist of hydrogen, with only the core consisting of helium, almost all of the hydrogen in the outer region of the star will be too cool to produce thermonuclear energy; it will not help maintain the stability of the star.

When that stage is reached, the gravitational contraction which had been halted by the onset of the thermonuclear reactions converting hydrogen into helium will resume. The central core of the star will be compressed to extremely high densities—about 10^4 grams per cubic centimeter, or about 1,000 times as dense as the element lead.

Gravitational energy released by this further contraction of the core will raise its temperature even more, to about 10^8 degrees Centigrade. As the core contracts, the outer regions of the star also fall inward, which releases further gravitational energy. The process heats the outer regions, causing them to expand in accordance with the gas laws. Therefore, the overall effect is one of increasing the radius of the star enormously even though the helium core is much smaller than it was before the renewal of gravitational contraction.

The enormous increase in radius causes the star to become what is known as a *red giant*. In extreme cases, the radius of the star will increase so much that its radius will become greater than that of the orbit of Mars—about 4×10^8 kilometers, or about 500 times the radius of the sun. Even though the rate of release of energy in the central core will be as great or even greater than before, the surface area will be so large that the rate of energy release per unit area will drop considerably. The surface of the star will therefore be cooler and will become red in color rather than the more common blue-white color of stars that are still shining by consumption of hydrogen fuel. Several of the brightest and most easily observable stars are red giants: Betelgeuse in Orion, Aldebaran in Taurus, and Antares in Scorpios.

When our sun becomes a red giant, the temperature of the earth will become so high that all the water of the oceans—and probably the carbon dioxide of the limestones as well—will be in the atmosphere. The surface of the earth will in some ways resemble the present surface of Venus—glowing at red heat under an atmospheric pressure hundreds of times greater than the present atmosphere of the earth and so opaque that sunlight will not reach the surface. It will not be a pleasant place for human life—or any life at all. On the other hand, if the human mind and spirit can surmount its present problems, there is no reason to believe that the gradual growth of the sun into a red giant billions of years from now will pose an insurmountable problem for whatever intelligent beings inhabit the earth in the distant future. One possible place to move would be to a satellite of Jupiter or Saturn; the temperature there may by that time resemble conditions found on the earth today.

THE FORMATION OF HEAVIER ELEMENTS

It is during the red giant stage of stellar evolution that we first find element formation other than fusion of hydrogen into helium. The difficulty

Figure 9. (Left) An artist's conception of the evolution of an average star such as our sun. At the lower right of this painting, the yellow sun is shown evolving into a red giant. The red giant may diminish gradually into a white dwarf, through the stages trailing downward on the right, or it may eventually explode into a supernova, which is depicted at the extreme left. The remnants of the supernova may then join the fate of a white dwarf, or become a pulsating neutron star (not shown).

Figure 10. What will the inhabitants of the earth do when the sun burns out billions of years from now? Perhaps our remote progeny will be as different in appearance from us as we are from our remote apelike ancestors, but that is something we never will know. Perhaps the only thing we can predict with any confidence is the dynamic surge of life itself which, if it still exists in that distant time, will respond with adaptive abilities to the challenge as it has done in the past.

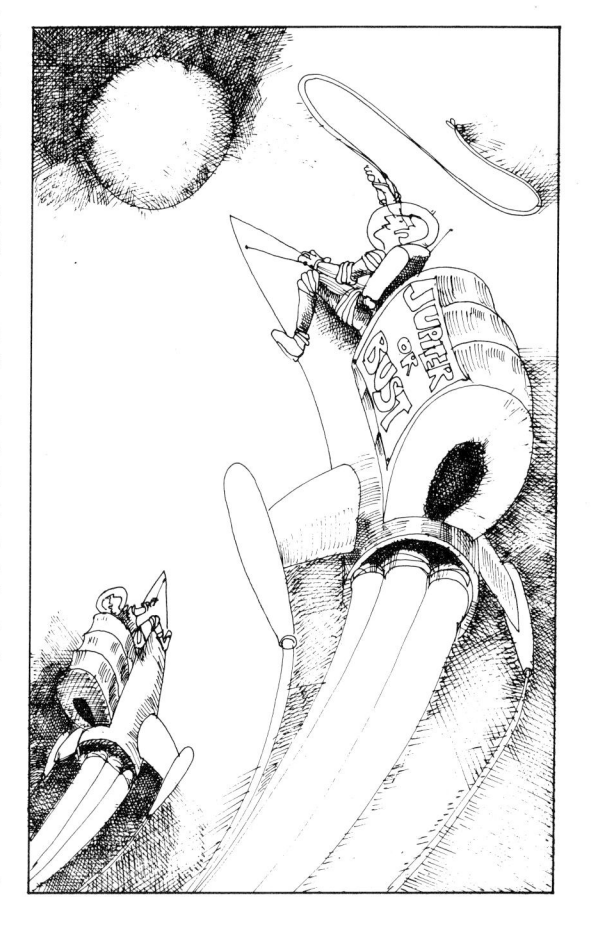

Figure 11. The galaxy M104 in Virgo, photographed through the 200-inch telescope at Mount Palomar. The halo surrounding the galaxy is made up primarily of red stars. Star birth is confined to the galactic plane. Our own Milky Way Galaxy probably looks quite similar to M104.

of forming elements heavier than helium during the primeval fireball of the big bang was in large part caused by the absence of nuclei of mass 5 and mass 8. For example, the reaction

Helium 4 + helium 4 $\longrightarrow$ beryllium 8

would not accomplish anything because beryllium 8 is so unstable that it would immediately disintegrate:

Beryllium 8 $\longrightarrow$ helium 4 + helium 4

In the interior of a red giant, however, the density of matter is 10^{10} times greater than it was in the primeval fireball. Furthermore, hundreds of millions of years are available, instead of the 1/2 hour that was available before the primeval fireball expanded and cooled. Under these different conditions, there would be a small but nevertheless significant chance that during the fleeting existence of the beryllium 8 nucleus, a *third* helium nucleus would collide with it:

Beryllium 8 + helium 4 $\longrightarrow$ carbon 12

the overall reaction being equivalent to:

(3) Helium 4 $\longrightarrow$ carbon 12

Like the fusion of hydrogen into helium, the fusion of helium into carbon produces energy—thereby giving the star a new lease on life. It will have obtained a new fuel to replace the consumed hydrogen. The star can therefore achieve a new stable structure and continue to shine, using helium as its fuel for long periods of time—perhaps hundreds of millions of years for a star the mass of our sun.

Like the first step in the fusion of hydrogen into helium, this reaction of three helium 4 isotopes into carbon 12 is very slow. It serves as a "bottleneck," permitting the star to conserve its fuel reserves.

The fusion of helium into carbon is also important to the formation of the elements because, as already mentioned, it bridges the gap of mass 5 and mass 8 and sets the stage for the formation of heavier elements. In fact, the formation of carbon 12 is almost immediately followed by the reaction

Carbon 12 + helium 4 $\longrightarrow$ oxygen 16

If any hydrogen nuclei from the outer regions of the star are mixed into this extremely hot mixture of carbon, oxygen, and helium, then isotopes such as carbon 13, nitrogen 14, oxygen 17, and oxygen 18 can also be produced. Observations of certain rather peculiar stars belonging to the general class of red giants indicate that neutrons are at least sometimes present in their cores, possibly produced by reactions such as

Carbon 13 + helium 4 $\longrightarrow$ oxygen 17 + neutron

In later-generation stars, formed from dust clouds already containing heavier elements, neutrons can be captured by heavier elements to cause further formation of isotopes. That might explain the existence of the short-lived (about 10^6-year half-life) element technetium in certain stars, which is being produced continually in the interior of the star by the reaction neutron + molybdenum 98 $\longrightarrow$ molybdenum 99 followed by the radioactive decay of molybdenum 99 into technetium 99.

Although the fusion of helium into carbon and oxygen will suffice to sustain the life of the star for some time, the time will be shorter than that required for the conversion of hydrogen into helium because the energy released in the reaction of three helium 4 into carbon 12 is only 1×10^{-5} erg per reaction, as compared to the 4×10^{-5} erg released by

(4) Hydrogen 1 $\longrightarrow$ helium 4

As before, the star's fuel eventually will become exhausted, further gravitational contraction will take place, and the core will become even more hot and dense. It is likely that further energy sources such as

Carbon 12 + carbon 12 $\longrightarrow$ magnesium 24

will then take over, and further element building such as

Oxygen 16 + helium 4 $\longrightarrow$ neon 20

and

Magnesium 24 + helium 4 $\longrightarrow$ silicon 28

will take place. But these energy sources are much less effective in prolonging the life of the star because their energy release per reaction is relatively small.

Once this stage of stellar evolution is reached, the theoretical problems associated with the details of its further evolution are so complex they are poorly understood. At this stage the residual life of the star is probably limited to a few days or hours. Nevertheless, it is during this time that most of the heavy elements are formed.

If a star is sufficiently small—say, about the mass of our sun—it may end its life peaceably. After all of its energy sources have been consumed and gravitational attraction is resumed, no new reactions will occur to

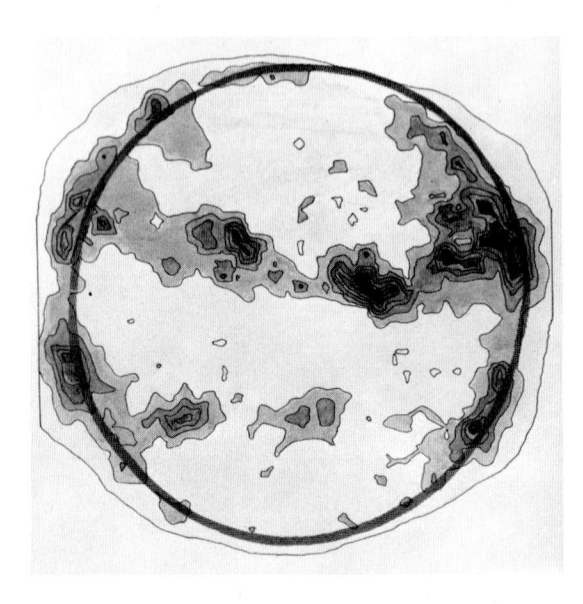

Figure 12. This *spectroheliogram* was synthesized from data picked up by the ultraviolet-sensitive sensors on NASA's Orbiting Solar Observatory IV. It shows the distribution of Magnesium 10 ions in the sun's atmosphere 100,000 miles above its surface. (The magnesium 10 has lost nine of its electrons because of the intense heat—about 1.5 million degrees Centigrade—at that altitude.)

stabilize the star with a denser and hotter core. Gravitational contraction will simply continue, compressing the matter so much that it will be able to bear the weight of the entire star by virtue of its own strength. Under those conditions, the density of the star will be about 10^6 grams per cubic centimeter. In other words, a teaspoonful of this material would weigh several tons.

If our sun ends its career in this way, and it is likely that it will, the radius of our sun will be about that of the earth even though its mass will be the same as it is today. A dense star of this kind, completely devoid of energy sources and shining only because of its residual high temperature, is known as a *white dwarf.* Many white dwarfs can be observed today. That final collapse probably represents the final stage of stellar evolution of most stars.

But not all. It is also known that many stars end their lives catastrophically. If a star is sufficiently massive after the energy resources of its core have been consumed, a new class of nuclear reactions may become dominant—reactions that *consume* energy rather than release it. The effect of that energy consumption is the opposite of the effect of energy release. Instead of stabilizing the further contraction of the star, the reactions will destabilize it and the material of its interior will collapse in a catastrophic free fall. Large amounts of gravitational energy will be released—sufficient to blow off the exterior part of the star. It is also possible that the relatively cool hydrogen still remaining in the outer envelope of the star will fall into the extremely hot center of the star. At these extremely high temperatures hydrogen will not release energy slowly, as was the case for the early stage of hydrogen fusion, but will react explosively. Most of the star's matter will be exploded into interstellar space.

At the high temperature of the explosion, much of the material will be exposed to an intense flood of protons and neutrons produced by the disintegration of heavier nuclei. It is during this *supernova* explosion that the remaining formation of the elements takes place, including massive nuclei such as uranium 238, plutonium 244, and californium 252. This is probably the most important mechanism that returns the heavier elements to interstellar space where they can mix with interstellar material to form new clouds of dust and gas, from which new stars are formed. Such a star was formed 4.6 billion years ago, and we call it our sun. Because heavier elements were present in the interstellar medium, having been formed in the cores of stars of earlier generations, our sun and solar system were formed with a full complement of elements. The formation of the complex minerals, organic compounds, and biological materials present on the earth was ensured.

The energy released in the explosion of a star as a supernova rivals the energy of the entire galaxy in which it is located. Consequently, it is not difficult to observe supernovas in other galaxies. In a typical galaxy, a supernova occurs about once every 100 years. They occur in our galaxy about once every 300 years; many more must be obscured by the opaque clouds of dust hiding much of our galaxy from our view.

The last supernova to be observed in our galaxy occurred in 1604 and was studied by both Johannes Kepler and Galileo. Only one generation earlier, in 1572, another had been observed by Tycho Brahe. We have not been so fortunate since then, but you do have a fair chance of observing a supernova during your lifetime.

A supernova bursts into full intensity in a few days and then generally fades from view after several months or years. But its remnants persist for thousands of years. The most famous supernova remnant is the Crab Nebula in Taurus, which was shown with the introductory essay to this book. It is visible not only optically; it is also a strong source of radio waves and x-rays. It is the remnant of a supernova described by Chinese astronomers in the year 1054. The remnants of the supernovas of 1572 and 1604 have also been identified as sources of radio waves.

What remains of a star after it has exploded as a supernova? If the residual core is sufficiently small, about the mass of the sun, it may contract, cool, and become an ordinary white dwarf. On the other hand, if it is significantly more massive, matter even of the density of white dwarfs will be insufficiently strong to support the weight of the massive remaining material and further compression will occur. A much denser body, called a *neutron star,* will form. The density of a neutron star will be about 10^{15} grams per cubic centimeter. At that density, the mass of the entire earth could be contained within an office building of moderate size.

Within the Crab Nebula is a faint star that emits regular, sharp bursts of light and radio waves every 1/30 of a second. Stars of this kind are called *pulsars.* The pulsar in the Crab Nebula is probably the neutron star that formed from the residual core of the supernova of 1054.

As our Galaxy grows older, more and more of its matter will become stored away in the form of white dwarfs and will be unavailable for the formation of stars of later generations. The dust and gas clouds of our Galaxy will become increasingly rare. Following this line of reasoning, the ultimate future of our Galaxy—and of the universe—should be a collection of cold, white dwarfs. The picture is just about as bleak as it would have been if the neutron had been a bit more massive and the entire chain of events—stellar evolution and element building—did not begin in the first place.

But this may not be the ultimate fate of our universe. Remember that the lifetime of typical stars such as the sun is about 10^{10} years, and the time required for our Galaxy and the other galaxies of the universe to come to this final stage will be many times this long. Looking back into the past, it has only been about 10^{10} years since the "big bang," the beginning of the present expansion of the universe. Why should we be able to see much further into the future than we can look back into the past? Times of the order of 1 to 2×10^{10} years may represent the time scale on which major, epochal changes take place in the structure of the entire universe. And until we understand much more than we do now about the fundamental laws governing the evolution of the universe, we can refrain from writing those last pages in our book of knowledge about it.

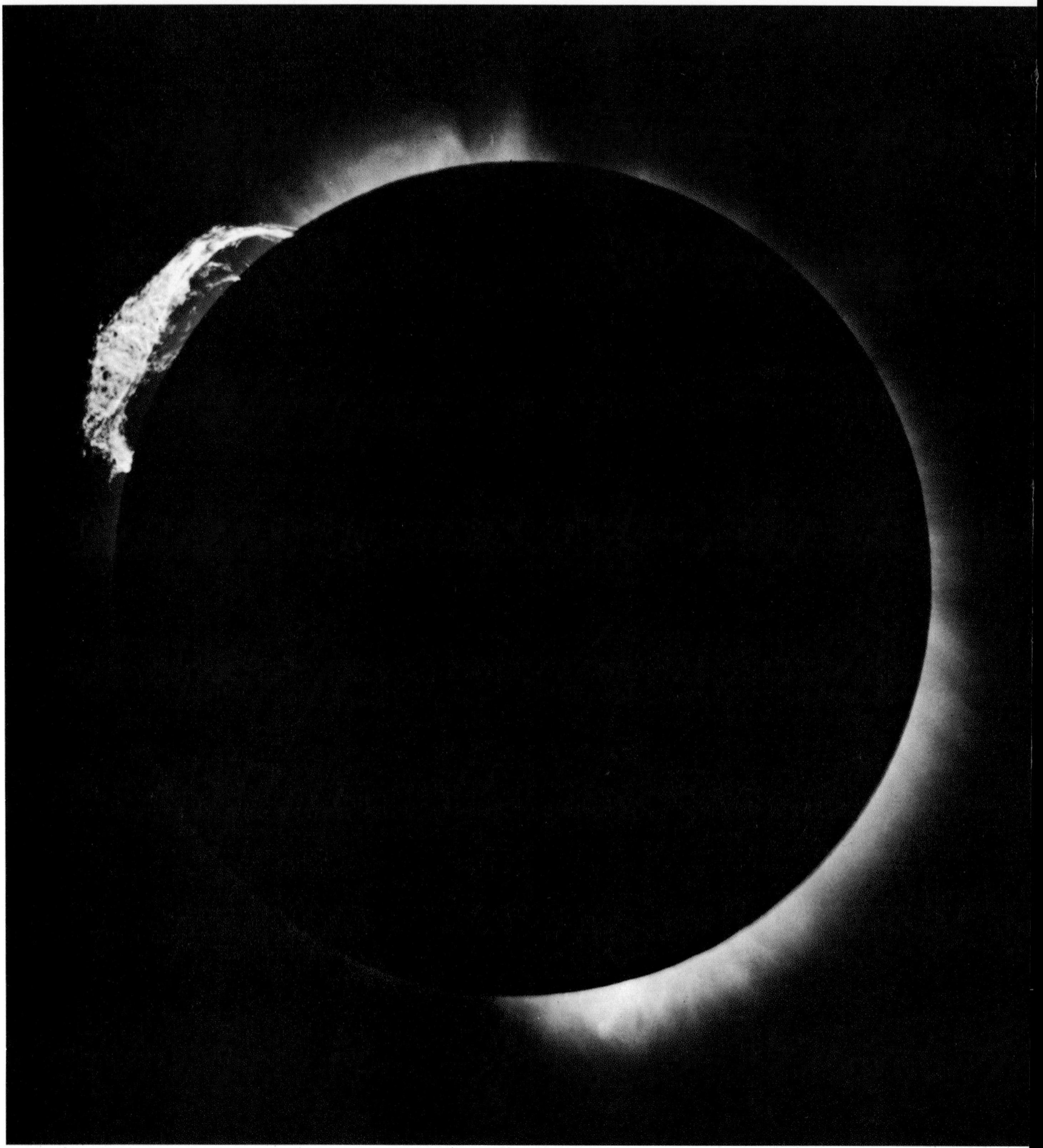

Our solar system is a vestigial by-product of the formation of the sun. Because the sun is a star, a complete understanding of how stars themselves form will enable us to understand how our solar system came into being. The preceding chapter mentioned that single- and multiple-star systems formed when a large cloud of dust and gas fragmented into thousands of smaller clouds, which eventually condensed into stellar bodies. This general scheme has been the basis for our thinking about the origin of the solar system since the time of Immanuel Kant—over 200 years ago—except for a brief interlude at the beginning of the present century. Then it was fashionable to think of the solar system as being formed as a consequence of the (highly improbable) close approach of the sun to another star.

During the last 20 years, we have learned a great deal about the origin and early history of the solar system through careful and detailed studies of meteorites. More recently, new insights have been gained from studies of lunar rocks brought to the earth and from geophysical instruments placed on the moon. As a result of this work the range of speculations concerning the origin and early history of the solar system has been narrowed considerably. And there is even some hope that the new speculations sprouting up in their place will have greater durability than their predecessors.

What is a meteorite? At least several times a day, a fragment of stone or iron weighing more than 1 kilogram collides with the earth. Its interplanetary velocity is about 10 kilometers per second, but as it nears the earth it encounters the pull of the earth's gravitational field and its velocity increases to about 15 kilometers per second as it enters our atmosphere. At this velocity, the fragment immediately flares into incandescence. Its surface melts and vaporizes, leaving behind a fiery, smoky trail. Someone on earth sees it as a brilliant light flashing across the sky and then exploding into fragments that fade in brilliance, and go out. He then hears loud explosions reminiscent of thunder or, in modern times, of sonic booms. For several minutes all is silent. Finally he hears one or more "whizzing" or "buzzing" sounds and, once the fragment strikes the ground, a rather anticlimactic thud.

That is the fall of a typical meteorite—a very useful space probe launched from an uncertain place in the solar system and brought to the earth by the random processes of nature at no expense to the taxpayer.

Thousands of meteorites fall to earth every year, and some are hundreds of kilograms in mass. Yet only about five are recovered each year and brought to the attention of scientists. Most fall into the ocean or sparsely populated areas and are not even observed. Even in fairly populous areas, the probability of recovery is negligible. No observed meteorite has ever been recovered in the most populous state of the United States—California—in part owing to the large areas of the state of low population density and the mountainous terrain. In contrast, five meteorites have been observed and subsequently recovered in the state of Kansas. There is absolutely no reason why a meteorite is more likely to fall in Kansas than in California, so the difference must be attributed to differences in the uniformity of population density and topography.

Interestingly, many of the meteorites that have been recovered have struck or penetrated the roofs of buildings. In the 1950s one passed

12
Early History of the Solar System

The solar prominence Anteater, photographed during the total solar eclipse of May 1919.

Figure 2. According to legend, a "thunderstone" (meteorite) fell during a battle between the people from Ensiheim and Battenheim and was taken as an omen. An old German woodcut depicts the scene.

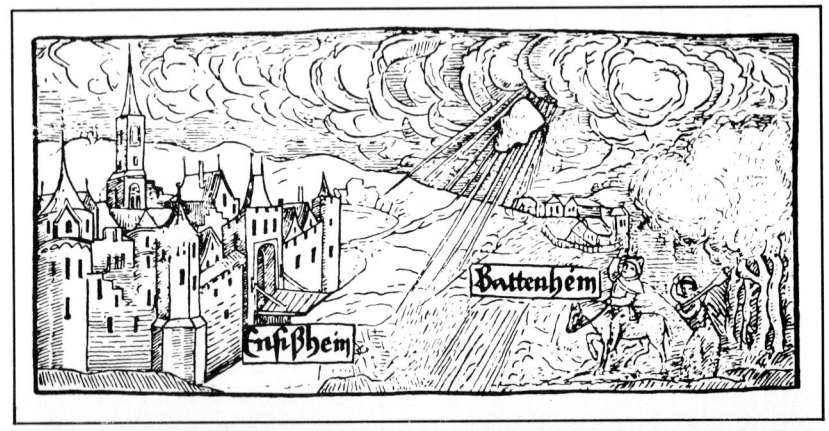

Figure 3. A *meteorite* is an interplanetary fragment that survives penetration of the earth's atmosphere and falls to the ground. A *meteor*, in contrast, would be completely destroyed in the atmosphere. This photograph of multiple flare-ups of a meteor was taken at midnight in July 1952.

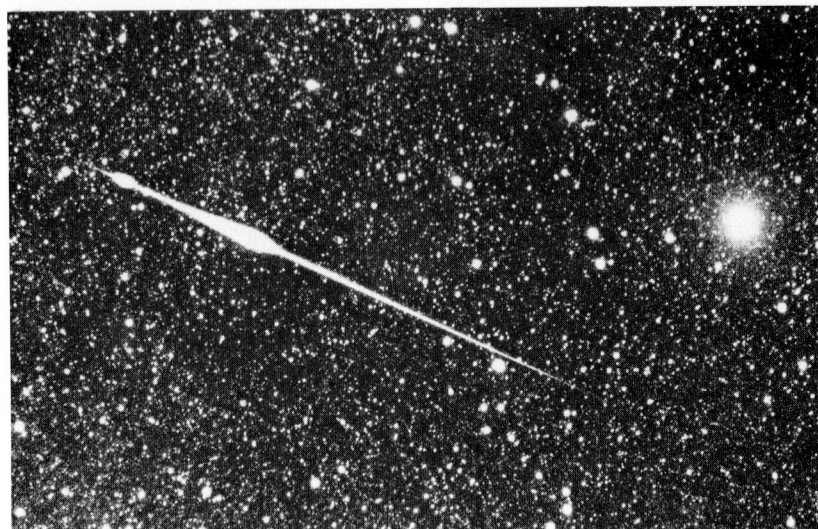

through the roof of a home in Sylacauga, Alabama, glanced off a radio, and struck Mrs. Hodge on the upper thigh as she was resting on a couch, causing a painful bruise. That is the only authenticated case of a person being injured by the fall of a meteorite, even though the probability of their recovery is definitely enhanced by the dramatic effect of having crashed into someone's attic.

The higher probability of capture by roofs was demonstrated a few years ago, when the Smithsonian Astrophysical Observatory established the Prairie Network of sky cameras, which patrol about 10^6 kilometers of the central United States every clear night. When photographed by these cameras, the trail of a meteorite is carefully measured and the point where the meteorite fell can be calculated within about 1 square kilometer. Despite this intensive effort, the only meteorite to be recovered by this network fell in Oklahoma in January 1970. In the same time, six meteorites were recovered after they struck houses and barns. After taking into consideration the fact that the Prairie Network fails to observe meteorites that fall in the daytime, the problems associated with cloudy weather at some of the stations, and the difficulty of finding a meteorite in a field of tall corn, we are led to the conclusion that the area of roofs is also about 10^6 square kilometers, corresponding to about 300 square meters, or a surprising 2,700 square feet per person. The explanation of this puzzling conclusion may be that most meteorites explode into a large number of fragments. The fragments are distributed over an area of about 50 square

214 Physical Science Today

kilometers, and only one of the fragments must strike a roof before something can be recovered and a search for other fragments initiated.

In addition to the meteorites that have been recovered following the observation of their fall, about a thousand have simply been found. In some cases these objects have been lying about the surface of the earth for several million years.

Now examination of the interior of the meteorites shows they are not all the same. Most are composed primarily of stony material consisting of silicate minerals quite similar to those found on earth. Others are pure iron-nickel alloys. A relatively small number contain comparable quantities of silicate minerals and metallic iron.

It seems probable that this meteoritic iron was the first metal prehistoric man used in the manufacture of tools. At least tools of meteoritic iron have been found that predate the discovery of the chemical technology necessary to transform ores into metallic form. Although the actual number of such meteoritic artifacts is small, the distribution of iron meteorite "finds" is much higher in the New World than in the Old World, corresponding to regions of the earth where other evidence suggests that the prehistoric population density was high. All of this suggests our prehistoric ancestors were quite efficient at recovering meteoritic iron, probably excelling our present ability to do so.

In contrast, stony meteorites are much more difficult to identify as uniquely interesting objects, unless they are actually observed to fall. While passing through the atmosphere at supersonic velocity they acquire a black "fusion crust," which makes them fairly easy to identify soon after their fall. Within a few years, however, the fusion crust chips off and they become ordinary-looking brown rocks. There are probably millions of such stony meteorites lying about on the surface of the earth that have not been recognized as bodies from interplanetary space.

THE RECORD OF THE METEORITES

During a meteorite's long flight, its interior remains at the low temperature of interplanetary space. If a meteorite is picked up immediately after its fall to earth, it may be somewhat warm to the touch because it retains some heat from atmospheric friction. But its brilliant flight through the earth's atmosphere lasts only a few seconds and its incandescent material disappears quickly as vapor or liquid, so the heat penetrates only a fraction of a millimeter below the surface. As a consequence, only a thin surface layer is chemically and physically altered by the flight through the atmosphere, and the chemical and mineralogical integrity of the interior of the meteorite is preserved for scientific study.

Meteorites would be interesting enough by virtue of being samples of matter from outside the earth. But their importance far exceeds this fact. During the past 20 years it has been shown that these fragments of interplanetary matter were formed during the formative period of the solar system—4.6 billion years ago—and that most have been essentially unaltered since then. To be sure, collisions in interplanetary space have chipped them down to their present modest size, but even the finest details of their mineral and chemical structure have been preserved as a record of the entire history of the solar system. It is even possible that some of these meteorites still contain in unaltered form the earliest solid matter to form in the solar nebula. How can we draw this astounding conclusion from such primeval matter?

In the preceding chapter you read that chemical elements and their isotopes are formed in the interior of stars. Except for the lightest elements, much of the formation of elements takes place during the last

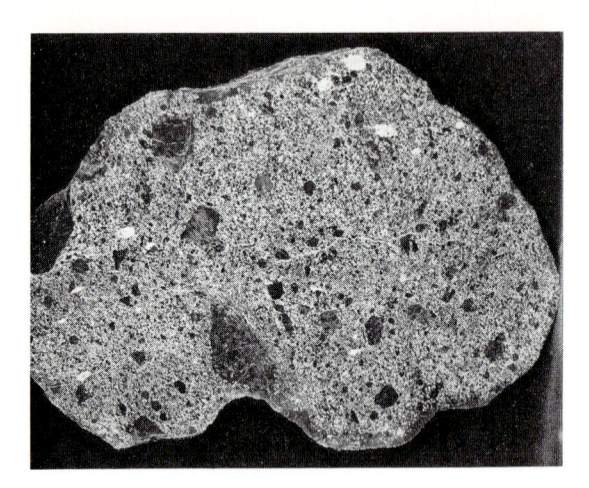

Figure 4. Santa Catharina (above), found in Brazil in 1875, is a typical iron meteorite. Kelly (above right) is a chondrite, a common class of meteorites. It was found in Colorado in 1937. Its colorful crystalline structure is exposed in the thin section (lower right) photographed in polarized light under a microscope.

stages of the life of a star. At this stage, many stars explode as supernovas. Nuclear reactions resulting in further isotope production take place during the explosion. These isotopes mix with isotopes previously produced and with hydrogen from the relatively surficial regions of the star. Eventually the isotopes are cast out into interstellar space, becoming the raw material from which new dust clouds and stars are formed.

Supernova explosions occur every 50 to 100 years in our Galaxy. At this rate, hundreds of millions of stars have exploded since our Galaxy was formed. The interval between these explosions is small compared to the age of our Galaxy, and we can think of the supernova process as a means of continuously supplying new isotopes to the interstellar medium.

Some of these newly formed isotopes are stable; some of them, such as uranium 238 and potassium 40, are radioactive. Long-lived radioactive isotopes have survived the 4.6 billion years since the solar system was formed, but shorter-lived radioactive isotopes were also produced. Some of these, such as carbon 11 and sodium 24, have very short lifetimes and disappear in times ranging from a fraction of a second to a few years. Others are sufficiently long-lived to survive for millions of years. Two important isotopes in this category are iodine 129, with a half-life of 17 million years, and plutonium 244, with a half-life of 80 million years. The half-life of these isotopes is long enough so that they were still present when the dust cloud formed around the sun, but it is not long enough to survive in significant quantities throughout the entire history of the solar system. Radioactive isotopes falling into this category—present when the sun and solar system were born but transformed eventually into stable isotopes—are known as *extinct radioactivities.*

In 1960, John Reynolds made a remarkable discovery. He measured the xenon isotopes in a particular stony meteorite, Richardton (named after Richardton, North Dakota, where it fell), and found that all but one of the nine isotopes of xenon were present in nearly the same proportions found in the more familiar xenon of the earth's atmosphere. But there was one conspicuous anomaly. The xenon from the Richardton meteorite contained much more xenon 129 than atmospheric xenon contains. Subsequent measurements on other meteorites confirmed his finding. After a series of careful experiments, there was only one plausible interpretation: the excess xenon 129 was the stable decay product of the now-extinct radioactive isotope iodine 129. When the meteorite was formed, it must have contained "live" iodine 129. Xenon is a gaseous element, and most of the xenon remained in the solar nebula when the stony material of the meteorite was formed. Iodine is less volatile, however, and it was incorporated in the minerals of which the meteorite is composed. After a few half-lives (tens of millions of years), the iodine 129 transformed into xenon 129 and remained trapped in the meteoritic minerals.

The fact that the meteorite was formed while there was still "live" iodine 129 available implies that the meteorite was formed early in the history of the solar system. A quantitative extension of the same line of reasoning shows that most stone meteorites were formed within about 100 million years after the collapse of the dust cloud from which the sun and solar system were to be created, and that since that time they have not been heated enough to drive off the xenon. In fact, much of this xenon 129 is still located in the *same* site within the meteoritic mineral previously occupied by the extinct iodine 129.

Later measurements demonstrated the existence of other anomalies in the relative abundances of the xenon isotopes. Some meteorites seemed to contain xenon isotopes in ratios that strongly suggested they were the products of nuclear fission. But there were several problems with this explanation.

Figure 5. The relative mass spectrum of isotopes of xenon from the Richardton meteorite looks similar to the spectrum found in the earth's atmosphere, *except* for the isotope xenon 129. The short, dashed lines show the measured levels in the atmosphere. The scale of xenon 136 to xenon 129 has been expanded.

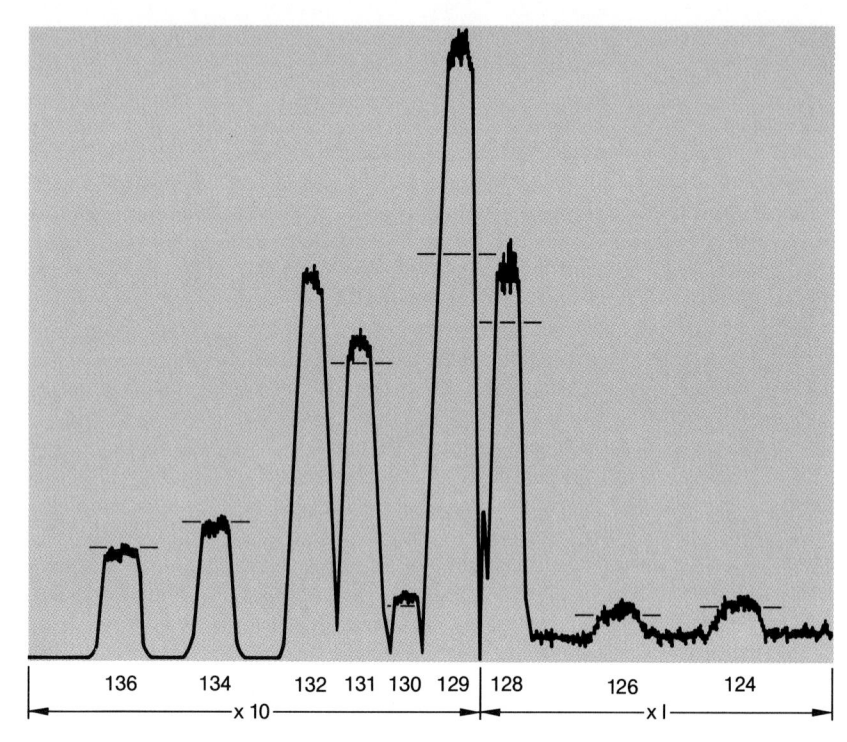

First of all, those meteorites contained far too little fissionable elements, such as uranium, to account for the fission-produced xenon. Furthermore, although the ratios of the xenon isotopes strongly resembled fission xenon, exact measurements of the fission xenon showed that the ratios did not exactly correspond to the ratios formed from any known fissionable isotope. This problem led to the hypothesis that the fission xenon was produced by *another* extinct radioactivity. The only likely candidate was an isotope of the transuranium element plutonium 244, with a half-life of 80 million years.

In 1971 this hypothesis was strikingly confirmed by Reynolds, who extracted extremely minute quantities of xenon generated by fission in a few milligrams of plutonium 244 that the Atomic Energy Commission had produced in a nuclear reactor. The isotopic composition of that xenon corresponded exactly to the mysterious fission xenon found in the meteorites. Not only were radioactive isotopes of "ordinary" elements such as iodine present in the solar nebula when the meteorite was formed; isotopes of transuranium elements were also present.

The discovery gives strong support to the hypothesis that many isotopes, both stable and radioactive, were formed in the explosion of supernovas. These violent events are the only events known to occur continuously in our Galaxy that give rise to conditions under which heavy elements can be synthesized. These measurements make clear that much of the matter we see—the paper of this book, the materials of the human body—was at one time deep in the interior of stars that exploded billions of years in the past.

THE AGE OF THE SOLAR SYSTEM

The stony meteorites were formed very early in the history of the solar system—within 100 million years of the time that the dust cloud began to collapse. If we could measure the age of the meteorites, then, we would have a good idea of the age of the solar system. Earlier chapters frequently mentioned, without justification, that the earth, sun, and solar system are 4.6 billion years old. Only in the past 15 years have we known this figure to be the approximate age of the solar system, and only in the last few years has it been known with any accuracy.

The measurement of the age of meteorites and thereby the solar system is an example of the contribution of nuclear physics toward our understanding of the universe. Relatively short-lived radioactive isotopes such as iodine 129 and plutonium 244 are used to measure the "formation interval" of the meteorites (the time between the beginning of the collapse of the dust cloud and the formation of the meteorites). In contrast, long-lived radio isotopes such as rubidium 87, potassium 40, uranium 238, and uranium 235 are used to measure the time that has elapsed since the meteorite was first formed.

The latter form of measurement has been made most accurately with the long-lived isotope, rubidium 87. This radioactive isotope has a half-life of 50 billion years, far greater than the age of the solar system and, for that matter, the "big bang" when the present epoch in the history of the universe began. Therefore, it is almost a stable isotope, and only a small fraction of the rubidium 87 ever formed has decayed into its stable daughter isotope of the element strontium. Although the quantity of strontium 87 produced by the decay of rubidium 87 is consequently very small, accurate techniques have been developed for measuring the slight increase in the abundance of strontium 87.

Strontium has three other stable isotopes: strontium 84, 86, and 88. Unlike strontium 87, these isotopes are not the daughter products of a long-lived parent and their abundance relative to one another does not change. Like other isotopes, they were produced together with stable strontium 87 and radioactive rubidium 87 in the interior of stars. They, too, were expelled into interstellar space by the explosion of supernovas and incorporated in the solar dust cloud. Because the dust cloud was well-mixed at the time the solar system first formed, the ratio of these isotopes to one another was uniform. In particular, the ratio of strontium 87 to strontium 86 had a certain value. Since that time, the amount of strontium 87 has slowly *increased* because of the radioactive transformation of rubidium 87 into that isotope. Because the abundance of the other strontium isotopes remains unchanged, this ratio has gradually increased over the entire history of the solar system. The increase is largest in bodies containing a relatively high proportion of rubidium; in bodies containing a low proportion of that element, the increase in this ratio is small, or even negligible.

Some meteorites contain very little rubidium and can be used to measure the ratio of strontium 87 to strontium 86 at the time the solar system was formed. The ratio turns out to be 0.6988. The chondrites, which are stony meteorites containing olivine or pyroxene, have relatively large amounts of rubidium; in these meteorites the decay of rubidium 87 has significantly increased the ratio. Frequently it is as high as 0.7600 and, in some cases, it is greater than 1.000. Laboratory measurements show that the increase in the ratio of strontium 87 to strontium 86 in stony meteorites is accurately proportional to their ratio of rubidium 87 to strontium 86 (Figure 6). Clearly all of the chondrites were formed at essentially the same time. On the basis of quantitative calculation of the data in Figure 6, together with knowledge of the decay rate of rubidium 87, that time was $4.55 \pm 0.10 \times 10^9$ years ago.

Another way to measure the age of a meteorite is to separate from it the individual minerals of which it is composed. Unlike a coarse-grained rock such as granite, which contains crystals large enough to be seen without a microscope, the individual mineral grains in a chondritic me-

Figure 6. Strontium 87/strontium 86 ratios are plotted against rubidium 87/strontium 86 ratios. Because the decay rate of rubidium 87 is known, the slope of the graph can be used to determine how long ago the chondrites were formed.

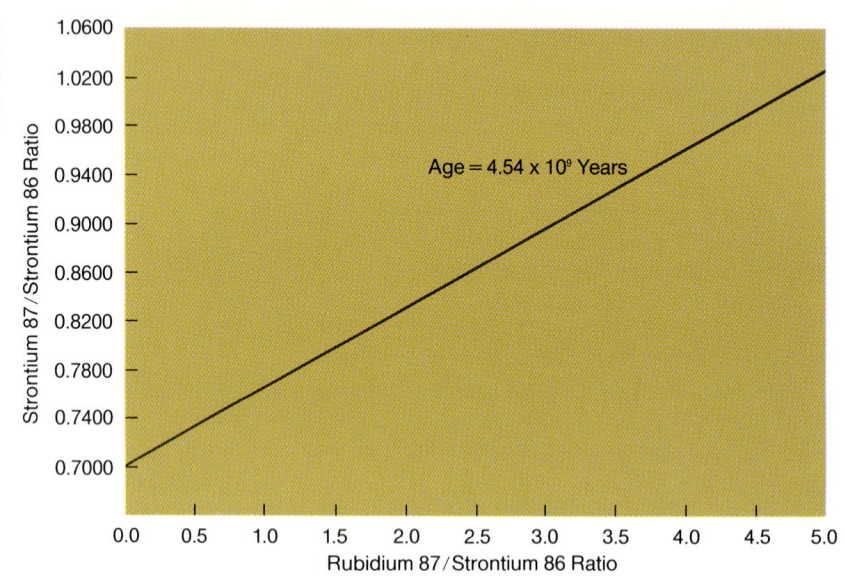

teorite are unusually small. Its component minerals—pyroxene, feldspar, olivine, and so on—have different ratios of rubidium 87 to strontium 86. Again, there is proportionality between this ratio and the ratio of strontium 87 to strontium 86. It is a remarkable fact that the "internal ages" of minerals separated from a single meteorite agree within experimental error with the "whole rock" ages of different meteorites. These results indicate that all of the chondrites were formed at nearly the same time, 4.55×10^9 years ago; and that their individual minerals formed at the same time and have not undergone any chemical alteration since then. These rocks are in the finest detail completely preserved relics of the solar system at the time of their formation.

One other thing. What if the meteorite Indarch, the mineral data for which is shown in Figure 7, contained a mineral having no trace of rubidium 87? If that were the case, excess strontium 87 could not have been generated in this mineral and its strontium 87/strontium 86 ratio would tell us what this ratio was at the time the meteorite formed. All of its minerals do contain at least some small amount of rubidium 87, so this simple measurement cannot be made. However, it isn't necessary to find such a mineral in order to find the answer.

Extrapolation of the line in Figure 7 down to zero value permits us to read off from the vertical scale exactly what this ratio would be if a rubidium-free mineral were present in the meteorite. The ratio in the meteorite when it was formed is 0.7005—not much higher than the 0.6998 found in meteorites containing almost no rubidium and having exactly undergone considerable chemical differentiation. But the difference is well outside experimental error. This means that the chondrites, otherwise primitive and undifferentiated meteorites, formed from the solar nebula at a time somewhat later than these more differentiated meteorites, during which time the strontium 87/strontium 86 ratio grew, due to the decay of rubidium 87, from the value of 0.6998 to 0.7005. The time required for this increase is about 100 million years. It's impossible to give an exact figure without the exact rubidium 87/strontium 86 ratio of the solar nebula for this time. But the important point is that these chondrites, which contained live iodine 129 and plutonium 244 at the time of their formation, did not appear long before the highly differentiated meteorites but were actually formed later.

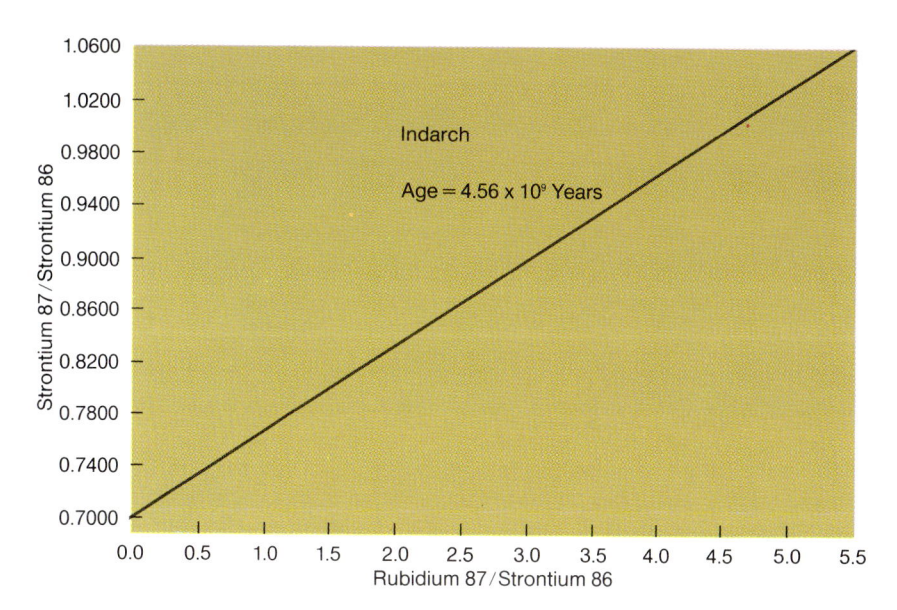

Figure 7. Indarch, an enstatite chondrite, is almost as old as the solar system. On the basis of the mineral data assembled in this graph, the age of this meteorite has been calculated to be 4.56 x 10⁹ years.

All the evidence based on the extinct radioactivities shows that the chondrites were formed very early in the history of the solar system and it applies with even greater force to differentiated bodies such as other meteorites and the earth's moon. The formation interval of the solar system, 4.6 billion years ago, was an extremely busy period. The planets, their satellites, and the parent bodies of the meteorites were not formed slowly in stages stretched out over hundreds of millions of years. It appears more likely that, once the formation of solid bodies began in the solar system, everything happened at once. Just how long it took — whether it was 10 million years, or 100 thousand years, or whatever — is a question that still has no clear answer. The closing section of this chapter presents the rather outrageous speculation that it was all over in a mere 1,000 years.

CHEMICAL COMPOSITION OF THE SOLAR NEBULA

All but 0.1 percent of the mass of the solar system is concentrated in the sun. If we could know the sun's chemical composition, we would know the composition of our solar system and its parent nebula; the remaining 0.1 percent would be negligible in the overall picture. But measuring the chemical composition of the sun isn't an easy thing to do.

The sun is a hot glowing body that emits light of nearly all wavelengths found in the visible spectrum as well as wavelengths far into the ultraviolet and infrared. But at certain wavelengths, solar radiation is greatly attenuated. As a result the solar spectrum contains dark lines, known as *Fraunhofer lines* (Figure 8). The significance of these lines is now understood. The atoms of various chemical elements, such as oxygen, calcium, and iron, have characteristic energies of vibration at which they strongly absorb light. The sharp absorption lines in the solar spectrum signal the presence of those atoms in the surface regions of the sun.

The characteristic wavelengths associated with each element can be measured in the laboratory, so it's possible to make a *qualitative* chemical analysis of the sun — at least for those elements with prominent absorption lines in the visible region of the electromagnetic spectrum. Now you might think that a strong absorption of a given spectral line indicates a

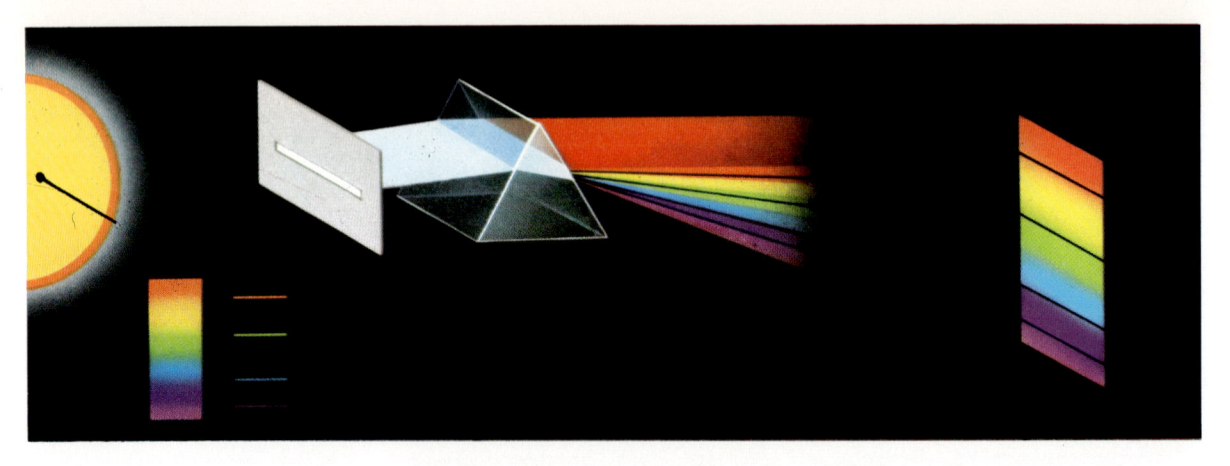

Figure 8. Fraunhofer lines are the dark absorption lines that appear during spectral analysis of low-pressure gases. They are named after J. von Fraunhofer, the first man to map the solar spectrum.

high abundance of a given element in the sun, and that the relative abundances of the elements could be determined easily. Unfortunately, the strength of the spectral absorption line depends on much more than the abundance of the chemical element producing it. The spectral line is a consequence of *the arrangement of electrons surrounding the nucleus of the atom.* And this arrangement varies greatly from one atom to another. So the strength of the spectral line depends not only on the concentration of the element, but on details of the structure of each individual type of atom—and on conditions such as pressure and temperature as well.

Spectral analysis is a familiar method of laboratory chemical analysis, although it is not free from difficulty. Artificial standards can be prepared in the laboratory by adding known amounts of the element under consideration to some material. The intensity of the spectral line in the material being analyzed can then be "bracketed" between the intensities of spectral lines of similar matter of known composition, and the concentration of the element in the material being analyzed can be determined by interpolation.

We cannot, of course, make up a number of "standard suns" of known composition and then compare the spectrum of the real sun to them. The relation between the abundance of an element in the real sun and the intensity of its spectral absorption line must be obtained entirely by physical theory. Even with this physical theory, however, the abundance of the chemical elements in the sun is only known imperfectly. For some elements, such as magnesium and calcium, the abundance is probably good to within a factor of about 2. For others, knowledge is quite sketchy. Ironically, the abundance of the element helium, first discovered in the sun, is one of the most poorly understood; the same thing is true for the most abundant element, hydrogen. For some of these elements, the abundance can be determined much more accurately in other stars and in gaseous nebulas. The fact that elements such as magnesium and calcium occur in those bodies as often as they do in the sun permits the plausible inference that the solar abundance of the more "difficult" elements, such as hydrogen, can be determined from their abundance in other stars and nebulas.

Subject to these qualifications, the solar abundance of the elements is given in Table 1, in which the element silicon has arbitrarily been assigned an abundance of 10^6. This table shows that most of the sun is composed of but a few elements. Hydrogen and helium are by far the most abundant. Lithium, beryllium, and boron are extremely rare, for if these elements were placed in the interior of even a relatively cool star they would be immediately transformed by thermonuclear reactions into

Table 1.
Comparison of Some Solar, Chondritic, and Crustal Atomic Abundances Based on Silicon = 10^6

Atomic Number	Element	Sun	Chondrites	Crust
1	Hydrogen	3×10^{10}	$<10^5$ (5.5×10^6)	1.4×10^5
2	Helium	(5×10^9)		
3	Lithium	0.29	50	290
4	Beryllium	7.2	0.6	31
5	Boron		6	93
6	Carbon	1.3×10^7	10^4 (8×10^5)	2,700
7	Nitrogen	3×10^6	800 (5×10^4)	330
8	Oxygen	2.9×10^7	3.5×10^6	2.9×10^6
10	Neon	(5×10^6)		
11	Sodium	6.3×10^4	4.38×10^4	1.2×10^5
12	Magnesium	8×10^5	9.1×10^5	8.8×10^4
13	Aluminum	5×10^4	9.5×10^4	3.0×10^5
14	Silicon	1×10^6	1×10^6	1×10^6
19	Potassium	1,600	3,160	6.4×10^4
20	Calcium	4.5×10^4	4.9×10^4	9.2×10^4
26	Iron	8×10^5	6×10^5	9.0×10^4

Note: Carbonaceous meteorite abundances are given in parentheses.

helium. Next there is a group of relatively abundant elements: carbon, nitrogen, oxygen, and neon. On the scale of silicon = 10^6, there are approximately 10^6 to 10^7 atoms of these elements present in the sun. Next in abundance are a group of elements that form solid oxides at temperatures less than several hundreds of degrees Centigrade: magnesium, silicon, iron, aluminum, sodium, nickel, and calcium, present in abundances of 10^5 to 10^6 atoms per 10^6 atoms of silicon. With the single exception of sulfur, all the remaining elements are much less abundant. Therefore, even though there are more than 80 elements possessing stable isotopes, all but about 15 are relatively rare.

This distribution is even more pronounced when solar abundances are compared to the abundance of the elements in the most common class of meteorites, the chondrites. The chondrites, being solid rocks, are nearly devoid of gaseous elements such as hydrogen, helium, and neon; they contain very little of the elements carbon, nitrogen, and oxygen because the most common compounds of these elements are gaseous, also.

With the exception of the more volatile elements, however, it turns out there is striking agreement between the abundance of the elements in the chondrites and in the sun (Figure 9). The agreement is even closer with a particular class of chondrites, the so-called type 1 carbonaceous chondrites, which show less extreme depletion of volatile elements.

The agreement between solar and chondritic abundances is surely no accident. Rather, it indicates the chondritic meteorites are good samples of the material of the solar system; they are depleted only in volatile elements. They must have originated in bodies that have undergone little chemical processing except for the loss of gaseous elements. The best candidates for the source of these meteorites are the small bodies found in the solar system—the comets and asteroids. The type 1 carbonaceous chondrites are probably derived from the asteroids—the minor planets moving in orbits in the vast region of interplanetary space between the orbit of Mars and the orbit of Jupiter.

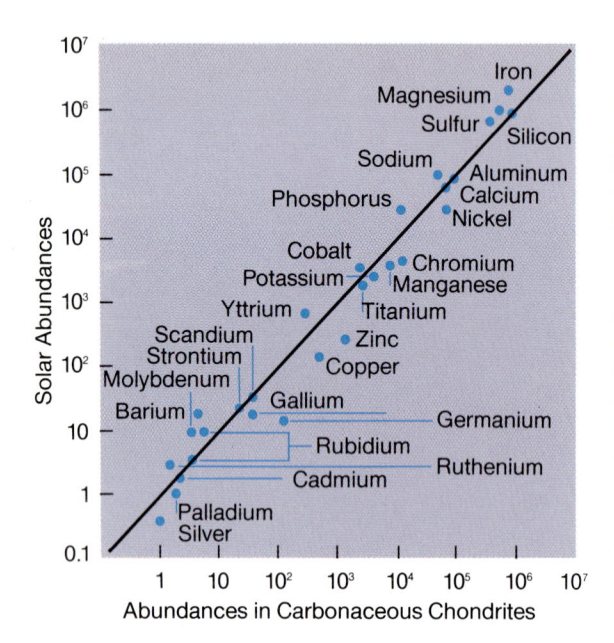

Figure 9. Why do we believe that chondritic abundances have anything to do with the average nonvolatile composition of the solar system? The reason is found in this amazing graph, which shows the one-to-one ratio of the nonvolatile abundances (all normalized to silicon = 10^6 atoms) in the sun and chondrites.

The striking agreement between solar and chondritic abundances for the more abundant nonvolatile elements makes it seem plausible that the rarer elements, such as rubidium and strontium, also have similar abundances in the chondrites and in the sun. The chondrites form the basis for our knowledge of the solar system abundance of many elements too rare to be measured accurately in the solar spectrum, but which can be accurately measured in the chondrites.

It should be emphasized that using meteorite data to learn the average composition of the solar system is *not* based on a frequently expressed belief that meteorites were at one time parts of an exploded planet as large as the earth. Many lines of evidence indicate that the meteorites were derived from small objects, probably comets and asteroids, which are quite unlike large planetary bodies such as the earth and the moon in their chemical history. The *only* reason for believing that chondritic abundances have anything to do with the average nonvolatile composition of the solar system is the remarkable agreement between the abundances of the major nonvolatile elements in the sun and chondrites.

EVIDENCE FROM THE MOON

Before the recent intensive study of the moon, most of our information regarding conditions in the early solar system was based on meteorite studies. The only planet we could study in detail was the earth, and the earth is so geologically dynamic that essentially all the record of its early history has been destroyed by volcanism, weathering, erosion, and similar geologic processes.

In comparison with the earth, the moon is a relatively inactive planetary body. Although we now know it is not completely "dead," the rate at which geologic processes occur on the moon is so much slower than on the earth that its early history has been only partially erased. Consequently, even though detailed study of the moon is only a few years old, we have already learned much more about the early history of the solar system — and the earth — than was possible from study of the earth alone.

Some of this evidence is presented in this section. Much more could be added, because the lunar record is currently being unraveled and new discoveries constantly require revision of conclusions made only a few months earlier. But this discussion is more in the nature of a "snapshot" of our current thinking.

Earlier, an explanation was given of the way the radioactive decay of rubidium 87 into stable strontium 87 in meteorites was used to measure the age of the meteorites and the solar system. In much the same way, rubidium-strontium measurements made on some of the rocks returned from the moon by the Apollo astronauts show that rock formation took place on the moon at this time — about 4.55×10^9 years ago, during the formation interval of the solar system.

These measurements show that at that time, the surface regions of the moon were greatly enriched in certain elements — for example, uranium, rubidium, potassium, barium, and the rare earths. In order for the surface regions to be enriched in these elements, deeper regions had to be depleted of the same elements. A quantitative extension of this qualitative line of reasoning leads to the conclusion that at least the outermost 200 kilometers of the moon melted during the formation of the solar system. Because the radius of the moon is 1,740 kilometers, this depth may seem shallow. But remember that the volume of a sphere in-

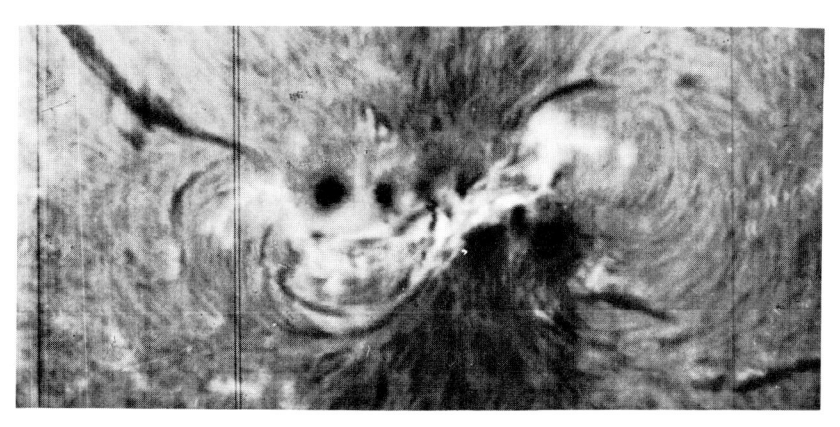

Figure 10. Periodically, the sun sends out charged particles into the solar system. Solar flares appear without warning and may last only a few minutes. This sunspot and flare were photographed in July 1946.

creases with the *cube* of its radius, and the outermost 200 kilometers represents about one-third the volume and mass of the entire moon. What was the source of energy that melted such a large mass of rock so quickly?

Before examining this question, consider some new evidence concerning the moon. The second manned lunar landing, Apollo 12, left on the lunar surface a sensitive device, known as a *magnetometer*, which measures the magnetic field at the surface of the moon. Unlike the earth, the moon has no strong, permanent magnetic field, so a compass would be of no value in lunar navigation. From time to time, however, the sun emits large quantities of charged particles called *solar flares*, which stream out into the solar system. The magnetic field accompanying these particles induces electric currents in the moon, which also produces a magnetic field in response to it. The lunar magnetometer measures both the direct magnetic field from the sun and the magnetic field induced in the moon. The way in which this total magnetic field changes with time forms the basis of calculations of electric conductivity deep within the interior of the moon.

The electric conductivity turns out to be surprisingly low. Unlike metals, the electric conductivity of rocky material increases with increasing temperature. Therefore the low electric conductivity of the deep lunar interior suggests that the present temperature of the central regions of the moon is relatively cool, about 800 degrees Centigrade, which is far below the melting point of rocks. The moon is sufficiently large that its central regions could not have cooled very much during the entire 4.6-billion-year history of the solar system. If the deeper regions of the moon had been melted, they would still be near the melting temperature today instead of at the much lower actual temperature. These data suggest that below a depth of about 300 kilometers, or half of its mass, the moon was never melted.

What about the region between a depth of 200 kilometers and 300 kilometers? Even if this region was not initially melted, the material just above it was melted so it must have become quite hot. Furthermore, according to the evidence presented above, this material as well as the material just below it was lunar material that was unmelted and had not yet been chemically fractionated. The material still contained radioactive elements that are important sources of heat over long periods of time —particularly potassium, uranium, and thorium. Following this line of reasoning we are led to a thermal model for the evolution of the moon. Slowly this region and the region just above it, which had been initially melted but later solidified, was heated to the melting point. Hundreds of millions of years after the formation of the moon, a second melting oc-

Figure 11. (Left) Astronaut Chárles Duke, Jr., pilot of the Apollo 16 lunar landing mission, stands beside the lunar roving vehicle at the Descartes landing site.

Figure 12. (Right) Astronauts collect samples of lunar rock for study and analysis in laboratories on earth. Here Charles Duke stands on the rim of a crater, which is 40 meters in diameter and 10 meters deep.

curred, creating basaltic rocks that spilled on to the lunar surface and caused the dark *mare* regions you can see even without a telescope. The basaltic rocks have been dated by the rubidium-strontium method at 3.6 and 3.3 billion years; some evidence indicates other mare basalts are 3.9 billion years old. All ages in between are probably represented by mare basalts, too, but sampling of the lunar rocks is not complete enough for that determination.

Other lunar workers are convinced that not only was the surface of the moon melted, but its deep interior was also melted. This belief is based on the fact that all of the rocks brought back from the moon are found to be magnetized. The most ready explanation for lunar magnetism is that at the time the rocks were formed the moon had a magnetic field similar to that of the earth's, but much weaker. The earth's magnetic field is produced by fluid motions in its molten iron core. Researchers believe that the moon once had a small fluid core, also, the motions of which acted as a dynamo and produced a magnetic field. We know the moon has no overall magnetic field like that of the earth at the present time. Therefore, if this hypothesis is correct, the lunar dynamo "turned off" some time in the last 3 billion years, probably because the fluid motions in the core slowed down. The question of whether the deep interior of the moon was ever melted or was always cold is hotly debated by lunar scientists. There are advocates of a hot moon, a cold moon, and a lukewarm moon. One of the latter school has referred to the moon as a "half-baked planet." Resolution of this disagreement is one of the major present problems in lunar science.

In any case, the chemical fractionation and the formation of igneous rocks on the moon was not confined to the initial instant of formation but persisted, probably at a diminishing rate, for at least the next 1.3 billion years. It is even possible that some lunar volcanism is still going on today. If so, it must be slight compared to that on the earth: if there were a Mount Etna or Kilauea on the moon it would have been recognized long ago. But there is some evidence for feeble geologic activity in the in-

Figure 13. The moon's craters are stark features of the lunar landscape. Early in lunar history, meteoritic impact formed the dark regions, or *maria*; later the impact basins filled with lava. Smaller meteorites formed the bright ray craters (left and middle right) later on.

terior of the moon. Very weak "moonquakes" are regularly measured by the seismic instruments the astronauts left on the moon. The moonquakes appear to be associated with brief increases in the water vapor content of the almost nonexistent lunar atmosphere. If this evidence is correct, the lunar interior still contains small amounts of water that are vented to the surface during moonquakes.

What kind of a lunar history must be postulated to fit all this evidence into a coherent scheme? The long-lived radioactivity provides a convenient explanation for the mare volcanism 3.6 to 3.3 billion years ago. But what explains the fact that the outer one-third of the moon melted during the formation of the solar system?

There is a process capable of producing a planet that is immediately hot near the surface. This process is called *gravitational heating*. Imagine a small planet that grows by sweeping up smaller bodies. The gravitational field of the small planet will be very weak, so the smaller bodies will not be accelerated much as they are swept up, and they will strike the surface at low velocity and produce negligible heating. As the planet grows larger and larger, its gravitational field will increase, and bodies will fall onto it at higher velocity. Upon striking the surface, their kinetic energy will be converted into heat. Eventually, when the planet has grown large enough, the kinetic energy of gravitational accretion will be sufficient to melt the incoming material, and beyond a certain radius the planet will be melted.

The actual process is more complicated than this simple picture. There is a competing mechanism that prevents the surface from heating up. A hot body radiates its energy into space, and if the planet grows slowly enough, its surface will be unable to heat up. The kinetic energy

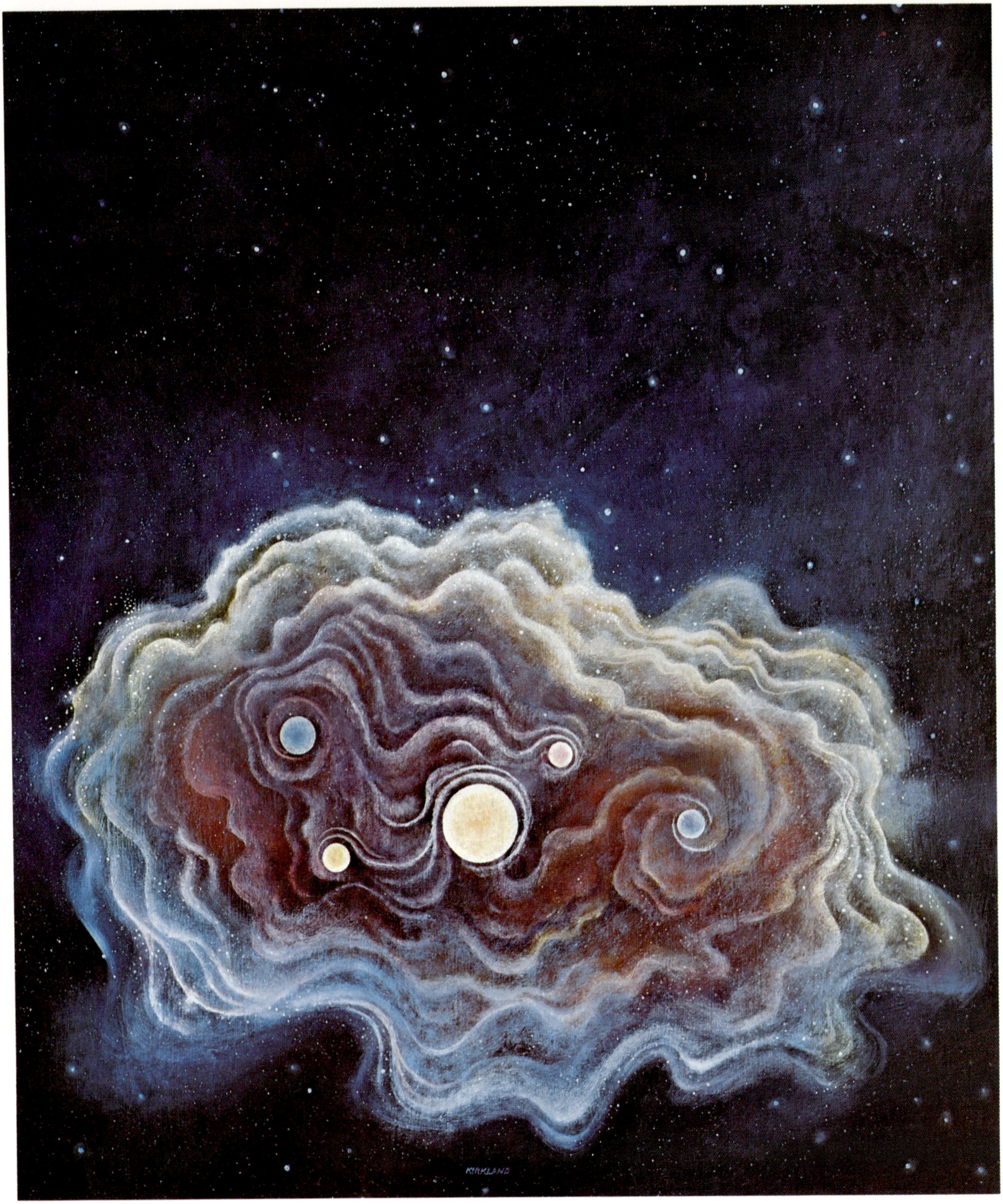
KIRKLAND

of the impacting body will be converted into heat, which will radiate into space without significantly heating the planetary surface. Only if the planet grows rapidly enough will it be possible for this "gravitational accretion" to cause a melted region near the surface.

How rapidly would the moon have to grow in order for melting to occur in the outermost 200 kilometers? The answer, which comes out of a detailed calculation of the physical theory involved, is 1,000 years. The moon must have grown on this extremely rapid time scale in order to explain what we have observed. Perhaps by stretching out various steps in the calculations the time scale for lunar formation could be extended to 10,000 years—but not much longer.

Now 1,000 or even 10,000 years is not very much time to build a moon. Consider what is required. As the solar nebula contracted, its rotation caused it to flatten out into a disk, and the gravitational energy of its contraction heated it to such a high temperature that at the distance of the earth and moon, all elements were gaseous. As contraction continued, most of the matter of the solar system went into the formation of the sun, leaving as a residue at least enough matter to form the planets.

According to older ideas about the formation of the planets, this residual matter cooled and various chemical compounds began to condense from it in the form of very fine crystals. Among the first crystals to form were those of metallic iron and magnesium silicates; at lower temperatures crystals of more volatile compounds such as sulfur compounds could form. Eventually, at distances well beyond the orbit of the earth, solid crystals of water, carbon dioxide, and ammonia would condense.

But it would take much more to form planets of the present size. Individual crystals would have to coagulate to form small bodies, about 1 meter in diameter, which would then collide with one another and grow into even larger bodies, eventually reaching the size of the planets observed today. The problem is that these small bodies would have a negligible gravitational field; upon colliding with one another they would not stick together but merely bounce off one another. In fact, at collision velocities of about 1 kilometer per second, which were likely to be present at this stage in the history of the solar system, the planetesimals would break each other apart when they collided and would become smaller, not larger. Older theories of the formation of the solar system never really got around this problem in a very satisfactory way. Clearly nature was somehow more clever in a way we have yet to figure out, otherwise the moon, the earth, and the reader would not be here.

Regardless of how this difficulty was overcome, it's clear that it was in fact overcome, and it's possible to proceed from that point and calculate the time scale for forming bodies such as the earth and moon. Some idea of this time scale may be obtained in the following way.

Suppose you look in on the formation process when the earth had grown to one-half its present radius. By that time it would have a significant gravitational field, and bodies colliding with it would at least stick to it. How long would it take for the earth to grow the rest of the way to its present size?

Imagine that the remaining matter of the earth was spread out in a ring and the "embryonic earth" was sweeping up the remaining matter as it orbited the sun. Suppose, further, that the residual matter of the solar nebula was spread out more or less uniformly in the inner solar system, and that the particles in this region were moving in orbits that could intersect the earth's orbit. This means they would be confined to a ring having a width of about one-tenth the distance of the earth from the sun, and a similar thickness. The distance of the earth from the sun is 1.5×10^{13}

Figure 14. With eye witnesses being as rare as they are, any rendition of the formation of our solar system is conjectural. But the event itself is fascinating and invites speculation; this is a graphic interpretation of the present theory that our solar system condensed out of an enormous galactic dust cloud that was probably a remnant of a nova explosion.

centimeters, so the width and thickness of this ring would be 1.5×10^{12} centimeters. Consequently, the earth would be sweeping out a volume equal to the circumference of its orbit (about 10^{14} centimeters) multiplied by the width and height of the band swept out, or $10^{14} \times 1.5 \times 10^{12} \times 1.5 \times 10^{12}$, or about 2×10^{38} cubic centimeters per year. The remaining portion or the earth's mass (about 5×10^{27} grams) would occupy this volume, causing the density (mass per unit volume) to be $5 \times 10^{27}/2 \times 10^{38} = 2.5 \times 10^{-11}$ gram per cubic centimeter. The cross-sectional area of the earth would be $\pi(R/2)^2$, where R is the present radius of the earth. The relative velocity of the earth and the matter it would be sweeping up will be called u. A reasonable value for u is 1 kilometer per second. The total mass swept up per year would be $\pi(R/2)^2 \times u \times$ density. Substituting in the radius of the earth (about 6.4×10^8 centimeters), $u = 1$ kilometer/second $= 3.15 \times 10^{12}$ centimeter per year, and density $= 2.5 \times 10^{-12}$ gram per cubic centimeter, as calculated above, you get the mass swept up per year as

Mass $= \pi(3.2^2 \times 10^{16}) \times 3.15 \times 10^{12} \times 2.5 \times 10^{-11}$

or about 2.5×10^{19} grams per year. Now that's certainly a lot of mass—25 trillion tons per year—but even at that rate it would take a long time to finish building the earth. At the rate of 2.5×10^{19} grams per year, 200 million years would be needed to sweep up the remaining matter to finish building the earth. Actually, it would not take *that* much time, because the earth's radius would increase as it grew larger and the cross-sectional area sweeping up the remaining mass would become greater. Even more importantly, as its gravitational field increased along with its radius and mass, it would "draw in" particles it would otherwise miss. An opposing effect would be the decrease in density of dust during the "sweeping-up" process. When these effects are put into the calculation, the time required can be reduced to about 10^7 years. In this theory, the time scale for the formation of the other bodies of the inner solar system—for example, the moon and Mars—would be similar. This timespan is much longer than the 10^3–10^4 years inferred from lunar studies. Furthermore, this calculation completely ignored the time required to get the earth to one-half the size of its present radius.

Clearly there is a gross discrepancy between the time scale of this older model of the formation of the solar system and the time scale indicated by recent lunar studies. At the present time, we are forced to seek alternatives to these simple models for the formation of the planets.

Recently, A. Cameron has provided at least some semi-quantitative ideas of how a new theory for the formation of the solar system may come about. According to Cameron's calculations, the gas and dust of the solar nebula dissipated into interstellar space within about 1,000 years after the formation of the sun. The giant planets Jupiter and Saturn, which are composed principally of hydrogen and helium, must have formed from the residual gas of the solar nebula so they must have formed within about 1,000 years. Rough calculations show that "gas giants" of this kind may form that rapidly, although the processes by which they form could be quite different from those described previously for the earth and moon. Once these giant planets were formed, violent processes must have occurred in the solar system—particularly within the orbit of Jupiter, which is about 5 times the distance of the earth from the sun, beyond Mars and the belt of asteroids. Any planetesimals "left over" from the formation of the giant planets would be subject to strong perturbations by Jupiter and would be hurled into high-velocity orbits, many of which would pass through the asteroid belt and inner solar system.

Their effect would be to smash up any planetesimals in this region that failed to grow to a radius perhaps as large as 100 kilometers.

Under such circumstances, the picture of the inner solar system would be greatly different from the one described earlier. It would no longer consist of a few lonely, embryonic planets moving through a sparse medium of dust.

Instead, great quantities of dust would be originating throughout the entire region of the inner solar system and would be carried inward along with the gas from which the sun is still forming. In this much denser medium, containing both gas and dust, relative velocities would be lower and planetesimals and embryonic planets could grow for a while. But their growth would almost always be cut short by a high-velocity collision with a body from the vicinity of Jupiter. This growth and destruction would compete rapidly. The few embryos that managed to escape destruction would be able to grow rapidly from the copious quantities of dust available, and they would have greater resistance to destruction because of their large size. Ultimately they could sweep up enough of the dust coming from the asteroid belt to form the inner planets. It is possible that, in this way, the earth and moon formed in 1,000 years.

This new picture for the rapid formation of the solar system is admittedly sketchy, and much work remains to be done before it may be regarded as satisfactory. A great number of complex problems in physics and chemistry must be solved before anything approaching a satisfactory model is achieved. Modern exploration of space has shown us the direction our thinking should take, and now we must get to work.

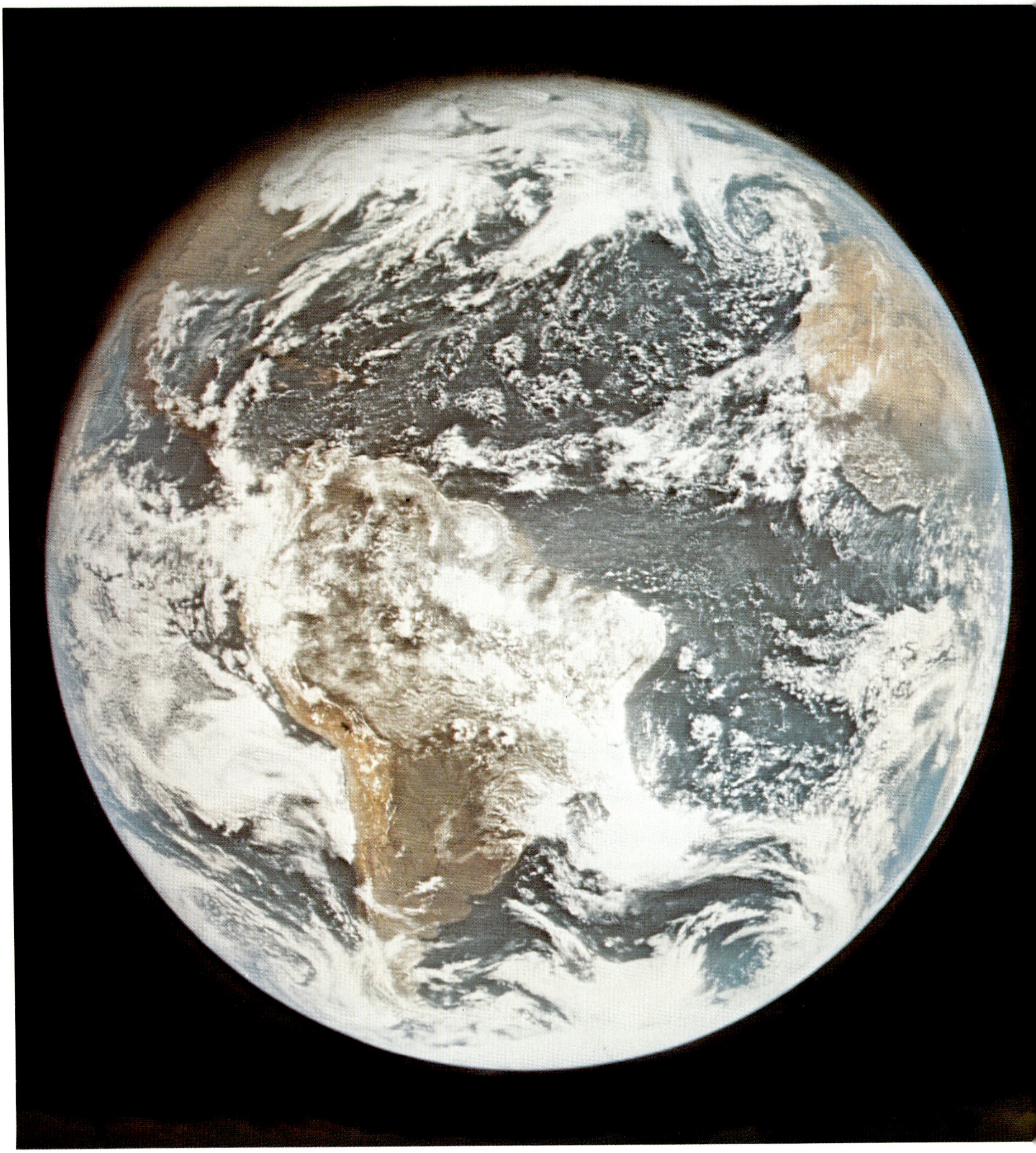

The earth is the third planet of the solar system, that collection of planets orbiting a quite ordinary star we call our sun. The sun, in its turn, is only one of billions of stars making up one galaxy in the collection of millions of galaxies in the universe. In comparison with the elusive phenomena beckoning to us from deep space, the earth often seems a little too familiar, a little too well understood, to kindle our imagination. We merely accept that it is the place where we were born and will live out our lives, and try to make the best of it.

And yet, what do we really know about our planet? That it's old; that it has an atmosphere and continents and oceans; that dinosaurs lived here once — perhaps those are the sorts of answers that frequently spring to mind, followed by the thought that in any case geologists know all about it. It is true that the centuries-old science of geology has solved many problems relating to the natural processes at work on the earth's surface, and in so doing it has unraveled much of the earth's history. And the sciences known as mineralogy and petrology have given many sound ideas about the nature of the earth's rocky material and the way various rocks came into being. So perhaps it may come as a surprise that truly fundamental questions about the whole earth system have, until now, defied solution.

The formulation of those questions started a few centuries ago, when geology first emerged as a scientific discipline. People began to wonder about the sequence of ancient environments, locked in the accumulation of sedimentary rocks. They methodically examined the warped and broken ribbons of earth materials — testimony to the titanic forces that gave rise to mountains. They studied igneous rocks cast out from the molten masses deep within the earth — remnants of the earth's volcanic past. Over time, their studies began to portray not the eternal earth that man and myth had created but a planet born in violence.

In the last two decades their observations started converging into a unified picture. As a result, we are now enjoying one of those rare moments in the history of science when the work of seemingly diverse fields comes together to fill in many seemingly unrelated shelves in our store of knowledge. Armed with the earlier views of field geologists, mineralogists, petrologists, and paleontologists, the earth scientist is beginning to realize the significance of the evidence being gathered by astronomers and planetary scientists in terms of our own planetary system. The conclusions now coming to his mind concern the origin of the third planet and the dynamic processes involved. And viable theories on the age of the continents and ocean basins, the structure and composition of the earth's interior — even the awesome mechanisms that warp the earth's crust — are at last within reach.

THE EARLY EARTH

Following its accretion during the violent events that took place in the origin of our solar system, the earth clearly underwent a process of evolution. Between 4.6 billion and 3.8 billion years ago, the earth's rocky crust began to accumulate. During this period, volatile substances that had been trapped beneath the early crust were driven out, or *degassed,* from the interior. This process — which is still going on, incidentally —

13
The Third Planet

The planet earth, photographed 35,700 kilometers above its surface.

Figure 2. (Far right) Geologic time scale. Before the advent of radiometric (or absolute) dating, geologists developed a relative time scale based on major units of sedimentary rocks and the distinct fossils associated with them. These stratigraphic units became incorporated into geologic systems which were often named after the area in which the fossils were found. These systems are the basic time-stratigraphic units of historical geology. They have since been correlated with absolute time scales by various methods of radioactive isotope decay.

The atmospheric composition during the pre-Cambrian differed drastically from what it is today. And yet, had it been otherwise, life as it now exists could not have evolved. Before 600 million years ago little oxygen was present in the atmosphere. It was only with the appearance of oxygen-producing plants (about 2 or 3 billion years ago) that the atmospheric composition began to change, thereby promoting diversification and radiation of life forms throughout the oceans and, later, the land. In this figure the time scale is distorted in that the entire pre-Cambrian period lasted 4 billion years.

From *Adventures in Earth History* by Preston Cloud.
W. H. Freeman Co. Copyright 1970

led to the formation of the proto-atmosphere and the proto-ocean. About 3.6 billion years ago the earth's surface began taking on the appearance of the planet we live on today; even its internal composition was about the same.

But it was a world of rock and water, enveloped by a totally different atmosphere. Hydrogen, oxygen, carbon, and nitrogen almost certainly were present, perhaps in a mixture of methane, ammonia, and water molecules. As concentrations of simple molecules built up in the atmosphere and oceans, more complex molecules became possible. What was the source of energy for the chemical reactions needed to produce them? Sunlight, perhaps—there was no barrier to ultraviolet radiation at that time—or perhaps lightning or even shock waves produced by falling meteors.

But after most of the hydrogen had drifted off, oxygen and ozone (O_3) molecules began to form in the upper atmosphere, where incoming ultraviolet radiation from the sun broke down the molecular bonds between the atoms of water vapor. It is possible that the appearance of the oxygen-ozone layer in the upper atmosphere contributed to the conditions necessary for the development of life on earth. That layer would provide an effective barrier against incoming solar radiation of longer wavelengths, which has a tendency to decompose complex organic molecules.

Rocks as old as perhaps 3.4 billion years have been found that contain organic molecules; rocks formed about 3.1 billion years ago contain what appears to have been bacteria and simple algae. The appearance of bacteria and algae represents a profound turning point in the history of the earth. Some scientists speculate that through energy transformation processes, particularly photosynthesis, these organisms totally changed the surface chemistry of the earth and laid the foundations for the eventual appearance of life as we know it.

By the beginning of the Cambrian Period of the Paleozoic Era, some 650 million years ago, the oceans were teeming with life. Within 250 million years, some forms of plants and animals had left the seas for terrestrial niches. Amphibians evolved, then reptiles, then mammals. Dinosaurs appeared and, mysteriously, disappeared from view. Most of the plant and animal species existing today developed from ancestral forms that appeared during the last 60 million years—a little more than 1 percent of the entire history of the earth. It was only a few million years ago that the hominid line evolved, eventually leading to modern man.

During this span of time, the evolution of life surely was linked to the processes shaping the physical appearance of the earth itself. Continents were inundated again and again by the seas; molten masses from the earth's interior were erupted on the surface or injected within the crust during times of ancient mountain building. The crust shifted restlessly.

What were the physical mechanisms underlying and paralleling the evolution of life? Until a few decades ago they were not well understood. Today, however, direct observation and experiment as well as indirect inference from other geophysical studies are providing new evidence. This chapter and the ones immediately following describe the nature of the evidence and present models of the earth and its components. In that descriptive context, it will be possible to review, finally, the processes that brought it all into being.

THE EARTH TODAY

The earth is one of the inner planets of the solar system. Three-fourths of its surface is covered by water and it still is enveloped in a gaseous

Geologic Succession					Approx Age × 10⁶ years	Age	Atmospheric Evolution	Radiometric Dating
Eon	Era	System-Period || Series-Epoch					
Phanerozoic	Cenozoic	Quaternary		Recent		Man	O₂ / N₂	Carbon-14
^	^	^		Pleistocene	3	^	^	^
^	^	Neogene		Pliocene	^	^	^	^
^	^	^		Miocene	22	^	^	^
^	^	Paleogene		Oligocene	^	Mammals	^	^
^	^	^		Eocene	^	^	^	^
^	^	^		Paleocene	62	^	^	^
^	Mesozoic	Cretaceous			130	Reptiles	^	Lead 206 / Lead 207 / Strontium / Argon
^	^	Jurassic			180	^	^	^
^	^	Triassic			230	^	^	^
^	Paleozoic	Permian			280	Amphibians	^	^
^	^	Carbon-iferous	Pennsylvanian		^	^	^	^
^	^	^	Mississippian		340	^	^	^
^	^	Devonian			400	Fishes	^	^
^	^	Silurian			450	^	^	^
^	^	Ordovician			500	Marine Invertebrate	^	^
^	^	Cambrian			570	^	^	^
^	^	Ediacarian			640	Multi and Uni-cellular Organisms	^	Uranium 238 / Uranium 235 / Rubidium / Potassium
Precambrian	Proterozoic	Upper			950	^	^	^
^	^	^			1350	^	^	^
^	^	Middle			1650	^	^	^
^	^	Lower			1800	Anaerobic Bacteria	CO₂, CO, H₂, HCl, S, SO₂, H₂O, CH₄, Ar, NH₃ etc. Primordial Atmosphere	^
^	^	Archean			2600	^	^	^
^	^	No Record			3600	^	^	^
^	^	^			4700	^	^	^

Figure 3. An old engraving tells the tragic story of collapsing buildings, fires, tsunami, and the engulfment of the marble quay that accompanied the Lisbon earthquake of 1755. At that time, the earthquake was variously attributed to the Devil, God's anger with his worshipers in the Lisbon churches, and the need to remind humanity of the flames of hellfire within the earth. But a few philosophers held different views. Jean Jacques Rousseau, for instance, regarded earthquakes as natural occurrences and pointed out that man cannot expect to prevent an earthquake by building cities and churches where one is likely to occur.

Figure 4. (Far right) Seismic waves give geophysicists information about the earth's interior. In (a), an earthquake caused by a sudden movement in the crust sends out shock waves. The *primary waves* (P) vibrate in the direction of propagation. The *secondary waves* (S, broken lines) travel only 60 percent as fast and vibrate from side to side. *Longitudinal waves* (L) travel around the crust in a belt. The shaded area shows the *shadow zone,* where only L waves are transmitted. Because internal waves are not transmitted through the shadow zone, geophysicists conclude that the outer core must be liquid. In (b), if seismic waves reach at least three widely scattered recording stations (in this case, California, Australia, and Japan), then the epicenter of the earthquake can be determined precisely (here, in the Aleutian Islands of Alaska).

atmosphere. Compared with Jupiter, which has a radius of about 68,800 kilometers, the earth has an average radius of about 6,400 kilometers. Its mass has been determined to be 5.98×10^{27} grams; other measurements show its volume to be 1.04×10^{21} cubic meters.

Once the total mass and the total volume of any object are known, then its density may be calculated. The earth's density turns out to be 5.5 times that of water; in other words, its specific gravity is 5.5 . The common rocks comprising the earth's surface, however, have specific gravities ranging from only about 2.5 to about 3.0 . Clearly, whatever makes up the interior of the earth must be denser than the rocks on the surface.

We know about the internal structure of our planet through geophysical evidence. For example, the paths of man-made satellites now orbiting the earth are perturbed slightly by variations in the earth's gravity field, and these perturbations may be analyzed to give an accurate measure of the earth's dimensions.

Other indirect evidence about the earth's structure comes from *seismology,* the branch of geophysics that studies earthquakes and the shock waves associated with them. Earthquakes are the characteristic signature of titanic geologic processes constantly at work in the earth. When giant blocks of the earth's crust shift suddenly along fracture surfaces (or *faults*), the abrupt movement triggers a group of elastic shock waves that radiate outward in all directions.

These seismic waves are of two basic types. *Primary waves* are longitudinal compressional waves, which means the transmitting medium is forced to move back and forth in the same direction as the wave propagation. Longitudinal compressional waves can be transmitted through solid, liquid, or gaseous matter.

Secondary waves are propagated only through solids. They travel in a transverse, or crosswise, motion; in other words, they force the transmitting medium to vibrate at right angles to the direction of wave propagation. One type of transverse wave is the "long wave," which is restricted to propagation around the earth's surface.

All seismic waves radiate outward from the source of the disturbance, or the *focus* of the earthquake. But only long waves are restricted to the earth's surface: primary waves and all other forms of secondary waves can travel through the earth along paths such as those depicted in Figure 4.

Most of what we know about the structure of the earth's interior comes from the study of the paths and velocities of primary and secondary waves. The possible routes of seismic waves from one earthquake focus are shown in Figure 4. As the waves travel into the earth, their paths curve—just as light waves bend, or refract, when they pass from air into water or a glass lens. As the waves travel deeper into the earth, their velocity increases. The velocity of the waves at various depths is determined by the elastic properties of the rock they pass through; waves seem to travel faster through more rigid materials. Although all rocks would be considered rigid when compared to, say, rubber, some are more rigid than others. We know that the density of the earth increases with depth. If only the density of the rocks increased, the velocity of the waves should *decrease*. Because the velocity *increases* instead, the rigidity of the rocks must increase even more rapidly than their density.

The waves generated by earthquakes are monitored throughout the world with delicate instruments called *seismographs.* If several seismic recording stations pick up tremors, the location and time of occurrence of the disturbance can be determined (see Figures 4 and 5).

The behavior of seismic waves as they travel through the earth is summarized in the velocity-versus-depth curve in Figure 6. You may

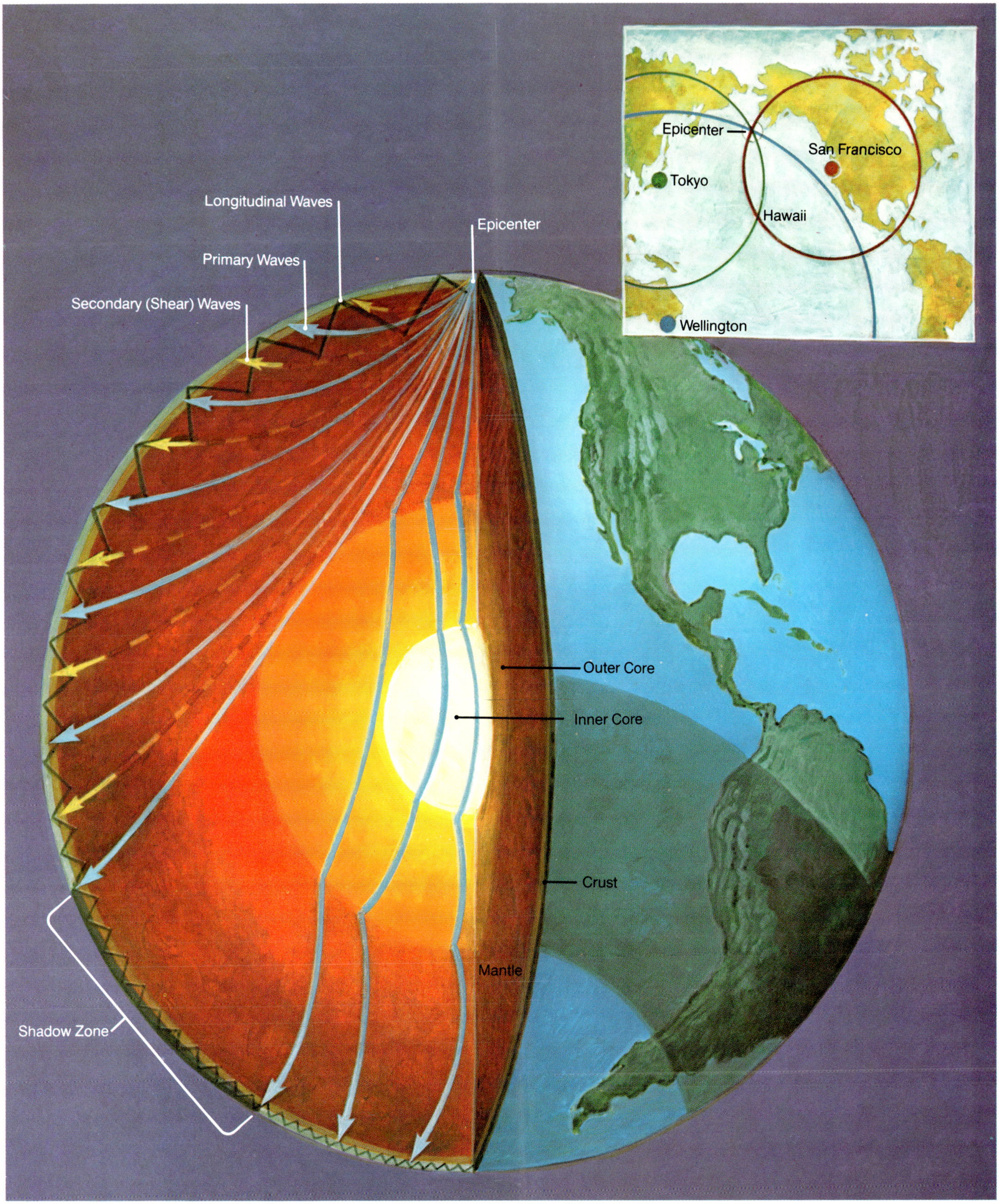

Epicenter
San Francisco
Tokyo
Hawaii
Wellington
Longitudinal Waves
Epicenter
Primary Waves
Secondary (Shear) Waves
Outer Core
Inner Core
Crust
Mantle
Shadow Zone

draw several conclusions about the interior of the earth on the basis of this curve. First, notice the abrupt increase in the velocities of both primary and secondary waves at a depth of about 30 kilometers. This boundary evidently marks a transition from materials of relatively low density into materials of relatively high density. It is known as the *Mohorovičić discontinuity* in honor of the man who discovered it. Below the Mohorovičić discontinuity, seismic velocities of both primary and secondary waves rise steadily. At a depth of about 3,000 kilometers, the velocity of primary waves drops abruptly and secondary waves do not continue at all.

Now there is one property of secondary waves that has not been mentioned before but is important to the discussion. Because of their crosswise motion, *secondary waves cannot be sustained in liquids;* only primary waves can. Below depths of 3,000 kilometers, then, the inner earth must be liquid.

On the other hand, after the abrupt decrease in velocity at a depth of 3,000 kilometers, primary wave velocities increase gradually to depths of about 5,000 kilometers, and then they seem to rise sharply (refer to Figure 6). It may be that the innermost part of the earth becomes solid again; we don't really know because secondary waves are damped out well before they can reach this region. The dimensions of the liquid portion of the core may be determined quite accurately, however, because the core casts a secondary wave *shadow* (see Figure 4). The liquid core effectively prevents secondary waves from reaching seismographic stations located within a certain region on the opposite side of the globe.

In summary, physical evidence alone indicates the earth has a structure that is concentric about its center. The deep interior, called the *core*, has a specific gravity ranging between 10 and 11. At least the outer 2,000 kilometers of the core is liquid. Encircling the core is the *mantle*, which has a specific gravity ranging from about 3.2 at its upper surface to perhaps 4 or 5 near the mantle-core boundary. Finally, the outer surface layer, called the *crust,* can be sampled and studied directly. The crust is composed of relatively low-density rock material. It ranges in thickness between about 30 kilometers and 35 kilometers in continental regions to as little as 5 kilometers in oceanic regions.

THE EARTH'S COMPOSITION

The composition of the solar system can be extrapolated from the composition of the sun. By far the most abundant elements in the sun are hydrogen and helium. These elements are followed by carbon, nitrogen, oxygen, and neon, which have relative abundances of 10^6 to 10^7 on a scale where the total silicon is 10^6. The next group of elements consists of magnesium, silicon, iron, aluminum, sodium, nickel, and calcium; they have abundances of 10^5 to 10^6. Presumably, all the objects in the solar system, including the earth, must be composed primarily of the abundant elements.

Now the planets are too dense to contain much hydrogen and helium; besides, virtually all of the existing hydrogen and helium is concentrated in the sun. Moreover, the densities of the earth and the other "terrestrial planets"—Mercury, Venus, and Mars—are high enough to preclude an abundance of the relatively light and volatile elements carbon, nitrogen, oxygen, and neon. These elements apparently escaped from the inner planets during their formation, but they may still be the major components of the planets beyond the asteroid belts. Probably *magnesium, silicon, iron, aluminum, sodium, calcium, and nickel* dominate the composition of the inner planets. *Oxygen*, too, must be included

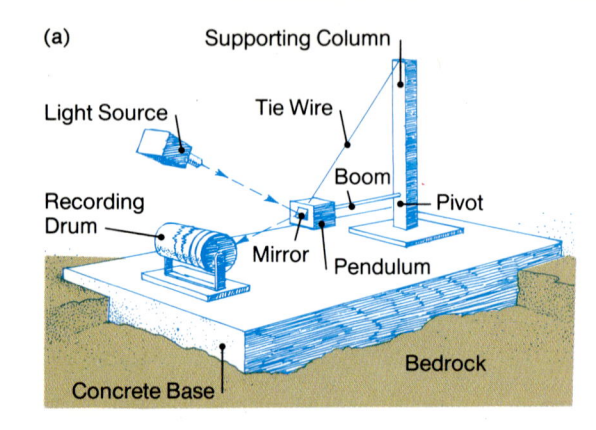

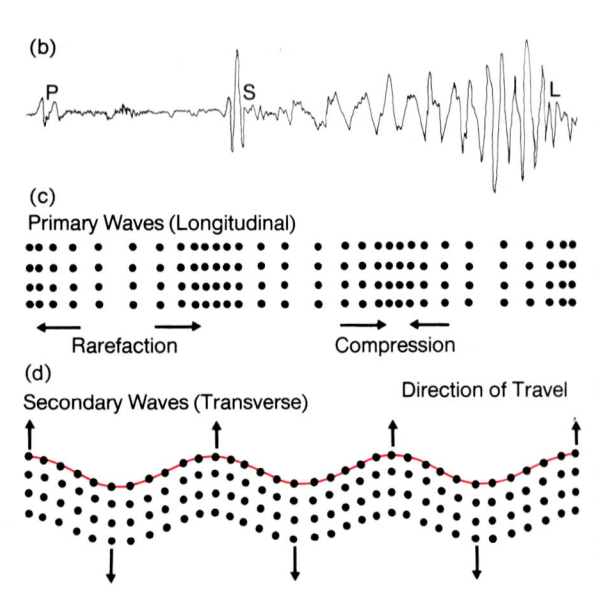

Figure 5. (a) The seismograph is an instrument for recording earthquakes. The inertia of a large reference mass *m* causes the apparatus to remain in place during an earthquake, but the rest of the apparatus, which is anchored to the earth, moves about. The jiggling motion is recorded by a light beam, which is reflected from a mirror (attached to the reference mass) onto a revolving drum covered with photographic paper. A typical recording is shown in (b). Modern seismographs use these same principles but are electric rather than mechanical. (c) This series of dots represents compression and expansion of matter during the passage of a *compressional* wave. Particle motion is in the same direction as the propagating wave. (d) This diagram shows displacements of particles in matter from their normal positions during the passage of a *transverse* wave. Particle motion is at right angles to the direction of propagation.

in the list because of its strong tendency to form chemical compounds with these six elements.

Both seismic velocities and the earth's moment of inertia indicate that much of the earth's mass is concentrated in its core; the attenuation of secondary waves indicates that the core is at least partly liquid. Of the dominant elements just listed, only iron and nickel are capable of making up such a dense, liquid core. Moreover, the idea that the earth's core is made of iron and nickel is substantiated by the existence of the earth's magnetic field. In fact, the magnetic field must be associated with motions in a conducting liquid such as molten iron.

The abundances of elements in the chondritic meteorites—especially in the carbonaceous chondrites—compares closely with other objects in the solar system. Could it be that the earth's overall composition is the same as that of the chondrites? If it were, and if most of the iron and nickel were in the core, then the crust and mantle would be composed merely of chondritic material without much iron and nickel. Unfortunately, the composition of the earth is a little more complicated than that.

The preceding chapter showed how the radioactive decay of rubidium 87 to strontium 86 may be used in determining the age of the solar system and in tracing the losses of volatile elements such as rubidium. Because of the radioactivity of rubidium 87, the ratio strontium 87/strontium 86 changes with time. That ratio has been determined for many samples of terrestrial materials, including the basaltic rocks of the ocean basins, widely believed to be direct partial melts of the upper mantle; the peridotites, rich in magnesium-silicates and possibly samples of the upper mantle; and most of the igneous, metamorphic, and sedimentary rocks of the continents. The values of the strontium 87/strontium 86 ratio in basalts and peridotites are too low to have evolved over the last 4.6 billion years in an earth that had a rubidium/strontium ratio like that of the chondrites. Compared to the mantle, the rocks of the earth's crust are enriched in rubidium, but even if the average composition of crust *and* mantle is considered, the rubidium/strontium ratio for the earth is substantially lower than it is for the chondrites. And it is highly unlikely, on the basis of chemical considerations, that the "missing" rubidium is in the earth's core.

If the abundance of rubidium in the earth is less than it is in the chondrites, you might suppose the earth also has lesser amounts of other relatively volatile elements such as potassium and sodium. And you would be right: the potassium/uranium ratio for the earth, for example, is lower than it is for the chondrites by a factor of about 10. It is this sort of evidence that suggests the early earth underwent some fractionation and loss of volatile elements.

If the earth's crust is enriched in alkali metals such as potassium and rubidium, the earth's mantle must be lacking in those elements. We are thus left with magnesium, calcium silicon, oxygen, and perhaps some iron as constituents of the mantle.

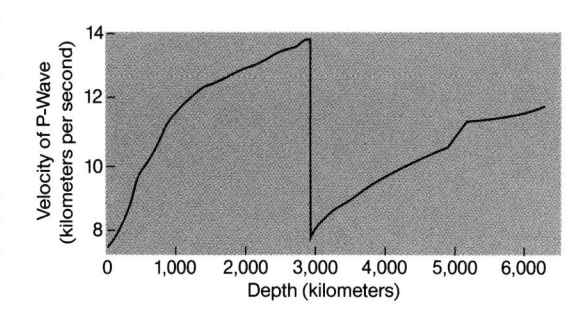

Figure 6. The velocity of a primary wave is plotted as a function of its depth of penetration into the earth. Such graphs are evidence of continued propagation through the liquid core, starting at about 2,900 kilometers. Secondary waves cannot propagate past this point.

FORMULATION OF AN EARTH MODEL

What do we get when we summarize these details of structure and composition? As far as we know, the earth looks something like the model depicted in Figure 7. Remember, however, that the model is based entirely on inference and indirect evidence. No one has been to the earth's core; no one has even drilled a hole into the upper part of the earth's mantle.

In this model, the earth consists of a crust of relatively low-density rocks extending to depths of 30 to 40 kilometers. The base of the crust

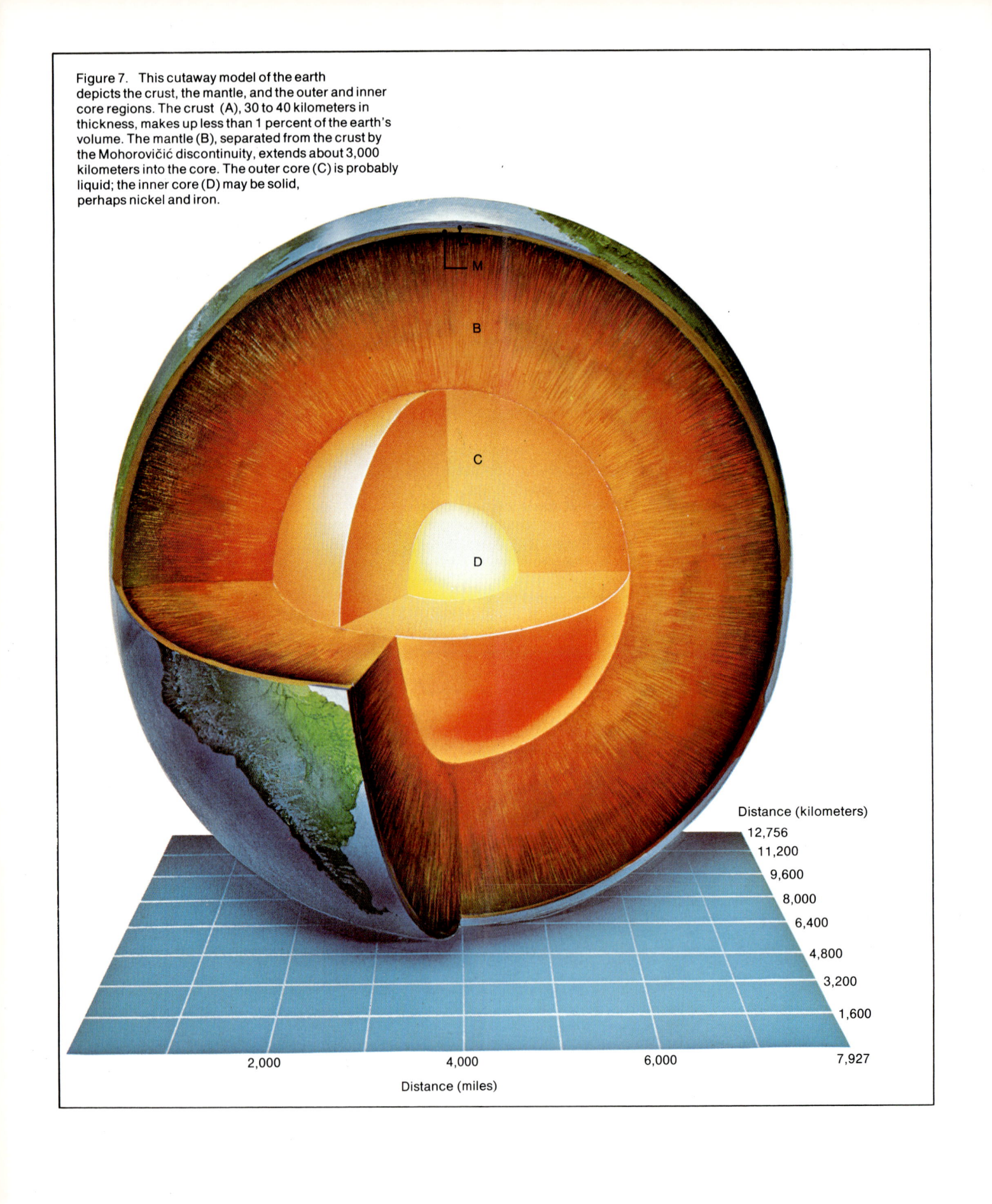

Figure 7. This cutaway model of the earth depicts the crust, the mantle, and the outer and inner core regions. The crust (A), 30 to 40 kilometers in thickness, makes up less than 1 percent of the earth's volume. The mantle (B), separated from the crust by the Mohorovičić discontinuity, extends about 3,000 kilometers into the core. The outer core (C) is probably liquid; the inner core (D) may be solid, perhaps nickel and iron.

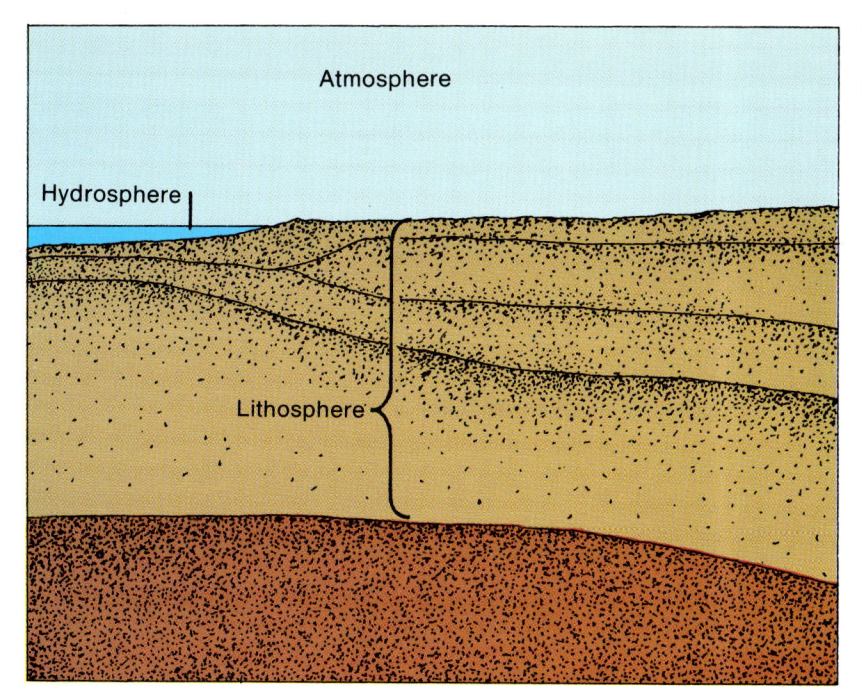

Figure 8. The earth model depicted in Figure 7 may be further embellished with such zones as the *lithosphere*, the *hydrosphere*, and the *atmosphere*. The lithosphere extends down through the crust and part of the mantle.

varies in depth: under some mountain ranges it is as deep as 60 kilometers and yet under the ocean basins it gets as shallow as 5 kilometers. This fundamental boundary, the Mohorovičić discontinuity, evidently separates the crust from the mantle. From the behavior of seismic waves, it would seem that the mantle is composed of much denser material than is found in the crust.

The mantle is probably composed of a collection of magnesium-rich silicate minerals. It extends to a depth of some 3,000 kilometers, where another abrupt boundary, or discontinuity, occurs. Below this boundary is the earth's core, which is believed to be liquid. From its density, and from inferences concerning the abundant elements available to make up the earth, the earth's core is believed to be molten iron-nickel alloy. From the behavior of seismic waves, it is plausible that the inner part of the core may be solid.

The distribution of materials within the earth is consistent with the theory that at one time the earth was in a molten state. In such a "melt," materials would separate on the basis of density, with heavy matter sinking inward and lighter matter floating to the surface. The core probably originated when heavy, reduced metal such as iron and nickel gravitated to the center of the earth and low-density materials fused out of the original material, rising to the surface to become crust, ocean, and atmosphere.

The constituents of the earth are often discussed in terms of broad categories that imply a change in relative density. The *lithosphere* designates the earth's solid matter. The *hydrosphere* encompasses the liquid waters of the earth—its oceans, lakes and rivers, polar caps, and water vapor. The *atmosphere* is the term used to designate the thin envelope of gases still trapped by the earth's gravity field.

THE NATURE OF THE EARTH'S CRUST

Having considered the gross composition and structure of the earth, it's possible now to examine the crust in greater detail. The entire crust

Calculating the Distance to the Focus of an Earthquake

A general rule of thumb is that for most rocks the velocity of secondary waves, or V_s, is about 3/5 that of the velocity of primary waves, or V_p:

$$V_s = 3/5\, V_p$$

From this relationship, it can be shown that

$$D = 5/3\, V_s\, \Delta t$$

where D equals the distance from the seismic recording station to the site, or focus, of the earthquake, and Δt equals the difference in arrival time of the primary waves and the secondary waves.

<u>Step 1.</u> Suppose that analysis of the seismogram of an earthquake shows the time of the first arrival of primary waves at the station was 10:03:58 and the first arrival of secondary waves was 10:05:17. If you assume the velocity of secondary waves between the station and the earthquake was 4.3 kilometers per second, how far away from the recording station was the earthquake?

The difference in arrival time, or Δt, is:

$$\Delta t = \begin{array}{r} 10:05:17 \\ -10:03:58 \\ \hline 1:19 \end{array} = 1\ \text{minute} + 19\ \text{seconds} = 79\ \text{seconds}$$

The distance to the earthquake is then:

$$D = 5/3 \times 4.3\ \text{kilometers/second} \times 79\ \text{seconds}$$
$$= 566.1\ \text{kilometers}$$

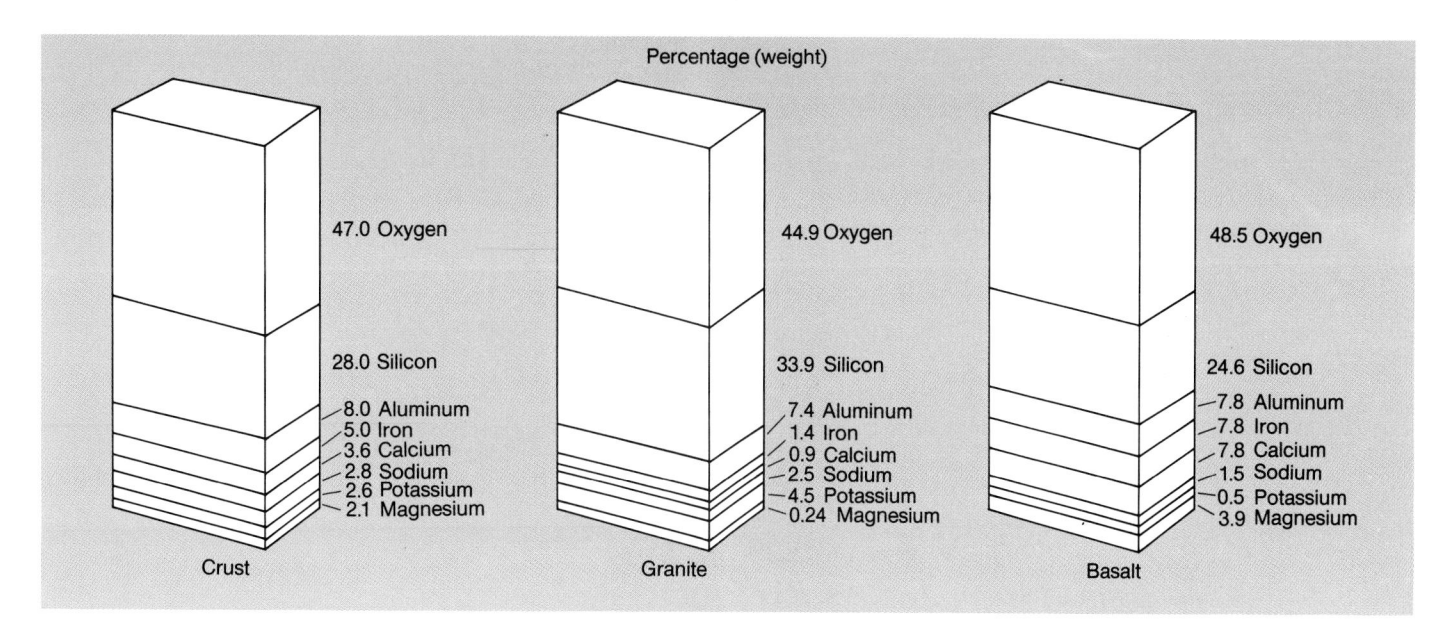

Crust	Granite	Basalt
47.0 Oxygen	44.9 Oxygen	48.5 Oxygen
28.0 Silicon	33.9 Silicon	24.6 Silicon
8.0 Aluminum	7.4 Aluminum	7.8 Aluminum
5.0 Iron	1.4 Iron	7.8 Iron
3.6 Calcium	0.9 Calcium	7.8 Calcium
2.8 Sodium	2.5 Sodium	1.5 Sodium
2.6 Potassium	4.5 Potassium	0.5 Potassium
2.1 Magnesium	0.24 Magnesium	3.9 Magnesium

Percentage (weight)

Figure 9. This illustration shows the eight most abundant elements in the earth's crust. Also shown are the relative abundances in two rock types: Granite and Basalt.

is composed of low-density materials, but it can be subdivided into two categories: *continental crust* and *oceanic crust*. Usually we think of the parts of the crust standing above sea level as "continents" and the parts below sea level as "ocean basins." But to the geologist, these simple definitions are not strictly true, for broad regions of almost all continents happen to be submerged in shallow sea water. The geologic distinction, which is based more on chemical composition and the nature of the rock types, is made between continental crust and oceanic crust.

The earth's crust is composed of rocks. Rocks may be defined as minerals, or aggregates of minerals. And minerals are chemical compounds that occur naturally in the materials of the earth. And like all chemical compounds, minerals must obey the rules of physical chemistry. Figure 9, which gives the chemical composition of the continental crust and two important rock types, shows that the entire chemical make-up of the continental crust may be described in terms of only 8 elements. All other known elements make up less than 1 percent of this part of the earth's crust.

The concentrations of potassium and sodium in the crust are greater than the concentrations of these elements in chondrites. Now if the total earth is depleted in these elements relative to chondrites, the continental crust must contain more alkali metals than the core and the mantle. And if the mantle is composed primarily of peridotite, then the continental crust is also enriched in silicon, oxygen, aluminum, and calcium. Conversely, crustal concentrations of iron, nickel, and magnesium are much lower than those deduced for the total earth, so these elements must be concentrated in the core and the mantle, as was discussed earlier.

The oceanic crust is composed almost entirely of the rock *basalt*. Although the continental crust is a much more complicated assemblage of rock types, the rock *granite* appears to predominate. When the chemical composition of granite and basalt are compared, it turns out that granite contains more silicon, oxygen, potassium, and sodium and less iron, magnesium, and calcium. These chemical differences correspond

Figure 10. Old and new mountains: the North American Appalachians (left) and the South American Andes, in different stages of orogenesis.

to a characteristic difference in rock types: granitic rocks are predominantly composed of light-colored, low-density minerals, whereas basalts are composed of dark-colored, high-density minerals.

What is the physical relationship of the crust to the underlying mantle? The oceanic crust is uniformly thin and rests directly on the upper part of the mantle. The continental crust is much more complex in its composition and structure. Because of the lower density of its rock materials, it generally stands higher than the oceanic crust. The exact nature of the lateral transition from continental to oceanic crust is not known for many parts of the earth, nor is the nature of the vertical transition from continental crust to upper mantle well understood. In some places the Mohorovičić discontinuity is sharply defined but in other places the transition is apparently gradual. The oceanic and continental crusts, together with a variable thickness of upper mantle material, constitute the *lithosphere*. All of the lithosphere lies above the low-velocity seismic zone.

Within the lithosphere the various parts of the crust are generally in *isostatic balance*—a kind of floating equilibrium. The higher parts of the crust, such as mountains, are buoyed up by thickened "roots" of lower density crustal material beneath them. These "roots" of low-density material displace higher density material of the mantle, or possibly the lower crust, and the resulting buoyant force supports the excess mass of the elevated crustal block. The exact nature of the material transfer necessary for isostatic compensation is not known, but it appears likely that the mechanism involves material movement in the low-velocity seismic zone.

The inference that continents are literally floating in isostatic balance on denser material is substantiated by the lack of notable *gravity anomalies* over continental regions. Positive gravity anomalies are values of the acceleration due to gravity that exceed the expected value for a particular region. In general, positive gravity anomalies represent the effect of excessive mass; thus, you might expect to find an anomalous value for gravity if the measurement were made over an extremely dense body of iron ore. The lack of substantial gravity anomalies over continents—even over high mountain ranges—indicates the excess mass of the higher elevations of continental areas must somehow be compensated for by the presence of low-density materials beneath them.

CONTINENTS AND OCEAN BASINS

The most impressive structural features of continental areas are mountain ranges. The term "mountain" is the name given to a variety of topographic features on the earth's surface. However, careful study of mountain ranges has revealed they share certain characteristics. Most of the

Figure 11. The northwest Pacific Ocean covers continents, trenches, sea mounts, and oceanic plains. The arcuate systems of adjacent mountain chains and island chains are clearly visible.
(Courtesy of the National Geographic Society.)

mountains of the world tend to occur in elongated ranges, which extend for hundreds, even thousands, of kilometers. Mountain ranges, particularly if they are geologically "young," tend to occur on the margins of continents. Rock materials within mountains commonly include thick sequences of strongly deformed sedimentary rocks, which display a variety of compressional structures such as folds and faults. Igneous activity—the introduction of molten material into the earth's crust—seems to be concentrated in mountainous areas. All mountains show evidence of vertical uplift. The erosion that follows such uplifting is responsible for the physiographic forms we call mountains.

All continents seem to be built around great regions of strongly deformed igneous and metamorphic rocks of great age. These regions are known as shields, or *continental shields,* and they are found on all continents. Shield regions commonly are located in the central portions of the continent, as they are in North America, with younger rocks appearing toward the outer margins. But there are exceptions to this arrangement.

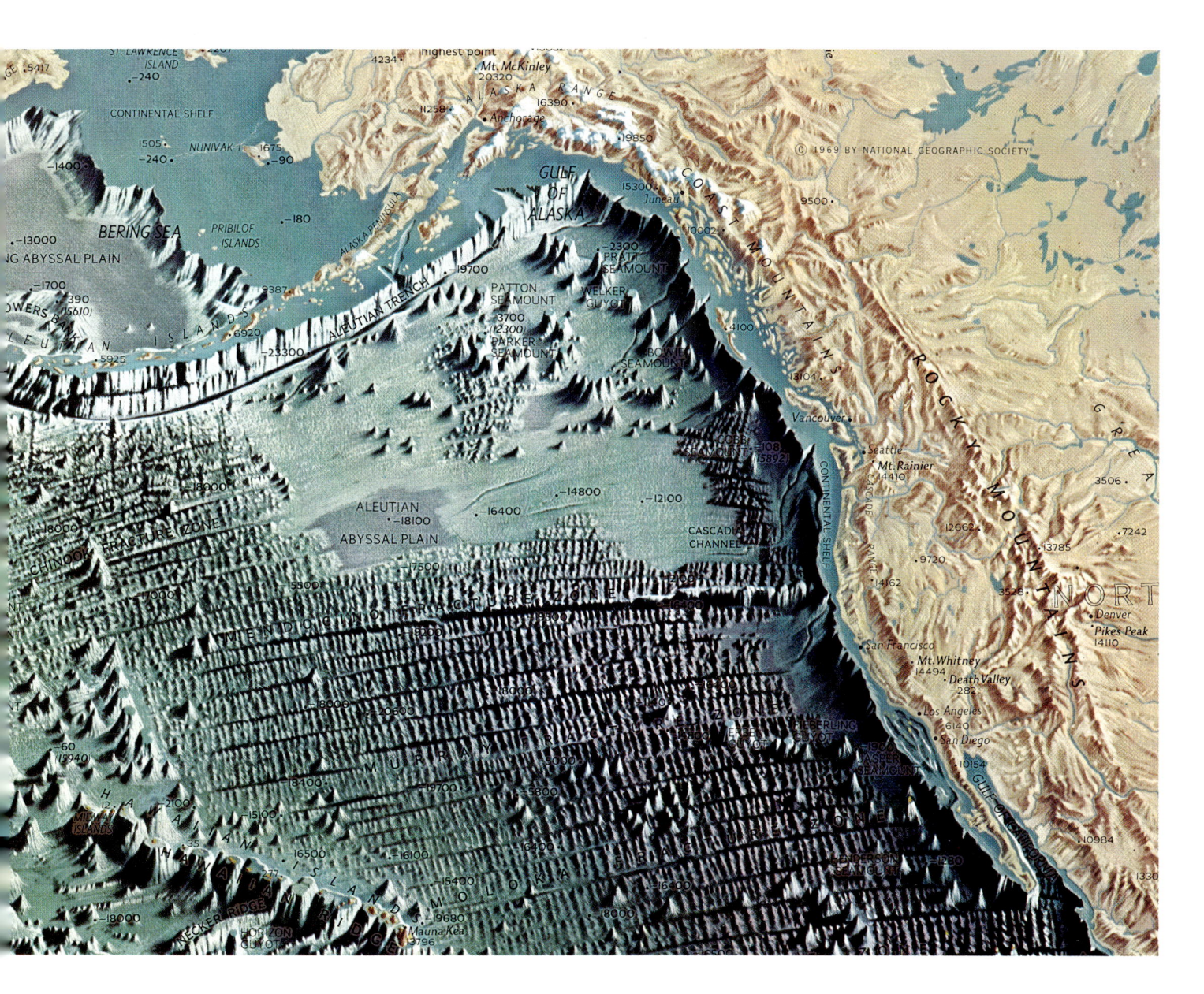

In Australia, for example, the oldest rocks are on the edge of the continent.

Surrounding the continental shields are regions of younger but still quite old crystalline rocks, both igneous and metamorphic, that have been termed *cratons*. The cratons, along with the shields, form the most stable parts of the continents. They are not subject to earthquake activity, as are the continental margins. Unlike the shields, cratonic areas typically are covered with several thousand feet of younger sedimentary rocks. In many cases, these sedimentary rocks have accumulated in intracratonic basins that lie within the craton. An example is the Illinois Basin on the North American continent where, during the Paleozoic era, some 3,000 meters of sedimentary rocks accumulated.

The ocean basins show much greater topographic extremes than the continents (see Figure 11). The features range from the *abyssal plains,* which are smooth, flat regions lying at depths ranging from 3,500 meters to 6,000 meters, to the *oceanic trenches,* which reach depths of 9,000 meters. In all oceans there are *mid-oceanic ridges* that rise above the level of the oceanic abyss; portions of those ridges may even rise above sea level. All of these great ridge systems are characterized by broad valleys along the center of the ridge, which evidently are of tensional, or *rift,* origin. Many oceanic ridges are studded with active volcanoes and are the loci of numerous earthquakes.

The northwest Pacific Ocean in the vicinity of the Aleutian Islands illustrates one of the most fascinating of all oceanic features—the great *island arc-oceanic trench systems* (see Figure 11). This feature commonly consists of an arcuate, or curved, chain of volcanic islands with an arcuate deep oceanic trench on the oceanward side. The margins of the entire Pacific Ocean Basin are festooned with island arcs, each with its corresponding trench, or with arcuate systems of mountains on some of the continents. Where mountains rather than trenches adjoin the ocean basin, the mountains themselves may be bordered on the oceanward side by a deep oceanic trench, as they are for the Andes Mountains of South America.

Detailed geophysical studies of the ocean basins have also shown the existence of a large number of fracture zones. The significance of the fracture zones, the mid-oceanic ridges, and the arcuate systems of islands and trenches is considered in a later chapter.

WHY STUDY THE EARTH?

Despite all we know about the third planet from observation and indirect evidence, a great deal remains unknown. Fundamental problems in the study of the earth concern its early history. How was our planet initially formed? If the earth accumulated through the accretion of solid particles during the consolidation of the early solar system, did this activity result in a young earth that was entirely molten? How did the earth achieve its differentiated structure, with a core apparently consisting of heavy, reduced metals, a mantle primarily of magnesium silicates, and a crust of less dense silicates? How old is the earth's crust? Why are the oldest rocks yet found on the earth's surface only about 3.5 billion years old if the earth itself is known to have formed about 4.6 billion years ago? Can this discrepancy be taken to mean that the earth had no crust during the first billion years of its existence? Or did it take that long for the earth to develop a stable crust?

Even a casual survey of the earth's features shows the crust has been deformed continually. Within mountain belts, rocks are wrenched into great, compressional folds. Giant faults have visibly dislocated rock masses. But what are the origins of the forces causing such deformation? What is the origin of the great chains of mountains found on every continent? What causes their enormous uplift?

Earlier in this chapter you read about the differences between the continents and the ocean basins. No one has satisfactorily explained why there should be a continental crust and an oceanic crust. And why should the continental crust be distributed the way it is? If it formed from the migration to the surface of low-melting, low-density compounds from an early molten earth, why isn't it now spread over the whole surface of the earth?

We study the earth because it's the planet we happen to live on, but we still understand amazingly little about its composition and processes. In a time of intense interest in the earth and its closest neighbor, the moon, exciting new ideas about the origin of the earth, moon, and the entire solar system are being proposed. Physics, chemistry, mathematics, biology, and geology are being integrated into a new earth science that promises to explain the secrets of the earth's structure, composition, and movements.

Many of the reasons for studying the earth relate to its history, but still others concern the earth's future and man's possible effect on it. To understand the significance of the depletion of resources such as oil and iron, we must learn about the structure, composition, and processes of the earth. To see the long-range effects of automobile exhaust and other pollutants on the air we breathe, we must know where air currents move and what they will carry. To understand what happens to effluents from industrial or domestic sources that flow into the waters of the earth, we must trace out the circulation of ocean waters as well as the characteristics of lakes, rivers, and streams. If we know how and where water moves underground we can find new sources of water for our growing populations. It is in the science of the earth that we may integrate the past, present, and possible future of our planet, and perhaps we may come to see how and where we fit into the interlocked picture.

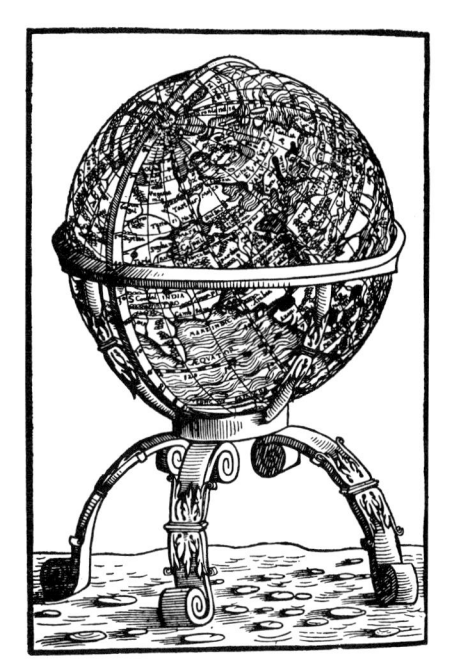

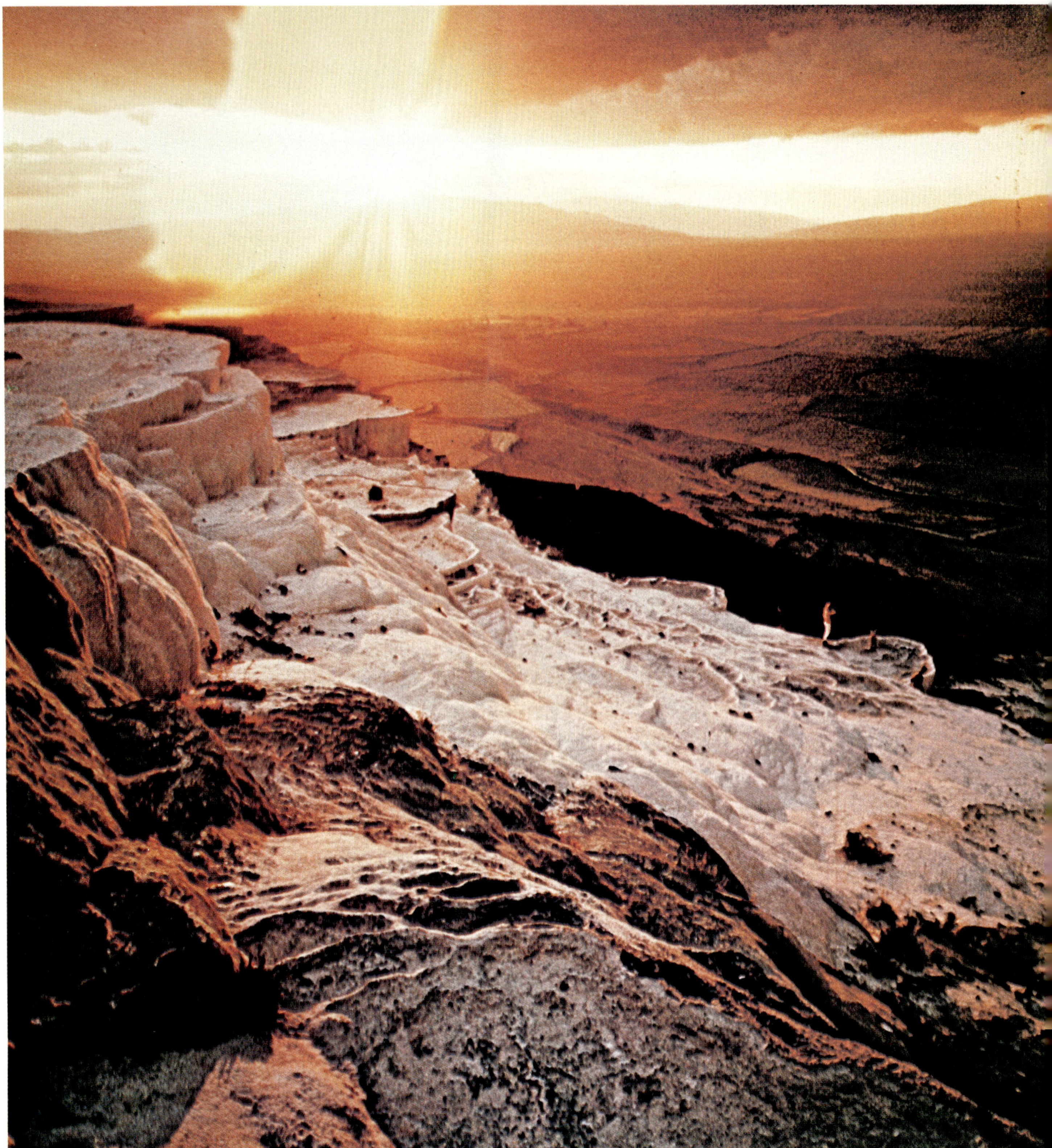

Essentially 99 percent of the earth's crust is composed of only 8 elements. In fact, most of what we see of the earth can be understood in terms of this handful of elements and the way they come together into chemical compounds, which we call *minerals.* The materials we know as *rocks* are simply aggregates of one or more mineral species. And yet, despite this seemingly limited number of component elements, a truly bewildering variety of rocks and minerals occur in nature—granite and basalt, sandstone and limestone, quartz and feldspars and micas, to name a few. How is it possible to sort through them all? First, you can examine them at the atomic level because virtually all of the characteristic properties of rocks and minerals depend on their internal atomic arrangement. Second, if you think of rocks as the major material units of the earth, you can simplify things by keeping in mind that they can be grouped according to their origin and general distribution. Then you have, basically, rocks of the continental crust, rocks of the oceanic crust, and rocks of the mantle.

CRYSTALLINE AND AMORPHOUS SOLIDS

Pick up a smooth beach pebble just washed over by the surf and you may see glistening bits of things of intricate geometric shapes locked inside its polished surface. Pick up a rock in the mountains and you may see the same shimmering bits, not polished smooth but jutting out in clearly discernible forms. These interesting features are the result of something called "crystallinity."

There are two kinds of solid materials, and a *crystalline* solid is one of them. The other kind is an *amorphous solid.* They are different chemically and physically because their internal atomic arrangement is different. In a crystalline structure, atoms are arranged in an orderly and repetitive way (see Figure 2). In fact, because the atoms occur in a periodic arrangement it's possible to predict the location of *any* atom in terms of its direction and distance from a given point in the structure. In other words, if you were to start out at a particular arrangement of atoms and travel along a straight line in any direction, you eventually would bump into an equivalent arrangement of atoms. Crystallographers have worked out a unique reference system for defining this point-to-point periodicity in a structure (see Figures 3 and 4).

The orderliness characteristic of all crystalline structures is missing in an amorphous structure, where the location of atoms is random with respect to both distance and direction. For that reason, the crystalline condition is often spoken of as an *ordered state,* and the amorphous condition as a *disordered state.* At the pressures and temperatures found in the earth's crust, the ordered state represents the lowest energy configuration for most chemical compounds. Accordingly, amorphous materials such as glass become ordered and crystalline with the passing of time in a process known as *devitrification.* Glass objects found in the tombs of Egyptian pharoahs have already begun the process.

Crystalline materials are noted for their directional, or *vector,* properties. An individual crystal is often characterized by external planar surfaces, which are known as *crystal faces.* Why do crystals take on the geometric shapes that they do? The answer is found in the mechanism

14
Materials of the Earth

Even the most common earth materials take on memorable configurations—limestone beds in Pamukkale, Turkey.

Figure 2. (Left) The crystalline nature of quartz is apparent in this photograph. Unlike the chunk of obsidian glass that appears above it, quartz is bounded by sets of regularly distributed planes called *crystal faces*. (Right) A two-dimensional sketch of the internal atomic arrangement of quartz, shown above, is contrasted with that of glass to point out that the different appearances of these two materials can be traced to differences at the atomic level.

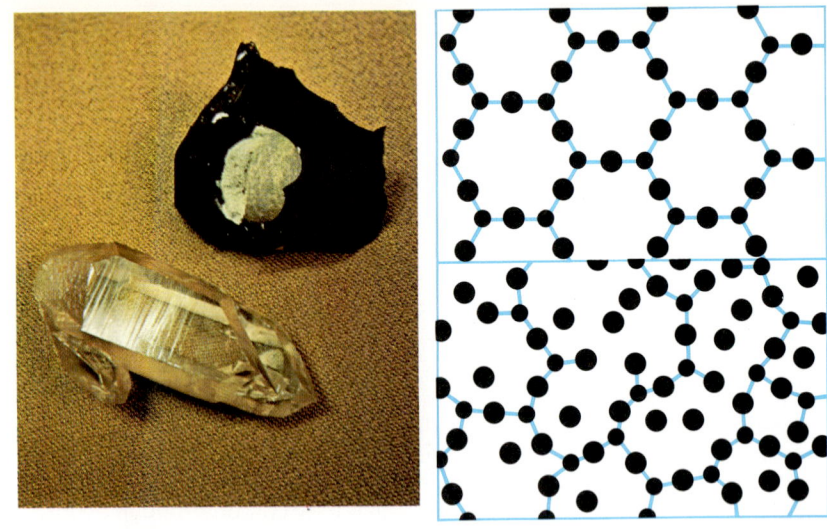

Figure 3. It is possible to define the basic repeating unit of a crystalline structure. When such units are put together in three dimensions, they form *crystal lattices*.

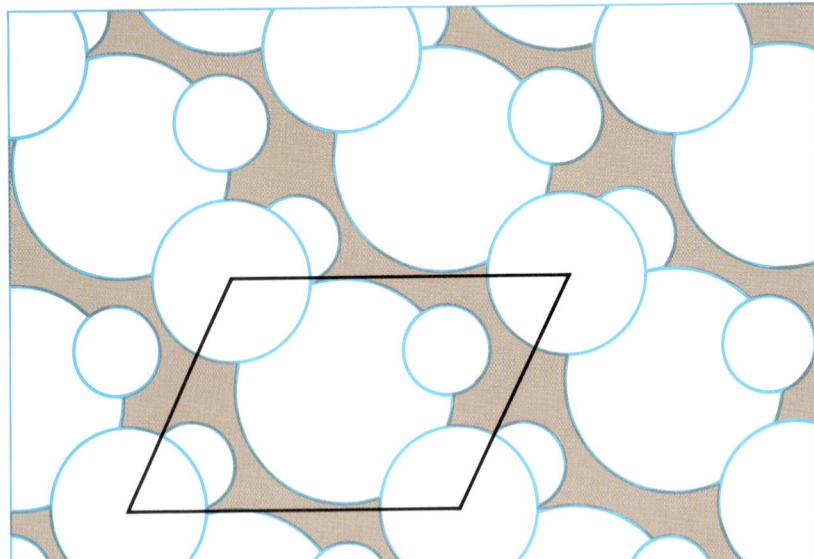

Figure 4. Examples of Bravais lattices. These lattices indicate translational direction, *not* atomic arrangement. A crystal structure patterned after a certain kind of lattice is the basic building block for the growth of a particular crystalline material.

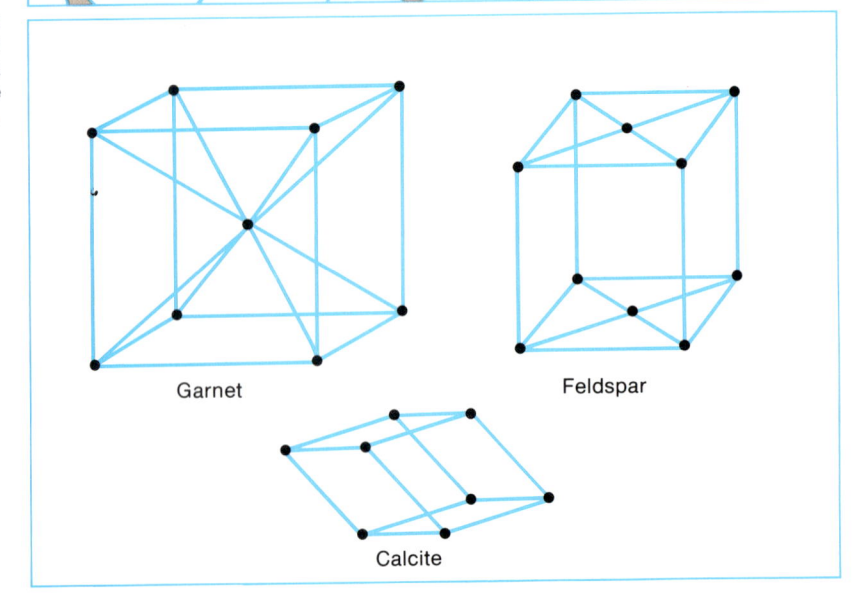

Garnet

Feldspar

Calcite

of crystal growth. Imagine a three-dimensional crystalline structure growing from a molten substance or from a solution. Because of thermodynamic considerations related to the crystal lattice and the chemical properties of the elements making up the crystalline compound, atoms will be added on more readily in some directions than in others. As the crystal grows, its external form will reflect these more favorable growth directions in external crystal faces.

Why do some rocks glisten in the sunlight? Most crystalline materials have another vector property, called *cleavage*. All this word means is that the material tends to break along planes where the weakest (or smallest) number of chemical bonds will be encountered. It is the cleavage planes, not the crystal faces, that reflect light from the sun.

The number and angular relationships of cleavage planes are often useful in identifying common minerals. Halite, for example, has three mutually perpendicular cleavages. If you break off a fragment of this mineral, you will end up with a lot of small cubes (Figure 6). Calcite also has three cleavage directions but the angular relationships produce rhombohedrons, not cubes.

Cleavage is a *breaking* phenomenon; crystal faces are *growth* phenomena. Some minerals, such as quartz, show a strong tendency to form crystal faces but do not show any tendency toward cleavage. Other materials, such as calcite, display both tendencies but the directions of the crystal faces differ from the directions of the cleavage planes. Both are related to the geometry of the internal atomic arrangement, however.

Some materials resist scratching or abrasion to a greater extent in one direction than they do in another. This vector property is called *differential hardness,* and it is related to directional variations in the strength of the chemical bonds between atoms.

Most crystalline materials also show directional variations when a beam of light is transmitted through them. The electromagnetic radiation interacts with the electric field that is created by charged atoms situated regularly through the crystal structure. As a result, the velocity of light varies with direction because the magnitude and spacing of the charge also varies with direction in most crystalline materials. This directional variation produces a double refraction of the sort illustrated in Figure 7.

Another vector property of crystalline materials is their ability to diffract x-rays. This property has been used successfully in exploring the internal structure of crystalline solids.

Because the internal atomic arrangement of amorphous solids is essentially random, these materials have no vector properties. Consequently, they can't develop crystal faces because there are no preferred growth directions. They can't display cleavages because they are just as easy (or difficult) to break apart in one direction as they are in any other. And they can't show directional variations in hardness or in the way they interact with light.

How can one kind of solid substance be so different from another? Consider what happens when a typical crystalline solid and a typical amorphous solid are heated. The crystalline solid seems to change very little as it is being heated, although there are some internal rearrangements going on at the atomic level. Then, when it reaches a certain temperature, it melts; it passes abruptly into the liquid state. Some crystalline solids do melt incongruently; in other words, a reaction that occurs at the melting temperature causes them to change into both a liquid and another solid of different composition.

Most crystalline solids have a definite melting point, however, and their changes of state are abrupt.

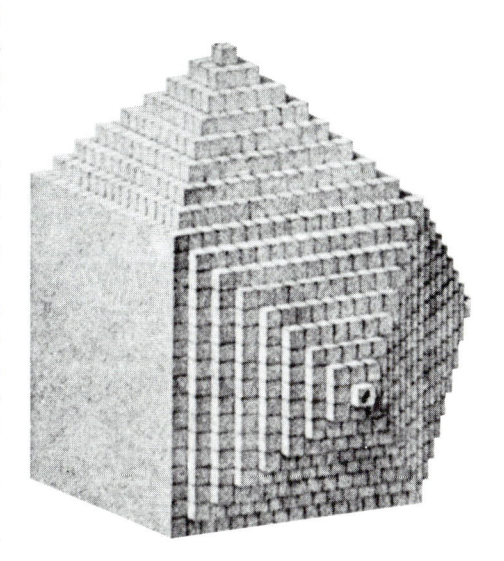

Figure 5. In the eighteenth century, Abbe René-Just Haun, a founder of modern crystallography, accidentally dropped a crystal which broke into pieces. He noticed that from the outside surface to the inside of the crystal, all the small particles of matter were arranged symmetrically. With the help of this structural diagram he elucidated his "law of diminution," according to which certain shapes decrease regularly in the number of particles that make them up. Crystals grow in this regular, symmetric way.

Figure 6. The cleavages along the planes of weakest atomic bonding in this halite crystal reflect light.

Figure 7. This photograph shows the double refraction in a crystal that is produced by directional variations when a beam of light is transmitted through it.

Amorphous solids exhibit markedly different behavior when they are heated. Glass, for example, merely becomes increasingly less resistant to deformation, or less *viscous,* as it is heated; it does not undergo an abrupt change of state. At some temperature its viscosity will have decreased to a point where it probably would be called a liquid. And yet, perhaps glass shouldn't be considered a solid in the first place—in many ways it is more like a "supercooled liquid." Even the random atomic structure of an amorphous material is a closer approximation to the disorder of liquids than to the highly ordered structures of crystalline materials. Thus, at low temperatures, amorphous substances show a curious blend of the strength and rigidity of solids and the random atomic arrangement of liquids.

IMPORTANT MINERAL GROUPS

About 2,500 naturally occurring minerals have been discovered in the earth's crust, and each one is a unique chemical compound with its own characteristic crystalline structure. Clearly it would be impossible to consider them all in this chapter. In this survey of earth materials, only two groups will be considered—the silicates, which represent virtually all rock-forming minerals; and the carbonates, which are important in interpreting the earth's chemical and biological history.

The Silicates

All of the silicate minerals are compounds of metals and negatively charged silicon-oxygen ion groups. In their characteristic bonding arrangement, four oxygen atoms are grouped around one silicon atom as if they were located at the apexes of a regular tetrahedron. There are several variations on this fundamental atomic arrangement. In some minerals the oxygen atoms are bonded only to a single silicon atom and are not shared with adjacent silicon atoms. These single tetrahedral groups are packed together in such a way that they form structural sites into which positively charged ions of metals such as iron, magnesium, and calcium may be bonded.

In other silicate minerals, the silicon-oxygen tetrahedral groups may be arranged in chains, in sheets, or in three-dimensional frameworks when adjacent silicon atoms share two or more oxygen atoms. Figure 8, which lists some of the more important rock-forming silicates, depicts these various structural arrangements.

The Carbonates

The carbonate minerals are compounds of positively charged ions of metals and a negatively charged ionic group composed of carbon and oxygen. In the carbonate group, three oxygen atoms are arranged in a plane around a carbon atom. The three oxygen atoms are spaced equally around the carbon atom, forming a bond angle of 120 degrees. Two important carbonate minerals, calcite and dolomite, are illustrated in Figure 9.

Figure 8. (Far right) The structures of rock-forming silicates are based on a variety of atomic arrangements of silicon-oxygen groups. (a) Olivine and (b) garnet have single tetrahedrons. (c) Augite has single tetrahedral chains and (d) hornblende has double tetrahedral chains. (e) Mica has tetrahedral sheets. (f) Feldspar and (g) quartz have three-dimensional tetrahedral frameworks.

The silicates and carbonates occur in different kinds of rocks. Years of careful study have shown there are three major material units that make up the crust of the earth. The first is *igneous rock,* which results from the crystallization of molten material known as magma. The second is *sedimentary rock,* which forms from decomposed or disin-

(a) Olivine

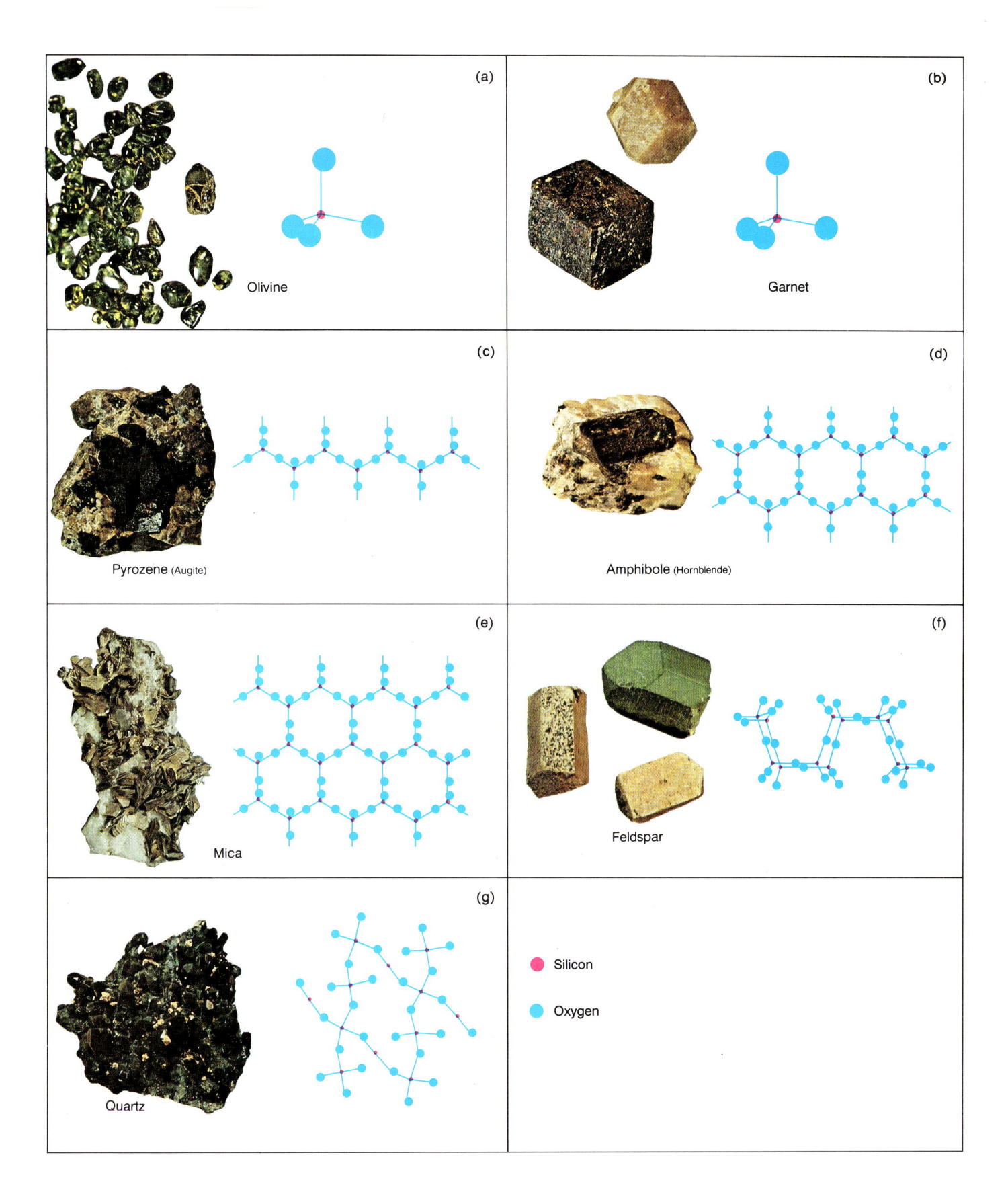

(b) Garnet

(c) Pyroxene (Augite)

(d) Amphibole (Hornblende)

(e) Mica

(f) Feldspar

(g) Quartz

● Silicon
● Oxygen

Figure 9. The structure of minerals in the carbonate group is based on planar carbon-oxygen groups. Calcite (above) and dolomite (below) are two well-known carbonate materials.

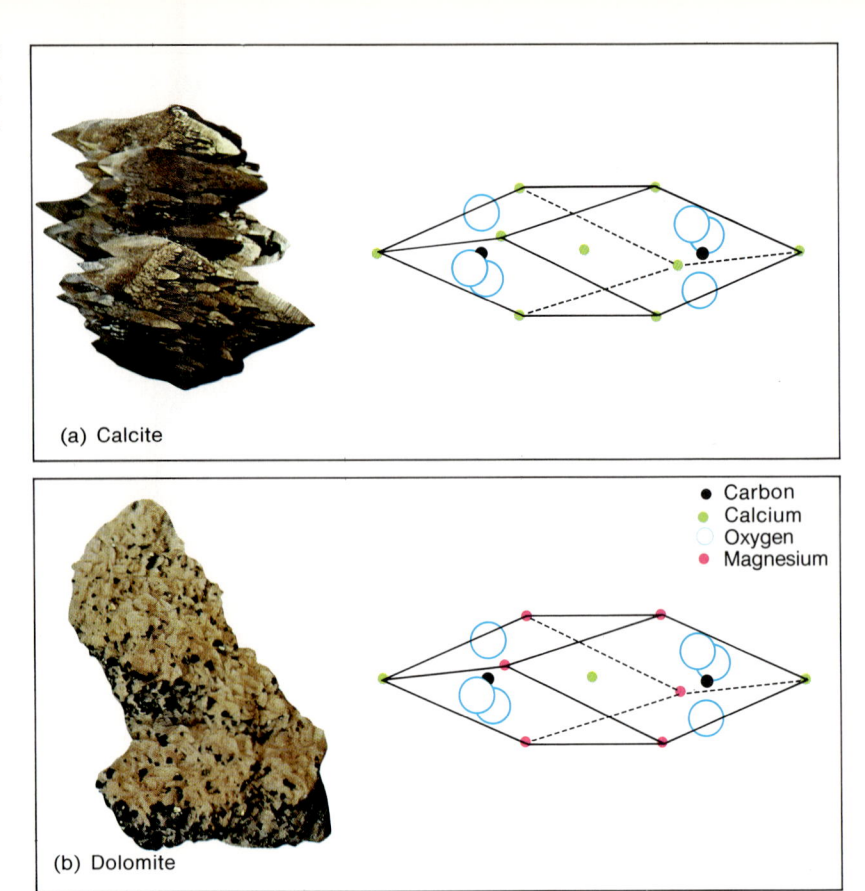

(a) Calcite

(b) Dolomite

- ● Carbon
- ● Calcium
- ○ Oxygen
- ● Magnesium

tegrated materials of preexisting rocks. The third is *metamorphic rock,* which results from the recrystallization of preexisting rocks due to changes in temperature, pressure, and the chemical environment.

FORMATION OF MAGMAS AND IGNEOUS ROCKS

If you have been fortunate enough (or unfortunate, depending on where you were standing) to witness a volcanic eruption you have seen the hot, surging lava. The fluid lava is *magma*—a body of silicate melt that originated within the crust or the upper part of the mantle. From time to time, magmas work their way up to the earth's surface, there to break through as volcanic eruptions. When that happens we are able to study directly the chemical composition and temperature of the magma. Still other bodies of magma originate in the earth but never make it to the surface. They crystallize within the earth's crust, and all we can learn about them is what we can get from the composition, texture, and geometry of the "frozen" magma—which is a form of rock.

Little is known about the origin of magmas. We do know that the earth is much hotter in its interior than it is on the surface, for an increase in temperature with depth can be measured in deep drill holes and in deep mines. This increase in temperature, which is known as the *geothermal gradient,* ranges between 10 degrees and 50 degrees Centigrade per kilometer, depending on where the measurement is taken. If this gradient were extrapolated all the way to the earth's center, however, temperatures would reach 60,000 to 300,000 degrees Centigrade! Such temperatures are not consistent with what we know of the densities and elastic properties of the earth's interior. Consequently, the rate of tem-

Figure 10. This hot, fluid substance is magma. The lava oozed out at the edge of Kilauea Iki in Hawaii.

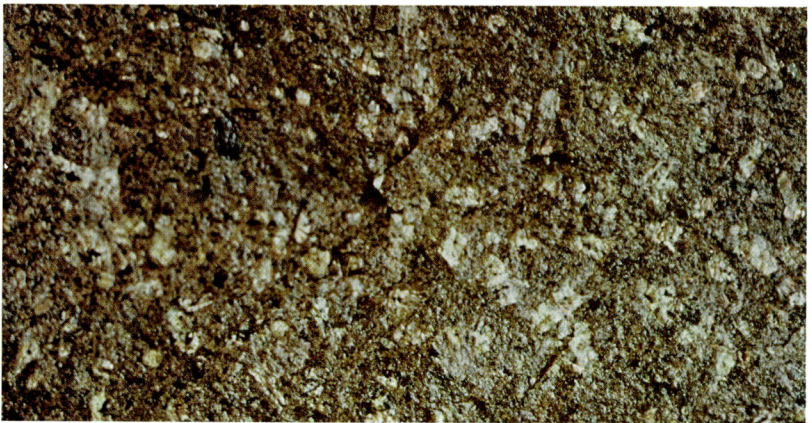

Figure 11. Porphyries form by crystallization of magmas that have been "seeded" with previously crystallized minerals. The large, earlier crystals are surrounded by the small grains that form during rapid crystallization.

perature increase must somehow diminish with depth. In general, the temperature at depth is less than the melting temperature of whatever rock materials make up the deeper parts of the earth. And if that is the case, the temperature at the base of the crust must be no more than 800 to 900 degrees Centigrade, for at higher temperatures the common rock types of the crust would melt.

Magmas are not generated everywhere in the earth's crust, nor are they generated all the time in any one place. They seem to be limited to the ocean basins, the mid-oceanic ridges, and those parts of the continents where young mountains are forming. It would seem that magmas are generated where the earth's crust is active, and where the temperature distribution temporarily rises above the melting temperature of the rock material that makes up the earth at a particular location.

Once a body of magma forms, it tends to rise into the earth's crust because it is mobile and less dense than the surrounding rocks. If the magma body is charged with gases and under considerable pressure the eruption may be spectacular, but other magmas flow quiescently out on the surface. Many, in fact, erupt beneath the waters of the oceans and never are observed.

Rocks produced by volcanism are called *volcanic,* or *extrusive,* igneous rocks. Because crystallization takes place so rapidly, they usually have tiny crystals. In fact, some volcanic lavas cool so rapidly that crystallization does not take place at all. The result is a volcanic glass such as the rock obsidian. Large quantities of gaseous substances are often dissolved in the magma. During a volcanic eruption, the sudden drop in pressure may cause these gases to boil off vigorously, producing bubbles in the congealing lava.

14 | Materials of the Earth

Figure 12. Intrusive igneous rock, which crystallizes within the earth's crust, forms from hot magma that never reaches the surface. This photograph shows an intrusive contact between granite (light-colored rock) and older volcanic rock.

Sometimes, certain minerals begin to crystallize from the magma while it is still within the earth's crust. If that happens, the eruption will carry the suspended crystals out along with the liquid lava. The resulting rock has an interesting texture, in that the larger crystals are surrounded by the rapidly crystallized material that was still liquid during the eruption. Such rocks are called *porphyries* (see Figure 11). The small crystal size of volcanic rocks is related not only to the rapid crystallization that results from the low temperatures on the earth's surface, but also to the rapid loss of dissolved gases. Dissolved gases in the melt tend to reduce its viscosity and depress its freezing temperature; if the gases are retained they tend to promote the growth of larger crystals, but if they are lost the crystals will be smaller.

If you have traveled through mountain regions you may have noticed a uniform, strongly crystalline rock with large, glistening crystals. Such rocks are commonly exposed where profound uplift of the crust, followed by erosion, has revealed some of the deeper crust. If you are willing to hunt around a bit, you may find a *contact,* where the uniform crystalline rock literally cuts through some earlier rock. These crystalline rocks are the *plutonic,* or *intrusive,* igneous rocks. They were formed from magmas that did not reach the earth's surface but instead crystallized within the crust.

Because these rocks did not reach the surface they crystallized slowly, for the rocks of the crust are poor heat conductors. Moreover, because they crystallize at depth and under pressure, intrusive magmas tend to retain their dissolved gases much longer. Consequently, intrusive rocks tend to have large crystals because their growth from the melt is promoted by slow cooling and the presence of dissolved gases. Granite is a typical intrusive rock.

Besides varying in their mode of origin, igneous rocks also vary in their chemical composition. Some magmas are rich in silicon dioxide, sodium, and potassium but are depleted in calcium, iron, and magnesium. Such magmas crystallize as light-colored rocks consisting of abundant light-colored minerals, such as quartz and feldspar. Other magmas are depleted in silicon dioxide, sodium, and potassium but are rich in iron, calcium, and magnesium. These rocks are usually dark in color because they are composed primarily of the darker minerals such

Table 1.
Composition of Chemically Equivalent Intrusive and Extrusive Rocks

Intrusive Rocks	Mineral Composition	Extrusive Rocks
Granite	Quartz, potassium and sodium feldspars, minor biotite, hornblende, and muscovite	Rhyolite
Syenite	Quartz less than 5 percent, potassium and sodium feldspars, minor biotite, hornblende, and muscovite	Trachyte
Diorite	Quartz usually absent, intermediate plagioclase feldspar, fairly abundant biotite, hornblende, and pyroxenes	Andesite
Gabbro	Quartz absent, calcium plagioclase feldspar abundant, abundant pyroxenes, hornblende, and olivine	Basalt
Peridotite	Feldspars usually absent, abundant olivine, pyroxenes, oxides of iron	No extrusive equivalent

as hornblende, pyroxenes, and olivine. Table 1 lists the common igneous rocks and compares the intrusive types with the chemically equivalent extrusive types.

FORMATION OF SEDIMENTARY ROCKS

Sedimentary rocks form because of the chemical and mechanical interactions of the earth's rocky crust with the atmosphere and the hydrosphere. Although sedimentary rocks are spread out laterally, they are quite thin in comparison with the total thickness of the earth's crust. Nevertheless, they are important for two reasons. First, deposits of sediments accumulate under the influence of the earth's gravitational field, so they are characteristically laid down in horizontal or nearly horizontal beds. In this way, beds of sedimentary rocks provide a frame of reference for the deformation of the earth's crust, and when we observe folded, broken, or otherwise deformed sedimentary rocks, we can often tell the extent of the deformation and the direction in which the forces acted. And second, the environment in which sedimentary deposits accumulate is also the arena of life on the earth. It is in sedimentary rocks that the record of life is preserved.

Sedimentary rocks form through a series of interrelated processes that occur on the earth's surface. These processes include *weathering*, the chemical and mechanical degradation of rock materials; *erosion*, the removal of this material from the site of weathering; *transportation*, the movement of eroded material through the agencies of running water, wind, or ice to some other locality; *deposition* of the transported material into a basin of sedimentation; and, finally, *lithification*.

Weathering

Weathering consists of a complex set of chemical reactions that occur when crystalline rocks are brought into prolonged contact with the gases of the atmosphere and the waters of the hydrosphere. In order to understand the nature of some of these reactions, consider what happens when granite, a typical crystalline, igneous rock, undergoes weathering. Assume it consists of potassium and sodium feldspars, quartz, and minor

Figure 13. (Above) A masterpiece of erosion, the Grand Canyon of the American Southwest is the geologist's delight. Down through the ages the Colorado River has relentlessly carved away at the earth to expose *sedimentary formations* that are silent testimony to the geologic processes continuously at work on the earth's surface. (Below) Rocks subjected to increased pressure, increased temperature, and slowly applied stress behave plastically even though they are brittle under normal conditions. An example of the resultant *folding* is this formation in England.

Figure 14. Chemical weathering has enlarged the joints of this granite at Pikes Peak, Colorado.

amounts of biotite and hornblende—minerals that originally crystallized at high temperatures from a silicate melt. When this assemblage of minerals is exposed to the chemically reactive gases of the atmosphere, to water, and to the low temperatures prevailing on the earth's surface, most are chemically unstable. The chemical reactions that occur tend to form *new* compounds that are stable under the physical and chemical conditions of the earth's surface—conditions that are grossly different from those of the crystallizing magma from which the granite was formed deep in the crust.

The processes of chemical weathering also accelerate mechanical disintegration. As the minerals of the weathering rock break down, the strength derived from their interlocked texture diminishes.

Erosion

Rocks often undergo erosion during the weathering process. Erosion represents the expenditure of energy directed against the rock mass by running water, wind, ice, or simply gravity. It results in the physical removal of the breakdown products from the site of weathering. Following erosion, sedimentary particles or material taken into solution may be transported away from the site of weathering by the same agencies. During transportation additional chemical breakdown may occur and the rock debris will undergo mechanical abrasion, particularly if the transporting mechanism is running water.

Transportation

If you examine sediment undergoing transportation, you will discover substantial changes in the size, shape, and mineralogical composition of the material that are a function of the extent of the transportation. Sediment near its source often will contain large proportions of relatively unstable minerals such as the feldspars and the micas, and all of the material may be angular and relatively large-grained. Conversely, material that has been transported for long distances will be small-grained

Figure 15. Most landforms are the result of water acting in the form of rivers and streams, but other erosional agents such as wind, ice, and waves are also important. (Above) Motueka River Valley in New Zealand was carved by a glacier and further eroded by running water. It is now occupied by a meandering river. (Left center) The wall of Red Rock Canyon in Central California displays the effects of water acting on different sediments in the same rock. (Below left) A sloping landscape north of Christ Church, New Zealand, is now being carved into a dendritic erosional pattern by rainfall and running water. (Right) The famous Half Dome in Yosemite Park, California, is a granitic intrusion that has eroded less than its surroundings. A glacier later worked to carve it into this remarkable configuration.

Figure 16. Further examples of landforms. (Above left) Wind and blowing sand have been the major forces shaping this arch in the arid southwestern United States. (Left center) Rainfall and gradual uplift have formed rolling fields near the coast of Dorsey, England. Coastal waves have cut the cliffs to form a secondary feature, analogous to the carved face of Yosemite's Half Dome. (Right) In Bryce Canyon National Park, Utah, bedded sediments in a limited drainage area have eroded to form spires. The spires are regular and steplike due to the variable composition of the sediments. (Below) Cathedral spires of soft pumice stand on the slopes of a collapsed volcano, now occupied by Crater Lake, in Oregon. Rainfall and wind have eroded the softest parts of the slopes. Because of the limited upstream drainage area, there is relatively little undercutting by running water.

Figure 17. In New Zealand, South Island, the sedimentary debris transported by the Godley River to Lake Tekapo has formed this delta.

because of the abrasive action of transportation, and the grains will tend to be much more rounded.

Moreover, the continued chemical processes that are active during transportation will tend to remove the chemically unstable minerals from the material. Thus sedimentary materials observed far from their sources will have very little feldspar or mica, and will consist primarily of clay minerals, and of quartz: the quartz of the original rock is so chemically stable that it persists even through long transportation, and the clays are the most common breakdown product of most of the other original minerals. Other components of the original rock, particularly the alkali metals sodium and potassium, tend to form soluble compounds and are transported away from the site of weathering in solution. These elements, especially the sodium, ultimately find their way to the sea.

Deposition

Transportation ends with deposition of the sedimentary debris at some locality. Such a locality is usually called a *basin of sedimentation,* although it may not really be a topographic basin. For example, if a rushing stream carries sediment down from a mountainous region, the stream will lose its ability to transport coarse material when it reaches the base of the mountain and flows out onto the adjacent plain. In such cases, a thick wedge of coarse sedimentary debris will be deposited along the base of the mountain range, forming an *alluvial fan*. If the stream flows instead into a lake, its velocity will be decreased and its transported sedimentary load will be deposited. But deposits in lakes and in alluvial fans represent only temporary basins: the real journey's end for sediment is the sea. As long as the continents remain above sea level, we can expect that running water, wind, and glacial ice will transport weathered rock debris to the edges of the continents, dumping the material into the shallow seas of the continental shelves. Where major rivers empty into the sea, great *deltas* are built as the sedimentary load is dumped at last.

Figure 18. Increased pressure, increased temperature, and cementation changed these sedimentary deposits to rock through the process of lithification.

Because deposition involves reduction in the velocity of the transporting medium, a lateral gradation in particle size appears in the final deposits. The coarsest particles are dropped first and the finer particles are carried farther into the basin; dissolved materials are carried farthest of all. As a consequence of the separation process, sand deposits are concentrated in relatively high-energy environments, such as in the beds of swiftly moving streams and along the seacoasts in the surf zone. Deposits of fine-grained clays, on the other hand, occur in low-energy environments, such as along the broad flood plains of streams where water moves sluggishly or in lagoons.

Lithification

After deposition, sediments become rocks through the processes of lithification. The first step is one of compaction and dehydration. It is generally brought about by the burial of the sediment beneath great thicknesses of subsequently deposited material. The resulting pressure tends to compact the sediment, driving out pore water and, with the higher temperatures encountered at depth, often causing dehydration of the clay minerals. Following compaction and dehydration, the sediment may also undergo cementation, during which the grains are cemented together by secondarily precipitated mineral matter such as silica, calcite, or various hydrated iron oxides.

Sediment that is mechanically derived, transported, and deposited is called *clastic,* or *detrital,* sediment. Such sediment is classified on the basis of particle size. Table 2 lists the lithified equivalents.

Figure 19. Vast, undulating formations of limestone lie beneath the ocean off the coast of Florida, southeast of the Bahama Islands, and in the Caribbean Sea. Here the camera looks out across the Great Bahama Bank.

Table 2.
General Classification of Sedimentary Particles

Sedimentary Particles	Sedimentary Rocks
Boulders, cobbles, pebbles	Conglomerate
Sand	Sandstone
Silt and Clay	Mudstone and Shale

Chemical Sedimentation

What happens to the part of weathered rock that goes into solution? This material is ultimately carried into lakes and seas, where it may be precipitated by chemical or, more commonly, biochemical processes to form a second major category of sedimentary rocks—the chemical sedimentary rocks. One example is limestone, composed primarily of the mineral calcite; dolostone, composed primarily of the mineral dolomite, is another, and both are abundant rock types.

Virtually all deposits of limestone are biochemical in origin. The accumulation of calcium carbonate in marine sedimentary basins is almost exclusively brought about by the life processes of marine organisms. These organisms secrete shells and other hard structures of calcium carbonate, which they extract from sea water. As these animals die, their calcite hard parts accumulate on the sea floor and form thick deposits. The remains are commonly broken up and moved about by waves and oceanic currents. To some extent, then, limestones of biochemical origin will have textural features similar to some of the clastic sedimentary rocks even though their mineral content differs.

The sedimentary deposits from the early Paleozoic include thick and widely distributed beds of dolostone. They contain fossils of marine organisms, but the origin of this carbonate rock has been something of a puzzle. We have not found any marine organisms that secrete hard structures of dolomite, so dolostone evidently is not of primary biochemical origin. Moreover, few existing environments are precipitating dolomite

The Cost of Mining Earth Materials

Suppose that the principal copper mineral in a mine is chalcopyrite and that the percentage of chalcopyrite in the ore being mined is 1.5 percent. If the price of metallic copper is $.50 per pound and the cost of processing and smelting the ore is $1.84 per ton, what is the maximum permissible cost per ton of mining the ore if the mining company is to make 5 percent profit?

Step 1. First you must calculate the percentage of metallic copper in chalcopyrite:

63.54 = atomic weight of copper (Cu)
55.85 = atomic weight of iron (Fe)
64.14 = atomic weight of 2 sulfur (S) atoms
$\overline{183.53}$ = molecular weight of chalcopyrite ($CuFeS_2$)

$$\text{Percentage of copper in chalcopyrite} = \frac{\text{atomic weight Cu}}{\text{molecular weight } CuFeS_2}$$

$$= \frac{63.54}{183.53}$$

$$= 0.346$$

$$= 34.6 \text{ percent}$$

Step 2. The percentage of copper in each unit of ore will be

$$1.5 \text{ percent} \times 34.6 \text{ percent} = .52 \text{ percent}$$

Then each ton of ore will yield 10.40 pounds of metallic copper, which will be worth

$$10.40 \text{ pounds} \times \$.50/\text{pound} = \$5.20$$

Step 3. If the mining firm is to have at least 5 percent profit, the cost of producing this copper must be no more than

$$\frac{\$5.20}{1.05} = \$4.95$$

A cost of $1.84 per ton has already been incurred for processing and smelting the ore. So

$$\$4.95 - \$1.84 = \$3.11$$

is the maximum allowable cost per ton for removing the ore from the mine.

inorganically. Perhaps dolostone is a secondary product resulting from the addition of magnesium to primary deposits of limestone.

FORMATION OF METAMORPHIC ROCKS

What happens when a mass of sedimentary rock—a body of shale for instance—is brought into an environment where the temperature and pressure are much higher than the environment in which the rock was formed? What happens to a large body of volcanic basalt when it is subjected to high pressures, mechanical deformation, and the presence of abundant water? What happens to a thick bed of biochemically derived limestone when it is subjected to high temperatures and pressures? The changes in rock bodies that occur as physical and chemical conditions are changed are called *metamorphism*.

As metamorphism proceeds, chemical reactions tend to reorganize the elemental material of the rock into new compounds that are stable under the new conditions. One of the first mineralogical reactions is *dehydration*, which involves the loss of water. An example is the breakdown of the clay minerals, which have substantial amounts of water held in their crystalline structures, and their conversion to micas, which have less water.

In a region surrounding a hot, igneous intrusive body, metamorphism may occur at high temperatures but low pressures. In such environments, low-density minerals with open crystal structures will form because such minerals are stable at high temperatures and low pressures. Conversely, in rock masses subjected to rapid burial but without a great temperature increase, extremely high-density minerals with closely packed crystal structures will form because they are stable under those conditions. Most commonly, however, metamorphic conditions include both high temperatures and high pressures. The new minerals that are stable under such conditions include most of the garnets, the feldspars, quartz, the micas, the amphiboles, and, at relatively high temperatures, the pyroxenes.

Metamorphism brings about some obvious changes in the physical appearance of rock masses. These changes result from *recrystallization*. This process involves not only the mineralogical changes just discussed; it also involves changes in the size, shape, and textural interrelationships of the mineral grains of the rock. Recrystallization of rocks undergoing metamorphism generally results in an increased grain size, so that metamorphic rocks are usually coarser in texture than their igneous or sedimentary precursors.

Furthermore, many minerals that are stable under metamorphic conditions have a tabular or sometimes elongate crystal form. Examples are the micas, with their tabular crystalline form, and certain amphiboles of the metamorphic rocks, which have extremely elongate, almost needlelike form. Because these minerals are formed under relatively high-pressure conditions and often under conditions of directional stress, it is common to find an orientation, or parallelism, of these minerals in metamorphic rocks. In the case of parallel orientation of platy or tabular minerals such as the micas, the rock is said to be *foliated*. Foliation may also result from the orientation of elongate minerals such that their long dimensions lie predominantly within one plane in the rock (see Figure 20).

Metamorphism is a process that occurs slowly within the earth's crust, and you might expect that the reactions producing the new mineral assemblages of the metamorphic rock have reached *chemical equilibrium*. If metamorphic processes are truly equilibrium processes, then for a given bulk chemical composition, metamorphism at the same pressure

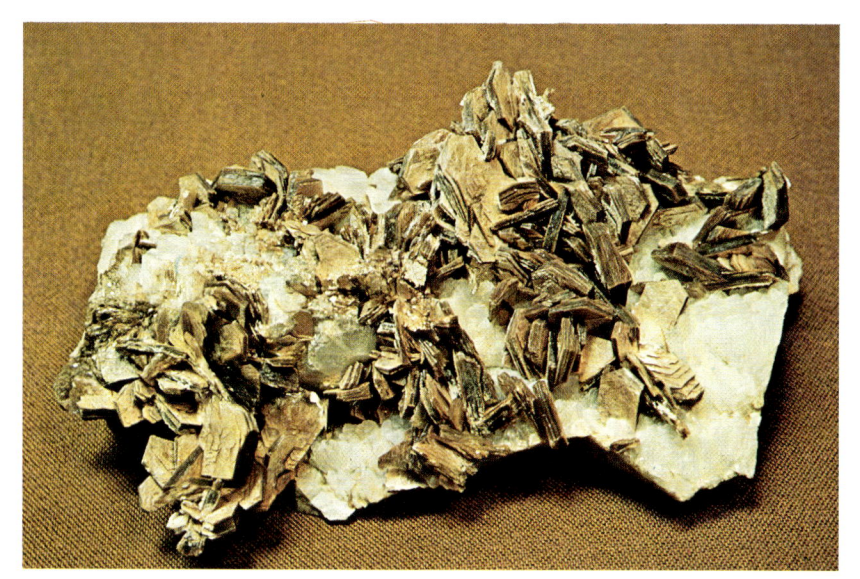

Figure 20. Layers of mica form the curious igneous rock known as pegmatite.

and temperature conditions should always produce the same mineral assemblage. Because the temperatures and pressures under which metamorphic minerals can form have been studied in great detail, it is often possible to equate rocks of grossly different chemical and mineralogical compositions to a common set of metamorphic conditions. Such rocks are said to belong to a specific *metamorphic facies*—a collection of rocks all formed under the same metamorphic conditions.

Although the principle of metamorphic facies provides a genetic way of classifying metamorphic rocks, it is not too useful in assigning common, descriptive names. A nongenetic, nonchemical scheme has evolved that is based on the structure and mineral composition of the rocks (see Table 3).

Table 3.
Common Metamorphic Rocks

Metamorphic Rock	Common Mineral Constituents	Commonly Derived From
Foliated Types:		
Slate	Clay minerals, chlorite, minor micas	Shale
Schist	Various platy minerals such as micas, graphite, talc, plus quartz and sodium, plagioclase, feldspar	Shale, basalt
Gneiss	Quartz, feldspars, garnet, micas, amphiboles, sometimes pyroxenes	Shale, granite
Nonfoliated Types:		
Quartzite	Quartz, sometimes minor muscovite and feldspar	Sandstone
Marble	Calcite or dolomite, plus minor calcium-silicate minerals	Limestone or dolostone

14 | Materials of the Earth

Weathering and Erosion

Volcanism

Sedimentation

Lithification

Sedimentary Rocks

Plutonic Intrusion

Metamorphism

Metamorphic Rocks

Melting

Magma Chamber

From the Mantle

Figure 21. The rock cycle, depicted schematically and anthropomorphically. All three types of rock—igneous, sedimentary, and metamorphic—can be uplifted to form mountains. All three can sink into magma, either to form underground crystallized rock or to return to the earth's surface during volcanic eruptions. Rock formed from volcanism or from uplift can be weathered and eroded into sediments. Sediments can be transformed into sedimentary rock or into metamorphic rock, which can again be uplifted or melted into magma in the continuing saga of our restless rocks.

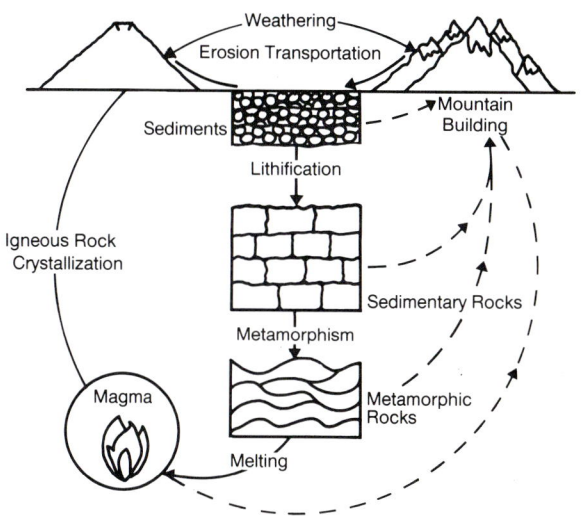

Slate is a rock that has been recrystallized to the point that it tends to split parallel to the new foliation, although this foliation may not be easy to see. *Schist* is a rock in which one of the tabular minerals, usually a mica, is the most abundant mineral. They have a kind of *closed foliation;* in other words, the platy minerals are tightly packed together in a parallel manner and are not separated by alternating bands of some other mineral. *Gneiss* is a rock with several different mineral constituents. It usually displays an open foliation, which is a layering or banding that segregates different minerals from one another.

Years ago, geologists recognized that metamorphic rocks seemed to occur in series. Within a given bulk chemical composition they found gradual transitions into rocks representing higher and higher pressure and temperature conditions. Occasionally they came across retrograde series representing conditions of falling temperature and pressure but these series were much less common because reaction rates are much slower under falling temperature. In fact, such a decrease in reaction rate, together with the almost total loss of water in relatively high-grade metamorphic rocks, probably contributes to the preservation of high-grade metamorphic rocks at the earth's surface.

A typical series progression might be described in terms of the metamorphism of a shale. During the earliest stages of metamorphism, the shale, which is predominantly composed of clay minerals, begins to lose its water. Its component clay minerals become somewhat recrystallized, forming micaceous minerals such as chlorite. As newly generated metamorphic minerals, the clays may be recrystallized at right angles to directional stresses. The new rock, which would be called a slate, possesses a well-defined rock cleavage that is parallel to the incipient foliation. The recrystallization of the clay minerals becomes complete at higher grades of metamorphism and the rock may then consist primarily of micas, sodium plagioclase feldspar, and quartz.

At a stage when the micas are first developing in appreciable quantities, the rock possesses a certain sheen along cleavage surfaces and could be called a phyllite. As metamorphism progresses and the micas become more abundant and better developed in large flakes, the rock will become a schist. At still higher grades of metamorphism, progressive dehydration of the micas yields potassium feldspar and aluminosilicates such as kyanite or sillimanite. Garnets may develop and the resulting rock will be a gneiss, with a mineral composition not unlike that of an igneous granite.

Because the original shale may have formed from the weathering products of a granite, the cycle is complete: from igneous rock it will have progressed through weathering, erosion, transportation, deposition, and lithification to sedimentary rock, and then finally through metamorphism back to a granite-like, although foliated, metamorphic rock.

THE ROCK CYCLE

Earlier in this chapter you read about the kinds of changes taking place in assemblages of high-temperature minerals, such as those characteristic of igneous rocks when they are exposed at low temperatures to the oxidizing and hydrating influence of the atmosphere and hydrosphere. Those changes represent a kind of metamorphism, although by convention they are classified as rock weathering, not metamorphic, processes. Similarly, at high-temperature rock metamorphism you might imagine rock masses reaching the melting temperature and bodies of silicate melt being formed.

At this stage the realm of metamorphism passes into the realm of the generation of magmas and the beginning of igneous activity. However, as in most natural processes the boundaries between these phenomena are gradational and vague. Thus at low temperatures, sedimentary rocks often undergo genuine mineralogical changes during the lithification process. Such changes are termed *diagenesis* but they are really the beginning of very low temperature metamorphism. Similarly, at high temperatures, rock masses undergoing high-grade metamorphism commonly show veinlike or podlike masses of crystalline rock of the composition of granite. Such masses are believed to be the result of partial melting. Clearly the relationships between the various kinds of rocks form a kind of cycle. This relationship, which is termed the rock cycle, is depicted in Figure 20 on the preceding pages. The diagram shows that magmas may reach the surface, forming volcanic rocks, or they may be implaced within the crust, forming plutonic rocks.

In their turn, volcanic rocks, or uplifted plutonic rocks, may undergo erosion and thus yield sediment. Upon lithification, this sediment forms sedimentary rocks, which may themselves be uplifted and eroded to produce a new or second cycle sediment. Once they are buried deep within the earth's crust, sedimentary rocks or volcanic rocks may undergo metamorphism because of the high temperatures and pressures and become metamorphic rocks. Finally, under extreme metamorphic conditions where temperatures reach the melting point, new bodies of magma may be formed. The rock cycle is complete.

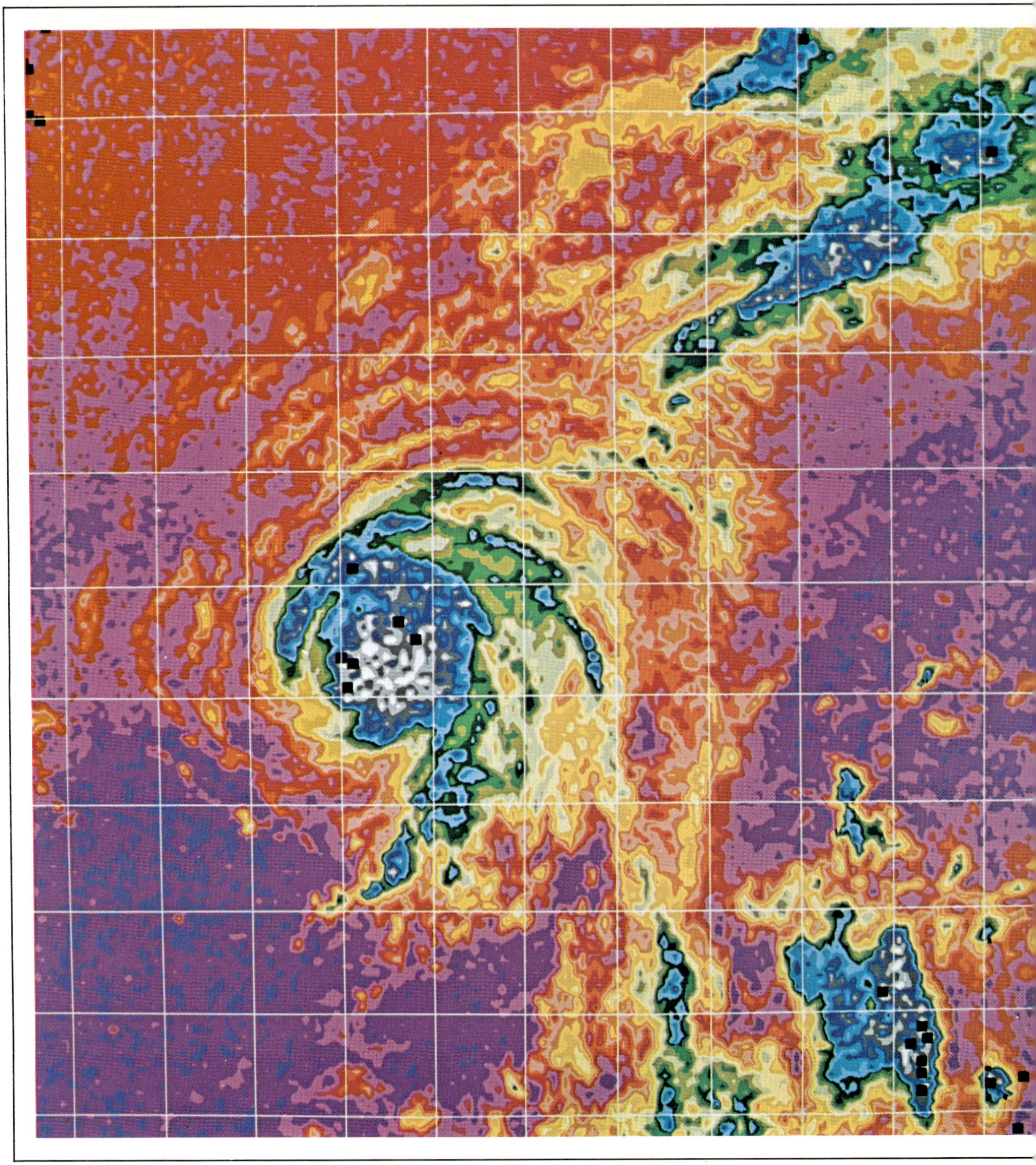

Eric Sloane, in *Look at the Sky,* conveys his immense delight with the sea of air we live in:

"The New England air that I breathe in as I write these lines was somewhere over Chicago yesterday. And that very same air which I breathe out will be about five hundred miles at sea by tomorrow. The air we live in is more like a mass of rushing rivers.

"The invisible stuff around us is amost never still. Clinging to the earth by its own weight, the sun may boil it up into dense mountains but it immediately seeks its own level exactly in the manner of water; then it cascades downhill as wind, and it settles into quiet pools of calm air. But always, somewhere, there is another air mass building up, or another air mass on the move.

"In my mind's eye I see the atmospheric sea as a gaseous symphony that gyrates and squirms over the surface of the earth . . . Destined never to be stagnant, the atmosphere is everlastingly stirred by giant bubbles of air that build up in still places to become what weathermen call air masses. These bubbles wobble for a while over their birthplaces and then, when they become full grown, they break away to roam across the land and crash into other kinds of air. And in their collisions, along their fronts, they create disturbances known as storms."

Many of us, like Sloane, look on the atmosphere as one of our most beautiful natural resources, and one of the most priceless. We grow up watching cloud shapes and wondering about storms in the night; we muse over a breeze we can "see" because of the pollutants it carries. And in an era of concern over our interaction with the environment, we occasionally ask: Are there limits on this resource, also? To what extent have man's activities affected this sea of air and its movements above the whole earth?

Consider the kind of episode that has caused us to ask these questions in the first place. In December 1952, London experienced a 5-day seige of atmospheric pollution of such magnitude that death rates ran many thousands, and serious illness many tens of thousands, above their normal levels. The London disaster resulted from a large mass of clear high-pressure air that stagnated over the city. Air temperatures near the ground were lower than air temperatures higher up, a condition known as a thermal inversion. The heavier cold air remained trapped near the ground. A fog developed, and gradually the output of thousands of soft coal fires in factories and homes throughout the city turned the London air into a sea of death.

Since that disaster, London has applied severe restrictions on the use of soft coal. Today Londoners enjoy 70 percent more sunny days in the winter than occurred 10 years ago as well as a threefold increase in visibility. Nevertheless, London has not solved its total air pollution problem. Even though it has removed the more serious threat of soft coal, it has joined the rest of the major cities of the world in facing a series of other air pollution difficulties.

Today when we think about the changing atmosphere, the first thing that usually comes to mind is this sort of growing local and regional air pollution problem that most urban areas now face on a regular basis. The effects are dramatic and serious; they demand immediate attention. These problems can, however, be brought under control.

15

The Changing Atmosphere

NASA computers painted this portrait of Hurricane Camille. The colors signify temperature changes with altitude.

Figure 2. London now has more clear days even though severe smog conditions such as those depicted in the photograph to the right recently prevailed. Londoners applied severe restrictions to the burning of soft coal and they have since enjoyed a 70 percent increase in clear weather. The technology exists to clean up much of atmospheric pollution, once economic, social, and political problems are worked out.

Even though the complex mechanisms of local air pollution are not fully understood, *the basic science needed to bring about the solution is within reach.* Fundamentally, local air pollution is not a problem of basic science but of politics, economics, and careful engineering.

On the other hand, how much is known about the effects that man's activities have on the global atmosphere? Here the problem becomes less straightforward. The physical laws that govern the motions and changes of the atmosphere have been known for a long time. But only in the last few years, with the advent of powerful digital computers, have scientists been able to begin exploring the details of how these fundamental laws interact to produce details of weather and climate.

In comparison to the incredible energies underlying normal weather patterns, the energy involved in man's activities, even through nuclear explosions, is small. It would seem, on the surface, that man is in no danger of inadvertently changing the weather. But the atmosphere is a complex system and no one knows how delicate its various balances may be. For example, consider what might happen if man were to succeed, however unintentionally, in lowering the mean temperature of the earth by a few degrees—say, something like 2 to 6 degrees Centigrade. That is probably all it would take to trigger a new Ice Age. On the average, natural fluctuations in the mean earth temperature have never varied more than about 1/2 degree Centigrade during the last century. Until now, man's effect on these temperature changes has probably been small. But evidence indicates that his effect is increasing as a result of the carbon dioxide, dust, and waste heat that are the by-products of his activities. Until the new computer models are significantly improved, no one can say how great the danger may be, or even whether or not a danger exists.

And what about deliberate assaults on the atmospheric system? Rainmakers in many cultures have had limited success. Today, however, it appears that man's technical ability to modify the weather will develop soon—and well before his social and political advancement will be ready to meet the challenges this modification will create. It may not be long before subtle pressures applied at crucial points will be enough to cause world-wide changes in the weather. Will we inherit unknown winds in spite of our knowledge of the atmosphere, or because of it?

THE ATMOSPHERE—A "STILL SHOT"

The atmosphere is a dynamic system, continually in motion and continually changing. How can you study a system that is time varying? One way is to look first at its static characteristics—the features that do not change appreciably, or at least rapidly, with time. Suppose you have a motion picture of the atmosphere and you are about ready to run the film through a projector. Pause, for a moment, and look at a single frame of that film, one that depicts a panoramic cross-section of the sea of air.

Atmospheric Pressure

The most noticeable characteristic of the atmosphere is that it is not homogeneous. Its density, for instance, changes with altitude. At the earth's surface, the mass of molecules in each cubic meter of air is about 1 kilogram. At an altitude of 16 kilometers, the mass of molecules is only 0.1 kilogram. At 32 kilometers, or twice that altitude, the density is 0.01 kilogram. At 48 kilometers it becomes 0.001 kilogram; at 64 kilometers, 0.0001 kilogram; and so on. Because the mass of the individual molecules in the air does not change, the air itself must be thinner at higher altitudes. And the values above show that, for every 16 kilometers above the earth, the number of molecules decreases by a factor of 10.

The distance 16 kilometers is characteristic of the way atmospheric density falls off with altitude. This distance is referred to as the *scale height* of the atmosphere. It tells you how much higher you would have to go to find a density one-tenth that found at the altitude where you happen to be at the moment. The value of scale height is not perfectly constant but changes slightly with altitude.

Why does the air get thinner at higher altitudes? Suppose you could take away the normal atmosphere and surround the earth with a uniform layer of air. Because of the earth's gravitational pull, the molecules in that air would begin to move down toward the earth's surface. All the molecules would be attracted to the earth; all would try to move toward it. Soon there would be many more molecules in a cubic meter of air near the surface than there would be at higher altitudes, and the pressure near the surface would become so great that no more molecules could squeeze in. Ultimately a state of equilibrium would be reached, and the atmosphere would end up looking like it does in Figure 3.

The pressure that the air exerts on 1 square inch at the earth's surface is the same as the weight of all the air you would find in a 1-square-inch column extending from the earth's surface up to the "top" of the atmosphere. This pressure is called the *atmospheric pressure*. At sea level it has a value of roughly 14.7 pounds per square inch. The mass of this column of air is the same as the mass of a 1-square-inch column of water 33.9 feet high, or a 1-square-inch column of mercury 76 centimeters high. A barometer, an instrument that measures the pressure of the atmosphere, often uses a column of mercury, so the pressure is sometimes expressed in centimeters of mercury.

In meteorology, however, pressure is usually expressed in a unit called a *millibar.* A pressure of 1 millibar is the same as a pressure of 0.0145 pound per square inch. Because pressure falls off with altitude and is a more important number for most meteorological purposes than altitude, height is often expressed in units of millibars. Sea level is just over 1,000 millibars (1,013 millibars for standard atmospheric pressure);

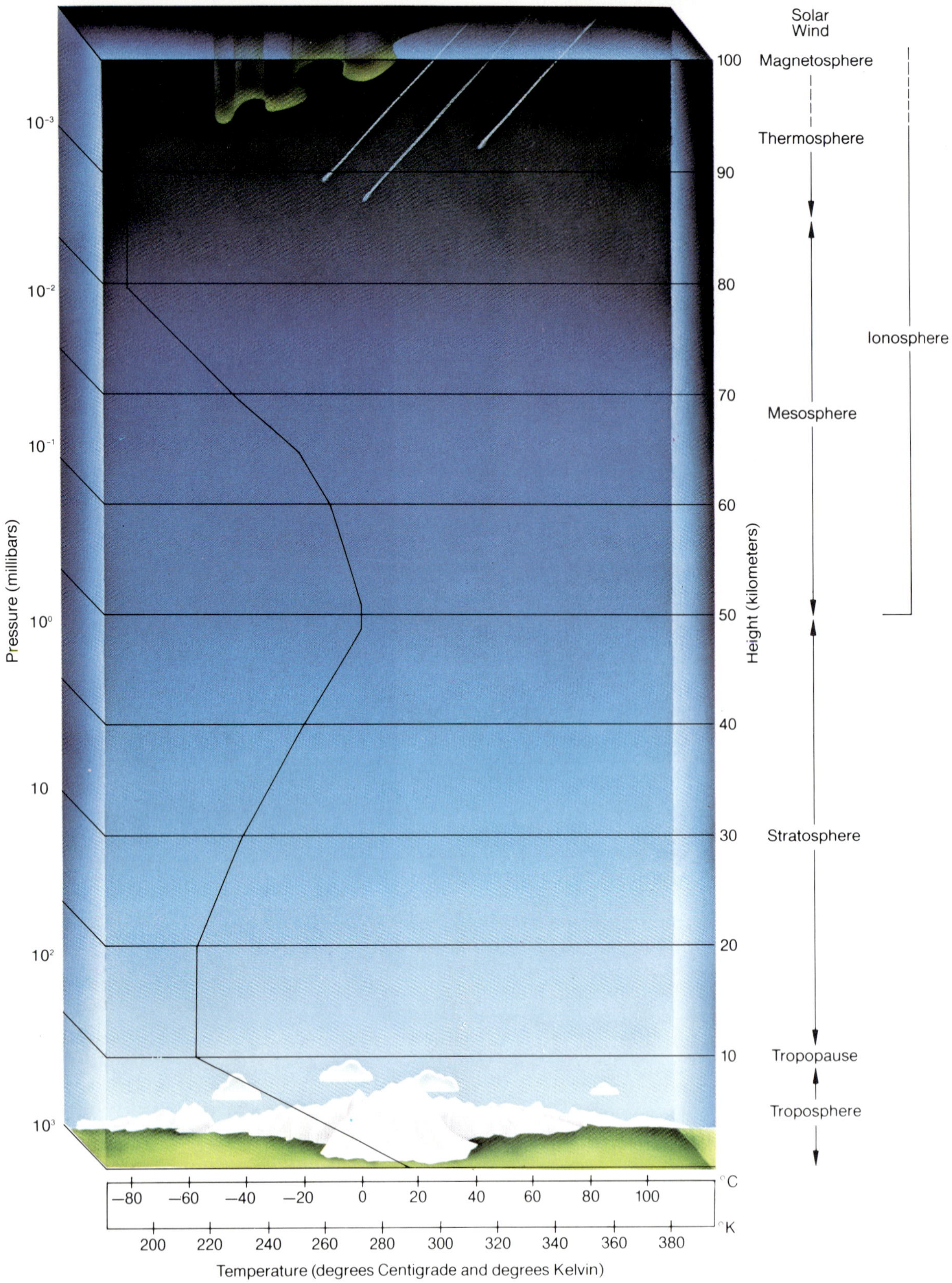

Solar Wind
Magnetosphere
Thermosphere
Ionosphere
Mesosphere
Stratosphere
Tropopause
Troposphere
Pressure (millibars)
10⁻³
10⁻²
10⁻¹
10⁰
10
10²
10³
Height (kilometers)
100
90
80
70
60
50
40
30
20
10
°C
−80 −60 −40 −20 0 20 40 60 80 100
°K
200 220 240 260 280 300 320 340 360 380
Temperature (degrees Centigrade and degrees Kelvin)

900 millibars is an altitude of about 3,000 feet; and 500 millibars is an altitude of about 18,000 feet.

If there are 14.7 pounds of pressure pushing in on every square inch of our bodies, why don't we collapse? The answer is that the stuff inside — the water, air, and tissue — is pushing out with exactly the same pressure as the air pushing in. There is no danger of collapsing from normal atmospheric pressure. On the other hand, if you were to step from a pressurized spacecraft into the vacuum of space, parts of your body would rupture because your internal pressure would be greater than that of your surroundings.

Regions of the Atmosphere

Figure 3 shows the temperature of the atmosphere at various altitudes. Approximately the first 10 kilometers is called the *troposphere*. Most of the normal weather patterns you observe occur in this region. The temperatures of the troposphere vary considerably, but they average about 290 degrees Kelvin (17 degrees Centigrade) at the surface and drop rather uniformly to just under 220 degrees Kelvin at an altitude of 10 kilometers. For meteorological purposes, the lowest kilometer or so of the troposphere is often distinguished from the main body of the troposphere. In this lower region, frictional forces from the earth's surface features have a significant effect on air motion. Wind velocities tend to decrease at this lower altitude.

The transitional region between the troposphere and the next region is called the *tropopause*. Above the tropopause is the *stratosphere*, which is characterized by a relatively constant temperature of about 220 degrees Kelvin, or −53 degrees Centigrade, in its lower portion. Such a region, where the temperature is constant, is said to be *isothermal*.

The values given in Figure 3 are for the *U.S. Standard Atmosphere*, a widely accepted description that is based on *average* values. If you actually were to go out and measure the altitude at which the isothermal part of the lower stratosphere begins, you would find it may range from about 8 kilometers in polar regions to perhaps 18 kilometers at the equator.

Above about 20 kilometers in the average atmosphere, the stratosphere no longer is isothermal but gradually increases in temperature to an altitude just under 50 kilometers. There are only limited weather patterns in the lower stratosphere because circulation patterns tend to be fixed. The only visible weather at this altitude is an occasional cirrus cloud. The air in the stratosphere is quite stable and is not exchanged rapidly with air in the lower atmosphere. For this reason, some scientists are fearful of the impact that a large fleet of high-flying SST aircraft may have on the stratosphere. Although aircraft fly efficiently in this region of thin, cloudless air, there is a possibility that accumulated effluents from many aircraft might upset the existing heat balance.

Above the stratosphere are two more regions, often referred to as the *mesosphere* and the *thermosphere*. The mesosphere is characterized by temperatures that decrease with altitude, whereas temperatures in the thermosphere begin to increase again.

The very high atmosphere (Figure 4) is also known by other names. This is a region where the sun's radiation breaks up the molecules into electrically charged ions and free electrons, which is a state of matter called the *plasma state*. Beginning at about 50 kilometers and extending several thousand kilometers is a region referred to as the *ionosphere*. The plasma it contains reflects lower frequency radio waves, which are useful in world-wide communications systems. Beyond, and overlapping the ionosphere, lies another region of plasma that is associated with the

Figure 3. (Far left) This drawing shows the structure of the earth's atmosphere.

Figure 4. The magnetosphere begins about 40,000 miles away from the earth in the direction of the sun and extends much farther in the opposite direction. The distortion is caused by the solar wind, which is made up of streams of atomic particles sent out in all directions from the sun. As the solar wind approaches the earth, it encounters the earth's magnetic field and a shock wave forms. On the side of the earth away from the sun, this shock wave gradually weakens until it no longer is detectable.

Table 1.
Components of the Atmosphere

Component	Percentage
Nitrogen	78.08
Oxygen	20.95
Argon	0.93
Carbon dioxide	0.03

earth's magnetic field—the *magnetosphere*. Beyond that lies the extremely tenuous plasma of interplanetary space. This plasma is an extension of the sun's atmosphere and is known as the *solar wind*.

Components of the Atmosphere

Of all the components of the atmosphere, nitrogen is the most prevalent. It represents 78 percent of the volume of "dry" air; oxygen makes up 21 percent (see Table 1). To these components, natural processes and the activities of man add various trace gases. Ozone, for example, is a molecule of three oxygen atoms that forms in the stratosphere under the influence of ultraviolet radiation from the sun. Carbon dioxide is another trace component, brought about by plant and animal respiration and the burning of fuel. Its concentration has changed significantly over time.

The percentages in Table 1 are given for "dry air," although most air in the lower atmosphere is not "dry." The amount of water vapor present in the air ranges from essentially none in some desert and polar areas to as much as 4 percent in humid, tropical regions. The maximum amount of water vapor that a certain volume of air can hold depends primarily on the air temperature.

The expression "relative humidity" refers to how much water vapor a volume of air is actually holding relative to how much it is capable of holding. When air is holding as much water vapor as it can for a given temperature and pressure, it is considered saturated. The "absolute humidity" is a measure of the mass of the water vapor contained in 1 cubic meter of air. For a given absolute humidity, the relative humidity of warm air will be lower than that of cold air—which is just another way of saying that warm air can hold a lot more water vapor than cold air.

Atmospheric Stability

Although atmospheric pressure changes over time and from place to place, it can be considered static at least for short periods of time. Suppose you take a parcel of air and prevent heat energy from flowing into it or out from it. If you expand that parcel into a large volume, its pressure will drop and so will its temperature. If, instead, you compress the parcel into a smaller volume, its pressure will increase and so will its temperature. A change of this sort, in which (ideally) heat energy is neither gained nor lost, is said to be *adiabatic*.

You know that pressure in the atmosphere decreases with altitude. Imagine a parcel of dry air at sea level (1,013 millibars). If you lift such a parcel through an atmosphere of dry air, allowing no heat to flow in or out, it should take on a specific volume and temperature at each altitude. The rate at which the temperature of a dry air parcel falls as that parcel is moved upward through the surrounding atmosphere is called the "dry-adiabatic lapse rate." This rate is equal to 1 degree Centigrade for every 100 meters, or 5.5 degrees Fahrenheit for every 1,000 feet. Of course, the atmosphere is not composed of "dry" air, so the actual value for the standard atmosphere turns out to be more like 3.6 degrees Fahrenheit for every 1,000 feet.

Whether or not the atmosphere is stable or unstable depends on this lapse rate. When the lapse rate for an atmosphere of unsaturated air is the same as the adiabatic lapse rate for a parcel of air within it, the overall atmosphere is in *neutral equilibrium*. Anywhere the parcel is moved, it stays put.

But consider what happens when the lapse rate for the surrounding atmosphere is less than the adiabatic lapse rate. When the parcel is moved down to a lower level and then released, it moves back up to its old level; when it is moved up, it will move back down. The overall atmosphere is in *stable equilibrium*. Cloud structures that develop in such a stable atmosphere are *stratiform*, or arranged in horizontal layers. Finally, when the atmospheric lapse rate turns out to be greater than the adiabatic lapse rate, the atmosphere is *unstable*. In such cases, an air parcel that is moved slightly upward and then released continues to rise; an air parcel that is moved slightly downward and then released continues to fall. Cloud structures developing in an unstable atmosphere are *cumuliform*. They tend to be towering structures with considerable internal turbulence.

These are all examples for unsaturated air. But air usually contains water vapor, and water vapor can condense or evaporate—changes that involve heat energy. Meteorologists get around the problem by defining both "moist" and "dry" adiabatic lapse rate. As air moves upward and is cooled, sooner or later its relative humidity reaches 100 percent and it becomes saturated. The same kind of lapse rate analysis continues to work for the saturated air but in this case it is called a "moist-adiabatic lapse rate."

DYNAMICS OF THE ATMOSPHERE

Now that you have looked at a still shot and have glimpsed the more-or-less static characteristics of the atmosphere, you can let the film run through the projector and watch Sloane's "gaseous symphony that gyrates and squirms over the surface of the earth."

If the atmosphere may be likened to a symphony, the sun may be considered the conductor of its movements. The energy emanating from the sun is the source of all weather patterns—from winds and clouds to the compression and expansion of enormous amounts of air. The structure of our atmosphere plays a part, as does its composition. But basically our weather begins far out in space, with the energy radiating from the surface of the sun.

The Balance of Heat Energy

The surface of the sun is extremely hot—about 6,000 degrees Centigrade. As a result, the photons it emits are most energetic. For some purposes,

Figure 5. In the broadest terms, clouds are cumuliform (above), stratiform (below), or some combination of the two. Cumuliform clouds are towering, turbulent structures created in an unstable atmosphere; stratiform clouds are characterized by streaming horizontal layers.

Figure 6. (Far right) The heat balance between the earth and incoming solar radiation. The earth's 30 percent albedo is due to direct reflection of visible sunlight, shown by the arrow on the left. The remaining 70 percent, which is absorbed by the atmosphere (20 percent) and the earth's surface (50 percent), is eventually reradiated into space in the form of infrared waves. Several other processes, depicted in the right-hand corner of the drawing, become involved due to complications in the energy transport mechanism.

photons are thought of as discrete, tiny packets of energy. But when you are considering an object that gives off countless numbers of photons, such as the sun, it's convenient to think of their combined effect as an electromagnetic wave.

There is also an "invisible" form of light found in the electromagnetic spectrum. Although most of the photons from the sun correspond to visible light, there are other, more energetic photons that can be described by electric and magnetic fields that oscillate more rapidly than they would for light. The energy they carry is called *ultraviolet energy*. Other photons carry far less energy than they would for light. This energy takes the form of heat, or *infrared energy*. Even slower oscillations represent waves of radio energy. This distribution in the intensity of the various forms of radiant energy given off by the sun is referred to as the *solar spectrum*.

When radiant energy from the sun enters the earth's atmosphere, a number of things happen. The more energetic photons, which correspond to the shortest (ultraviolet) wavelengths, collide with oxygen molecules high in the atmosphere. Their energy tears these molecules apart, changing them to free atoms. At a somewhat lower level, these free oxygen atoms encounter oxygen molecules and react to form ozone; ozone in turn absorbs photons of somewhat longer ultraviolet wavelength. Although these processes involve a mere 1 to 3 percent of the total incoming energy, that percentage represents the most intense ultraviolet energy from the sun. Without this protective barrier of absorbing gas, life on earth, as we know it, could not exist.

Just as most of the very short wavelength ultraviolet energy is absorbed by the atmosphere, so is much of the long wavelength infrared energy from the sun absorbed. However, different processes are involved that have to do with water vapor at rather low altitudes. On the other hand, little of the visible light energy from the sun is absorbed by the atmosphere. Some is reflected off clouds and particles and some continues on down to the earth's surface, where it is either reflected or absorbed. Because little of the visible light from the sun is absorbed by the atmosphere, the atmosphere is often said to have an *optical window*.

A brilliant layer of white clouds or snow reflects most of the solar energy that reaches it, whereas a black asphalt road reflects very little. All of the earth's features—great expanses of clouds, green stretches of wilderness, brown deserts, blue waters—reflect intermediate amounts of light energy. The amount of radiant energy reflected by all these processes is known as the *planetary albedo*. Until recently, scientists believed the value of the average earth albedo was 35 percent. But now spacecraft measurements have shown the actual value is close to 30 percent. Planned satellite experiments will moniter the value of the albedo with great precision because even minor changes in the reflectivity of the earth—for instance, the changes brought about by the contrails of large fleets of high flying aircraft—could significantly alter the global heat balance.

Altogether, as Figure 6 indicates, about 70 percent of the incoming radiant energy actually gets absorbed by the earth and its atmosphere. Surely if the earth and its atmosphere are absorbing that much energy from the sun, there must be some mechanism that works to maintain a long-term heat balance; otherwise the earth would have burned up long ago. In fact, an *equal amount* of energy must be reradiated back into space.

This reradiation is not a simple process. Unlike the sun, the earth is relatively cool, with an effective "black body" temperature of about 245 degrees Kelvin (−28 degrees Centigrade). As a result, almost all the

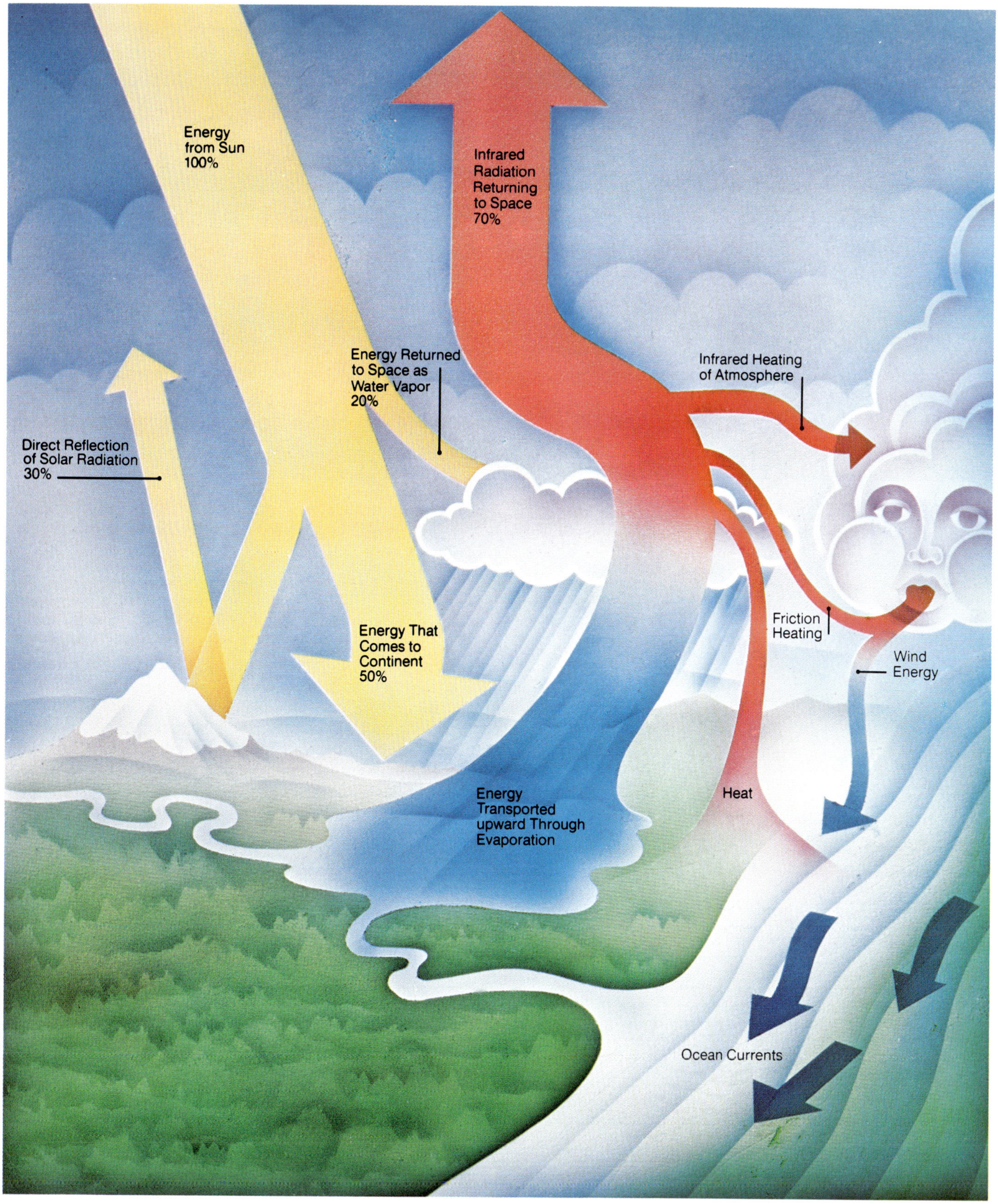

Energy from Sun 100%
Infrared Radiation Returning to Space 70%
Energy Returned to Space as Water Vapor 20%
Infrared Heating of Atmosphere
Direct Reflection of Solar Radiation 30%
Friction Heating
Energy That Comes to Continent 50%
Wind Energy
Energy Transported upward Through Evaporation
Heat
Ocean Currents

Hadley Cell Model

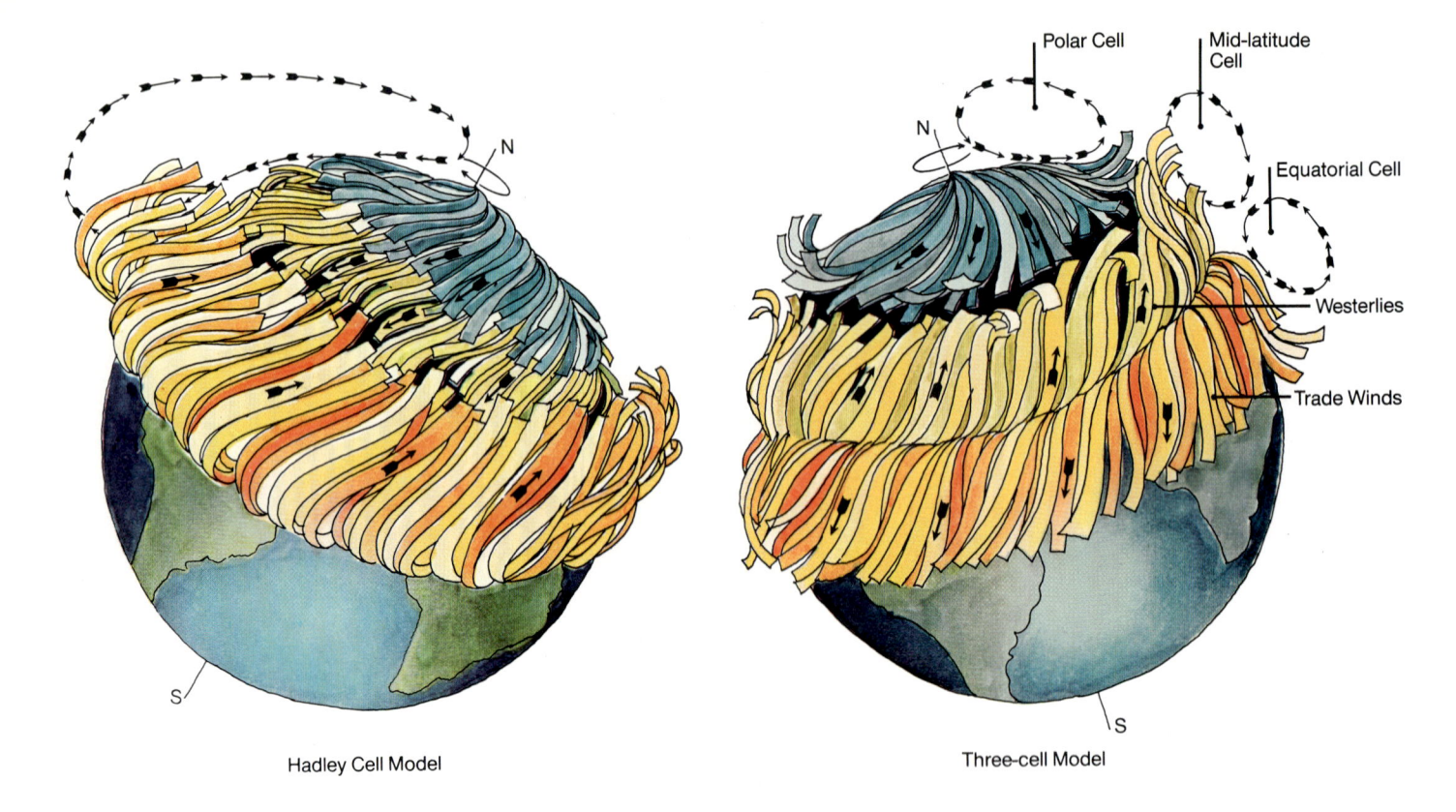

Three-cell Model

Figure 7. (Below) A large surface area near the poles intercepts the same amount of sunlight as a smaller area near the equator. That is why much more heat per unit area is absorbed at the equator.

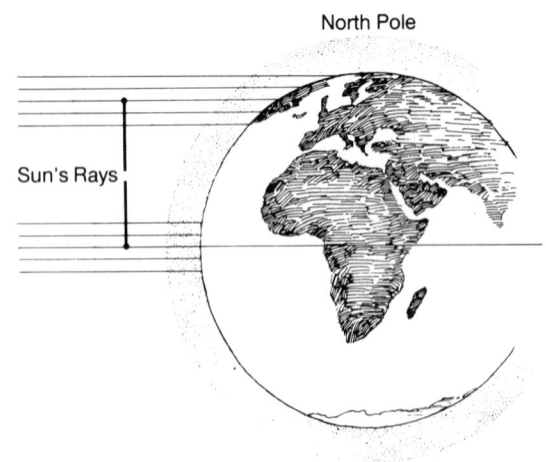

energy radiated into space by the earth system is in the form of infrared energy.

Just as infrared energy trying to get down to the earth's surface is blocked by the "opaque" atmosphere, so infrared energy trying to escape back into space gets blocked on the way up. This entrapment at low altitudes is sometimes called the *greenhouse effect,* although the actual mechanism is different from that of the florist's greenhouse.

This trapped heat energy evaporates water from the oceans and land. The moist air rises and moves to higher altitudes. Eventually its temperature falls, its relative humidity rises, and it soon becomes saturated. If conditions are right, the water vapor it carries will begin to condense into droplets. As water vapor condenses, heat energy is released. Because it is now high in the atmosphere, this energy is able to escape into space and the global heat balance is maintained.

The Distribution of Heat Energy

An atmospheric symphony limited to the ups and downs of water vapor would be a little tiresome, perhaps even a little dreadful: without any horizontal movement of heat over the earth's surface, the temperature difference from pole to equator would average hundreds, not tens, of degrees. Fortunately there is more to the atmosphere than the simple vertical movement of heat energy.

Horizontal heat movements exist because the heat energy that is balanced between incoming and outgoing radiation is not distributed evenly across the earth's surface. Because the earth is a sphere, 1 square kilometer of surface area in northern and southern latitudes gets less solar energy than 1 square kilometer at the equator (Figure 7). Because

Physical Science Today

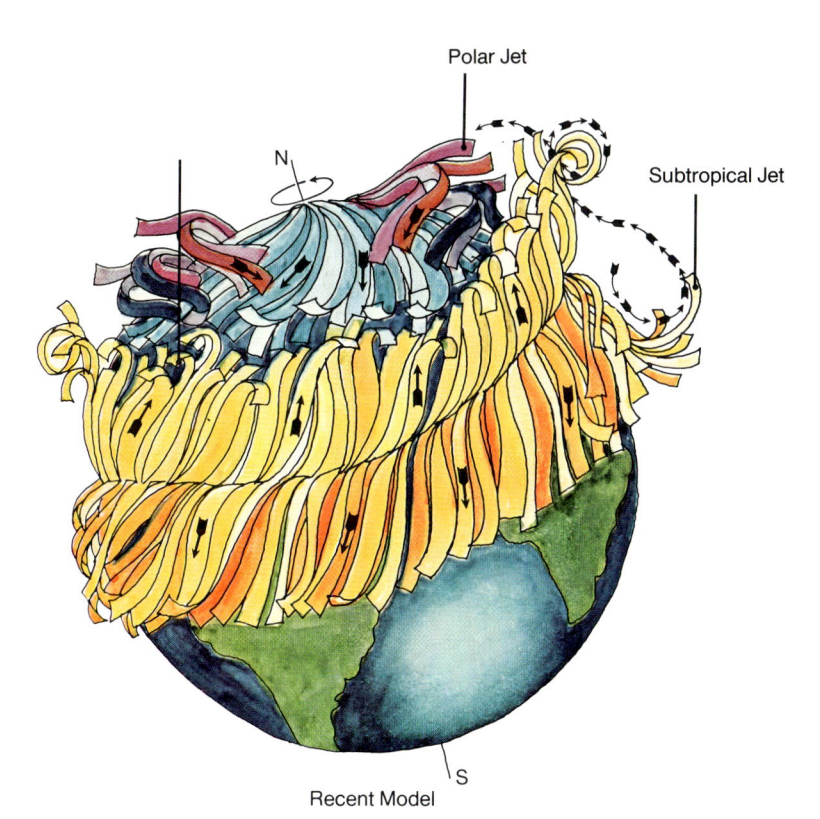

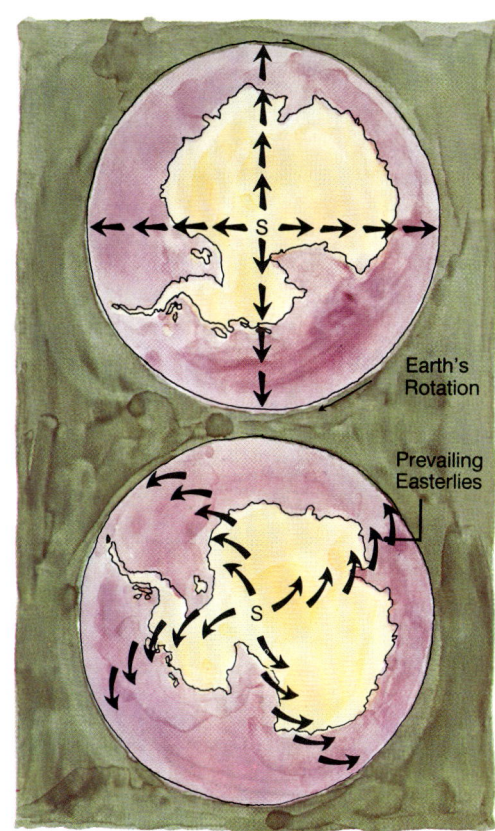

of the earth's inclination on its axis the intensity of incoming solar radiation at any one place varies, depending on the season. Complicating these variables is the presence (or absence) of clouds, atmospheric dust, and moisture in the air in various regions of the world. Moreover, incoming radiant energy heats land surfaces much more rapidly than it does water surfaces, and the distribution of land and water surfaces throughout the world is anything but uniform. These and other variables give rise to air motions that distribute heat around the earth's surface.

The seventeenth-century meteorologist George Hadley was the first to propose a model of how this distribution works. He suggested that warm, moist air rises at the equator and then moves toward the poles in a large, rotating pattern (Figure 8a). Today this pattern is called a *Hadley cell*.

One thing that complicates such a model is the *Coriolis force*, which is not really a force at all but an effect of the earth's rotation. Suppose you are riding one of the outer horses on a merry-go-round and you throw an opened bottle of ink to someone riding a horse on the other side. A spectator standing on the ground will see the ink bottle travel in a straight line. But once the merry-go-round stops and you've been put to work cleaning up the mess, you'll find that the trail of ink drops running across the platform is curved. The reason is simple: the merry-go-round continued to turn as the bottle flew through the air. Even though the path of ink drops on the platform implies there was some kind of force acting to curve the path, the bottle itself moved in a straight line.

Such an effect occurs on the rotating earth. The effect is strongest near the poles and is weak near the equator. What implications does this effect hold for the atmosphere? For one thing, in the absence of other forces a straight north-south motion is impossible. In the Northern

Figure 8. (Above, left) In the original Hadley cell model of atmospheric circulation (a), tropical air rises and moves poleward; the cold polar air sinks and moves to the equator. If global circulation were that simple, however, winds could reach speeds of 830 kilometers per hour! Interactions of the winds with the earth's surface were taken into account in a later model (b). A more realistic model of atmospheric circulation in use today (c) also takes into account the interactions between regions and explains the phenomenon we call jet streams.

Figure 9. (Above) The Coriolis effect causes an apparent deviation of moving particles in a rotating reference frame. The deviation is perpendicular to particle velocity as well as to the axis of rotation.

15 | The Changing Atmosphere

Hemisphere the movement of air will appear to curve to the right; in the Southern Hemisphere, to the left.

In the presence of the Coriolis effect the great long cells that Hadley proposed become unstable and break up into several cells in each hemisphere. Recognizing this, scientists propose a somewhat more accurate three-cell model. It contains an equatorial and a polar cell that rotate in the same way as the original Hadley cell and a mid-latitude cell that rotates in the opposite direction (see Figure 8 b). This model predicts several features of the earth's atmosphere, including the westerlies of mid-northern latitudes and the trade winds of the tropics. But even this model does not explain many of the most important features of weather and climate; and it is entirely useless for day-to-day weather forecasting.

A more recent model, depicted in Figure 8 (c), comes closer to describing the global circulation pattern. It includes an interface, or *front,* between two large air masses—the polar air mass and the tropical air mass. Rotating cells are still included, particularly at lower latitudes, but several high-level wind systems, or *jet streams,* have been added as well as a winter break in the tropopause. This model agrees fairly well with observational data and probably describes at least the gross properties of general atmospheric circulation.

FRONTS, HIGHS, LOWS, AND TONIGHT'S WEATHER

Before considering how better models of the atmosphere might be constructed, pause to look at some of the regional weather effects that all such models must explain. Even though specific details are different through global space and time, these basic effects remain much the same.

Consider the motions called *winds.* The force that drives them is usually a pressure difference between one region and another. Contours of constant pressure, called *isobars,* can be plotted in much the same way that lines of constant potential energy are plotted. The steepness of the slope represents the rate at which pressure changes perpendicularly across these isobars. This slope is called the *pressure gradient.*

Once air begins to move, the Coriolis effect deflects it toward the right in the Northern Hemisphere, or toward the left in the Southern Hemisphere. The amplitude of this effect increases as the velocity increases. Finally an equilibrium state is reached in which the pressure gradient force and the Coriolis effect are balanced exactly and the wind flows parallel to the isobars. When that happens the moving air is called a *geostrophic wind.*

Close to the ground, however, the wind flow generally does not quite parallel the isobars. Here a third force—*friction* with the earth's surface—becomes important. It is only at higher altitudes that ground friction forces are unimportant and geostrophic flow is regularly observed. And even at higher altitudes, sudden changes in the value of the pressure gradient and other, more subtle effects may create an air flow that is not perfectly parallel with the isobars.

Figure 8(c) depicts two large parcels of air—the polar and the tropical air masses. Within the two air masses are subparcels of warmer and colder air. As these air masses move about they encounter one another, and fronts develop where there are abrupt changes in temperature and pressure. Cold air typically moves more rapidly than warm air. When a region of cold air overtakes a region of warm air, a *cold front* develops. Cold air is more dense than the warm air it replaces, so the cold front tends to plow below the warm air. The warm air that is lifted becomes cooled and large cloud structures usually develop. These clouds fre-

Figure 10. In the photograph to the far right, several cold air masses are descending from the pole and ploughing under warmer air masses to form cold *fronts*. These fronts generally parallel the *isobars,* the contour lines shown in the diagram beneath the photograph. Because of the Coriolis effect, both cold and warm air masses tend to move toward the right through the stages depicted in the four sketches in the far right margin. The overall effect is a counterclockwise cyclonic storm system.

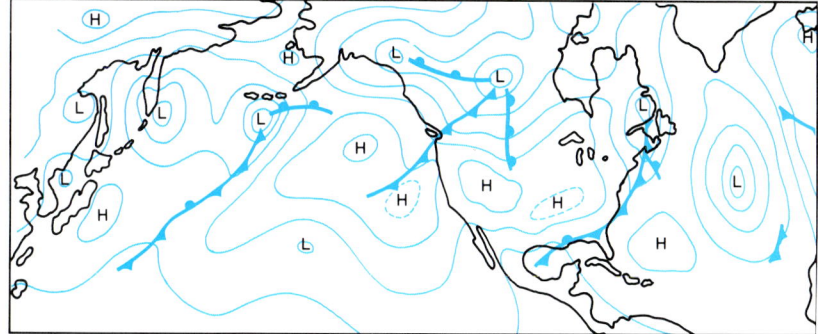

quently are confined to the region near the front. A *warm front* develops when a warm air mass overruns a colder mass. Usually a warm front slopes more gradually than a cold front, and cloud structures may extend well ahead of it.

Cold and warm fronts are shown in Figure 10. These are traveling fronts, and wind generally moves parallel to the lines of constant pressure. Across the front there is a rapid change in wind direction, called a *wind shear.* For fronts in the Northern Hemisphere, this change is always in a counterclockwise, or cyclonic, direction.

Sometimes there is no change of pressure across the front and wind blows roughly parallel to it. This *stationary front,* as it is called, frequently develops between the cold polar air mass and the warmer tropical air mass.

This front may develop kinks as it interacts with the global atmospheric circulation processes, such as the *jet stream,* which occur aloft. The forces at work in this process generate a low-pressure area at the kink, and a pair of moving fronts develops. A cold front moves down from the polar air mass and a warm front moves up from the tropical air mass. In the Northern Hemisphere these fronts move in a counterclockwise direction around the low-pressure area. The cold front usually moves more rapidly than the warm front and soon overtakes it, producing an *occluded front.*

Figure 11 shows this process in a large, low-pressure area with ground level winds spiraling about in a counterclockwise fashion. The wind near the surface is not perfectly geostrophic but gradually spirals inward. Such a large-scale weather system centered on a low-pressure area is known as a *cyclone.*

High-pressure areas produce an outward-flowing spiral of ground level air which, in the Northern Hemisphere, travels in a clockwise direction. These structures, which frequently are not as distinct as the cyclone structures, are called *anticyclones.* It is the gradual march of these large

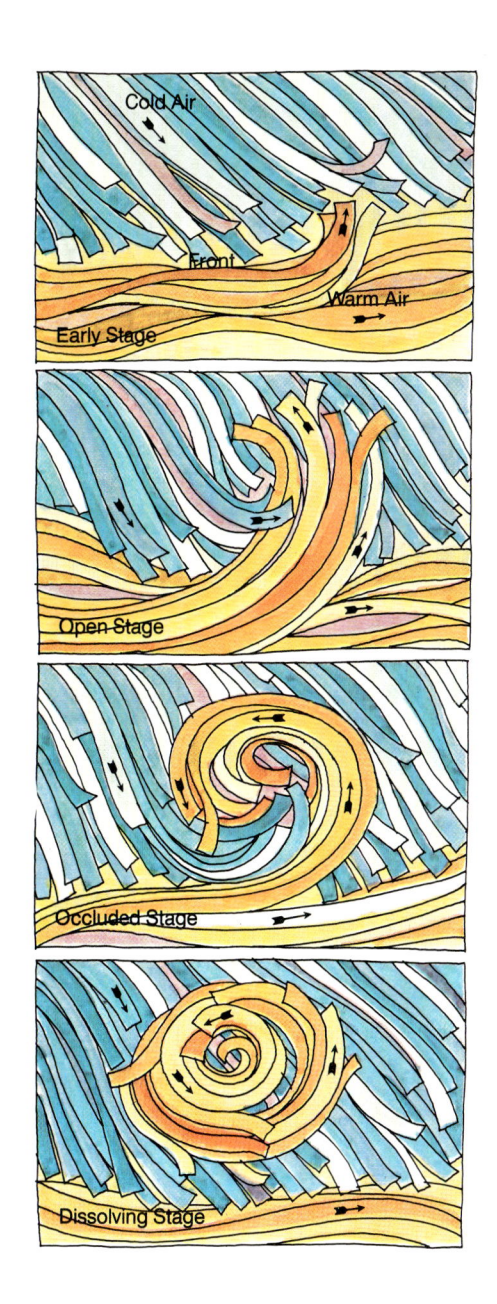

15 | The Changing Atmosphere

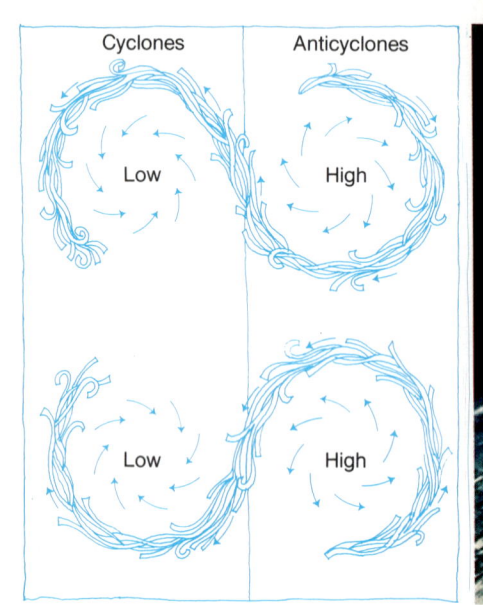

Figure 11. *Cyclones,* or cyclonic storm systems, result from the convergence of air toward a low-pressure center. In the Northern Hemisphere, they rotate counterclockwise. In the Southern Hemisphere, they rotate clockwise because the Coriolis effect is reversed. *Anticyclones*, resulting from the movement of air away from a high-pressure center, follow rotation patterns exactly opposite the patterns of cyclonic storm systems. Hurricane Gladys, a severe tropical cyclone, is shown in the photograph to the right.

low-pressure cyclones and high-pressure anticyclones in a general northwest to southeast direction that produces most of our weather on the North American continent.

A MODEL FOR THE ATMOSPHERE

By studying the individual weather patterns of past years, the meteorologist has learned how to make some educated guesses about what the weather will be like in the next day or so. What happens when he wants to predict detailed behavior of the weather for periods of several weeks, or the general behavior of climate for many years? For such tasks he must have an atmospheric model that is much more precise than the ones just considered.

Most of the physical laws that govern the workings of the atmosphere have been known for many years, and they can be expressed in a set of mathematical equations. These are known as the "primitive equations" and, despite what their name may imply, they are exceedingly difficult to solve. They cannot be solved by hand calculations and even the largest computers have trouble with them.

Before the primitive equations can be used to predict future weather conditions, present weather conditions must be described precisely. A few years ago weather stations provided only part of the picture. For example, they had little information from the regions of the earth covered by water—which meant the weather above roughly 70 percent of the earth's surface was virtually unknown. Today, with improved ground stations and a number of sophisticated satellite stations, the information is becoming more detailed.

To predict weather, detailed measurements that have been gathered throughout the world are put into a computer along with a program that will solve the primitive equations. Imagine that you have detailed information on the world-wide weather at 8 A.M. on June 14. If you want an exceptionally detailed prediction, you find the largest computer available, load it up with your information, push the start button, and go home. Twenty-four hours later you come back to excellent weather predictions: the computer has modeled all of the weather between June 14 at 8:00

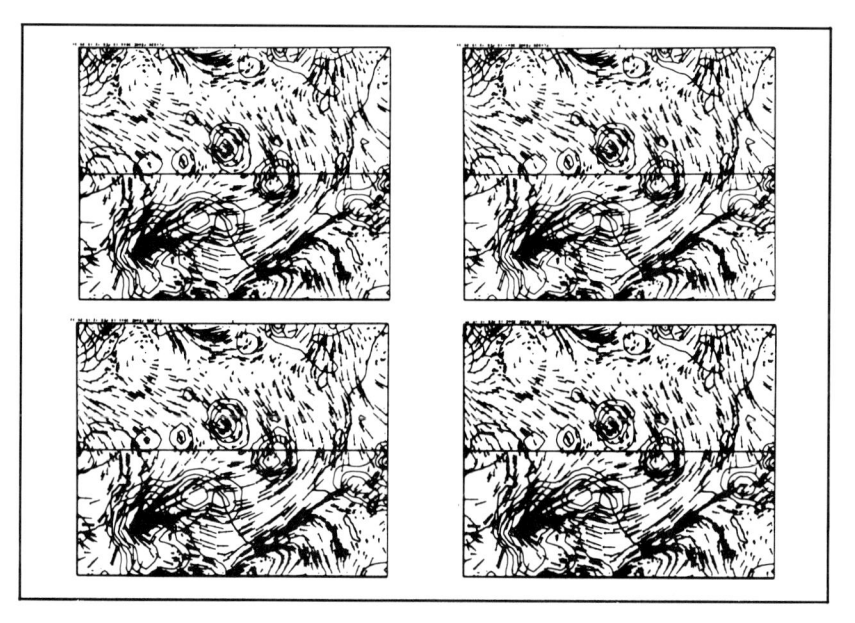

Figure 12. This map of a monsoon is based on a global computer model of the atmosphere that simulates seasonal variations in climate. The Asian monsoon dominates the exchange of air masses between the Northern and Southern hemispheres. The outermost precipitation contour represents 0.2 centimeter of rainfall per day; successive contours represent 0.5, 1.0, 2.0, 5.0, and 10.0 centimeters per day. The length of the wind arrows represents 24-hour displacements.

A.M. and June 14 at 8:30 A.M. with great precision. Go away for another 24 hours and when you come back you will have a prediction through 9:00 A.M. on June 14.

You realize, then, that the computer is taking longer than real time to solve the problem! Its predictions are generally accurate, but they are only available *after* the actual weather has already occurred. Of course, if you are satisfied with far less detail in the prediction, you can get results at real time or even faster than real time.

Clearly a lot more work has to be done. The United States has instituted a major national research program to improve atmospheric modeling capabilities. Each year scientists are collecting more accurate data, building bigger computers, and writing more sophisticated computer programs. This work is part of an international cooperative effort that goes by the name of GARP (Global Atmospheric Research Program).

This research is leading toward the realization of accurate weather prediction. Eventually global weather scanning will be perfected. A network of satellites and ground stations will transmit continuous information to computers and will help us make accurate predictions for periods of weeks or longer. Knowledge of this sort might help save billions of dollars in crops that could be harvested before an imminent hurricane. Moreover, construction, transportation, and other losses from storms and floods might be lessened. These benefits alone might well exceed the initial cost of scientific exploration programs. But the value of these programs must also be weighed against the inestimable cost of human life that is incurred in the wake of unexpected storms.

CONTROLLING THE WEATHER

It is only a matter of time before man will have the ability to control and alter at least some major aspects of climate and weather. Major weather systems contain incredible amounts of energy. It would be out of the question to use "brute strength" to deflect a hurricane from its course, once it is established. But it may be possible to apply crucial leverage in the heat engines that drive hurricanes. Research efforts in seeding hurricanes are in fact attempts to remove the driving heat energy by forcing precipi-

Figure 13. We have not yet thought through all the implications of controlling the weather of our planet. Nature indiscriminately brings rain that is essential to crops, and floods that create insufferable conditions for much of the world's population. We would like to improve on nature; the question is, have we reached a stage where we can do better?

tation before a storm reaches hurricane levels. It might one day become entirely feasible to shift whole weather systems by manipulating them through such subtle controls.

The question to ask concerns the desirability of doing so. A hurricane is an awesome, destructive force. At the same time, hurricanes pump tremendous quantities of moist air into the interior of the United States, providing much of the rainfall needed in the vast agricultural regions of the country. Would it be possible to suppress hurricanes without turning off the important rain they bring? Would there be other long-range effects of such monumental manipulation? The answers do not yet exist. Consider a further ramification of the issue. If it were possible to control major storm systems while providing adequate rain for the United States—but only at a cost of disturbing the weather patterns of other nations—under what conditions would we be justified in proceeding with such modification?

At the moment there is no agency for consistent long-range planning on such matters, nor is there any way of coordinating long-range planning with other peoples of the world. We have not yet developed rational ways of performing the enormous value judgments involved in such decisions. Moreover, little experience is available to serve as the basis for evaluating the relative social and ecological costs and benefits of such proposed changes. As with so many other fields of science, the power this knowledge brings is growing far faster than our ability to make intelligent and wise decisions. The most challenging and exciting problems in the years ahead may not lie in the field of science itself but in that interface where man tries to evolve enough understanding to control the science and technology he has created.

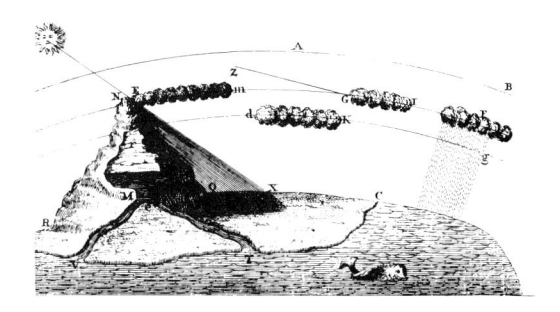

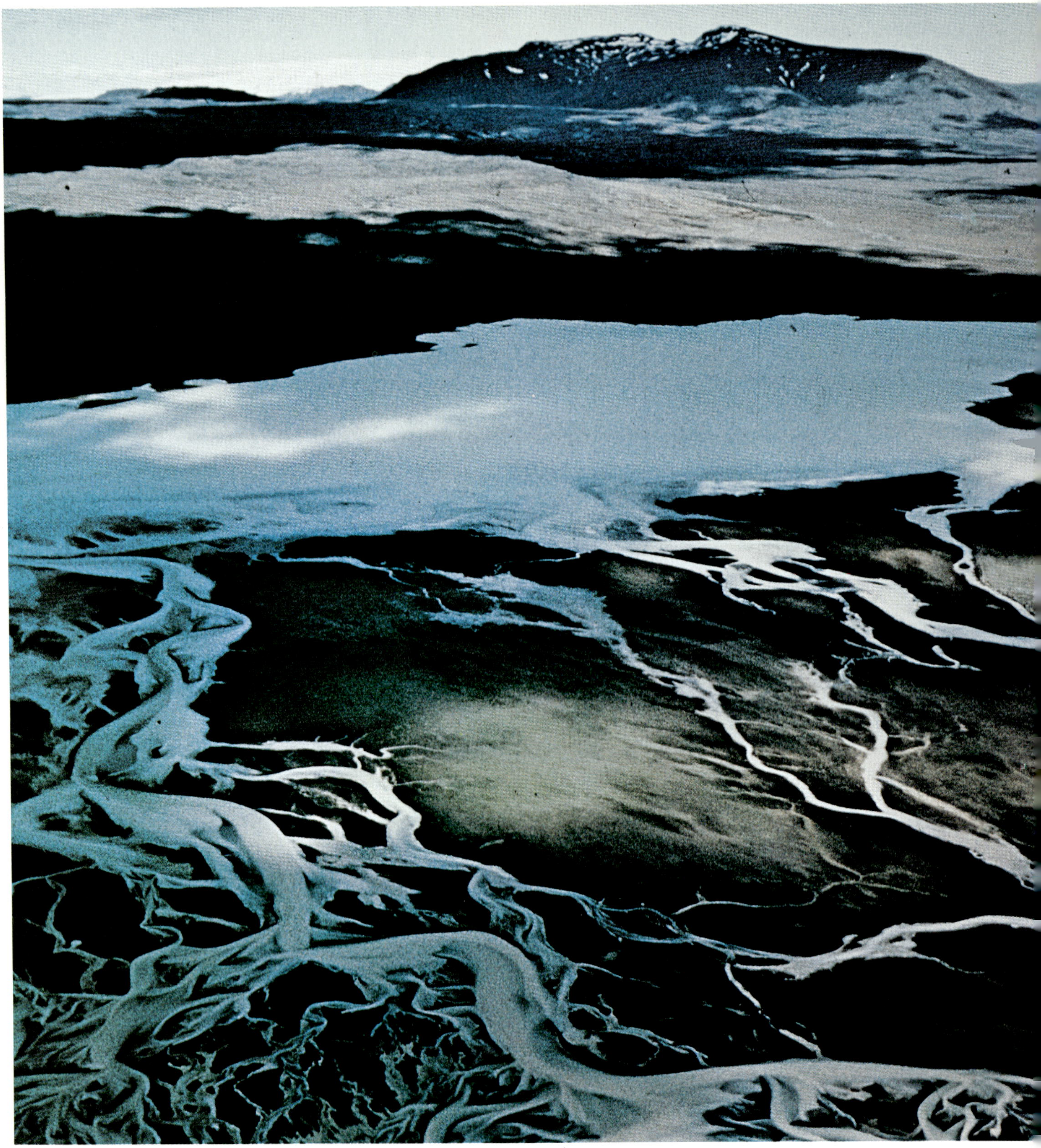

That unique chemical compound known as water exists all around you. As a liquid it is the lifeblood of the biosphere, and you would be hard-pressed to survive without it even for a few days. As a vapor it is a most active component of the surrounding atmosphere; as a liquid and a solid it forms the hydrosphere—the world's ocean and ice caps, rivers and glaciers, lakes and aquifers. And in all three states it infiltrates, covers, and carves the rocks of the lithosphere.

This compound pervades the natural world to such an extent that it's difficult to define where one sphere ends and another begins. The vast hydrosphere is not an isolated system: it is bound to the atmosphere by the waters exchanged between them. Its ocean dominates the water of regions throughout the world; its waters play a major role in maintaining the heat balance between the earth and the sun. Even the movement of water across the continents is not random but is directed by features of the lithosphere which, in turn, has characteristics largely determined by flows of water in the past.

Its disposition on and above the face of the earth has been likened to an immense plumbing system, complete with "reservoirs" of various sorts of stagnant waters, "canals," and "aqueducts" in which water travels rather swiftly. Each of the reservoirs is the subject of a specialized science. The ocean is the realm of oceanography; the lakes, limnology; and the ice caps, glaciology. The relatively rapid movement of water in the atmosphere, in rivers, and in shallow depths of the soil is the particular concern of a science known as hydrology.

In following this model of a global water system, the hydrologist postulates that every droplet of water travels a route that takes it, by way of some of the water reservoirs, back to the largest reservoir of all—the ocean. He knows, of course, that all water droplets do not follow a single common route, and that their excursions away from the ocean may last a few hours or many thousands of years. In fact, most of the water starting out on the cycle from the ocean falls back immediately as precipitation, so that the cycle is, in effect, short-circuited. In other cases, water starting out at the same time eventually may become side-tracked and locked for millennia in glacial or ground water storage. There are, in sum, over-simplifications in this model, but it does serve to put the parts in focus for a survey of the whole system. Using Figure 2 as the focusing point, this chapter explores the earthbound features of that system—the world ocean; the rivers, lakes, and ground water; and the solid water that makes up the world's ice fields.

THE WORLD OCEAN

Water covers three-quarters of the earth's surface, and all but about 2 percent of it is found in the ocean. The proportion of water to land is 4 to 1 in the Southern Hemisphere and about 1.5 to 1 in the Northern. To say that the average depth of the ocean is close to 4 kilometers is a somewhat empty description unless you think of it in terms of its horizontal dimensions as well, which range from 5,000 to 15,000 kilometers. Relatively speaking, that's about as thin and wide as the page these words are printed on.

Just as a continent has distinctive topographical zones, so does the ocean. The *coastal zone* is most familiar to us; ecologists know its salt

16
Waters of the Earth

A temporary holding station in the earth's water cycle: an interior lake fed by a melting glacier in Iceland.

Carried in Atmosphere from Ocean to Continent 7%
Precipitation on Continent 23%
Precipitation to Ocean 77%
Evaporation from Ocean 84%
Evaporation from Continent 16%
Runoff
Surface Flow to Ocean 7%
Ground Water

marshes to be the richest biological producing grounds in the world. The *continental shelf,* home of most of the world's fisheries, drops off gradually from the coast until it meets the *continental slope.* The shelf is of widely varying width but is generally narrower than 100 kilometers. The *deep ocean bottom,* or the abyssal zone, is by far the most extensive zone, complete with its own extensive mountains, valleys, and plains. In fact, the Mid-Oceanic Ridge is the largest single feature of the earth's surface.

Each of these zones is enriched by the sedimentary material derived from the weathering of rocks and soil on land. Clays are washed down to the ocean by rivers; dust particles are blown out over the ocean from deserts. Once in the ocean these materials may be carried by currents until finally they settle, very slowly, to the ocean bottom. Other material, the discarded skeletons on tiny plants and animals that live in the surface waters of the ocean, also falls to the sea floor and accumulates as sediment. These skeletons, composed primarily of calcium carbonate or silica, may be mixed with land-derived mud or, in places far from shore, may form relatively pure deposits called *oozes.* In the deepest ocean waters most skeletons are dissolved by the sea water. Here, the land-derived mud, which has accumulated slowly, again dominates.

What effect do these deposits have on the immense amount of water contained in the ocean basins? The most immediate and well-known effect is its saltiness, or *salinity.* Salinity is the measure of how many grams of dissolved salts are present in 1,000 grams of sea water. As you might expect, some parts of the ocean are less salty than others depending on the flow of fresh water from, say, melting ice into certain regions; other parts have high salinity because of intense evaporation that leaves the salts behind. But surprisingly enough, even though the total concentration varies from place to place and from depth to depth, the relative proportions of the abundant elements—sodium, chloride, magnesium, calcium—are constant in sea water throughout the world. Evidently sea water has become well-mixed over geologic time, not only within a region, but from one region of the ocean to another.

The Oceanic Currents

Within the boundaries of the ocean all water is in ceaseless, complex motion. Often the rate of motion is too slow to be measured directly, but it is enough to transport water, dissolved substances, and suspended

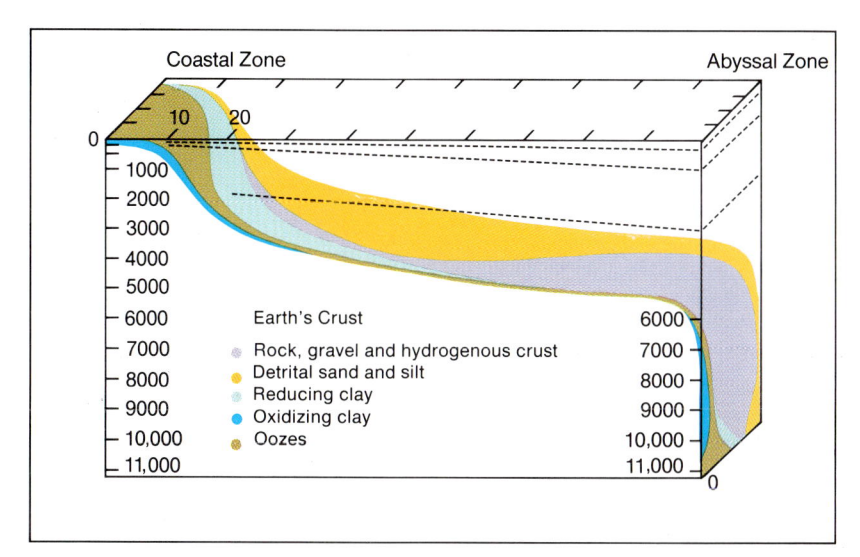

Figure 2. The hydrologic cycle (left) is a model that is used to explain the movement of the earth's water. In this drawing, the percentages given for the amounts of water always in transit over land and sea equal percentages of the total rainfall. The mean annual rainfall is 85.7 centimeters over the entire surface of the earth.

Figure 3. This drawing shows the predominance of land-derived sediment in shallow water and in the deepest parts of the ocean. Organic oozes are major deposits over much of the ocean floor. Their geographic distribution is variable; only the effect of depth is shown here. The distance between the front and back planes of this drawing represents the percent composition at a given depth. The curve of intersection between the colored surface and the front plane represents the percent of ocean floor above a given depth, in meters. The surface therefore is a profile of the ocean floor.

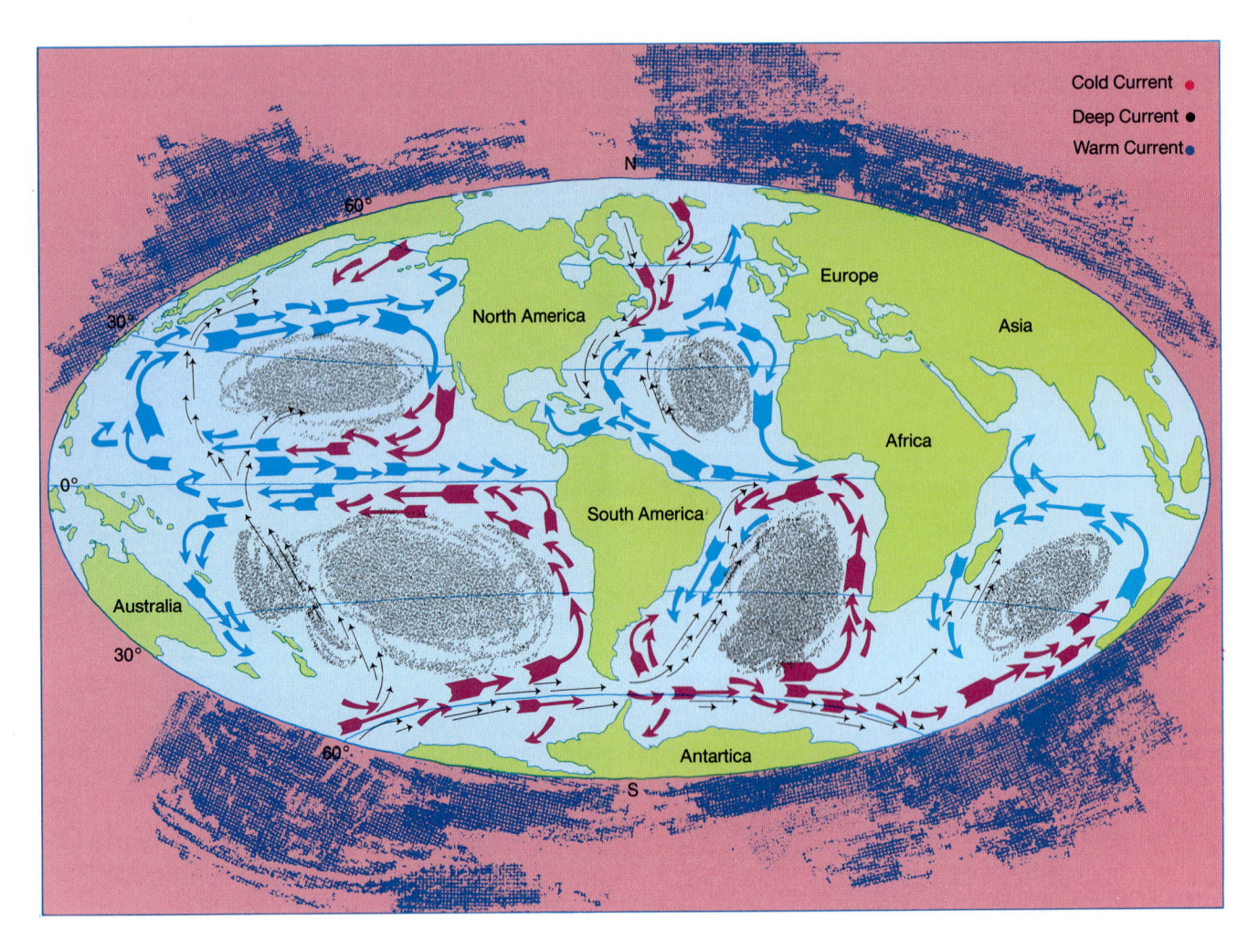

Figure 4. Major ocean currents. Notice the pattern of strong surface currents on the equator and near the perimeter of each ocean basin and the relatively still, current-free areas between them. Warm currents move from low to high latitude, warming the surrounding area; cold currents move the opposite direction. The preliminary mapping of deep ocean currents is based on recent research. All these currents are long range and all involve the movement of cold water.

particles. The primary transport agents are called oceanic currents. *Surface currents* include such familiar flows as the Gulf Stream. Knowledge of these surface movements began accumulating centuries ago, with the first explorations of the high seas, and their positions and speeds have been accurately charted.

Deep currents, on the other hand, follow quite different patterns. Investigations of these deeper movements began only in the last 100 years and knowledge of their courses still is incomplete. But today oceanographers can determine the history of a water sample—where it has come from, when it was last at the sea surface, and occasionally how fast it has traveled—from such region-specific variables as temperature and, to a lesser extent, salinity.

How do surface currents originate? As winds move across the surface water, they create a stress that pushes it along. The piling up of water in turn creates pressure differences within the water itself, which leads to further motion. Because the resulting "surface" currents extend several hundreds of meters into the water, they clearly are not the result of wind action alone.

Most of the movements of the surface currents resemble the movement of the air above them, however, and for this reason the surface currents are also referred to as *wind-driven circulation.* In temperate

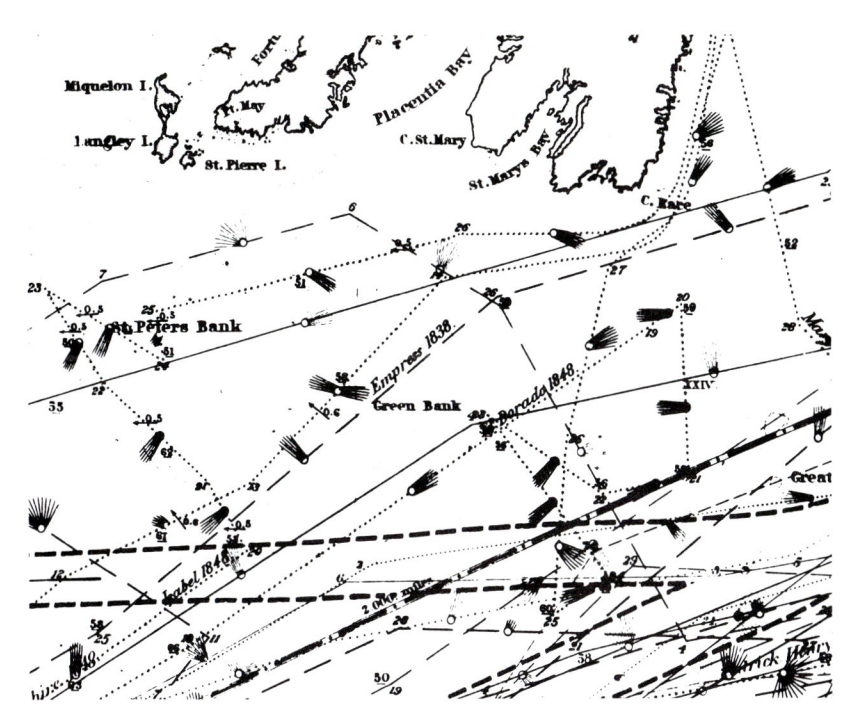

Figure 5. (Left) Captain Matthew Maury drew this wind and current chart in 1849. Man's dependence on the seas for trade and exploration hastened not only the completion of accurate wind and current charts but also the development of scientifically sound instruments for navigation.

latitudes these currents travel from the west; in the tropics they travel from the east, paralleling the westerlies and the trade winds, respectively. But currents as well as winds are weak throughout much of the tropics. As in atmospheric circulation, streams of water travel clockwise about the centers of such regions in the Northern Hemisphere and counterclockwise in the Southern Hemisphere. These oceanic movements are called *gyres.* Moreover, just as it complicates atmospheric circulation, so does the Coriolis effect influence the direction of the movement of surface water.

The deep ocean currents are not driven directly by wind; rather, they are propelled by differences in sea water density, which are caused by variations in temperature and salinity. In a column of stable sea water, the density of the water increases with depth. If the density of a parcel of water changes with respect to water of the same depth—if it increases, for example, because of a decrease in temperature or an increase in salinity—then vertical and horizontal flow occurs. The deep ocean currents created by this process are called *thermohaline circulations* ("thermo" = "heat"; "haline" = "salt").

Sea water of the highest density lies at the bottom of the deep oceans and is continually replenished by cold surface water that sinks around Antarctica and the Arctic North Atlantic. After it sinks, this water migrates slowly toward the equator, warming as it goes and slowly rising again toward the surface. This global mixing—from the surface at the poles, to the bottom, and then toward the equator—apparently requires many centuries, so that some deep ocean water is well over a thousand years "old." This age has been determined from carbon 14 measurements. In other words, it has been more than a thousand years since it has been exposed to atmospheric exchange at the surface.

Increasingly accurate measurements have shown how slowly the oceans move, which may come as a surprise to those who have felt the force of large waves whipped up during storms at sea. The sluggishness of the ocean may even be something of a consolation to those of us who worry about man's activities suddenly disrupting life on earth. But it also

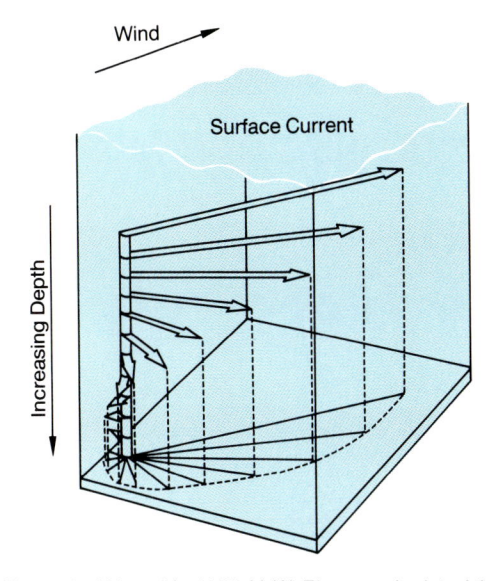

Figure 6. (Above) In 1905, V. W. Ekman calculated the effect of wind blowing steadily over an ocean of infinite depth, extent, and uniform eddy viscosity. He found that the surface layer is driven at an angle 45 degrees to the right of the wind direction in the Northern Hemisphere and 45 degrees to the left of it in the Southern Hemisphere. The successively deeper layers of water move more and more to the right until, at a given depth, the direction in which the water moves is opposite to what it is at the surface. Associated with the rapid decay of velocity with depth in the "Ekman spiral" is an upwelling and sinking of masses of water, occurrences that actually have been observed off coastlines.

16 | Waters of the Earth

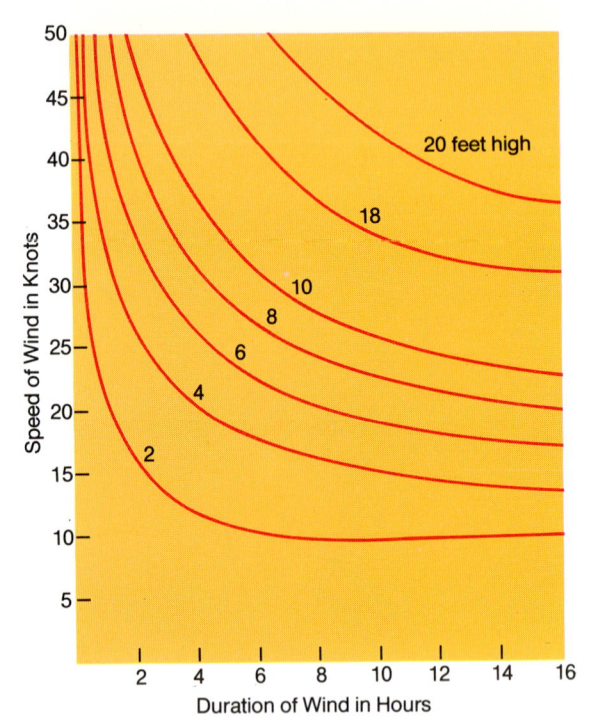

Figure 7. Wave height varies with the strength of the wind present. These calculations are based on a distance from the windward shore that is great enough to preclude the influence of other factors on wave height.

impresses us that any inadvertent changes will not be corrected by natural processes for many centuries.

Waves and Tides

In addition to the circulation of the ocean, there are periodic motions known as waves and tides. Although waves transmit great amounts of energy and cause violent motions, they do little transporting. During wave motion, water molecules move up and down and may oscillate from side to side, but they drift only slightly in the direction of wave propagation. Surface waves are created at the interface between sea water and air, two fluids of differing density. When the ocean is relatively calm, a wind of steady force and direction will ripple its surface for quite a distance. These *wind waves* persist long after the wind that nurtured them has died down. As the waves move farther and farther away from the wind stress that created them, the length between their peaks and troughs gradually becomes extended until *swells* are formed, which are waves of longer periodicity. The greater the wavelength, the greater will be the velocity of these swells. In fact, a swell that breaks on the California coast may have been generated by distant storms 10,000 miles away in the southwest Pacific.

The ocean is not always calm, however, and more than one wind moves across its surface with varying degrees of intensity. When waves of differing direction, wavelength, and amplitude come together, they produce what is known as a *sea*. On the open sea, particularly in an area where waves are being generated, the surface usually becomes chaotic as waves of different amplitude and direction interfere with one another. Only when waves come into shallow coastal water is their energy sorted out into regular trains of crests and troughs. As a wave moves in toward the shore, the water molecules caught up in its motion encounter the resistance of the land. When the resistance below becomes great enough, the faster moving crest of the wave breaks away and water thunders up on the shore, much to the insatiable delight of surfers, sandpipers, and other creatures tied in one way or another to the spectacle.

There are two other types of ocean waves, both of very long wavelength and both almost invisible on the open sea. These motions are known as tidal waves, and tides. "Tidal waves" have nothing whatever to do with the tides but are episodic disturbances of the water's surface. Usually they are caused by a volcanic explosion or an earthquake displacement of the sea floor. They are better known by their Japanese name, *tsunamis*. Earthquake-prone Japan is particularly afflicted by *tsunamis*, as are many low-lying islands throughout the Pacific. Few coasts are completely free from danger, for *tsunamis* can travel across entire ocean basins with little energy dissipation. Although barely noticeable on the open sea, these waves travel 500 miles an hour, becoming huge walls of water as they cross shallow continental shelves.

The true tides have some of the characteristics of "tidal waves" but they are predictable and regular, seldom destructive. The energy for the regular daily or twice-daily rise and fall of the tides comes from the gravitational force of the moon and, to a lesser extent, the sun. As the earth rotates, the pull of the moon tends to create a bulge of water on the earth directly "beneath" it, and on the opposite side of the globe. Any shoreline on the earth will spin through these bulges and experience a "high tide" twice a day.

But the bulges of water really travel about the ocean as waves of a regular, 12–1/2-hour periodicity. Because they are waves they are affected by the depth of water over which they travel and by the features of the land

Figure 9. Earthquakes can trigger tsunami, shock waves that can be large enough to inundate low-lying coastal areas thousands of miles across the ocean from the source of the seismic disturbance.

they break against. And the associated movement of water is not easy to define. The *range of tide* — the distance between the level of low and high tide — changes from place to place in an unpredictable manner. The world-wide average is about 3 or 4 meters (10 or 13 feet), but within some embayments such as the Bay of Fundy in eastern Canada the tidal range is greater than 20 meters. At any given place, however, changes with time are predictable because they are caused by predictable changes in the gravitational pulls of the moon and sun as those bodies move relative to the earth.

Even where tidal ranges are great enough to be obvious, visible wave forms are seldom evident. Nevertheless, when the wave of a high tide enters certain estuaries the shoaling of the bottom may cause the wave to steepen, and even break. This genuine tidal wave is called a *bore* and the more impressive examples of it, as in the Seine estuary of France, would be quite destructive if they were not so predictable.

Upwelling and the Distribution of Nutrients

All living things in the ocean as well as on land are part of a system whereby photosynthesis in green plants produces oxygen at exactly the same rate that respiration of both plants and animals consumes it. In photosynthesis, carbon dioxide and water combine to form living plant material, giving off oxygen as a by-product. Photosynthesizing occurs only where there is light. In the oceans the process is restricted not only to daytime hours but also to the upper 100 meters, or 327 feet, of even the clearest water. In turbid water, light may not penetrate more than a few meters, especially if water pollution is severe. The intensity of light is a factor in how much photosynthesis can take place. For that reason, even though deep water is rich in all the chemical substances needed in the process, it does not support green plant growth.

In surface ocean water, however, photosynthesis proceeds so vigorously that it reaches a state where the supply of chemical substances is the limiting factor. Very little of the light is actually converted into the energy stored in green plant cells, but normally all of the available nitrogen or phosphorous is used and limits the standing crop of plant material.

Thus, oceanographers regard nitrogen and phosphorous as "limiting nutrients."

The biomass of the surface layer is based on phytoplankton and zooplankton, microscopic plants and animals. The phytoplankton quickly use up the nitrogen and phosphorous present in the surface layer. When the phytoplankton die or are eaten by zooplankton, which also die, the resulting *detritus* sinks through the water column. Bacterial action causes the breakdown of these organic compounds to the original phosphate, nitrate, and bicarbonate constituents. It is for this reason that the deeper waters are richer in nutrients.

Not all of the organic remains are broken down directly by bacteria. Some of it is food for creatures of the abyss that can survive and propagate even in the deepest oceanic trenches, subsisting only on the remains floating down to them. The detritus that is inedible and cannot be broken down by bacteria ends up on the ocean floor as the organic sediment known as ooze. The main contributors to ooze are diatoms, coccolithopheres, foraminifera and radiolaria—long names for tiny plants and animals.

Although there is some vertical mixing everywhere in the ocean, in some locations this mixing is much more pronounced. Along the equator, for example, in both the Atlantic and Pacific oceans, there is a westward-moving current. Because of the earth's rotation, the water in such a current is spun away from the equator and there is a slow, poleward motion superimposed on the westward flow. Surface water is dragged out of a narrow band along the equator, and cold water rises from below to take its place. The process is known as *upwelling*. Because the deep water has higher concentrations of nutrient minerals, the areas where upwelling occurs are particularly productive, supporting the densest pastures of phytoplankton and herds of zooplankton. Upwelling also occurs adjacent to coasts where prevailing offshore winds blow surface water away from the land and deeper water rises. One such region lies off the coast of Peru.

Man's Use and Misuse of the Ocean

With increasing use of the oceans, abuses seem to be becoming more and more common. Oil and petroleum products, for example, are spilled into the oceanic environment on the order of megatons each year. Dredge spoils and industrial wastes add more than 40 million tons each year to the nearshore areas of the United States alone. These waste products clog the feeding mechanisms of minute organisms and directly poison many species. The much smaller addition of toxic elements such as mercury, arsenic, lead, and the chlorinated hydrocarbons is possibly a more critical problem in nearshore areas. These chemicals become concentrated as they move up the marine food chains until they become harmful to a life process, such as reproduction, respiration, or feeding. Even if the marine organism itself isn't affected, man may be harmed when he ingests the concentrated chemical.

A more subtle problem of industrial use of the ocean is the indirect effect of sand and gravel dredging, such as that taking place in the North Sea. When the sand and gravel is loaded on barges to be transported elsewhere, the "fines," or the lightest fraction of bottom material, is returned to the sea. Fragile fish eggs are covered by the redeposition of these "fines" and are prevented from hatching. Unfortunately, the areas off the coasts of eastern England where the most sand and gravel are being removed are major spawning grounds. This sort of problem involves international control and coordination that is difficult to achieve. Moreover, inadequate research and insufficient funds and technology hamper the control of ocean pollution. But it is not only a scientific problem. It has economic and political implications that will continue to grow as human population increases.

RIVER, LAKE, AND GROUND WATER

The world ocean is an incredibly complex body of water that mixes vertically as well as horizontally. In contrast, when rainwater falls on the continent it moves in a single-minded downhill direction between topographic divides into drainage basins. The exit drain of any such basin is a single stream. Most streams, as they move downhill, receive tributary streams that contribute to their flow. Each tributary has its own basin at its beginning, but all of these make up part of the basins that feed the flow of water called rivers.

Drainage basins are convenient units for studying the continental part of the hydrologic cycle. Their boundaries are easily defined and both ingoing and outgoing water and energy can be measured. Changes in water quality as water passes through the basin can be monitored with sure knowledge that the effects of any modification will show up downstream. Drainage basins, because they usually occur in nested hierarchies —that is, with smaller tributary basins as components of larger basins— can be studied on whatever scale is appropriate (see Figure 10).

Precipitation, in the form of snow or rain, represents the only significant supply of water into drainage basins. Of the rain falling on the con-

Figure 10. An example of nested drainage basins in the United States. The Continental Divide separates basins draining to the Atlantic from those draining to the Pacific. The arrows indicate the flow of water from higher to lower elevations.

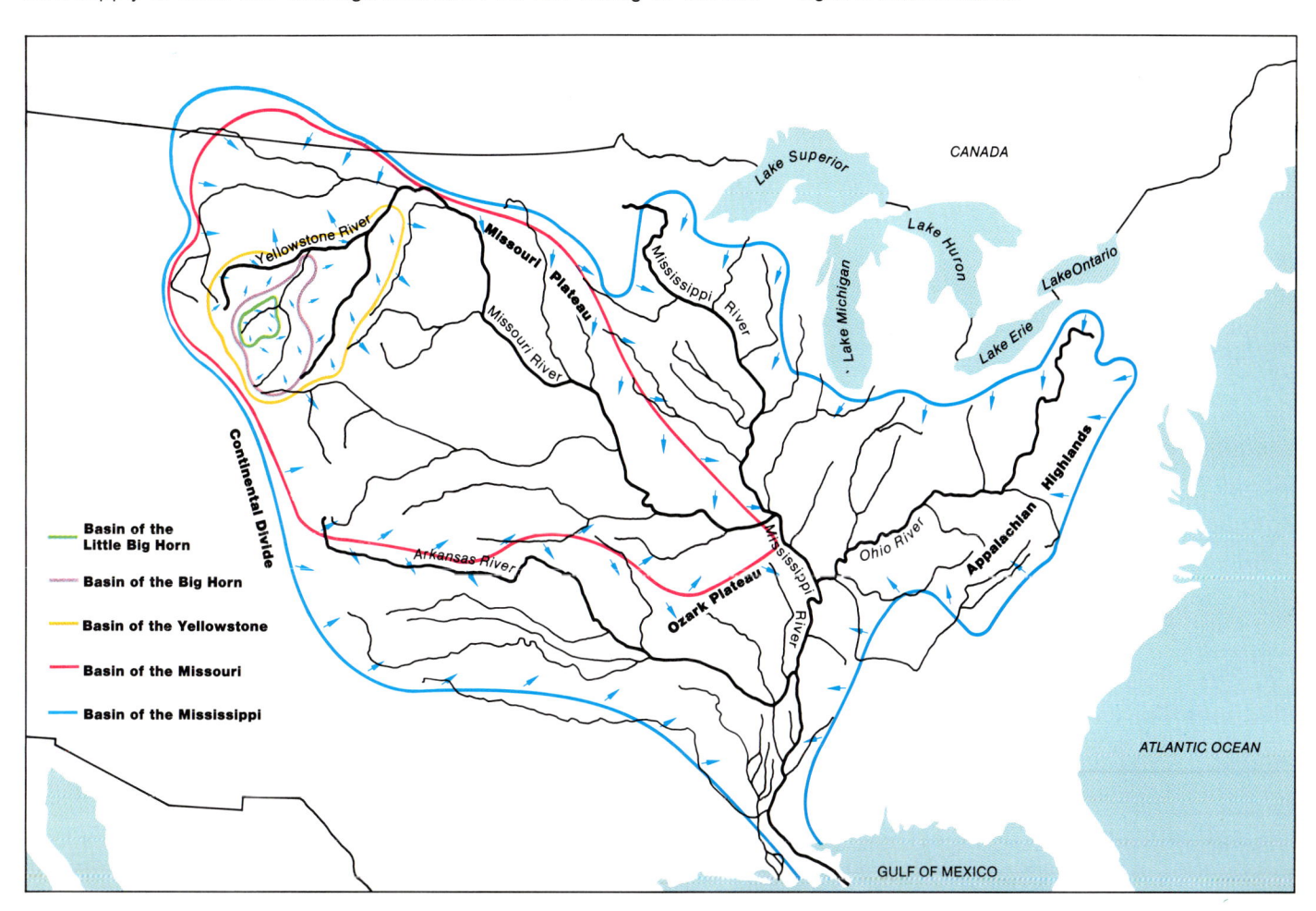

tinental surface, very little turns directly into running water on its way to the sea. At the beginning of a rainstorm in a humid, vegetated region, most of the rain is intercepted by the leaves of trees and grasses. If the storm is brief, the rainwater goes no further but is retained on plant shoots until it evaporates into the atmosphere. During a prolonged rainfall the retention capacity of vegetation is soon exceeded, and water is transmitted to the soil in drips from the foliage or in streams down stems. At this stage, trees are like saturated, leaky umbrellas providing ineffective shelter from the rain.

Dry soil plays a similar role. Acting at first like a sponge, it soaks up water and forms thin films, which are held around soil particles by surface tension forces. Plants absorb this soil water through their roots and later lose it to the atmosphere by transpiration. When water fills all the gaps in the surface soil, the soil begins to transmit water to deeper layers. Unlike vegetation, soil has a maximum rate at which it can transmit water. This limit is called the *infiltration capacity.* If water arrives on the soil surface at a faster rate than the infiltration capacity, it will accumulate and begin to flow downhill. The runoff starts as a thin sheet but is soon guided by the terrain into rivulets and channeled flow.

Variations in the duration and intensity of rainfall, the height and density of vegetation, and the type of soil present mean that there can be no typical disposition of rainwater. Even at the same location there may be marked seasonal variations. In the winter, deciduous trees have no leaves to intercept rain. If the ground is frozen, then filtering through the soil, or *percolation,* is impossible and most of the rain accumulates directly as runoff. Because the main controlling factors always involve the vegetation and the fragile soil, man has drastically altered the processes of runoff and percolation in inadvertent modifications that have seldom been beneficial and usually lead to accelerated erosion.

Although some river water originates from direct runoff, this source usually accounts for only a small percentage of the total discharge. In most humid regions, streams and rivers flow permanently even though rainstorms in the basins are episodic, occurring at intervals of a few days or a few weeks. Permanent rivers tap a permanent reservoir of fresh water that is periodically replenished following rains.

Ground water makes up this permanent but unseen reservoir. Above the depth where rock pressure prohibits voids, all rocks and sediments of the earth's crust have intergranular spaces and cracks in them. Most of these spaces eventually fill up with water percolating down from the soil

Figure 11. This drawing shows how water collects beneath the earth's surface between saturated soil and rock layers. This ground water supplies wells and springs.

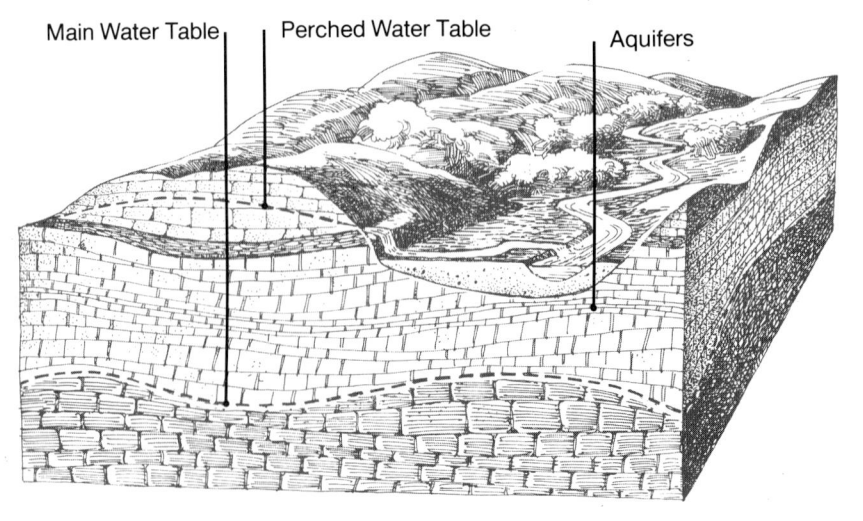

Figure 12. These limestone caves illustrate the effect of water on sandy, sedimentary rock.

above. The seeping water may join water that was retained in sedimentary rocks from the original accumulation of the sediment. The amount of water held in this reservoir can only be estimated until more information is gained from deep drilling research.

Ground water extends upward to merge with the soil water at only a few places in a "typical" drainage basin. Where the surface of the ground water reservoir — the *water table* — intersects the land surface, stored water can be released from the land as *springs*. Rock layers that supply water in this fashion are called **aquifers**. Between the level of the water table and the soil water over most of the basin, there is a layer of rock or sediment that is not permanently saturated and that transmits water after rainstorms. Water percolates down until it arrives at the water table; it then moves laterally until it is tapped by a spring or seeps into a river. After an extremely heavy rain, the water table gradually rises as more rock becomes saturated. The rise causes an increased outflow from springs.

It usually takes some time for ground water to thread its way through tiny passageways within rocks. But its speed depends on the character of the rock. In coarse sandstone or gravel, the movement can be more rapid than in clays and shales, which have smaller pore spaces. These dense rocks have a greater total volume of pores, however, which increases their storage capacity. Ground water may move rapidly through limestone because normally ground and surface water is slightly acidic and dissolves the carbonate rock. Joints may become enlarged to huge caverns and tunnels in which entire underground river systems occasionally develop. Powerful springs appear where such rivers emerge on the surface of the

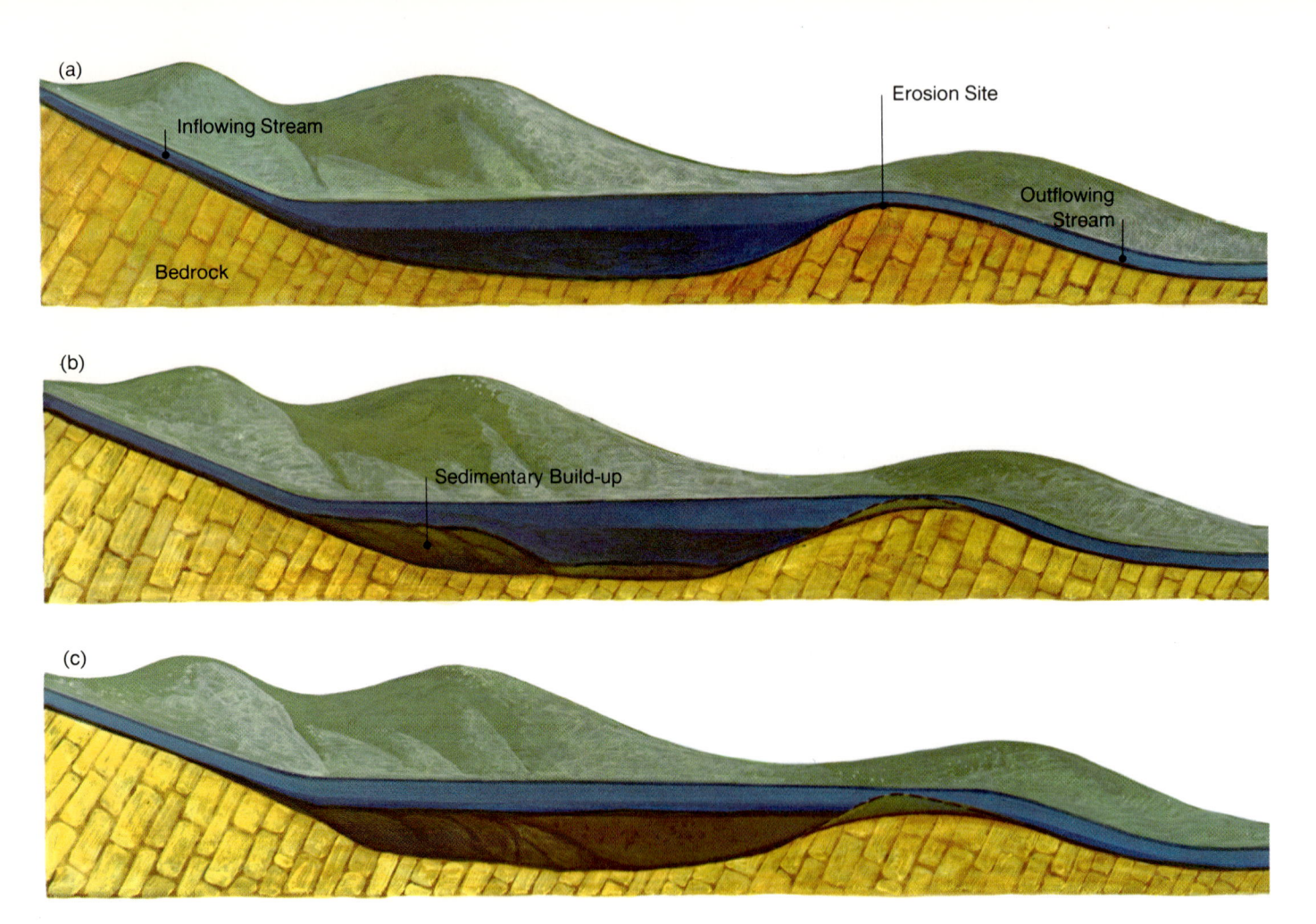

Figure 13. Life history of a lake. In the first stage, the lake occupies a hollow, which is supplied and drained by streams. During the second stage, the lake begins to get filled with sediment deposited from inflowing river water and the spillover point becomes lowered because of erosion. As a result, the lake gets shallower. In the final stage the lake becomes completely drained.

earth. Such springs are characteristic of limestone country, where rivers may be almost fully grown at their "source." In almost all other types of rock or soil, the permanent source, or *headwaters,* of a river is usually an insignificant patch of marshy ground supplied by ground water. From such sources, water begins moving downhill in an inconsequential trickle.

The flow of surface water in channels erodes the surface of the continents and maintains a downward slope from river sources to their estuaries. But other agents, such as vertical earth movements and erosion by ice or wind, can disrupt this natural orderliness by forming closed depressions. In humid climates these depressions form lake basins. Rivers flow in until the depression is filled and then the water spills over at its lowest point to continue its downhill journey. Lake basins are temporary features in humid lands because the inflowing river eventually fills them with its deposits of sand and mud while the outflowing river continues to erode the spillover point (see Figure 13). In arid regions, evaporation from a lake surface may be so great that it takes all of the inflowing river water to counteract loss by evaporation. Under truly arid conditions, all of the water is periodically evaporated away and the only trace left of it is a solid salt pan. In many parts of the world, river water reenters the hydrologic cycle by evaporation to the atmosphere rather than by flowing into the ocean.

A good way to examine the hydrologic effects of an interruption of the flow of a river with a lake is to look at artificially created reservoirs.

Figure 14. These folds in Malaspina Glacier in Alaska record the migration of water to the edges of the ice sheet. Such migration may take thousands of years.

Man, having the greatest collective thirst of all land animals, has taken steps to increase his fresh water supply and stabilize its flow by constructing his own storage reservoirs. By damping its fluctuations there can be a steady and predictable supply for drinking, irrigation, and power generation and less danger of disastrous flooding after heavy storms. The larger the volume of the lake or reservoir relative to the volume of water passing through it, the more efficiently it performs this function. The hydrologic characterization of any drainage basin can be deduced by observing the time variation flow in the river that drains it.

SOLID WATER

The ice and snows of high mountain areas contribute supplies of water to some lowland drainage basins. But in the Arctic and Antarctic, there are huge areas in which water is perpetually frozen. In fact, more than 75 percent of the fresh water within the earth's hydrologic cycle is locked in the polar ice caps.

The polar continental regions drain entirely by streams of slow-moving ice, and distinct drainage basins, although present, are difficult to define. Ice moves slowly downhill under the influence of gravity, but the relative immobility of solids ensures that, where water enters the solid state, it accumulates as a mass. How long any individual parcel of water is stored in this form depends mainly on the length of time between thaws—the time when the temperature rises enough to return the water into the liquid state. In temperate latitudes, where the most snow falls, thawing may be diurnal or seasonal. Therefore, the most noticeable effect of storage is on the network of rivers. In polar areas, where thawing is slight and superficial, masses of ice accumulate and the duration of icy storage becomes the time required for migration to the edge of the ice sheet—and it may take many thousands of years.

Most of the Arctic is covered with ice that has been formed by the freezing of sea water, a separation process that purifies the water almost as efficiently as does evaporation. Frozen seawater is less dense than liquid water and floats above it. It does not build up to great thicknesses, however, because it isn't strong enough to support its own weight. Whenever there is any tendency for sea ice to build up to excessive thicknesses, the vertical pressure of its own weight is counteracted by its buoyancy

16 | Waters of the Earth

Figure 15. (Far right) The above maps show the present position of the north and south magnetic poles relative to the north and south geographic poles. The large map illustrates the maximum spread of glaciation during the Pleistocene; the solid line defines the probable southern limit of pack ice, and the shading depicts glaciated land areas.

and the ice is effectively squeezed by these two forces into a thin, broad sheet. Even at the center of the Arctic Ocean, the ice thickness averages a mere 3 meters, or about 9 feet. When ice accumulates over solid continental material, much greater masses of ice can be supported. The average thickness in the center of both the Antarctic and Greenland ice sheets is more than 3 kilometers. In fact, the weight of the overlying solid water is so great that the plastic lithosphere has buckled beneath these immense loads and the thickest ice occupies hollows that it has impressed into the rocky basements.

The ice impressed into hollows beneath the Antarctic ice sheet is presently quite stagnant. Without benefit of the perspective that geology offers, the glaciologist might conclude that the water had been removed permanently from the hydrologic cycle. Such a state of permanence is never achieved on the dynamic face of the earth. A mere 20,000 years ago the hollows now occupied by the Great Lakes in the United States were filled by equally stagnant bodies of frozen water that had been created beneath a 3-kilometer-thick ice sheet. In a short period of rising temperatures between 15,000 and 6,000 years ago, the margins of the ice sheet retreated northward at a spectacular rate (Figure 15). The previously stagnant solid water was released and huge floods of meltwater created extensive lakes over the Great Plains. This grand thaw brought to a close one of several glacial periods of the most recent Ice Age, the Pleistocene.

Many of the effects of these glacial periods still affect the development of twentieth-century man. While land ice still covered higher latitudes, the atmospheric circulation was disturbed and much of the tropics underwent heavy rainfall. Today, shallow drilling is being used to tap the ground water beneath such unlikely areas as the Sahara Desert. That ground water was formed during the Pleistocene Epoch. As the ice sheets melted and their load on the subjacent rocks was released, the lithosphere bowed up. This vertical rise in the land surface is still continuing over much of northern Canada and Europe. Contemporary uplift at rates of 1 meter per century pose some problems in harbor construction but also provide splendid opportunities for land reclamation along the seashore.

The most practical consequences of former ice covers are related to the movement of the ice. Depending on its velocity and the resistance of rocks encountered, moving ice scraped the rock surface bare of soil and dug hollows that became lake basins in much of northern Canada. Great spreads of silt, sand, and gravel were deposited that blanket the bedrock. They form the raw material for the fertile soils of the American Midwest and provide the main source of building materials and shallow ground water over much of the Northern Hemisphere.

The direct uses of ice are limited. Most of this huge potential supply of fresh water is located in inhospitable areas remote from consumption centers. But as fresh water becomes more scarce and more expensive to transport, we can anticipate schemes to bring the polar ice and thirsty people together. John Isaacs of the Scripps Institution of Oceanography has suggested towing a small Antarctic iceberg into the major ocean current system that sweeps toward the equator along the west coast of South America. After about 1 year, he predicts, the iceberg could be maneuvered across the equator and towed out of the current system at Los Angeles to provide almost 200 million dollars worth of fresh water for Southern California.

A HISTORY OF THE HYDROSPHERE

There are fundamental and intriguing questions left unanswered by the descriptive explanations of the hydrosphere's internal workings. Where did all the earth's water come from? Why does our planet have so much

North Pole

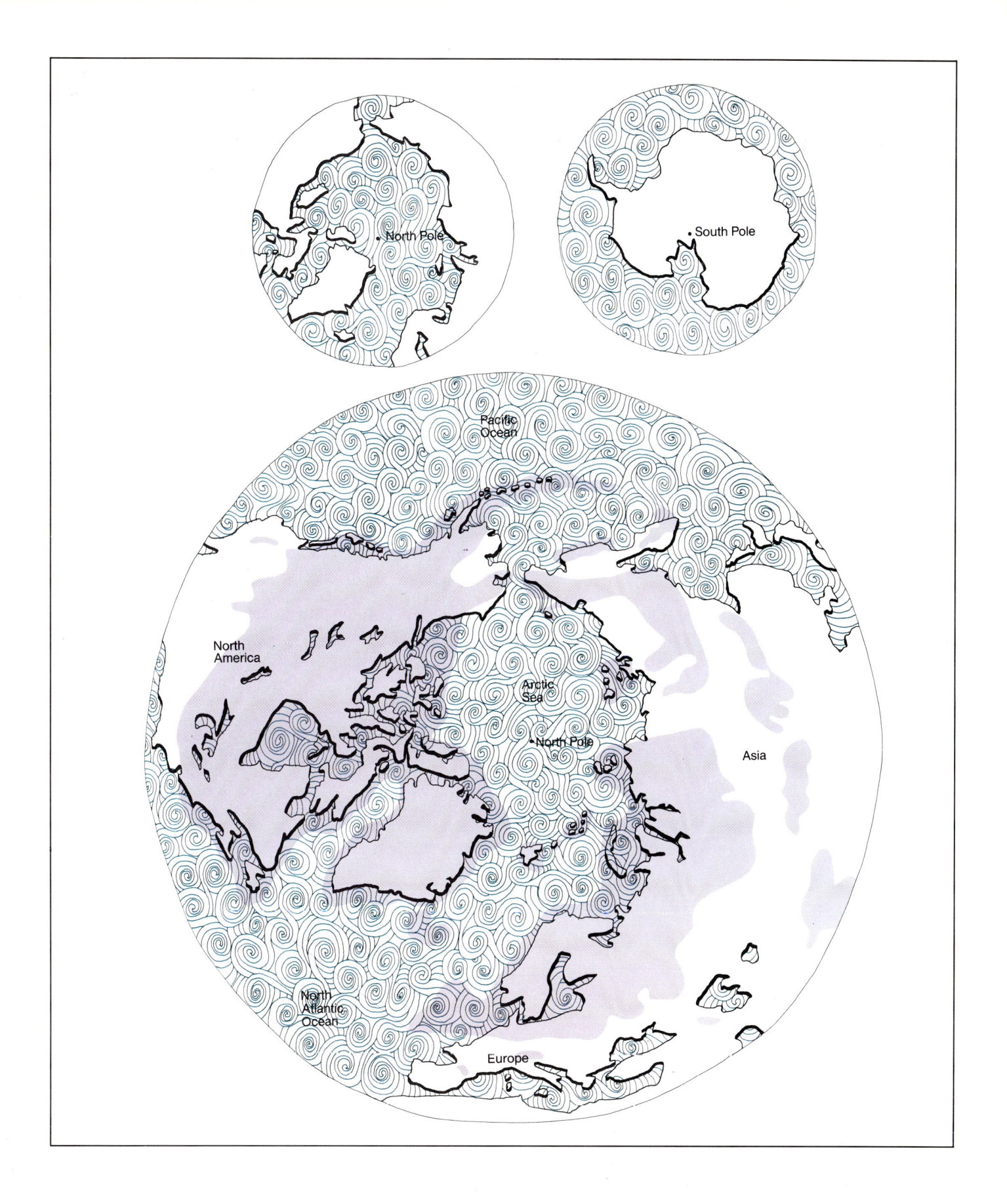

South Pole

Pacific Ocean

North America

Arctic Sea

North Pole

Asia

North Atlantic Ocean

Europe

Are the Seacoast Cities in Trouble?

Many times in the last ten million years, large parts of the continents of the earth have been buried under huge chunks of ice — ice that has advanced and retreated over the years. One consequence of this motion has been a corresponding rising and falling of sea level: the more ice that forms on the earth, the less water there is in the seas. Suppose all the ice now covering the earth melted. How would that affect the sea level?

Step 1. Geochemists estimate the total amount of ice currently on the earth to be 9,528,249.11 cubic kilometers, or roughly 9.5×10^6 cubic kilometers. If this ice melted, however, its volume would decrease, because ice has a volume 4 percent greater than the volume of an equivalent mass of water, due to the greater distance between its intermolecular bonds. The volume of the melted ice would therefore be:

$$104 \text{ percent} \times \text{volume of water} = \text{volume of ice}$$

$$\text{Volume of water} = \frac{\text{volume of ice}}{1.04}$$

$$= \frac{9.5 \times 10^6 \text{ cubic kilometers}}{1.04}$$

$$= 9.1 \times 10^6 \text{ cubic kilometers}$$

Step 2. The terrestrial seas cover about 3.6×10^8 square kilometers. What would happen if over 9 million cubic kilometers of melted ice were spread over the surface of the seas? To determine how much the sea level would be raised, divide the volume of the melted ice by the total surface area of the terrestrial seas:

$$\text{Increase in sea level} = \frac{\text{volume of melted ice}}{\text{area of seas}}$$

$$= \frac{9.1 \times 10^6 \text{ cubic kilometers}}{3.6 \times 10^8 \text{ square kilometers}}$$

$$= 2.5 \times 10^{-2} \text{ kilometer}$$

$$= 0.025 \text{ kilometer}$$

$$= 25 \text{ meters}$$

So if all the ice on the earth melted, approximately 25 meters would be added to the present sea level. Although the continents now covered with ice would rise slightly if the pressure exerted by the weight of the ice were removed, North America would not rise significantly; New York, Seattle, San Diego, and other coastal cities would be submerged. Fortunately, there is no evidence that such an event is likely to occur.

more water than its neighbors in space? Is the volume of the ocean steadily increasing or decreasing? Plausible answers, although probably not final answers, are now available for most of these questions.

The hot rocks of the mantle have been the source of water throughout the 4.6 billion year history of the earth. Molecules of water are locked up in them, and when melting occurs, these molecules are released as vapor. New water vapor may thereby be introduced into the atmospheric or oceanic stage of the hydrologic cycle during a volcanic eruption or it may condense deep in the crust as an addition to the ground water reservoir. It was in this manner that the surface waters of the earth accreted rapidly at an early stage in its history. Quite extensive seas, at least, can be inferred from certain properties of rocks that are 3.5 billion years old. Probably the amount of water at the earth's surface has increased slowly since this time, although some water molecules do return to the mantle.

The share of the hydrosphere's water that is stored in the ocean has always been large, but it has fluctuated dramatically during some brief periods of the earth's history in response to vast glaciations. Ice Ages, even when broadly defined as periods when extensive bodies of frozen water accumulate at the poles, are rare in the history of the earth. As read from the geologic record, they recur infrequently at intervals of several hundred million years. Probably one requisite for an Ice Age is a chance configuration of spreading oceans and drifting continents, which effectively insulates the polar areas by hampering convective heat exchange with warmer latitudes. The present distribution of continents is an ideal pattern for an Ice Age, and in fact we may now be experiencing only a mild "interglacial" period of the Pleistocene.

The other chemicals of the hydrosphere, the dissolved components that give the sea its saltiness, also come more or less directly from the hot rocks of the mantle. Some components, notably the chloride ion, escape in volatile compounds from molten rock in the same way that water does. These components reach the surface in volcanic gases and mineral springs. Most of the other common substances—sodium, magnesium, and potassium, for example—are delivered to the ocean after chemical weathering of igneous rocks on the continents or the ocean floor. It was once thought that the saltiness of the ocean had increased constantly throughout its history. But there are also ways in which minerals are taken out of the hydrosphere and returned to the lithosphere or the underlying mantle. In fact, it is now generally believed that there is a natural balance in which the rate of addition of each type of ion is equalled by the rate at which it is extracted from the ocean so that the salinity remains constant.

The most obvious way in which salt is removed from the ocean is by complete evaporation of local arms of the sea. This is happening today in some desert areas and has been recorded from the past by thick layers of rock salt within continental rocks. But recent revolutionary advances in geology, dealt with more fully in the next chapter, have shown that oceanic sediments are riding across the ocean floor and traveling down a "conveyor belt" to the depths of oceanic trenches, there to be incorporated into the lower lithosphere or even the mantle.

This descent into the depths of the earth could be considered one stage of a grand hydrologic cycle, in which molecules of water and salt are dragged into the mantle to become incorporated in nearby molten rocks. The same rocks then melt and rise to the surface during volcanic activity, where the volatile components mix and dissolve in the hydrosphere's water and the rest stay behind to be weathered—again by that pervasive chemical, water.

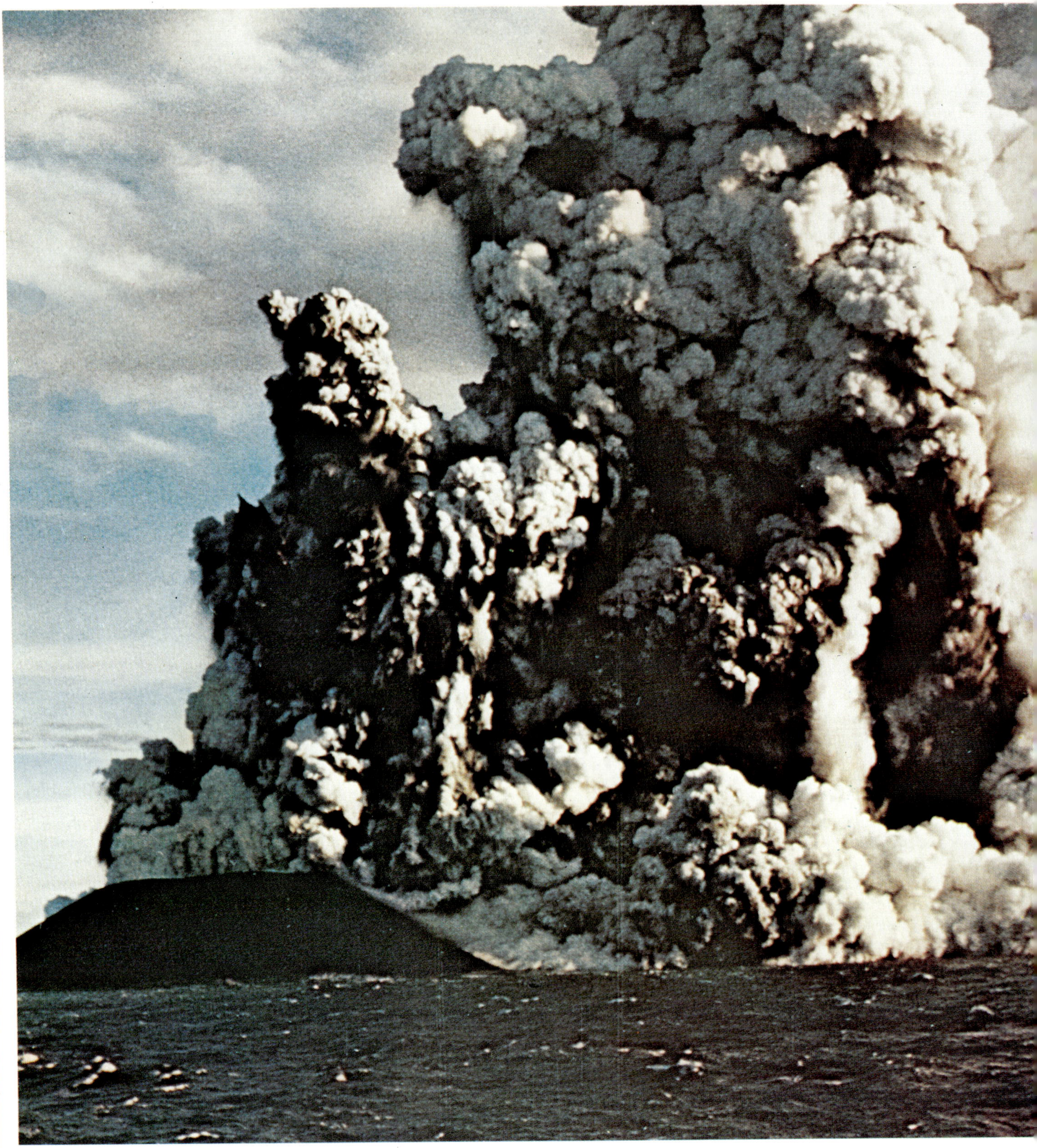

Often, when we are faced with the discomforting reality of the transience of life, we seek solace in the enduring beauty of our environment, in its "eternal hills" and "everlasting rocks." However, if geologic study over the past 200 years has taught us anything at all it is that the features of the earth's crust—its mountains, hills, valleys, and even the rocks themselves—are insufferably temporal. Rocks succumb to weathering, decomposing into smaller and smaller bits; or they may be changed by metamorphism or even melting. Early geologists correctly perceived the landscape features of the earth were, for the most part, the work of erosional agents. And as surely as the erosional power of a swiftly flowing stream carves out a valley in hard rock, continuation of that process destroys the entire valley. Hills are simply the remnants of upland areas between stream valleys, and if erosion continues the hills themselves are ultimately eroded away.

The geologic record abounds with evidence for the transience of the surrounding countryside. Look at sedimentary rocks and you see evidence for the filling of old river channels. Look at a geologic map showing the areal distribution of various rock types and you may see the drainage patterns of rivers that flowed hundreds of millions of years ago; sometimes you see old land surfaces, complete with hills and valleys, buried under later sediments. Look, finally, at the distribution of deformed and metamorphosed rocks on the continents and you see the crumbled spines of ancient mountain ranges.

But if erosion and deposition go on continuously, why do continents still exist? What has kept the seas from filling up with the sedimentary debris of the weathered land? The answer lies in *tectonism,* the continued vertical uplifting and attendant deformation of the continents by forces originating in the great heat energy of the earth's interior. The evidence for tectonism is all about us—in the young mountains with their markedly deformed rocks, in the persisting elevation of continents above sea level, in the great canyons cut by streams over continually rising land surfaces, in the wave-cut terraces and beaches formed at sea level but now elevated hundreds of feet above the surf.

The landforms called mountains are the most spectacular signs of tectonism. Every mountain range contains extensive evidence of massive rock deformation in its folded and broken strata. Its rugged peaks and great, sweeping valleys result from profound erosion following thousands of feet of vertical uplift. Often, uplift follows a period of great compressional deformation.

For many years one of the most perplexing questions in geology has been why mountains exist or, more fundamentally, what forces cause the vast deformations of the earth's crust. Compressional deformation, with its folding and faulting, implies shortening of the crust—but by what mechanism? And how can the continents keep on being uplifted and eroded, uplifted and eroded, and so on through all of recorded geologic history? The continents are not infinitely thick, so where does the constant supply of rock material come from? And more puzzling yet, where does all the eroded material go, since it has evidently not yet filled up the oceans? The answer to all these questions lies in the relationship between the continents and the ocean basins.

17
The Restless Earth

The island Surtsey is born, November 1963.
(From *Surtr,* a giant of subterranean fire in Icelandic mythology.)

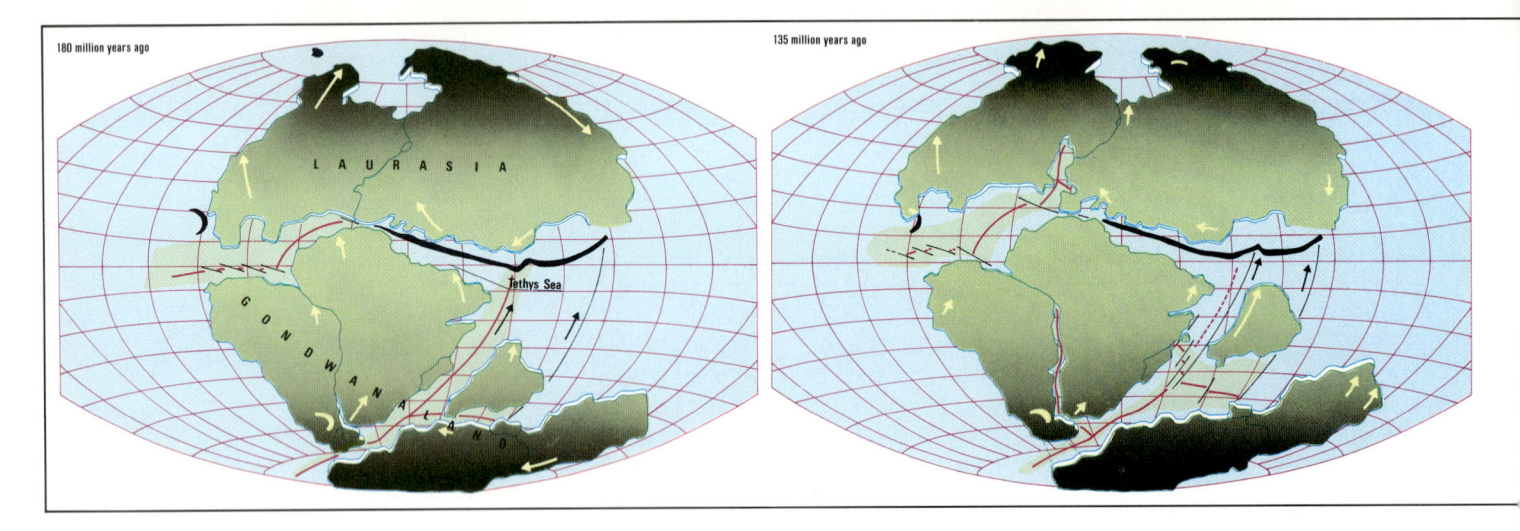

CONTINENTAL DRIFT

For many years one of the principal tenets of geology was that the size, shape, and distribution of continents and ocean basins were the same now as they had been throughout all of the earth's history. But in 1924 the German scientist Alfred Wegner, following an earlier idea of Alexander du Toit, proposed a radical new idea which he called the "Theory of Continental Drift." He suggested that all of the earth's continents had once been part of a super-continent. About 150 million years ago, according to Wegner's theory, that great super-continent broke up and the pieces began to drift into their present positions on the globe.

The evidence for the theory of continental drift was impressive. Wegner first called attention to the surprising "fit" of the continents one against the other—especially the fit of South America against Africa when the intervening South Atlantic Ocean is removed. With certain allowances for some submergence of the continental shelf regions of North America and Europe, those two land masses would also fit together. In addition, *structural trends* (the orientation of folds in mountain chains of the same age) were continuous in many cases across the oceans—both in North America and Europe and in South America and Africa. (More recently, major boundaries between rocks of different age in South America and Africa have been shown to be continuous if the continents are restored to their "pre-drift" configurations.)

Other evidence came from the distribution of certain kinds of fossils. For example, fossils of the *Glossopteris flora,* a genus of creeping land plant, were found in rocks in New England and in the British Isles. Earlier, paleobotanists had explained this phenomenon by invoking a now-subsided "land bridge" between the continents. But Wegner's hypothesis maintained that the distribution was to be expected, for at the time the plants lived, North America and Europe were contiguous parts of a giant continent. Furthermore, fossils of amphibians that lived during the Permian (about 200 million years ago) were found in southern Africa and also in southern South America. Some of these fossil animals are so similar they are classified into the same *species.* How could that similarity be the result of parallel evolution on opposite sides of an intervening ocean? And surely those amphibians were incapable of migrating across thousands of miles of ocean waters. According to the Wegner hypothesis, South America and Africa were contiguous land masses when these creatures lived and evolved.

Figure 2. Fossils of *Glossopteris flora* from New England *and* the British Isles, and fossils of the same kinds of amphibians from southern Africa *and* southern South America surely could not be dismissed as parallel evolution. The existence of these identical fossils on opposite sides of the ocean is cited as evidence for the theory of continental drift.

312

Physical Science Today

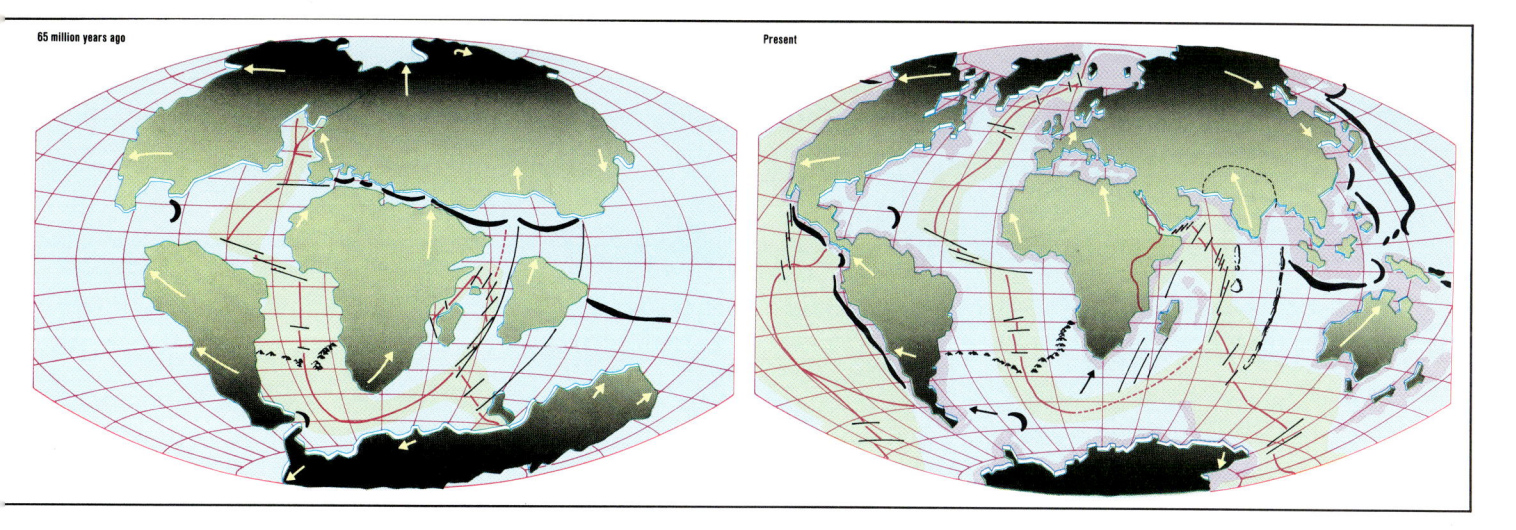

Finally, the Wegner hypothesis offered an explanation for the fact that young mountains and their associated volcanic and seismic belts are commonly located on the edges of continents. Prime examples are the coastal ranges of western North America and the great Andean chain of South America. Those mountains could have been formed as rocks crumpled during the westward drift of the American continents following their separation from Europe and Africa. The location of other young mountains in Europe and Africa fitted nicely with the proposed movements of these continents during the period of continental drifting.

During the 1920s and early 1930s, the Wegner hypothesis was largely rejected by geophysicists on the grounds that *a source of energy large enough to cause continental masses to drift across the earth could not exist.*

EVIDENCE FROM OCEANOGRAPHY

Following World War II the science of oceanography developed rapidly, perhaps under the impetus of extended naval operations in the South Pacific. Several institutions, most notably the Lamont Dougherty Geological Observatory of Columbia University and the Scripps Institution of Oceanography of the University of California, began operating oceanographic research vessels. Large-scale topographic and geologic mapping of the sea floors was undertaken along with geophysical investigations. And the results shed light on the relationship between the continents and the ocean basins. First, dredging of the sea floor revealed that the ocean basins are composed almost entirely of basalt; somewhat later radiometric age measurements showed that virtually all of the basalt was formed within the last 150 million years. Second, *no* sedimentary rocks were found in the ocean basins that carried fossils older than the Jurassic (about 150 million years ago). It had been hoped that coring samples of sedimentary rocks taken from the sea floor might yield sequences of sediment with preserved fossils from the entire history of the earth, but such was not the case. The ocean floors are apparently relatively young features.

Extensive mapping of the ocean floors also pointed to topographic and structural diversity. As you read earlier, great ridges lie in all of the ocean basins and the margins of the oceans are festooned with great arcuate island chains, all of predominantly volcanic origin and each accompanied by a deep oceanic trench lying on the convex and ocean-

Figure 3. Postulated continental drift movements 180 million years ago (far left) to the present (right). The present sequence begins after the single land mass of Pangea broke up into Laurasia and Gondwanaland. Progressive rifting and plate evolution led to the creation of the Atlantic Ocean and the northward drift of India, which eventually collided with Asia. Australia began to separate from Antarctica and also drifted northward. Drift is still continuing and the global picture will probably be significantly altered over the ensuing hundreds of millions of years of geologic time.

Compression Forces

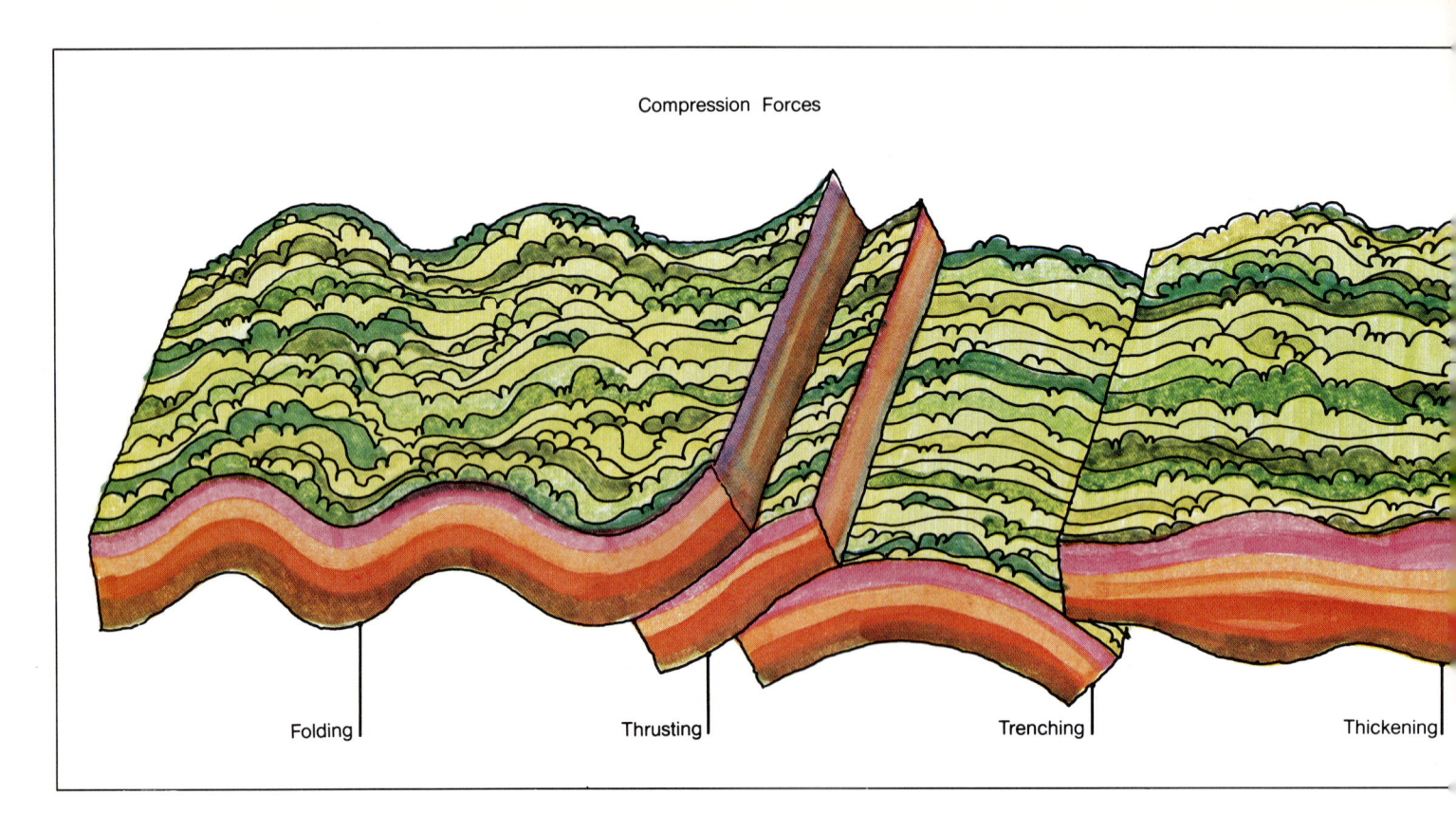

| Folding | Thrusting | Trenching | Thickening |

ward side. The structure of the mid-oceanic ridges, moreover, strongly suggests they are tension features: each is characterized by a central *graben,* or a down-faulted block. Such down-faulted blocks are characteristic of those regions where the earth's crust has been stretched apart, allowing a central block to subside.

EVIDENCE FROM GEOPHYSICS

Seagoing geophysicists have studied the magnetism of the sea floor, its gravitation anomalies, and its seismicity. And now, other interesting features must be integrated into the study of the continents and ocean basins. The oceanic crust is uniformly thin, usually between 3 and 10 kilometers; the continental crust is much thicker. The island arcs and their associated trenches are the sites of earthquakes, as are the mid-oceanic ridges, but most of the rest of the ocean basins is relatively free of seismic activity. The actual location (or foci) of the disturbance that causes the earthquake beneath the island arcs is particularly revealing. As Figure 5 shows, earthquake foci beneath island arcs become deeper *toward* the continents, which suggests there may be some kind of great fault plane sloping down from the islands toward the adjacent continent. Finally, gravity surveys show that the trenches have negative gravity anomalies. In some way they are associated with deficiencies in mass in the crust or the subjacent mantle.

PALEOMAGNETISM

Rocks provide a record of past orientations of the earth's magnetic field because tiny particles of iron-bearing minerals tend to become aligned with the prevailing field during either magmatic crystallization or sedi-

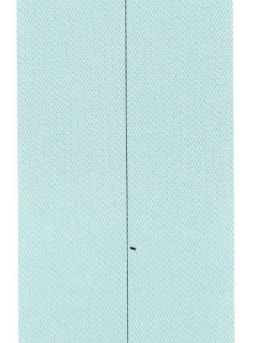

Physical Science Today

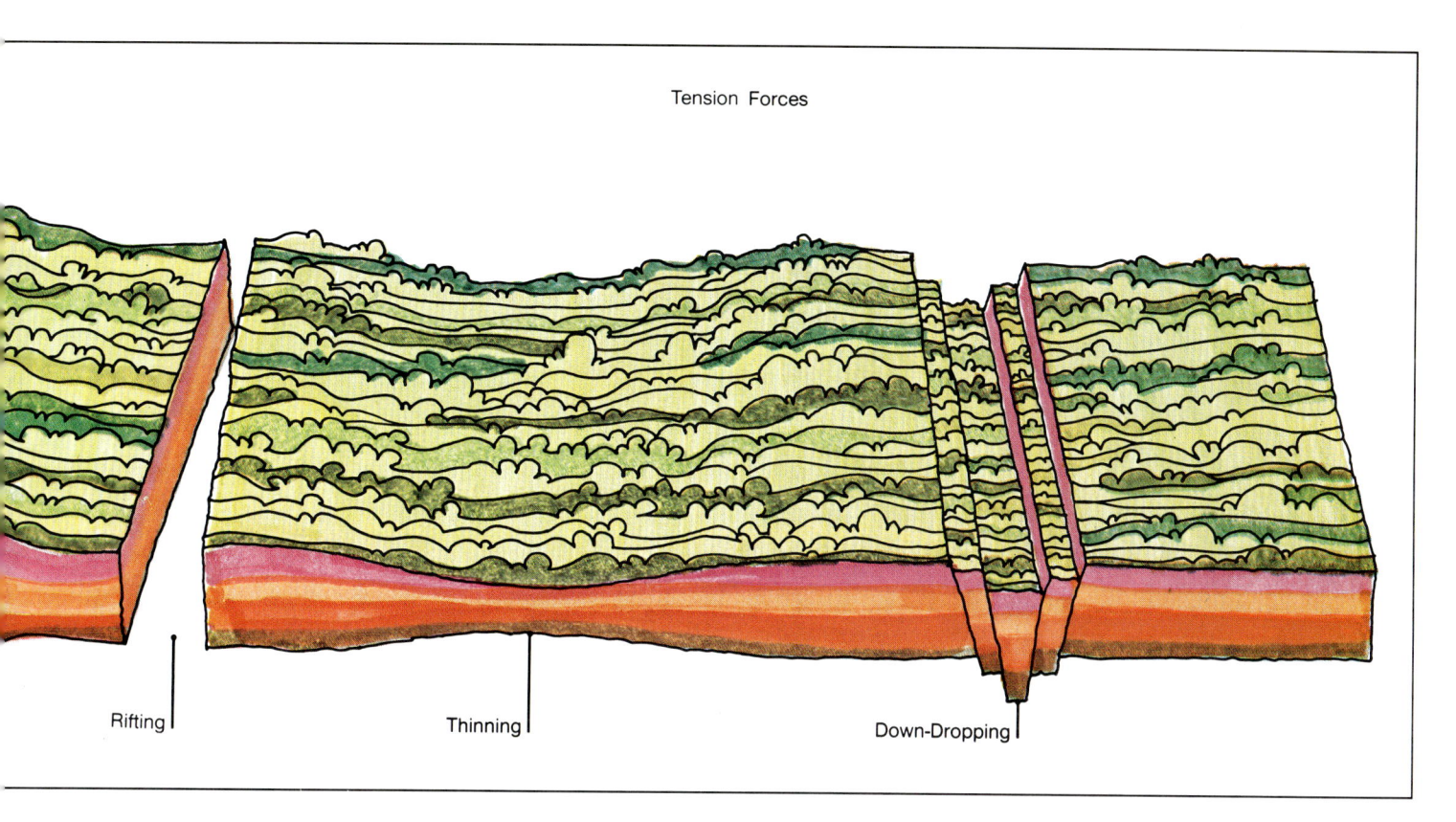

mentation. The study of so-called *remanent magnetism* has been significant in attempts to reconstruct the earth's past. Specifically, the orientation of remanent magnetism in rocks of known age indicates the direction of the earth's magnetic poles at the time the rock formed—provided that any structural changes such as folding or faulting are compensated for. The remanent magnetism in rocks formed during the last 10 million or 20 million years essentially parallels the present magnetic field, but the magnetic orientations in older rocks differ markedly from the present position of the poles. The difference is so great that rocks of Precambrian age (greater than 650 million years ago) may show pole positions in the middle of what is now the Pacific Ocean! These measurements have been interpreted to imply that the earth's magnetic poles have "wandered" in the past.

However, it is difficult to reconcile this view with the apparent coincidence of the magnetic axis of the earth with its rotational axis. It is also difficult to explain away the great physical difficulties encountered in any theory that proposes some mechanism for allowing the earth to shift its rotational axis. An alternative explanation is that the land masses—the continents—have moved with respect to the rotational and magnetics axes during the earth's history. In all but the most recent rocks, then, there is an *apparent* change in the position of the magnetic poles brought about by a *real* change in the position of the rocks after their magnetic orientation was "frozen in." The poles didn't wander; the rocks did.

Investigations into the remanent magnetization of rocks showed that some rocks appeared to have their magnetic polarity reversed. In other words, the direction of their north magnetic pole pointed in the direction of the present south magnetic pole of the earth. At first the phenomenon was believed to be the result of some strange physical or chemical event

Figure 4. (Far left) Ship coring is a method used to obtain samples from the ocean floor. This drawing depicts the drilling vessel Glomar Challenger 5,486 meters above the ocean floor. (The line width has been greatly exaggerated so that it can be seen.)

Figure 5. (Above) The response of crustal plates to compression and tension accounts for most of the earth's geologic features. Rifting usually occurs in the ocean floor where the crust falls into the mantle.

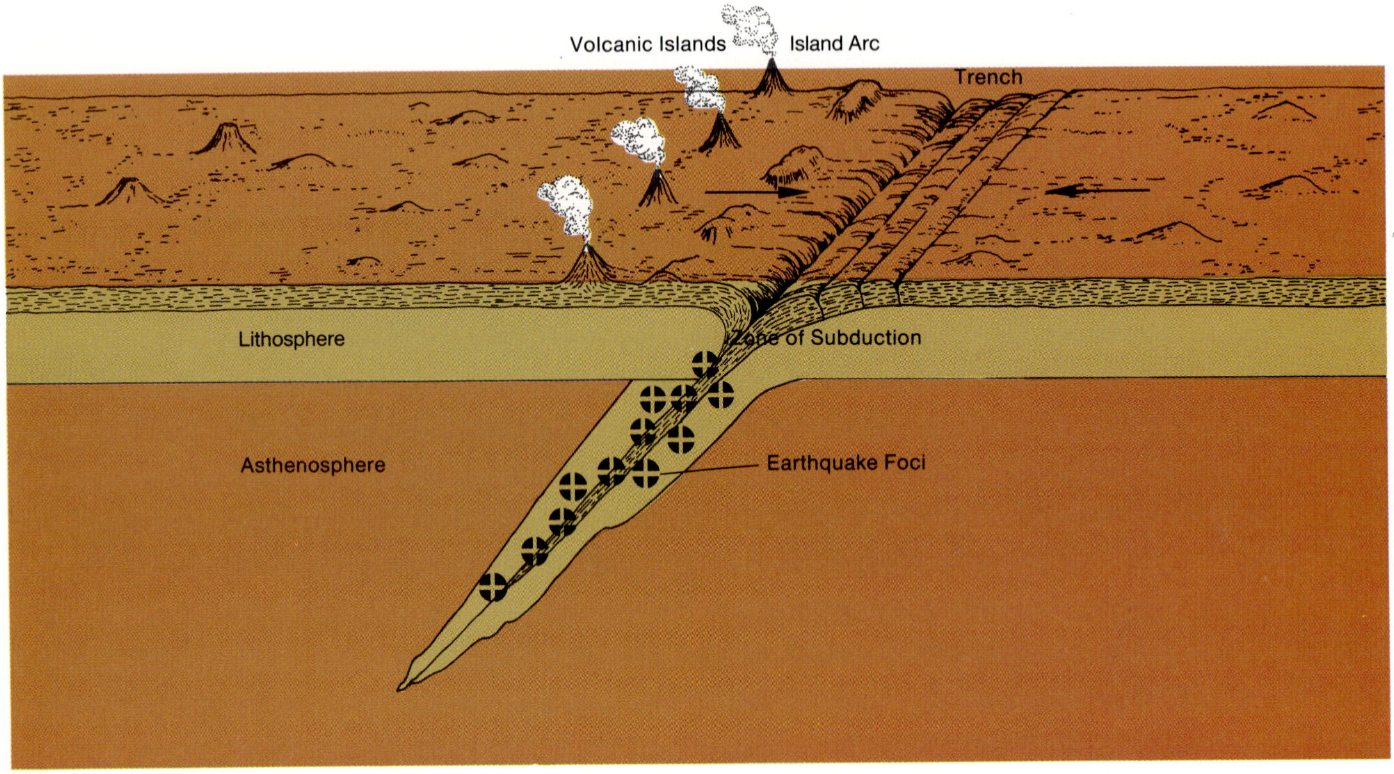
Volcanic Islands
Island Arc
Trench
Lithosphere
Zone of Subduction
Asthenosphere
Earthquake Foci

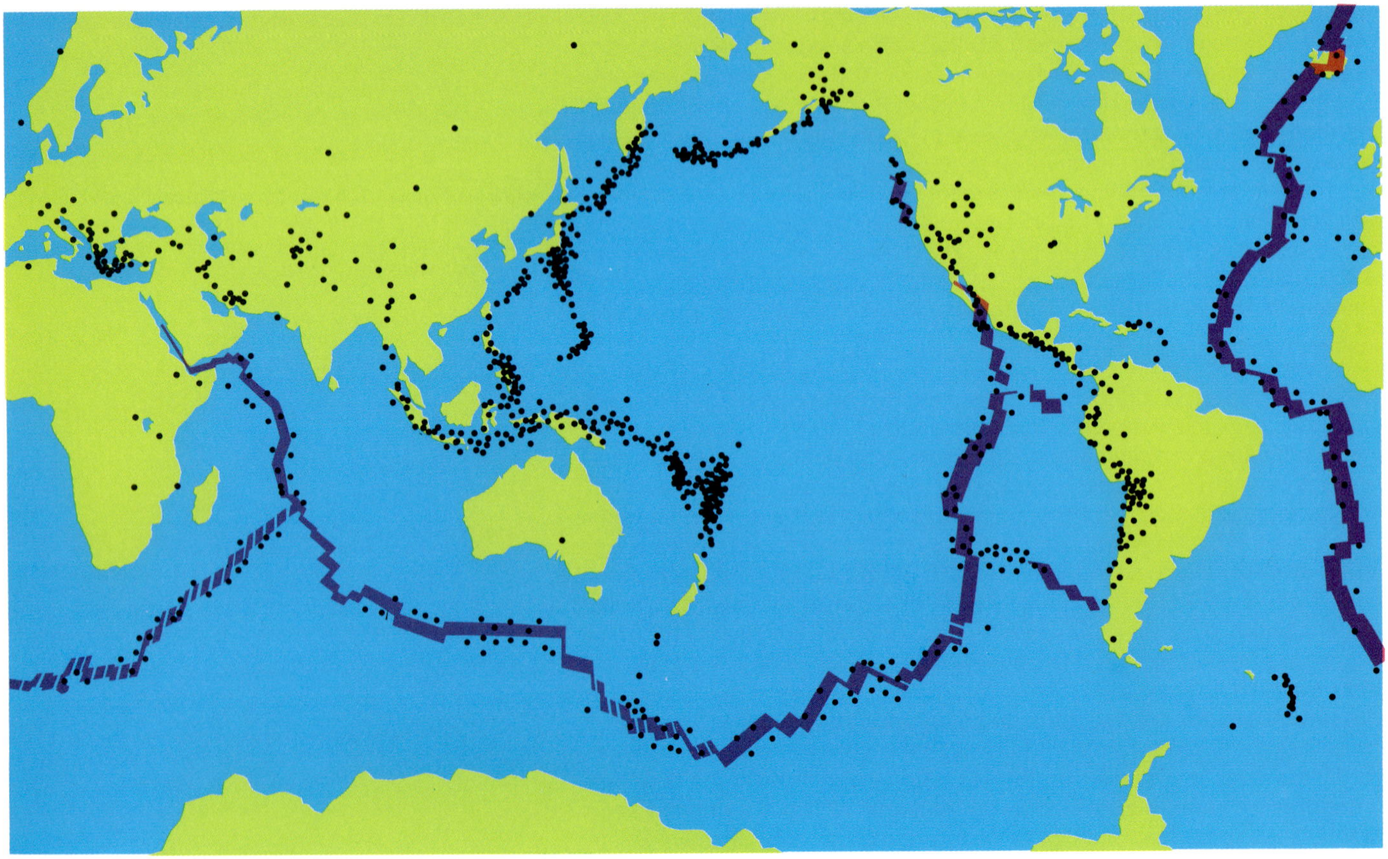

in the history of the rock, but none of the proposed mechanisms seemed adequate to explain it. As more and more rocks were measured they, too, displayed reversed polarity. It became evident that they were actually records of the earth's history when the magnetic field *was* reversed. It is now well established that the earth's magnetic field does reverse itself periodically. Although the mechanism causing the reversal has not been identified, the times during which the magnetic field has been reversed are well known for the most recent 20 million or 30 million years of earth's history. The magnetic reversal time scale has been worked out on the basis of radiometric methods used to date rocks whose magnetic directions have been measured.

SEA FLOOR SPREADING

Magnetic surveys of the sea floor turned up some curious anomalies. A *magnetic anomaly* is a measured value of the earth's magnetic field, at a particular locality, that is greater or less than the value that would be expected if you assume a uniform geomagnetic field. The pattern of magnetic anomalies mapped on the sea floor in many places are elongated, or "stripe-like" (see Figure 6).

The magnetic anomalies near the Rejkannes Ridge, which is a part of the mid-Atlantic ridge system, are both positive and negative and they are symmetrically distributed on either side of the ridge. In 1965 F. J. Vine and his associates correlated the linear magnetic anomalies about the Rejkannes Ridge on a one-to-one basis with the known times of normal and reversed magnetic field polarity. Vine proposed a radical mechanism to explain the relationship. He suggested the sea floor literally spreads away from the mid-oceanic ridges, and that the basalt of the sea floor is like a magnetic tape recorder that records variations in the earth's magnetic field through time. The correlation of the magnetic anomaly patterns with the magnetic reversal time scale occurs because each newly created "belt" of basaltic sea floor "freezes in" the polarity of the earth's magnetic field at the time the basalt is erupted. Comparison of the width of the magnetic belts with the magnetic reversal time-scale indicates a spreading rate of a few centimeters each year. That rate would account for the opening of the Atlantic Ocean as well as the drifting apart of North and South America from Europe and Africa since Jurassic times, about 150 million years ago.

The mechanism of sea floor spreading apparently involves fusion of the upper part of the mantle beneath the mid-oceanic ridges, followed by volcanism on and near the ridges themselves. This process creates new basalt all along great segments of the ridges, on a time scale of several hundred thousand to perhaps a million years. Continued rifting and spreading causes the tensional structures of the ridges and brings on renewed melting and volcanism in the vicinity of the ridge. In fact, the ridges are topographically high because they are built up by repeated eruptions of basalt. There is now evidence of higher heat flow beneath the ridges, and it seems probable that all the volcanic and spreading activity is driven by thermal convection in the earth's mantle.

As the oceanic crust spreads away from the mid-oceanic ridges, it carries with it a "memory" of the intensity of the earth's magnetic field at the time it was erupted and cooled. At the same time, volcanism forms new oceanic crust along the ridges. Sometimes actual volcanic mountains that form along the ridges are dragged away by the spreading; old volcanos can be found today on the older sea floor, sometimes hundreds of kilometers from the ridge. This model for the spreading of the sea floor suggests that the most recently formed basalt should be found along the

Figure 6. (Far left) The above drawing shows how the collision of plates produces sites of seismicity. The faster collisions produce trenches. When collisions are slower, buckling can occur and mountain ranges form. Volcanic activity characterizes the mid-oceanic ridges, which have been formed by rifts in the ocean floor. (Below) The black dots mark the locations of seismic activity throughout the world. The maximum density of earthquake foci occurs along the crests of ridges (thick black lines), on parts of the fracture zone (thin black line), and along deep trenches (hatched bands).

Direction of Earth's — Magnetic Field

Figure 7. This series of drawings is based on the magnetic stripe pattern from the Reykjannes Ridge south of Iceland. This magnetic record, which lies on the ocean floor, accumulated when layers of crust were successively displaced. Each new layer became magnetized along the direction of the earth's magnetic field. New layers, fed from the molten mantle, surfaced at the ridges. As spreading continued, each successive layer separated into two bands, one on each side of the ridge. The paleomagnetic record locked in this striped pattern helped confirm the theory of sea floor spreading.

mid-oceanic ridges, with older basalt being found farther and farther away. To the extent that radiometric age determinations have been made on the basalt of the sea floor, that is exactly the distribution being found.

What happens to "old" sea floor if new basalt is continually generated at the ridges? Because of their seismicity, the deep oceanic trenches and their associated volcanic island arcs were thought to be related somehow to relative movements of the ocean basins and the continental masses. In fact, the planes of earthquake foci dipping toward the continents under the island arcs had been interpreted as great thrust faults where the continents were riding up and over the ocean floor. But later interpretations of the nature of earthquake movements beneath the island arcs suggested that great slabs of cooler crust and upper mantle must be descending along those planes. This interpretation accounted for the fact that the old oceanic crust that was spreading away from the mid-oceanic ridge systems was being consumed, somewhere.

The island arc-trench systems are now considered to be the sites where great slabs of lithosphere (the upper mantle above the low-velocity zone as well as thin basaltic crust) descend into the mantle. Numerous observations are consistent with this interpretation.

For one thing, all the trenches show negative gravity anomalies, which implies they lie above a region with some kind of mass deficiency. These negative anomalies correlate well with the notion of a great slab of lithosphere descending into the mantle, there to be transformed back into mantle material.

Second, the island arcs and mountain arcs are active volcanically as well as seismically. Presumably the magmas are generated by the partial melting of the descending lithospheric slab and perhaps other material that is dragged down with it.

Third, in many parts of the world the ophiolite suite of rocks—the periodotites and gabbros, the basalt and minor deep sea sediment—is being found where there is reason to believe a trench-arc system once existed. This characteristic suite of rocks may be actual pieces of the old oceanic crust and upper mantle that were broken off and literally plastered against the edges of the continents or against the island arc instead of being dragged down under it.

Finally, in parts of the world where there is either a trench-arc system or evidence of an ancient one, belts of "blueschists" occur. Blueschists are rocks metamorphosed under the unusual combination of extremely high pressures but relatively low temperatures. This sort of metamorphism probably occurs when sedimentary and volcanic materials accumulate in deep oceanic trenches and are then dragged down rapidly along with the descending lithospheric slab.

PLATE TECTONICS

The theory of continental drift and the new evidence for spreading of the sea floor has brought about a new theory of the tectonics of the earth's outer crust. "Plate tectonics," as this theory is called, promises to account for many of the enigmatic features of the earth's crust. The new theory maintains that the outer part of the earth consists of a small number of semi-rigid plates that are in motion relative to one another. These plates constitute the *lithosphere* of the earth—the continental and oceanic crusts together with that portion of the mantle above a depth of about 100 kilometers. At that depth the velocity of seismic waves is lower. Sometimes called the *asthenosphere,* it is a zone of plastic yielding and may be partially melted.

The boundaries of lithospheric plates may be of three main types, although there evidently are many variations. At *accreting plate bound-*

(a)

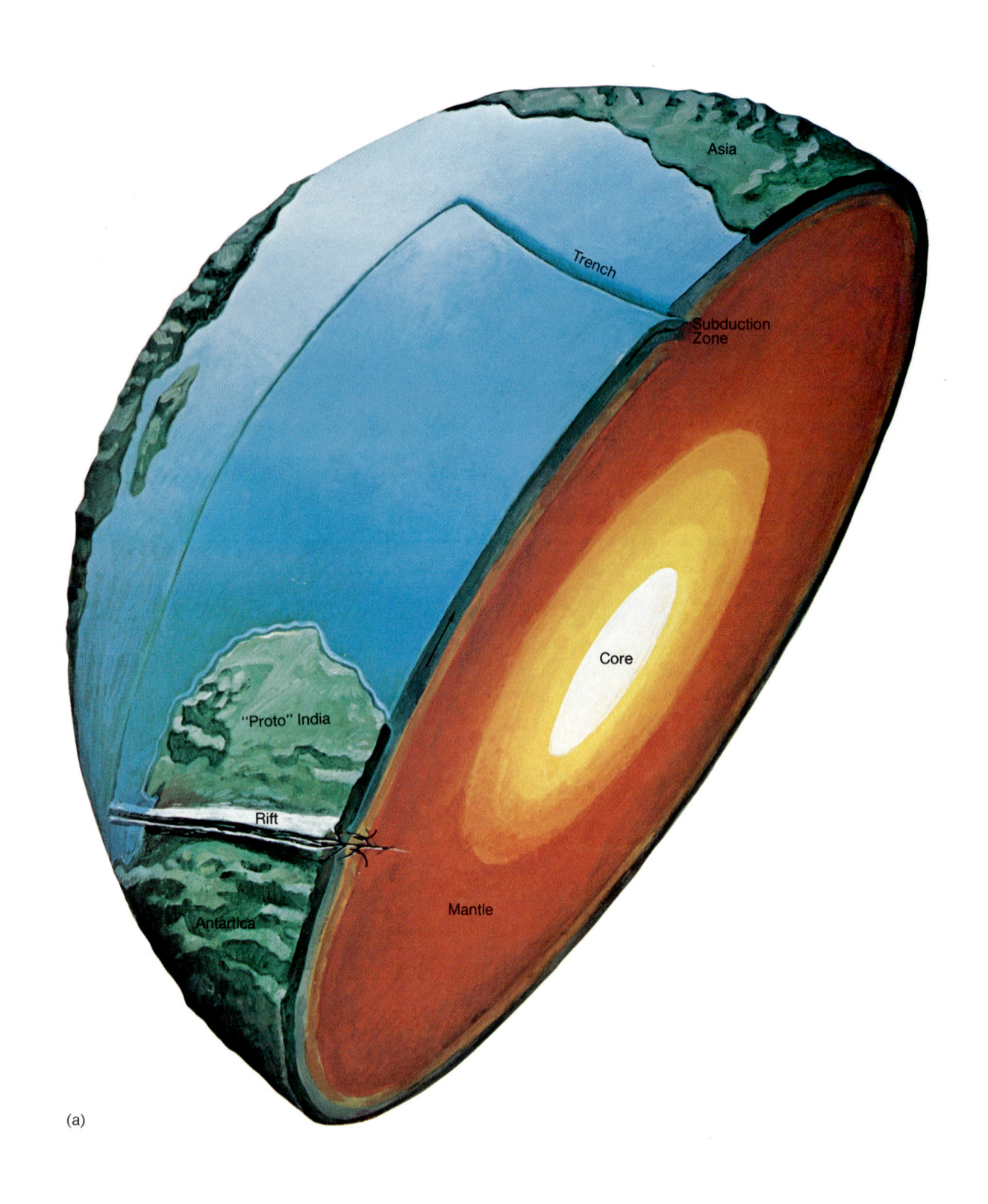

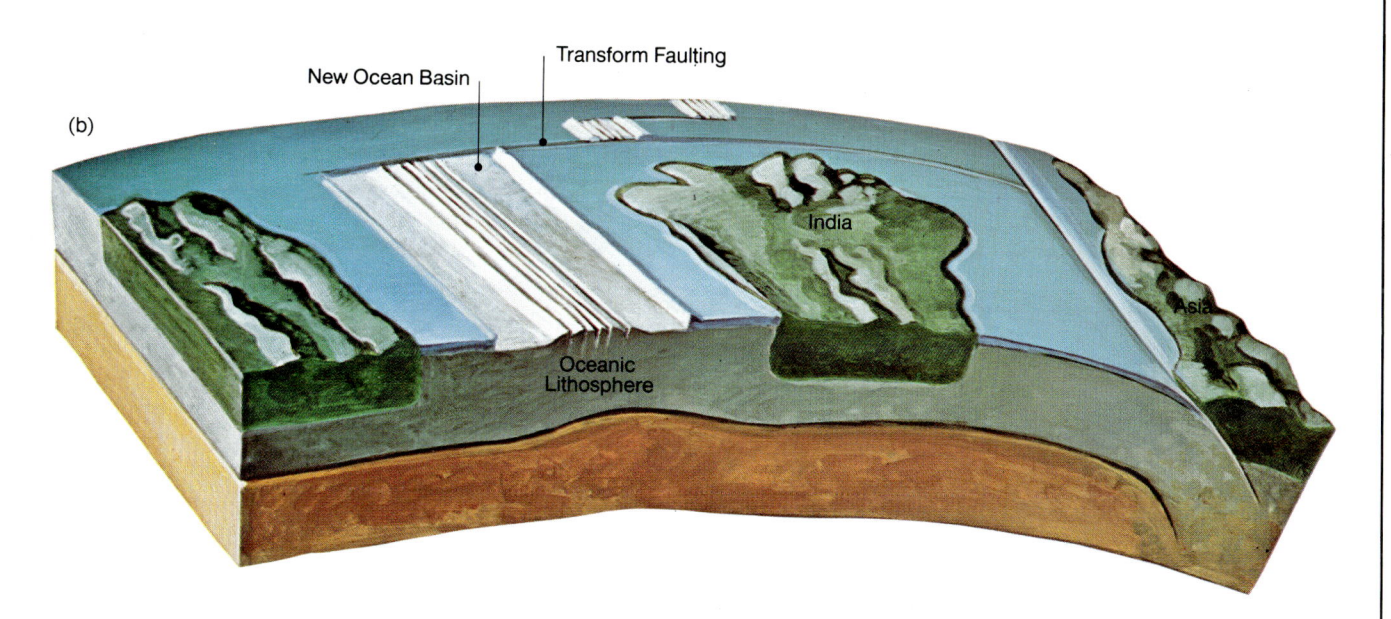

Figure 8. According to plate tectonic theory, India was part of a large land mass near the South Pole during early Mesozoic time. This continent, which was originally one plate, now commonly referred to as "Gondwanaland," included Antarctica, Australia, India, Africa, and South America. During the Jurassic period, India began to split away and to drift northward as part of a newly formed plate, shown in (a). At the trailing edge of the plate, a rift formed on the Gondwana continent, and basalt rose up from the mantle to fill the gap. At the leading edge of the plate, a subduction zone and trench formed, which has been obliterated by subsequent drift. In (b) the formation of a new oceanic basin is indicated. Its history of spreading is recorded in the magnetic stripes that are now observed in the modern Indian Ocean. Finally, in (c), India collided with the Asiatic land mass, underthrusting its southern margin to form the uplifted Himalaya Mountains.

Continental Drift

<u>Step 1.</u> Suppose that samples of basalt dredged from the sea floor on both sides of the Mid-Atlantic Ridge at a distance of 600 kilometers from the ridge crest are dated at 3.7×10^7 years. What has been the rate of sea floor spreading in centimeters per year?

The rate in kilometers per million years is

$$600 \text{ kilometers} / 3.7 \times 10^7 \text{ years} = 16.2 \text{ kilometers}/10^6 \text{ years}$$

If you convert 16.2 kilometers to 16.2×10^5 centimeters, you can calculate the rate in centimeters per year:

$$\frac{16.2 \times 10^5 \text{ centimeters}}{10^6 \text{ years}} = 1.62 \text{ centimeters/year}$$

<u>Step 2.</u> Would this rate of spreading account for the separation of North America and Europe and the opening of the Atlantic Ocean since Jurassic time some 150 million years ago? This time of initiation of continental drifting is strongly indicated by the geologic evidence.

The Mid-Atlantic Ridge is about equidistant from North America and Europe: the distance from the ridge to the edge of the continental shelf of either continent is about 2,500 kilometers. Thus a spreading rate of 1.62 centimeters per year would bring about this separation in

$$\frac{2,500 \times 10^5 \text{ centimeters}}{1.62 \text{ centimeters/year}} = 1,543 \times 10^5 \text{ years} \approx 154 \times 10^6 \text{ years}$$

that is, in 154 million years.

aries two plates pull apart, and fusion of the mantle produces new basaltic crust through volcanism. Each plate "accretes" (adds to) its area at such localities, which are marked by mid-oceanic ridges.

Consuming plate boundaries, in contrast, are marked by island arcs, mountain arcs, and trenches. Here, two plates move and collide, and the overriden plate descends into the mantle where it is transformed and partially melted. Some of the molten material may include sediment from the continent, if the trench is adjacent to the continental edge, and it may return to the surface as volcanic rocks of intermediate composition. The volcanism builds the island arc or the mountain arc, forming new continental crust. The region where a crustal plate descends into the mantle is termed a "zone of subduction."

At *lateral slippage boundaries,* the plates slip and slide not vertically but laterally against one another. These boundaries may result in enormous faults, such as the San Andreas fault of California, along which hundreds of kilometers of displacement have occurred.

Plate tectonics explains many of the most puzzling geologic problems—including the puzzle of how continents have drifted apart. Continents are parts of the lithospheric plates, and move with them. When plate motion was initiated along the site of the present mid-Atlantic Ridge system about 150 million years ago, the part of the old supercontinent that was riding on the western plate rifted away and became the North American and South American plates. The separation of those plates and the generation of new basaltic sea floor along the ridge created the Atlantic Ocean.

There is evidence that such processes are occurring today in the opening of the Red Sea, the Persian Gulf, and perhaps the Great Rift Zone of East Africa. These areas exhibit all of the features we would expect to find along the boundaries of a newly initiated plate separation. Volcanism, earthquakes, boundaries marked by tensional faults and subsided blocks are represented and, in the case of the Red Sea and the Persian Gulf, land masses have been separated by newly formed basaltic sea floor.

Figure 9. Aerial radar image of the San Andreas fault. Lateral slippage of crustal plates has created this fault in California. Earthquakes have occurred periodically along the fault.

Figure 10. (Below) This satellite photograph, looking east from Africa across the Red Sea and Gulf of Aden to the Arabian peninsula, shows the rifting that is further evidence of continental drift.

MOUNTAIN BUILDING

As our understanding of present-day tectonics increases, we are able to look at older geologic features with greater insight. For example, a great belt of basalt extending from the Lake Superior region of Minnesota and Wisconsin all the way into central Kansas separates older granitic rocks both to the east and to the west. This belt is extremely narrow relative to its length, and the basalt is known to have been formed over 1 billion years ago. It is now looked on by most earth scientists as a fossil rift zone, along which plate separation and an accreting plate boundary were initiated. Apparently the rifting did not proceed very far, for the basaltic belt is narrow and a major ocean basin never formed.

And what of the origin of mountains? Mountains occur in great linear belts that often show arcuate structures when viewed on a broad regional basis. Frequently young mountain belts lie along the margins of the continents, where they are essentially in contact with the sea floor. These belts include the coastal ranges of North America, the Andes Mountains of South America, and the Alps, the Carpathian, the Pyrenees, and the Atlas mountains. Sometimes mountains lie within the continental plates but such mountains are always old. The Urals of Central Russia are an example. The Rocky Mountains of the west-central United States appear to be anomalous and do not fit the plate tectonic model in a simple way.

According to plate tectonic theory, mountains are formed at consuming (colliding) plate boundaries in six stages of *orogenesis,* or mountain building.

First stage. The initiation of plate motion produces a consuming plate boundary, and a trench forms.

Second stage. Melting along the descending lithospheric plate causes volcanism and an island arc or a mountain arc results. These structures are formed almost entirely of volcanic rocks.

Third stage. Erosion of the island or mountain arc, and of the nearby continent, transports sediments to the ocean. In and adjacent to the trench and the mountain or island arc, great thicknesses of "greywacke" sediments and volcanic rocks accumulate. Greywacke is an immature sediment in which the processes of weathering and segregation of components is not much advanced because the distance between the site of derivation (the volcanic arc) and the site of deposition (the trench and its vicinity) is small.

Between an island arc and the continent, well-sorted, mature sediments may accumulate. The sedimentary material may have been derived from weathering and erosion of the continent and may have been transported some distance from the place where they were first derived. Such sediments are typically deposited in shallow seas between the continent and the arc and are usually limestone, sandstone, and shale.

Fourth stage. Continued plate movement causes compressional deformation of the accumulated sediment. Usually the greywacke sediment is deformed first; it lies on the oceanward side of the island arc. Later the mature sediments that accumulate between the arc and the continental edge will become deformed.

Fifth stage. As the compression of sedimentary rocks and the volcanic rocks continues, the whole rock mass of the arc thickens and deforms plastically. Metamorphism and deep-seated igneous activity result as the deformed mass is pushed deeper into the upper mantle, where temperatures are higher and where the distribution of heat in the region has been disturbed by the consumption of the descending slab and the deformation. The vertical uplift that accompanies the thickening and deformation of the rock mass accelerates erosion—and the landform we call "mountains" is formed.

Sixth stage. The metamorphism and plutonism produce new continental crust, which is essentially welded on to the old continental plate.

Figure 11. At the later stages of mountain building, continued plate movement compresses masses of rock, thickening and deforming them.

In this way continents can accrete and grow.

This outline of the mountain-forming processes is simplified, and probably somewhat oversimplified at that. But almost all young mountains show evidence of most of these processes. As in all natural phenomena, variations occur in the basic patterns and we must study each mountain system individually if we are to completely understand its evolution. The important contribution of plate tectonics is that it provides a fundamental mechanism from which our observations can begin. For the first time the enormous compressional deformation that occurs during mountain-building seems understandable.

If you look in detail at older mountain systems, you will find many similarities with the younger systems in terms of structure and rock distribution. For example, most older mountains show arcuate structural patterns that often bend away from the older part of the continent. The convex side of these arcs often incorporates a belt of highly deformed greywacke, presumably the deposits of the trench and the associated continental slope. Just inside the greywacke belt there is usually a belt of volcanic, plutonic, or metamorphic rocks—the remains of the old volcanic arc. Finally, on the concave side of the igneous-metamorphic complex there is often a third belt of moderately to intensely deformed sedimentary rocks that make up the typical "shelf" assemblage: sandstone, shale, and limestone. Because these patterns are found in most of the older mountains of the earth, it seems likely they formed through processes involving colliding plate margins.

Consider the implications of one of the most thoroughly studied North American mountain ranges, the Appalachians. These mountains are made of deformed rocks about 250–600 million years old and they show most of the patterns discussed earlier. The deformed greywacke belt lies to the east. The patterns disclose plate movements more-or-less perpendicular to the present margin of North America *before* the most recent opening of the Atlantic Ocean, which began only about 150 million years ago. In other words, at one time there must have been an opening of a "proto-Atlantic Ocean," with a consuming plate margin somewhere in what is now western New England. But the "proto-Atlantic" must have closed up again. Following this episode, the present Atlantic Ocean began

Figure 12. A simplified history of the Appalachian Mountains, showing the stages leading to the accretion of "new" continental crust. (a) Sediments were deposited along the edge of a stable continental margin. (b) Compressional folding began, which signaled the development of a consuming plate junction. The resulting uplift led to thrusting, faulting, and deposition of sediments in inland basins. (c) This stage was followed by further uplifting, faulting, and igneous intrusion. (d) Finally, a mature landscape emerged with a new continental shelf along its margin. The approximate locations of features are shown as they exist today in the Appalachians.

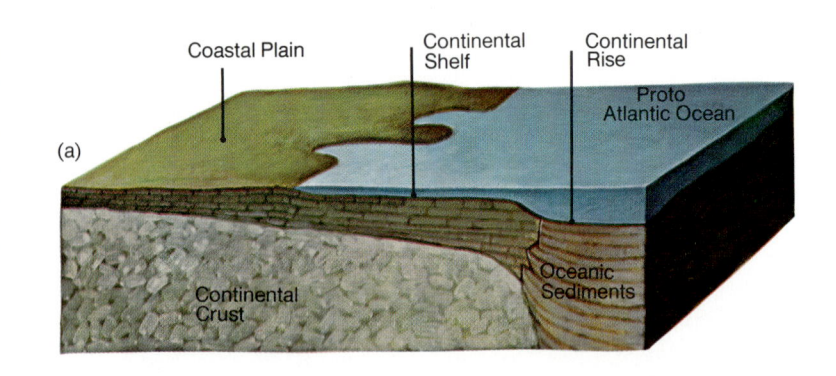

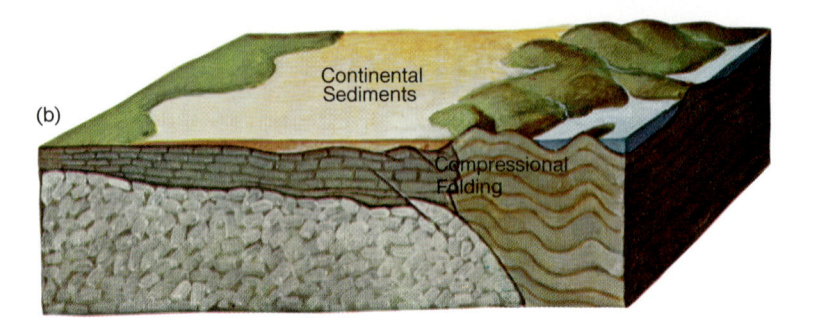

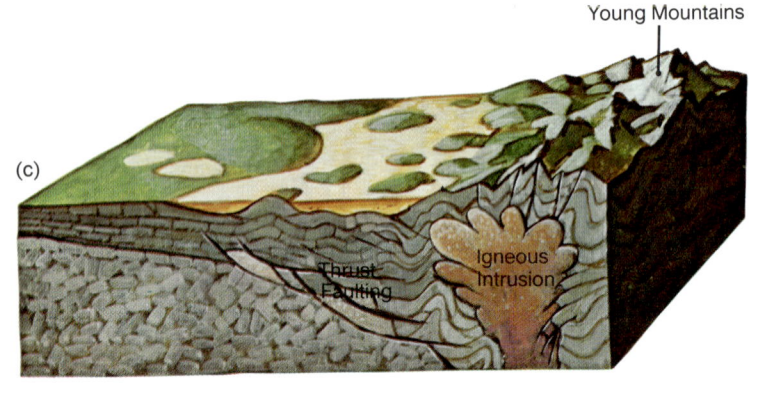

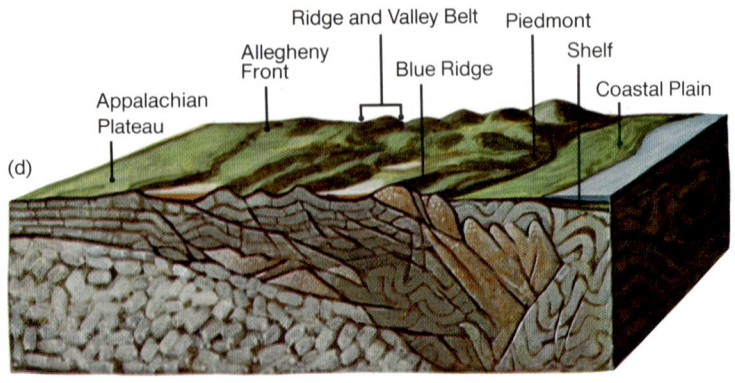

to open. Now the consuming plate boundary is somewhere near the western edge of North America and the whole continent is coupled onto the westward-moving plate.

Prior to the plate tectonic interpretation, Appalachian geology left many questions unanswered. For example, many of the sedimentary rock units thicken to the east, as if the source had been in that direction. But where was the source? There is nothing there *now* but the Atlantic Ocean basin. Earlier geologists invoked a hypothetical land mass called "Appalachia" which, after supplying the needed sediments, conveniently subsided into the Atlantic. Oceanographic research, however, has not turned up any evidence of that "foundered" land mass.

Furthermore, Appalachian folds are commonly overturned toward the west and the movement on major thrust faults is from east to west; surely the stresses had come from the east. But from where? In terms of the plate tectonic model, the stresses came from the spreading of a proto-Atlantic sea floor, away from a center of rifting and spreading. The sediments thickening to the east were derived from the erosion of a volcanic island or mountain arc. That ancient arc is now represented by the igneous and metamorphic complex of the Piedmont zone, and parts of it may even be on the other side of the Atlantic, rifted away by the most recent opening of that ocean.

To say the earth is restless, then, may be something of an understatement. Its landscapes are incessantly worked by the surging waters of the seas, carved by thousands of rivers and streams, and polished by the work of the winds. The rocks of the earth are constantly transformed through weathering, erosion, sedimentation, metamorphism, and melting. And even the crust itself does not stay in place but moves and jostles about, crumpling here and stretching there, driven by the internal heat energy of the earth.

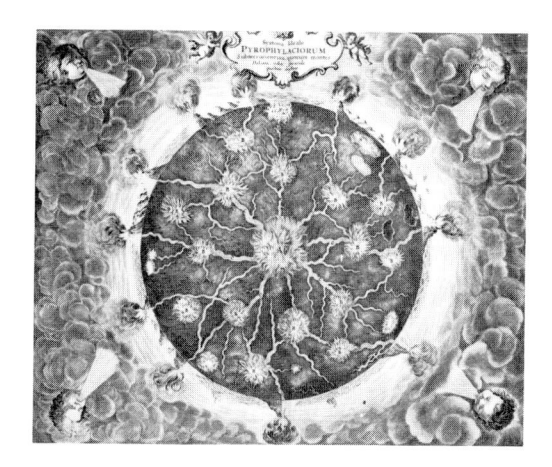

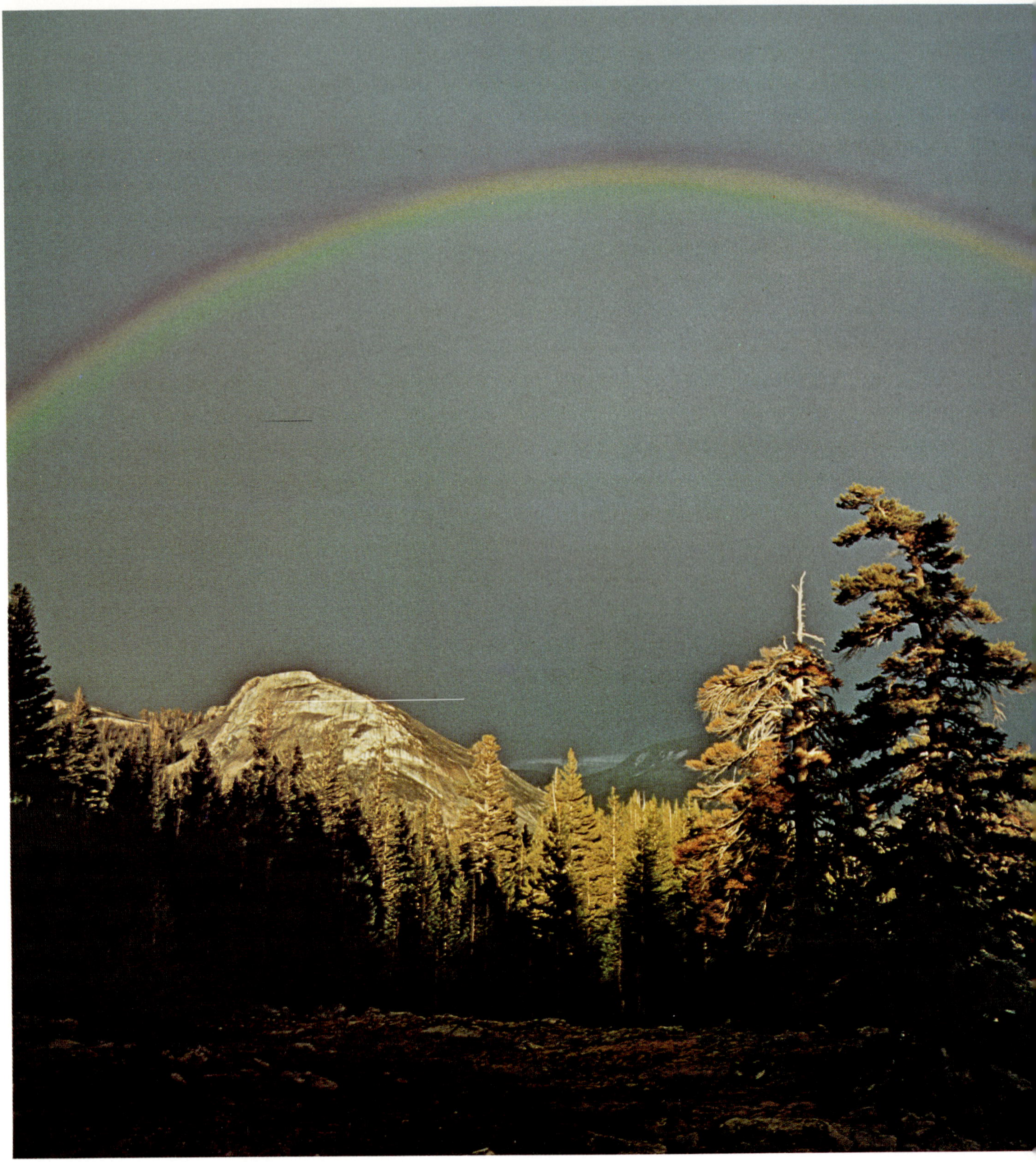

18
A Nice Place to Visit...

To the natural scientist intent on studying the complex interactions of real physical systems, the outdoor laboratory called planet earth is most assuredly a great place to visit. However, there is an implicit second half to the title of this chapter—most of us automatically assume that it includes the words, "... but I wouldn't want to live there." And perhaps at one time it might have been looked upon as a clever way of saying we really don't have much choice about it. For better or worse, the planet earth is the only habitat available to us and our fellow creatures. Unfortunately, things look like they are about to get worse before they get better, and today people are more likely to be thinking, "It's a nice place to visit ... can we keep it that way?"

Through our growing technical capability, we now have access to powerful tools that can be used to build a richer, more comfortable, and more diverse way of life for peoples all over the world. But all powerful tools carry a potential for damage. And suddenly we're worried about it; we want to learn how to assess what impact we're having on the whole earth system. Invariably we find we must turn to the natural scientists for guidance. These are the people who have probed and poked about in, on, and above the surface of the earth, locating metals and fuels, charting winds and ocean currents, recording the magnetic dipole field. As a result, they are the ones who are in a position to tell us what we know, and what we don't know, about the resources of our planet and how those resources may be evaluated in terms of survival, fun, and profit.

THE STUDY FOR SURVIVAL

The message from the environmental movement during the last few decades is that we're in trouble. Often it's not quite clear just why we're in trouble, however. We hear about all kinds of serious local pollution problems, and occasionally we hear about global problems involving such substances as DDT, mercury, and carbon dioxide. And we hear a great many cries that we are perilously close to running out of natural resources. Put it all together and it begins to sound like we eventually may not even have "A Nice Place to Visit."

The difficulty comes in trying to assess the various threats to the earth system. Everyone agrees that we cannot sustain growth and expansion forever on a finite earth with finite resources. But there is very little agreement on just how close we are to being in trouble and about just how finite our current and potential resources really are.

With some important exceptions, serious students of earth resources argue that we will not run out of any basic raw materials in the next several hundred years, or that if we do, we will be able to replace them with other resources and technologies. Even if these estimates are overly optimistic, the argument goes, the crises that progressive resource shortages would bring about probably would be gradual and at first merely inconvenient. Does that mean we are safe from a major earth system crisis for at least six or eight generations? Certainly not. Going without two automobiles, air conditioning, and aluminum beverage cans probably wouldn't hurt people. Living at a subsistence level in tents and eating fish meal might be unpleasant but it would not necessarily mean

Figure 2. A horse density of about 2,000 horses per square mile over the entire United States would be needed to support our society at its current level of energy use. This calculation is based on each horse working an average horse day of 12 hours and expending 10^7 joules of energy per day.

the end of us. But eating poisoned food, breathing poisoned air, or coping with arctic or desert climatic conditions is an entirely different story.

Perhaps the potential crisis facing our earth system is really twofold. It could be argued that there is a long-term crisis posed by limited space and materials, which could be called a *resource crisis*. And there is the potentially rapid failure of one or more of the systems vital to human life, which could be called a *life support systems crisis*. But there is not very good agreement today on the imminence, or indeed even on the possible existence, of these two sorts of earth system crises. In fact, several fundamental intellectual arguments are raging on this subject among the people who study long-term growth patterns, resource utilization, and the earth-ocean-atmosphere system.

TECHNOLOGICAL INNOVATION AND SUBSTITUTION

If we manage to avoid damaging the earth's life support systems, how soon will we run out of the raw materials necessary to sustain the high standard of living enjoyed in the developed world—and desired by much of the rest of the world? The question is not simple, although you will find many prepared to offer simple-minded answers. Obviously some resources are limited. Conventional reserves of natural gas will probably be adequate for only a few decades, for example, and reserves of coal are expected to last for only a few centuries. You could go around from resource to resource, looking at current utilization rates for known reserves and projecting their expected duration. And when you were done you would probably predict many shortages in the not-too-distant future. The difficulty lies in trying to identify which, if any, of those predicted shortages will actually reach crisis proportions.

The unknown variables are technological innovation and substitution. If we tried to support the United States today with the energy of horses alone, we'd have a crisis. The current United States consumption of energy is about 6×10^{19} joules per year. Theoretically speaking, a good horse can put out something like 10^7 joules in a 12-hour day, so it would take at least 8×10^9 horses to support the United States at its current level of energy use. That works out to a horse population of about 2,000 horses per square mile for the entire country.

The lesson is obvious: projections of long-term requirements based on current technology can be misleading. Nevertheless, we must not assume that technology will *automatically* come along to prevent a projected crisis. The important intellectual argument mentioned earlier centers on whether or not technology will *always* be developed in time to "save" us.

Few serious scientists would argue that technology cannot be developed to meet foreseeable crises. The arguments have to do with the time scales and the constraints imposed on their development. Most of our social systems—our economy, our use of energy, and so on—are growing in an exponential, or "compound interest," way. Today we consume twice as much electric energy as we did 10 years ago. In another 8 or 10 years we will consume twice as much again, so that by decades the growth rate increases from 1 to 2 to 4 to 8 to 16, and so on. Will we always be able to develop the next level of technology in the time we have available without totally unacceptable side-effects? Some respected scientists say yes and urge us to push ahead at full speed. Others (perhaps a larger number) say there may be problems and urge a slowing-down or even an end to growth in order to gain enough time to come up with the necessary solutions. A few urge an actual reversal of the growth trend—a return to a less technical society with a lower standard of living on the grounds that we have already excceeded the tolerance of the earth system. The latter view is probably more widely held by nonscientists in the environmental movement than by professional physical scientists. Finally, until quite recently a majority of people (including scientists) probably hadn't thought much about the problem at all.

A CASE STUDY: OUR USE OF ENERGY

There are so many ways in which man's activities interact with and are potentially limited by the earth system that any attempt at a comprehensive review would fill a book in itself. But examining a single aspect of the problem should give you an idea of the types of interactions that arise and some of the ways in which knowledge of the earth system can be important.

Our consumption of electric power has been doubling every 10 years. Consider how this activity may interact with the earth system and how our knowledge of earth and atmospheric science is important in evaluating the interactions. Today most of our electric power comes from burning fossil fuels—coal, oil, and gas. Hydroelectric power presently supplies only about 15 percent of our electricity, and our growing demand for electric energy is such that hydroelectric power will become even less significant as time goes on. In the future, nuclear energy may provide a major portion of our power but today it supplies only a small percentage of our total requirements. Moreover, because of opposition by environmental groups it is a technology that is being applied more slowly than had been predicted. For the next several decades, a major portion of our expanding needs will continue to be met with fossil fuels.

Locating fossil fuels, extracting them from the earth, and projecting the size of the fuel reserves involve a knowledge of earth science. Once a fuel is found and extracted, it must be burned. The combustion process releases oxides of sulfur and nitrogen as well as carbon dioxide and particles. In the burning process, energy is converted into steam which, in turn, is converted into electric energy through steam turbines. There is a fundamental rule of thermodynamics that limits the efficiency of such systems. It is true that modern steam electric plants based on the burning of fossil fuels are much more efficient than earlier designs. In 1910, for example, it took more than 2 kilograms of coal to produce 1 kilowatt-hour of electricity; today it takes less than 1/2 kilogram. Yet even with this improved efficiency, modern plants still eject almost 60 percent of the total energy as waste heat into the atmosphere, the waterways, and the world

**A Case Study: The Aswan Dam—
An Expensive Response to the
Need for Power**

In 1954 the decision was made to build the largest rock fill dam in the world near a small, earlier dam at Aswan, Egypt. The completion of this dam, one of man's greatest engineering feats, brought many economic rewards—a vast increase in arable land, good flood control, and a huge output of electricity. But there are many inherent environmental problems in a dam of this size—including nutrient and silt holdback and a buildup of toxic mineral salts. The future will tell if the problems and costs will outweigh the advantages.

Although the project initially cost 1,000 million dollars, once the power station near Kalabsha (see map below) is in full operation the power-output will be about 1,000 million kilowatt hours. In addition to providing power, the High Dam at Aswan should allow the growth of two crops annually in Upper Egypt, open up more than a million acres of new farm land, and eliminate floods. The filling of Lake Nasser just above Aswan is slowed by water losses through evaporation and underground seepage. But as the lake water rises, people residing in villages along the river have been forced to evacuate their homes. More than 100,000 people have been resettled in Nasser City, Kom Ombo, and Kalabsha. Moreover, before dam construction could begin, 40 million dollars had to be spent to raise the 3,200-year-old temple of Ramses II (shown in the photograph to the left).

Dam construction has damaged offshore fisheries, but it is hoped that stocking coastal lakes and Lake Nasser will increase the annual catch. Coastal erosion near the delta will need control because the river will carry less silt. Flooding was a natural way to fertilize the soil each season. Now fertility of delta soil must be maintained by chemical fertilizers. Because accumulated mineral salts may cause soil sterility, drainage canals will be installed in irrigated areas.

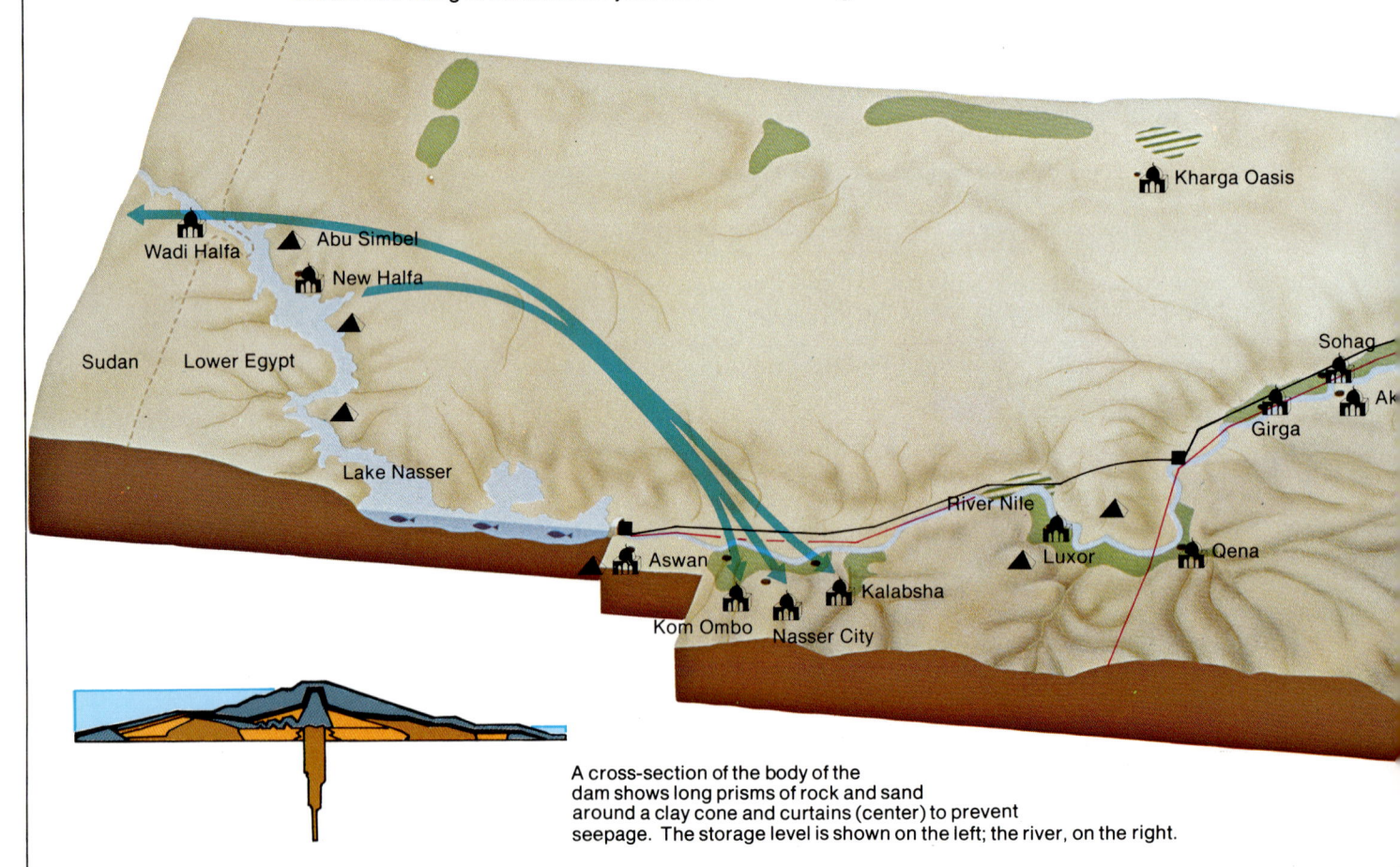

A cross-section of the body of the dam shows long prisms of rock and sand around a clay cone and curtains (center) to prevent seepage. The storage level is shown on the left; the river, on the right.

Building the dam.

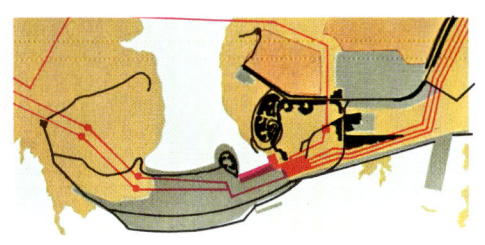

Plan of the dam site.

Hydroelectric power drives the new large pumps needed to handle irrigation flow.

For at least 6,000 years people irrigated with wells and the Archimedian screw.

Throughout history the delta of the Nile has been naturally fertile in contrast to the Sinai desert beyond the Gulf of Suez. The dam promises to bring fertility to the arid upper Nile region.

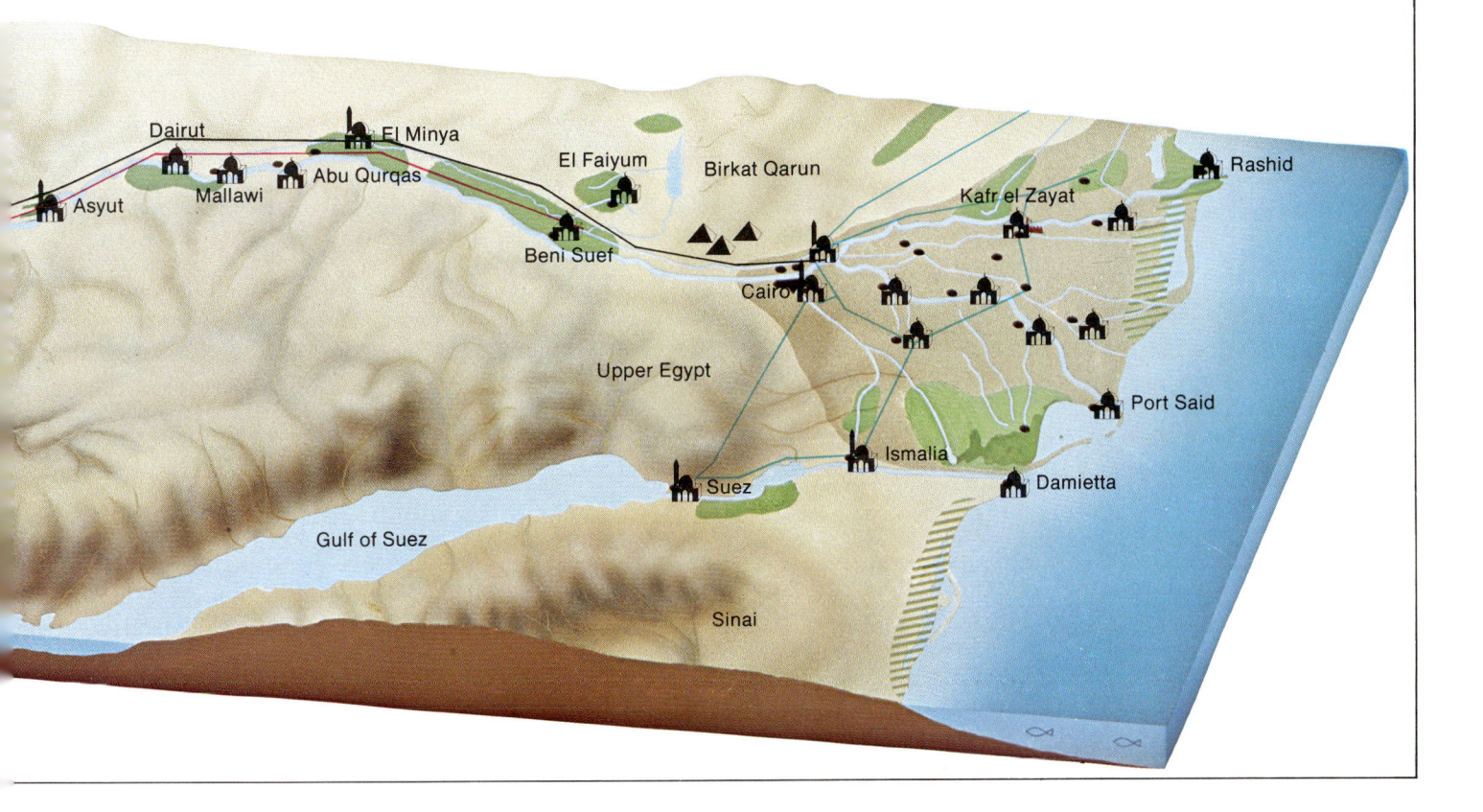

ocean. Because of the fundamental thermodynamic limitation on these plants, there is little that can be done to improve their efficiency as long as they continue to run at their current operating temperatures.

What are the effects of the combustion by-products and waste heat on the earth system? It is likely that combustion by-products may have some influence on weather and climate. Of course, weather and climate have changed naturally without any help from man since the earliest days of the atmosphere and they continue to change today. Fluctuations in solar radiation, the earth's orbital parameters, and the location of the continents are all thought to have had long-term effects on climate. Variations in oceanic circulation as well as volcanic activities are also known to have produced natural climatic change.

Until now, our impact on global weather and climate appears to have been slight. However, because of the burning of fossil fuels the carbon dioxide concentration in the atmosphere over the last century has increased 15 percent, although it is difficult to measure this concentration change to high accuracy. Modern measurements tend to show that the carbon dioxide concentration is growing at about 0.2 percent a year. Not all of the carbon dioxide we inject into the atmosphere remains there: as with many atmospheric trace gases, carbon dioxide is absorbed by the ocean. About one-half of all the carbon dioxide we inject into the atmosphere ultimately ends up in the ocean. The absorption rate depends critically on such factors as water temperature, gas concentration, and surface roughness.

Scientists predict there will be an 18 percent increase over existing levels of carbon dioxide by the year 2000. If nothing else changes, that increase will probably raise the mean temperature of the earth by about 0.5 degree Kelvin because of the effect carbon dioxide has on the global heat balance. And changes in the mean temperature of the earth by even a few degrees have the potential to influence the climate significantly.

The carbon dioxide concentration is not the only thing that is changing. The changing concentrations of dust and other small particles (called *particulates*) in the atmosphere are also affecting the surface temperature of the earth. But most of the high-altitude particulates seem to come from natural sources, primarily volcanic eruptions.

From 1900 to 1940, the mean temperature of the earth appears to have increased about 0.5 degree Kelvin. Since 1940, however, it has dropped about 0.2 degree Kelvin. Some scientists believe that the increase came from carbon dioxide and that the decline has been brought about by the particulates from recent volcanic eruptions. But there is by no means agreement on this theory or even on the fact that the overall change is man-made. Particulates very close to the earth's surface may actually have a heating effect, but both the distribution and the effect of particulates throughout the atmosphere are still not fully known.

Nevertheless, a reasonable conclusion is that our use of fossil fuels on a massive scale does hold the potential for bringing about significant changes in global atmospheric conditions. Until now, at least, we have been fortunate and have not blundered too seriously. The ocean and atmosphere are a highly complex, and possibly delicate, multiple feedback system and scientists are actively working to model this system with giant digital computers. But until the models are worked out, only educated guesses can be made about the effects of carbon dioxide particulates and similar factors on a global scale. The best guess at the moment is that there may be a problem but it probably is not yet serious. Nevertheless, we will undoubtedly feel more comfortable when further study allows us to remove some of the "educated guesses."

What about local weather patterns? In this area we know the effects of our activities have been appreciable. Clearing woodlands to make fields, plowing large expanses of land, irrigating with large quantities of water, building huge reservoirs—all such activities appear to have had some climatological effects.

When air is cooled, its relative humidity increases, and if cooling continues, it ultimately becomes saturated. However, condensation into droplets does not immediately begin merely because the air has become saturated. For condensation to occur, appropriate cloud *condensation nuclei* such as dust, smoke, sea salt, or various chemicals must be present. The physics of the condensation process is only partly understood. For the most part, condensation occurs at about the saturation point, but some *hygroscopic nuclei* have a chemical affinity for water and may begin to produce condensation at relative humidities as low as 80 percent. Naturally occurring condensation nuclei are vital in the production of clouds and precipitation.

There is not much direct evidence that specifically links particulates from power plants to local weather changes. There is, however, a collection of data that does indicate man-made particulates can have significant effects. Recent measurements in the state of Washington have established that pulp and paper mills, as well as large saw mills and aluminum smelters, are significant sources of cloud condensation nuclei. Sodium sulfate is a major effluent of pulp mills and it forms efficient hygroscopic nuclei. Scientists at the University of Washington have compared the statewide precipitation patterns for the period between 1929 and 1946 with the patterns for the period between 1947 and 1966. Their comparisons show as much as a 30 percent increase in precipitation levels in some areas of the state that are close to, or downwind from, major industrial sources of cloud condensation nuclei. The evidence strongly suggests that we have inadvertently caused parts of the rainy Pacific Northwest to get a good deal more rainy.

Not all particulates introduced into the atmosphere are good condensation nuclei. Nor does the generation of efficient nuclei imply that increased precipitation will follow. For example, consider what happens during the sugar cane harvest in the Bandaberg district along the coast of Queensland in Australia. Large numbers of condensation nuclei ap-

Figure 3. How little we know about the workings of our atmosphere becomes clear when we try to explain why particulates introduced into the atmosphere of the Pacific Northwest cause rain, and why particulates introduced into the atmosphere above Hawaiian cane fields cause drought.

Figure 4. By using infrared color film, it is possible to "see" the intense heat energy being radiated from concentrated urban areas.

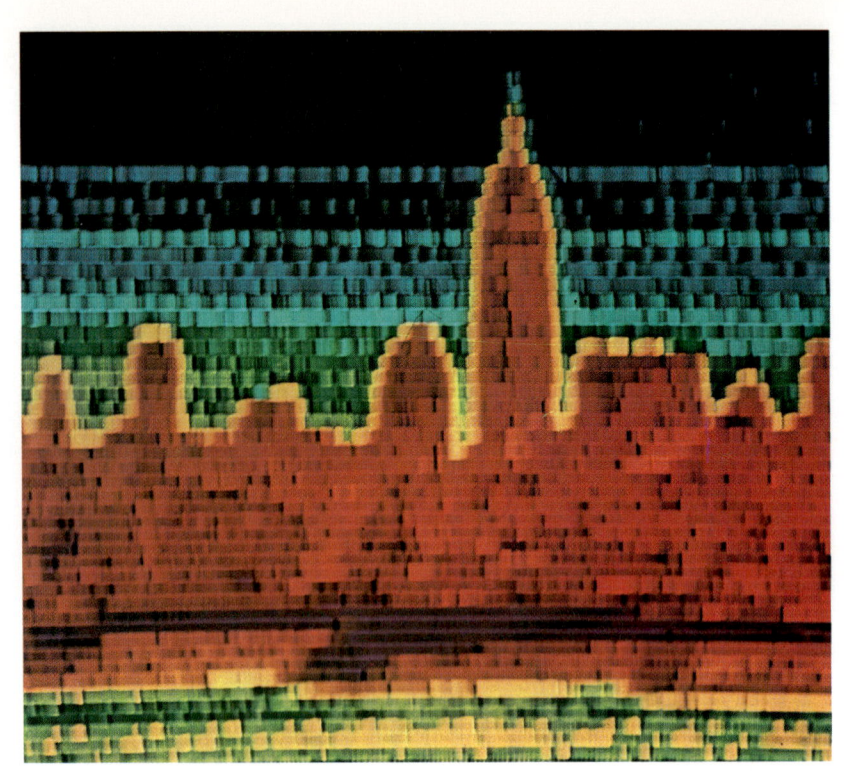

parently are introduced by the widespread burning of sugar cane during the harvest season. Careful studies of precipitation patterns, involving about 60 years of data, suggest there is a correlation between the nuclei generated by burning and the precipitation in the area downwind from this burning. In Queensland, however, the correlation does not point to an *increase* but rather to a *decrease* in precipitation rates in those areas. The hypothesis is that the great quantities of nuclei introduced by the cane burning leads to a smaller average droplet size upon condensation, and thus to a lower probability of actual precipitation. Similar results have been observed for smoke from cane fires in Hawaii.

Another effect of power generation through the burning of fossil fuels is the introduction of large quantities of the oxides of sulfur and nitrogen into the atmosphere. Electric power generation is one of the major sources of sulfur dioxide pollution in the United States. An active control program is now under way. New reserves of fuel with a low sulfur content are being sought, methods for extracting sulfur from fuel are being improved, and devices for removing sulfur from flue gases have been developed. Because it is linked to the combustion process, the production of nitrogen oxides is more difficult to control. While nitrogen oxide pollution from power plants is not as serious as the sulfur oxide problem, major control efforts are now planned in this area, also.

What about some of the other side-effects of energy use? We all know about air pollution, but few of us are aware of the vast "thermal pollution" or "thermal enrichment" that we are generating as we convert and use energy. With today's technology, essentially all the energy we use —about 6×10^6 joules per day in the United States alone—sooner or later ends up in the oceans and atmosphere. Our major cities sit in self-created "heat islands" in which the average temperatures are several degrees higher than temperatures in the surrounding countryside. It's bad enough that cities generate a considerable amount of heat through the concentrated use of energy. Compounding the problem is the interference with

Is the Waste Heat of the U.S. at the Danger Level?

When man uses energy, it ultimately ends up in the environment as waste heat. If this waste becomes a significant proportion of the solar energy input to the earth — say, one or two percent — the global heat balance could be disrupted, possibly resulting in drastic climatic changes. How would you determine for the United States the current percentage of waste heat in the atmosphere?

Step 1. The earth receives about 1.82×10^{24} ergs of solar energy each second, assuming there is no absorption of energy by the atmosphere. The United States covers approximately 1.8 percent of the earth's surface, so it receives about 1.8 percent of the total solar energy reaching the earth:

$$\text{Energy} = (.018) \times (1.82 \times 10^{24} \text{ ergs/second})$$
$$= 3.3 \times 10^{22} \text{ ergs/second}$$

This is equivalent to 3.3×10^{12} kilowatts.

Step 2. If there are approximately 2.0×10^8 people in the United States, the average amount of solar energy received per person is:

$$\text{Energy per person} = \frac{\text{energy received by United States}}{\text{population of United States}}$$

$$= \frac{3.3 \times 10^{12} \text{ kilowatts}}{2.0 \times 10^8 \text{ persons}}$$

$$= 1.7 \times 10^4 \text{ kilowatts/person}$$

Step 3. Each person in the United States presently uses an average of about 10 kilowatts of energy per second. To find the percentage of atmospheric energy that turns up as waste heat, divide the amount used per person by the amount received per person:

$$\text{Percentage of waste heat} = \frac{10 \text{ kilowatts/person}}{1.7 \times 10^4 \text{ kilowatts/person}}$$

$$= \frac{10 \text{ kilowatts}}{17 \times 10^3 \text{ kilowatts}}$$

$$= 0.588 \times 10^{-3}$$

$$\approx 0.0006$$

$$\approx 0.06 \text{ percent}$$

The percentage of energy that goes into the atmosphere as waste heat from the United States is less than 1 percent. But selected cities are already at several percent, and if the present growth of energy consumption continues, the percentage for the entire country will rise to a few percent within 100 to 150 years.

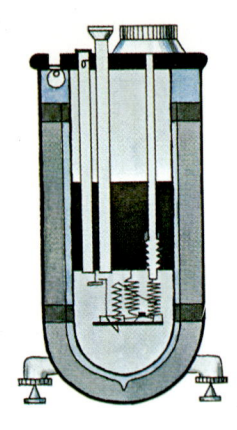

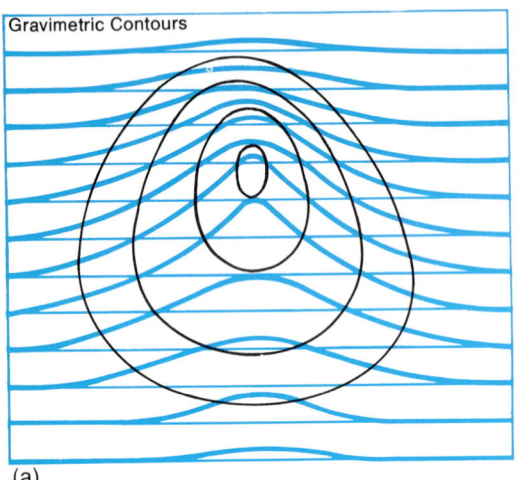

Figure 5. How do we prospect for minerals today? In *gravimetric surveys* (a, above), the Worden gravimeter measures distortions caused by variations in a gravity field. The microscopic motions of a sensitive quartz mechanism, enclosed in the vacuum flask, are magnified and read against a scale through the eyepiece. If known distortions caused by the sun, moon, latitude, and height above sea level (h) are allowed for, a localized anomaly may disclose the presence of a mineral deposit, such as a salt dome. When the regional field is plotted, such anomalies appear as "kinks" which can be used to map out the position of the disturbance. In *seismic surveys* (b, above right) shock waves are sent into the crust by vehicle-mounted vibrators, explosives, or airguns for surveys at sea. These waves are then reflected or refracted by abrupt changes in strata, which monitoring equipment records. The record from one survey shows a fault and a dome. In *magnetic surveys* (c, far right), an airborne magnetometer measures the strength and direction of the earth's magnetic field. For example, two strongly magnetized projections of igneous rock yield two clear peaks in the recording. A plot of the regional magnetic contours then reveals potential magnetic deposits, such as iron ore.

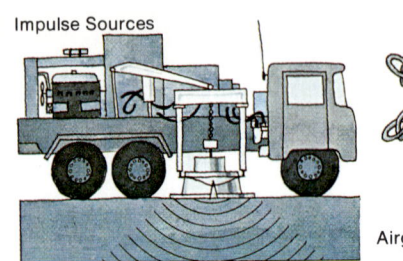

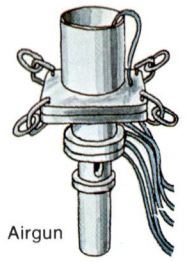

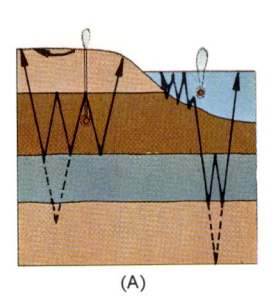

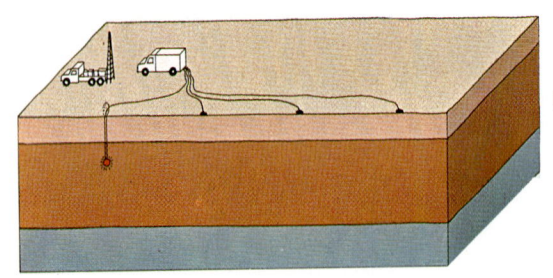

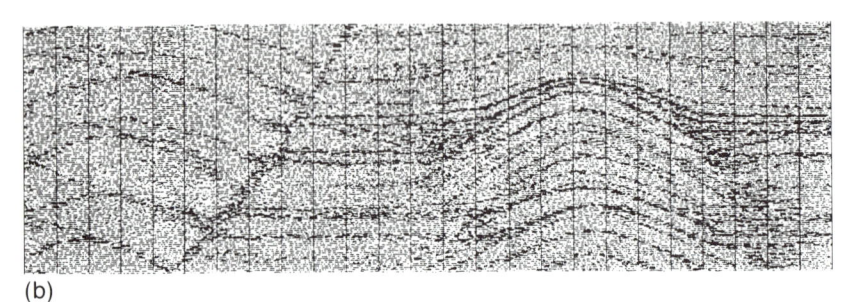

evaporative cooling, the principle mechanism for removing heat from the earth. Evaporative cooling works less effectively in the city because precipitation runs off quickly through storm drains. Moreover, the rampant spread of asphalt paving lowers the albedo and the surface roughness from buildings produces turbulence and slows ventilating winds.

In sum, energy consumption is increasing much faster than the population. Even if we stabilize population growth and solve all of the atmospheric problems, the oceans and atmosphere may still be affected by the heat energy we release. Oceanic and atmospheric circulation play a predominant role in climate, and such circulation is driven by heat energy. If we begin to produce waste heat on a global scale at levels approaching even a small percentage of the solar input, significant climatological effects can be expected. In some parts of the United States, waste heat output has become significant when compared to the level of the solar energy input. We don't know just how serious this problem may become. Once again an answer will require careful computer modeling that can lead to a reasonable development policy for use of energy. Current models suggest that we have no serious short-term problem in this area, but there are no definitive answers, especially with respect to future growth.

THE STUDY FOR BENEFITS AND USE

Obviously a knowledge of the earth has yielded great benefits to mankind. Continuing with examples in terms of our energy resources, it is clear that without a knowledge of modern geology we would not have the coal, oil, and gas needed to fuel the great majority of today's electric power stations as well as the uranium that will power a growing portion of stations in the near future. Indeed, there has been an important two-way benefit. While earth science has contributed immensely to the energy industry, the energy industry has prompted and has often funded the development of many of the techniques of earth science.

Sometimes a study of the fundamental physical processes of the earth system unexpectedly will solve problems in seemingly unrelated areas. For example, who would have guessed that research in global tectonics would have an impact on our energy problem? And who would have guessed that an understanding of the heat balance and spectral characteristics of the earth's atmosphere would prompt ideas about new ways to produce electric power? In fact, fundamental ideas from both geology and meteorology appear to be of major significance in new power generation schemes that involve geothermal and solar power.

Geothermal Power

Geothermal power in a sense is an "old" technology. People have always used the warm waters of hot springs for bathing and for their supposed curative powers. In modern times, hot earth waters and naturally occurring steam have been used ingeniously in several regions. At the beginning of this century, Italians began to produce electricity with steam from the Larderello geothermal field in Italy—a region that was described in Roman documents 2,100 years ago and that more recently served to inspire Dante's vision of hell. By the beginning of World War II, this natural steam field was producing 135 megawatts of electricity. The facilities were destroyed during the war but they have since been rebuilt. Today they generate about 400 megawatts, which is enough power to support a small city.

Geothermal water and steam has been used extensively in Iceland. Much of the capital city of Reykjavik is heated with this natural energy. In more recent years, natural steam has been applied to the production of modest quantities of electricity in such widely scattered regions as New Zealand, Mexico, parts of the Soviet Union, Hungary, and Japan. Other countries showing interest in this field include Algeria, Chile, Colombia, Czechoslovakia, El Salvador, Ethiopia, France, Indonesia, Kenya, the Philippines, Taiwan, Turkey, and Yugoslavia.

If geothermal energy is already being so widely explored or used on at least a small scale, why is it referred to as a "new" power generation technology? The answer is simple. At its current level of production, geothermal electricity must be viewed as an interesting but insignificant power source. But developments in our knowledge of geology and global tectonics have now led some earth scientists to believe that the total earth "reserves" of extractable geothermal energy may be large enough to support a significant portion of our total energy needs.

You read earlier that the continents are not fixed but drift about the globe as part of a great moving system of tectonic plates. In the subduction zones or sinks, material is being carried deep into the earth's interior; in the spreading zones or ridges, material is being carried upward from lower depths toward the surface.

Most of the world's major spreading ridges lie in the deep ocean but the East Pacific Rise comes up through the Gulf of California. Several such regions have been located in which hot materials are being transported upward toward the surface. At the head of the Gulf of California, Mexico has begun the development of a geothermal steam field at Cerro Prieto that is expected to have 75 megawatts of installed electric generating capacity. Near San Francisco, a steam field known as The Geysers is

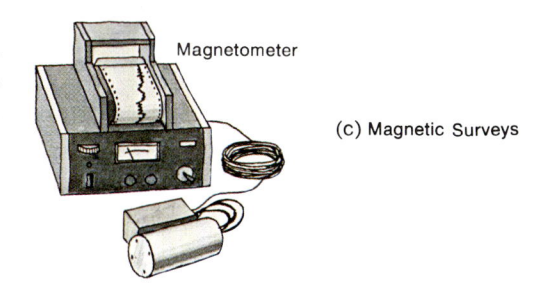

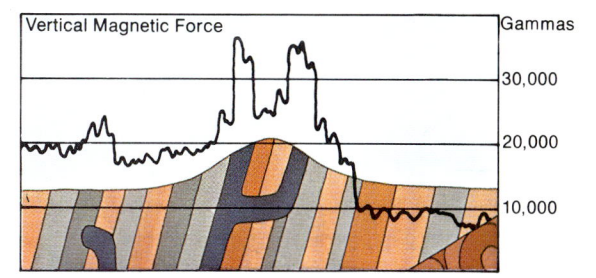

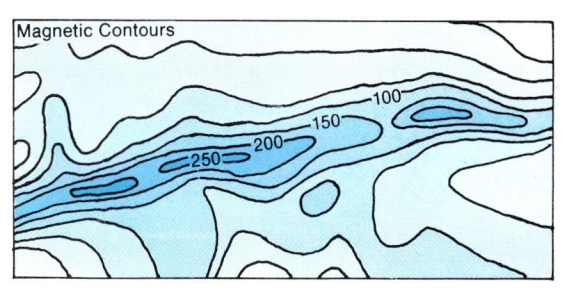

(c) Magnetic Surveys

Figure 6. Geothermal power may be one energy source for the future. The wells in this sulfur bank in Sonoma County, California produce about 800,000 pounds of steam pressure every hour.

now producing about 200 megawatts of electric power and may be producing as much as 600 megawatts by 1975.

Studies in the Imperial Valley area in south central California tend to confirm the existence of a major geothermal reserve that may ultimately allow a generating capacity of 30,000 megawatts, which is just slightly less than California's current level of total electric energy consumption. These studies have made use of all of the most modern methods of geologic exploration, including infrared photography, gravitimetric studies, studies of heat flow and temperature gradients, measurements of seismic activity, of geochemistry, and of electrical conductivity and magnetism.

As our understanding of the earth system grows we may find it increasingly easy to locate other major geothermal reserves and plan their development. It seems unlikely that geothermal energy will come to supply all of our energy needs, but further knowledge of earth science may allow this energy source to become a major part of our total energy picture.

Of course, nothing comes without a price tag and it is wise to inquire about some of the side effects, both positive and negative, that may result from the development of geothermal power. The hot water or steam available from geothermal sources is not the same as the steam in a conventional steam electric plant. Almost inevitably there are large amounts of dissolved or suspended materials, particularly salts, in the geothermal steam. Early experimental plants released large quantities of concentrated brine into the local environment to the point that surface waters and land could no longer be used. This situation is widely recognized as intolerable, and now essentially all development proposals call for the re-injection of the spent brine into the deep earth layers from which it came.

Electric energy may not be the only benefit from the development of geothermal power. The Imperial Valley geothermal field may produce more than twice as much distilled water from geothermal brine and sea water than will be carried into Southern California under the California State Water Plan near the end of the century. This enormous quantity of water — up to 5 or 6 million acre feet each year — along with the large quan-

Figure 7. This artist's conception of the Meinel's solar power "ranch" shows banks of lenses collecting solar energy in tubes. The pressurized gas that flows through the tubes conducts the collected heat to the power plant in the background. There, the heated gas transfers its energy to large tanks of molten salt, which store the energy until it is needed.

tities of electric power that would become available would clearly have a major impact on the development pattern of the southwestern United States. Concerned planners and environmentalists are already working to see that a technology capable of bringing clean power and abundant water to Southern California does not also bring environmental disaster through uncontrolled or unplanned industrial, agricultural, and residential growth.

Solar Power

Nuclear fission is the process that splits heavy molecules such as uranium 235 and releases large quantities of energy. This is the process that generates the power in a nuclear reactor, and in a very real sense geothermal power is power from the earth's natural nuclear fission reactor. Another kind of reaction, known as nuclear fusion, releases energy by fusing very light molecules together into heavier ones. Man has yet to devise a scheme for producing a controlled fusion reaction, but we may be producing significant quantities of electricity from fusion energy before we actually build our first fusion reactors. The source of this energy will be the enormous fusion reactor we call the sun.

Most of the energy we use right now comes from the sun — through plant energy stored by photosynthesis, through much older plant energy in the form of fossil fuels, and through hydroelectric power produced by water carried aloft by the solar-driven atmospheric heat engine. However, with the exception of photo cells (which work well but are a thousand times too expensive for commercial electric energy production), no reasonable technology now exists for converting incoming sunlight directly into electric power.

Numerous proposals have been suggested for making direct use of solar energy but none have stood the test of economic reality. Today several highly respected scientific teams are taking a new look at this

Figure 8. How can oil be transported from Alaska's oil-rich North Slope to an energy-hungry nation? Environmentalists put a temporary stop to a proposed pipeline, fearing that it would play havoc with the ecology of the countryside by thawing the permafrost on either side of the pipeline (which may have to be heated to keep the oil flowing) and by interfering with the migratory routes of animals. Recently the project has been resumed.

problem. What they are finding offers hope that within a few decades reasonably priced solar power may become a reality.

Projections suggest that by the year 1990 the United States will have roughly a million megawatts of installed electric generating capacity, or about 3 times the total 1970 installed capacity. If this capacity were to be met entirely with solar power (which of course it will not be), it would require a collecting surface equivalent to a square about 75 miles on a side. This comes to about 5,600 square miles—a lot of land, but only something like 1 percent of the amount of land used in farms, and less than twice the amount of land that has already been disturbed in the United States in the strip mining of coal.

One particularly interesting proposal for solar power generation has been made by scientists at the University of Arizona. Their idea involves capturing solar energy in almost precisely the same way that energy is captured by the earth's atmosphere. The energy from the sun appears largely in the optical portion of the electromagnetic spectrum; in other words, as visible light. The atmosphere is "transparent" at those frequencies and the energy reaches the earth without difficulty. On the other hand, most of the energy reradiated from the earth is in the form of infrared energy, which means it is of much lower frequency. At the lower frequencies the atmosphere is not transparent, so energy is trapped and escapes only after being transported to high altitudes by water vapor.

Why not use a similar mechanism in a solar energy system? A solar energy collector could be designed to include a specially treated window that, like the atmosphere, will let focused sunlight pass through but will act as a mirror for infrared energy. Energy could enter as sunlight and could be used to heat a fluid. The infrared energy reradiated from the very hot fluid would not escape out of the system because at that lower frequency the special "window" is not a window but a mirror. The energy would be trapped, which would allow a high conversion efficiency for the system. The heated fluid could then be transported through pipes to a heat exchanger, where it could be used to boil water in a conventional electric steam plant. Of course, days are not always sunny and there is no sun at night, which means that some energy must be stored for later use. Current proposals include schemes for storing energy directly in the form of heat and for using electricity to produce hydrogen from water, in which case the stored energy could be recovered easily by burning the hydrogen.

Nuclear Power

The possibilities of geothermal power and solar energy are two examples of how a knowledge of the earth system leads to new ways of meeting the growing demands for energy. Recently earth science has begun to play an increasingly important role in the area of nuclear energy. As a popular phrase suggests, "nuclear power is clean power"—at least as far as air pollution is concerned. But waste by-products *are* produced by nuclear reactors. Once the fuel has finished its useful life in a modern reactor, it is removed and stored on site for a few months and then shipped to a fuel reprocessing plant where the cladding is stripped off the individual fuel pellets and the spent fuel is dissolved in acid. The materials that can be recycled, such as plutonium and any unused uranium fuel, are extracted by means of chemical processes. The remaining waste material contains the radioactive fission by-products. These by-products remain extremely radioactive for thousands of years.

Until now, such "high-level wastes" have been stored in concentrated liquid form in large "tank farms." The tanks usually consist of a steel inner tank resting in a steel tray designed to catch unexpected leaks. A reinforced concrete outer tank surrounds the inner tank and the full assembly is placed below ground level. Because of the high level of radioactivity, the material must be continually cooled. Today many millions of gallons of high-level waste are in storage. Most of it comes from military sources but a growing amount is being added because of the generation of electric power through nuclear energy.

Responsible scientists have been concerned for many years about the long-term safety of these waste materials. The fact that the wastes are in a liquid form is disturbing to them because liquids are difficult to contain. And the long storage times required are also disturbing. As we look back over human history we find little to encourage a belief that any social institution can be expected to persist without major disruptions for periods of thousands of years. One would feel much better if the maintenance of nuclear waste materials could somehow be taken out of human hands.

Recently there has been rapid progress in the development of technologies that will convert the liquid wastes into solid form. Preliminary methods have already been developed for large-scale use and improved processes that will fuse the waste into a ceramic-like material of very low solubility. These methods are expected to be perfected in the near future.

Once the material is in solid form, the problem then is to dispose of it in a fashion that removes it from the realm of human activities. Today, at least, this means placement in some highly stable geologic formation. After a thorough study of the possibilities, scientists working for the United States Atomic Energy Commission concluded that the best storage site was bedded salt deposits. Their conclusion was confirmed by a review committee of the National Academy of Sciences National Research Council in late 1970.

The idea goes along the following lines: Find a good deposit of bedded salt in an area of tectonic stability, in which both the overlying and underlying rock structures have integrity against the intrusion of water that could dissolve the salt. Then carefully build a mine down into the deposit. Package the waste in long thin cylindrical containers. Bore holes in the salt, place the containers in the holes, and fill the holes with salt.

Salt is a good heat conductor and will readily carry away the heat generated by the waste material. Salt has great compressive strength; it also exhibits slow "plastic flow" at moderate temperatures so it will form a sealed cell around the waste container. Although sites should have low earthquake probability in any case, any fractures that might occur would be self-healing. As long as water is not introduced into the deposit by man or by completely unexpected geological developments, the waste material will remain safely sealed.

The largest appropriate deposits of bedded salt lie in an area in the state of Kansas. At one time it was proposed that an experimental disposal site be established in an unused salt mine near Lyons, Kansas. Through a series of complicated interactions, this proposal mushroomed into a major scientific and political affair. The result was that the state legislature in Kansas barred the storage of radioactive wastes anywhere in the state. While there is no agreement among members of the scientific community, prominent scientists both in Kansas and elsewhere feel that the Lyons site was a poor choice. A number of deep wells have been drilled down from the surface; the location of some of the older wells is not known. A nearby salt mine is still being actively worked. This mine uses hydraulic "solution" mining in its operations and thus introduces water in the area.

Figure 9. Geologists study the earth because it is essential that we take stock of our resources and that we learn about the dynamic processes, such as earthquakes, at work on the earth. But they also study the earth because it is an adventuresome way to live.

There is still a widespread belief in the scientific community (including scientists in Kansas) that bedded salt disposal is the best solution we have today for the storage of nuclear wastes. How things will actually get worked out is anybody's guess. But the episode clearly has been one more instance in which a knowledge of the earth can help us to meet our energy needs—and perhaps an instance where a more thorough application of this knowledge might have avoided a major political problem.

THE STUDY FOR CHALLENGE

The preceding sections of this chapter outlined many practical reasons why we should and must study the earth system. Most of the examples were limited to the topic of energy use, and yet any number of similar topics would lead to equally compelling reasons. But there is still another factor that urges many people to take up the study of the earth that is quite apart from the serious nature of the subject. Earth science happens to be fun.

Most of us know one or two people who take great delight in spending hour after hour putting together 1,000-piece picture puzzles of a mallard duck circling above clumps of cattails as the sun sets over a remarkably uniform-looking swamp. Why do they do it? For the challenge, of course. Yet one suspects that if the only pleasure derived was that fleeting glow of success as the last piece fell into place, they probably would have given up their puzzle-solving long ago. The very act of *doing* a puzzle must provide enjoyment and meaning.

The motivations of the earth scientists are probably not that different. Undoubtedly they relish the momentary thrill of finally piecing together thousands of pieces of data into a new theory, a new level of understanding. But it can be a long way between such exciting moments and there must be some other motivation that keeps them going. One suspects that, just like picture-puzzle aficionados, the earth scientists get enormous pleasure and meaning out of the simple day-to-day acts of *doing* earth science.

In our time much of modern earth, ocean, and atmospheric science is done in laboratories, offices, and computer centers. Yet today, as always, most earth scientists from time to time find themselves wandering off to remote regions of the globe to run the experiments and gather the data that form the foundation of their work. It can be an adventuresome way of life—

. . . Outside the temperature stands at fifteen below. The night is clear and the snow squeaked under your feet as you shed your cross-country skis and fumbled with thick-gloved hands at the door of the equipment shelter. The chart recorders have new paper now; you've changed the film in the all-sky cameras; you've put new tape on the tape recorders. The timing circuits had drifted off and you've reset them. WWV just wasn't there but the signal from Dominion Observatory Canada was loud and clear. You make a few final notes in the log, give the equipment a final check, and then don your parka for the long run down to camp. There's a snow-cat in camp that you could have used to get up here, of course. But on a night like this, with just a sliver of a moon and a vast, silent display of a shimmering aurora, who in his right mind wants to ride? Your gloves back on, you step out of the equipment hut, the freezing air nipping the inside of your nostrils. You clamp on your skis and push off. The crew should have hot coffee ready by the time you're back . . .

Such vignettes could go on and on, to the arctic tundra, to an observatory in the high *altiplano* of Peru, to the deep waters off continental shelves, to the slopes of a Hawaiian volcano, to the center of the Arabian Desert. You might almost begin to reach the conclusion that men and women study earth science because it is a way to mix intellectual challenge with a life style of adventure and fun. You might reach such a conclusion—and you might not be all that wrong.

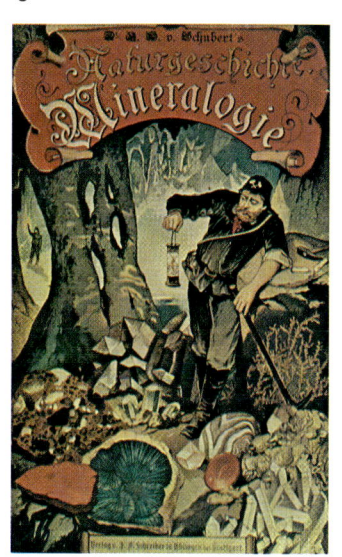

The Beginning and the End; An Imaginative Essay

We stand here on the surface of the earth and look out at the heavens. The object closest to us in space is the moon, which we see without difficulty. Its average distance from the earth is some 380,000 kilometers and, after some 6,000 years of what we call civilization, we have managed to reach it.

The farthest known object is a quasar thought to be about 5 billion trillion times as far from us as the moon is. It can be seen only in a high-powered telescope by someone who knows where to look. We are probably quite safe in saying man will never reach it.

Time passes for us here on the surface of the earth and few of us have any reasonable chance of living a hundred years. But the earth has already been in existence nearly 50 million times longer than the longest lifetime, and the rest of the universe has been in existence for much longer still. With our consciousness trapped in a point of space and at a moment of time, we have nevertheless managed to evolve, to our own satisfaction, a plausible picture of a universe enormously large and enormously durable.

But curiosity never rests. It pushes always at the boundaries of the known. If the universe is enormous, where does it end, if it ends at all? And when and how will it end, if it ends at all? And when and how did it begin, if it began at all?

The point where we begin probing the nature of the universe in space and time is at one observed fact. All the spectral lines of the distant galaxies are displaced toward the red. Indeed, the dimmer the galaxies and, therefore, the farther away (we suspect) they are, the greater this "red shift."

The most logical way of interpreting the red shift is to suppose it is evidence of the recession of the light source. All the distant galaxies, in other words, are receding from us; and the farther away from us they are to begin with, the more rapidly they recede.

To put it more simply, the universe seems to be expanding.

If that is the case, the simplest speculation we can advance is that the universe will continue to expand and that the galaxies will continue to move farther and farther from one another. Of course, we must modify this view by realizing that galaxies tend to exist in clusters and that, within the clusters, gravitational attraction predominates over the tendency toward expansion.

But the clusters separate with time and the average distance between them grows increasingly. The separations eventually would be so immense, if this expansion were to continue indefinitely, that no foreseeable trick of instrumentation would enable one cluster to be visible to the questing minds located on some planet with another cluster.

This is the kind of end we can expect for an "indefinitely expanding universe." It is not an *absolute* end, for the galactic clusters would continue to exist and continue to drift apart for perhaps all eternity. However, when a state is reached such that there is no further perceptible change to some observer, we can call it an

346

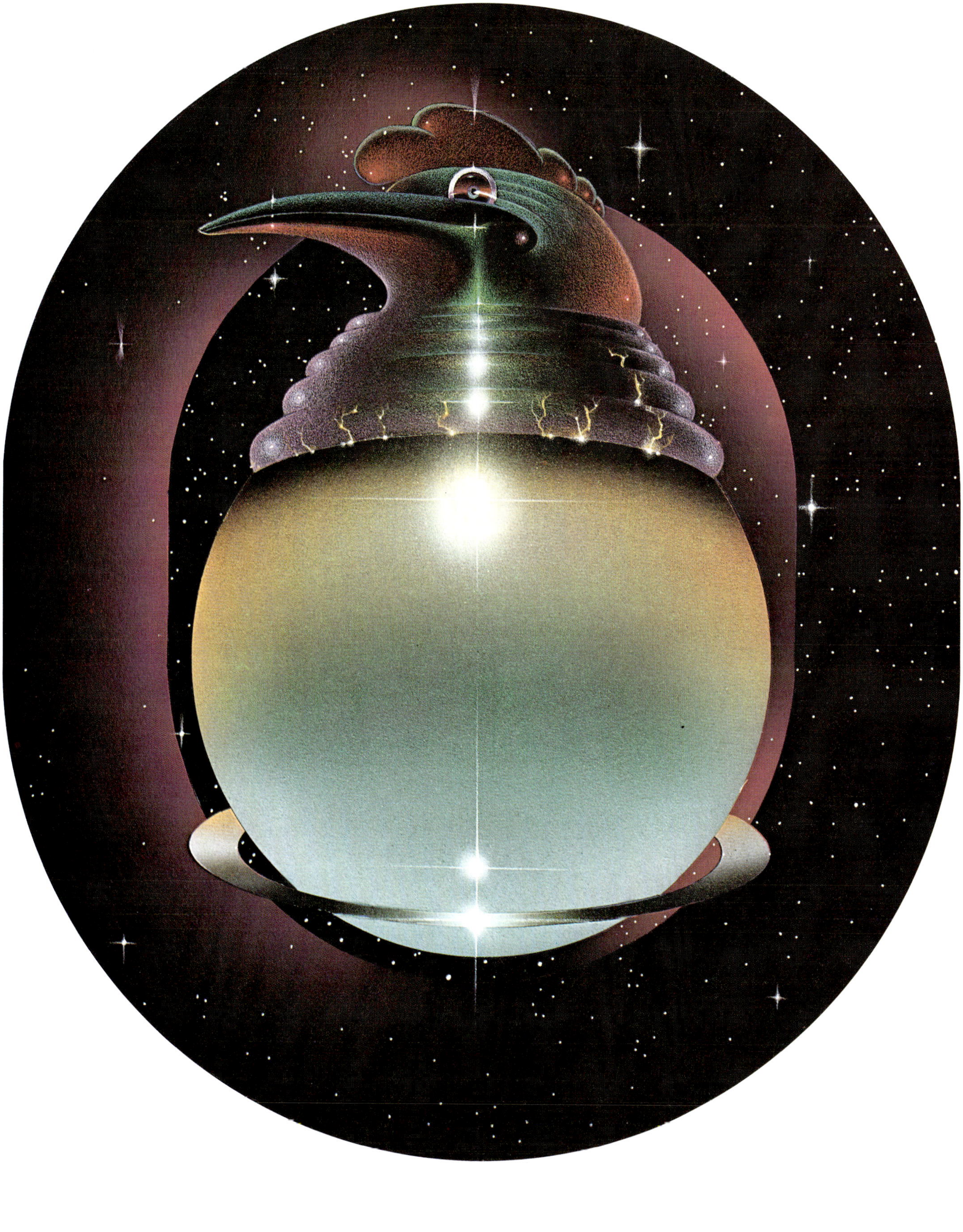

Figure 2. The Indefinitely Expanding Universe (Future)

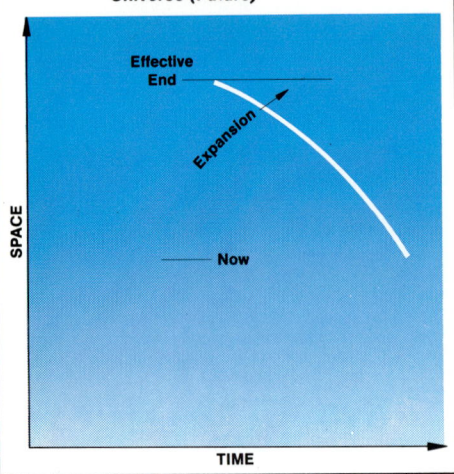

Figure 3. The Indefinitely Expanding Universe (Past and Future)

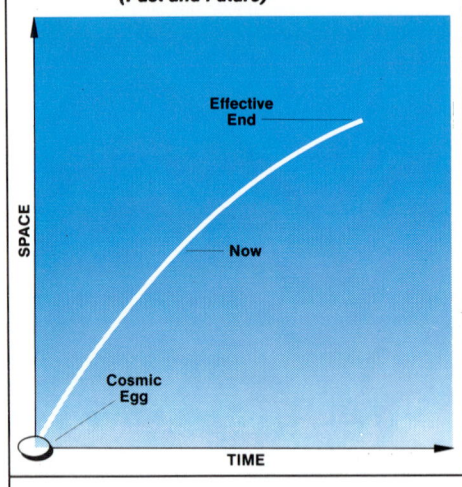

"effective end." In this essay I deal chiefly with effective ends rather than absolute ones.

At the effective end of an indefinitely expanding universe, we would be surrounded by nothing detectable but the twenty or so galaxies of our own local cluster. Beyond that, as far as we could see with any instruments within reason, would be nothingness, except for intergalactic gas and dust. Figure 2 is a simple diagrammatic expression of the picture posed by indefinite expansion.

But will we see even the galaxies of our own cluster? Will they remain essentially unchanged during the slow course of the billions of years during which all the other galactic clusters will drift away from us beyond the unseeable horizon?

Well, everything we see, everything we do, is a manifestation of energy transfer. And the "first law of thermodynamics," which is believed to be valid through all of space and through all of time, says that the total energy of the universe is constant. This is also referred to as the "law of conservation of energy."

If the total energy of the universe must remain constant, then surely the galaxies will always exist. Ah, but this first law also states that energy, even though it can be neither created nor destroyed, can change its form. And the "second law of thermodynamics"—also believed to be valid through all of space and time—dictates something about the nature of the change. The second law states that less and less energy can be converted into work as time progresses. Another way of putting it is that a quantity called "entropy" (which represents evened-out energy that can't be made use of) steadily increases with time, tending always toward a maximum.

What this means is that the universe, and every part of it, is running down. We on earth mask this fact because we are always rebuilding things. When water runs downhill, we can pump it uphill. We can reverse the process, however, only by drawing upon other energy sources that have not yet run down—primarily the energy of solar radiation. But the sun's radiation depends on the continuing consumption of its hydrogen. That hydrogen supply will someday run out. The sun will expand to a red giant and then collapse to a small and extremely dense "white dwarf."

The white dwarfs may fade out slowly or collapse further to still smaller and still denser "neutron stars." In the end, perhaps before all the other galactic clusters have drifted completely away or perhaps afterward, all the stars of all the galaxies may dwindle into dense pygmies. Then, only neutron stars and thin interstellar gas would exist—and exist, and exist—and that would be the effective end of an indefinitely expanding universe in which entropy is indefinitely increasing. And from the earth (if we imagine ourselves and it to still exist) we would see nothing at all except, perhaps, for a tiny glow from the dot of light that was once our sun.

If that is the far future, what about the far past? Suppose we run the film backward, so to speak, and look into the past; suppose we carry the curve of Figure 2 to the left, before the "now" point in the other direction.

Moving forward in time, we have an expanding universe. Moving backward in time, we have a contracting universe. As we probe deeper and deeper into the past, we can envision galaxies coming closer together and entropy decreasing steadily toward a minimum.

If we go back far enough in time, we can imagine all the matter of the universe contracting to the point of being collected into a single dense ball—the "cosmic egg." If we think about that point in time where the cosmic egg exists, we can think of it undergoing a cataclysmic explosion. It is this "big bang" which would have started the universe on its career of expansion, an expansion that continues today.

The cosmic egg represents an "effective beginning" for the universe.

Conceivably it might have existed for eons before the explosion, for all eternity perhaps, but it is at the time of the explosion that we can imagine changes beginning to take place. And it is the onset of change that offers us an effective beginning (see Figure 3).

If we picture the universe now as beginning at the cosmic egg (with minimum entropy) and ending at a scattering of neutron stars (with maximum entropy), it would seem that the beginning is the less satisfactory of the two extremes. After all, how did the cosmic egg come into existence? How did matter get to be squeezed so tightly together in the first place? Surely a cosmic egg is a highly unstable arrangement; everyone who speculates about it invariably deals with its vast explosion. If it is so unstable, why should it exist in the first place? If, on the other hand, it existed because it *was* stable, then why did it become so unstable that it exploded at some particular moment in time?

These are difficulties that have led some astronomers to picture a universe in which the cosmic egg never existed at all. They cannot deny that the universe is expanding and that the galactic clusters are continually separating. They suggest, though, that new matter forms in the universe and collects together as the clusters separate. This new matter would make up new galactic clusters between the separating old ones.

If we imagine ourselves watching such a universe from some point within it, old galaxies would be drifting away continually beyond the horizon but new ones would always be forming in the foreground. The fine details of what we see—the

Figure 4. The Steady-State Universe

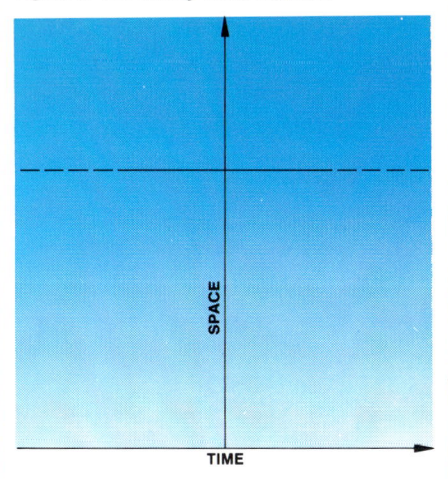

individual galaxies, their shapes, their positions—could change, but the overall view would not.

In such a "continuous creation universe" or "steady-state universe," size would be infinite and duration eternal, and the part we could see would never change. There would be no beginning and no end. Or if we consider lack of change as the mark of an effective beginning and an effective end, then the steady-state universe would be all beginning and end and nothing else (see Figure 4).

At the present moment, however, we must eliminate the steady-state universe as a possibility. There is some reasonable evidence we can draw on (the radio wave background that exists in every direction, for instance) which seems to be enough of an indication that there was indeed a big bang in the past and that the cosmic egg did at one time exist, even if only momentarily. Which means we are stuck with its paradoxes.

Suppose we tackled the problem from the other direction. Suppose we ask ourselves what sort of a beginning of the universe would be a stable one. What would a universe be like before any of its present organization existed? What form could we accept both as ultra-primitive and ultra-stable, so that we could imagine its form as having existed for an indefinite period before changing into our universe of today?

One possibility, perhaps the easiest to envisage, is that the universe existed to begin with as an exceedingly thin gas made up of the simplest possible atoms—those of hydrogen. This is, in essence, the universe that exists today in the space between the clusters of galaxies. We are supposing, then, that the universe began in a state like that of intergalactic space, which would certainly be stable over enormous periods of time.

Such an exceedingly thin gas would, however, be subject to its own vastly diffuse gravitational field. Because of gravitational influences, random motions of the atoms would be a little more likely to bring them together than to separate them. Once they were together, interatomic forces might tend to keep them together.

Any clustering of atoms would serve to intensify the gravitational field in that neighborhood and make the gathering of still more atoms somewhat more likely. At some point, the general tendency for the gas to contract (whatever the details—uniformly, in clusters, with or without turbulence) would become measurable. That would represent an effective beginning of the universe.

Any degree of contraction would intensify the gravitational field and hasten further contraction. For many billions of years, perhaps, the universe would continue to contract at a steadily increasing rate. The contraction would heat the universe generally and produce a higher and higher temperature in the matter composing it as that matter compressed into a smaller and smaller volume. The temperature rise would exert an increasing expansive effect that would counter the gravitational contraction and begin to slow it down.

The inertia of the contraction would carry the universe past the point where the temperature effect would just balance the gravitational in-pull, if the universe were not contracting. The universal contraction would continue to some minimum volume, represented by the enormously dense cosmic egg, to the point where the expansive effect of temperature would finally gain control. The substance of the universe would then be pushed outward, faster and faster, until the accelerating force became the big bang.

The general picture is rather like that of an elastic ball dropping from a height onto a hard surface, compressing itself, and then having the force of the compression reversed so that it hurls up again in a high-flying bounce.

In this view, the universe would begin at a point of general emptiness, with matter

Figure 5. The Hyperbolic Universe

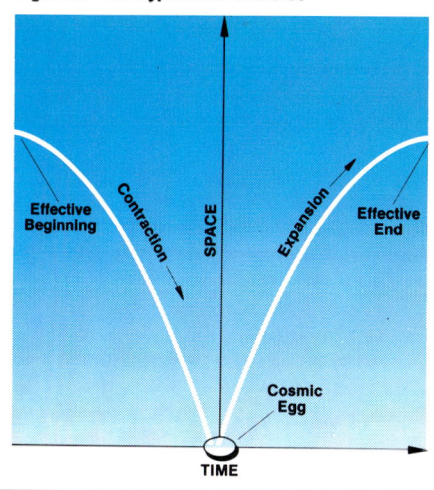

distributed as single hydrogen atoms. It also would end at a point of general emptiness with matter distributed in dense clots as neutron stars. The cosmic egg would be the unstable and momentary midpoint. This is a "hyperbolic universe" (see Figure 5).

But now it is the ending that raises a question. The beginning, with its thin scattering of hydrogen atoms, is simple. But the end with its thin scattering of neutron stars seems less so. In the end there would be regions of extremely intense gravitational fields. Could they, growing large enough, reverse the expansive effect? If expansion overcame contraction at the time of the cosmic egg, is it possible that contraction will overcome expansion at the time of the neutron-star scattering?

If the answer is yes, is our present state of universal expansion to be succeeded by another contraction, somewhat similar to that which preceded the formation of the cosmic egg? And can we envisage the eventual formation of another cosmic egg?

The picture is a troublesome one. A neutron star universe would be close to entropy maximum, whereas the cosmic egg would be close to entropy minimum. The second law of thermodynamics says that entropy must increase with time; events must go from cosmic egg to neutron star.

Ah, but the second law has been observed and tested only within the context of an expanding universe, and we have no right to insist on its applicability to a contracting one.

Suppose the universe were contracting. The radiation emitted by the galaxies, as viewed from the center, would undergo a shift toward the violet and therefore gain in energy in the direction of contraction. The extent of that "violet-shift" (the opposite of the galactic red shift so prominent in an expanding universe) would increase as the velocity of approach increased with accelerating contraction. The energy poured into the center of the universe by the intensifying gravitational field would be compressed and itself intensified. Under the lash of that energy, the downhill changes taking place in an expanding universe would be reversed. The universe would wind itself up again.

Of course, once the cosmic egg formed, it would explode once more and we would go through the cycle again, and again, and again. Instead of a hyperbolic universe, once-in-and-once-out, we would have an "oscillating universe," or a "pulsating universe"—one expanding, contracting, expanding, contracting, and so on indefinitely.

But would each period of the cycle be exactly like the one before? Could it be, for instance, that in the process of expansion a great deal of matter would be turned into energy within the various stars? This energy, in the form of photons, neutrinos, gravitons, and so on, might be viewed as escaping and streaking outward forever. The matter that would end up as neutron stars, at the point where contraction would begin again, might therefore be considerably less than the mass of the original cosmic egg. When the cosmic egg formed again, it would be smaller than the one before and would explode less violently. The next period would then be shorter and result in a still smaller cosmic egg, and so on indefinitely.

Finally, a cosmic egg would form that would be too small to store enough energy, in forming, to bring about an explosion once more. Perhaps it would just pulsate in a gradually dying tremor, itself no larger than a single moderately large neutron star, and that would be the effective end of such a "damped-oscillating universe."

And what about the beginning of such a universe? If we run the evolution backward, we might envision that the cosmic egg formed at each compression-state of the cycle is larger than the one before and the period between successive

Figure 6. The Damped-Oscillating Universe

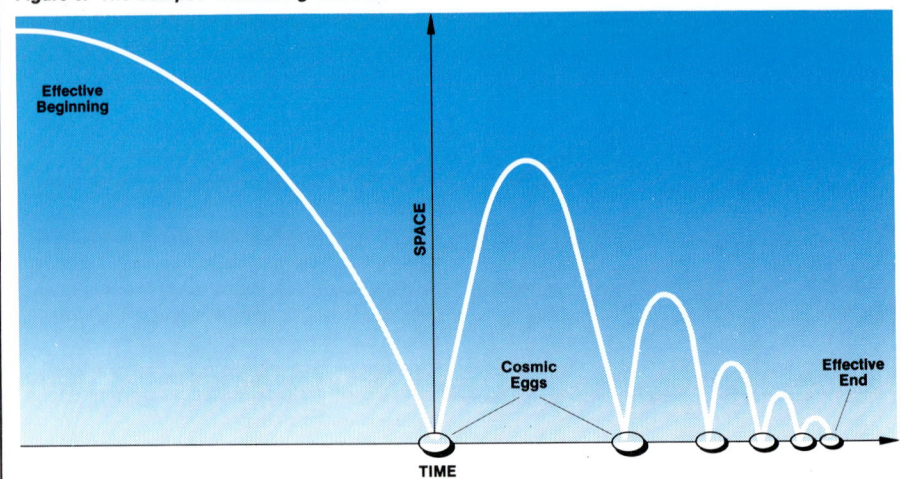

cosmic eggs is longer. Eventually, in this early dawn of existence, we would have a cosmic egg so huge and an explosion so vast that gravitation would never manage to overcome the outward thrust. Because we would in this case be traveling *backward* in time, this means there would be a *first* compression.

Going back to the effective beginning of a damped-oscillating universe, then, we would have a thin distribution of matter, not in the form of neutron stars (which might imply a previous cosmic egg) but in the form of hydrogen gas. In other words, the damped-oscillating universe could be pictured as beginning in the same way the hyperbolic universe begins; but instead of forming one cosmic egg it would form a whole series of ever smaller ones, and many bounces in place of one (see Figure 6).

Yet such a damped-oscillating universe doesn't look entirely reasonable in the context of Albert Einstein's theory of relativity. According to relativistic notions, space is curved by an amount depending on the nature of the general gravitational field of the universe. Photons and neutrinos escaping from the stars and traveling in "straight lines" are not really traveling in straight lines in the Euclidean sense. They are not streaking outward forever, never to return. They are, instead, following a very gentle curved path, the curvature of which is, in fact, what is meant by the notion that "the universe is curved." In this sense, radiation does not leave the universe and is not lost to it.

In the contracting phase of a universe, the gravitational field would intensify and the curvature of space-time generally would grow more marked. As the galaxies approached one another, the photons and neutrinos would spiral inward, too, following sharper and sharper curves.

The result may well be that there is *no* loss of matter in the course of a cycle, and that each cosmic egg is just as large as the one before.

If that be so, we have a "steady-oscillating universe" in which each period is equal to the one before and the one after, and the number of periods can be viewed as infinitely great (see Figure 7). Such a universe, like the steady-state universe of Figure 4, would be without beginning or end. But it would not be changeless, merely periodic.

Each period would begin and end with a cosmic egg, expanding out of one and contracting into one. Or you can say, with equanimity, that each period could begin and end with a thinly spread-out volume of neutron stars, contracting into the intermediate stage of a cosmic egg and expanding out of it again.

Of course, we are only justified in picturing an unending number of pulsations in the steady-oscillating universe if we think of time as proceeding uniformly and endlessly. If we suppose each period between cosmic eggs lasts a trillion years (just to use a round number), then we could say that a trillion years after the most recent cosmic egg there will be another one and a trillion years after that, still another one, and so on.

Are we, however, justified in thinking of time in this way?

At the subatomic level, the direction of time is arbitrary. This means that all the laws of particle behavior are the same whether time moves "forward" or "backward." (At least no one has yet found any concrete evidence of a process that can run only in one direction.) This means that if we are a subatomic particle living among other subatomic particles, we would have no way of telling whether time were moving forward or backward. We would merely define one direction as forward and then the other direction would be backward.

In our own universe, at the macroscopic level of organization, we consider time as moving in one direction only. Because entropy is steadily increasing, all changes seem to be of one general kind. Objects suspended in midair always drop; objects hotter than their surroundings always cool off. We call this apparent motion of time "forward."

If we run a movie film backward, we can see a broken vase reassemble itself out of its pieces and fly up from the floor to the table; we can see a diver rise from the water to the diving board while the water un-splashes and gathers together. We know at once that things don't happen that way in what we call "real life." That is what would happen if time flowed "backward."

But in a contracting universe, all the galactic clusters would be coming together and the universe would be winding up, with entropy steadily *decreasing*. The film, so to speak, *would* be running backward. The events taking place are the ones that would take place if time moved in the reverse direction.

In that case, we can imagine ourselves starting at a cosmic egg (any cosmic egg) in a steady-oscillating universe such as that shown in Figure 7. The universe would expand for half a trillion years with time moving forward and then contract for half a trillion years with time moving backward. It would end at the same time as it began.

Instead of imagining an infinite number of pulsations occurring over an eternity, we might imagine a single expansion, followed by a single contraction, over a time of half a trillion years and back (Figure 8).

The picture at its simplest might be one of a single universe existing in four dimensions—the three spatial dimensions and the fourth dimension, time. We might imagine it to be a "hypersolid," beginning as a small volume filled with cosmic egg at time-zero, and ending, coronet-fashion, as an enormous volume containing a thin scattering of neutron stars at time-half-a-trillion-years.

Everything in such a fixed universe would be, in the larger sense, permanent. Somehow, though, our consciousness would progress steadily forward through the hypersolid and at each instant of time we would be conscious of a different three-dimensional cross-section. It is this progress-of-consciousness forward at a fixed speed, so to speak, that would give us the illusion of change and time.

If we could then imagine a consciousness greater and less evanescent than our own—sweeping forward from cosmic egg to maximum expansion, then sweeping back to cosmic egg again—we would have the illusion of a universe endlessly pulsating but always repeating itself in exactly the same way at every stage.

And yet the notion of a fixed universe does not quite fit the fundamental perception known as the Heisenberg uncertainty principle. If the same uncertainty principle that has been found to explain events on the submicroscopic level were to apply also on this super-macroscopic scale, then it would limit the precision with which any measurement can be made or any effect predicted. If we consider the state of the universe at any instant of time, there is more than one state possible at some other instant of time in the future.

As the consciousness we are imagining sweeps backward through time and then forward again, it may, this second time, pass through a somewhat different universe of equal probability. In fact, even though it sweeps back and forth an infinite number of times, it may in each sweep never quite repeat the details of the previous sweep, or any previous sweep in that direction.

We might imagine (in this somewhat science-fictional fashion) that an unending number of universes exist in the time interval between time-zero and time-half-a-trillion years, each different in detail from all the rest and separated by something more subtle than time and space. Instead of envisioning a steady-oscillating universe

Figure 7. The Steady-Oscillating Universe

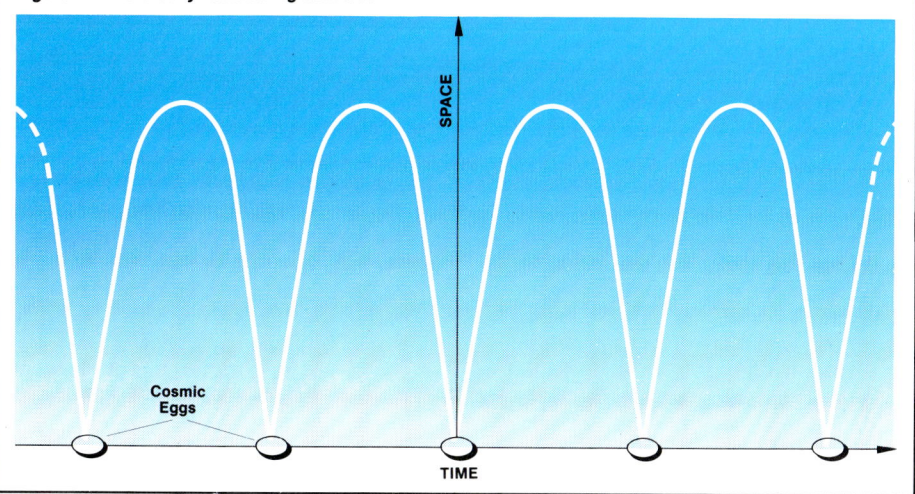

Figure 8. The Steady-Oscillating Universe (Bidirectional Time)

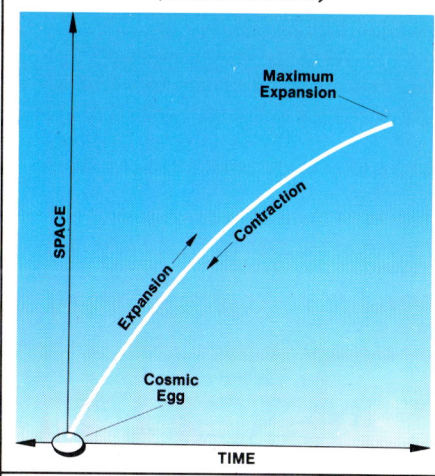

unending in time, we might just as well imagine a steady-oscillating universe unending in probability states.

But in either case, whether it oscillates equally and unendingly in time or in probability states, what is unending is the material content of the universe. Whether the matter is compressed into a cosmic egg or thinned out into separate neutron stars, it is always there. We are surely entitled to move on, in our questioning, from the matter of the beginning and end of the organization of the universe (which is what we've really been doing so far) to the beginning and end of the matter that makes it up.

Actually, this problem is not as difficult as it might sound. Matter is *not* eternal. We can make it begin and end in the laboratory.

Every particle has its "antiparticle." A particle and its antiparticle are identical in all respects except for electric charge and the relative directions of spin and magnetic field. Thus a proton carries a positive electric charge of a certain fixed size; and an antiproton carries a negative electric charge of the same fixed size. It differs from a proton in no other respect, except that its magnetic field is also reversed with respect to the direction of its spin.

Particles can combine to form nuclei, atoms, molecules—and the corresponding antiparticles can do the same. Ordinary particles can combine to form ordinary matter—and antiparticles can combine to form antimatter.

If ordinary matter is considered as existing by itself, it cannot be entirely destroyed. A small fraction of it (about 1 percent) can be converted into energy if the particles are particularly well-packed. This energy formation through better packing is the source of the radiation of the sun and of stars like it.

The same would be true of antimatter if that were considered as existing by itself. In an antiuniverse, antistars would also radiate energy produced through better packing.

Particles of matter can, however, combine with those antiparticles of antimatter analogous to themselves to produce "mutual annihilation." All the matter in both sets of particles would no longer exist—as matter. It would be converted to energy in the form of energetic photons. And photons, the constituent particles of radiation such as visible light, are their own antiparticles. They are equally at home in both the universe and the antiuniverse.

The reverse is also true. Energetic photons can be converted to matter—but never to matter alone, and never to antimatter alone. A particular quantity of energy will form particles of matter of a particular type in particular numbers, but it will inevitably form the same number of the equivalent antiparticles of antimatter.

If the matter of the universe is not to be considered as having existed eternally, then the logical alternative is to suppose it was formed out of energy. But if matter was formed out of energy, an equivalent amount of antimatter must also have been formed.

And in that case, where is the antimatter? Our planet is composed entirely of matter, with no significant quantities of antimatter at all. The moon is matter, too; the entire solar system is. In fact, as nearly as we can tell from the cosmic ray particles that reach us from outer space, our entire Galaxy is matter—perhaps even our entire universe.

Where is the antimatter, then?

It is not likely that once the particles and antiparticles formed they would remain well-mixed. In such a state they would interact after a comparatively short interval of time and undergo mutual annihilation back into energy. If matter and antimatter were to form out of those particles and antiparticles, and if they were to continue to exist, they would have to be separated from the moment of formation and they would have to continue to remain separated by distances great enough to avoid significant amounts of interaction over extended periods.

If the matter within our Galaxy were partly antimatter, we would probably be aware of radiation from various directions arising from the occasional mutual annihilations that would be inevitable. We can be reasonably sure that our Galaxy is almost entirely matter, if only because such radiation is *not* detected. It may also be that any galaxy, or even any cluster of galaxies, is entirely matter—or entirely antimatter. It may be, however, that the universe consists of clusters of galaxies and anticlusters of antigalaxies.

We can't be sure which would be which—except that our own is matter by definition. The photons and gravitons that galaxies emit would be the same whether those galaxies were composed of matter or of antimatter. To be sure, galaxies would emit neutrons, whereas antigalaxies would emit antineutrons; but both neutrons and antineutrons are extremely difficult to detect. If two galaxies were in contact or near-contact and the pair gave rise to unusual quantities of radiation, we could conclude that one was a galaxy and one an antigalaxy but we would not be able to tell which was which.

But then, it may be that all the galaxies of the universe are matter and that a similar enormous collection of antigalaxies exists somewhere far beyond the range of detection by our instruments. Together they might form a separate "antiuniverse."

How could the separation of matter and antimatter into galaxies and antigalaxies, or into a universe and an antiuniverse, come about?

For one thing, charged particles and antiparticles react in opposite fashion to a given electromagnetic field. If, in the process of explosion, the cosmic egg produced a huge electromagnetic field, all charged particles would swerve off in one direction and the corresponding antiparticles would swerve off in another. Then, too, there is a gravitational attraction between all particles of matter and, presumably, between all antiparticles of antimatter. This might mean there is gravitational repulsion between matter and antimatter—though, to be sure, no such effect has yet been detected in the laboratory. (Nor has such an effect been ruled out in the laboratory, either.)

It is useless to try working out the details; whether the separation occurred because of electromagnetic forces, gravitational forces, or both; whether it took place in one grand sweep, or little by little with numerous interactions and annihilations en route; whether it resulted in galaxies and antigalaxies or a universe and an antiuniverse; whether the separation took place in the course of seconds or of millions of years.

In fact, the only way we can avoid the assumption of eternally existing matter in the universe is to assume it arose out of energy. In that case, there is antimatter somewhere, too. And the neatest solution, however it came about, is to suppose there is a universe and an antiuniverse.

If we do this, we can perhaps avoid a second difficulty I have so far skirted around.

If we imagine a contracting universe made up of matter alone and collapsing into a cosmic egg, we may have trouble thinking up something to stop the collapse. Individual stars can collapse into white dwarfs, say, 10,000 kilometers across; or beyond that into neutron stars, say, 10 kilometers across; or beyond that into black holes in which the volume can shrink to zero while the density and gravitational intensity rise to indefinitely large figures.

The larger the shrinking star, the more likely it is to plunge past the intermediate stages and into the black hole ultimate. The black hole has a gravitational field so intense that nothing can ever get out again, not even massless particles such as those of light and other radiation. We know of no mechanism that would make a black hole explode.

It seems quite likely that a star, large but not enormously large, contains enough matter to make black hole formation possible. Certainly the entire universe contains enough matter for the purpose. Well, then, assuming the universe can contract to form a cosmic egg, what would keep it from plunging past the cosmic egg to form the hugest black hole conceivable and ending all and everything in that fashion? The expansive properties of ordinary matter, under that compression, would not seem enough to overcome the potentiality of black hole formation.

But suppose, instead, that as the universe contracts, galaxies and antigalaxies are forced closer and closer together. Or suppose that as the universe contracts, the antiuniverse contracts, also; and that as they travel back through time they come closer and closer together. In either case, as contraction proceeds the incidence of interaction and mutual annihilation increases. The mutual annihilation at a steadily increasing rate not only would produce far more radiation and therefore expansive tendency than could be produced by any contraction involving matter alone (or antimatter alone), it also would destroy mass and therefore lower the gravitational intensity underlying the contraction.

As a result, the contraction may well stop short of black hole formation and produce instead a kind of cosmic egg that is pure energy and not matter, antimatter, or a mixture of the two. Now we have a steady-oscillating universe beginning as an energy egg and producing two universes (see Figure 9). Conditions in the antiuniverse would be no different in essentials than those in the universe. An antiman in an antiuniverse would be perfectly justified in defining his own as a universe and calling ours an antiuniverse.

Figure 9. The Steady-Oscillating Double-Universe

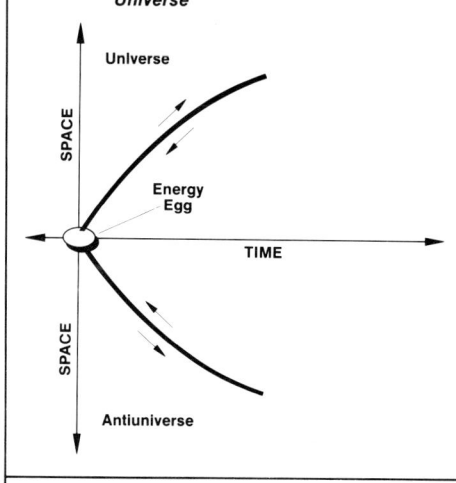

Does the double universe of Figure 9 now answer all questions? Unfortunately, it doesn't. We can still ask ourselves: Where does all the energy, out of which the universe and the antiuniverse arose, come from? Must we view that energy as constant and eternal, without beginning or end?

To be sure, that is exactly what the law of conservation of energy asks us to do; but how far can we trust that law? To what extent must we consider it a generalization applying only to the local conditions of our own universe in its present state?

To see what I mean, suppose we consider a law similar to it—the law of conservation of momentum, where the momentum of any object is the product of its mass and velocity. By that law, momentum, like energy, can neither be created nor destroyed.

And yet, a bullet at rest in a rifle has zero momentum with respect to the earth, whereas the same bullet in flight has a great deal of momentum. Has not the pulling of the trigger succeeded in creating momentum out of nothing?

No! While the bullet speeds in one direction the rifle has, through recoil, moved in the opposite direction. If we define momentum in one direction as positive and in the other as negative, then we will have $+x$ momentum for the bullet and $-x$ for the rifle. The two will add to zero, so that the zero momentum before the firing will equal the zero momentum after the firing if, in both cases, rifle and bullet are considered together. The whole "system" can neither gain nor lose momentum even though individual parts of a system can.

The same is true of "angular momentum." Something that is not rotating will begin rotating in one direction if another part of the system undergoes an equivalent rotation in the opposite sense.

There is also such a thing as a law of conservation of electric charge. A proton with its positive electric charge cannot be created out of energy (which has no charge) unless an antiproton, with a negative electric charge, is also created. The two charges, positive and negative, are equal in size and add up to zero, so that no *net* electric charge has been created.

Can this sort of thing be applied to energy, too? Can there be "positive energy" and "negative energy" and can equal quantities of both be created out of "zero-energy," just as equal quantities of positive and negative momentum can be created out of zero momentum and as equal quantities of positive and negative electric charge can be created out of zero electric charge?

Let's suppose the existence of two opposite kinds of energy, then. What might the consequences be?

Let's begin by remembering that all the known particles making up the universe represent some form of energy. In that sense, everything we can detect in the universe, without exception, is energy in one form or another arranged in some form of interrelationship. If the universe contained no energy at all, there would be nothing left (as far as we know) that we could detect. Nor would we exist or anything that (as far as we know) could serve as detector.

We can express this by saying that positive energy and negative energy would interact to produce *nothing!*

But if that is so, then the reverse ought to be true. A quantity of nothing might suddenly become equal quantities of positive energy and negative energy. The two states of existence—nothing on one side, equivalent amounts of positive and negative energy on the other—might have equal probabilities of existence. For that reason, nothing could give rise to double energy on a purely random basis.

In fact, if we imagine a field of nothingness, random effects might be continually producing bubbles of negative and positive energy, always in equivalent amounts. Some bubbles might be larger than others and it would seem reasonable to suppose that, the larger the bubbles, the less frequently they form.

When the bubble-pair is formed, it may be that there is some tendency for the two to separate—a mutual repulsion. The larger the bubble-pair, the farther they move apart and the longer it takes them to come back together and interact—turning back into the nothing from which they came.

Suppose that the quantity of positive energy formed is roughly equal to all the energy present in the universe and antiuniverse represented in Figure 9. There would then be enough negative energy produced to form a negative universe and a negative antiuniverse of equal size.

Moreover, it's easy to assume that the bubble of positive energy formed is the equivalent of the energy-egg mentioned earlier (a cosmic egg made up of energy only). It explodes to form the universe and antiuniverse, which expand and then contract again in a period of (say) a trillion years.

Meanwhile the bubble of negative energy formed is an antienergy-egg. It explodes to form the negative universe and the negative antiuniverse, which expand and then contract again in the same period of (say) a trillion years.

When the contraction stage is completed, the energy-egg and the antienergy-egg are formed again. Perhaps they explode again and, after another expansion-contraction cycle, form and explode. They may do so an indefinite number of times. Or else, after a few expansions and contractions, or even after only one, the energy-egg and the antienergy-egg may interact and sink to nothingness again.

In what way, though, are the positive and negative universes different and what keeps them separate? Here we are in the realm of pure fantasy, for we have nothing to guide us.

We can reason, however, that the positive and negative universes must be identical except in some key respect in which they are opposites, and that these opposites must have no internal effect. Just

Figure 10. The Steady-Oscillating Quadruple-Universe

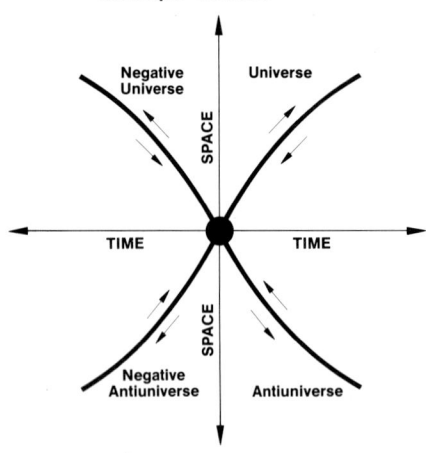

as an antiuniverse would seem perfectly normal to an antiman, a negative universe would seem perfectly normal to a negative man; and a negative antiuniverse would seem perfectly normal to a negative antiman.

In antimatter, the electric charge on a particular antiparticle is the opposite of that on the analogous particle. In negative matter, it might be hypothesized, the direction of time is reversed, as compared with that in positive matter.

In the universe and antiuniverse alike, time goes in that direction we arbitrarily call "forward" during the period of expansion and "backward" during the period of contraction. In the negative universe and negative antiuniverse alike, time would go "backward" during the period of expansion and "forward" during the period of contraction. This difference in time direction could not be detected from within a universe where the difference affects all things alike, including the detector.

It is only when universe and negative universe are compared from outside by an observer who is a member of neither one that an effect would be detected. That effect might be that the time interval between the two is continually increasing as long as both are in the expanding stage, and continually decreasing as long as both are in the contracting stage.

Just as the universe is separated from the antiuniverse (and the negative universe from the negative antiuniverse) in space, so the two positive universes are separated from the two negative universes in time. The result of this speculation is the "quadruple-universe" shown diagrammatically in Figure 10.

In such a case, we can finally imagine an absolute beginning and end to all aspects of the universe. There is an absolute beginning with the quadruple universe is created out of nothing; and there is an absolute end when the quadruple universe sinks back into nothing. In between, the quadruple universe may oscillate once, twice, or an indefinite number of times.

Once the quadruple universe sinks back into nothingness, it (or another like it) may begin again. In fact, it would be logical to suppose that it must begin again. Random factors must be producing bubble-pairs of energy here and there in nothingness continually.

Indeed, numerous such bubble pairs may exist now. If the nothingness is unending (and why should it not be?) there may well be an infinite number of quadruple universes existing along with ours and separated from ours by something more subtle than time or space.

And yet, just as I feel that I have evolved a scheme in which all questions are answered, in which there is an absolute beginning and an absolute end, I find that I can question still further.

What is this nothingness out of which all the quadruple universes arise and into which all sink?

Is it really nothingness or is it something other than nothing which we confuse with true nothingness only because we don't know how to detect or study it? Is it something more subtle than energy, but just as real and just as other-than-nothing?

And in that case, can it exist in opposite varieties so that the quadruple universe is really an octuple universe sinking back to a "nothingness" that is really a still more subtle other-than-nothing?

Can we allow ourselves to wonder whether everything that exists is one of an infinite number of such objects that arises out of something more fundamental than itself which is, in its turn, one of an infinite number of such objects that arises out of something still more fundamental, which is itself one of an infinite number of such objects that—

And so on, and so on, and so on—
Without beginning and without
END

Isaac Asimov

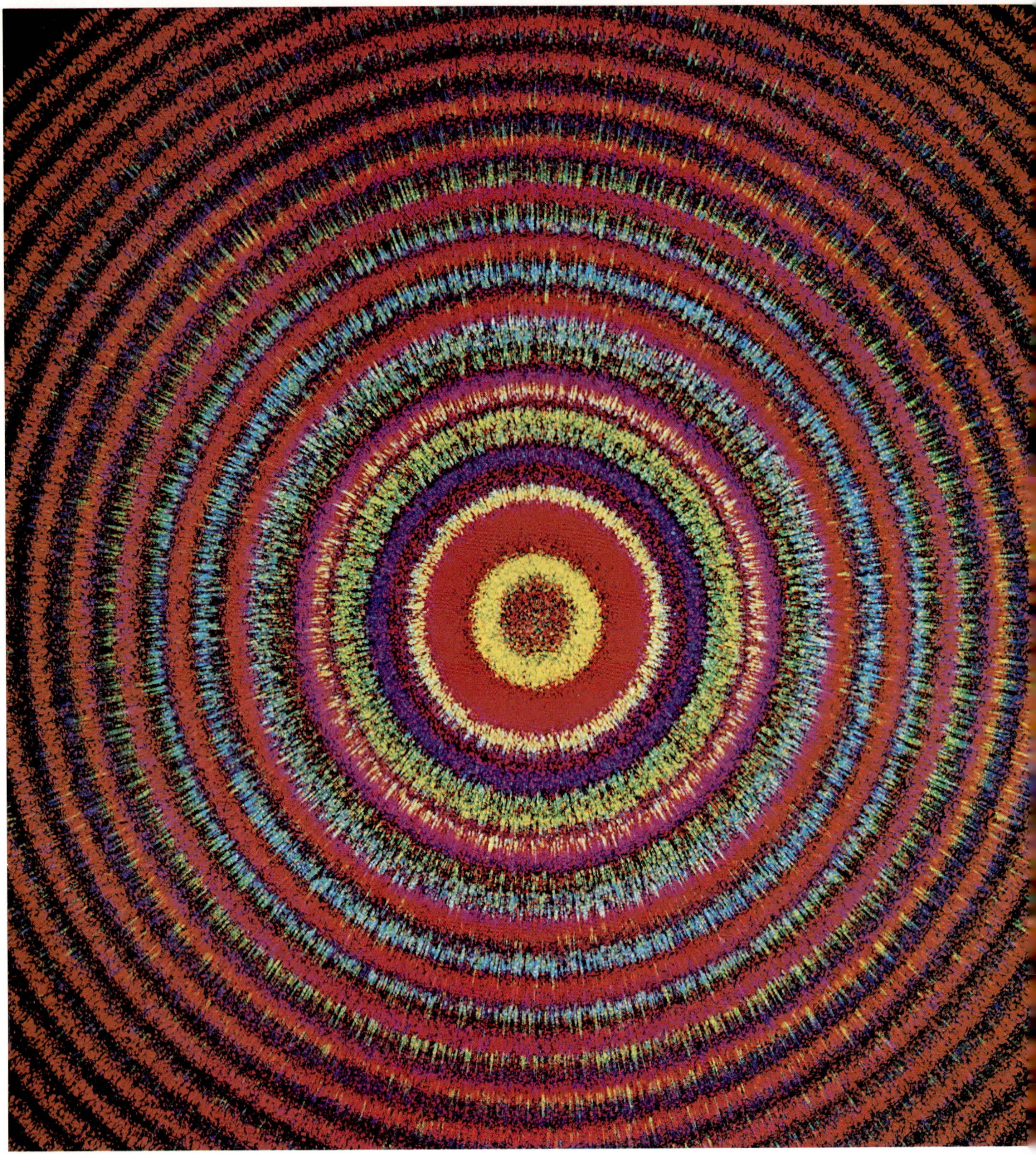

In earlier chapters you looked at some simple models of the atom. Each model was invented to explain a particular set of observations. John Dalton's chemical atom, with its hooks and eyes, was helpful in thinking about how atoms joined together in chemical combination. In the kinetic theory of gases the atom was thought of as a tiny featureless billiard ball, continually moving, colliding, bouncing.

Each model expressed some aspect of how atoms behave. If you were interested in pressure or Brownian motion, you used the billiard-ball model, and there was no need to worry about the hooks and eyes of the chemist's model. Each model was valuable in its own range of application. The simple assumptions used in the kinetic theory model allowed such phenomena as pressure, temperature, and the speed of atoms to be understood. Definite predictions could be made on the basis of the model, which agreed very well with experiment. But the kinetic theory model had nothing at all to say about how atoms bound themselves together in molecules. At the end of the nineteenth century, there was no unified model for the atom that explained all its features.

Worse yet, there were many features that no model explained at all. In 1859, when Robert Bunsen and Gustav Kirchhoff began using the light emitted by atoms as a method of chemical analysis, they saw that each element had its own characteristic spectrum. During the late nineteenth century, experimenters measured the wavelength of a vast number of spectral lines with great precision. But no one could think of a model that explained why atoms emitted light only in definite colors.

Physicists realized that the interpretation of atomic spectra might be the key to understanding the structure of the atom. Kinetic theory had nothing to do with the inner construction of atoms, and observations from chemistry were too descriptive and complex to be related directly to atomic structure. Atomic spectra, on the other hand, involved *single* atoms, and the careful measurements of spectral wavelengths were solid data that would be a stringent test for theory to explain. The atoms were sending out abundant information about their structure. At the end of the nineteenth century, the problem was to decode the message.

ELECTRON, PROTON, NEUTRON: THE BUILDING BLOCKS

The modern picture of the atom was developed in the brief period from 1895 to 1925. It was a time of intense activity and discovery that resulted in a model capable of explaining not only the origin of atomic spectra but the details of chemical combination as well. It even explained why gas molecules behave like billiard balls. However, because so many different lines of investigation contributed to that model, it would be confusing to discuss them in strict historical order. It's easier to understand how the modern atomic model was developed by taking one aspect of it at a time.

Thomson and the Search for Electrons

The word "atom" comes from the Greek word for "indivisible," and for a long time chemical atoms seemed to be truly indivisible. Finally, in 1897, J. J. Thomson showed conclusively that particles could be split off from

19

The Quantum Atom

Why do atoms emit light only in definite colors? The interpretation of atomic spectra was a turning point in understanding the structure of the atom.

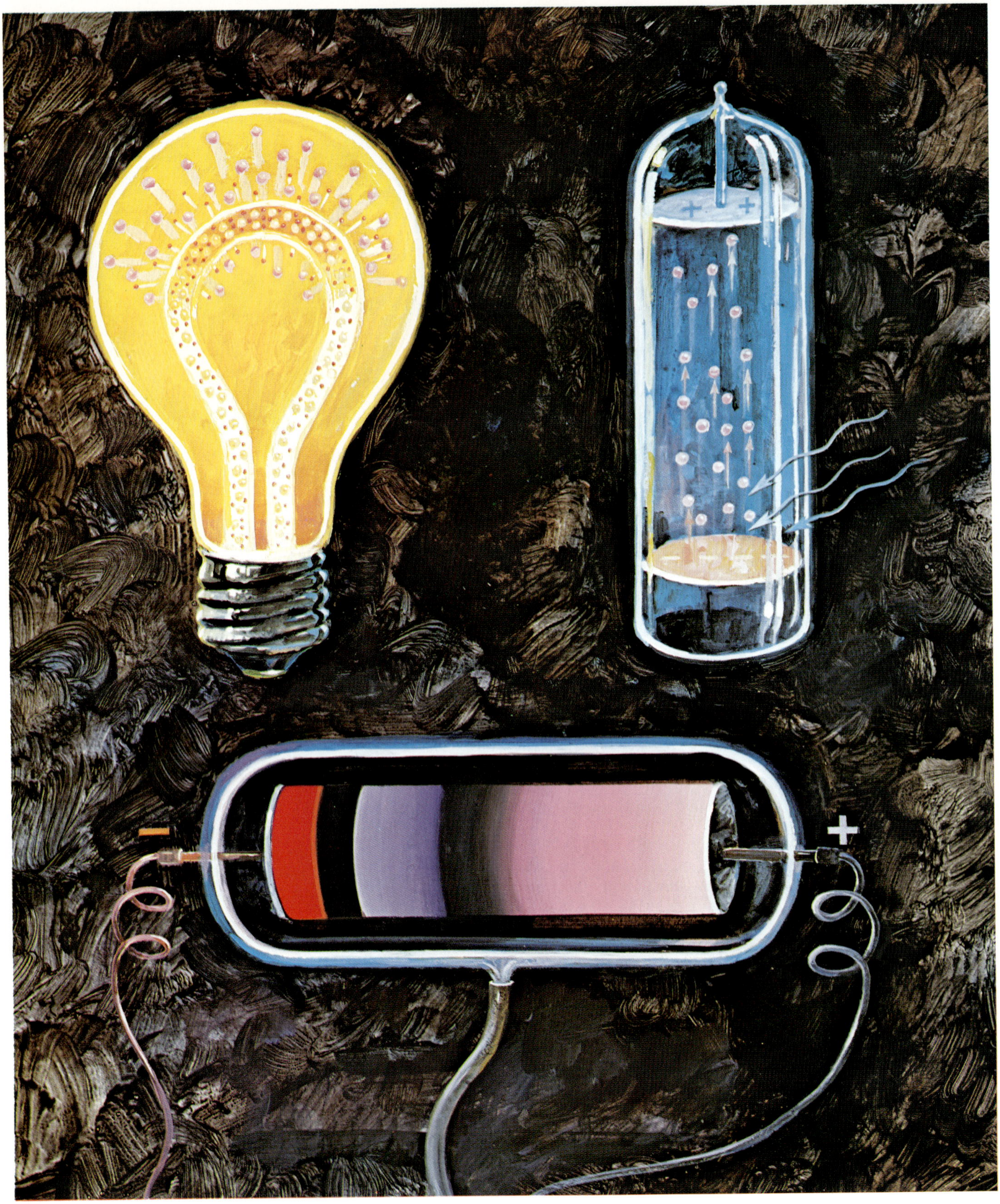

atoms. The particles, called electrons, have a negative electric charge and a mass much smaller than the mass of an atom.

There are many ways of releasing electrons from matter. For example, if matter is heated enough, it emits large numbers of electrons. Electrons can also be released from matter when light shines on it; the electrons are ejected particularly easily when ultraviolet light shines on sodium or potassium metal. A third way of releasing electrons is to pass electricity through a gas at low pressure.

In all three methods energy is transferred to electrons. When matter is heated, the increase in thermal energy gives some of the electrons enough speed to overcome the forces that bind them to the matter. When light shines on a metal, for example, a photon of light may give its energy to an electron, and if the forces holding the energetic electron in the metal are not too strong it may escape completely. When the electric forces produced by, say, a high-voltage battery are used to accelerate charged particles in a gas, the particles collide with neutral atoms and knock out electrons. We know electrons must be bound to atoms because it takes energy to release them.

Thomson used the third method to release electrons. In his apparatus, which was similar to a short length of a neon sign tube, the gas was held at a low pressure so the light would be barely visible. And yet Thomson observed a fluorescent glow in the glass at one end of the tube. It was as if particles were moving from the negative end of the tube toward the positive end and striking the glass. How did he know that the glowing streams in his electric discharge tube were charged particles and not ultraviolet light or x-rays? By bringing a magnet or electrically charged particles near the tube he was able to deflect the streams, which proved they were particles with mass and electric charge.

You know from earlier chapters that the effect of a force on a mass is to produce a change in its momentum. When Thomson brought electric charges near the stream of electrons, the force exerted on the electrons pulled them sideways. Now if he had known the strength of the force, he would have been able to find the mass of the electrons by measuring their sideways momentum. By Coulomb's law, the strength of the force is proportional to the *electric charge* carried by each electron. But the charge of an electron had not yet been measured.

Thomson could not find the mass directly; all he could measure was the ratio of charge to mass, which came out to be about 1.76×10^{11} coulombs per kilogram. Interestingly enough, no matter what gas he used in his apparatus the negatively charged particles *always* had a charge/mass ratio of 1.76×10^{11}. When the charge/mass ratio of the negatively charged particles emitted from hot matter was measured, it turned out to have the same value. The charged particles ejected from metals by photons had the same ratio as well. It was clear that electrons were present in every kind of matter. Electrons had to be one of the universal building blocks of the atom.

The charge/mass ratio of an electron is thousands of times larger than the charge/mass ratio of a charged atom. What could account for the difference? If the electric charge on an electron is about the same as the electric charge on a charged atom, the mass of the electron must be very small compared to the mass of an atom. The problem was to find a way to measure the charge directly.

Millikan's Oil Drops

It isn't difficult to measure the charge/mass ratio of a particle. Robert Millikan in the United States reasoned that the charge of an electron or a

Figure 2. (Far left) Electrons are emitted when light bulb filaments are heated, when photons hit a metal (upper right), and when electricity is passed through the rarified air of a discharge tube.

Figure 3. J. J. Thomson (1856-1910), the physicist who determined the charge-mass ratio of the electron.

19 | The Quantum Atom

Figure 4. In Thomson's electric discharge tube, the stream of electrons produced at the left passes through slits and deflected by positively and negatively charged plates.

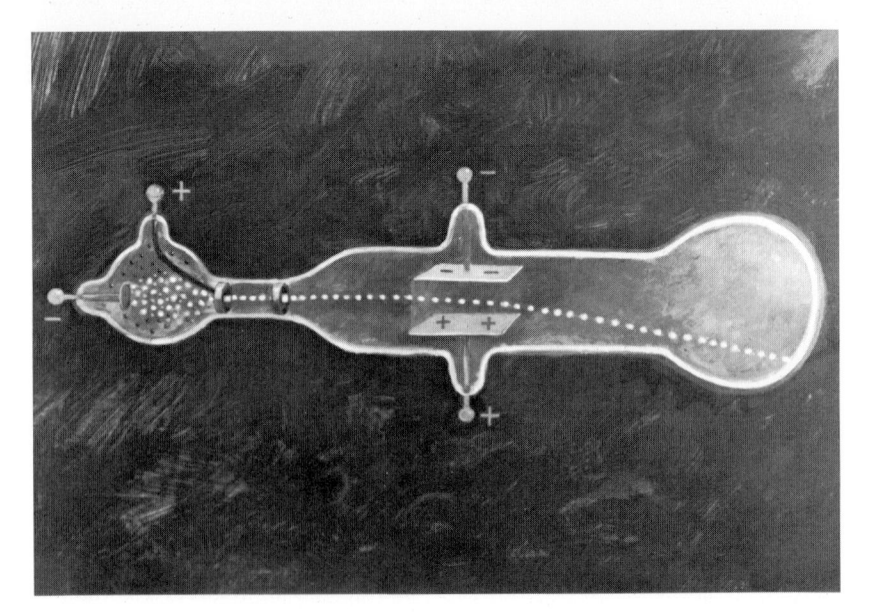

Figure 5. Robert Millikan, the scientist who devised a method of detecting the charge on an electron.

charged atom could be found if they were attached to a large particle of known mass and the charge/mass ratio measured. The spray from an atomizer is ideal for the purpose. The droplets are about 10^{-4} centimeter in diameter, and if they are allowed to enter a properly illuminated box their motion can be followed through a short-range telescope. They look like bright stars of light.

One of the advantages of using such small droplets is that they fall slowly in still air. You know that a Ping-Pong ball doesn't fall in air with the acceleration due to gravity; it's retarded by air resistance. In fact, if a Ping-Pong ball is allowed to fall far enough it will reach a steady speed because the force of air resistance increases with speed. As the speed of a falling Ping-Pong ball increases, so does the air resistance, until finally the retarding force is equal in strength to the weight force pulling the Ping-Pong ball down. When the forces are in balance the Ping-Pong ball moves at a steady terminal speed.

The force of air resistance is relatively large on drops of the size Millikan studied. Their terminal speed of fall is only a fraction of a centimeter per second, which means that a single drop can be followed through a telescope for a long time. The theory of air resistance shows the terminal speed of a spherical drop is proportional to the square of its diameter. By timing the rate of fall of a particular drop, Millikan could infer the size, hence the mass, of the drop.

When an electric force is applied to the drops, the motion of some of them changes, which means they are electrically charged. A charged drop can even be made to rise by manipulating the direction and strength of the electric force. By letting it rise and fall, you can study a drop for hours. Sometimes a drop changes its motion abruptly while the electric force is acting because of a sudden change in the amount of charge carried by the drop. Millikan's method is so sensitive that the departure of a single electron or the addition of a charged atom from the air noticeably affects the motion of a drop.

Millikan found out that none of the drops had a charge of less than 1.6×10^{-19} coulomb. Furthermore, the charge carried by a drop was always a small integer multiple of 1.6×10^{-19} coulomb. For example, some drops had a charge of 3.2×10^{-19} coulomb, or twice the basic charge. Millikan concluded that electric charge comes in units of 1.6×10^{-19}

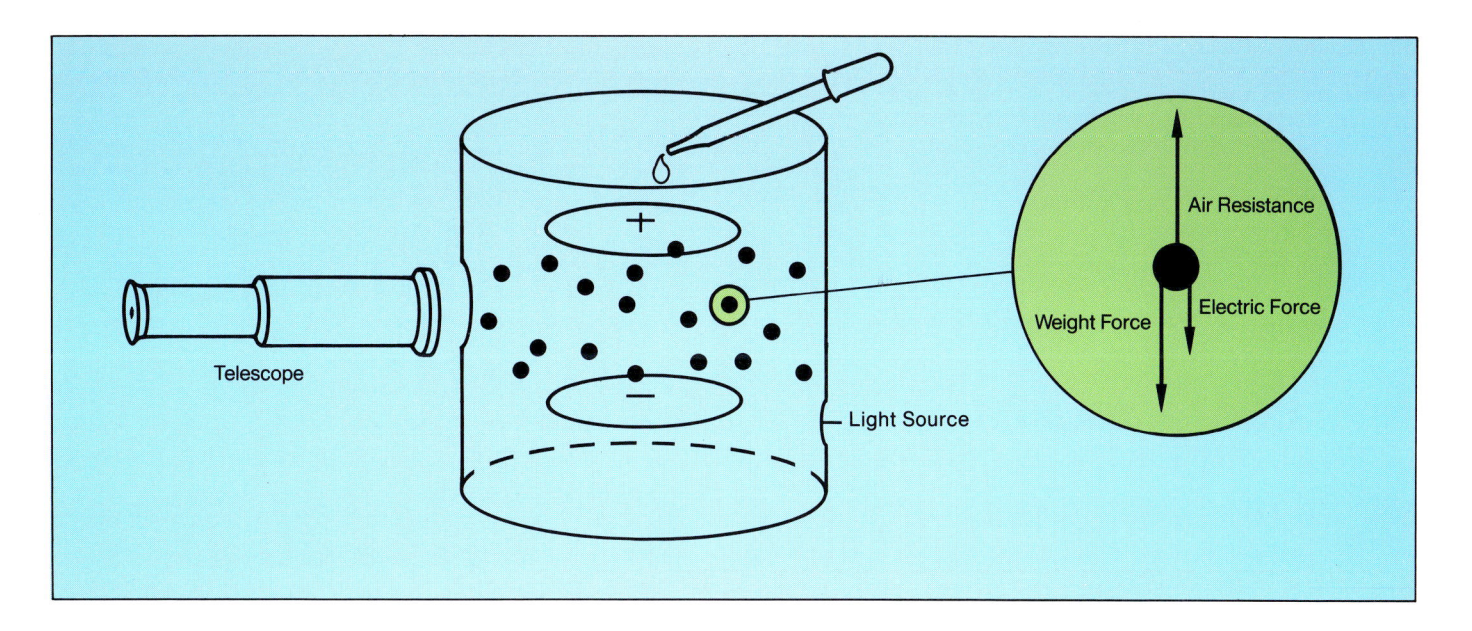

coulomb and that the magnitude of charge on an electron is exactly this unit. An electron has a charge/mass ratio of 1.76×10^{11} coulombs per kilogram, so its mass comes out to be 9.1×10^{-31} kilogram. That is about 1,840 times smaller than the mass of hydrogen, the lightest atom.

Clearly electrons are part of an atom but they have only a small fraction of the atom's mass. Usually atoms are electrically neutral, with equal amounts of negative and positive charge. How the positive charge and the mass of an atom were arranged remained a puzzle. Thomson thought the positive charge and the mass were smeared out uniformly over the atomic volume, with electrons dotting it here and there like raisins in a cake. Thomson's model had a certain degree of stability, but it explained little and was soon discarded in light of new experiments by Ernest Rutherford.

Rutherford Finds the Nucleus

Rutherford, the son of a New Zealand potato farmer, came to England on scholarship and worked with Thomson. One of the best experimenters of the twentieth century, he introduced a way of studying matter that has become one of the most important tools of physics.

Rutherford decided to study matter by shooting particles at it. He knew from his work in radioactivity that some heavy atoms such as radium spontaneously emit fast charged particles, called *alpha rays,* which have the same atomic mass as helium. Because the speed of these particles exceeds 10^9 centimeters per second, they are able to penetrate 6 or 7 centimeters of air before they lose all their energy and come to a stop.

When an alpha ray passes through matter, it is hardly deflected at all by collisions with electrons. Alpha rays are thousands of times heavier than electrons, and when an alpha ray hits an electron it is like a bowling ball hitting a marble. The electrons go flying off, but the alpha ray is barely disturbed.

Rutherford and his coworkers let a beam of alpha rays hit thin sheets of metal. To see what was happening, they used a special phosphor that glowed momentarily when struck by an alpha ray. By moving their detector, they found that most of the alpha rays passed through the thin metal foil with little deflection. But some of the alpha rays hitting the foil

Figure 6. In Millikan's experiment, charged oil drops are made to rise and fall by manipulating the electric force produced by charged plates. The electric force can be either up or down. The air resistance always opposes the motion of the drop. At terminal speed, the upward forces on the drop balance the downward forces.

Figure 7. (a) In Thomson's model of an atom, the positive charge would be spread out and could not strongly deflect an alpha ray. (b) In Rutherford's model of an atom, the positive charge was highly concentrated and could exert strong forces but only within a small region of the total volume of the atom.

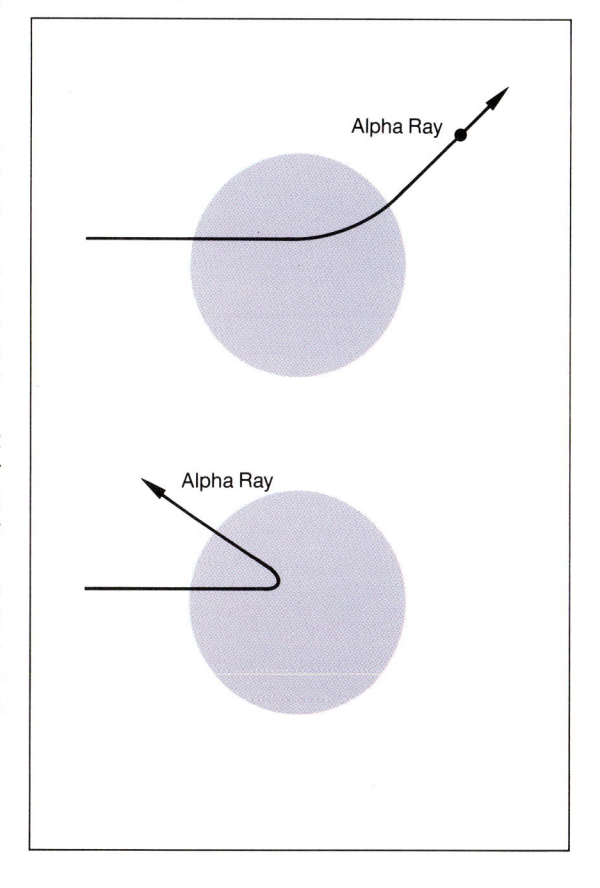

19 | The Quantum Atom

Figure 8. In Rutherford's model, alpha rays passing far from the nucleus are not deflected much. The closer the alpha ray comes to the nucleus, the stronger the repulsive force and the more it is deflected.

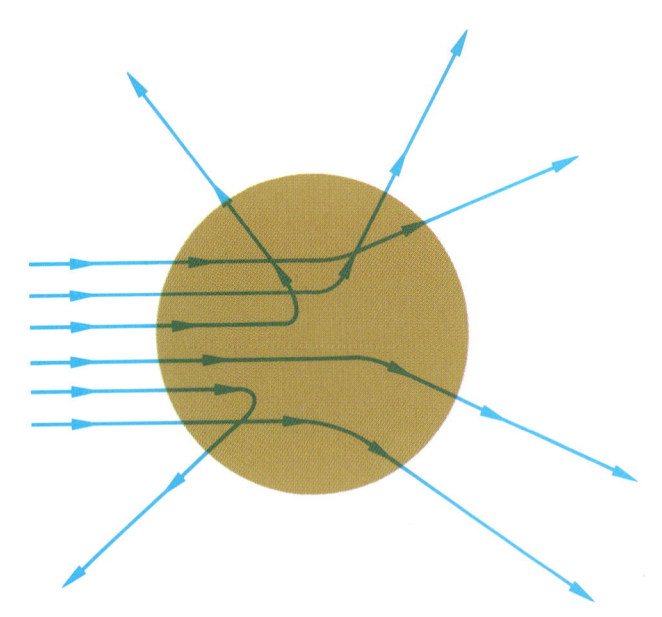

Figure 9. The English physicist Ernest Rutherford (1871-1937), who developed a method of studying the charge distribution of the heavy nucleus of an atom.

were deflected at large angles—and a few even bounced straight back! Rutherford was amazed. It was as if cannon balls had bounced back from a sheet of paper.

Strong forces are needed to deflect a fast-moving alpha ray through such large angles. According to Coulomb's law, the force between two bits of charge is strongest when they are close together. But in Thomson's "raisin cake" model of the atom, the charge was spread out and the alpha ray could never be close to more than a small part of the total positive charge. The Thomson atom could only deflect an alpha ray slightly, and it would take many small deflections with many atoms to add up to a large net deflection. Because Rutherford's metal foils were too thin to allow that many deflections to occur, Thomson's model had to be incorrect.

Rutherford suggested that the positive charge in an atom is concentrated in a small volume, called the *nucleus*. An alpha ray passing near the nucleus experiences a powerful force at small separations and is deflected strongly. A single encounter would be enough to swing an alpha ray through the large angles he had observed.

The alpha ray scattering experiments had some additional consequences. According to Coulomb's law, the electric force between the alpha ray and the nucleus is proportional to the amount of charge in the nucleus, and the strength of the force determines how many alpha rays will be scattered at large angles. Rutherford was able to determine the amount of charge in the nucleus of some elements by carefully comparing his experimental results with the theory based on his model. It became clear that the number of units of charge in the nucleus was its atomic number. And it is by atomic number that elements are listed in the periodic table.

Dmitri Mendeleev set up his periodic table by listing the elements more or less by atomic weight, with an eye to their chemical properties. But atomic weight is only incidentally involved in the chemical properties of the elements: the important factor is atomic number. Moreover, because atoms are electrically neutral and because charge comes in units of 1.6×10^{-19} coulomb, the atomic number is equivalent to the number of electrons.

Rutherford's experiments also showed that the nucleus is small in comparison with the size of the atom. Alpha rays are positively charged, like the nucleus. Because like charges repel, an alpha ray of a given speed can penetrate only to within a certain distance of the nucleus. An alpha ray with enough speed can come close enough to the nucleus to "touch" it—close enough so that the nuclear forces act on the alpha ray. The nuclear force and the electric force deflect the alpha ray in different ways, and by looking for this difference the size of the nucleus can be estimated.

The radius of even the heaviest nucleus is less than 10^{-12} centimeter. That is really small compared to the typical size of an atom, which is 10^{-8} centimeter. The size of an atom must therefore be determined by the electrons, and it must be the electrons that engage in chemical reaction. As far as chemistry is concerned, the nucleus is simply a point deep inside the atom.

THE ENERGIES OF ATOMS AND PHOTONS

The Rutherford model of the atom may be summarized in the following way:

1. The atom consists of a dense, central nucleus surrounded by electrons.

2. The nucleus is less than 10^{-12} centimeter in size and contains all the atom's positive charge and most of its mass.

3. In a neutral atom, the number of electrons equals the number of units of positive charge.

4. According to Coulomb's law, the electrons are attracted to the nucleus and repelled by one another.

All the particles and all the forces are specified in this model. It would seem there is nothing left to do but apply Newton's laws to find out the motion of the electron. But you can tell already what the motion is like. Coulomb's law of electric force between two charges and the law of gravitational attraction between two masses both vary as the inverse square of the distance. The motion of an electron around a proton, according to Newtonian mechanics, would be just like the motion of a planet around the sun. The electron would have to orbit around the proton in circles or ellipses.

This picture of the atom as a miniature solar system seems simple and esthetically appealing. But there are several reasons why it cannot be correct. First of all, there is nothing in this model that tells why the size of an atom is 10^{-8} centimeter. Suppose the sun had only one planet; what would be the size of the solar system? The planet might be close to the sun, like the planet Mercury. Or the planet might be 100 times farther away, like Pluto. The inverse square force allows orbits of *any* size, and the solar system model of the atom has no way of explaining why the billions and billions of hydrogen atoms are identical in size and structure. If all the electrons in different hydrogen atoms followed different orbits, how could all the light emitted by hydrogen atoms have the same frequencies?

Another strong argument against the solar system model is that such a system could not exist for any appreciable length of time. An electric charge radiates electromagnetic energy if it is accelerated. For example, a radio transmitter sends out its signal by making electrons move up and down its antenna. Any particle moving under the action of a net force must be accelerating. Because an electron circling a nucleus has an acceleration due to the electric force, it would radiate energy so fast that it would careen into the nucleus after only about 10^{-12} second. But atoms in the universe are at the very least 10^{17} seconds old!

A principle of physical science may work well in one range but fail in another. Newton's laws do a fine job of predicting the motions of the planets, but that does not prove they can be applied to the motion of

Figure 10. Max Planck (1858-1947).

Figure 11. (a) An atom may emit a photon by changing its energy from E_1 to E_2. (b) The three photons have related energies: the energy of A equals the sum of the energies of B and C.

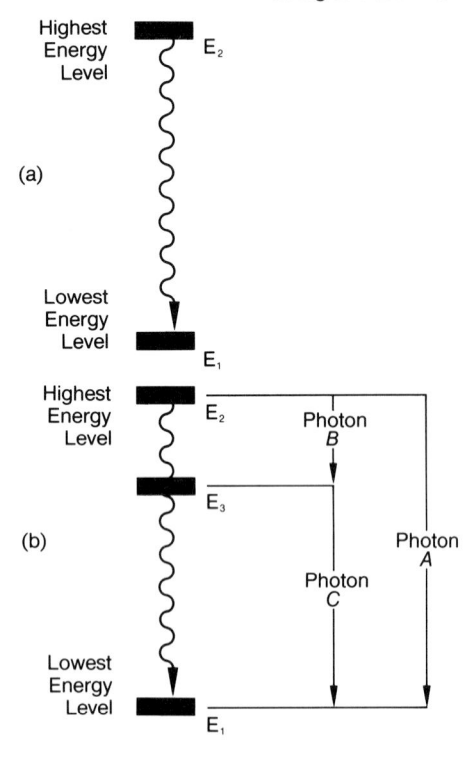

electrons in atoms. For example, relativity principles tell you that Newton's laws fail for particles moving at speeds comparable to the speed of light. It wouldn't be too surprising if they failed on the atomic scale as well. Only experiment can decide.

To find out how the atom is put together, turn from Newton's laws to a more fundamental concept—*energy*. Energy comes in many forms. A battery has chemical energy, a stretched rubber band has mechanical energy, sunlight has radiant energy, a hot poker has thermal energy. If all forms of energy in the universe are taken into account, the total energy remains the same; it is *conserved*. A block sliding across a table loses mechanical energy as it slows down but it generates an equivalent amount of heat energy as it rubs against the table.

Light is a form of energy. And when an atom gives off light some of its energy must be given off, too. If you examine the light an atom emits or absorbs, you can tell how much the energy of the atom has changed. Photons are tiny packets of light energy. They are like particles in many ways and can even be counted one by one in suitable detectors. Your eyes are almost sensitive enough to detect a single photon, but in ordinary daylight more than 10^{14} photons enter each one of your eyes in each second, so you are not aware of the particulate nature of light energy.

The Photoelectric Effect

When light falls on a metal, the incoming energy of a photon is converted to the kinetic energy of an electron. If an electron receives enough energy from a photon it may even leave the metal, provided it is heading in the right direction. By measuring the kinetic energy of the electrons that are emitted, the energy of the photons can be inferred.

Experiments on this photoelectric effect reveal two important properties of light. If the light falling on the metal is made dimmer, the number of electrons released from the metal decreases proportionately—but there is no change in the energy of each electron that does come out. You know, from the photon picture of light, that a dim light sends off less photons than a bright light. If there are more photons there are more electrons emitted, but there is no change in the energy of each individual electron.

What *does* change the electron's energy is the color of the incoming light. A photon of blue light gives an electron more energy; a photon of red light gives it less. In fact, the energy of a photon is directly proportional to its frequency. This relationship is expressed as $E = h\nu$. What it means, in the metric system, is this:

Energy of photon = $(6.6 \times 10^{-34}) \times$ frequency

where the energy of a photon is expressed in joules and the frequency in oscillations per second. For example, red light has a frequency of 5×10^{14}, so the energy of a photon of red light is 3.3×10^{-19} joule. (That isn't much energy on an everyday scale, but it's enough to make an electron move at 1/300 the speed of light.) The constant of proportionality, 6.6×10^{-34} joule-second, is called *Planck's constant* after the man who found a special case of the energy-frequency relation. A few years later, Albert Einstein showed that it applied to photons in general. Planck's constant can be determined experimentally from the photoelectric effect.

The joule, however, is too large a unit of energy to be useful for describing atomic events. Instead, a much smaller unit of energy, called the *electron volt,* is used:

1 electron volt = 1.6×10^{-9} joule

A photon of red light with energy of 3.3×10^{-19} joule, then, has an energy of about 2 electron volts.

You may now put together what you have learned so far about photons and atoms. The spectrum of light emitted by atoms in a hot flame does not contain all colors; it contains only certain colors that are characteristic of the elements that happen to be present in the sample being studied. In other words, only photons of certain special energies are emitted by a given kind of atom. And in accordance with the law of conservation of energy, you would have to conclude that the energy of an atom can change only by fixed amounts:

Energy of emitted photon = change in energy of photon

or

Frequency of emitted photon = $\dfrac{\text{change in energy of atom, in joules}}{6.6 \times 10^{-34}}$

if you use the energy-frequency relation for photons.

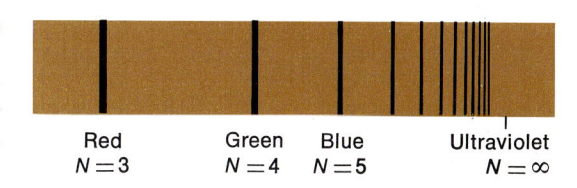

Red	Green	Blue	Ultraviolet
$N=3$	$N=4$	$N=5$	$N=\infty$

Figure 12. The Balmer spectrum for atomic hydrogen.

Energy Levels and the Balmer Formula

The energy relations of an atom can be expressed in simple diagrams by plotting energy on a vertical scale. Figure 11(a) shows two allowed energies, or *energy levels,* in an atom. The atom may emit a photon by changing its energy from the higher energy level to the lower one.

Energy levels are not just another way of cataloging atomic spectra. The concept of energy levels is one of the key ideas of physics on the atomic and subatomic scale. Here are some of the reasons. Suppose there are three energy levels in an atom, as shown in Figure 11(b). If the atom is in the highest energy level, it might emit photon *A* and go down to the lowest energy level. Or it might emit photon *B* and go to the middle level, then emit photon *C* and end up at the lowest energy level. But there is a definite relation between the energies of photons *A*, *B*, and *C*: the energy of *B* plus the energy of *C* must equal the energy of *A*. Many examples of this *combination principle* can be found among the experimentally measured spectral lines for a given element.

Hydrogen, the simplest atom, has a simple spectrum. In 1885, J. J. Balmer, a Swiss schoolteacher, found by guesswork a formula that fits well with the photon energies from the visible spectrum of hydrogen. If photon energy is expressed in electron volts, Balmer's formula becomes

Photon energy = $13.595 \left(\frac{1}{4} - \frac{1}{N^2} \right)$

or, approximately,

Photon energy = $13.6 \left(\frac{1}{4} - \frac{1}{N^2} \right)$

where *N* is any whole number greater than 2.

The lowest photon energy, for $N = 3$, turns out to be 1.89 electron volts. It corresponds to a red spectral line. The next highest photon energy, for $N = 4$, is 2.55 electron volts and corresponds to a blue spectral line. The first few spectral lines are widely separated in color. But consider, say, the photons for $N = 100$ and $N = 101$. Their energies are nearly identical. In fact, the photon energy according to Balmer's formula never exceeds $13.6/4 = 3.40$ electron volts. *All the Balmer spectral lines—and there are an infinite number of them—are crowded between 1.89 and 3.40 electron volts.* But the illustration of the hydrogen spectrum in Figure 12 shows that is exactly what happens: the lines for large *N* are faint but obviously numerous and they crowd up near a limiting value.

Balmer's formula is strongly suggestive of the relation between photon energy and the change in the atom's energy:

Photon energy = $13.6 \left(\frac{1}{4} - \frac{1}{N^2} \right)$
   = change in atom's energy

You get Balmer's formula if you assume the lowest energy level has an energy of $-13.6/4 = -3.4$ electron volts and the higher energy levels possible have energies given by $-13.6/N^2$. Negative energies may seem inconvenient to work with. But notice that when *N* is very large, the highest energy level is close to zero. This corresponds with the idea that if the electron in hydrogen is far from the proton, the electric force is weak and the electron's kinetic and potential energies are nearly zero. According to this argument, *the electron is close to the proton when N is a small number.*

Figure 13 is a diagram of the energy levels of hydrogen inferred from Balmer's formula. Some of the photon energies predicted by Balmer's formula are shown by solid vertical lines. As you might expect from the combination principle, there are also photons with energies corresponding to the dashed lines in the diagram. Each set of photon energies obeys a formula like Balmer's. For example, for the set with lower energy level $N = 3$ (Figure 13), the photon energies are given by

Photon energy = $13.6 \left(\frac{1}{9} - \frac{1}{N^2} \right)$

J. R. Rydberg generalized Balmer's formula to:

Photon energy = $13.6 \left(\frac{1}{n^2} - \frac{1}{N^2} \right)$

In this formula, *n* is the number for the lower energy level involved in an energy change, and *N* is the number for the upper level. When $n = 2$, you

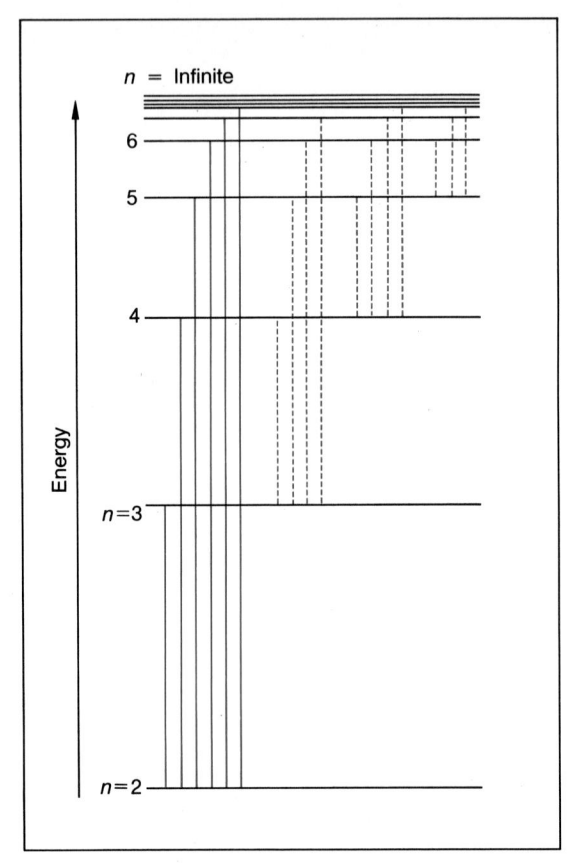

Figure 13. Some of the energy levels of hydrogen are inferred from the Balmer formula. The solid lines correspond to the photon energies described by Balmer's formula. The dashed lines are expected photon energies, which have also been found.

have Balmer's formula. When $n = 1$, you have the photon energies for the part of the hydrogen spectrum lying in the ultraviolet. The formula accounts for the entire optical spectrum of hydrogen.

According to Rydberg's formula, all the energy levels in the hydrogen atom are given by the expression

Energy of any level $= -\dfrac{13.6}{n^2}$ electron volts

One remarkable feature of this expression is the appearance of sets of whole numbers in a subject as complex as atomic spectra. Another feature of utmost importance is the implication that *the energy of an atom is never less than a certain minimum value.* For hydrogen, the minimum energy is −13.6 electron volts. Here is an insight into the stability of atoms. If there is a lowest possible energy, an atom cannot continuously lose energy. The most it can do is reach its state of minimum energy.

The energy level formula for hydrogen (Figure 14) shows there is a considerable gap between the lowest energy level, at −13.6 electron volts ($n = 1$), and the next highest energy level, at −3.4 electron volts ($n = 2$). That is a gap of 10.2 electron volts, which is a sizable amount of energy on the atomic scale. Suppose a hydrogen atom is in its lowest energy state and another particle collides with it. What will happen? The hydrogen atom has no energy states between −13.6 and −3.4 electron volts. It cannot absorb an amount of energy less than 10.2 electron volts. If the collision is not forceful, the hydrogen atom will not absorb any energy—it will simply bounce away, more perfectly than the best billiard ball. That explains why the billiard ball model of the atom works so well in the kinetic theory of gases. The average kinetic energy of a gas atom at room temperature is only about 0.02 electron volt. When two atoms in a gas collide, there can be no transfer of energy and they bounce back.

All atoms and molecules have a large energy gap between their lowest energy level and their next highest level. This gap is a significant factor in guaranteeing the stability of living things. The genetic code that determines your genetic make-up and the regulation of your body chemistry is embodied in long, complex molecules. These molecules are continually bombarded by other molecules in ceaseless thermal motion. If such collisions could easily transfer energy, the regulatory molecules would be disrupted and would no longer function properly.

Ground Levels and Excited Levels

When an atom's energy has its lowest possible value, the atom is said to be in its *ground level.* If it has a higher energy, it is said to be in an *excited level.* Usually an atom does not stay in an excited level for long; instead, it emits photons until it has returned to its ground level. For most excited levels, this process takes an average of about 10^{-7} or 10^{-8} second.

To make isolated atoms give off their characteristic spectra, energy has to be pumped in. If a gas is hot enough or if electricity is passed through it, its atoms make energetic collisions and transfer energy, which momentarily raises them to an excited level. An atom soon radiates away the excess energy and returns to its ground level. A highly energetic collision can transfer enough energy to an atom to rip an electron completely away. The minimum energy needed to release an electron from an atom in its ground level is called the *ionization potential* of the atom. The ionization potential of hydrogen is 13.6 electron volts because the ground level of hydrogen is 13.6 electron volts below zero energy.

Atoms may also absorb energy from photons. But to be absorbed, the photon must have exactly the right amount of energy to raise the atom

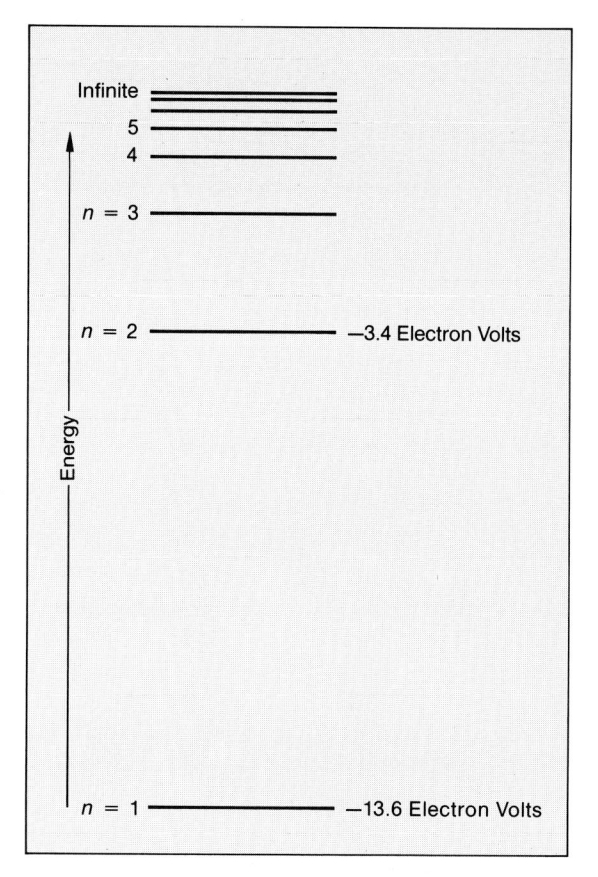

Figure 14. Energy levels of the hydrogen atom. Notice the gap between the lowest level, $n = 1$, and the next highest level, $n = 2$.

Figure 15. Atoms are fussy about accepting electrons. Here a hydrogen atom refuses 9.5 electron volts of energy but accepts 10.2 electron volts.

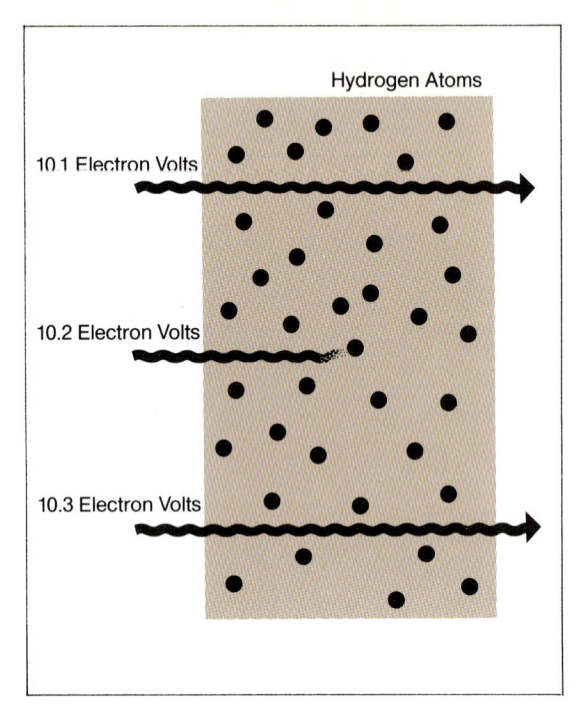

Figure 16. Hydrogen atoms are highly selective about the photon energies they will absorb. In fact, the photon energy must be within about 10^{-6} electron volt of the correct energy in order to be absorbed.

from one energy level to another. Even in a hot gas, however, atoms are in their ground levels most of the time. So the only photons that normally have a reasonable chance of being absorbed are the ones with energies that correspond to starting at the ground level $n = 1$ and going to an excited level. For example, the first excited level in hydrogen is 10.2 electron volts above the ground level, so a 10.2-electron-volt photon is the lowest energy photon that can be absorbed by a hydrogen atom in its ground level. A 10.1-electron-volt photon or a 10.3-electron-volt photon cannot be absorbed. On the other hand, a photon with any energy above 13.6 electron volts can be absorbed because that is enough energy to ionize a hydrogen atom, and the unbound electron can take away any energy over 13.6 electron volts.

To sum up, there are many reasons to believe that atoms have only certain allowed energies. The direct lines of evidence come from experiments on the energy of emitted and absorbed photons and from measurement of the energy transfers in collisions between an atom and another particle. You can say the energy of an atom is *quantized*—from the Latin word "quantus," which means "how much." With one or two extensions, the idea of energy levels is enough to explain the main features of the periodic table, as you will now see.

THE PERIODIC TABLE AND ATOMIC STRUCTURE

Hydrogen, with only one electron, is the simplest atom. Its energy levels are so regular and so orderly that they can be expressed accurately by a simple formula, as you have seen. Other atoms are much more complicated. When there are several electrons in an atom, each electron is repelled by the other electrons and is at the same time attracted by the nucleus. The motion of each electron greatly affects the motion of the others. The energy level structure is too complex for a simple mathematical formula to describe, but in any case, energy levels can be determined by experiment.

There is a close connection between the energy level picture of the atom and the behavior of the chemical elements described by the periodic table. Many of the elements have strongly contrasting chemical behavior. Describing such clear-cut differences should not require an accurate knowledge of the energy levels. But it does require a more general picture of energy levels than the simple hydrogen atom provides.

Bohr's Quantum Atom

Consider the idealized energy levels shown in Figure 17 that illustrate the main features found for atoms more complex than hydrogen. The overall pattern of energy levels is not too different from the one for hydrogen. There are levels labeled by n, a whole number in a set starting with $n = 1$. Now n is called the *principal quantum number*. But you can see from Figure 17 that there is generally more than one energy level for each value of n. In fact, there is one energy level for $n = 1$, two energy levels for $n = 2$, and so on. Generally speaking, the energy separations are small between levels having the same value of n. In hydrogen, the separations are too small to show in the sketches. Energy levels having the same value of n have nearly the same value of energy, and it is natural to group them together. But when n is large, the grouping becomes less distinct.

How do you keep track of all the energy levels? There is another label that is used. The new label, called ℓ, is any whole number in a set starting with 0. For example, the three energy levels corresponding to $n = 3$ are labeled $\ell = 0$, $\ell = 1$, and $\ell = 2$ respectively. Here, ℓ is the

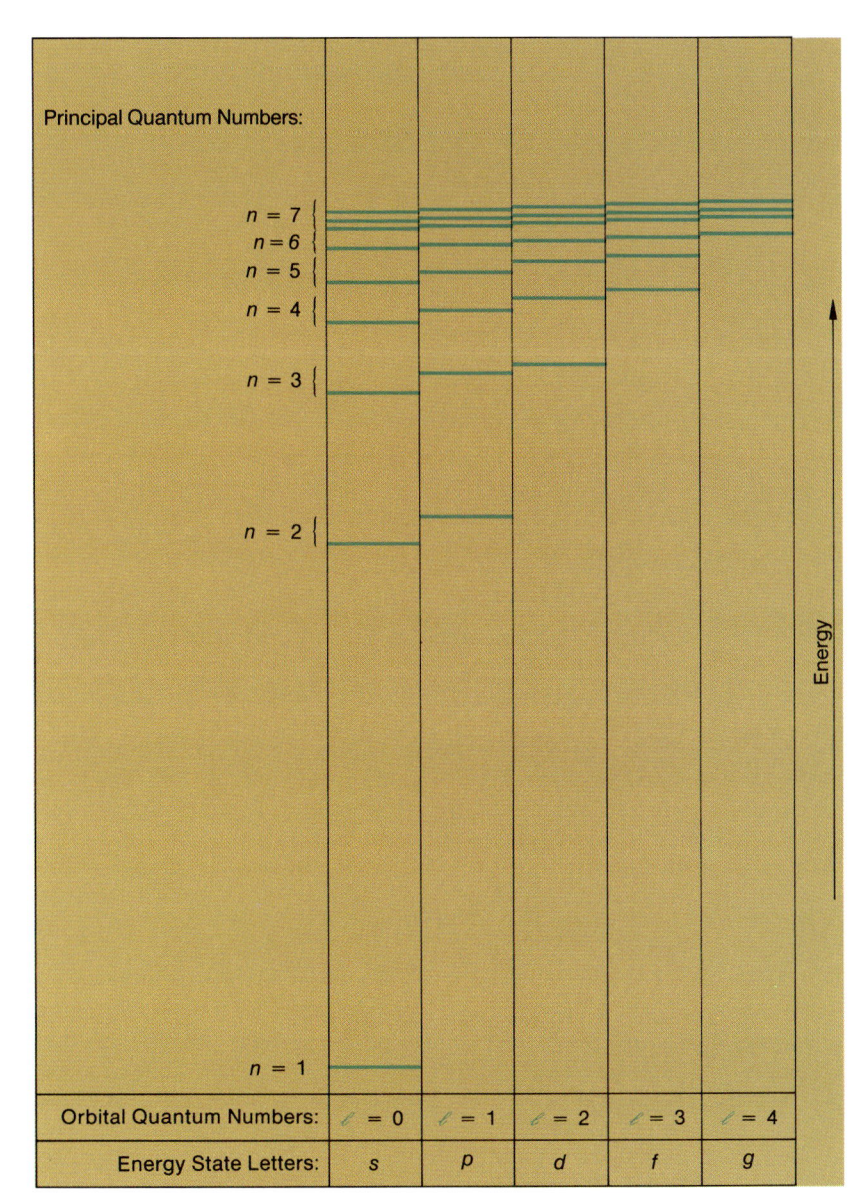

Figure 17. This drawing shows the idealized energy levels and two corresponding systems of notation used for labeling them.

orbital quantum number. According to *another* system of labeling, the ℓ numbers in turn are replaced by letters (see Figure 17). In this system, the lowest energy level $n = 1$ ($\ell = 0$) is called the 1s level. Similarly, the two energy levels corresponding to $n = 2$ ($\ell = 0$ and $\ell = 1$) are called 2s and 2p levels. In hydrogen, the 2s and 2p levels have nearly the same energy.

To see how energy levels are related to the periodic table, you can use a building-up method devised in 1921 by Niels Bohr, a Danish physicist who had studied with Rutherford. Bohr started off by picturing a completely neutral atom. He then imagined that the nuclear charge was increased by 1 unit and an electron was added to form the atom of the next higher atomic number. With each added electron the energy levels change in this model, but the general features you saw in Figure 17 stay about the same.

You know that an atom has a strong tendency to emit photons until it reaches its lowest energy. Consequently, you might tend to believe that

Figure 18. (a) An s level contains two hidden states that show up when a magnet is brought nearby. A p level contains six hidden states. (b) Here, s and p levels are shown split by a magnetic field. An s level can contain two electrons, one for each hidden state. A p level can contain six electrons.

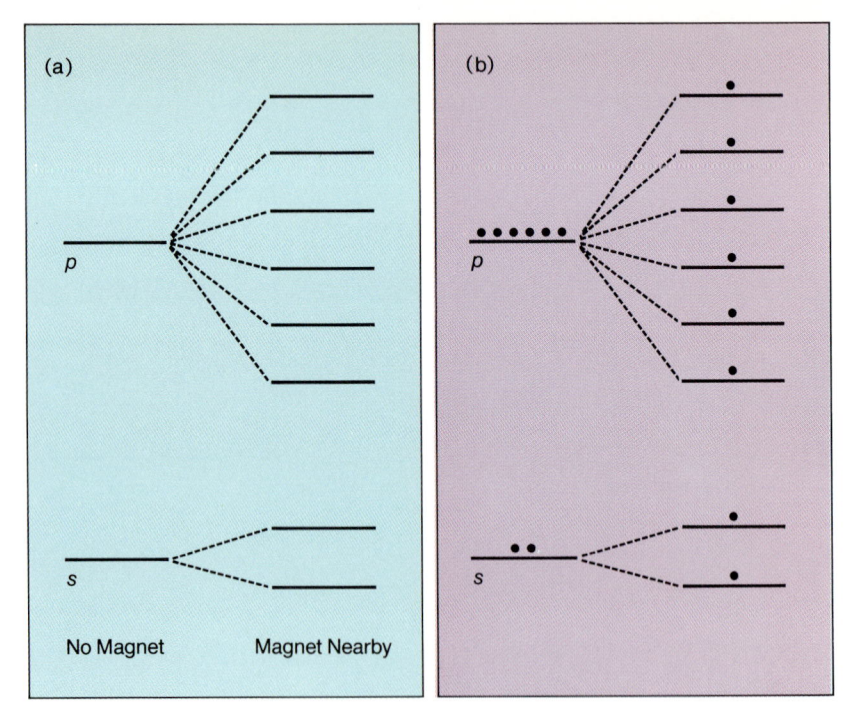

the lowest energy could be achieved by putting each successive electron in the 1s level because the 1s level has the lowest energy. But that gives no variation in behavior that can explain the periodic table. Some additional ideas need to be taken into account.

The Pauli Principle and the Shell Model

Distinguishable configurations of an atom can have equal energies. You know that the 2s and 2p levels in the hydrogen atom have nearly the same energy. Similarly, there are additional "hidden" energy levels that don't show up in the *idealized* energy level diagram (Figure 17). These hidden levels, or states, follow simple rules: Every s level, regardless of its n value, actually consists of two states. Every p level consists of six states, and every d level consists of ten states. The number of hidden levels depends only on ℓ, not on n. A simple general formula is:

Number of hidden levels = $(4 \times \ell) + 2$

For example, $\ell = 2$ in a d state, so the formula says that there are hidden levels, or states.

The hidden levels can be studied experimentally. The energy of an electron in an atom is due mainly to its kinetic energy of motion and its potential energy due to the electric force. If a magnet is brought near an atom, the electrons have, in addition, *magnetic potential energy.* As Figure 18 indicates, the hidden levels are split apart by the magnetic energy. The splitting is not large. For an s state, the splitting is about 4×10^{-5} electron volt with a moderately strong magnet. This energy corresponds to a photon with a frequency of about 10^{10}. That is not a visible photon; rather, it corresponds to high-frequency radio waves such as the ones used in radar. With radio techniques, the emission and absorption of such photons can be measured to show the existence of the hidden levels.

You can say that every s level contains two states, and that every *p* level contains six states. In 1925 Wolfgang Pauli in Germany found a new principle governing the behavior of atoms:

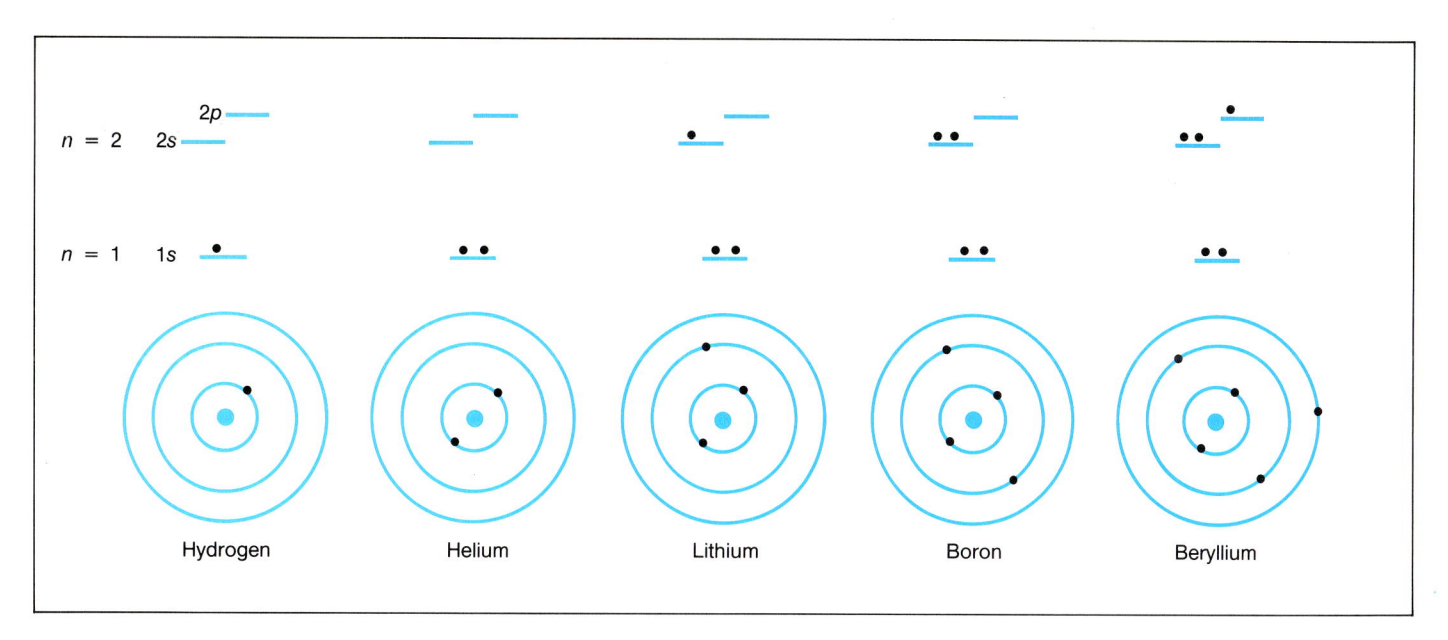

■ **The Pauli Exclusion Principle.** There can be no more than one electron in each state of an atom.

This principle means that an *s* level cannot contain more than two electrons, because an *s* level consists of two separate states. Similarly, a *p* level has the capacity to hold no more than six electrons.

When an electron is close to the nucleus, it is strongly bound by the electric force and it takes considerable energy to tear it away. You can think of electrons in $n = 1$ energy levels as being close to the nucleus. Electrons in $n = 2$ energy levels are farther away from the nucleus; they are not as tightly bound to it. The electrons in a complex atom tend to form a succession of layers, or *shells*. Each shell is connected with a particular value of n. The innermost shell, called the *K shell,* has $n = 1$. The next shell, called the *L shell,* has $n = 2$. And the shell for $n = 3$ is called the *M shell,* and so on.

Periodic Shells, Periodic Table

Now go back with these new ideas to Bohr's building-up method and look at the first few elements, starting with hydrogen. The idea is to think of electrons being added successively, arranged to give the lowest energy without violating the Pauli principle:

Hydrogen (one electron): The single electron goes in one of the 1*s* states to give the lowest possible energy.

Helium (two electrons): The second electron can go in the second 1*s* state. This fills the *K* shell. At helium, the two 1*s* states are fully occupied.

Lithium (three electrons): Because the 1*s* states are both filled, the third electron must be put in a 2*s* state.

Boron (four electrons): The fourth electron can go into the second 2*s* state, filling them.

Beryllium (five electrons): The next electron must go into a 2*p* state.

Figure 19. A schematic diagram of the electronic structure of the first five elements. (The energies and atomic sizes are not to scale.) This simple shell structure model can be used to explain the principles underlying the periodic table of the elements, shown in detail in the foldout chart at the back of this book.

19 | The Quantum Atom

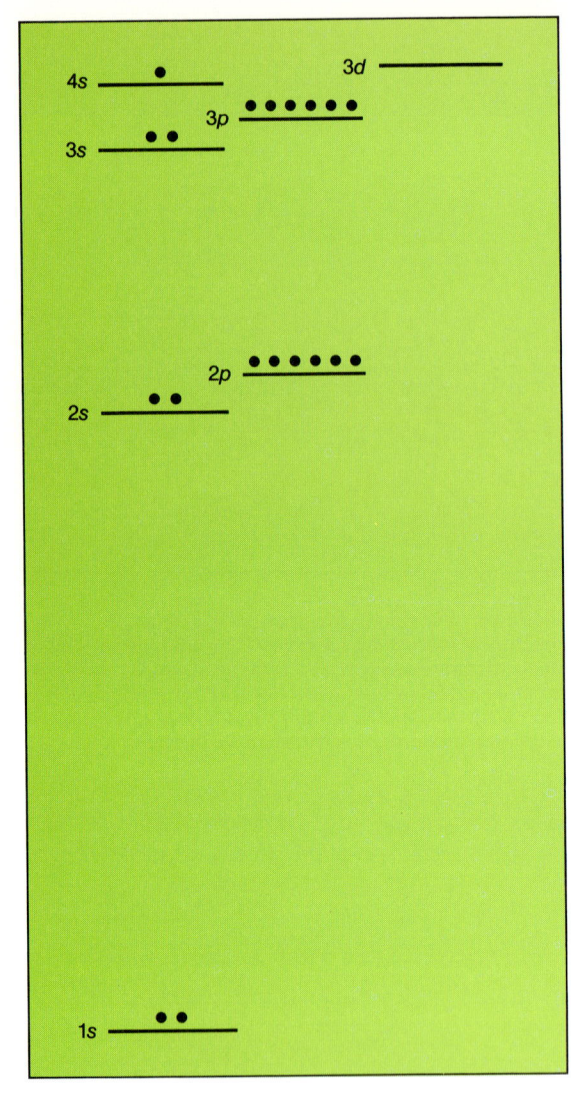

Figure 20. In potassium, atomic number 19, the nineteenth electron goes into the 4s level instead of the 3d level because of slight shifts that make the 4s level somewhat lower in energy.

The 2p states can accept six electrons and become completely filled when there are a total of ten electrons in the atom. The element with atomic number 10 is neon. Its L shell is completely filled. For the next element, fluorine, the eleventh electron must go into the 3s state.

The periodic filling of shells suggests the structure of the periodic table. Atoms with electrons only in completely filled shells are highly symmetrical and can exert very little force on other atoms, which accounts for the chemical inertness of noble gases such as helium and neon.

The third electron in lithium is in a 2s state and lies outside a closed shell. It doesn't take much energy to remove a 2s electron, and lithium is chemically active, with valence 1. Beryllium has a complete 2s state. But it certainly doesn't act like a noble gas, primarily because it takes very little energy to move one of the 2s electrons to a 2p state. That breaks up the symmetry and makes beryllium chemically active. And you can go on in this way to explain the chemical behavior of all the lighter elements.

The systematic chemical behavior starts to break down at potassium, with atomic number 19. Potassium has completely filled K and L shells, containing a total of ten electrons. The 3s and 3p states are also filled, for eight more. The nineteenth electron should go into a 3d state, if the idealized energy level diagram of Figure 17 is correct. Instead, the electron goes into a 4s state, which has the effect of making potassium a reactive metal, like lithium. The levels are not filled in order for the following reason: when n is larger than 2 or 3, the energy levels for different n are *so close together* that small details in the structure of the atom can slightly rearrange the order of the levels. In potassium, a 4s state has a lower energy than the 3d states and is filled first.

A similar thing happens with iron, nickel, and cobalt. In these atoms, the 4s states are filled with their complement of two electrons. The 3d states are not completely filled. The 3d states are successively filled in going from iron to nickel to cobalt. The process is shown schematically on the foldout chart at the back of this book. A similar thing happens for the rare earth elements, which start with lanthanum, atomic number 57. An inner shell has somewhat less energy than the outer shell after a few electrons have populated the outer shell. But chemical properties depend mainly on the outermost electrons because they are the first to interact when two atoms approach each other. The rare earth elements all have two electrons in the 6s level and differ only in the electron configuration of inner shells. Therefore, all rare earth elements behave about the same chemically.

The building-up method cannot be applied with accuracy to the complex atoms having high atomic numbers unless the energy level structures are known in detail. But this general discussion shows how the concepts of energy levels and shells can account for chemical properties.

WAVES AND PARTICLES

You have seen how the energy level model of the atom can be developed from experimental results by concentrating on energy. One of the surprising features of the model is its fundamental dependence on quantum numbers. How do simple numbers such as 1 and 2 and 3 come to play a role in the dynamics of an atom?

In 1924 Louis de Broglie, a young French physicist, began thinking about the double nature of light. Light acts like particles, but it also acts like waves. And the two pictures are connected by Planck's constant. In the metric system,

The Dual Nature of Light	
Particle	Wave
Photons have energy	Waves have frequency
Photon energy = 6.6×10^{-34} × wave frequency	

De Broglie wondered if "ordinary" particles, such as electrons, might have a dual nature as well and act like waves in some way. With the help of relative theory, de Broglie found that if particles do indeed have a wave aspect, the two pictures should be connected by Planck's constant. In the metric system,

The Dual Nature of Matter	
Wave	Particle
Waves have wavelength	Particles have momentum
Wavelength for a particle = $\dfrac{6.6 \times 10^{-34}}{\text{momentum}}$	

Figure 21. The French physicist Louis de Broglie theorized that every moving particle has the properties of a wave of wavelength given by $\lambda = h/mv$, where h is Planck's constant, m is the mass of the particle, and v is its velocity.

Consider what the wavelength would be for an electron moving with 10 electron volts of kinetic energy. The speed of the electron is 2×10^6 meters per second, and because its mass is approximately 10^{-30} kilogram, its momentum in the metric system is 2×10^{-24} kilogram meter per second. The wavelength comes out to be 3×10^{-10} meter, or 3×10^{-8} centimeter. But 10 electron volts is roughly the kinetic energy of the electron in a hydrogen atom. And 10^{-8} centimeter is roughly the size of a hydrogen atom. On the atomic scale, then, the wave nature of an electron cannot be neglected. That is one reason why Newton's laws, which deal only with particles, fail for atoms.

On the other hand, a wavelength of 10^{-8} centimeter is small on the everyday scale. The electrons in a television tube act like particles for all practical purposes. As a rough rule of thumb, you can't notice the wave aspect of particles or of light unless you are concerned with distances the size of a wavelength. The beam from a flashlight seems to go in perfectly straight lines. The wavelength of visible light is 5×10^{-5} centimeter, which is extremely small compared to ordinary objects. A sound wave easily goes around corners and through doorways because the wavelength of sound is 30 centimeters or so.

De Broglie was able to calculate the quantized energy levels of the hydrogen atom by combining his wave ideas with the solar system model. Recall that in the solar system model the electron travels around the nucleus in circles or ellipses. The closer the electron gets to the nucleus, the more the electron gets bound to it and the more negative its total energy becomes. One of the troubles with the solar system model is that the electron could be in any orbit. The model can't explain why only certain orbits and energies are allowed.

Figure 22. This wave pattern was one of many that de Broglie used to explain quantized energies and orbits. The pattern shown corresponds to energy level $n = 6$.

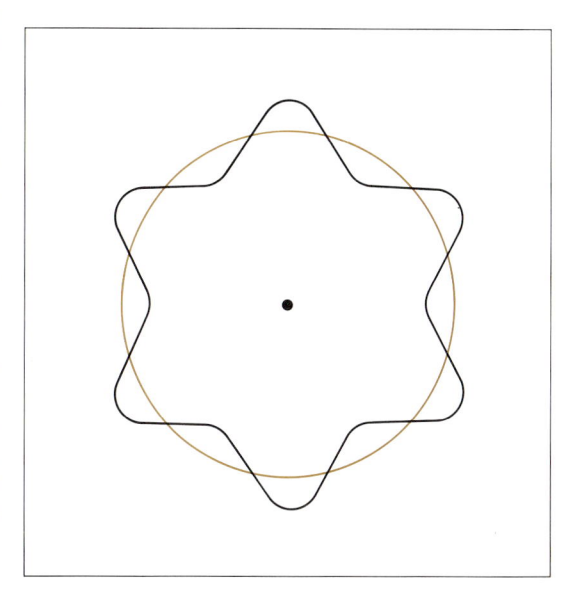

19 | The Quantum Atom 373

Relating Energy Levels to Wavelengths

When hydrogen atoms are excited by collisions, as they are in the atmosphere of the sun, photons are emitted in the transition from the $n=2$ energy level to the $n=1$ energy level. What is the wavelength of the radiation emitted? You can determine the wavelength by using the generalized Balmer formula.

<u>Step 1.</u> For the principal quantum number n, the energy level of the hydrogen atom is given by the relation Energy = 13.6 electron volts/n^2. Then the energy of the atom in the $n=1$ level is -13.6 electron volts, and the energy of the atom in the $n=2$ level is -13.6 electron volts/$(2)^2$, or -3.4 electron volts. In the transition of the atom from the $n=2$ level to the $n=1$ level, the excess energy (10.2 electron volts) is carried away by the emitted photon.

<u>Step 2.</u> To find the wavelength of a 10.2 electron volt photon, first find the frequency from the energy-frequency relation for photons:

$$\text{Energy} = (6.63 \times 10^{-34}) \times \text{frequency}$$

where the energy is expressed in joules. One electron volt is equal to 1.6×10^{-19} joule, so a 10.2 electron volt photon has an energy of 16.3×10^{-19} joule. Its frequency is then:

$$\text{Frequency} = \frac{\text{energy}}{6.63 \times 10^{-34}}$$

$$= \frac{16.3 \times 10^{-19}}{6.63 \times 10^{-34}}$$

$$= 2.46 \times 10^{15} \text{ cycles/second}$$

<u>Step 3.</u> For any light wave, the wavelength is determined by dividing the speed of light by the frequency of the wave. Then the wavelength for a photon emitted in the transition from the $n=2$ energy level to the $n=1$ energy level is:

$$\text{Wavelength} = \frac{3 \times 10^8 \text{ meters/second}}{2.46 \times 10^{15} \text{ cycles/second}}$$

$$= 1.22 \times 10^{-7} \text{ meter}$$

for each cycle. The shortest wavelength the eye can perceive is about 4×10^{-7} meter — the wavelength of violet light. Wavelengths shorter than this are called ultraviolet light. The sun contains a large amount of hydrogen and emits many ultraviolet photons with a wavelength of 1.22×10^{-7} meter. Such radiation is harmful to living cells; fortunately, most of the hydrogen photons are absorbed by oxygen molecules high in the earth's atmosphere.

Even in an excited state, an electron orbits a million times or so before a photon is emitted. De Broglie said the energy states of an atom were so stable they must correspond to a steady wave pattern, as Figure 22 illustrates. In his model, a steady wave pattern can occur only when a whole number of wavelengths fits around the orbit. The wavelength of an electron depends on its speed, but the speed depends on the size of the orbit. *So a steady wave pattern will fit only into orbits that have just the right size.* And when these ideas are put into mathematical form, they predict the energy levels of hydrogen.

The smallest orbit, closest to the nucleus, has only one complete wave. The next largest has two complete waves in its pattern. This is the significance of the principal quantum number n: *it is the number of complete waves around the orbit.* You can see how whole numbers become involved in the structure of the atom.

What does the size of the atom depend on? The size of the unit electric charge determines the relative strength of the electric force between the electron and the nucleus. The mass of the electron comes into the kinetic energy. And Planck's constant sets the scale of wavelengths. The size of the atom, and the values of the energy levels, are determined by these three fundamental constants.

De Broglie's atomic model was soon replaced by a more complete wave-particle theory devised by Erwin Schroedinger and by Werner Heisenberg in Germany. This theory does not need classical orbits to explain electron waves. It has been remarkably successful in accounting for even the finest details of atomic and molecular physics. The remaining unsolved problems in the structure of the atom appear to be connected only with calculational difficulties. The principles all seem to be known.

WAVES AND UNCERTAINTIES

You saw in the last section that an electron of moderate energy has a wavelength of about 10^{-8} centimeter. That is comparable to the spacing between the atoms in crystals, and crystals can in fact serve as diffraction gratings for electron waves. In 1927, Clinton Davisson and Lester Germer in the United States set out to study the way a beam of electrons bounced off a nickel crystal. De Broglie's idea that particles could act like waves had not been widely accepted at the time, and Davisson and Germer were surprised when they found that the electrons reflected strongly at certain angles. But their results were easy to explain with the picture of elec-

Figure 23. (Far left) Erwin Shroedinger. (Above left) Werner Heisenberg, flanked by Enrico Fermi and Wolfgang Pauli. These men contributed to the profound transformation of physics that shook our perceptions of space, time, and matter, for the more they probed the subatomic world, the more they realized that they would have to break away from traditional concepts of reality.

Figure 24. (Below) An electron with a long wavelength cannot be localized.

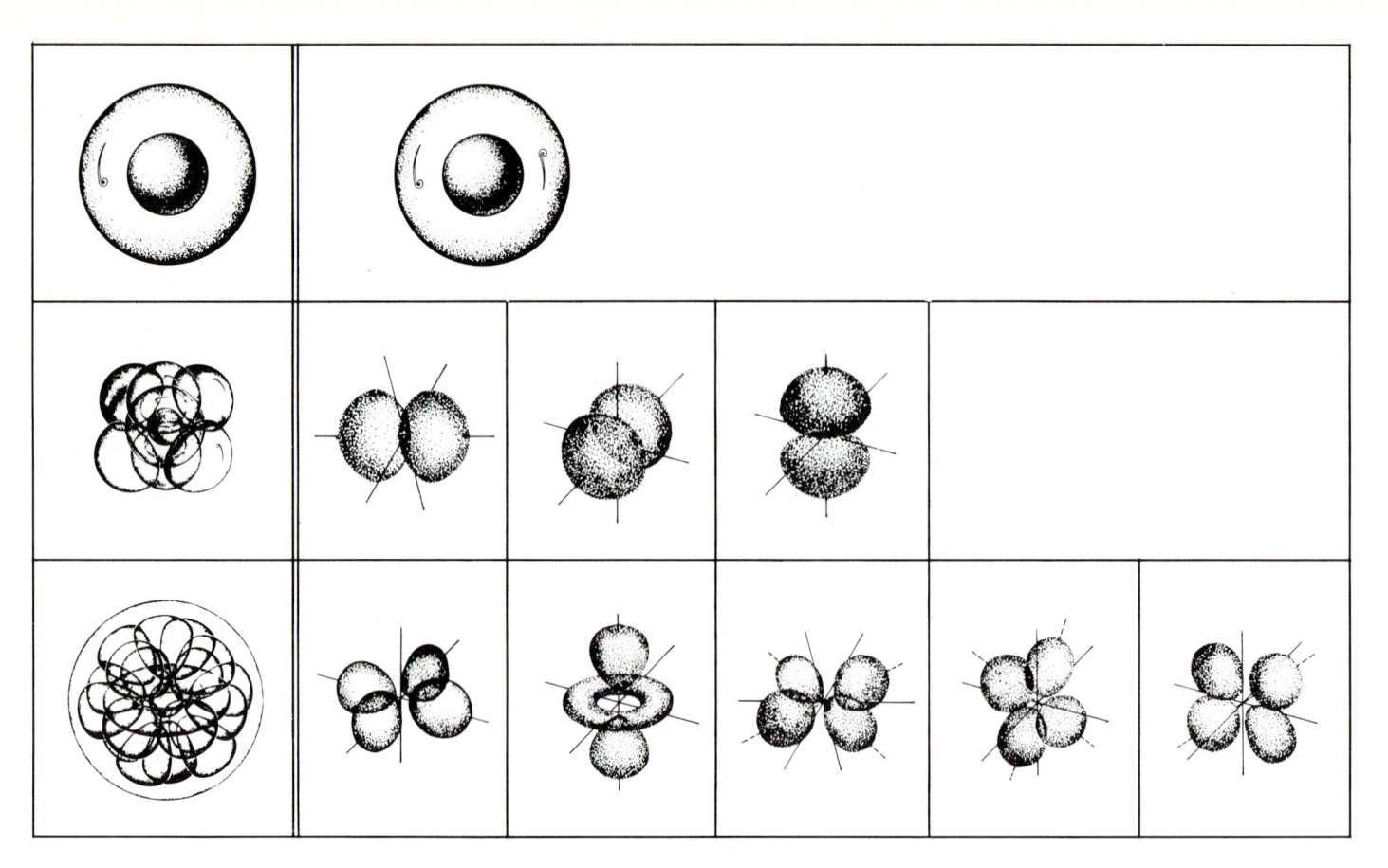

Figure 25. Electron "clouds" for the hydrogen atom in the ground state and in two excited states, showing the one shape characteristic of the s state (above), the three shapes characteristic of a p state (center), and the five shapes characteristic of a d state (below). The size of the clouds varies with principal quantum number.

Figure 26. (Below) Electrons can be described as spinning balls of charge that act like tiny magnets.

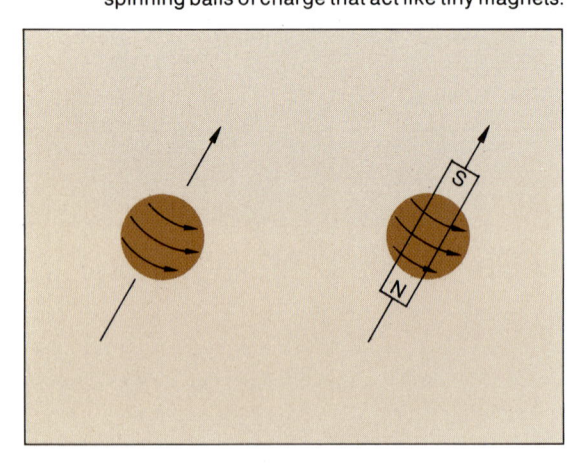

tron waves diffracting from the crystal. The wave nature of protons, neutrons, atoms, and molecules has also been demonstrated experimentally.

The wave nature of matter makes it impossible to apply the ideas of classical mechanics on the atomic scale. Classical mechanics is concerned with motion, as in the motion of a falling body or the earth around the sun. The ultimate goal of classical mechanics is to tell the position and speed of a particle, given the forces acting on it. But waves are more diffuse than particles. Consider an electron moving very slowly. Its momentum is so small that its de Broglie wavelength is large. A large wavelength means the electron is "spread out" — you can't pin down its exact *location* even if its *speed* is accurately known. Heisenberg showed that, in general,

Uncertainty in position × uncertainty in speed

is never smaller than 6.6×10^{-34} divided by the mass of the particle. This is a fundamental limitation on our knowledge. It stems from the wave nature of matter, not from imperfections in measuring instruments.

ELECTRON CLOUDS

In the modern theory of the atom, a definite orbit for the electron cannot be given. The electron waves are like a diffuse cloud smeared out over the atom, and all you can specify is the density of electrons at any point in the atom.

Figure 26 shows typical electron clouds for the hydrogen atom. You can see that the shape of the cloud depends on the quantum numbers n and ℓ. Look at the s states, for example. The radius of the thick part of the electron cloud increases with n; here is the shell structure of the atom again. When n is small, the electron spends most of its time near the nucleus.

The clouds for different ℓ values have characteristic shapes. For s states, $\ell = 0$, there is only one shape: a spherically symmetric cloud. There are three shapes for a p state, $\ell = 1$, and each of the three clouds is dumbbell-shaped. An electron in a p state has an orientation: the electron "sticks out" in two directions. Similarly, there are five shapes for d state clouds. An isolated atom is spherically symmetric, however. Electrons divide their time between the available shapes, averaging out the distribution. When atoms combine into molecules, the directed nature of the shapes is important in establishing chemical bonds, as you will see in the next chapter.

You know that an s level contains two states, and that a p level contains six states. These numbers are simply double the number of cloud shapes, so it seems that each cloud shape has the capacity to contain two electrons. But according to the Pauli principle the two electrons must differ in some way; otherwise you would have two electrons in one state.

The additional factor of difference was discovered by George Uhlenbeck and Samuel Goudsmit in 1925. They pointed out that the detailed structure of energy levels could be explained by assuming the electrons were *spinning*. Spinning charges are similar to tiny bar magnets, as shown in Figure 26. The magnetism of the electron was just what was needed to account for the behavior of an atom brought near a magnet.

But Uhlenbeck and Goudsmit's idea also showed how the two electrons in a charge cloud could be different. One electron has its magnetism pointing along one direction, and the other electron has its magnetism pointing along the opposite direction. The electrons are *paired* and the magnetic effects of one of the electrons cancel the magnetic effects of the other.

The contemporary model of the atoms was not built by arguing logically from a complete set of physical principles. As you have seen, the first models were incomplete and restricted in applicability. But the attempts to apply the models, and the experimental and theoretical work the attempts fostered, led to the discovery of new principles. And the new principles helped make later models more accurate and more universal. Ideas were important in pointing out new directions, and careful measurement and calculation were important to test the ideas. Physical science pulls itself up by its boots, creating knowledge and correcting and refining as it goes.

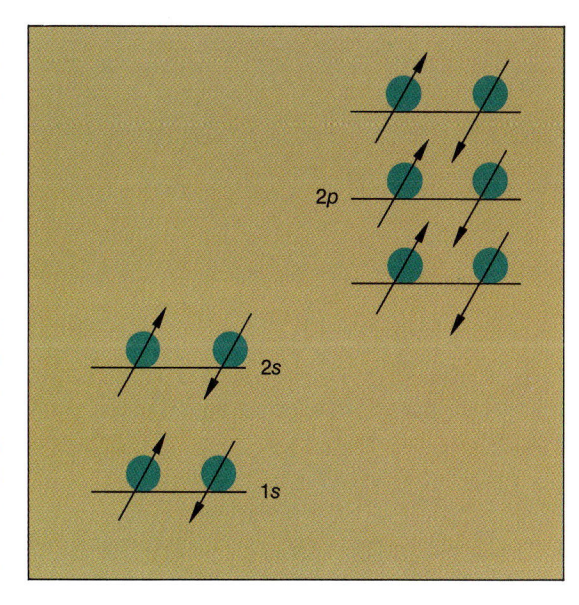

Figure 27. Electrons in filled levels pair up with their spins in opposite directions.

19 | The Quantum Atom

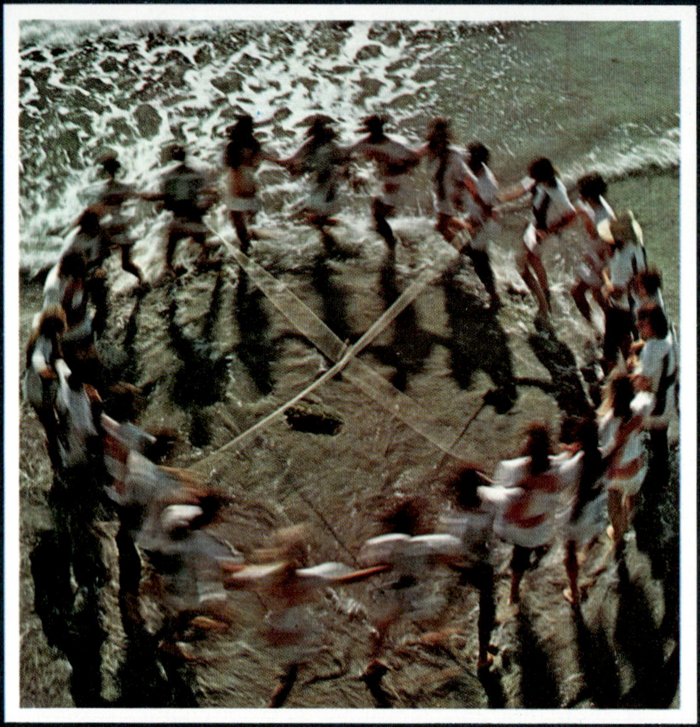

How can it happen that the rich diversity of the material world—its rocks and its seas, its grasses and its people—is created from a mere 100 kinds of atoms? How can it be that only about 20 different kinds of atoms are enough to make most of our material world? The earth's core is probably made of molten iron and nickel and its rocky crust is primarily silicon, oxygen, chlorine, carbon, aluminum, magnesium, calcium, iron, sodium, and potassium. Plants and animals (including man) consist almost totally of carbon, hydrogen, oxygen, nitrogen and, to a lesser degree, iron, chlorine, iodine, sulfur, and phosphorus. But that's just about the whole of it: the other 80 elements occur in much smaller amounts. How does this mere handful of building blocks come together to form the myriad substances of our world?

How atoms come together is a central question for chemistry. In 1661, Robert Boyle took the first step toward describing the combination of atoms by suggesting that "corpuscles" of a metal meeting with "corpuscles" of another kind of element may be more likely to join together than to remain joined in their original elemental grouping. The new form from such combinations would be as much an entity as the two kinds of "corpuscles" before they joined together. Under the spur of his suggestion, analyses of the components of various substances were undertaken and the quantitative results of such analyses eventually led John Dalton in 1801 to propose a persuasive picture of matter based on the idea of molecules. Dalton suggested that a few atoms of elementary constituents joined together to form a molecule, which he considered to be the smallest possible piece of any substance. An enormous number of identical molecules make up bulk material. Each molecule has the characteristics of the bulk material rather than of the component elements. Molecules of sugar had the properties of sugar—not of the carbon, hydrogen, and oxygen of which it was composed.

IONIC BONDS

Many materials do exist in the molecular form postulated by Dalton. But many other substances do not consist of molecules at all. For example, if you were to look at the structure of a crystal of sodium chloride, you would find that each sodium (Na) atom in the crystal is surrounded by six chlorine (Cl) atoms at equal distances. Furthermore, each of these chlorine atoms is surrounded in turn by six sodium atoms. This structure persists throughout the whole of the crystal of sodium chloride (Figure 2). It is not possible to assign any combination of sodium and chlorine atoms to particular molecular groupings. In fact, further experimentation would show you that the atoms in this crystal are electrically charged: the sodium atom has a +1 charge and the chlorine atom, a −1 charge. Such charged atoms are known as *ions* and the formation of the crystal of sodium chloride is the result of the attraction between the positive and negative ions.

If sodium chloride is heated to a sufficiently high temperature, pairs of sodium and chloride ions go into the gaseous state. In the solid form of this substance, each Na^+ ion is surrounded by six Cl^- ions and vice versa; in the gas, however, a single positive ion is associated intimately

20
Bonds Between Atoms

Atoms of carbon and hydrogen come together into a molecule.

Figure 2. (Below) Because of the way that sodium and chlorine ions are arranged in crystalline salt, each ion is associated with six of the other species and there is no way to "divide" the material into distinguishable molecules.

Figure 3. (Right) For a quantitative calculation, it is helpful to imagine the formation of a pair of associated ions in sodium chloride as taking place in three stages: (a) the expulsion of an electron from a sodium atom, (b) the acquisition of an electron by a chlorine atom, and (c) the approach of the resulting ions to each other.

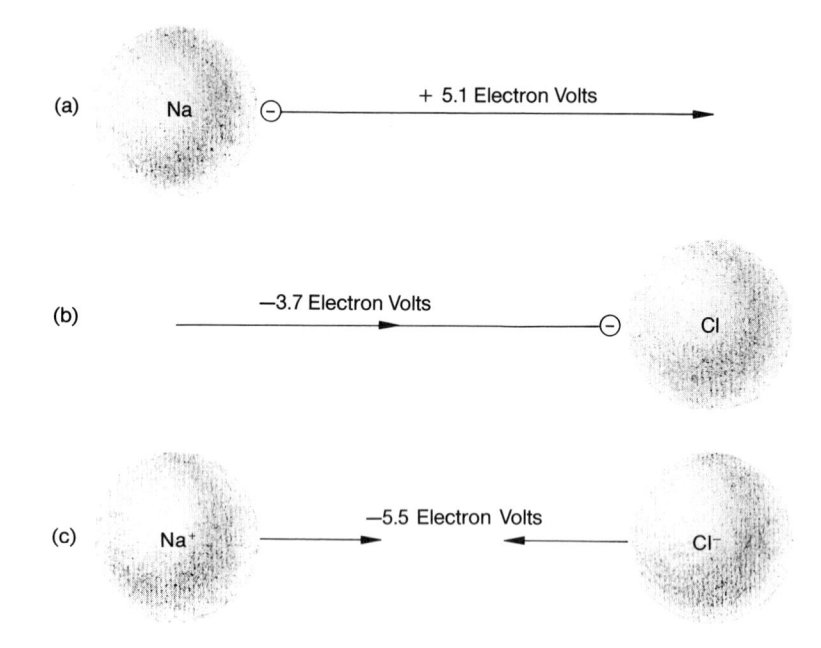

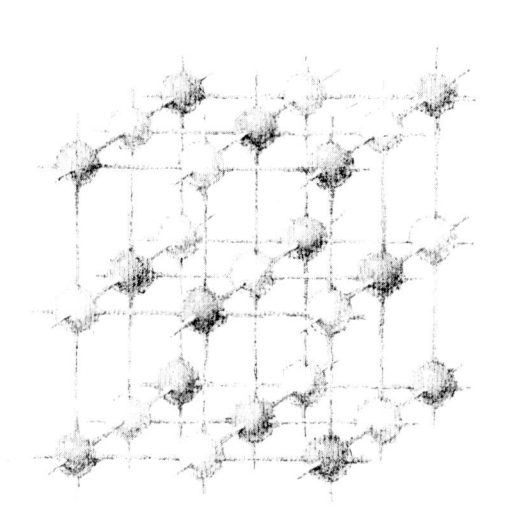

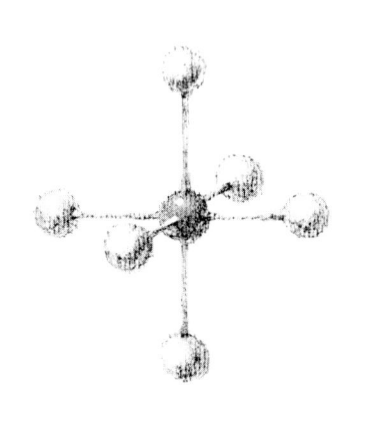

(a) Na ⊖ → + 5.1 Electron Volts

(b) −3.7 Electron Volts → ⊖ Cl

(c) Na⁺ → −5.5 Electron Volts ← Cl⁻

with a single negative ion. The attraction between a positive ion and a negative ion is known as an *ionic bond.*

What energy changes are involved in the formation of this sodium chloride ion pair from the atoms? The energy needed to remove an electron from a gaseous sodium atom to form a Na⁺ ion (known as the *ionization energy*) has a value of 5.1 electron volts. When an electron is added to a gaseous chlorine atoms to form a Cl⁻ ion, energy is released. The amount of this energy, known as the *electron affinity energy,* is −3.7 electron volts. When the attraction of opposite charges brings the Na⁺ ion and the Cl⁻ ion together in the gaseous state, 5.5 electron volts of energy is released (Figure 3). In summary, a total of 5.1 electron volts is required whereas 3.7 + 5.5 electron volts is released in the overall process of forming the ion pair from the atoms. Expressed in another way, (3.7 + 5.5) − 5.1 = 4.1 electron volts. The net release of 4.1 electron volts of energy indicates that the sodium chloride ion pair is in a more stable energy state than the individual gaseous sodium and chloride atoms.

In contrast to the situation for sodium chloride ion pairs, the ion pairs of carbon (C) positive ions and the chlorine (Cl) negative ions do *not* form. The energy needed to remove an electron from a carbon atom is relatively high, and the electron affinity energy from the formation of the Cl⁻ ion together with the combination energy of the ion pair is insufficient to overcome the large ionization energy of carbon. As a result, the overall reaction to form C⁺Cl⁻ does not occur spontaneously. But why do sodium atoms ionize with so much less energy than carbon atoms, and why do chlorine atoms release energy once they acquire an electron? These are questions of fundamental importance in understanding the formation of ionic bonds.

To understand why certain types of ions form in compounds that exist through ionic bonding, you must turn to the electron configurations of the cloud shapes, or *orbitals* of the atoms. In the preceding chapter you read that the sodium atom has two electrons in the 1s orbital, two electrons in the 2s orbital, and six electrons in the 2p orbitals, which means it has completed *K* and *L* shells. In the *M* shell, there is a single

Fuels and Bonds

By measuring bond energies — the energy released when chemical bonds are formed — you can estimate the amount of energy available from fuels. Take the example of coal or charcoal burning in air to form carbon dioxide.

<u>Step 1.</u> The carbon dioxide molecule consists of one atom of carbon joined to two atoms of oxygen by double bonds. The bond energy of a carbon-oxygen double bond is approximately 160 kilocalories per mole. If you start with free oxygen atoms and free carbon atoms and let them react to form carbon dioxide, then the amount of energy released, or the heat of formation, for every mole of carbon dioxide formed is:

$$\text{Heat of formation} = \text{number of bonds} \times \text{energy per bond}$$
$$= 2 \times 160 \text{ kilocalories/mole}$$
$$= 320 \text{ kilocalories/mole}$$

The two oxygen atoms in carbon dioxide interact very little, and their bond energy is negligible.

<u>Step 2.</u> Normally, however, you begin with solid carbon and gaseous oxygen in its usual diatomic form rather than with free atoms. It takes energy to remove a carbon atom from solid carbon, and it takes energy to tear apart an oxygen molecule. To remove a carbon atom from solid carbon requires 171 kilocalories per mole, and to dissociate an oxygen molecule requires 117 kilocalories per mole. These energies have to be supplied from the heat of formation, 320 kilocalories per mole. Hence the net energy output from burning charcoal is:

$$\text{Net energy release} = \text{heat of formation} - \text{energies needed to break existing bonds}$$
$$= 320 \text{ kilocalories/mole} - 171 \text{ kilocalories/mole}$$
$$- 117 \text{ kilocalories/mole}$$
$$= 32 \text{ kilocalories/mole}$$

<u>Step 3.</u> How would you express the energy output in terms of the amount of carbon burned?

For every mole of carbon dioxide formed, one mole of carbon is burned, because each carbon dioxide molecule contains one atom of carbon. The relative atomic weight of carbon is 12, so one mole of carbon weighs 12 grams. The energy output per gram of carbon burned is therefore:

$$\text{Energy release (per gram)} = \frac{\text{energy release (per mole)}}{\text{mass of carbon (per mole)}}$$
$$= \frac{32 \text{ kilocalories/mole}}{12 \text{ grams/mole}}$$
$$= 2.7 \text{ kilocalories/gram}$$

Positive Ion		Rare Gas Atom		Negative Ion	
Lithium	0.60	Helium	0.93		
Sodium	0.95	Neon	1.31	Fluorine	1.36
Potassium	1.33	Argon	1.74	Chlorine	1.81
Rubidium	1.48	Krypton	1.89	Bromine	1.95
Cesium	1.69	Xenon	2.09	Iodine	2.16

Figure 4. The radius of a positive ion is smaller than the radius of the rare gas atom having the same number of electrons. The radius of a negative ion is larger. The radii are given in angstrom units (1 angstrom = 10^{-8} centimeter).

electron in the 3s orbital. If sodium were to gain an electron, thereby having a -1 charge, the 3s orbital would be filled but the M shell would still be incomplete because the 3p orbitals would be vacant. However, if sodium *loses* an electron from the 3s orbital, the resulting positive ion has an electronic configuration of completed K and L shells and a totally empty M shell. This is the same electronic configuration possessed by the atoms of the unusually stable element neon. The stability of this filled shell configuration is reflected by neon's lack of any chemical reactivity.

The chlorine atom has the following configuration: two electrons in the 1s orbital; two electrons in the 2s orbital; six electrons in the 2p orbital; two electrons in the 3s orbital; and five electrons in the 3p orbital. If chlorine were to lose electrons in order to acquire the configuration of an inert gas atom, it would have to lose all the electrons in the M shell. To do so would mean that it would have a +7 charge. However, the energy required to lose seven electrons would be exceedingly high. On the other hand, chlorine can complete the M shell by acquiring only one additional electron. The configuration with completely filled K, L, and M shells corresponds to the atoms of the inert gas argon. That is why sodium tends to form an ion with a +1 charge and chlorine tends to form an ion with a -1 charge. The formation of these ions results in filled shells that are highly stable electronic configurations. If carbon loses an electron, it has a +1 charge. Its electronic configuration is that of the boron atom, which does not represent any particular stability. Consequently, the energy needed to remove an electron is quite large.

There is a difference between ions having the same electronic configuration as an atom of an inert gas and the actual atom of that gas. Positive sodium ions have the same electronic configuration as neon atoms; both have ten electrons about the nucleus. However, a sodium nucleus has a charge of +11, and it attracts the ten electrons somewhat more than a neon nucleus, charge +10, and pulls them in a little closer. As a result, the positive sodium ion has a smaller radius than the atomic radius of neon. Conversely, the negative chlorine ion has eighteen electrons and a nucleus of +17 charge, whereas argon has eighteen electrons surrounding a nucleus with a +18 charge. The surplus negative charge in the negative chlorine ion causes an expansion in the electron cloud, so the radius is larger than it is for the argon atom. Figure 4 compares the radii of positive and negative ions with the radii of the inert gas atoms of the same electronic configuration. This figure shows that positive sodium ions are approximately half as large as negative chlorine ions. In the formation of a crystal of sodium chloride, negative chloride ions occupy the bulk of the crystal's volume and the positive ions may be viewed as fitting into the space between the bulkier negative ions. Such a configuration is shown in Figure 5.

COVALENT BONDS

Chlorine gas consists of pairs of chlorine atoms. If you look at the formation of the ion pair Na⁻ Cl⁺ just as you did for the formation of an ion pair of Cl⁺ Cl⁻, you will see that the chlorine ion pair is not favored energetically. The energy needed to move an electron from a gaseous chlorine atom is 13 electron volts. The energy released when a chlorine atom acquires an electron, thus forming a Cl⁻ ion, is only 3.7 electron volts. If you assume that the combination of a positive and a negative chlorine ion releases about the same amount of energy as the combination of a positive and a negative sodium ion (5.5 electron volts), it follows that

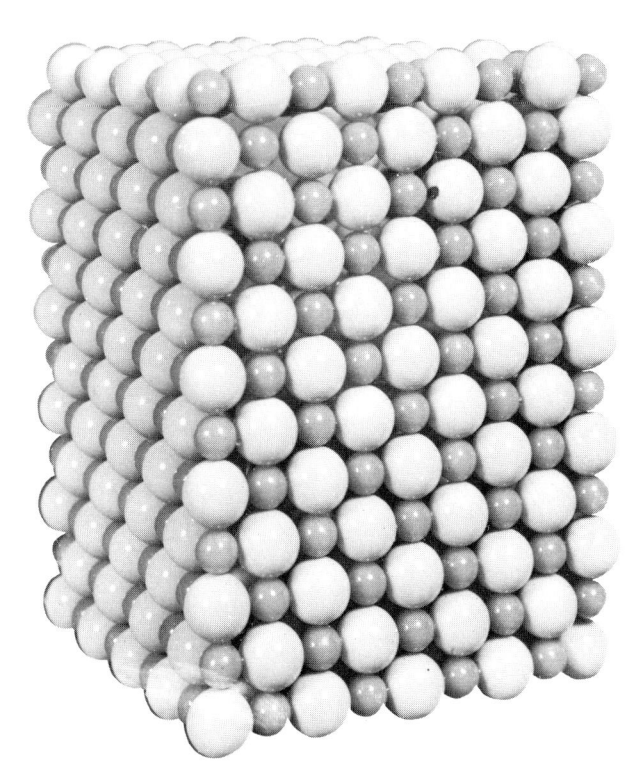

Figure 5. A "packing model" of sodium chloride is another way of depicting the arrangement of Figure 2. The smaller sodium ions are tucked in among the larger chlorine ions.

the overall energy for the formation of the Cl$^+$ and Cl$^-$ ions from the gaseous atoms would then require 3.8 electron volts. In other words, $13 - (3.7 + 5.5) = 3.8$. Therefore, the ion pair is at a higher energy state than the separate atoms and is not favored energetically.

We know from experiments, however, that chlorine atoms do exist in the gaseous state in the form of a combination of two atoms per unit and that energy is released when the pair of atoms combine. Because the result cannot be an ion pair, it would be safe to assume that the pair of chlorine atoms exists as a molecule with a molecular bond, not an ionic bond, holding them together. Molecular bonds are called *covalent bonds*.

Studies of the nature of the covalent bond indicate that the two atoms held together by such a bond *share* a pair of electrons. The attraction to electrons must be identical for two chlorine atoms, so it's reasonable to suppose that when these chlorine atoms approach each other, electrons would not be given up or acquired. In the case of sodium and chlorine atoms, the tendency to attract electrons is much stronger for the chloride atom whereas sodium gives up electrons readily. As a result, they do not share electrons equally and sodium completely gives up an electron to the chlorine atom. In the case of the chlorine *molecule,* however, electrons would have to be shared equally by the two identical atoms. *A covalent bond is defined as one formed between two atoms in which a pair of electrons are shared more or less equally by the two atoms.*

Does the formation of covalent bonds fit the model used to explain the formation of ionic bonds? As you have learned, chlorine atoms can complete the *M* shell by acquiring one additional electron. Now assume that the two chlorine atoms must share the electron pair of the covalent bond. If each chlorine atom donates one electron to the pair, and if you count both electrons of the bonding pair for each chlorine atom, the result would be that each chlorine atom would have its original seven

Figure 6. The formation of a covalent bond by two chlorine atoms gives completed *M* shells for both chlorine atoms.

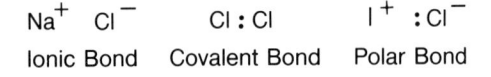

Figure 7. Ionic bonds exist between ions with no sharing of electrons. Pure covalent bonds hold atoms together by equal sharing of a pair of electrons. Polar bonds involve unequal sharing of the pair of electrons, which causes one atom to be slightly positive and the other slightly negative.

electrons in the *M* shell *plus* the extra electron that is shared by forming a covalent bond with the other chlorine atom. In Figure 6, the outer electrons of each chlorine atom are shown with the bonding pair of the covalent bonds located between the two chlorine atoms. The total number of electrons surrounding each chlorine atom in Figure 6 would be eight. By sharing electrons in covalent bonds, each chlorine atom seemingly acquires the electron configuration of the atoms of the inert gas argon. Thus, by sharing electrons in covalent bonds, atoms again may acquire stable, filled electronic shell configurations.

INTERMEDIATE POLAR BONDS

If there is a large difference in the tendency of two atoms to gain or lose electrons, then electrons would be lost by one atom and gained by another to form ions, and the two atoms would be attracted to each other by ionic bonding. However, if the two atoms have *equal* tendencies to lose or gain electrons, neither atom can take electrons from the other. The result would be that the two atoms would form molecules with covalent bonds in which the electrons would be shared. As you might expect, there are many pairs of atoms in nature that are found between these two extremes. For such pairs, there is a difference between the two atoms in the ability to gain and lose electrons, but this difference is not large enough to cause an actual gain or loss of electrons to form ions. Such atoms form intermediate bonds in which the electrons are not completely shared as they are in the case of chlorine molecules, nor are they completely gained or lost as in the case of the sodium chloride ion pair. Rather, for such pairs, bonds are formed in which the electrons are shared but not equally so. The atom that has the greater tendency to gain electrons is more likely to attract the shared pair of electrons than the atom that has less tendency to attract electrons. *Such unequal sharing is called polar bonding* (Figure 7).

What happens when polar bonds are formed and the sharing of the electron pair is unequal? The atom having the greater attraction to electrons acquires a slight excess negative charge because the electron cloud density is somewhat higher around that atom. Conversely, the atom with the lower attraction for electrons acquires a slight positive charge because it loses some electron cloud density to its bonded partner. In a diatomic molecule having a polar bond, the result is a positive center on one atom and a negative center on the other. Such diatomic molecules are known as *dipoles*. When placed in an electric field, they line up the way compass needles line up along the earth's magnetic field. The extent of the unequal sharing of charge is measured by the *dipole moment*, which is a measure of the tendency of the molecule to line up in an electric field.

Fluorine atoms have a stronger attraction for electrons than do chlorine atoms because the fluorine atom is smaller than the chlorine atom and its electrons are closer to the nucleus. In following this line of reasoning, you would expect that the attraction for electrons would decrease, in order, through the five members of the halide family: fluorine to chlorine to bromine to iodine to astatine. If we consider the gaseous molecules formed by these elements with hydrogen, all of the molecules are polar because all of these halide elements attract electrons more than hydrogen atoms. When these molecules are placed in electric fields, the dipole moment of the molecules can be measured. The molecule hydrogen fluoride (HF) has about twice as large a dipole moment as the molecule hydrogen chloride (HCl), approximately 3 times that of hydrogen bromide (HBr), and about 8 times that of hydrogen iodide (HI). Remember that the dipole moment measures the polarity of the bond.

If we were to conclude there were only ionic bonds and pure covalent bonds, it would be difficult to fit these dipole moment measurements into a logical scheme; given the gradual variation in properties of atoms, we would expect that there *should* be a gradual difference in the relative attraction of electrons between different pairs of atoms. It would be unreasonable to expect that the electrons would have to be either equally shared, as in a pure covalent bond, or totally swapped, as in an ionic bond.

Molecules with dipole moments attract one another because the positive end of one molecule is attracted to the negative end of an adjacent molecule. That is the origin of the attractive force holding molecules together in liquids and solids. As the temperature increases in a molecular solid held together by dipole attraction, the molecular vibrations tend to disrupt the structure until, at a sufficiently high temperature, melting occurs. Further increases in the temperature of the liquid cause even more violent vibrations of the molecules until eventually the dipole attraction holding the molecules together is overcome and boiling occurs as individual molecules escape.

DISPERSION FORCES

Not only do molecules have permanent dipole moments due to the unequal sharing of electrons of covalent bonds; they also have *fluctuating* dipole moments. Indeed, even isolated atoms that have no permanent dipoles have such temporary dipoles. In such cases, the electrons are moving about the atom in a complicated fashion, leading to their description as "electron clouds." And such electron clouds can fluctuate in their average position about the nucleus in such a way that, at any specific moment, there may be a slight excess of the electron cloud on one side of the nucleus compared to the opposite side. When that happens, the side with the excess electron cloud density will be negative, and the other side will be positive. For that particular instant, the atom has a dipole moment. However, that dipole is quickly destroyed as the electron cloud oscillates into a new momentary pattern, establishing a different dipole. So the dipole rapidly oscillates in direction as the electron cloud fluctuates from one side of the atom to the other (Figure 8).

The fluctuating dipole moment of an atom can lead to a steady attractive force between separated atoms. To see how, suppose atom *A* is momentarily negative on the side adjacent to atom *B*. This causes the positively charged nucleus of *B* to be attracted toward *A*, and the negative charge cloud of *B* to be pushed away from *A*. In this way, atom *B* becomes momentarily polarized, and is attracted toward *A*. As Figure 9 shows, the force is also attractive when the far side of atom *A* is negative. Such an attractive force is responsible for the existence of the inert gases in the liquid and solid state. Even though the individual atoms of inert gases are uncharged and unpolarized on a time average, at any instant they do have a dipolar nature due to this oscillatory behavior. This oscillating dipole attraction is known as the *dispersion force*. The dispersion force is much weaker than the forces holding atoms together in ionic and covalent bonds, and it is even weaker than the dipole attraction in molecules with polar bonds. However, it does add to the effect of the stronger forces in all these situations. When a neutral molecule is large and contains many oscillating electronic cores about the nuclei, the force of attraction with neighboring molecules becomes large enough to hold molecules

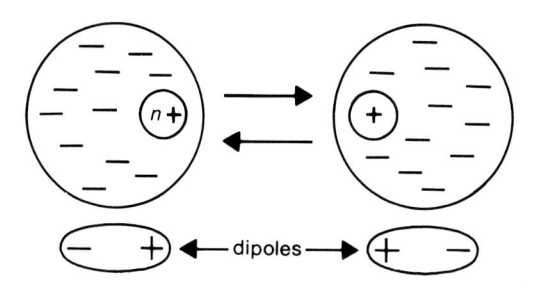

Figure 8. This sketch depicts the formation of temporary, oscillating dipoles by movement of the electron cloud about the nucleus.

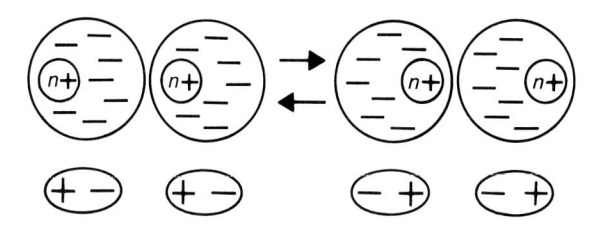

Figure 9. The fluctuating motion of the electrons in atom *A* induces a dipole moment in atom *B*. The net force is always attractive.

together even at room temperature. Paraffin wax used in candles and for sealing jars of jam is a familiar example of a solid held together by dispersion forces.

POLYATOMIC MOLECULES AND BONDING

Frequently, the formation of a single covalent bond in a diatomic molecule is inadequate to provide enough shared electrons to fill the electronic shells for the bonded atoms. For example, nitrogen has an outer electronic configuration of two electrons in the s orbital and three electrons in the $2p$ orbitals. If it forms a molecule with hydrogen with one electron in the $1s$ orbital, the diatomic molecule would provide nitrogen with only six total electrons in the L shell. Because nitrogen requires eight electrons to complete this shell, it must seek to share two more electrons. As a result of the stability associated with the filling of electronic shells, the molecule formed between nitrogen and hydrogen is actually NH_3, ammonia, in which a nitrogen atom is bonded by three covalent bonds to three hydrogen atoms. In an ammonia molecule, the nitrogen atom has its five original electrons plus the three new electrons it shares in the covalent bonds, which gives nitrogen a filled L shell. Each of the three hydrogen atoms has a completed K shell by virtue of the shared pair of electrons of the covalent bond, giving them the electronic configuration of the inert gas helium.

Following this line of reasoning, you can understand why oxygen, which has six electrons in the L shell, forms the water molecule H_2O, whereas carbon, which has four electrons in the L shell, forms the methane molecule CH_4.

There still seems to be a problem in explaining why some diatomic molecules are formed. For example, nitrogen exists as the diatomic molecule N_2. If these two atoms are sharing a single pair of electrons, then each atom could count only six electrons in its outer L shell. This puzzle was solved in the case of ammonia when it was recognized that nitrogen bonds with three hydrogen atoms rather than a single atom. But in nitrogen gas, we know that only two atoms are bonded together in each molecule. The idea that the completion of electronic shells establishes the stability of the molecule is still valid if we allow not a single pair of electrons to be shared between the two nitrogens but rather *three* pairs of electrons, forming three covalent bonds. That is known as a *triple bond* (Figure 10).

In the case of the carbon dioxide molecule, CO_2, the bonding must also be explained in terms of multiple sharing of electrons between the adjacent atoms. Carbon has four electrons in the L shell and therefore must share four more if it is to have a completed shell. The oxygen atoms have six electrons in the outer L shell and must share two more. If two covalent bonds exist between each carbon and oxygen pair, then the carbon shares two additional electrons from each oxygen atom and each oxygen shares two additional electrons from the carbon atom. The result is that the carbon atom shares the necessary four additional electrons from the two oxygen atoms while each oxygen shares the necessary two additional electrons with the carbon atom. The double bonds existing between the carbon and oxygen atoms provide enough shared electrons to complete the electronic shells of the carbon and oxygen atoms.

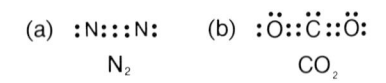

(a) :N:::N: (b) :Ö::C::Ö:
     N_2 CO_2

Figure 10. (a) This drawing shows the formation of a triple bond in a nitrogen molecule. (b) Two double bonds in carbon dioxide give the atoms filled L shells.

EXPLANATION OF THE ATTRACTION IN COVALENT BONDS

To understand why the sharing of a pair of electrons in a covalent bond holds atoms together in molecules, consider the hydrogen molecule

H_2. The hydrogen molecule consists of two nuclei and two electrons. The two positive nuclei repel one another, as do the two electrons. However, each nucleus attracts the two electrons. The fact that the hydrogen molecule exists must mean that the attraction of the two nuclei for the two electrons is greater than the repulsion between nuclei and between electrons. When two helium atoms are brought close together they do not form a molecule. If a molecule such as He_2 were to form, there would be two nuclei of +2 charge—each would repel the other and the four electrons that could serve to attract the nuclei would also have a repulsion force between them. Because He_2 does *not* form, you may conclude that the repulsive forces exceed the attractive forces.

If the attraction of the nuclei to the shared bonding electrons is greater than the repulsion between the nuclei, it is reasonable to expect that these electrons would most likely be found between the two nuclei. You read earlier that it is not possible to define exactly a position for an electron in an atom. The probability of electron distribution must be taken into account, which leads to a consideration of electron clouds or electronic charge densities. Therefore, the hydrogen molecule must be described as one in which the electronic charge density is highest in the region between the two nuclei and lower elsewhere. In contrast, in an isolated hydrogen atom the electronic charge density is equal on *all* sides of the nucleus. Therefore, the formation of a covalent bond results in a localization of electronic charge density in the region between the nuclei of the bonded atoms. If you draw a cross-section of the electronic charge density about the nucleus, you get the results shown in Figure 11(a) for the isolated hydrogen atom. When two such atoms are brought closer together and a covalent bond forms, the electronic distribution of the two atoms overlap and increases in the region between the two nuclei, as shown in Figure 11(b). Notice that in describing how two atoms are attracted and held together in a covalent bond, the same basic form of attraction that exists between ions—the electrostatic force—is used. Nuclei are held together by covalent bonds because of the attraction of the positive nuclei and the negative charge of the bonding electrons.

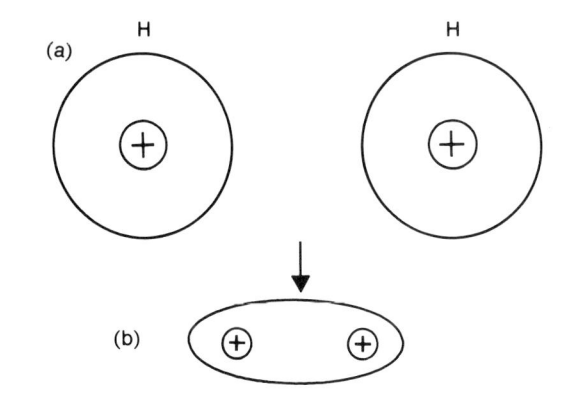

Figure 11. (a) Isolated hydrogen atoms have symmetric charge distribution about their nuclei. (b) To understand the bonding between two hydrogen atoms, you must take into account the attraction of both nuclei for the two electrons as well as the repulsion between the nuclei and between the electrons. In the end, you find that the hydrogen molecule can be formed with a configuration allowing the two electrons to have an increased charge distribution between the two nuclei.

MOLECULAR BONDING GEOMETRY

In polyatomic molecules, the atoms either must arrange themselves randomly about one another with no set pattern or they must always tend to be arranged in the same pattern with the same geometry. A study of the shapes and geometries of molecules quickly shows that the latter is indeed the case. When a molecule such as ammonia forms, the three hydrogen atoms are bonded to the nitrogen atom in such a way that a pyramid is formed with the nitrogen at the apex. In similar fashion, when the methane molecule CH_4 forms, the four hydrogen atoms are always found to be grouped tetrahedrally and at equal distances about the central carbon atom. In the molecule of water, the two hydrogen-oxygen bonds make an angle of approximately 104 degrees (see Figure 12). Many of the properties of molecules may be understood once you know the geometry of the atoms in the molecules.

A relatively simple molecule is that of hydrogen sulfide, H_2S, in which the two hydrogen-sulfur bonds are at an angle approximately 92 degrees apart. What is the experimental evidence that leads us to believe that hydrogen sulfide has a bent shape (Figure 12) rather than a configuration in which the two hydrogen atoms and the sulfur atom are along the same line? It can be readily ascertained that hydrogen sulfide has a dipole moment, which indicates the existence of an unequal distribution of charge in the molecule. If the molecule were to have a linear

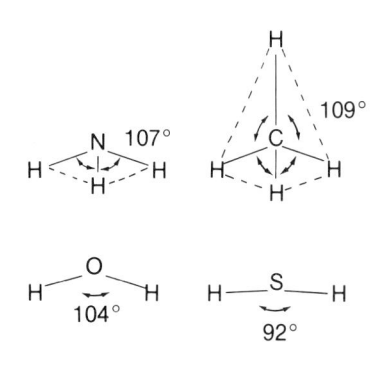

Figure 12. The shape of ammonia (pyramidal), methane (tetrahedral), water (bent), and hydrogen sulfide (bent).

20 | Bonds Between Atoms

shape with the two hydrogen atoms on opposite sides of the sulfur atom, then the polarity of the two covalent bonds would be in a direction that would oppose and cancel each other, giving a molecule with zero dipole moment. Moreover, the magnitude of the dipole moment is larger than would be expected from the unequal distribution of charge between a single hydrogen-sulfur bond. The measured dipole moment for hydrogen sulfide can only be explained if you presume that the molecule has a bent shape, so that the unequal distribution of electronic charge in the two bonds reinforces each other and produces a larger net dipole moment than either bond would produce by itself. Determination of the position of the hydrogen atoms around the sulfur atom by other physical means confirms the bent shape of the molecule.

It's possible to understand the formation of a molecule in terms of the stability of the closed shell configurations. In the case of the hydrogen sulfide molecule, the sharing of pairs of electrons with hydrogen produces an electronic configuration about the sulfur nucleus that is identical to the electronic configuration of the inert gas argon. If this explanation is valid, the theory should also explain the bent structure of the hydrogen sulfide molecule. The electrons in the outer orbitals of the sulfur atom taking part in the formation of the covalent bonds are in the $3p$ orbitals. In the preceding chapter you read that the p orbitals have directional properties and are at right angles to one another. Therefore, if the hydrogen sulfide molecule forms by the sharing of electron pairs in covalent bonds using two of the $3p$ orbitals of sulfur, the hydrogen-sulfur bonds should be at right angles to one another. The experimental angle of 92 degrees is sufficiently close to the right angle value of 90 degrees to justify this explanation.

The molecule phosphine, PH_3, has a pyramidal shape with the three hydrogen-phosphorus bonds at 91 degrees with respect to one another. In phosphorus there would be three electrons in the $3p$ orbitals, which would be shared with the electrons in the $1s$ orbitals of hydrogen. The result would be the formation of three covalent bonds using the three $3p$ orbitals of the phosphorus atom. Because these $3p$ orbitals are at 90 degrees to one another, the hydrogen atoms in such a molecule would form the base of a pyramid with the phosphorus atom, and the angle between them would be 90 degrees. Experimentally, the angles between the bonds are found to be 91 degrees. This agreement is further evidence that the directional properties of the orbitals involved in the formation of the covalent bonds determine the geometry of the molecule.

Can this model explain the geometry of other molecules such as methane? The methane molecule, CH_4, has a tetrahedral geometry with the carbon atom in the center of the tetrahedron and angles of 109 degrees between the hydrogen-carbon bonds. The carbon atom has two electrons in the $2s$ orbitals and two electrons in the $2p$ orbitals. You might expect to find the p orbitals used with the formation of a molecule between the carbon atom and two hydrogen atoms if the covalent bonds have to be formed by sharing one electron from hydrogen and one electron from carbon for each bond. However, that arrangement does not give carbon a filled L shell.

But consider what would happen if the s orbital and the p orbitals of carbon were involved in the formation of methane. The $2s$ orbital and the three $2p$ orbitals would combine to form four new *hybrid* orbitals. When the properties of these hybrid orbitals are actually calculated, it turns out that they are indeed identical, with equal distances between the hydrogen and carbon nuclei for all four bonds and with equal bond angles. In fact, the geometry would be that of a tetrahedral configuration, or 109-degree angles. This geometry conforms to the experimental observations for the structure of methane, and it provides evidence for the proposal that hybrid orbitals form from the original s and p orbitals.

By considering the geometry of atomic orbitals directly or the geometry of hybrid orbitals formed by combining atomic orbitals, it's possible to understand the molecular geometry of even the most complex molecules. This application of the theory of quantum mechanics has been immensely valuable to chemists because it provides them with great insights into molecular behavior. It has also enabled chemists to predict the properties of hitherto-unknown molecules and has led to the synthesis in the laboratory of a wide variety of new materials, including the plastics we use every day. At present, it is also contributing greatly to our understanding of the behavior of large biological molecules in living organisms.

HYDROCARBONS

Millions of different chemical compounds are known. That is an impressive number, and it can be accounted for by examining carbon atoms, which have the ability to join to one another in an incredible variety of chains and rings. Carbon is not unique in this ability; germanium and a few other elements will also form chains. Unlike carbon chains, however, such chains are limited to lengths of two to six atoms. In contrast, chains of thousands of carbon atoms are common in large biological molecules. The importance of the chain and ring compounds of carbon is reflected by the fact that they constitute a whole distinct area of chemistry with its own special name: *organic chemistry.* In organic molecules, a variety of other atoms such as hydrogen, nitrogen, and oxygen are connected to the backbone chain or ring of carbon atoms.

The compounds formed when hydrogen atoms are connected to the carbon chains are known as *hydrocarbons.* Of the hydrocarbons, the simplest member is methane, which has only one carbon atom with four hydrogens connected to it in a tetrahedral structure. As the number of atoms increase in the chain, the compounds methane, propane, butane, pentane, and so on are produced in succession (Figure 13). Nor do chains have to be linear: they can have side branches yielding different arrangements of the same number of carbon and hydrogen atoms. Such compounds are called *isomers.* There are still other hydrocarbons in which the carbon atoms exist in closed rings that may or may not have side chains of carbon atoms. In some cases, these rings will be fused together to form a multi-ring system. A group of hydrocarbons that have been discovered recently have rings linked through each other. They are called the *catenanes.*

These complex molecules should be viewed in three-dimensional space rather than in planes. Because of the tetrahedral orientation of the carbon atom's hybrid bonding orbitals, they do not lie in straight lines or planes. Even with the fixed angles between the bonding orbitals and the fixed distances between the nuclei of the bonded pairs of atoms, flexibility exists in the shapes of the molecules because there can be *rotation* about a bond. As Figure 14 suggests, a group of atoms may revolve about the bond, connecting it with another group when the atoms do not interfere with one another in this rotation. In molten, long-chain hydrocarbons such as melted paraffin, the molecules flex, coil, and move about. In some cases, however, the bonding in the molecular structure is so arranged that portions of the molecule can't revolve around the

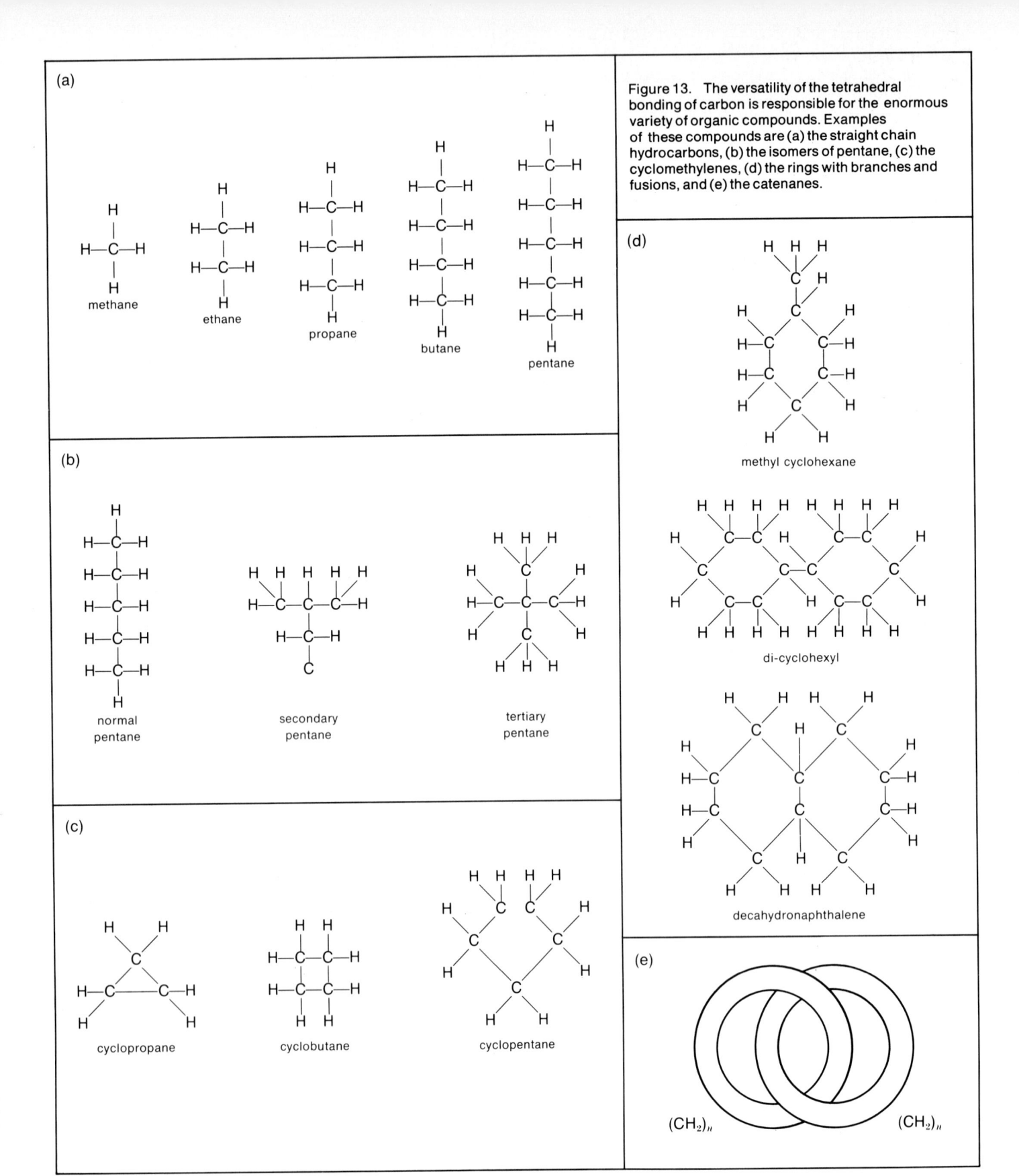

Figure 13. The versatility of the tetrahedral bonding of carbon is responsible for the enormous variety of organic compounds. Examples of these compounds are (a) the straight chain hydrocarbons, (b) the isomers of pentane, (c) the cyclomethylenes, (d) the rings with branches and fusions, and (e) the catenanes.

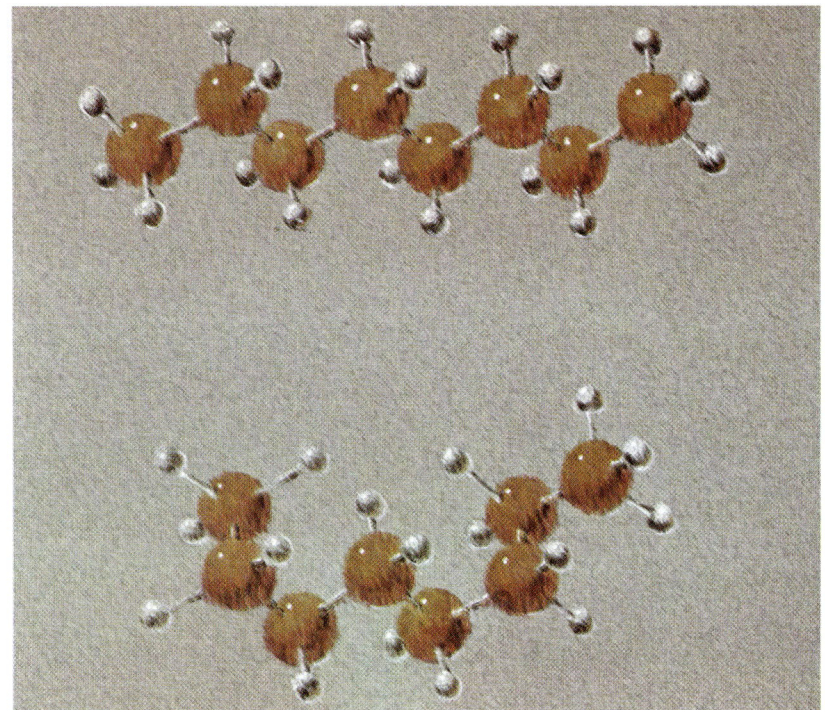

bonds. For example, the hydrocarbon adamantane behaves as if it were a rigid molecule because the bonding does not permit rotation (Figure 15).

NAMES FOR ORGANIC COMPOUNDS

Organic compounds that are not made up solely of carbon and hydrogen atoms have structures analogous to parent hydrocarbon structures. In such compounds, some of the hydrogen atoms in the parent hydrocarbon structure are considered to be "replaced" by other groups of atoms, even though that might not be the procedure used to synthesize the compound in the laboratory. This way of thinking is reflected in the names that have been given to compounds. It has actually reduced the confusion of having separate names for the almost infinite number of organic compounds. In fact, it even lets you know the structure of the molecule on the basis of the name alone.

The nomenclature system is as follows. The substituent groups replacing the hydrogen atoms in organic molecules are named with a *number* that tells which carbon atom in the chain they happen to be bonded to. For example, in the compound shown in Figure 16, the bromine atom is connected to the third carbon atom of the chain; and the *hydroxy,* or OH group, is connected to the first atom in the chain. This compound would be named 1-hydroxy-3-bromo-hexane.

Figure 14. (Above left) Normal octane is an example of how a molecule changes shape when one part rotates about a bond to the other part.

Figure 15. (Above right) In adamantane, the bonding holds all the carbon atoms in relatively fixed positions.

```
    H  H  H  Br H  H                          H  H  H
    |  |  |  |  |  |                          |  |  |
H - C- C- C- C- C- C- OH              H - C - C - C - NH
    |  |  |  |  |  |                          |  |  |
    H  H  H  H  H  H                          H  H  H
```
1-hydroxy 3-bromo hexane   1-amino propane or propylamine

Figure 16. (Left) Organic compounds are named to reflect their structure. For example, 1-hydroxy, 3-bromohexane contains one OH radical and a bromine atom bonded to the *third* atom of six carbon atoms. As another example, propylamine could also be called 1-amino propane.

What about shorter chains? Rather than tacking on the hydrocarbon name at the end, chemists frequently replace the "-ane" of the hydrocarbon chain name with a "-yl" and use the chain name as a prefix. For example, the compound with the amino, NH_2, group at the end of the

Figure 17. Several functional groups commonly occur in organic molecules.

(a) —CH₃   —CH₂—CH₃   —OH   —NH₂   —NO₂   —COOH
     methyl   ethyl   hydroxy amino   nitro   carboxy

(b) CH₃—CH₂—CH₂—CH₂—CH₃
     1   2   3   4   5

chain of three carbon atoms is more commonly known as propylamine rather than 1-amino-propane. Such common names are persistent for older organic compounds. No one speaks of the compound based on ethane and the hydroxy group as 1-hydroxy-ethane but rather uses the common name eth*yl* alcohol. Figure 17 shows the structures of some common groups together with their names.

RIGHT HAND, LEFT HAND, AND LIFE

Suppose you make a compound in the laboratory with a molecule that has four different substituents. They might be either single atoms or groups of atoms replacing the four hydrogen atoms in methane. Call the substituents *W*, *X*, *Y*, and *Z*. Possible structures are shown in Figure 18.

Now in three dimensions, the structure can take two distinguishable forms with the *same* bond lengths and angles. One form is the mirror-image of the other but neither can be turned in space so that it looks the same as the other. In order to make them the same, you would need to interchange two of the substituents (they could be any two) on one or the other of the molecules. Their relation is like that of a right-hand glove and a left-hand glove.

The two compounds will melt and boil at the same temperatures. They will undergo the same chemical reactions with most reagents. Unless there is some reagent that would react differently to the two mirror-image forms, you would have no way of separating them.

Louis Pasteur obtained tartaric acid as a by-product of making wine from grapes and he discovered that most of it was right-handed. His test was to *polarize* a beam of light into a plane by sending it through special crystals and then to send the light through a solution of the tartaric acid. He observed that the plane of polarization of the emerging beam was rotated clockwise. He was able to isolate a small amount of acid that had no effect on the light; in other words, it was optically inactive. Allowing the mold *Penicillium glaucum* to grow on a solution of that optically inactive acid, he found that the remaining acid rotated polarized light in the *opposite* way. Finally, he prepared a fourth form of the acid that was optically inactive but differed in its crystalline shape from the other inactive form.

Looking at the molecular structure of tartaric acid in three dimensions (Figure 19), you can see the structural origin of these differences. The two ends of the molecule are alike, and each end has one carbon atom about which the four substituents differ as *W*, *X*, *Y*, and *Z*. If the substituents are arranged in the same way about both atoms, that arrangement will be either right- or left-handed and the effects of the two ends on polarized light will add together. If the resulting left- and right-handed acids are mixed together in equal proportion, their effects

Figure 18. When four different groups are bonded to a carbon atom, the resulting molecule can take two different forms that are mirror images of each other.

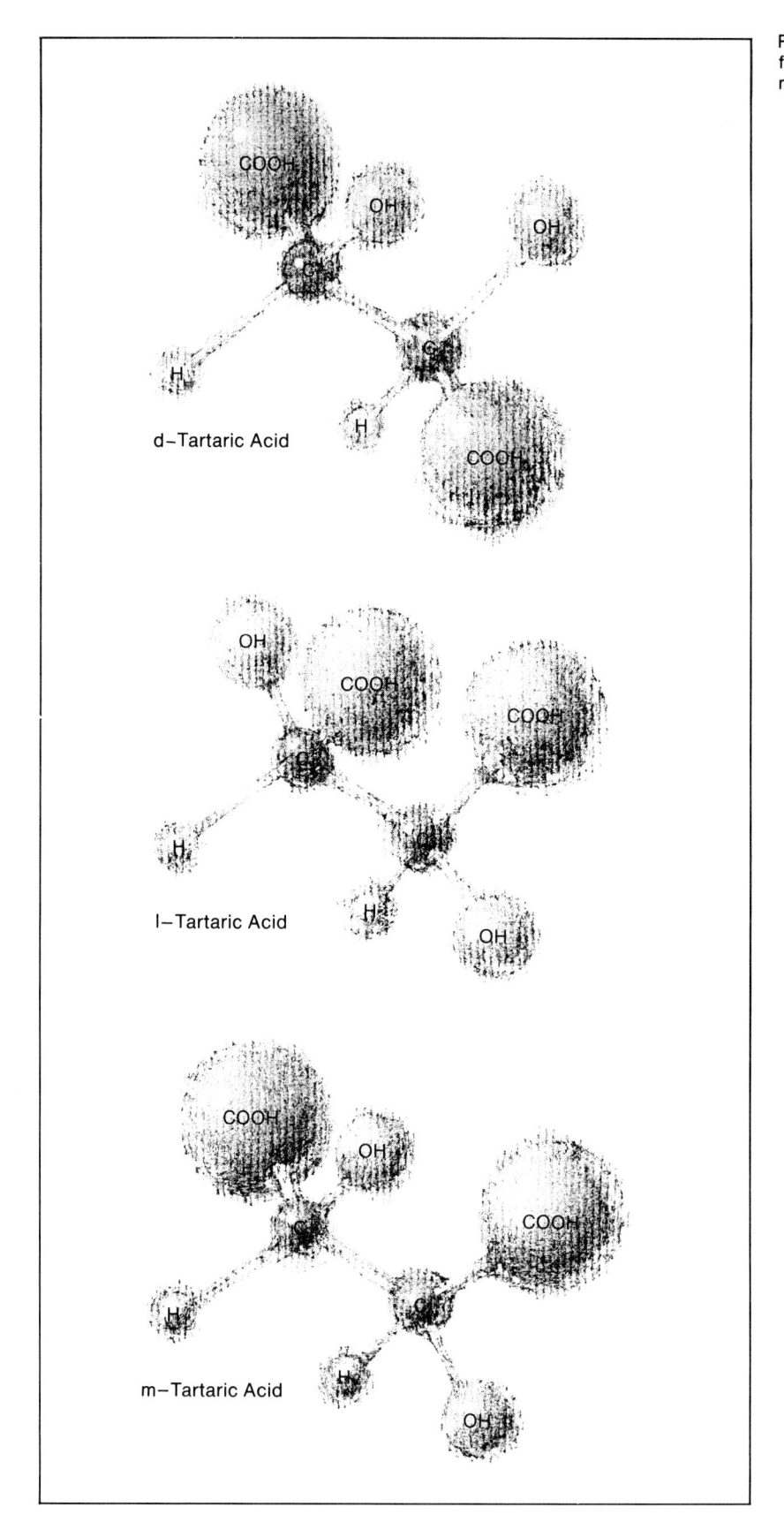

Figure 19. Tartaric acid occurs in two mirror-image forms and in another form whose two ends are mirror images of each other.

d–Tartaric Acid

l–Tartaric Acid

m–Tartaric Acid

Figure 20. The amino acids from which a protein chain is made have a carbon atom with four different groups attached. That is why they can occur in distinguishable mirror-image forms.

will cancel. And finally, if one end of the molecule has the right-handed arrangement and the other end the left-handed, their effects will cancel within the molecule itself.

Now both of Pasteur's optically active acids were products of living organisms—the right-handed from the growth of grapes and the left-handed from the growth of a mold that destroyed the right-handed ingredient of a 50–50 mixture. So far, the only means found for synthesizing optically active compounds in any but 50–50 mixtures employ either a life process or a product that such a process has already made optically active. How is this feat accomplished?

In the first place, much of living matter turns out to be made of optically active molecules. The proteins in our flesh and muscle, the molecules of deoxyribonucleic acid (DNA) that bear our genetic code—all are made largely from amino acids with the general structure shown in Figure 21. When the proteins are broken down chemically to liberate these acids for study, they are all found to rotate the plane of polarization of light in the same direction. Moreover, the enzymes that catalyze the production of these proteins in the body have the same property. In the simile of the organic chemist Emil Fischer, "The enzyme fits the substrate as the key fits the lock."

As one result, an organism may show quite different responses to the forms of an optically active molecule. Sometimes one form neither smells nor tastes the same as the other. Drugs and medicines offering right and left alternatives may be effective in one form and inert in the other.

It isn't difficult to imagine ways by which the right-left distinction may be propagated from generation to generation of organisms. But how did the distinction arise in the first place? The question probes deeply into the unknown origin of life.

UNSATURATED HYDROCARBONS

So far, you've read about hydrocarbons in which each of the carbon atoms are bonded to other atoms by a single bond. There are also a large number of organic molecules in which the carbon atoms are bonded to each other by double and even triple bonds. For example, the molecule ethylene, CH_2CH_2, has a double bond connecting the two carbon atoms. The molecule acetylene, C_2H_2, has two carbon atoms associated by a triple bond. The structures for ethylene and acetylene are shown in Figure 21. The multiple bonding prevents rotation, so these molecules are not as flexible as molecules with single bonds. An interesting chemical argument supported the idea of a double bond in such compounds as ethylene. All bonds may be visualized as rigid connections between the atoms. If only one bond connects the two carbon atoms, each CH_2 group can rotate freely about the bond. But if the pair of carbon atoms is connected by *two* bonds, the CH_2 groups can no longer rotate about the connection. Using such a picture, you would expect to find only one sort of dichloroethane with one chlorine atom attached to each carbon atom, because the single bond allows the CH_2Cl groups to rotate freely. Conversely, you would expect to find two distinguishable dichloroethylenes and in fact you do (Figure 22).

The molecule benzene is made of six carbon and six hydrogen atoms. In 1865 Friedrich Kekulé offered the first helpful suggestion for describing it. He also described a classic instance of the mental processes often responsible for intellectual discovery:

"I was sitting, writing at my textbook; but the work did not progress; my thoughts were elsewhere. I turned my chair to the fire and dozed. Again, the atoms were gambolling before my eyes. My mental eye, rendered more acute by repeated visions of the kind, could now distinguish larger structures of manifold conformation: long rows, sometimes more closely fitted together, all twining, and twisting in snakelike motion. But look! What was that? One of the snakes had seized hold of its own tail and the form whirled mockingly before my eyes. As if by a flash of lightning I awoke, and this time also I spent the rest of the night in working out the consequences of the hypothesis."

But placing the carbon atoms in a ring still caused conceptual trouble. It seemed natural to accommodate the four bonds of carbon by writing single and double bonds alternately around the ring, as shown in Figure 23. Then there should be two distinguishable dichlorobenzenes with the chlorine atoms attached to adjacent carbon atoms in both of them, as in Figure 23. But only one "orthodichlorobenzene" is ever found.

The benzene problem may be tackled in the following terms. Accept the accumulated evidence that the carbon atoms are connected in a

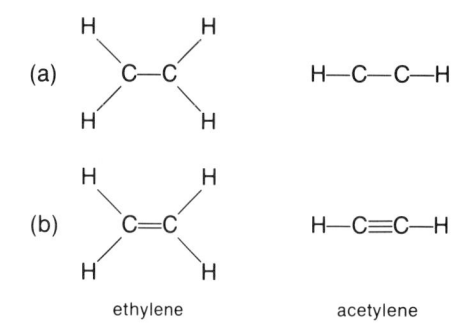

Figure 21. Ethylene and acetylene have multiple bonds between the carbon atoms. They can't be rotated to form molecules of different shapes.

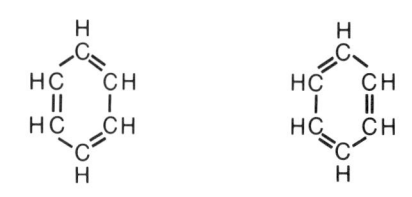

Figure 22. If it were possible to explain benzene according to the satisfactory explanations of the other hydrocarbons, then there would have to be two distinguishable dichlorobenzenes. But before 1865 only one was found and the structure of benzene remained a puzzle.

Figure 23. The "Kekule structure" of benzene (a) would make possible two distinguishable orthodichlorobenzenes (b) with a double bond and a single bond between the chlorine-bearing atoms.

Figure 24. Cis-trans-isomerism occurs because of restricted rotation about double bonds. Cis-dichloroethylene and trans-dichloroethylene are distinct chemical substances. In 1,2-dichloroethane, there is free rotation about the carbon-carbon bond.

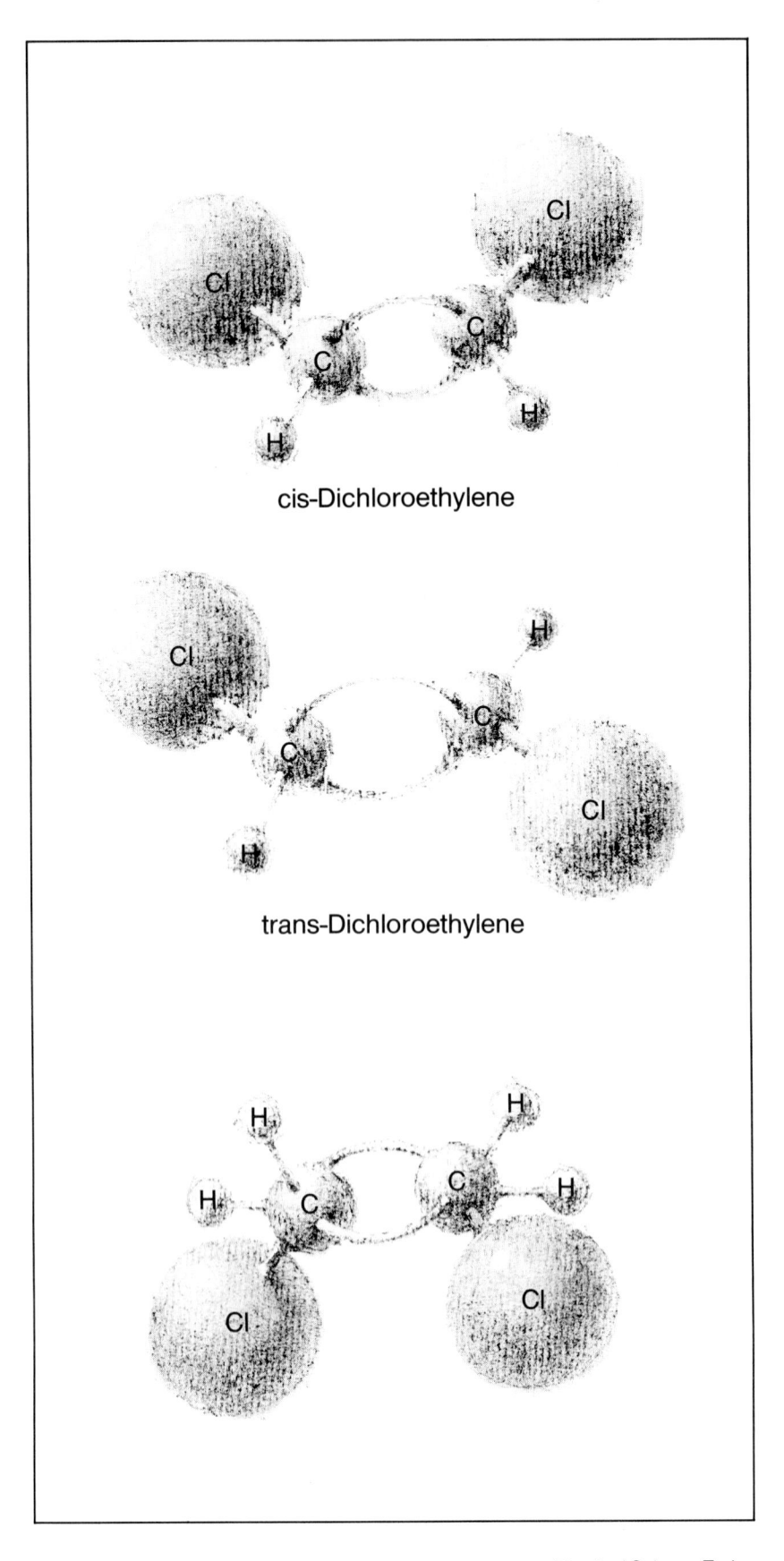

cis-Dichloroethylene

trans-Dichloroethylene

regular plane hexagonal ring. Accept the idea that each carbon atom is bonded to each of its two neighbors by a single localized bond. Then you may visualize an arrangement as shown in Figure 25, where each carbon atom still has one unshared electron in its outer shell. You may now ask what those six electrons might do if they roamed over the whole ring, with no specific pairing of these electrons. Calculations show that it is more favorable energetically for the electrons to roam around the ring in new delocalized shared orbitals than to stay in localized bonds between the atoms (see Figure 26).

METALS

Metals present the extreme case of delocalized bonding. Each atom contributes one or more electrons to a great pool of electrons, all roaming throughout the metal and threading their ways between the ionic cores of its atoms. If you wish to visualize the forces that hold the atoms together, the picture of the force in the single bond of the hydrogen molecule must be expanded considerably. The ions in the metal are able to remain close to one another despite the mutual repulsions of their positive charges because of the neutralizing effect of the negatively charged electronic pool that permeates the entire metal. It is as if the network of positive ions were immersed and held together by a "sea" of electrons.

This model of metals provides a picture of their most conspicuous distinguishing property: their ready conduction of electricity. The conductivity of gold is 10^{23} times greater than that of most nonmetallic materials. The electrons wandering in the electronic pool are quite free to drift under the urging of an electric field imposed from outside, unlike the electrons bound to atoms or molecules.

This account of interatomic bonding forms the basis of our understanding of molecular structure. All bonds between atoms, seemingly so diverse, are actually due to the same electrostatic forces, the interactions between positive nuclei and negative electrons. The diversity of these bonds arises only from the numerous ways in which the electrons locate themselves around and between the nuclei. The bonding structures found in nature are completely consistent with the allowed electron cloud configurations predicted by wave mechanics. And it is the principles of wave mechanics that may ultimately provide the answers to our questions concerning not only the various states of matter, but even the nature of life itself.

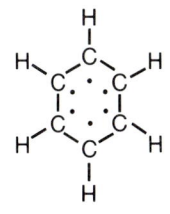

Figure 25. If the carbon atoms in benzene are connected by single bonds, six electrons are left free to roam over the ring. The structure is unique, and the "Kekulé paradox" is solved.

Figure 26. The "best" structure of benzene has the carbon atoms joined by single bonds, with the remaining six electrons in a new type of delocalized orbital that is shared by all six carbon atoms in the ring.

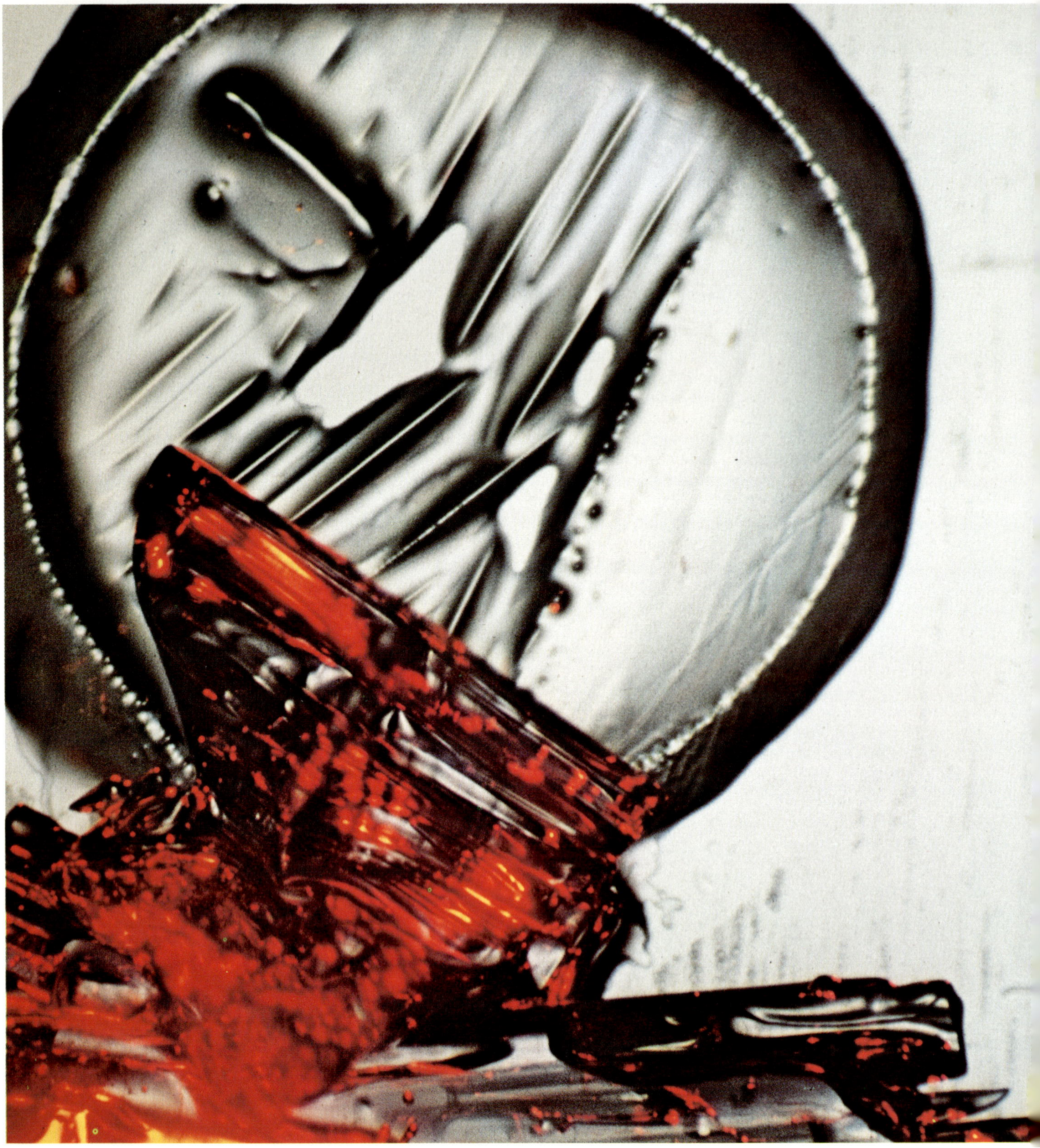

21
States of Matter

The camera captures matter as it changes state—a photomicrograph of a melted and partially refrozen ice crystal.

Receiving a prize for his brilliant work on absorption, the late Irving Langmuir began his acceptance address, "We distinguish three states of matter: solid, liquid, and gas." It is a classic distinction. But Dr. Langmuir went on to show that his own work had brought to light collections of atoms behaving like a solid or liquid or gas *in two dimensions.*

Since his time many other collections of atoms have been found that don't fit a previously familiar category: a recent book bears the provocative title *Seven States of Matter.* Perhaps that diversity is not surprising in light of our present understanding that all matter is made of electrons and nuclei whose interaction gives rise to the complexities of chemistry described in the preceding chapter. Indeed it is more surprising that in ordinary life the three familiar states of matter can be sharply distinguished at all. How do we do it?

RESTLESSNESS

In order to pursue the distinction, examine mercury, which can take the form of gas, liquid, or solid. According to the classic model of a gas, accepted ever since Daniel Bernoulli described it in 1738, you can visualize the atoms of mercury in a mercury-vapor lamp as rushing about at high speeds quite independently, affecting one another little except during the instants when they collide. Cooling the gas reduces their speeds, and their mutual attractions can bring them together into a liquid.

In the liquid state the atoms continue to move. They can pass around one another readily, permitting the liquid to flow. But the gas can be compressed easily into a smaller volume and the liquid cannot. The liquid resists compression so vigorously that mutual attractions alone must hold the atoms in a densely inhabited community. Nevertheless, the impacts of its neighbors occasionally will kick an atom out of the community. In fact, if liquid mercury happens to escape from its container, a chemical test-paper waved in the surrounding atmosphere will reveal a lingering, recognizable trace of the accident because of these disengaged atoms.

Cooled still further, the pool of mercury suddenly freezes to a solid and will no longer flow. This sudden transition occurs at a fixed definite temperature, −39 degrees Centigrade. The volume of mercury changes little. As a solid its atoms are not much more densely packed, but they can no longer roll past one another freely. The atom at one place in a solid is almost certainly the atom that was there earlier and that will be there later. The atom is still moving; its motion merely has retired to vibration about a fixed position. And it still receives vigorous kicks from other vibrating atoms. The fact prompted the physicist Max Born to observe, "It is odd to think that there is a word for something which strictly speaking does not exist, namely, rest."

CRYSTALLINITY

As long as its temperature remains below −39 degrees Centigrade, mercury is solid and behaves much like many other metals. Examined under

Figure 2. The grain boundaries separating the crystals that compose it often appear on brass door handles that have been etched by numerous contacts with sweaty hands.

a microscope, the solid mercury exhibits a grainy structure. Most metals conceal a similar graininess, which they exhibit when their surfaces are etched with an acid. This grainy structure can often be seen on old cast brass door handles that have been cumulatively etched by the sweat from human hands (Figure 2). To the modern metallographer, the sizes and shapes of the grains speak of the composition of a metallic alloy, of whether its components have segregated, and of how much reliance to put on its strength. The grains in other solids, especially in rocks, are more conspicuous. Grains of clear quartz, pink feldspar, and black mica come together to produce the handsome texture of granite.

To most of us, a crystal is a transparent bit of solid with flat faces meeting at sharp angles. Minerals such as quartz often appear naturally in that form; sugar can be crystallized into "rock candy." Toward the end of the eighteenth century, people found that the angles at which crystal faces meet are characteristic of its composition and are independent of its size and shape. They began to think of a crystal as being composed of tiny building blocks, all the same and stacked in regular arrays, the shape of which is determined by the crystal's substance.

If these imagined building blocks are in fact all alike, the atoms composing them must be arranged identically in each—and crystals must consist of *regular, repetitive arrays of atoms.* The implication was so powerful that in 1912 it directed a celebrated experiment. Were x-rays, then just discovered by W. C. Roentgen, a stream of particles or were they a kind of light? Guessing that they were light with an unusually short wavelength, and that the repetitive array of atoms in a crystal would form a "grating" spaced suitably for diffracting that light, W. Friedrich, P. Knipping, and Max von Laue focused the rays on a crystal of blue vitriol. They observed the traces of diffracted rays on a photographic plate.

By one stroke the experiment confirmed two well-advised hunches of the sort that scientists call "educated guesses." Von Laue himself wrote of it later, "Here the wave theory of x-rays and the atomic theory of crystals come together, one of those surprising events to which physics owes its powers of conviction."

Reading from left to right, five investigators of the structure of matter:

Figure 3. Daniel Bernoulli, the Swiss mathematician who described the random motion of gas molecules in 1738.

Figure 4. W. C. Roentgen, who discovered x-rays in 1901.

Figure 5. Max von Laue, the first scientist to observe diffraction patterns by radiating a crystal of blue vitriol with x-rays.

Figure 6. William Bragg and his son Laurence who studied the "reflection" of x-rays from crystal planes.

Almost at once William and Lawrence Bragg, father and son, proposed a simplified and powerful way of thinking about how crystals diffract x-rays. They looked upon the presumed regular arrangements of atoms as if the various crystal planes on which the atoms fall were semi-transparent mirrors. An incoming beam of x-rays is reflected a little from each "mirror," and a strong, cumulative reflection occurs when all the little contributions interfere constructively with one another. Figure 7 suggests how a single arrangement of atoms can be analyzed into many families of parallel mirrors, and how constructive interference from any family will depend on the angle that it makes to the incoming beam, on the spacing of the mirrors, and on the wavelength of the x-rays.

This picture of x-ray diffraction and the refinements that have followed form the experimental basis of modern crystallography. Measurements of the angles of reflected beams and of their relative intensities are put together to portray the way the atoms are arranged in a crystal. The crystallographer J. D. Bernal says the required inference resembles in many ways the solution of crossword puzzles: "The atoms provide the letters and the intensities the clues."

DIFFRACTION

Does von Laue's experiment alone assure us that x-rays are not a stream of particles? No; today that assurance needs and has received other support. Diffraction is characteristic of any wavelike phenomenon, and you know that a moving particle has wavelike properties. Again, crystalline gratings were central to the discovery. In 1927, Clinton Davisson and Lester Germer observed the diffraction of a beam of electrons bouncing from a crystal of nickel into an electrometer. At about the same time, George Thomson saw on a photographic plate the traces of diffracted electrons shot through the many tiny crystals that make up a gold foil.

Today it is sometimes advantageous to study the structures of crystals by diffracting a beam of particles. The number of electrons in an atom measures its scattering power for x-rays, and therefore the

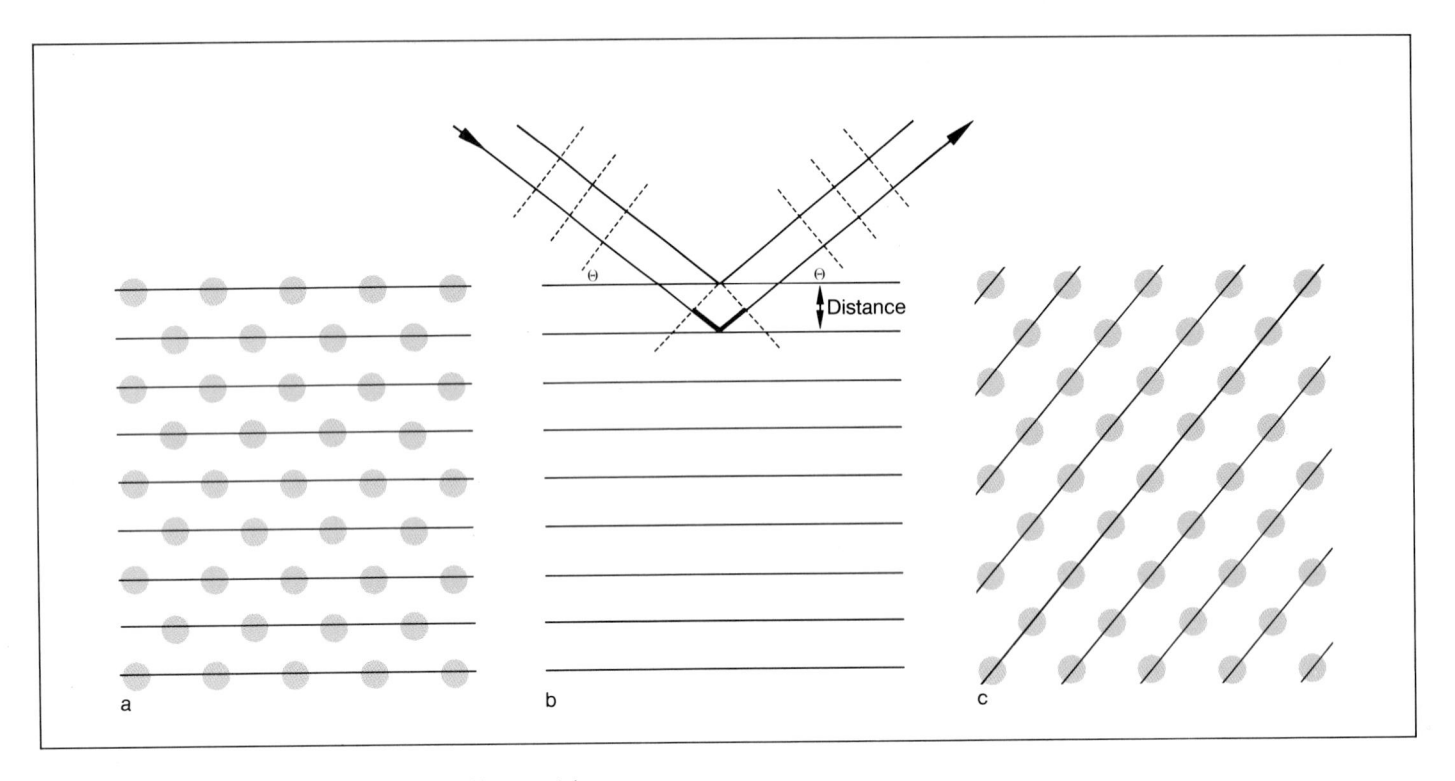

Figure 7. The atoms regularly arranged in a crystal can be allocated to families of parallel, equally spaced planes in many ways. Each family will afford a strongly reflected x-ray beam when Bragg's law, $n\lambda = 2d \sin\theta$, is satisfied.

intensity of diffracted x-ray beams declines to the vanishing point for hydrogen atoms. In such a case, diffracting a beam of neutrons is sometimes useful. Unlike x-rays and neutrons, electrons bear an electric charge, and an incident beam of electrons may not penetrate far into a crystal because the beam is repelled by the electrons already present in the atoms. Hence electron diffraction is particularly useful for studying the atomic arrangements near the surface of a crystal. Electrons projected at low velocities are now divulging much about layers of foreign materials absorbed on crystalline surfaces.

The product of all these studies is a large body of rapidly increasing knowledge about how the atoms are arranged in and on crystals. X-ray diffraction has confirmed that almost all solids consist of crystals, big or little, jumbled together. Moreover, it has revealed a patchy orderliness in liquids and glasses, which had long been thought to be utterly disorderly. It has shown that the atoms in plastics often form arrays that are orderly in one dimension but disorderly in the other two. It has confirmed that the atoms in a solid are vibrating and offers a rough measure

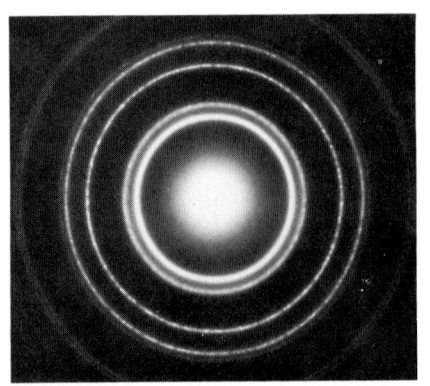

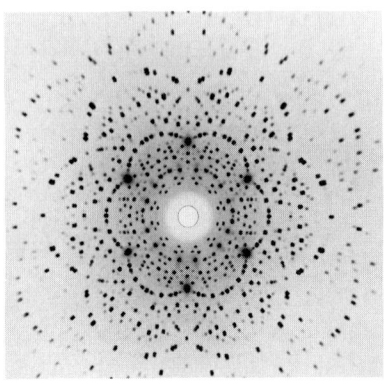

Figure 8. The crystal ordering in silver foil shows comparable symmetries when examined by electron diffraction (left) and by x-ray diffraction (right).

Diffracted Electrons

At what velocity must you project electrons in order to observe their diffraction by a crystal? To find out, use de Broglie's relation between the wavelength and the momentum of a moving particle: $p = h/\lambda$, where p is the momentum mv, λ is the wavelength, and h is Planck's constant, 6.6×10^{-34} joule-second. Substituting $\lambda = 3.0 \times 10^{-8}$ centimeter, a typical atomic spacing in a crystal, and $m = 9.1 \times 10^{-28}$ gram for the mass of an electron, you find:

$$p = h/\lambda$$
$$mv = h/\lambda$$
$$v = h/\lambda m$$

$$= \frac{6.6 \times 10^{-34} \text{ joule-second}}{(3.0 \times 10^{-8} \text{ centimeter}) \times (9.1 \times 10^{-28} \text{ gram})}$$

$$= \frac{6.6 \times 10^{-34} \text{ joule-second}}{2.7 \times 10^{-35} \text{ gram-centimeter}}$$

$$= \frac{6.6 \times 10^{27} \text{ gram-(centimeter/second)}^2 \text{-second}}{2.7 \times 10^{-35} \text{ gram-centimeter}}$$

$$= 2.4 \times 10^8 \text{ centimeters/second}$$

This is equivalent to about 10 million miles per hour. An "electron gun" will confer this velocity on a beam of electrons if the voltage of the gun is a few hundred volts.

Figure 9. Modern methods of examining crystals by x-rays provide contour maps of electron density to aid in the inference of crystal structure.

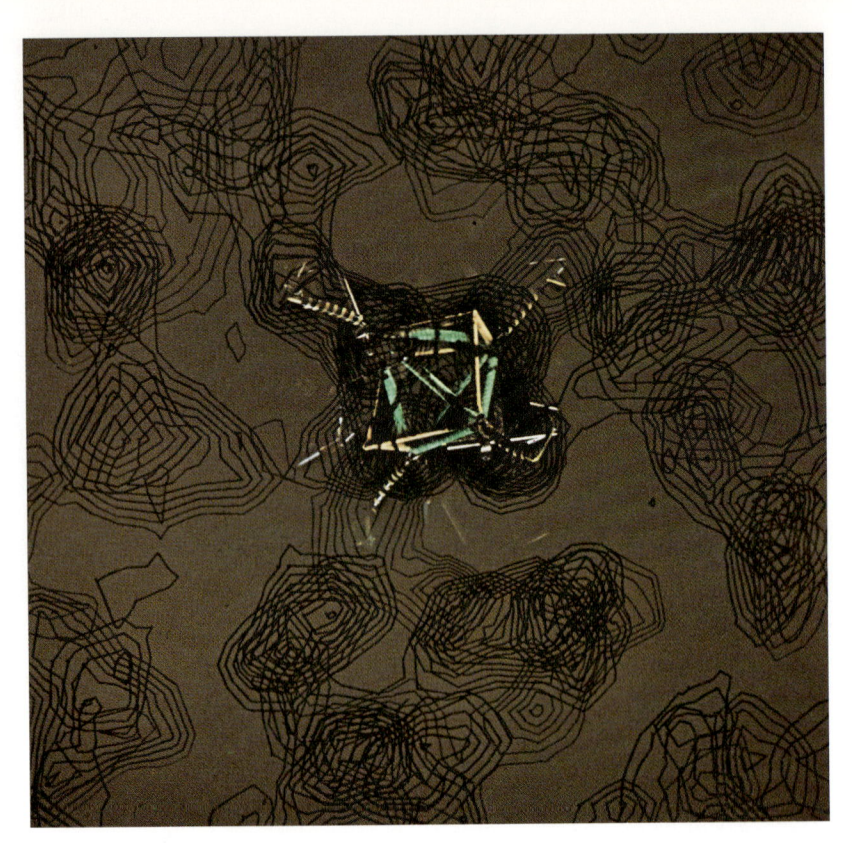

of their vibrational amplitudes. Even gases diffract electrons, yielding information about the lengths of the interatomic bonds in their molecules.

Inorganic materials provided the focus for the early studies of crystal structures. The Braggs themselves gave particular attention to minerals, and specialists are still busy exploring the complexities of the feldspars and micas. In contrast, organic materials yield less readily to analysis because the x-ray scattering powers of carbon, nitrogen, and oxygen are similar and feeble. More recent methods provide data that can be plotted as contour maps of electron density. Cumulative experience with such maps, combined with knowledge of the habits of organic molecules, enable an investigator to fix upon a probable crystal structure. Chemistry and x-ray crystallography in fact reciprocate. Dorothy Hodgkin, for example, determined the interconnection of the atoms in an organic molecule, penicillin, by studying the diffraction of x-rays from its crystals. James Watson and Francis Crick also used diffraction techniques to reveal the even more complicated structures of the molecules of deoxyribonucleic acid (DNA), the evolutionary key to heredity in all life forms.

SOLID, LIQUID, AND GAS

Now stand back from detail for a moment to examine some consequences of the experiments and insights that broadly distinguish solids, liquids, and gases. All three states are characterized by atomic restlessness, increasing in the order solid-liquid-gas. Because temperature is a rough measure of restlessness, an increase in temperature can convert a solid to a liquid, and a liquid to a gas, but it cannot achieve the opposite conversion, other things (such as pressure) being equal.

A liquid is like a solid, and unlike a gas, because it has a definite *volume.* In both solids and liquids, volume is controlled by the effective sizes of the atoms. They attract one another until they come together as close as they can, as long as their thermal restlessness is not enough to knock them apart. On the other hand, a liquid is like a gas, and unlike a solid, in its ability to *flow.* When a liquid flows its atoms must move past one another, but they need not get much farther apart. In other words, in a liquid interatomic *angles* are relaxed without necessarily relaxing interatomic *distances.*

Because melting a solid to a liquid or vaporizing a liquid to a gas increases the atoms' restlessness, they both require inputs of heat. The increased vigor of the atomic motions represents an increase in the kinetic energies of the atoms, and those additional energies represent the added heat. Thus at the freezing point of water, a piece of ice will melt to a pool of water at the same temperature only if it receives 80 calories of heat per gram. At its boiling point, water will vaporize to steam at the same temperature only when supplied with 538 calories per gram. Here are clear exceptions to the earlier statement that temperature is a rough measure of restlessness: during melting, the restlessness increases while the temperature stays the same.

Because a liquid is intermediate in these ways between a solid and a gas, a quantitative theory of liquids might draw on the theories of gases and of solids, but it has turned out to be especially difficult to construct such a theory. For gases, the picture of tiny particles rushing about independently except when they collide can be quantified. The theory can then be modified to take account of the sizes of the particles and the small effects of attraction and repulsion between them. For solids, the picture of particles at rest, ranked in orderly arrays, can form a broad basis for distinctions and calculations and it can then be modified to take account of the atomic vibrations. But the atoms in a liquid, closely associated but constantly changing their allegiance, resist any simple quantitative description.

One upshot of the distinction between gas, liquid, and solid is the fact that gases are all very much alike whereas solids have great diversity. Of course gases are chemically diverse: carbon monoxide is poisonous, oxygen is benign. But physically they are remarkably alike. For example, they all obey approximately the same *perfect gas law,* $p \times V = RT$, which connects their pressure p, volume V, and absolute temperature T by a constant R, which is the same for the same number of molecules, whatever their species.

Solids have the same chemical diversity but they also have the diversity arising from different kinds of *orderliness* in the arrangements of their atoms. The disorder of the perfect gases is a single thing—disorder is disorder. But the order of solids is many things. And liquids, in their diversity, are intermediate between gases and solids.

ALLOTROPY

The solid forms of carbon illustrate dramatically how diversity may arise from order. Soot consists largely of carbon, and the soft, shiny, black crystals of graphite are almost pure carbon. And more than a century ago Humphry Davy put a diamond in an atmosphere of oxygen, focused the sun's rays on it, and burned it to show that a diamond, too, is almost pure carbon.

Studies of these materials with x-rays have shown how the atoms of carbon are arranged in a diamond (see Figure 11). Each atom is immediately surrounded by four others at the corners of a regular tetrahedron, much as a carbon atom in a molecule of methane is surrounded by four hydrogen atoms. The four bonds connecting each atom with its

Figure 10. Disregarding immediate financial losses, Humphry Davy (1778–1829) burned diamond into graphite to prove that diamonds are made up of carbon.

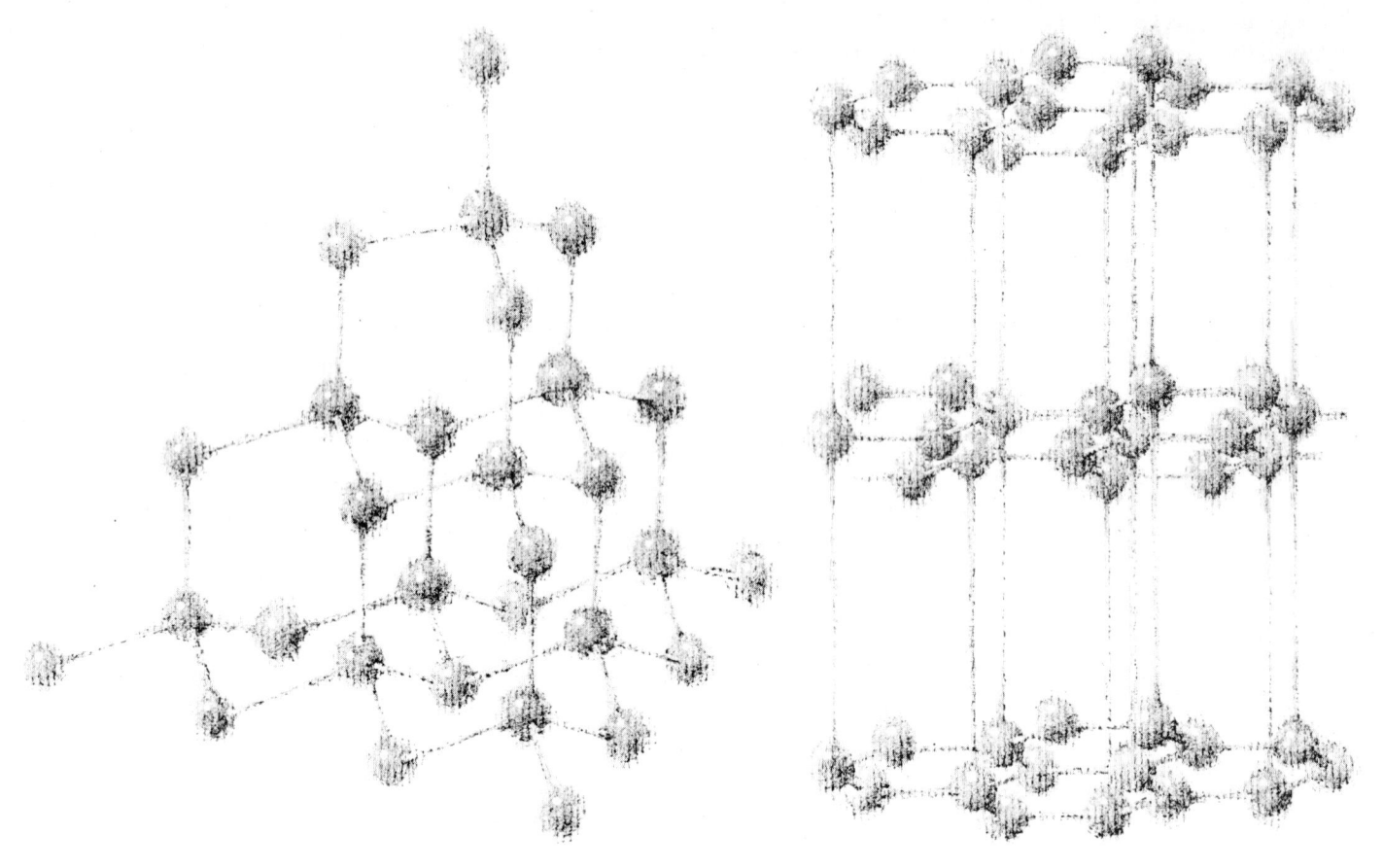

Figure 11. (Above) The carbon atoms in diamond are arranged in such a way that each atom simultaneously forms the center of a regular tetrahedron and an apex of each overlapping tetrahedron. This diagram shows the *average* location of the centers of the vibrating atoms.

Figure 12. (Above right) The planar structure of graphite is due to the bonding of each carbon atom to three other carbon atoms. The forces between adjacent planes of carbon atoms are weak, making graphite extremely soft.

neighbors are equally strong and give to a diamond its celebrated hardness.

In graphite the carbon atoms are arranged in plane parallel sheets, as shown in Figure 12. The neighbors within each sheet lie close to one another, but adjacent sheets are much farther apart. Accordingly, the bonds between the sheets are weaker: the crystals are soft and come apart readily between the sheets.

The occurrence of a single substance in several different solid forms is called *allotropy*. The word is usually reserved for elements, such as carbon; the word *polymorphism* is used for compounds. Tin provides another interesting example of allotropy. In a cold climate, a piece of tin may become afflicted by "tin pest" and crumble to dust. The atoms of the familiar metallic "white tin" shift their positions, adopting the diamond structure and forming the nonmetallic "gray tin."

The different allotropic forms of a material are stable in different ranges of temperature and pressure. When those temperatures or pressures change, passing through the dividing line between the ranges, the material will tend to change to the newly stable configuration. For tin at ordinary pressure, the dividing point is at 18 degrees Centigrade. You can easily observe a beautiful example of such a transformation in mercuric iodide. The brilliant red crystals will crumble to a brilliant yellow dust when you warm them through the dividing point of 126 degrees Centigrade. (If you do this experiment, be careful not to breathe the toxic vapor.)

When you lower the temperature again, the yellow form of mercuric iodide may not change spontaneously back to the red. You may have to touch the yellow with a bit of the red in order to initiate the reversal. This reluctance to change may remind you that liquid water may often be "supercooled" below its freezing point but ice cannot be "superheated" above its melting point.

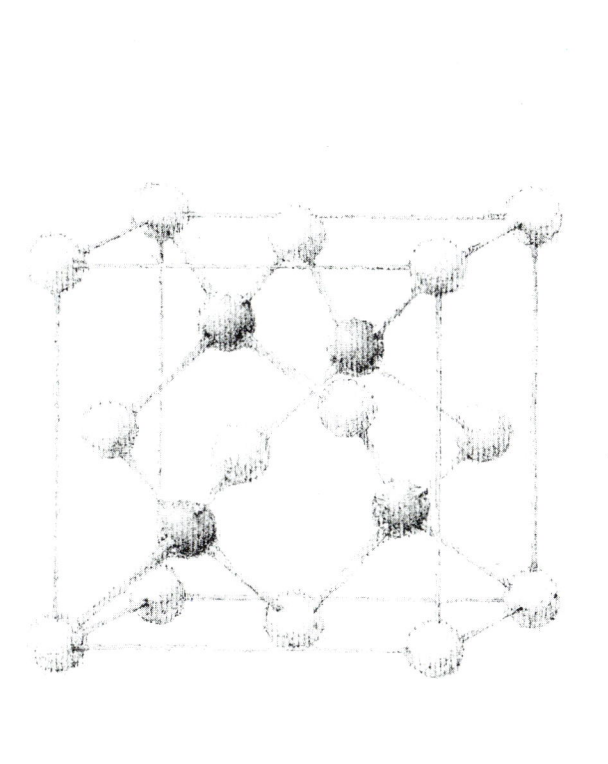

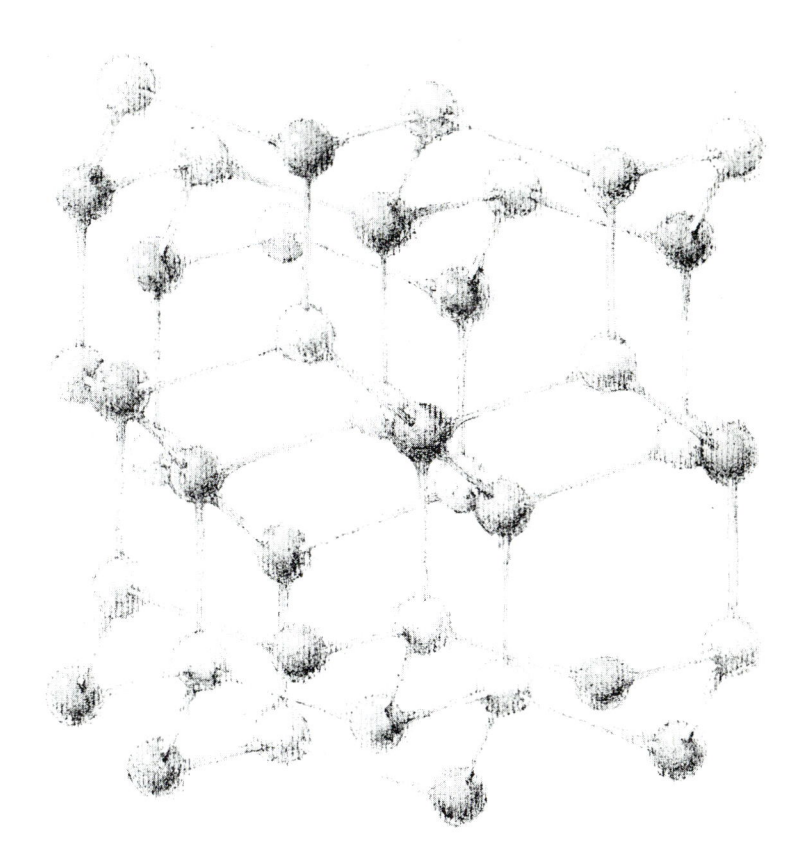

The world contains many relatively unstable solids that cannot find their way back to their more stable configurations. Under ordinary conditions, the stable configuration of calcium carbonate is found in the mineral calcite. But another form, aragonite, is not uncommon, and for some reason oysters prefer that structure for the calcium carbonate that they deposit on pearls and on the inner surfaces of their shells.

In a similar dualism, zinc sulfide forms the minerals sphalerite and wurtzite with the different crystal structures shown in Figure 13. In both structures, each atom is surrounded tetrahedrally by four of the other kind of atom: in fact, the sphalerite structure is the diamond structure, but for two different kinds of atoms. Then are there such things as "diamonds" with the wurtzite structure? The question points back to the fascinating allotropy of carbon.

Heating a diamond turns it black and it never clears again: its atoms will have jiggled into their more stable configuration, graphite. The dividing line between the stable ranges is not yet completely known, and earnest study of the geologic circumstances in which diamonds occur has not yet revealed how they were produced. After many years, efforts to make diamonds artificially did meet success in a process that subjects carbon simultaneously to high pressure and high temperature. The resulting diamonds lack the high quality of gems, but they *are* diamonds.

More recently, tiny diamondlike crystals of carbon were found in meteorites, and at about the same time scientists produced similar crystals by subjecting carbon to sudden and intense shock. X-ray exami-

Figure 13. The structure of sphalerite (left) and wurtzite (right).

Figure 14. (Below) The argonite form of calcium carbonate makes up the internal structure of the mother-of-pearl used in this Toltec artifact.

21 | States of Matter

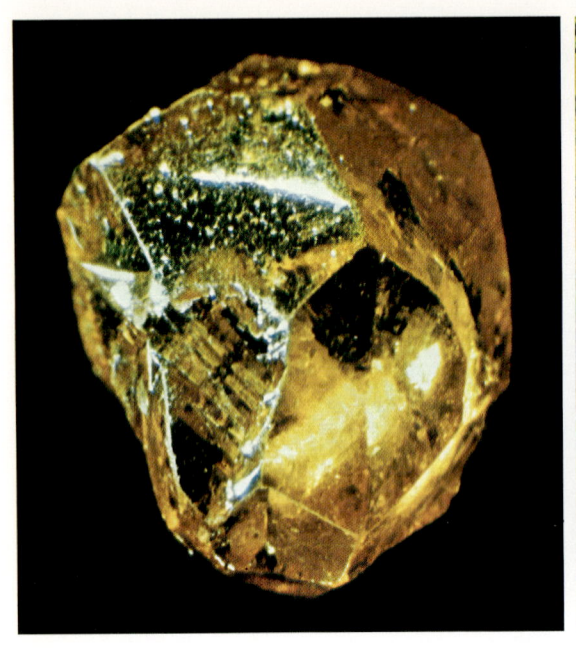

Figure 15. (Left) An African diamond in its natural state. (Right) A photomicrograph taken in polarized light reveals the triangular repeating patterns of a diamond.

Figure 16. (Below) A crystal made of positive (solid) and negative (open) ions conducts electricity by virtue of vacancies that permit the ions to move toward the appropriate electrodes on the crystal's surface.

nation of crystals from both sources showed the wurtzite structure! Because any meteorite examined by man has felt the shock of hitting the earth, meteoric "diamonds" may have arisen from "shocked" carbon. Today the rocks brought back from the moon are eagerly examined for similar "diamonds," which might clarify the role of meteorites in the moon's geology.

The ranges of stability of different allotropic forms depend not only on temperature but also on pressure. Studies of allotropy under high pressures first became possible in this century. Some of the most curious results relate to *water.* Several different forms of solid water—five according to one investigator, six according to another—appear in different ranges of temperature and pressure. The familiar form of ice that we make in the freezer or see in icicles is unique in occupying more space than the liquid; all the others are denser than water.

In *Cat's Cradle,* Kurt Vonnegut, Jr. makes dramatic use of water's versatility and of the fact that allotropic forms of solids may appear only when they are initiated, or "seeded." In his story, all the water in the world is frozen with seeds of a new "Ice-IX," which destroys the world. The scientist would say, "Improbable," because we have had a lot of water around for a long time, and the random fluctuations among its molecules have never produced such seeds. But he might be loath to say, "Impossible."

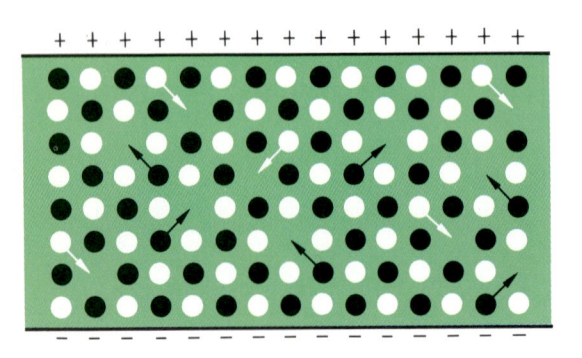

IMPERFECTIONS

Rubbed between the fingers, graphite feels greasy, and it has in fact been valued as an ingredient of lubricants. To explain this property, scientists had lived comfortably with the idea that the binding between the parallel sheets of atoms in graphite is so small that the sheets slip easily along one another. They received a rude shock when they tested graphite for use as a lubricant in outer space. In a vacuum the "lubricant" soon became abrasive. It turned out that normally the layers of atoms slip easily because impurities have crept between them, but these same impurities creep *out* again in a world without pressure. The experience emphasizes that the behavior of solids often depends not on their orderly perfection but on their *imperfection.*

Imperfections are ubiquitous in solids. An interesting theorem of thermodynamics proves that anything must be soluble in anything else. But, like most of thermodynamics, the theorem does not say *how* soluble. If two solids have the same crystal structure and if their atoms are about the same size, the two solids may be soluble in each other in all proportions. Silver and gold are one conspicuous example. Pure chromium oxide does not have the same crystal structure as pure aluminum oxide; but chromium, with the same valence as aluminum, can occupy some of the aluminum sites in a crystal of corundum, producing ruby. And gallium, different from germanium in both crystal structure and valence, can replace it in small amounts to yield crystals that are vital to a transistor.

The theorem can even be applied to a crystal of *nothing-at-all.* Imagine a crystal of sodium chloride and, beside it, a crystal in which nothing-at-all occupies the sites of the ions in sodium chloride. The two objects are somewhat soluble in each other: a little sodium chloride escapes from the crystal into the vacuum, leaving vacancies in some of the ionic sites in the crystal. These vacancies make possible a little electric conduction through that "electrically insulating" crystal. In a perfect crystal, the ions would be packed so tightly that they could not move past one another. If there is a vacancy, an electric field can push an adjacent ion into it, leaving a vacancy where the ion used to be. Then another ion can be pushed into that vacancy, and so on. Conduction takes place by a relay-race of ions, while vacancies move in the opposite direction.

DISLOCATIONS

Such imperfections as the replacement of one species of atom by another or by a vacancy are often lumped together in the category "point defects," because they afflict a crystal at isolated places in its structure. More recently, "line defects" have been given attention. They were first imagined as a way of explaining the plasticity of metals, and later experiments gave unquestioned evidence of them.

When a bottlecap is stamped out of sheet iron in a punch press, the metal is "plastically deformed." The atoms in its component crystals slide past one another along slip planes, which are determined by its crystal structure. If a crystal is without defects, all the atoms in a slip plane must slip at the same time in order to keep out of one another's way. But the force required to shift a plane of atoms by one interatomic distance can be calculated, and experiment shows that slippage actually occurs under forces no larger than one-thousandth of the calculated requirement.

The dilemma was resolved by guessing that only one line of atoms must move at one time—that a plane of atoms moves piecemeal, line by line. And a type of line defect called an *edge dislocation* was invented for freeing lines of atoms.

You can visualize an edge dislocation by imagining that you cut a crystal perpendicular to its slip plane and only partly through it, remove one layer of atoms, and finally close the crystal again. Then there is a misfit of the atoms in the neighborhood of the line where your cut stopped. Figure 18 shows the misfit in a plane perpendicular to that line. When the crystal feels the shearing forces illustrated, its atoms tend to slide so that the region of misfit moves to the left, and soon half the crystal has shifted over the other half by one atomic spacing.

In this simple form, the theory suggests that all dislocations could be swept out of a crystal by shearing it, leaving none to facilitate further

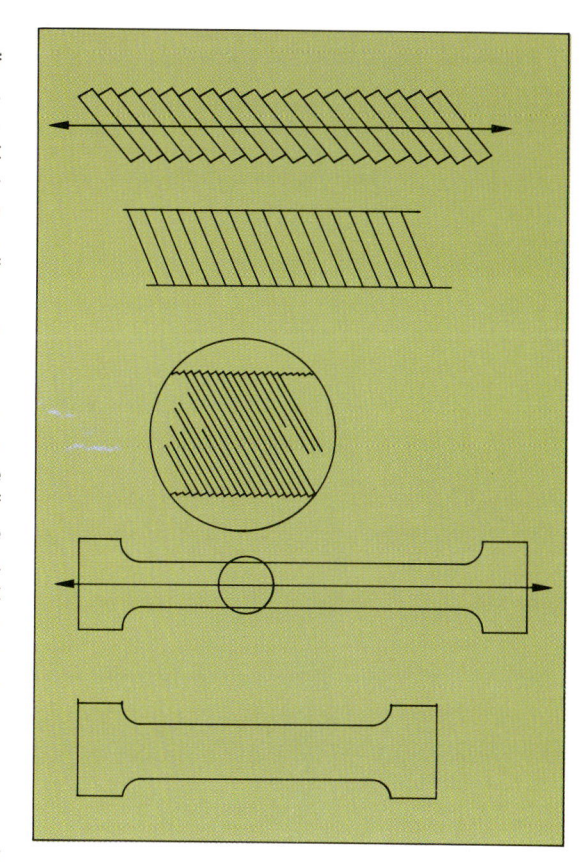

Figure 17. When a bar of metal is stretched, its component crystals behave like packs of playing cards and suffer glide along slip planes.

Figure 18. An edge-dislocation facilitates slip in a crystal by freeing lines of atoms (perpendicular to the paper) to shift their relative positions. Notice that the shearing forces represented by the arrows shift the light atoms by one interatomic spacing in each successive frame. Bottle tops are metal sheets that can be deformed plastically because of these edge dislocations.

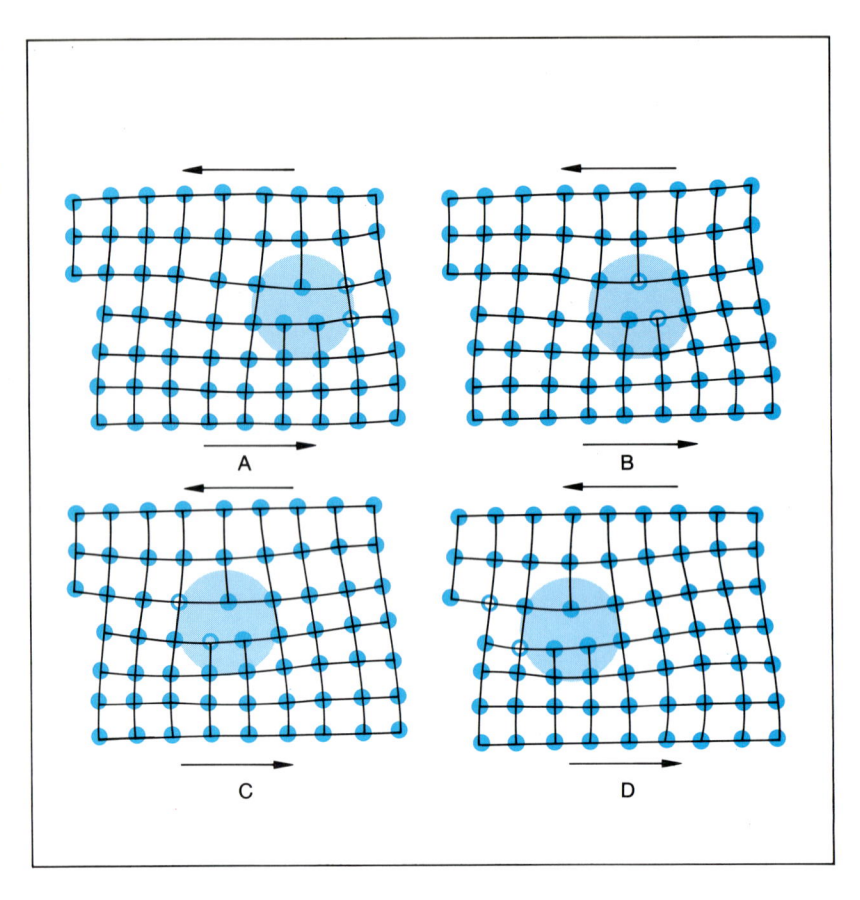

plastic flow. But the complete theory of dislocations also recognizes *screw dislocations,* which can be visualized by cutting, shifting, and closing a crystal (Figure 19). A combination of dislocations, edge and screw, can provide a line of misfit that wanders through a crystal in any direction, and it can also provide a "dislocation source" that perpetuates itself and cannot be swept out as a crystal glides.

A dislocation can close on itself within a crystal, but it comes to an end only at the crystal's surface. Visible evidence of edge dislocations has appeared in etch pits on a surface where two crystals abut with a very small difference in their orientation. Such a low-angle grain boundary can be visualized as a succession of closely spaced edge dislocations. Etching the surface there produces a regularly spaced row of pits, visible microscopically. Screw dislocations have also been seen, in phase contrast microscopy, on the naturally grown surfaces of crystals.

Dislocation theory can explain many diverse and puzzling occurrences such as the "work hardening" of metals, familiar to anyone who has tried to make a hole in soft lead with a twist drill. The theory has explained why crystals sometimes grow under circumstances that don't fit previous theories. It has even tackled with some success the difficult problem of the mechanism by which solids come apart when they are broken.

HOW SOLIDS ARE HELD TOGETHER

Considering the fracture of solids may raise a converse question: Why do solids hold together in the first place? You have already read the answer to that question in the preceding chapter. Solids cohere by the attractions

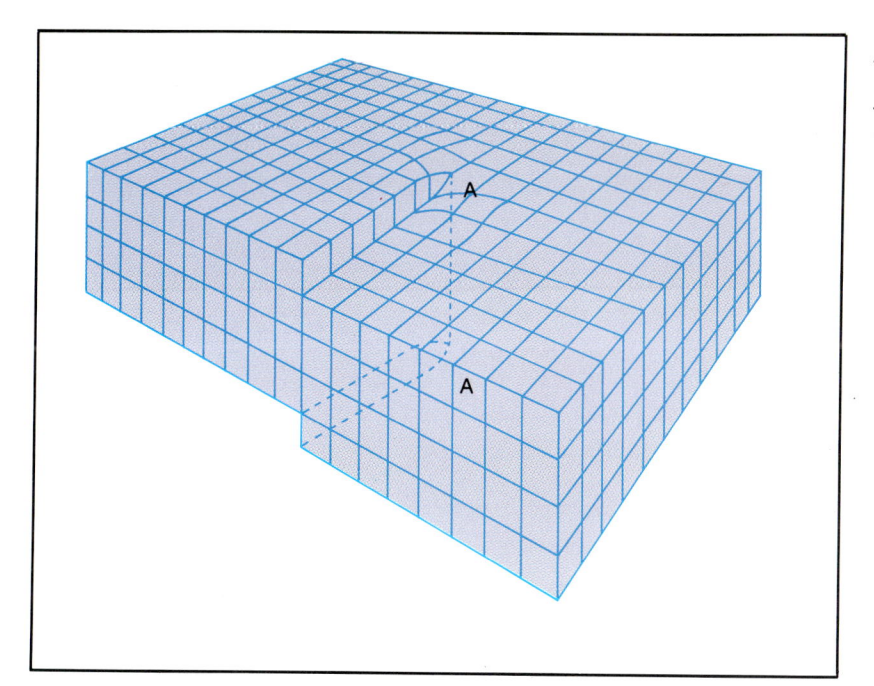

Figure 19. A screw-dislocation can be visualized as formed by cutting a crystal as far as the line $\overline{AA}$, shifting its abutting parts by one interatomic spacing, and finally closing it again. Such dislocations aid us in understanding how solids fracture.

between their negatively charged electrons and their positively charged atomic nuclei—the same attractions that hold atoms together in molecules. Indeed, a solid piece of matter can be regarded as a "giant" molecule. Accordingly, all the bonding mechanisms described earlier can be found in various solids, and they are often invoked to classify solids roughly into four types: ionic, covalent, molecular, and metallic.

In common salt, you have learned, each sodium atom has lost one electron to a chlorine atom, filling the outermost occupied electronic shell of the chlorine atom and leaving a filled shell in the sodium atom. You know that each resulting positive sodium ion gets as close as it can to its neighboring negative chloride ion. Because each ion is pulled strongly by its close neighbors and pushed less strongly by its distant neighbors, the solid holds together.

In a diamond, on the other hand, each atom shares a covalent bond with an immediately neighboring atom. Each bond between a pair of atoms may be visualized as formed by a pair of electrons, with each atom contributing one electron. The electrons in any pair owe special allegiance to the bonded pair of atoms. As they course from one atom of a pair to the other and back, the electrons spend time between the two paired atoms. Their negative charge helps compensate the repulsion of the nuclei, and the two atoms are pulled toward each other in a covalent bond.

The solids formed at low temperatures by neon and argon exemplify the third extreme. Imagine the electrons in each atom rushing about to form a cloud of negative charge around the nucleus. The center of gravity of the cloud, and therefore its center of charge, oscillates a little about the center of the nucleus. Instantaneously the two centers of balancing charge are not at exactly the same place: the atom behaves like an oscillating electric dipole. Two electric dipoles can either attract or repel each other according to how they face each other. In solid argon and other materials, the atomic dipoles adjust their oscillations and so provide net attractions between the atoms, called "dispersion forces."

The resulting cohesive forces are relatively weak; the solids are fragile and readily evaporate into gas. The dispersion forces are always

Figure 20. In a protein, the NH and CO groups in a chain are closely spaced, and the chains are propped apart by various large side groups.

Figure 21. Arnold Sommerfeld described a new organization of electron states that satisfactorily explained the gross electronic structure of metals.

present in cohesion, but in most solids they are masked by the stronger ionic, metallic, or covalent mechanisms. The electric charges of electrons in sodium chloride, silver, and carbon still oscillate and add their small contribution to bind the atoms together.

Rarely do any of these mechanisms remain wholly unalloyed with the others, as you read in the preceding chapter. And in fact, much of the solid world is held together by bonds that are mixed in a different and more elementary sense: atoms are held into aggregates by one kind of bond and the aggregates are held to one another by a different kind. Within the giant molecules that form the organic world, the atoms are connected by strong covalent bonds. Those molecules cohere to form wax, fat, and muscle primarily by the interaction of their electric dipoles—not only the oscillating dipoles providing dispersion forces, but also stronger dipoles fixed in the molecules.

In the long-chain molecules of a protein, the net charges are closely spaced and the strength of their interaction in neighboring molecules varies subtly because the chains bear neutral side groups of various sizes, separating the charges by various distances. The plastics chemist can engineer a molecule partly by choosing the spacings of its charges and the sizes of its side groups. His final cookery may deliberately cross-link adjacent molecules by covalent bonds.

Silicon, in the same group as carbon in the periodic table, crystallizes with the diamond structure. It might be expected to proliferate "organic" compounds by bonding to itself in rings and chains, as carbon does. Instead, and in a wholly different way, silicon proliferates the silicate minerals that dominate the earth's crust. In the geologic world, two silicon atoms never bond directly to each other; invariably one oxygen atom intervenes. The —O—Si—O—Si— chain plays somewhat the same role in inorganic material as the —C—C— chain plays in organic material.

These chains extend throughout the silicate minerals in such a way that each silicon atom is bonded to four oxygen atoms. Whenever one of the four oxygen atoms attached to a silicon atom is not engaged in bridging between two silicon atoms, it can readily accommodate one more electron to fill its outer shell. In minerals, one or another metal atom supplies that electron and thus becomes a positive ion. Accordingly, it's possible to think of the silicate minerals as made of negatively charged silicate "skeletons," often large and complex, within which the silicon-oxygen bonding is covalent. Tucked among these charged skeletons are positive metal ions, bonding the skeletons into solids by strong ionic bonds.

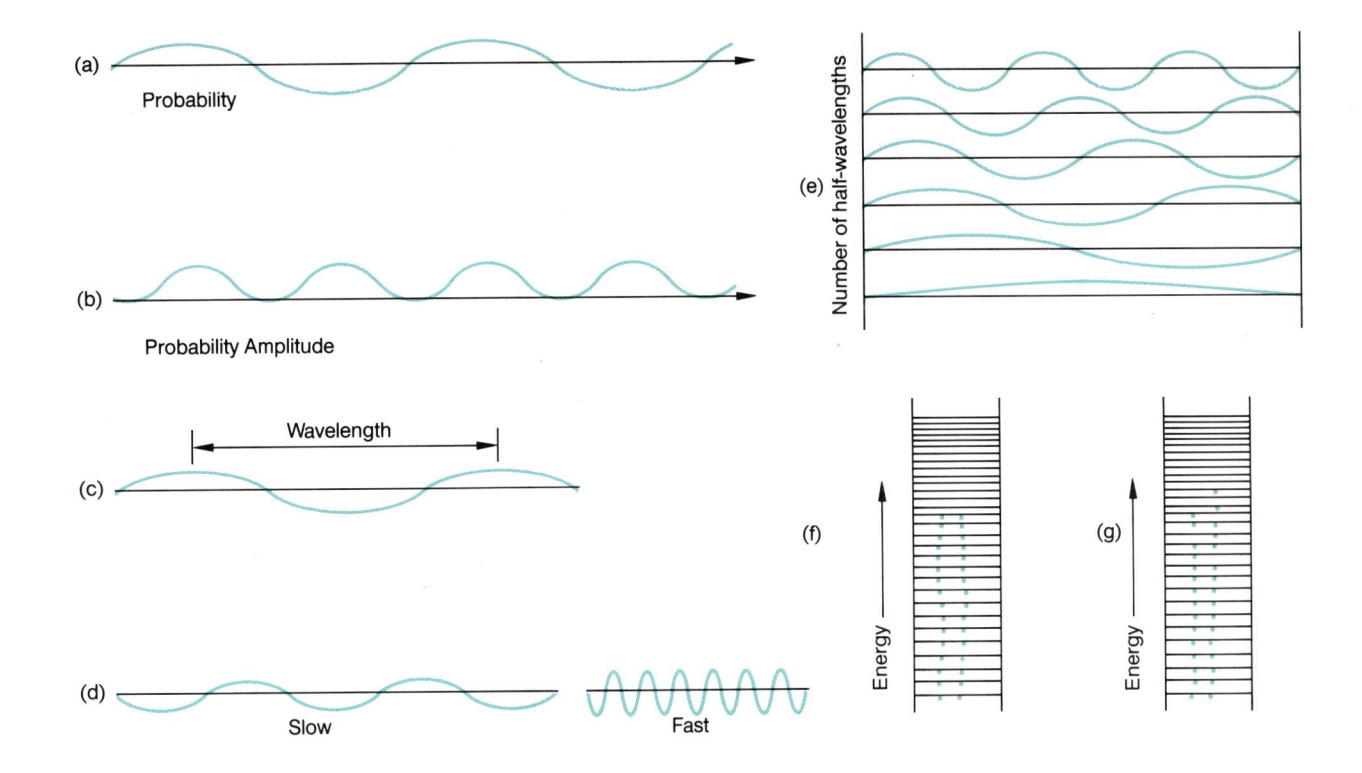

ELECTRONS IN METALS

A different picture of cohesion is needed to describe metals such as copper or silver. Metals behave as if one electron or so detaches itself from every atom. The remaining positive ions swim in a sea of roving electrons whose negative charge holds the ions together. Such metallic bonds are indiscriminate of number and direction; each ion collects around it as many ions as it can. All are attracted toward one another by the electrons threading between them. In their characteristic crystal structures, the metals resemble stacked cannonballs, with each sphere immediately surrounded by as many spheres as it can collect. The free electrons can move along distances through the metal when urged by an electric field, and this is how metals conduct electricity.

As with atoms and molecules, the observed behavior of a metal can be explained only by examining the expected behavior of its electrons. In earlier chapters, you saw that when electrons are bound in atoms, the possible states form a discrete set which does not merge continuously with one another. If its state is to change, an electron must make a "quantum jump" from one state to the other. In addition, if there are several electrons in the same situation — in the same atom, for example — electrons tend to be paired, one with spin "up" and the other with spin "down."

The same restrictions apply to the electron confined to a solid. Arnold Sommerfeld put these ideas together in a simple way to describe a metal. He pictured a piece of metal as a box containing nothing but

Figure 22. (a) One of the postulates of quantum wave mechanics is that the wave nature of an electron with a given energy can be represented by a sinusoidal wave in space. (b) This description is used in quantitative calculations to assess the probability of finding the electron at a given point by forming the square of the wave amplitude at that point. (c) Another attribute of this wave, as postulated by de Broglie, is that the momentum of such a wave is given by $p = h/\lambda$, where h is Planck's constant. (d) A slow electron has a long wavelength, and a fast electron a short wavelength. (e) Because nature is continuous, and because there are no electrons outside the box, it makes sense that the probability of finding the electron very close to the walls should be zero. (f) The spin of the electron enables two, but only two, electrons to adopt the same state of motion at any given time; a large number of electrons fill up their permitted states from the lowest to the highest energy level. (g) At ordinary temperatures, only those electrons near the highest energy level can gain extra energy and move to nearby (empty) states of energy.

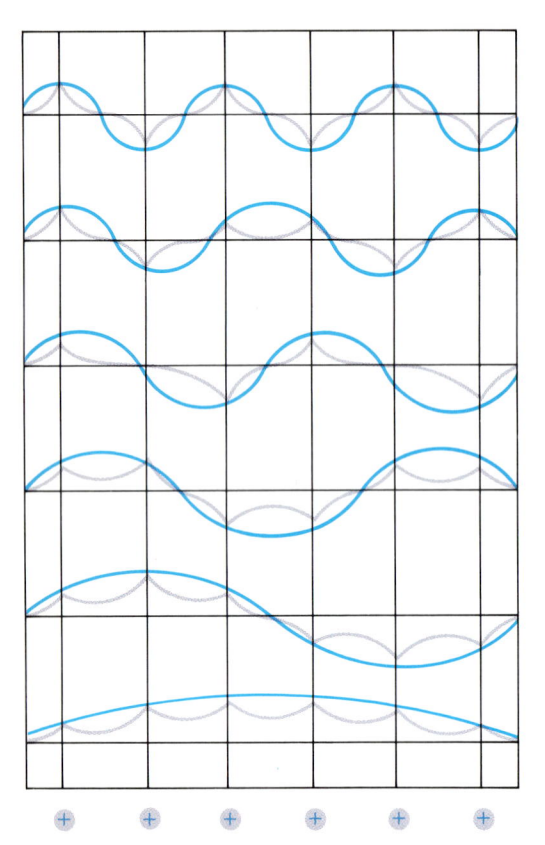

Figure 23. The sea of electrons as described by Sommerfeld is not without structure; the periodic lattice of positive ions causes periodic clustering of the negative sea due to electrostatic attraction.

probability waves representing the electrons (Figure 22). The walls of the box represent the positive ions in the metal, and prevent the electrons from escaping. But the walls confine the electrons, so they must also restrict the possible electronic states. In particular, the amplitudes of the permitted states must go smoothly to zero at the walls of the box because the electrons are never to be found outside, and only waves with selected wavelengths will accomplish that (Figure 22e).

What the electrons will do in the box can be understood by imagining that they are thrown into it one by one. The lowest energy level is the one with the longest wavelength, because it has the lowest velocity and therefore the lowest kinetic energy. According to the Pauli principle this level can be occupied by only two electrons. The next pair of electrons must occupy the level of next higher energy, and so on. The successive energies are so close to each other that no experiment could detect the difference, and therefore the whole distribution of energies is often considered quasi-continuous. Because there are a great many electrons in the piece of metal, the last pair is forced into a level with a very short wavelength and a correspondingly high energy (Figure 22f).

The great triumph of this simple model was that it demonstrated why the free electrons in a metal contribute negligibly to its heat capacity. Six calories of heat is required to raise by 1 degree Centigrade the temperature of 1 mole of a solid element. And it makes little difference what the element is. The amount of heat required for a gas is 3 calories. In a metal, where one electron is detached from each atom and roves like the atoms in a gas, the required heat should be 9 calories, 6 for the ions and 3 for the "electron gas," but it is still only 6.

To see why the electron gas in a metal has almost no heat capacity, imagine what the electrons must do in order to absorb heat. Because heat is a form of energy, the electrons must increase their kinetic energies. To accomplish that, they must find permissible states of shorter wavelength that are not already occupied by pairs of electrons. Such states are those with wavelengths shorter than the shortest already occupied. But the agitation available to excite the electrons is sufficient only to excite the tiny fraction that already have energies nearly large enough for them to attain those unoccupied high-energy states (Figure 22g). The contribution of the free electrons to the heat capacity of a metal is therefore nearly zero.

ELECTRONS IN A CRYSTAL

The simplicity and success of Sommerfeld's model invite refinements and extensions, and an obvious refinement is to take the ion cores of the metallic atoms into account more carefully. The cores change the foregoing picture of the electronic states in two respects, one expected and the other perhaps unexpected. You would expect to see changes in the shapes of the waves describing the states of the electrons. Because each ion core is positively charged, an electron is attracted to it. You would find any electron likely to be near rather than far from an ion core, and, accordingly, the waves develop peaks around these cores. Again as you might expect, the peaks look somewhat like those for the electron waves in separated atoms.

Without a deeper experience in physics, however, you might *not* expect that the permissible states now sort themselves into groups. Each group contains the same number of permitted states as the number of ion cores that were put in the box. The energies belonging to the states within each group are as closely spaced as they were in the ion-free box.

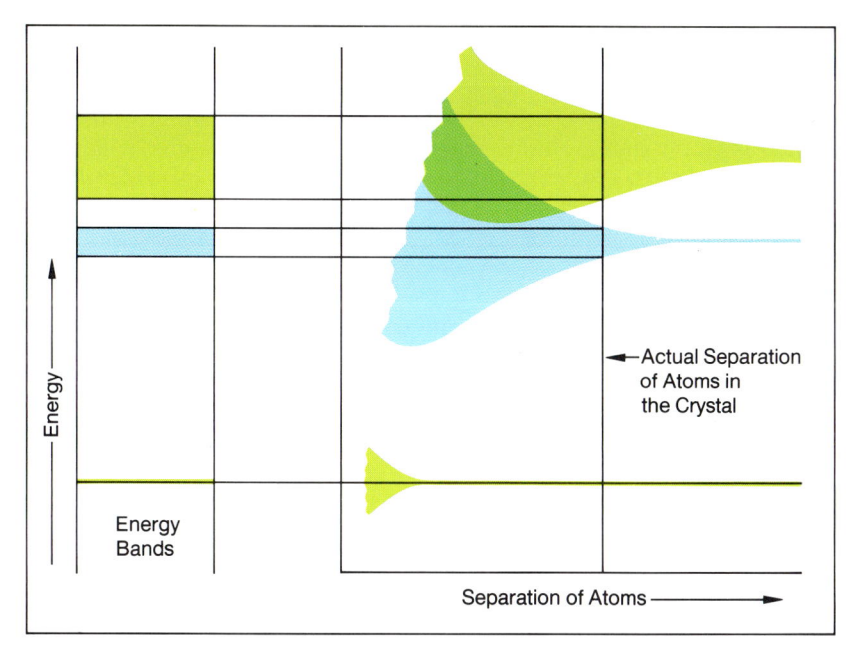

Figure 24. In an atom free from influence by other atoms, the electrons have definite energy states. In a crystal, the atoms influence one another and the electrons occupy groups, or *bands,* of states centered about the original energy levels. At the distance between atoms in a crystal, these bands may or may not overlap. Within each band, pairs of electrons tend to fill the lowest available energy states.

But that is not necessarily true from group to group. In some cases, the band of permitted energies in a group may overlap the band in another group; in others, an energy gap may separate two adjacent groups.

A helpful way of looking at these groups of electronic states is to imagine what would happen if you performed a "thought experiment." Pull all the atoms in a solid to great distances from one another and then slowly push them back toward one another. Each separated atom would start by offering an electron the states characteristic of its species, as you noticed when you studied the electronic structures of atoms in an earlier chapter. When Avogadro's number of atoms are brought back together, an atomic state on each atom combines with like states to yield Avogadro's number of states in a group. The total number of states remains the same, but their energies change as the atoms are brought together.

In a metal, the unpaired electron that each atom contributed to the electron gas was its most energetic electron when the atom was isolated. These electrons form the group with the highest energy. Because pairing is now possible, the group will accommodate twice as many electrons as it is offered, and the band of energies permitted in the group of highest energy is only half-filled. The behavior of those electrons then looks much like their behavior in Sommerfeld's box.

CONDUCTORS AND INSULATORS

A broad view of how electrons behave in solids arises in examining other groups of electronic states and other ways in which they become occupied. For example, the states of lower energy in the isolated atoms give rise to groups of lower energy states in the metal, separated by energy gaps from the half-filled group. In the atoms, those states are occupied by pairs of electrons and therefore the corresponding groups are fully occupied by the electrons in the ion cores in the metals, because no further pairing can occur.

When a group of states is fully occupied and an energy gap separates it from the group next higher in energy, an electron may have great difficulty acquiring enough energy to jump from its state in the lower

Figure 25. Conductors, insulators, and semiconductors have distinguishable band structures. They differ in the disposition of their bands of permitted energies and in the extent to which their electrons fill those bands.

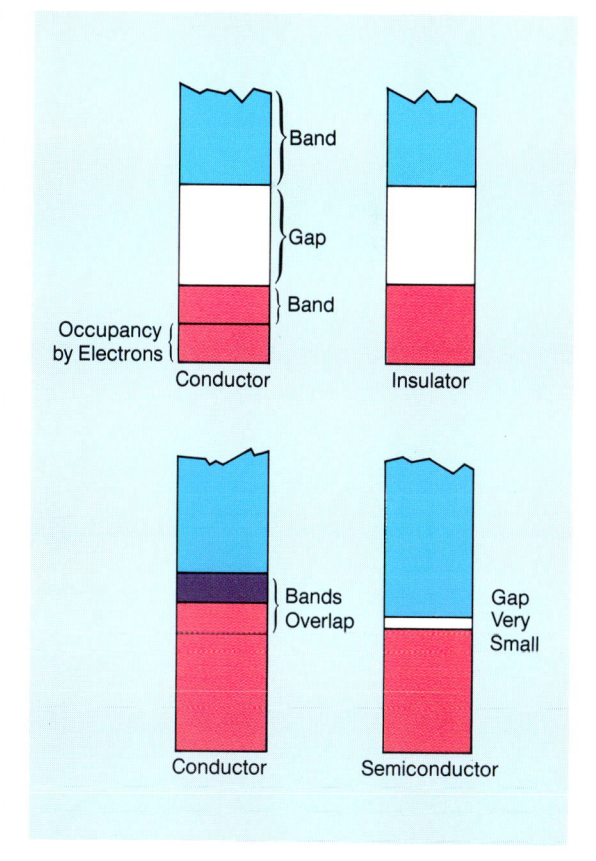

21 | States of Matter   415

group to a state in the higher group, even if the higher group is entirely unoccupied. Here lies the difference between good conductors of electricity such as metals, and good insulators such as sodium chloride. In order to drift through a solid under the urging of an electric field, an electron must be able to accept the little addition of kinetic energy entailed in its drift; if no state of higher energy is accessible, the drift is impossible. In insulators, the highest group is separated from the lower filled groups by a considerable energy gap.

If the energy gap is very small, however, the heat vibrations of the atoms may be able to kick a few electrons into states in the unoccupied group. This is what happens in the semiconductors, such as silicon and germanium, so important in the technology of electronics. The conductivities of these two solids are much less than those of true metals because relatively few electrons have been promoted to states that permit them to drift in an electric field. In even sharper contrast with the metals, the conductivities of semiconductors increase with temperature because more electrons are promoted when they are hot.

Why do the conductivities of metals *decrease* when they are hot? The question has an interesting answer, which gives further credibility to the wavelike nature of a moving electron. If you picture an electron being pushed by an electric field as a wave attempting to drift through the metal, you can use the same ideas that you used earlier in thinking about the diffraction of x-rays by a crystal. The pushed electron wave advances until it encounters a part of the crystal with a spacing of atoms that is just right to reflect it. It then moves off in another direction, only to be reflected again by another such encounter. This incessant scattering impedes its drift, and the electric conductivity of the metal will be lower if the scattering is more frequent. The higher the temperature of the metal, the further its vibrating atoms depart from their normal spacings and the more often an electron wave confronts a spacing that is instantaneously the right spacing to reflect it.

From this picture of *electronic resistance,* it might be expected that the resistance of a metal might approach zero at the absolute zero temperature. And so it does—if the metal is pure. But in many metals—in lead, for example—a more surprising thing happens. At a critical temperature, low but finite, the resistance drops to zero abruptly: the metal becomes a *superconductor.* For lead, the critical temperature is 7 degrees Kelvin.

Until recently, the explanation of superconductivity was the largest problem remaining to be solved by the wave-mechanical theory of solids. It is clear now that below the critical temperature, a cooperation of atomic vibrations with the electron waves replaces the contest carried on at higher temperatures. But theorists are still working on the details of the theory, and experimenters are contributing new data to check it. A vigorous search for superconductors with higher critical temperatures progresses: the highest found so far is at 21 degrees Kelvin in an alloy of niobium, aluminum, and germanium. Here is one of the truly exciting fields of contemporary physical research.

PLASMAS

In the 1930s Irving Langmuir, the man whose two-dimensional states of matter opened this chapter, examined an electric discharge with a probe, recognized some unusual properties in a region of the gas just above the cathode, and coined the name *plasma* to denote them. Since then, further studies of the gas in that region, and of such seemingly different phenomena as the aurora borealis ("northern lights"), have revealed what is sometimes called "the fourth state of matter."

A plasma consists of a disordered collection of ionized atoms — of ion cores and the electrons that have left them. You have just read that a metal is also a collection of ion cores and roaming electrons. To be sure, a metallic collection is ordered and a plasma is disordered, but a plasma differs from a metal in a more important way. In a metal the electrons are in *thermal equilibrium* with the ion cores; in a plasma they are not.

If you heat a particular place on a piece of matter — gas, liquid, or solid — the piece conducts the heat to all its parts so that all the parts rapidly reach the same temperature. The atoms in the hotter part, moving more energetically on the average, kick their cooler neighbors into more vigorous motion; soon the average restlessness becomes the same throughout the piece of matter.

The atoms in most pieces of matter are able to accomplish these exchanges of kinetic energy because they are more or less tightly coupled to one another. Because the atoms are farther apart in a gas than in liquids and solids, gases are our best thermal insulators when they are prevented, by enclosure in little pockets, from transferring heat by moving bodily. In a metal, the roaming electrons interact intimately with the ion cores, and the tight coupling, combined with their high mobility, makes metals our best thermal conductors. The electrons rapidly interchange kinetic energy with the ion cores.

In a plasma, however, the coupling between electrons and ions is relatively loose. The electrons are sufficiently far from the ions to reduce the strength of their interaction, and they move sufficiently rapidly to reduce the time in which any one electron can interact with any one ion. So if the plasma is subjected to some outside influence that primarily affects only the ions or only the electrons, the effects of that influence may be slow to reach the other constituent. When energy is pumped into a plasma at low pressure by an electric field, for example, it is primarily the electrons that are accelerated, and they transfer their energy to the ions so slowly that the plasma must be regarded as a mixture of two gases at two different temperatures. In the *ionosphere,* the ionized region of the earth's atmosphere, the temperature of the ions and atoms is quite stable whereas the temperature of the electrons has wide fluctuations.

The effects of a magnetic field on a plasma become important in explaining many phenomena in the atmosphere and in the stellar world. These effects result from the fundamental fact that the motion of a charged particle in a magnetic field is altered from its otherwise straight line. If the motion of the charged particle is at right angles to the field, the particle will trace a helical (spiral) path whose axis lies along the field. At very high field strengths, the helix becomes so tightly coiled that it approximates to a straight line along the field. Thus magnetic fields of suitable configuration and sufficient strength, operating on sufficiently light particles moving sufficiently fast, might confine the particles to a limited space, much as the walls of a container would confine them. This possibility is the basis for some present attempts to achieve thermonuclear fusion by confining a plasma at temperatures higher than any confining walls would tolerate.

Recognition of the unique properties of this fourth state of matter has already formed the basis of new technological devices such as the "magnetohydrodynamic electric generator," and no doubt more will follow. Meanwhile the relevant theoretical and experimental work, complex and fascinating, continues to advance. Whether it can lead to controlled nuclear fusion as a usable source of energy still remains to be seen.

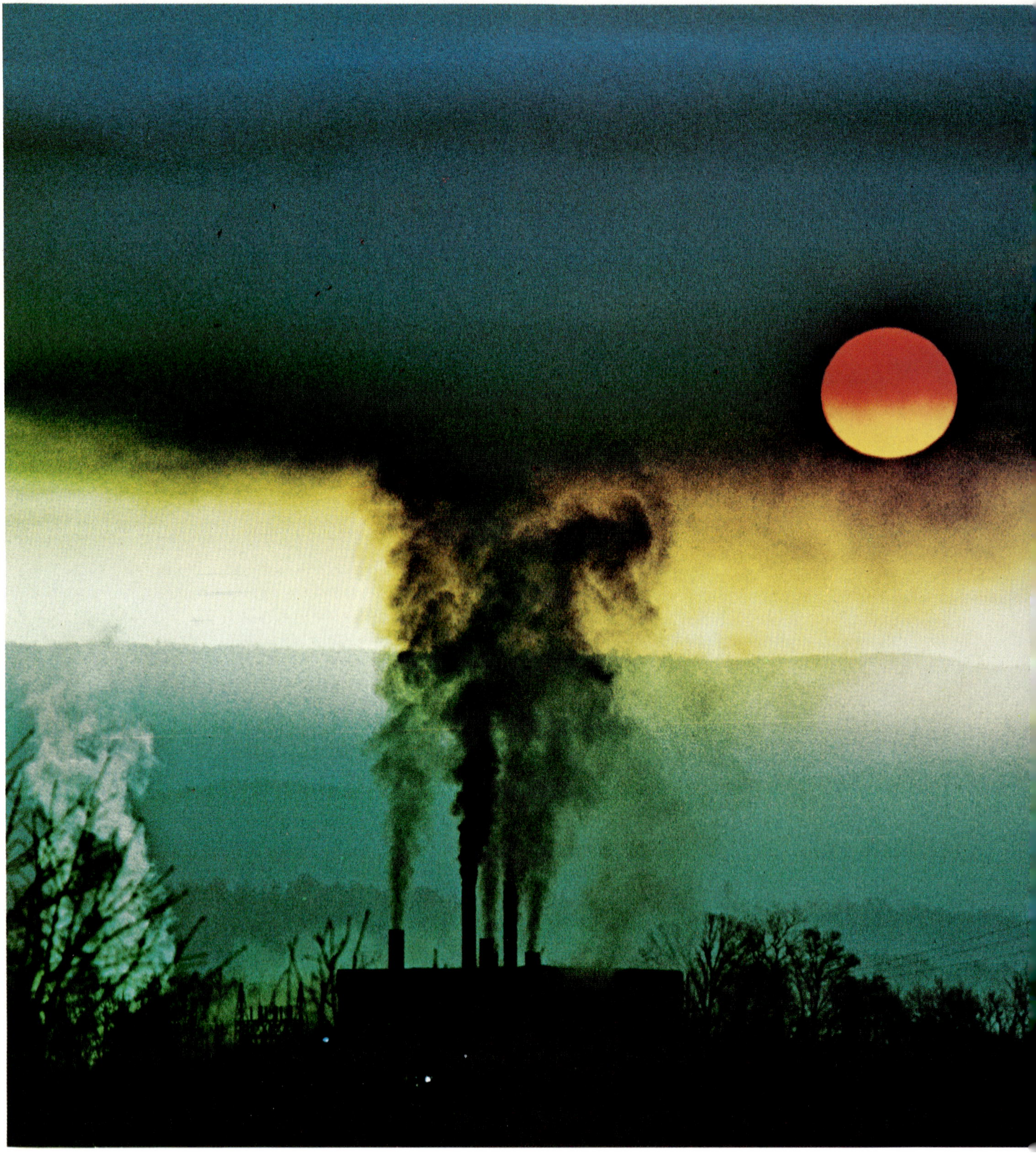

There are millions of known chemical reactions—no one knows precisely how many there are. Some of the reactions take years to occur; others take place in less than one trillionth of a second. Some liberate enormous quantities of heat, some absorb heat, and a great many occur with very little heat change. How can all these diverse processes be thought about in manageable ways?

Begin with a definition of a chemical reaction. When substances interact with one another in such a way that they form new substances, we say chemical reaction has occurred. Typically, the bonds between atoms are broken or rearranged during the course of a chemical reaction. And in accordance with the law of conservation of matter, mass can be neither created nor destroyed during this process. In other words, the stuff that results from a chemical reaction always has the same total mass as all the stuff you start out with. All of it can be accounted for.

Consider the reaction of oxygen with hydrogen to produce water. The equation describing this reaction is

$$2H_2 + O_2 \longrightarrow 2H_2O$$

If you mix gaseous hydrogen and oxygen together very carefully and then discharge an electric spark in the gaseous mixture, the two substances will react violently with one another and form water. The equation above indicates the newly created water molecules contain *exactly* the same number of hydrogen and oxygen atoms as were contained in the hydrogen and oxygen that were consumed. The amount of hydrogen that will react with a given amount of oxygen is rigidly fixed.

In fact, *any* substance that goes into making a given compound must combine in a definite, fixed proportion with other substances. That statement is known as *the law of definite proportions*. If there is an excess of oxygen in the reaction mixture, it will remain after the reaction is over; similarly, if there is an excess of hydrogen it will remain after all of the oxygen has been used up.

When that happens, you end up with a mixture of liquid (the water) and gas (the excess oxygen or excess hydrogen). Such a mixture is called a *heterogeneous* substance because the properties of its constituents are not all the same. In contrast, hydrogen alone (or oxygen alone) is considered a *homogeneous* substance because the properties of all the atoms or molecules comprising it are identical. Water, too, is a homogeneous substance: each one of the newly created water molecules is like all the others.

BALANCED EQUATIONS

The relationship between the amounts of the reactants and the products involved in a chemical change is called the *stoichiometry* of the reaction. Chemical equations are exact statements of the stoichiometric amounts of materials consumed and produced in a chemical reaction. These equations must be balanced; that is, the correct formulas of the compounds must be used, and the numbers of each kind of atom must be the same on both sides of the equation.

Often you can tell whether or not an equation is balanced merely by looking at it. Consider the equation given earlier. Theory tells you that

22
Chemical Change

Paradoxically, many of our more stunning sunsets exist because of our more offensive modifications to the delicate chemical balance of the atmosphere.

Figure 2. The element sodium is an extremely reactive metal. It reacts violently with water to form sodium hydroxide, with the evolution of gaseous hydrogen.

hydrogen molecules should be diatomic (H_2) and that each molecule of the common form of oxygen also contains two atoms (O_2). The formula for water is also consistent with theory because common hydrogen can form one bond and oxygen can form two bonds. But if you know that hydrogen and oxygen react to form water, why can't you simply write

$$H_2 + O_2 \longrightarrow H_2O$$

The answer is that the equation is unbalanced: there are two oxygen atoms on the left-hand side and only one oxygen atom on the right-hand side. Then why not write

$$H_2 + O_2 \longrightarrow 2H_2O_2$$

That may be a balanced equation, but it has nothing to do with the reaction that produces water: H_2O_2 is the formula for hydrogen peroxide. In balancing equations, you can't change the *formulas* of the compounds.

The way to proceed is to first double the number of water molecules. That will make the total number of oxygen atoms the same on both sides of the equation:

$$H_2 + O_2 \longrightarrow 2H_2O$$

Next, double the number of hydrogen molecules,

$$2H_2 + O_2 \longrightarrow 2H_2O$$

and you have, finally, a balanced equation.

Sodium hydroxide is another substance that can be used to illustrate how chemical equations are balanced. The element sodium is represented only by the symbol Na; there is no subscript number to indicate how many atoms a molecule of sodium contains. Nevertheless, sodium is not monatomic. At room temperature, sodium is a solid crystal that is like a continuous, giant molecule: there is no way of determining exactly which atom belongs to which molecule. All solid metals display this characteristic so it's not possible to assign simple molecular formulas to them.

Instead, metals are always represented in equations by their elemental symbol.

The element sodium is an extremely reactive metal. When brought into contact with water it reacts vigorously to form sodium hydroxide with the evolution of gaseous hydrogen. The formula for sodium hydroxide is NaOH. The reaction might be thought of as a simple substitution, with sodium replacing one of the hydrogen atoms of water. But this picture is not quite accurate: water is a covalent, molecular compound, and sodium hydroxide is an ionic solid containing sodium ions (Na$^+$) and hydroxide ions (OH$^-$). The reaction itself proceeds in the following manner:

Na + H$_2$O $\longrightarrow$ NaOH + H$_2$

Now, the hydrogen atoms occur on the left-hand side only in multiples of two. In writing a balanced equation, then, you must include a like number of hydrogen atoms on the right-hand side. First, double NaOH so there is an even number of hydrogen atoms in the products:

Na + H$_2$O $\longrightarrow$ 2NaOH + H$_2$

Now double the sodium and balance the number of hydrogen atoms by doubling the number of water molecules:

2Na + 2H$_2$O $\longrightarrow$ 2NaOH + H$_2$

which gives you a balanced equation for the reaction.

CHEMICAL STRUCTURE AND REACTIVITY

By itself, a chemical equation will not enable you to predict what the result of a reaction will be, nor will it tell you what conditions will cause the reaction in the first place. It only states the formulas of known reactants and their products. On the other hand, if you know enough about chemical behavior you may often be able to use existing equations to extrapolate how substances will react under different conditions.

For example, the electronic theory of molecular structure suggests a compound of the formula CH$_4$ should exist. And in fact the compound methane has this predicted composition and structure. It is found in nature as part of the mixture called natural gas. But what are the chemical transformations underlying methane?

Methane can undergo a number of reactions. For example, it burns in oxygen, producing carbon dioxide and water:

CH$_4$ + O$_2$ $\longrightarrow$ CO$_2$ + H$_2$O

A balanced equation may be written for this reaction:

CH$_4$ + O$_2$ $\longrightarrow$ CO$_2$ + 2H$_2$O

When a substance interacts chemically with oxygen, and when this reaction proceeds with the rapid release of heat and light energy, we say that *combustion* has taken place. Combustion of methane is important because it is the main reaction that occurs when natural gas is burned to produce heat or electric power. The reaction also shows some interesting characteristics. Air and natural gas can be mixed without the occurrence of any reaction. If the reaction is initiated with a lighted match or an electric spark, however, combustion occurs rapidly and may even lead to an explosion.

Chemists talk about reactions that are known to occur and those that might occur. Reasoning often indicates that unknown, but potentially useful, reactions should be possible. If you know methane reacts

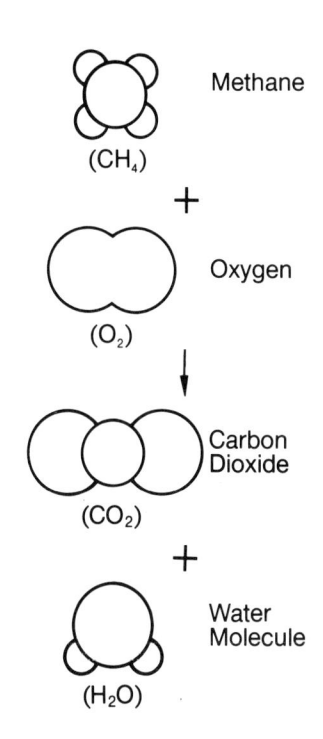

Figure 3. The combustion of methane in oxygen produces carbon dioxide and water.

Figure 4. An example of an *exothermic* reaction (the energy released in the fire) and an *endothermic* reaction (energy consumed through the use of the fire extinguisher).

with oxygen, for example, you might suspect a similar compound, such as ethane (H_3C—CH_3), would also react with oxygen. Furthermore, theoretical considerations may also enable you to infer which reaction conditions are likely to allow the new reaction to proceed smoothly. For instance, if you know how to make methane burn smoothly without exploding, you might be able to predict how to do the same with ethane. Alternatively, such considerations may also lead you to conclude that a hypothetical reaction is impossible. Both kinds of considerations are valuable in attempts to control nature by controlling chemical change.

ENERGY CHANGES IN CHEMICAL REACTIONS

When considering compounds, you have to ask two fundamental questions. First, will a reaction occur among the compounds if a pathway is opened up? And second, will the reaction occur rapidly enough to be significant? In the reaction of methane with oxygen, the answer to the first question is yes, but you would not get the answer merely by mixing methane and oxygen to see if they react. The answer to the second question is also yes, but again you would not find the answer merely by mixing the reactants, for without a triggering spark or flame, the reaction will not occur rapidly enough to be significant.

The first question involves chemical energetics. Energy changes occur during chemical reactions. In some cases, reactions produce energy that is delivered to the surroundings; in other cases, reactions occur only if they are driven by some external energy source. Either way, there is a transfer of energy at the electron level. For example, in a heat-producing, or *exothermic,* reaction, part of the potential energy of the electrons is transformed as they move into new energy levels. The transfer may be accompanied by violent molecular motion. A heat-consuming,

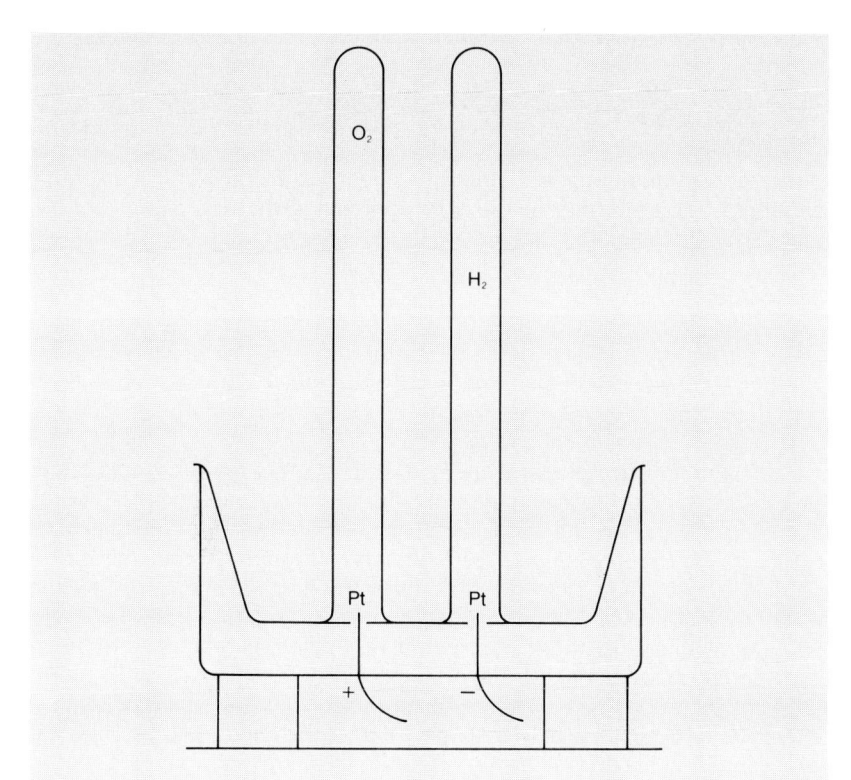

Figure 5. In the electrolysis of water, an electric current passes through a dilute solution of sulfuric acid in water. The sulfuric acid acts as an electrolyte. An amount of electric energy equal to 68.32 kilocalories per mole of water is used to chemically change water into one mole of hydrogen gas and one-half mole of oxygen gas.

or *endothermic,* reaction involves a chemical change in which energy in the form of heat is lost.

Electrolysis is one means of providing external energy. Electric conduction through a gas or liquid can occur when positive and negative ions are made to move through the substance. For example, when an electric current is passed through a water solution containing some ionized solute, the water decomposes into hydrogen and oxygen. The chemical change may be stated as:

$$2H_2O \xrightarrow{electrolysis} 2H_2 + O_2$$

To give a more definite statement about this reaction, you can specify the conditions needed to bring it about. They are often included in equations by notes written above or below the arrow:

$$\text{Reactants} \xrightarrow[\text{conditions}]{\text{conditions}} \text{products}$$

For the reaction given above, you could write:

$$136.64 \text{ kilocalories} + 2H_2O \xrightarrow[\text{1 atmosphere}]{25°C} 2H_2O + O_2$$
ELECTRIC ENERGY

Here, the pressure of the reactants is 1 atmosphere and the temperature at which reaction occurs is 25 degrees Centigrade.

When a mixture of hydrogen and oxygen is ignited with a flame or an electric spark, the reaction proceeds rapidly and a large amount of energy—68.32 kilocalories per mole—is given off. (A mole is merely Avogadro's number; it is a weight equal to the molecular weight of a substance.) A complete description of this reaction would include the

Figure 6. (Far right) The cycle of life. Oxygen-breathing, plant-eating animals release carbon dioxide (CO_2) to the air. When animals die and decompose, nitrogen is extracted from protein and is used with CO in chemical reactions, triggered by solar energy, to enable the plant to grow. The most important growth process of plants is photosynthesis. The green chlorophyll in plants absorbs red and orange wavelengths of sunlight and acts as a catalyst to add the energy needed to convert carbon dioxide and water into carbohydrates and oxygen. Oxygen is then released into the air, completing the life cycle. It is estimated that in every second, 5 million calories of energy are stored as carbohydrates by plants on this planet. The combustion of plants reverses the process and energy is released in a form that is used by animals for fuel. The equation for this latter process is:

$$6CO_2 + 6H_2O \xrightarrow[\text{Chlorophyll}]{\text{Sunlight}} C_6H_{12}O_6 + 6O_2$$

(Carbon Dioxide) (Water) (Glucose) (Oxygen)

temperature, the pressure of the reacting gases, and the fact that liquid water, rather than steam, is produced:

$$2H_2 + O_2 \xrightarrow[\text{1 atmosphere}]{25°C} 2H_2O + 136.64 \text{ kilocalories}$$

GAS   GAS   LIQUID

Under other conditions, the amount of heat produced would be slightly different. The important things to remember are that the amount of liberated heat is large, and that the amount of heat produced (or consumed) during a reaction varies with the conditions under which a reaction is carried out.

From these two examples, you can see that the energy imparted to water during electrolysis can be stored in the form of hydrogen and oxygen and then released again if those two substances are allowed to react with each other to form water. The stored energy is the energy of chemical combination, or *chemical energy*.

The chemical storage of energy is important to many aspects of our society; batteries, for example, represent a widespread portable source of stored energy. But the chemical storage of energy is fundamental even to life itself. Green plants absorb sunlight. Through the mechanism of *photosynthesis,* they convert solar energy to chemical energy by causing water to react with carbon dioxide, thereby producing carbohydrates. The chemical energy that carbohydrates represent is then used to maintain the life process in plants and animals. When we burn coal or petroleum products such as gasoline or natural gas, we are using chemical energy stored millions of years ago by photosynthetic plants. High explosives such as dynamite consist of high-energy compounds which, when detonated, decompose rapidly and release a large amount of energy. Providing enough fuel to thrust a spacecraft out of the earth's gravity field means the maximum amount of chemical energy must be stored in the minimum weight of propellant. Of course, when the chemical energy is converted to mechanical thrust, the propellant must be burned at a rate that is controllable.

REACTION RATES

In the regulation of chemical reactions, the rate of reaction is as significant as the underlying energetics. The rate of a reaction is a measure of how fast reactants disappear and products appear. Some reactions are so rapid that their rates are measured on time scales of a millionth of a second or less; others are so slow that only the study of geologic ages will indicate whether or not chemical change is occurring at all. Energy relationships indicate that most metallic elements should react with atmosphere oxygen to form oxides. Oxidation is an exothermic reaction, so you might expect that reaction would occur spontaneously and that metals could not exist in contact with the atmosphere. Some metals such as sodium, potassium, and calcium do react rapidly with air. But often the oxidation of a metal is slow. The rate at which iron rusts is immeasurably slow in dry climates but speeds up in the presence of moisture. That is why steel tools can be protected against rusting by a thin coating of grease; the film keeps water from contacting the metal. Oxidation of iron can be made extremely rapid. If fine steel wool is held in the flame of an oxygen torch, the metal bursts into flame and burns spectacularly.

Rates of reactions vary enormously. They are sensitive to conditions such as temperature, pressure, the concentrations of reactants, the physical condition of the reactants (gas, liquid, solid, solution, and so forth), and the presence or absence of something called a catalyst. A

Animals

O_2

Food Eaten

CO_2

Organic Nitrogen

Green Plants

Ultraviolet Light From the Sun

O_2

CO_2

Photosynthesis

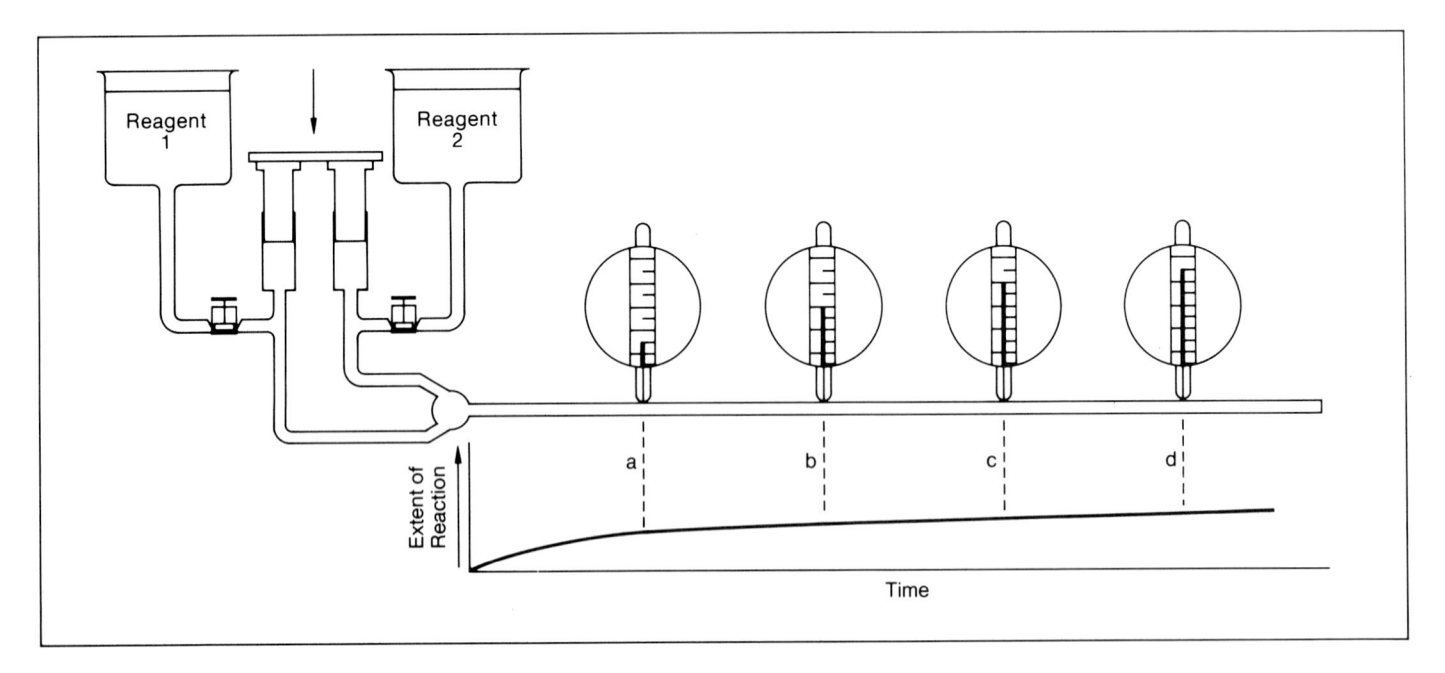

Figure 7. The rate of a reaction is a measure of how fast reactants disappear and products appear.

catalyst is a material that increases the rate of a chemical reaction without being consumed by the reaction. It is a powerful tool for controlling chemical change. The function of the catalyst is well understood in some reactions and poorly understood in others, but chemists are convinced that catalysis is never a matter of remote control.

Reactants always become bound to catalytic substances at some stage in the reaction. The interaction is temporary and merely helps the reactant pass through some awkward configuration that is required to change reactants into products. The role of platinum in accelerating the reaction of hydrogen and oxygen is an example of a catalytic effect. The chemical reactions in most living organisms take place at atmospheric pressure and moderate temperatures. Each reaction is speeded up by its own catalyst, called an enzyme. Numerous enzymes are involved in the digestion process, for instance. If you had to rely only on your stomach acids to break down food, it would take years to digest each meal.

An important industrial use of catalysis is in the breaking up, or *cracking,* of crude oil molecules into gasoline. Gasoline represents only a minor fraction of petroleum, and cracking is a way to increase the yield. A catalyst is used to initiate the complicated sequence of reactions that break down the hydrocarbons of petroleum into smaller molecules.

TYPES OF REACTIONS

There are between 5 million and 10 million known chemical compounds and all can be involved in numerous chemical reactions. Obviously, no one scientist knows all the reactions; and if he did, his knowledge would be out of date tomorrow because new reactions are discovered each day. Studying chemical reactions merely by memorizing facts must ultimately lead to failure and frustration. What is needed is a generalized picture of chemical reactivity. If the overall view is well designed, specific details about individual reactions or small groups of reactions can be fitted neatly into the big picture. However, the details should not be ignored. A theoretical overview unrelated to facts, and bare facts unattached to any kind of theory, are equally worthless. Each reaction presented here is representative of a certain type of reaction.

Examination of a balanced chemical equation provides a logical basis for classifying reaction types. You look at the equation and describe the reaction in terms of the changes in chemical bonding that have occurred. This means that you must not only be able to write the correct equation for the reaction; you must also know the structures of the compounds involved. Five categories of reactions are discussed here: substitution, acid-base, addition and elimination, isomerization, and oxidation-reduction.

Substitution Reactions

Substitution reactions always involve replacement of an atom or a group of atoms in a molecule by another atom or group. An example of such a reaction is the substitution of chlorine for bromine bonded to carbon:

$$\begin{array}{c} H \\ H-C-Br + Cl^- \longrightarrow H-C-Cl + Br^- \\ H \end{array}$$

In substitution reactions, the reactants may be either molecules or ions. In most cases one of the reactants is considered to be the *substrate* at which the substitution occurs, and one the *entering group*. In the reaction

$$\begin{array}{c} \text{LEAVING GROUP} \\ H \\ H-C-Br + Cl^- \longrightarrow H-C-Cl + Br^- \\ H \\ \text{SUBSTRATE} \quad \text{ENTERING GROUP} \end{array}$$

the CH_3Br is the substrate and Cl^- is the entering group that replaces Br^-. The substance that is replaced—in this case, Br^-—is called the *leaving group*.

Generally, the entering group possesses at least one pair of electrons that are available to form a bond with the substrate molecule. In the above example, Cl^- has four lone pairs,

$:\ddot{\underset{..}{Cl}}:^-$

and can make a bond to the carbon to help push away the Br^- group.

These details of the reaction are referred to as a *reaction mechanism*. The displacement of Br^- by Cl^- occurs because the Cl^- partially bonds to the CH_3Br substrate, thereby "easing out" the Br^-. This description helps point out that a good entering group in a substitution reaction should have *an available pair of electrons* in order to make a bond with a substrate molecule.

Substitution reactions are of vital importance in our bodies. The digestion process consists primarily of a water molecule substituting for a much larger sugar, fat, or protein molecule. Our lungs serve the essential function of transporting oxygen (O_2) into our bloodstream and carbon dioxide (CO_2) out of it. But other small, gaseous molecules can tag along to a small degree in this process. If they are inert to chemical reactions in the body, then there is no problem. But if they are reactive chemicals, they often are dangerous. For example, carbon monoxide (CO) replaces the oxygen (O_2) bound to iron atoms in hemoglobin molecules in the blood and thus reduces the amount of oxygen reaching the cells of the body. It is this reduction of oxygen that decreases the alertness of the drivers in rush-hour traffic jams.

Acid-Base Reactions: Proton Transfer

Acids are hydrogen-containing, nonmetallic substances that are corrosive because of their great chemical reactivity. Some acids, such as concentrated hydrochloric acid, are so strong that they may cause painful burns; others, such as carbonic acid, are weak enough to be used in soft drinks. *Bases,* in contrast, are hydroxides of metal that are capable of neutralizing acids. We use a weak base, sodium bicarbonate, to neutralize excess stomach acid. We use a strong base, sodium hydroxide, to clear out drain pipes. Most acid-base reactions are substitution reactions in which a hydrogen ion (a proton) is transferred from one basic substance to another.

When acids are dissolved in water they form ions. For example, hydrofluoric acid ionizes in water to produce hydrogen ions (protons):

$$HF \xrightarrow{H_2O} H^+ + F^-$$

ACID   HYDROGEN ION   NEGATIVELY CHARGED ION

When we measure the acidity of a solution, we are measuring the concentration of hydrogen ions. However, the best evidence we have shows that the hydrogen ion (proton) does not actually exist in water. The ionization of an acid in water is better illustrated by the following reaction:

$$HF + H_2O \longrightarrow H_3O^+ + F^-$$

ACID   WATER   HYDRONIUM ION   NEGATIVELY CHARGED ION

When separating into simple ions, the proton becomes attached to a water molecule (H_3O^+) rather than floating about freely in the solution. The ion H_3O^+ is called the *hydronium ion.*

When bases are dissolved in water, they, too, dissociate into ions as the following example for sodium hydroxide illustrates:

$$NaOH \longrightarrow Na^+ + OH^-$$

OH^- is the hydroxide ion that characteristically emerges when bases are dissolved in water, just as H_3O^+ is the hydronium ion that characteristically emerges when acids are dissolved in water. Both produce an excess of negative ions and positive ions, respectively, in the solution. When both kinds of ions are present, they have a strong tendency to recombine into water:

$$H_3O^+ + OH^- \longrightarrow H2H_2O$$

The characteristic properties of both acid and base are destroyed, or neutralized.

A scale called the *pH scale* has been devised to compare the acidity or basicity of solutions. On this scale, a solution with equal concentration of hydrogen ions and hydroxides (very low concentrations of each) has a pH of 7, which is the neutral pH region. The greater the hydrogen ion

Figure 8. pH is the negative of the logarithm of hydronium ion concentration (in moles per liter).

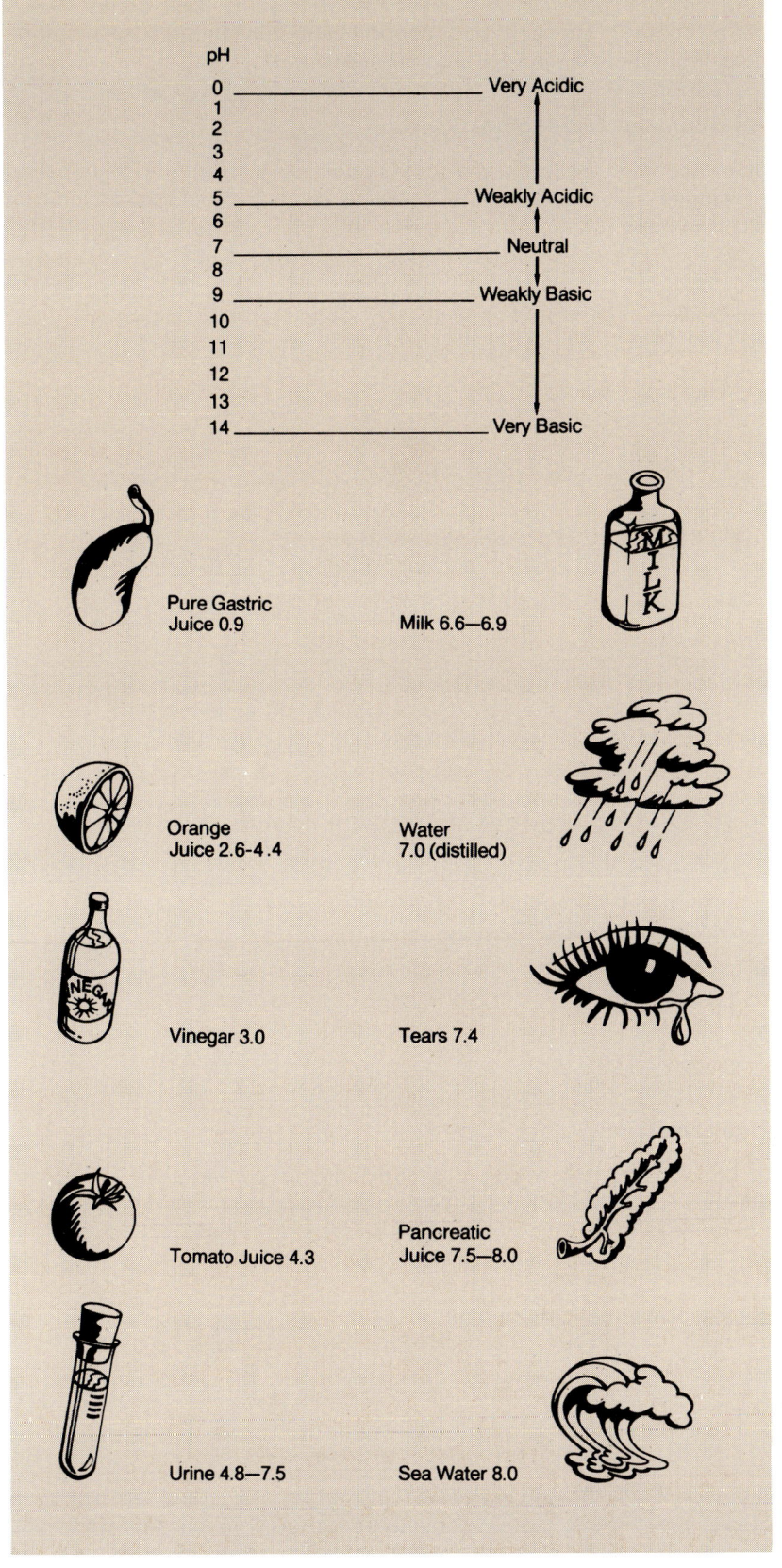

pH	
0	Very Acidic
1	
2	
3	
4	
5	Weakly Acidic
6	
7	Neutral
8	
9	Weakly Basic
10	
11	
12	
13	
14	Very Basic

Pure Gastric Juice 0.9

Milk 6.6—6.9

Orange Juice 2.6-4.4

Water 7.0 (distilled)

Vinegar 3.0

Tears 7.4

Tomato Juice 4.3

Pancreatic Juice 7.5—8.0

Urine 4.8—7.5

Sea Water 8.0

22 | Chemical Change

concentration becomes, the lower the pH reading. Soft drinks have a pH of about 3; stomach acid, about 1; and concentrated hydrochloric acid, −1. Numbers above 7 are considered basic.

Addition and Elimination

In an addition reaction, a molecule is built up through the combination of simpler substances. In the following examples, the dashes indicate electron pairs.

$$H_2C=CH_2 + H-Cl \longrightarrow H-CH_2-CH_2-Cl$$

ETHYLENE   HYDROGEN CHLORIDE   ETHYL CHLORIDE
(C_2H_4)   (HCl)   (CH_3CH_2Cl)

$$SF_4 + F_2 \longrightarrow SF_6$$

SULFUR TETRAFLUORIDE   FLUORINE   SULFUR HEXAFLUORIDE
(SF_4)   (F_2)   (SF_6)

Elimination is the opposite of addition. In an elimination reaction, some molecule is always broken down into smaller units:

$$CH_3CH_2Cl + KOH \longrightarrow H_2C=CH_2 + H_2O + KCl$$

ETHYL CHLORIDE   POTASSIUM HYDROXIDE   ETHYLENE   WATER   POTASSIUM CHLORIDE

or, as another example,

$$XeF_6 \longrightarrow Xe + 3F_2$$

XENON FLUORIDE   XENON   FLUORINE

Isomerization

In an isomerization reaction, a compound is converted to an *isomer,* which is a substance having the same composition as the original compound but a different structure. For example, n-butane,

$H_3C-CH_2-CH_2-CH_3$

can be converted into i-butane:

$$H_3C-CH(CH_3)-CH_3$$
(with CH₃ above and CH₃ below the central C-H)

Although both compounds have four carbon atoms and ten hydrogen atoms, they have different chemical properties. Isomerization reactions appear deceptively simple, but the actual mechanisms of change are usually complicated. In the reaction shown above, the conversion from one form to the other involves two or more steps in which the molecules are actually taken apart and reassembled.

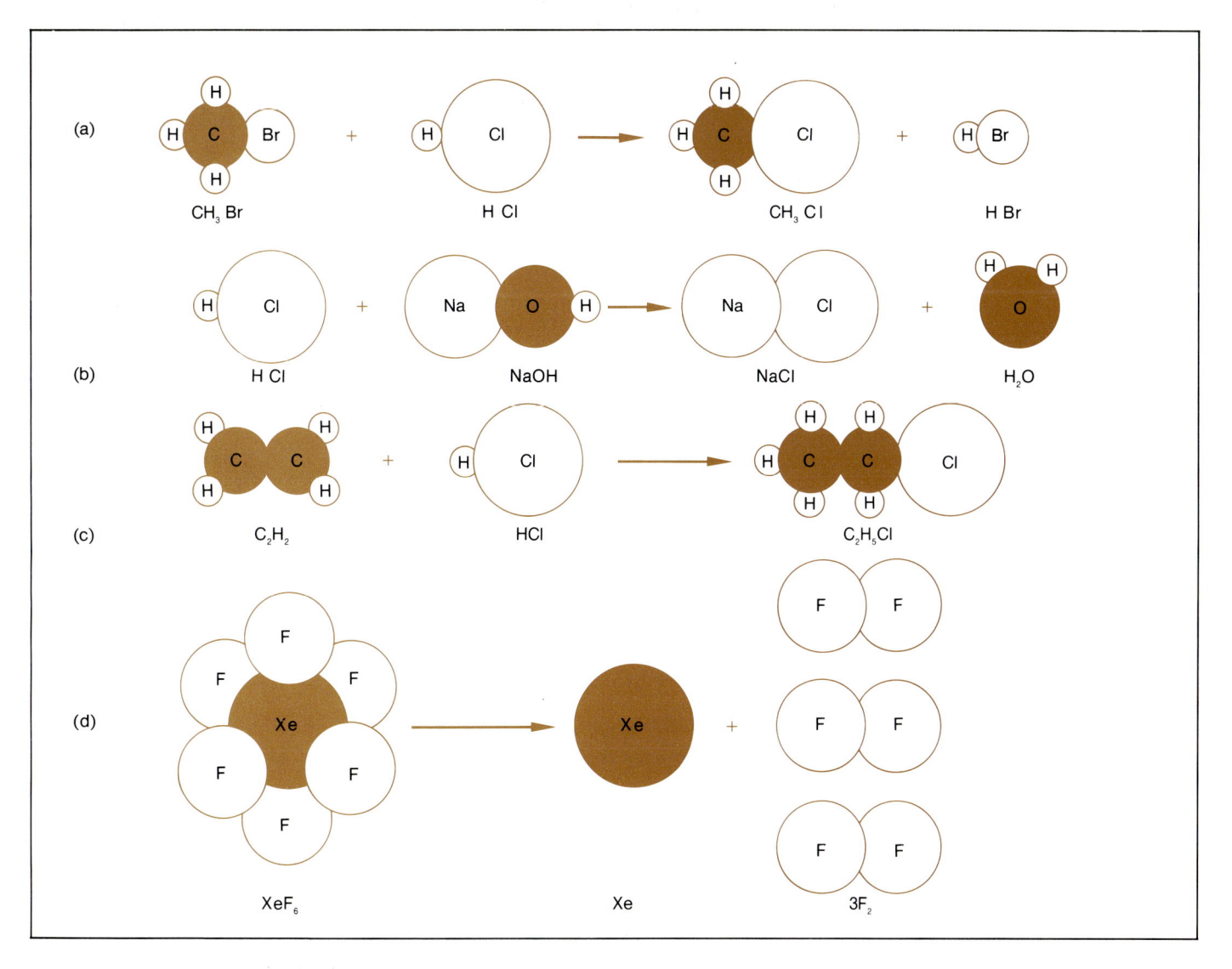

Figure 9. Examples of four kinds of chemical reactions: (a) substitution, (b) acid-base, (c) addition, and (d) elimination.

Oxidation-Reduction

Historically, the term *oxidation* meant "reaction with oxygen." It was applied to the process of combustion, both fast and slow, of various materials exposed to atmospheric oxygen. However, most of the changes caused by reaction with oxygen can be accomplished by treatment with other substances, which came to be called *oxidizing agents*. Development of the electronic theory of molecular structure brought further insight into the structural changes that occur in oxidation reactions. When a molecule or ion is oxidized, it loses electrons. The chemical oxidizing agent gains electrons in the process and is itself reduced. Reactions involving oxidation of one reactant and reduction of the other reactant are commonly called *redox reactions*.

A simple example of a redox reaction is the reduction of ferric ion by stannous ion to form ferrous ion and stannic ion. Here the super-

22 | Chemical Change

script numbers are used to indicate the difference in the number of electrons between the charged atom and the neutral atom. (For example, Fe^{3+} has three fewer electrons than the neutral atom, Fe.)

$$Fe^{3+} + Sn^{2+} \longrightarrow Fe^{2+} + Sn^{4+}$$
FERRIC ION   STANNOUS ION   FERROUS ION   STANNIC ION

In this reaction, Sn^{2+} loses electrons to Fe^{3+} ions. A redox equation is balanced by making the electron loss of the reducing agent exactly equal to the electron gain of the oxidizing agent. The reduction of Fe^{3+} may be written

Fe^{3+} + one electron $\longrightarrow Fe^{2+}$

and the oxidation of Sn^{2+}

Sn^{2+} − two electrons $\longrightarrow Sn^{4+}$

It thus takes two Fe^{3+} ions to oxidize a single Sn^{2+} ion and balance the electrons:

$2Fe^{3+} + Sn^{2+} \longrightarrow 2Fe^{2+} + Sn^{4+}$

Oxidation reactions are responsible for food spoilage—that's what causes apples to turn brown and milk to sour. Chemicals are added to foods as preservatives because they will oxidize before the food will. Packaged bread remains fresh longer than home-baked bread because the former contains preservatives, or "antioxidants," as they are called on the labels.

CHEMICAL CHANGES IN THE ENVIRONMENT

Chemical change is studied best in a controlled laboratory situation. There, compounds can be measured precisely and mixed together carefully in preparation for possible reactions. Unfortunately, pressed by a rapidly expanding population and technology, we increasingly use our environment in the same way a chemist uses a reaction flask.

The essential difference between man and his world and the chemist and his flask is that the chemist does not stand inside his flask while a new reaction proceeds. If a reaction occurs in which the reaction products are harmful to life, he can simply stop up the flask and throw it away. But we are mixing together new combinations of compounds under new conditions that are difficult to predict and control. We do not have the alternative of simply discarding a river or the air we breathe if chemical reactions result in unpleasant or harmful effects.

While it is unlikely that we will poison ourselves off the planet, it is certain that unexpected chemical reactions can cause great discomfort to many individuals. The following two examples illustrate how chemical reactions in the natural environment affect us. In the first case, the new chemical reaction was discovered in a laboratory at about the same time it was noticed in a lake. In the second, the reaction products were breathed by numerous people for years before even the starting materials for the reaction were known.

Mercury Reactions

Organic mercury salts found in living materials are extremely toxic. Seeds, for instance, are commonly coated with organic mercury salts to prevent fungi from growing on them. People ingesting these compounds are

Understanding the Chemistry of Chocolate Chip Cookies

½ cup butter... Butter consists of long-chain fatty acids such as linolenic acid: $H_3CCH_2CH=CHCH_2CH=CHCH_2CH=CH(CH_2)_7COOH$. Fats with no double bonds (C=C) are saturated with hydrogen; those with more than one double bond are polyunsaturated.

6 tablespoons brown sugar
6 tablespoons white sugar ... Cane sugar, or sucrose, is a double sugar of glucose + fructose = sucrose

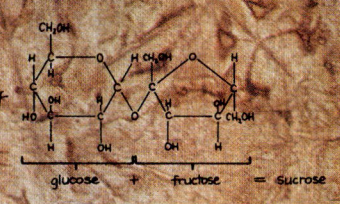

Add 1 beaten egg... An egg is mostly protein, or long chains of amino acids: $(-\overset{O}{\underset{}{C}}-\overset{R}{\underset{}{CH}}-\overset{H}{\underset{}{N}}-)_n$ When heated it coagulates quickly and permanently.

Sift 1 cup flour... Flour is a starch made of repeating glucose units in long branch chains. It is soluble in water.

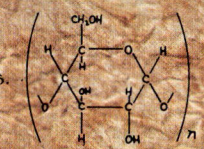

½ teaspoon baking soda... Sodium bicarbonate ($NaHCO_3$) forms bubbles of carbon dioxide when heated in acid.

½ teaspoon salt... Sodium chloride ($NaCl$) is very soluble in water.

Add a few drops hot water... Water (H_2O) evaporates when heated.

1 package chocolate chips... The flavor of chocolate is due to a mixture of compounds from the dried seeds of the cocoa plant.

2 teaspoons vanilla... Vanilla is a solution of

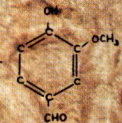

Bake at 375° for 10-12 minutes

Flour molecules form a thick solution when mixed with the other ingredients. The beaten egg, which contains trapped air bubbles, makes the batter lighter. How do the other ingredients affect the batter? The baking soda accepts protons and forms the unstable compound carbonic acid, which decomposes to carbon dioxide (CO_2) and water. As the batter is heated, the carbon dioxide bubbles expand, rise, and want to escape into the air. But the oven heat causes the water molecules in the batter to evaporate and the protein molecules in the egg to coagulate, which thicken the batter. Many of the carbon dioxide bubbles become trapped, and the cookies rise.

Figure 10. Rusting is the slow oxidation of iron into iron oxide. These rusted objects in a California ghost town show the effect of oxidation over time.

poisoned because the organic component in the compound provides the passport for transporting the mercury across cell membranes. Once inside living tissues the mercury diffuses quickly to nerve cells, causing severe damage such as paralysis.

Inorganic mercury salts found in rocks and minerals are also poisons because they form complexes with the sulfhydryl groups and act on the enzymatic systems in the body. But it takes a lot of them to damage the human organism because inorganic mercuric salts are not absorbed easily through the digestive tract. Rarely is the concentration high enough for the salts to pass through the membranes into cells where they can attack critical enzymes.

Aware of the toxicity of organic mercury salts, we have been cautious in the disposal of compounds containing them. But we have not been as cautious about returning the metallic mercury and the inorganic salts to the land because we did not know of any way these compounds could change into the organic form and become harmful to living organisms. So these "less toxic" mercury compounds have been spread on the earth's surface by discarded electonic parts and paints, which contain mercury, and by the wastes of chlorine-alkali plants.

Then in the 1960s a new reaction was discovered. High concentrations of methyl mercury were found in fish in the Great Lakes. The only major input of mercury into these lakes was inorganic industrial and metallic mercury. So the only way to explain the appearance of mercury in the flesh of sea life was that the layer of mercury on the lake bottom reacted with the organisms in the sludge to produce the highly toxic organic compounds.

Simultaneously, the same unexpected reaction was discovered in a chemistry laboratory, but it was too late—the damage was done. The mercury was already in the water; the reactions were initiated, and there was no way to stop them. For years to come, even if no more pollutants are added to the water, the mercury that now spreads over the lake and river bottoms will slowly undergo the chemical change into substances that are toxic to fish, and ultimately, to us if we eat the fish.

The Chemistry of Smog

The sunny skies of the Los Angeles basin have served as a reaction vessel for another novel chemical reaction. In this case, the reaction products affected millions of people for several hours each day. Because the air stung people's eyes like smoke and looked hazy like fog, it was called "smog."

For some time we did not understand smog. Initially we blamed it on smoky factories and backyard incinerators. Strict controls on these two offenders cleared the skies a bit and improved the garbage collecting business, but the smog continued to get worse.

The first component of smog to be identified by chemists was *ozone* (O_3). It is a strong oxidizing agent that affects many plants and causes rubber to crack. It may also account for a number of the unique discomforts of smog, but its effects on man are not yet clearly understood.

The amount of ozone in the air became a standard for how smoggy a day was. When the ozone concentration reached 0.5 parts per million (one molecule of ozone for every 2 million molecules of the gases of air), a smog alert was called and people were asked to keep their physical activity to a minimum.

But there is a lot of air in the Los Angeles basin and 0.5 parts per million adds up to an enormous amount of ozone. Where did it come from? The search for an answer to this question led eventually to the notion that ozone was created in the atmosphere by some sort of chemical reaction. The starting materials had to be chemicals emitted daily into the atmosphere in enormous quantities. The two pollutants were found to be gasoline (euphemistically called "hydrocarbons") and the oxides of nitrogen — both from automobile exhaust.

The thick brown smog is formed by the following chemical reactions (see Figure 11). The oxides of nitrogen form when the nitrogen from the air (N_2) is oxidized by oxygen (O_2) at the high temperatures of the combustion chamber in the automobile engine. The hydrocarbons (long molecules with repeating —CH_2— units) result from the incomplete burning of the gasoline. These pollutants from the exhaust become trapped in the cool ocean air of Los Angeles that is held down by a blanket of hot air from the desert. A *photochemical reaction* (that is, a reaction requiring light) takes place to initiate the complicated set of smog reactions.

The brown gas of the polluted sky, which is nitrogen dioxide (NO_2), absorbs ultraviolet light from the sun. It decomposes, forming an oxygen atom (O) as one of its products. The oxygen atom can then combine with an oxygen molecule (O_2) in the air to form ozone (O_3). Two nitrogen oxide molecules (2NO) can also react with an oxygen molecule. In this way nitrogen dioxide (NO_2) is regenerated to repeat the first reaction and form more oxygen atoms.

If a hydrocarbon is present, the oxygen atom can extract a hydrogen atom from it, leaving what is called a hydrocarbon radical. A *radical* is an atom or molecule with an upaired electron that forms when a bond that consists of a pair of electrons is split in half. A radical is quick to react with another molecule; if it reacts with a stable molecule, one of the reaction products is also a radical. The type of reaction that propagates itself is called a *chain reaction*. The smog reaction is a chain reaction.

The hydrocarbon radical can react with other hydrocarbons, with an oxygen molecule, with a nitrogen oxide molecule, and so forth. A number of different reactions can occur, eventually forming the particu-

Sunlight
↓

Warm Air
——— Inversion Layer ———
Cool Air

NO_2 + sunlight → NO + O

$2NO + O_2 \rightarrow 2NO_2$

$O + O_2 \rightarrow O_3$

Chain Reaction
$\begin{cases} O + RH \rightarrow R\cdot + \cdot OH \\ \cdot OH + RH \rightarrow R\cdot + H_2O \\ R\cdot + O_2 \rightarrow RO_2\cdot \\ RO_2\cdot + NO \rightarrow RO\cdot + NO_2 \\ RO_2\cdot + O_2 \rightarrow O_3 + RO\cdot \\ RO_2\cdot + NO_2 \rightarrow \text{peroxyacyl nitrates} \end{cases}$

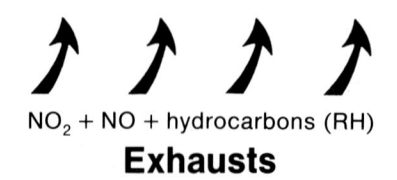

↑ ↑ ↑ ↑
NO_2 + NO + hydrocarbons (RH)
Exhausts

late haze that reduces our visibility and the organic nitrates that severely irritate our eyes.

If two radicals combine, a stable molecule is the only product. Because no radicals are produced to continue the chain reaction, the reactions producing a stable molecule are referred to as *chain termination reactions.* So, as the sun sets and the oxygen atoms are no longer being created out of stable nitrogen dioxide molecules, the smog reaction is terminated for the night.

We now understand the chemical changes that produce smog. As in any chemical reaction, the smog reactions could be stopped by removing any one of its special reactants or the conditions for reaction. Because any notion of altering the weather conditions that trap the hydrocarbons is an implausible solution, the easiest solution would be to stop the flow of hydrocarbons and oxides of nitrogen into the atmosphere.

Smog elimination is no longer a puzzle to scientists, but it remains a social and political dilemma. It would be expensive and inconvenient to curtail or stop automobile use because that mode of transportation is vital to the lifestyle not only of southern California, but of urban centers throughout the world. Although smog is unhealthy and annoying, it is not deadly. The Los Angeles community has chosen to live with smog for the present and to phase it out by gradual, albeit costly, modifications in the automobile.

Figure 11. (Far left) One result of smog reactions is the haze over downtown Los Angeles, the city with the dubious distinction of putting up with incredible levels of smog.

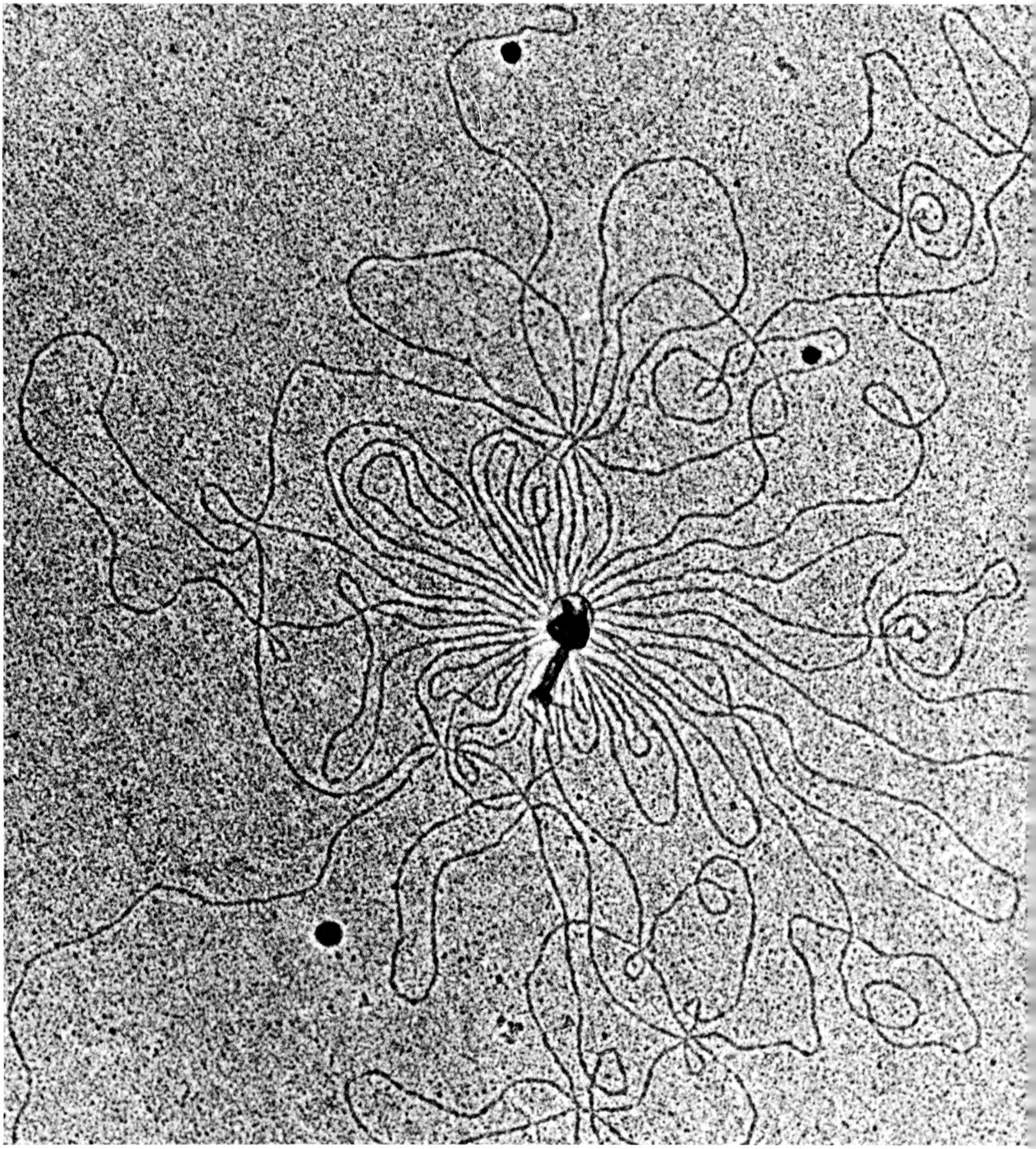

23
Giant Molecules

E. coli bacteriophage, magnified 90,000 times, are molecules that have come together into the phenomenon of life.

What do the following have in common: the resemblance of a baby to his father, the strength of a nylon rope, the digestion of lunch in the intestine, and the infection of a tobacco plant by a virus? Answer: these are a few of the many phenomena around us that depend on giant molecules. In fact, exaggerating only slightly, it can be said that most of today's medical science and industrial chemistry and nearly all of current biochemistry and molecular biology are devoted to the study of giant molecules.

Giant molecules include DNA and RNA—the information-storing and information-carrying molecules—and proteins, which serve both as the engines of chemical change in all living systems and as the pillars and posts of biological structure. They also include the synthetic fibers and polymers that pervade our industrial culture. Scientific studies of these molecules over the past 30 years have led to an understanding of the patterns of arrangement of atoms in such substances, and of how the resulting structures are related to their functions. These studies also have forced scientists to realize that viruses and perhaps even more complex organisms may be considered as giant molecules or, in essence, that the dividing line between living beings and inert materials is so blurred as to have lost all meaning.

POLYMERS FROM MONOMERS

All giant molecules are polymers. They are formed from small molecules that are called "monomers," which are linked end to end by repeating chemical bonds of the same type. An example is polyvinyl chloride, a plastic material that is found in rubberized sheetings and fabrics. Here the monomer is vinyl chloride, which may be written as

$$CH_2 = CH$$
$$\quad\quad\;\;|$$
$$\quad\quad\;Cl$$

During polymerization, the monomer units are joined when the electrons in the double bond move to form linkages between the units, giving

$$... CH_2-CH-CH_2-CH-CH_2-CH ...$$
$$\quad\quad\;\;|\quad\quad\quad\;\;|\quad\quad\quad\;\;|$$
$$\quad\quad\;Cl\quad\quad\;\;Cl\quad\quad\;\;Cl$$

This structure may be represented by the more compact notation $(-CH_2CHCl-)_n$, where the atoms in parentheses are the repeating unit of the polymer. The subscript n indicates that numerous units are linked together.

You can think of a polymer as a very long sentence composed of words—the monomeric units. The "letters" making up the monomeric words are the atoms. Although there are 100 or so kinds of atoms in the natural world, fewer than 10 are found in most polymers; only 6 are found in most biological polymers (see Table 1).

In a simple polymer such as polyvinyl chloride, the words of the sentence are all the same. Most biological polymers, in contrast, have a variety of words in the polymer sentence even though each monomer bonds to the next in the same way. In DNA, for example, there are four

Table 1.
Some Types of Polymers

Polymer	Monomer	Chemical Structure of Monomer	Types of Atoms	Number of Monomer Types	Function of Polymer
Polyvinyl chloride	Vinyl chloride	$(-CH_2-CH-)_n$ with Cl side chain	H, C, Cl	1	Synthetic plastics
Deoxyribonucleic acid (DNA)	Nucleotide	$(-Sugar-P-O-P-O-P-O-)_n$ with Nucleic acid side chain	H, C, N, O, P	4	Information storage
Proteins	Amino acids	$(-N(H)-C(H)(side\ chain)-C(=O)-)_n$	H, C, N, O, S	20	Enzymes, structural molecules, antibiotics, hormones

Source: D. Eisenberg, 1972.

different "words" called *nucleotides*. Each nucleotide consists of a triphosphate group, a sugar called deoxyribose, and one of four nucleic acid bases. The DNA polymer can be represented by:

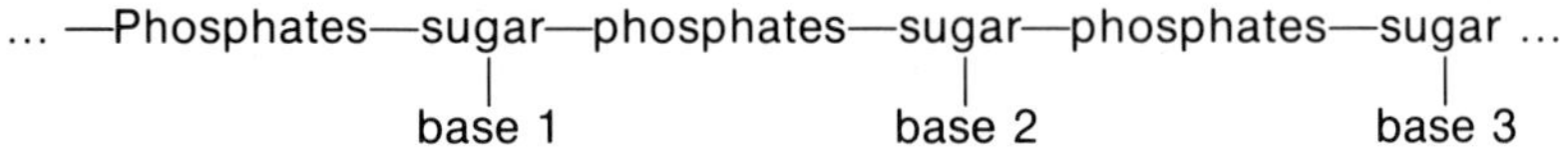

... —Phosphates—sugar—phosphates—sugar—phosphates—sugar ...
                | | |
              base 1   base 2   base 3

Notice that the sugar-phosphate *backbone* of the polymer has a repeated bonding of the identical sugar-phosphate groups. But each *side-chain* base can be any of the four, so the polymer as a whole does not have the monotonous repeat of polyvinyl chloride. The possibility for varied order of the bases is what makes DNA useful for storing information, rather than for making rubberized sheetings.

In proteins, the backbone is even simpler. It can be represented by the following structure:

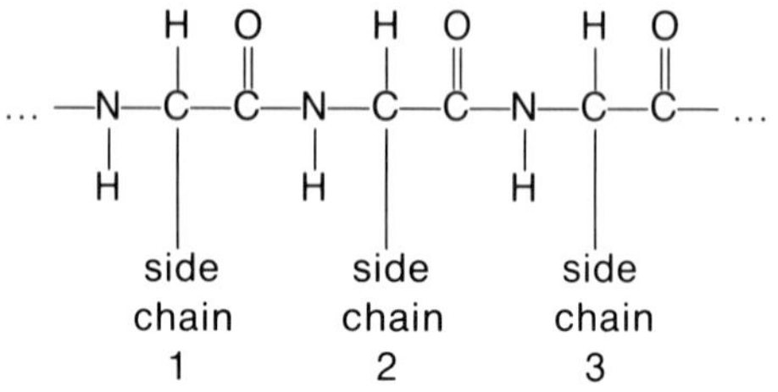

$$\cdots -N(H)-C(H)(side\ chain\ 1)-C(=O)-N(H)-C(H)(side\ chain\ 2)-C(=O)-N(H)-C(H)(side\ chain\ 3)-C(=O)- \cdots$$

or, in more compact notation,

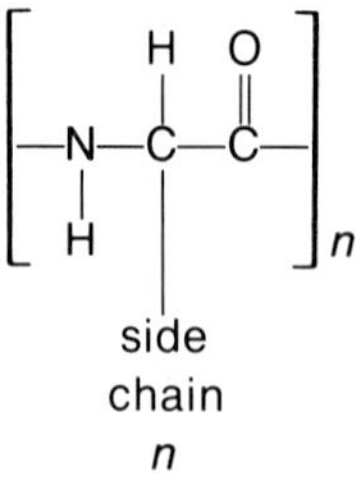

$$\left[-N(H)-C(H)(side\ chain)-C(=O)- \right]_n$$

Each unit like the one depicted above is called an *amino acid* or, more strictly speaking, an *amino acid residue.* There are twenty different kinds of amino acids, and it is in their side chains that they differ from one another. The amino acid *glycine,* for instance, has a side chain consisting only of a hydrogen atom. Or consider the side chain of *alanine,* shown in Figure 2. It is a —CH_3 group. Figure 2 also shows the side chain of *lysine,*

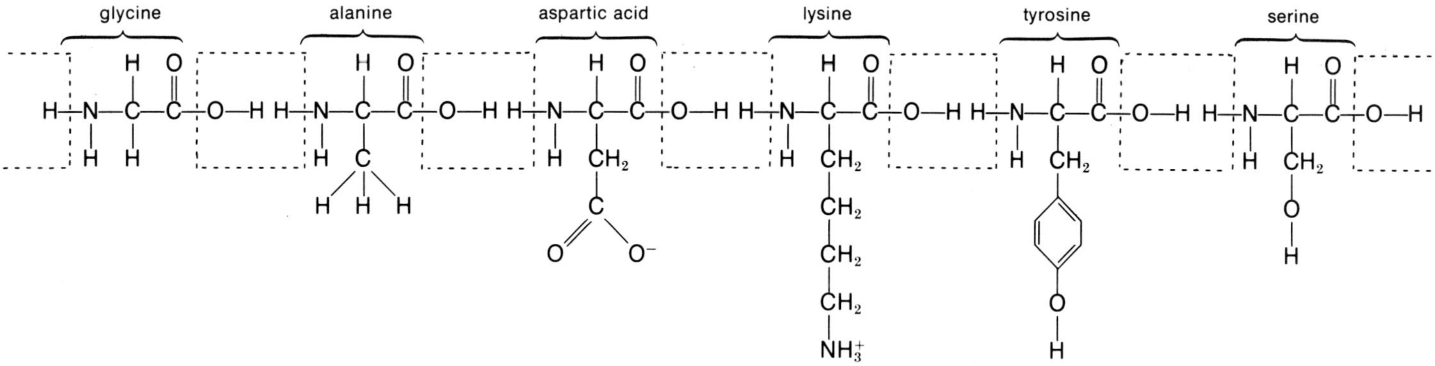

Figure 2. Amino acid structures. The atoms surrounded by dashes form the water molecules that are removed when the amino acids are polymerized into a protein.

which includes a positively charged amino group. Clearly an enormous variety of proteins can be formed when the twenty different amino acids are polymerized in various orders. Nature has taken full advantage of this variety in evolving protein molecules of widely differing functions.

An important property of some polymer strands is their ability to form bonds with similar strands lying side to side with them. An example is *nylon 6*. In this case the monomer is aminocaproic acid, $H_2N-CH_2-CH_2-CH_2-CH_2-CH_2-COOH$, and it polymerizes by the removal of water:

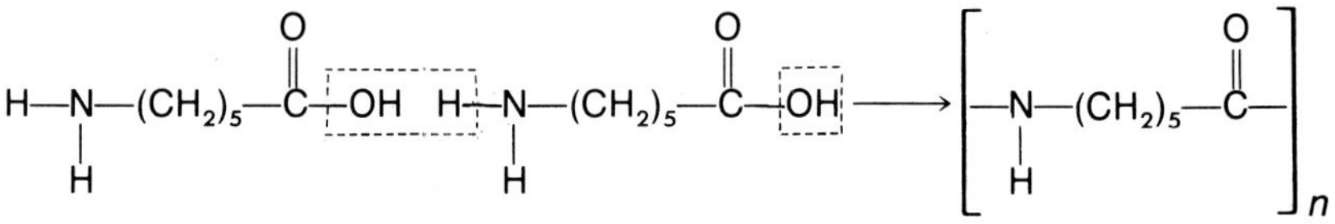

Here, the atoms that are removed during polymerization are enclosed in dashes, and the bracket with the subscript n indicates that many of the enclosed units are linked together. The C=O groups protruding from a strand of nylon are somewhat negatively charged, and the N—H groups are somewhat positively charged. These groups therefore tend to attract one another, forming *hydrogen bonds*,

$C=O^- \ldots H^+-N$

As these hydrogen bonds form, they tend to order the strands of nylon side by side (Figure 3).

The energy of interaction of a hydrogen bond is only about one-twentieth that of the covalent bonds between atoms of a strand, yet the energy is sufficient to cause the chains to fall into order. The formation of the hydrogen bonds also increases the melting point of the nylon. In short, the nature of the monomer units of a polymer can determine how one strand of a polymer bonds to another, and this bonding can affect the properties of the polymer.

A polymer can also bend back and bond to itself. Such repeated bending and bonding will build up a three-dimensional structure. Enzymes, which are discussed later in this chapter, have such a structure.

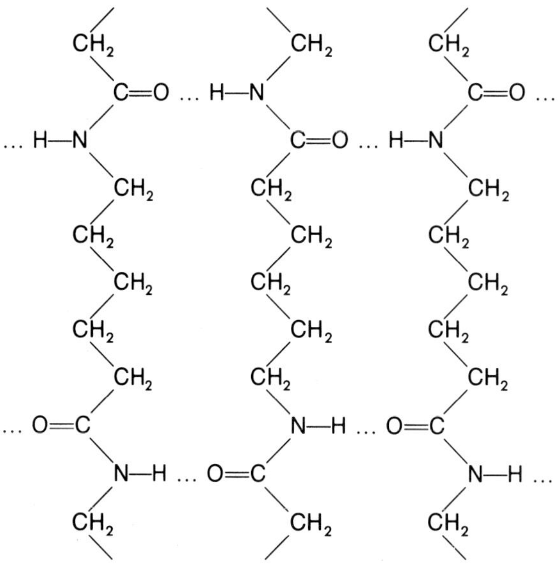

THE SIZES OF GIANT MOLECULES

Proteins and DNA are enormously larger in size than the small molecules of, say, water and oxygen. Nevertheless, they are still tiny compared to

Figure 3. How strand of nylon form bonds with similar strands.

23 | Giant Molecules

Figure 4. (Right) Molecules of the enzyme glutamine synthatase, seen through an electron microscope, are magnified 500,000 times. Each molecule has twelve globular polymer chains arranged in two hexagonal rings.

Figure 5. (Below) The molecular structure of the four nucleic acid bases of DNA is shown, as well as the pairing of T to A and C to G. Note the hydrogen bonds that cause the pairing.

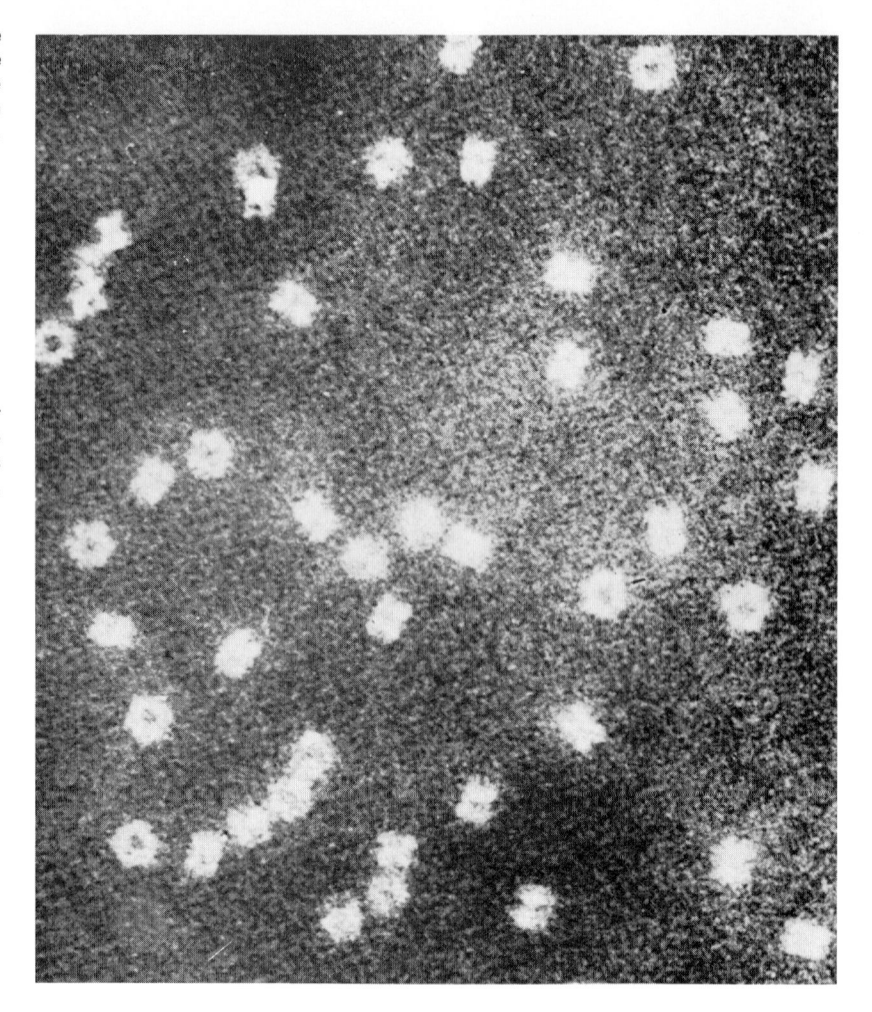

anything that can be seen with a light microscope as opposed to an electron microscope. Most enzymes have molecular weights between 10,000 and 1,000,000. They are thus between 600 and 60,000 times heavier than a water molecule, and yet a penny weighs more than a million, million, million large enzyme molecules.

The enzyme glutamine synthetase, for example, has a molecular weight of about 600,000. It is known to consist of twelve identical protein chains, each with a molecular weight of about 50,000. Because the average molecular weight of an amino acid is about 100, there are roughly 500 amino acids in each protein chain. These chains are folded into spherically shaped globs that are stacked together like two hexagons on top of each other (see Figure 4). The dimensions of glutamine synthetase molecules have been measured in the electron microscope. Two million of these molecules stacked edge to edge would be an inch long. Expressed another way, if the glutamine synthetase molecule were expanded to the size of a beach ball and if the beach ball were expanded by the same ratio, the ball would be as large as the moon.

The smallest viruses are about 10 times larger than large enzyme molecules. The polio virus has a molecular weight of about 6,000,000, and the tobacco mosaic virus has a molecular weight of about 40,000,000.

The sizes of DNA molecules depend on the complexity of the organism from which they are taken. DNA is usually in the form of a long, double-stranded polymer. The double strand of DNA from the bacterium *E. coli* has a molecular weight of about 2,800,000,000. And when it is fully extended to about one-twentieth of an inch, it is roughly 1,000 times as long as the bacterial cell in which it is normally tightly folded.

DNA: INFORMATION STORAGE AND REPLICATION

Suppose you wish to make a replica of a bronze statue. You would probably start by making a plaster cast of the statue, and then you would use this *template* to cast a replica of the statue in bronze. For a biological cell to divide into two identical cells, the proteins and other of its components must be duplicated. It turns out that nature uses the device of a template in cell division. The template is a molecule of DNA, which replicates itself. This replication process is the essential step in cell duplication because the DNA molecule carries a list—in coded form—of the proteins and other components that must be in the new cell.

To understand how the DNA template works, you must know something about the nucleic acid bases that are attached to the sugar-phosphate backbone of the DNA polymer, which was depicted earlier. DNA contains four kinds of bases: thymine (T), adenine (A), cytosine (C), and guanine (G). These bases are attracted to one another because of the way their atoms are arranged. This attraction is similar to that of two strands of nylon, in that the attraction arises from hydrogen bonding between the bases. But unlike nylon, the attraction in DNA involves four different bases, and each one pairs only with one other base. The base A pairs to base T, and vice versa; base C pairs to base G, and vice versa. Figure 5 depicts the two kinds of pairing and the hydrogen bonds that give rise to them.

Once you know about this characteristic pairing of nucleic acid bases, you can easily see how a molecule of DNA might be replicated. The molecule consists of a sugar-phosphate backbone, with one of the four bases extending from each sugar. Think of this molecule as the original bronze statue. Now bring up an assortment of nucleic acid-sugar-phosphate groups—the nucleotides—and you will end up with something that looks like this:

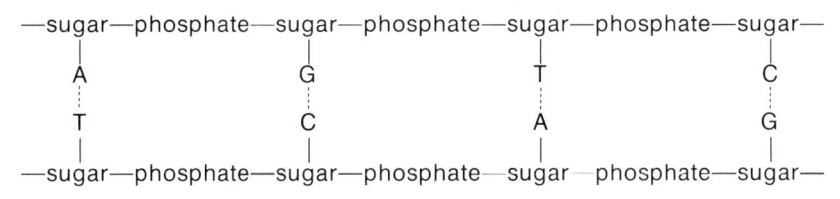

ORIGINAL
STRAND

—sugar—phosphate—sugar—phosphate—sugar—phosphate—sugar—
    | | | |
    A   G   T   C
    T   C   A   G
    | | | |
—sugar—phosphate—sugar—phosphate—sugar—phosphate—sugar—

COMPLEMENTARY
STRAND

These groups may be thought of as the plaster for the template. Each nucleotide base will pair to its usual mate, as shown schematically above. As this pairing takes place, an enzyme called *DNA polymerase* forms the polymer bonds between sugar-phosphate groups on this second strand. The second strand is complementary to the first; in other words, just like the plaster cast, it can act as a template to reproduce the original. In order for this to happen, the complementary strand must separate from the original strand, and more nucleotide bases must be brought up to pair with the complementary strand. These bases will then be polymerized into a strand that is a replica of the original (see Figure 6).

James Watson and Francis Crick, the scientists who in 1953 first conceived of this mode of replication of DNA, also found how the DNA is

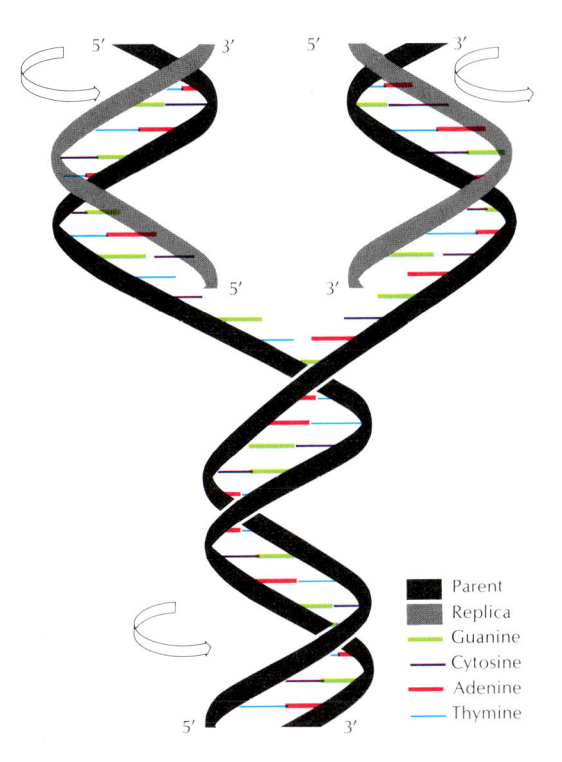

Figure 6. During replication, the double-stranded helical structure opens and each original strand acts as a template for a new strand. T always pairs with A, and G with C.

Figure 7. In the β-pleated sheet of silk, the side chains actually protrude above and below the plane of the sheet. Sheets adhere to one another because these chains nestle together. Here, the side chains of all amino acids are indicated by the letter R.

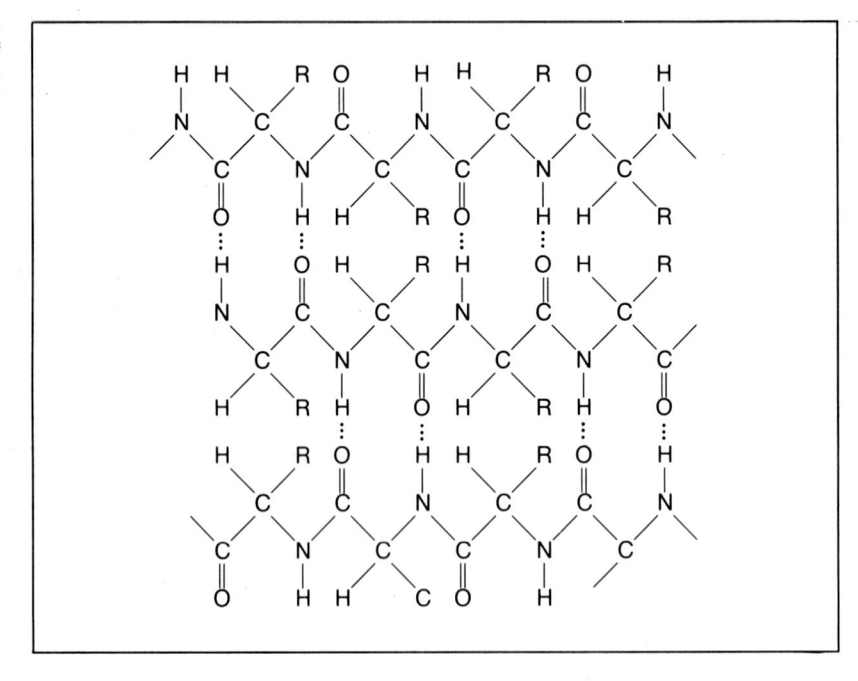

coiled. A DNA strand is normally paired to its complementary strand. The double strand is then twisted, forming a double helix (Figure 6). The pairs of nucleic acid bases are in the center and the two sugar-phosphate backbones run down the outside. This coiled structure is soluble in the aqueous medium of a cell, because the relatively insoluble bases are sequestered in the center and the highly soluble sugars and phosphates are located on the outside.

How is genetic information encoded in DNA molecules? The code is actually in the *sequence* of nucleic acid bases. Each sequence of three bases on the DNA polymer is a code word for one amino acid. It has been discovered, for example, that the sequence of bases TTT on DNA corresponds to the amino acid lysine. Each sequence of several hundred bases is a code message for a particular protein. Much of the machinery of a living cell is devoted to translation of the base sequence of DNA molecules into proteins. Many of these proteins are the enzymes that catalyze all the chemical reactions of the cell. Therefore, the information that a new cell receives is primarily a "list" of the enzymes it should make. Which enzymes are present and which order they appear in determine the details of the future history of the cell.

A *genetic mutation* is an incorrect DNA message; its consequence is a defective or a missing enzyme. What causes mutation? One cause is the accidental elimination of one nucleic acid base during replication of a segment of DNA. Then, when the genetic code of this segment is translated into protein, the wrong set of three nucleic acid bases will occur at that point and the wrong amino acid will be added to the corresponding protein. Indeed, all amino acids after that one will be wrong—the code units made of three bases each will be read starting one base too soon. The result will be a non-sense enzyme and, depending on the importance of that enzyme to the cell, the results will be debilitating or disastrous.

FIBROUS PROTEINS

The most abundant proteins in our body are the fibrous proteins that serve as the pillars, posts, and plasterwork of our biological structure.

These proteins include *collagen* in our tendons; *α-keratin* in our skin, hair, and nails; and *actin* and *myosin* from our muscle. In contrast with the compactly folded enzymes, which will be considered later, these proteins are extended and are often made up of a small number of the twenty different kinds of amino acids. Most importantly, they are regularly *crosslinked* by hydrogen bonds. It is this feature that provides the stability necessary for their structural roles.

The *β-keratin* structure of silk is a relatively simple example of a fibrous protein. Not surprisingly, there are elements of similarity between the structure of silk and that of nylon. Both are formed from extended polymer chains that are crosslinked to neighboring chains by hydrogen bonds. In silk, the basic polymer chain is more complicated than that of nylon (it is a repeat of a unit of six amino acids), but the mode of cross-linking is similar. It also consists of N—H ... O=C hydrogen bonds. In silk, the basic amino acid chain is hydrogen bonded to two chains on either side of it; these two chains run in the direction opposite that of the basic chain. Similarly, the outer chains are hydrogen bonded to other chains, and so on. This bonding leads to a sheet structure, called the *β-pleated* sheet. A small section of it is shown in Figure 7. This hydrogen bonded sheet has the combination of strength and flexibility required of a fabric.

The protein of hair and wool is *α-keratin*. It, too, is stabilized by N—H ... O=C hydrogen bonds, but here the hydrogen bonds are between different amino acids of the same polymer chain. This type of bonding is possible in a helical folding of the protein chain. In *α-keratin*, the helix is called the *α-helix* (or *alpha*-helix). It is a single helix of amino acids, in contrast to the double helix of nucleotides in DNA. In the *α-helix* (Figure 8), the C=O group of each amino forms a C=O ... H—N hydrogen bond, with the amino acid four points ahead of it in the protein backbone. The elasticity of wool has its basis in this hydrogen-bonded pair. It is clear from Figure 8 that when a fiber of wool is stretched slightly, some of the helical hydrogen bonds will be broken; their tendency to re-form makes the fiber snap back.

ENZYMES: OUR PROTEIN CATALYSTS

As you read in the preceding chapter, catalysts are substances that increase the rates of chemical reactions without themselves being consumed in the reactions. Catalysts are essential to the thousands of chemical reactions that occur in living beings. Without catalysts, biochemical reactions would occur over the course of days rather than the ten-thousandths of a second it actually takes.

The biochemical catalysts designed by nature are enzymes. All enzymes are globular proteins—they are polymers of amino acids that are folded into compact, globular shapes. Some contain several globular protein chains, clustered together by hydrogen bonds or other forces. An example is the enzyme glutamine synthetase (Figure 4), which consists of twelve identical protein chains.

How do enzymes act as catalysts? Today this is an active subject for research in many laboratories, but a number of general points are clear. Suppose the catalyzed reaction is represented in the following way:

$$A + B \xrightarrow{\text{enzyme}} C + D$$

This formula means that *A* is reacted with *B* in the presence of the enzyme to form *C* and *D*. Here, *A* and *B* are the *substrates* of the enzyme and

Figure 8. The alpha helix. Amino acid side chains are indicated by R, and hydrogen bonds by dots.

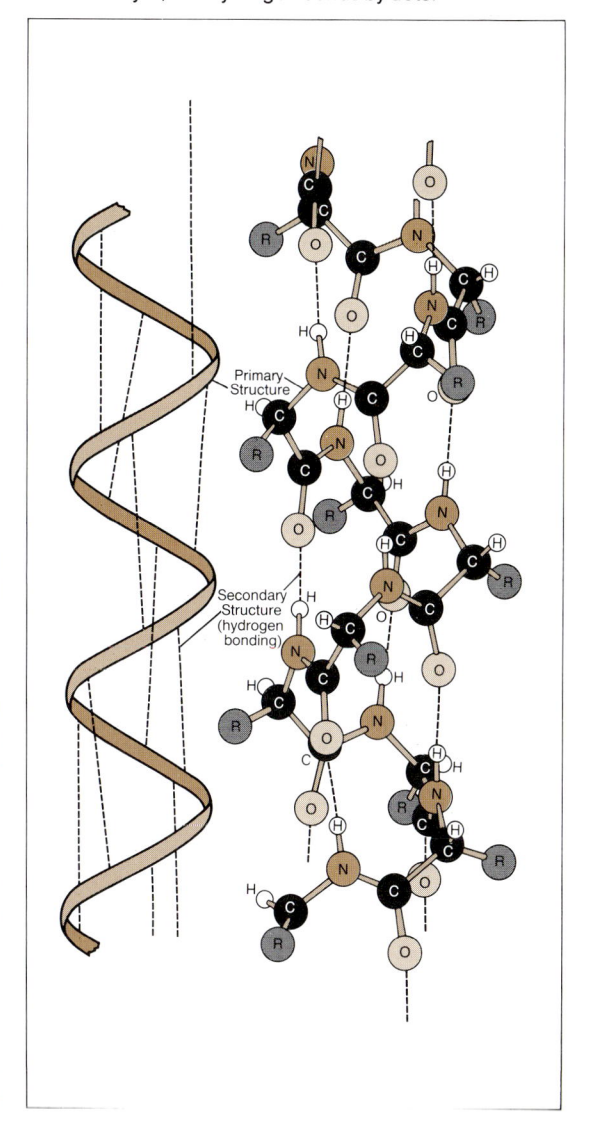

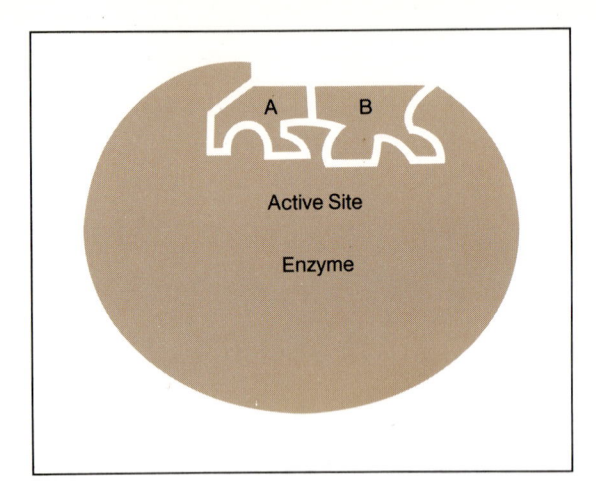

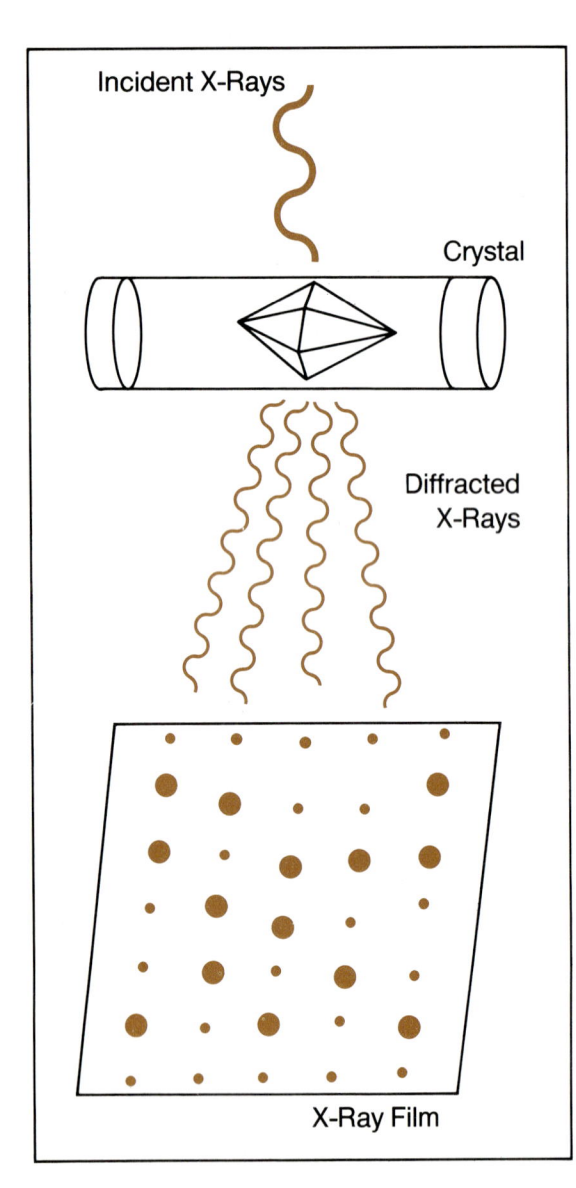

C and *D* are the *products.* In any living system, all the enzymes, substrates, and products are surrounded by water.

Most biochemists believe that catalysis by enzymes involves the following steps:

1. *First, the enzyme meets the substrates in the solution and binds them to form an enzyme-substrate complex. In this complex, the enzyme holds the substrates in a mutual orientation that is favorable for reaction. Because the substrates are usually molecules with molecular weights that are small compared to the molecular weight of the enzyme, it is clear that catalysis takes place at a relatively small region of the enzyme. This region is called the* active site, *or* catalytic site *(see Figure 9).*

2. *As the enzyme-substrate complex is being formed, side chains of some amino acids in the active site of the enzyme come into contact with the substrates. The electric charges—primarily the capacity for hydrogen bonding of these side chains—in some way facilitates the conversion of A and B to C and D.*

3. *The enzyme then releases the products C and D and is prepared to bind new molecules of A and B. If the enzyme were not returned to its original state at the end of the reaction, it would not be a true catalyst; it would be "consumed" in the reaction.*

You can think of the enzyme as a tiny machine. The substrates are the raw materials from which it makes products. The fabrication of products involves holding the raw materials in proximity and applying special forces. When the products leave, the machine is in its original configuration, ready to receive more raw materials.

A great deal of evidence suggests that the catalytic activity of an enzyme depends on the precise arrangement in three-dimensional space of its protein backbone, and the attached side chains. One line of experiments that suggests this dependency exists involves the unfolding, or *denaturing,* of enzymes. In these studies, the speed of enzymatic catalysis is followed as the water solution containing the enzyme and substrate is heated. As the temperature approaches 50 degrees or 60 degrees Centigrade, catalysis slows and then stops. At that point the heat is great enough to break apart the hydrogen bonds and the other forces that hold the polymer chain in its active, three-dimensional form. In other words, the active site is disrupted by heat because hydrogen bonds and other forces are overwhelmed by thermal vibrations, and the polymer unwinds. It should be noted, however, that these temperatures are not sufficient to break the much stronger covalent bonds between amino acids in the polymer backbone. In short, catalytic activity is lost because the backbone *unwinds,* not because it is broken.

There is an interesting sidelight to such denaturation experiments. If the hot enzyme solution is cooled slowly, enzymatic activity returns. Apparently the unwound polymer chain can rewind by itself into its active configuration, which indicates that the unwound enzyme "knows" what its three-dimensional shape should be. It seems that the three-dimensional structure is encoded in the linear sequence of amino acids of the backbone. One of the central questions of molecular biology today is how the knowledge of three-dimensional structure of a protein is contained in its amino acid sequence.

THE STRUCTURE AND FUNCTION OF SOMETHING CHYMOTRYPSIN CALLED

The three-dimensional structures of enzymes can be determined by an x-ray diffraction study of crystals of the enzyme. A crystal is placed in a

Figure 9. (Far left, above) A highly schematic drawing of an enzyme-substrate complex. The substrates are held in position to react with one another.

Figure 10. (Far left, below) The essentials of an x-ray diffraction experiment on an enzyme crystal. The spots on the x-ray photographic film are related (indirectly) to the positions of atoms of the enzyme molecules in the crystal.

Figure 11. (Left) A representation of the folding of the protein backbone of the enzyme chymotrypsin. The active site is in the region of serine (His 57), one of the amino acids participating in catalysis. The start (NH_3^+) and finish (CO_2^-) of each of the three polymer chains (A, B, and C) are shown. The five —S—S— linkages are indicated by the symbol ⌐⌐. The numbers next to these symbols show the amino acid involved in the linkage in terms of its order in the chain.

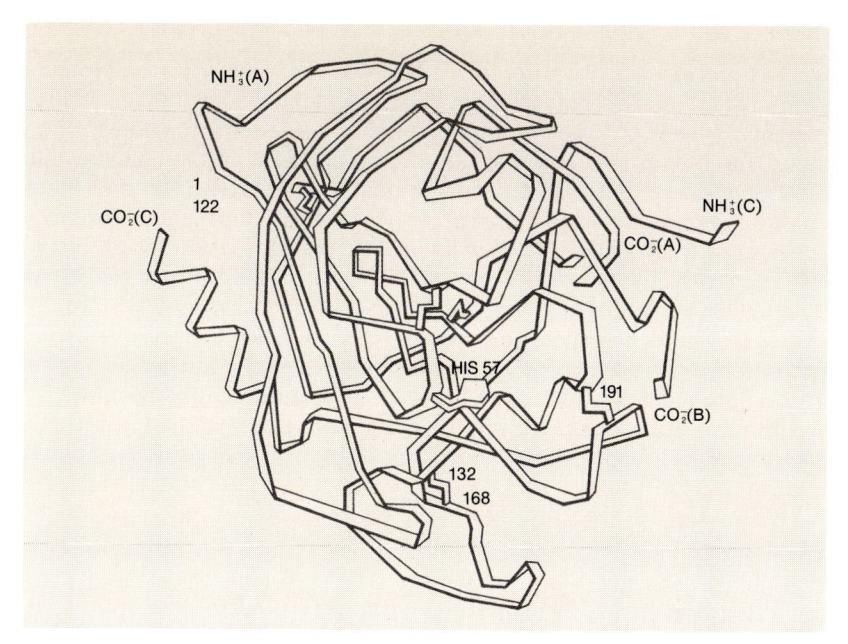

beam of x-rays, and the x-rays that are diffracted are recorded as black spots on a piece of photographic film. This pattern of spots is related indirectly to the separation of atoms in the enzyme. Computer processing of a large number of such x-ray diffraction photographs can yield a picture of the enzyme that shows the relative positions of all the constituent atoms (see Figure 10). The procedure is laborious, but by 1972, such experiments had been carried out for about thirty enzyme molecules.

The enzyme *chymotrypsin* is one whose three-dimensional structure is known. Consider the structure as an example of how enzymes fold, and of how the folding relates to their function. Chymotrypsin is secreted into the small intestine, where it aids in the cleavage of proteins into amino acids. The amino acids can then be absorbed by the intestine and used for synthesizing the new proteins that are required. Chymotrypsin does not cleave just any bond between amino acids; it cleaves best those bonds just beyond an amino acid with a bulky side chain such as tyrosine:

The three-dimensional structure of chymotrypsin is shown in Figure 11. As with other enzymes, the folding of the polypeptide backbone is so complex that it is remembered in detail only by the scientists who have

Figure 12. (Far right) A representation of a part of a tobacco mosaic virus. The vertebra-like RNA polymer can be seen coiling out the top. Each nucleotide is represented by a bump. The RNA binds to a groove in the lobe-shaped protein monomer. There are 2,200 protein monomers in one virus.

worked closely with the structure. The important features of the folding of the 241 amino acid polymers are that the molecule is a compact, slightly flattened sphere of about 40×10^{-8} centimeter—about 1/15,000,000 of an inch. The folding is so compact that there are only thirteen slots within the enzyme large enough to accommodate small molecules, and these contain water molecules. The polymer backbone of this molecule actually consists of three amino acid chains rather than the usual single chain. These chains arise from a single chain that is cleaved in two places by other enzyme molecules. The three chains adhere to one another through hydrogen bonds and through five —S—S— linkages, each formed from two nearby amino acids having sulfur-containing side chains.

There are many hydrogen bonds that crosslink the molecule to itself, thereby ensuring stability in the rugged and somewhat unpredictable environment of the intestine. These bonds are of three types:

1. β-pleated sheet type hydrogen bonding is found between many of the backbone chains that run parallel to one another for short stretches.

2. α-helical type hydrogen bonding occurs. The main α-helical section lies at the end of the amino acid chain (the corkscrew section on the left-hand side of Figure 11).

3. There are small numbers of many other types of hydrogen bonds between side chains and backbone, and from the backbone to itself.

The amino acid side chains involved in binding and cleaving the substrate are clustered together in the active site. This site is a small depression near the enzyme surface; it is visible in Figure 11. The amino acids here include serine, which is located 195 positions away from the first amino acid of the backbone; and histidine, which is 57 positions away from the start. The side chains of these amino acids are the ones that participate in the cleavage of the protein substrate. Other amino acids in this region account for the preference of chymotrypsin to cleave proteins near a bulky amino acid such as tyrosine: they bind only to bulky side chains; and when they bind, they position the bond to be cleaved next to the catalytic side chains of serine and histidine.

With only a few amino acids in the active site, what role do the other hundreds of other amino acids of the enzyme serve? The answer probably is that they form the framework for the few amino acids that precisely position and cleave the substrate. Because some flexibility of the enzyme is often required for catalysis in addition to the precise positioning, the framework must be more elaborate than a few amino acids.

From similar studies on thirty or so other globular proteins, many details of enzymatic catalysis are starting to emerge. Studies of defective molecules have, in the case of hemoglobin, revealed the nature of a molecular disease. It seems likely that within a few years this sort of knowledge should enable doctors to remedy enzymatic defects by various forms of chemical therapy.

VIRUSES AS GIANT MOLECULES

Viruses are considered by many to be the simplest organisms that can be termed "living." Even the smallest viruses can reproduce themselves—one of the chief characteristics of living beings—although they require a host biological cell to help. The larger viruses have increasingly complex structures and patterns of metabolism, until the largest are distinguished from simple bacteria only on the basis of virtually arbitrary definitions.

1

2

3

4

5

6

7

What has surprised us in recent years is that at least some of the simplest viruses seem to *self-assemble*. In other words, if the component proteins and DNA (or RNA) polymer are mixed in a test tube under the right conditions, they spontaneously form an infective virus. Such viruses have all the characteristics of a virus that is formed naturally in a host biological cell.

Simple viruses consist only of a nucleotide polymer core and a protein coat. In some viruses the core is DNA and in others it is the closely related polymer RNA. The only difference between these polymers is in the sugar of the sugar-phosphate backbone. The protein coat is built from globular proteins. In some cases the coat contains many copies of only one type of protein.

An example of a simple virus is *tobacco mosaic virus* (TMV). The structure of TMV is depicted in Figure 12. The core is a single helical polymer of RNA, some 6,600 monomers in length. The protein coat consists of lobe-shaped, globular proteins, which are bonded to one another and are wound helically around the RNA. There are 2,200 identical copies of the coat protein in the helical coat, each containing 158 amino acids. The entire virus is about as long as 1,000 water molecules and has a molecular weight of 40 million.

When a virus enters a cell of a tobacco leaf, its RNA subverts the biochemical machinery of its host cell, directing it to make new copies of its *own* RNA and coat protein. The virus thereby reproduces itself with the aid of the host. The coat protein alone does not cause these effects; the RNA does. Thus the RNA is the infectious agent. Because the biochemical machinery of the tobacco cell has been taken over by the invader, the host cell is damaged.

When the RNA and coat protein molecules of RNA are mixed in a test tube under the proper conditions of *pH* and salt concentration, they spontaneously form a virus. This process was discovered in 1955. Since that time, intensive study has given us many insights into this self-assembly. The first step seems to be the formation of a two-layer disk made up of thirty-four coat protein molecules. The disk binds to one end of the RNA, and the RNA lies in a groove in the disk (see Figure 12). As the RNA binds, it causes the disk to split into a "lockwasher" (Figure 12). The lockwasher starts the helix of protein molecules. The virus particle grows by addition of new disks, which join the existing helix by binding to the RNA and becoming lockwashers. Growth stops when the end of the RNA strand is reached.

How does self-assembly occur? There are no enzymes present that can align and form covalent bonds between the protein units or between the protein and RNA. There is no template to order the units that go into the virus. Self-assembly takes place because the protein units tend to form strong, non-covalent bonds with one another. Some of these bonds are undoubtedly hydrogen bonds; others are different. These bonds are at proper angles to lead to the disk. The RNA tends to form a strong non-covalent bond with the protein in the groove, and this new interaction changes the protein-protein force, thereby making a lockwasher. In short, the structure of the virus is encoded in the tendency of the protein unit to form certain bonds.

Self-assembly of a virus can be viewed as a second level in the programming of structural information in an amino acid sequence. You saw the first level in the denaturation experiments described in the preceding section: the linear sequence of amino acids in a protein determines the three-dimensional structure of the protein. Now you see that the three-dimensional structure of proteins and RNA can determine the form of aggregation of the macromolecules. Could it be that even larger biological structures self-assemble from components? That is another basic question that is currently under active investigation.

The experiments on viruses have even broader implications. It is now clear that the "living" or "nearly living" viruses are simply aggregates of proteins with RNA or DNA. The proteins and nucleotide polymers are themselves merely polymers of simpler monomers. Once the monomers have been linked together, the resulting polymers fold spontaneously into three-dimensional structures; then these structures aggregate spontaneously to form viruses ready to infect cells.

From such experiments, it is clear that these "living" organisms are controlled by no new or unusual principles. Indeed, the controlling principles are the same ones that influence the simplest organic and inorganic chemical reactions. There is no clear dividing line, then, between "inanimate" and "living"; there is only a gradual increase in the sophistication of structure and in the complexity of chemical reactions. At some level of complexity we refer to an aggregate as a "giant molecule," and at some higher level we think of it as being "alive."

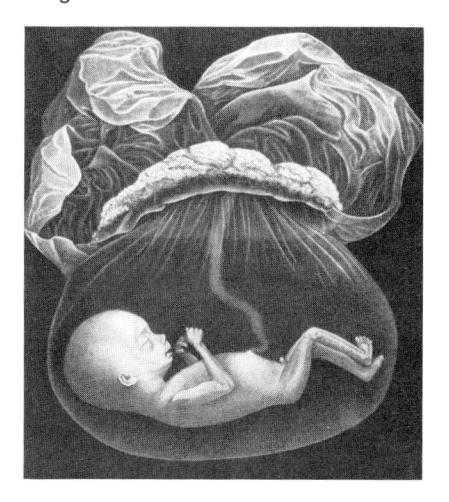

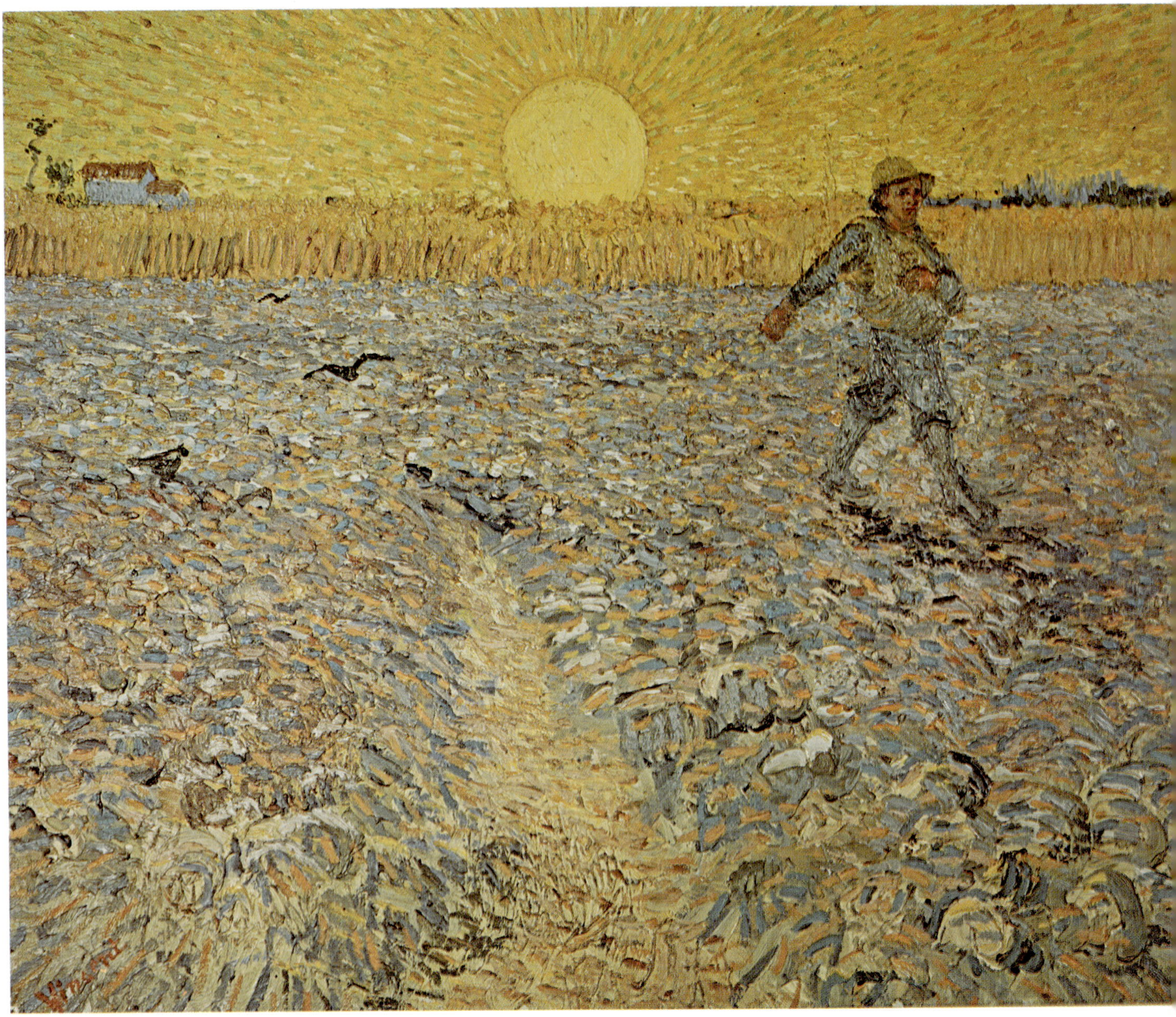

The Sower
Vincent Van Gogh (1853-1890)

So much of our understanding of the world around us is based on our perceptions of it, either directly or from models of abstract theories.

In portraying the interplay between his observations and his feelings, Van Gogh produced incredibly vibrant images of our dynamic world. *The Sower* evokes a sense of the stuff and substance of the universe precisely because the artist, in his way, has penetrated the mysteries of the interaction of light and matter. Science, in contrast, would explain that interaction in terms of atoms, molecules, and the scattering of photons. Which is the more meaningful interpretation?

Substance sends out messages that science interprets in one way, and art in another. The adventurous person—whether scientist or artist—can learn from both interpretations. Each one broadens our intuitive base by providing us with abstract as well as visual models of reality. The scientist as well as the artist seeks to apply this intuition to the building of models consistent with detailed observations that hopefully will explain the nature of the world, even the nature of life itself.

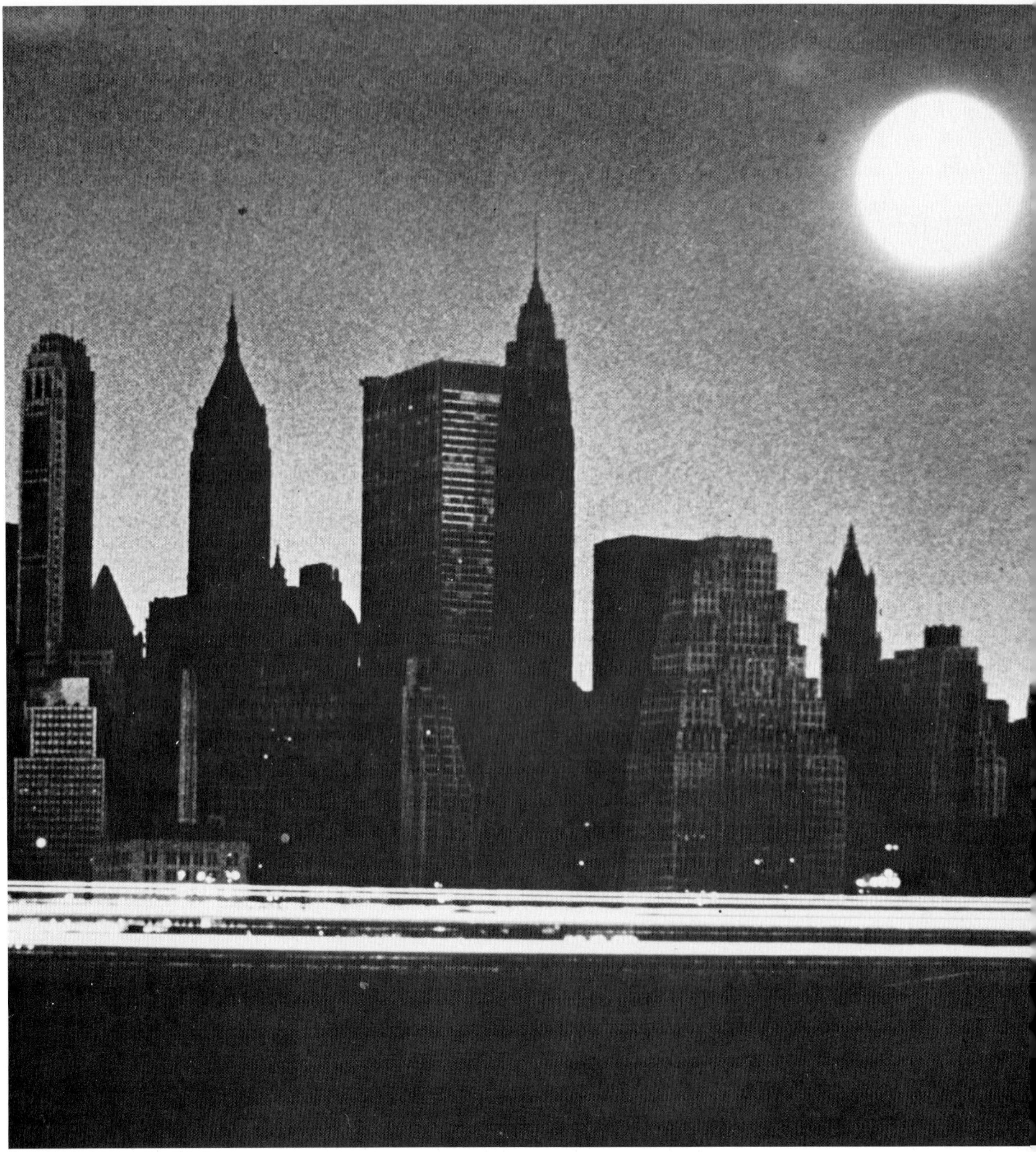

In the nineteenth century a few enterprising people planned to harness the power of Niagara Falls by converting its thundering power into mechanical energy with great, churning water wheels. An intricate network of pulleys and belts would then whisk all the mechanical energy to busy factories and workshops. The plan never was carried out. Even though they had the right idea for generating power (what, after all, are the giant turbines harnessing the Falls today, if not glorified water wheels?), they had too many problems with their scheme for transmitting and distributing the power.

By the turn of the century, a scheme for electric power transmission had been perfected and large amounts of power could be moved efficiently. Electric energy could be transmitted as a current through wires and then converted into mechanical energy, heat, and light far from the places where power was generated. With electric power transmission to stimulate its growth, the relatively simple level of human activity would mushroom into the complex society we know today. Efforts would no longer be confined by the limits of man's muscular power or that of his draft animals; man could move far from water wheels and natural sources of fuel. With electric power transmission, wind mills and pulleys would soon become remnants of the past in a world that would use electricity to run its radio transmitters, electrocardiographs, and electric guitars.

ELECTRIC CHARGE

Electricity can be made to do useful work because electric and magnetic forces can push or pull on things. The ability of an object to produce or respond to electric force is measured by its electric charge. Like mass, "charge" is a fundamental property of matter.

How big is a charge? Today it's described in terms of the *coulomb*. Unlike "meters" or "miles," the "coulomb" is an unfamiliar term because you seldom have need to talk about the massive amount of electric charge it represents. When you scuff your shoes on a thick nylon rug and then reach out and touch someone, the charge that passes in the spark is only a tiny fraction of a coulomb — a mere 10^{-6} coulomb or so. A coulomb is more descriptive of titanic electric storms than of everyday experience; a lightning flash carries about 20 coulombs of charge from a cloud to the earth.

There are two kinds of electric charge, called positive and negative. Two charges of the same kind, say, two positive charges, repel each other; a positive and a negative charge attract each other. An object that contains equal amounts of negative and positive charge is said to have *zero net charge*. It is electrically *neutral*. The force between two electric charges is directly proportional to how big they are. It also depends on how close they are: the smaller the distance between them, the greater the force. By measuring the charge in coulombs, you can write Coulomb's law for the electric force between two stationary charges as:

$$\text{Electric force} = (9.0 \times 10^9) \frac{\text{charge in } A \times \text{charge in } B}{(\text{distance } \overline{AB})^2}$$

Here, the charges would be expressed in coulombs; the force, in newtons. (The newton is the unit of force needed to accelerate 1 kilogram in 1 meter

24
The Original Charge Account

How essential electric energy is to our civilization becomes clear when it suddenly is not available—a recent blackout in New York City.

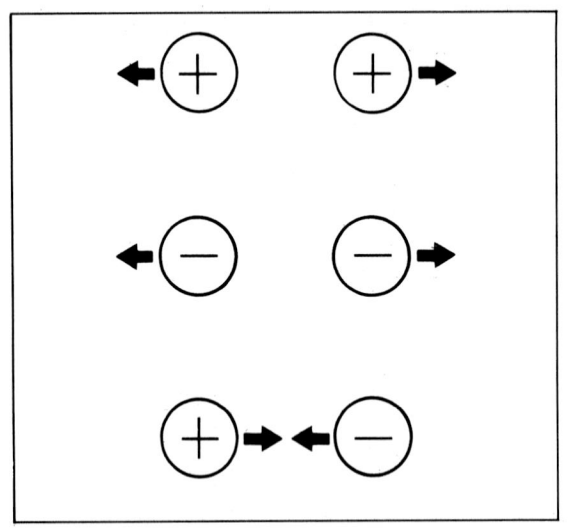

Figure 2. Like charges repel; unlike charges attract.

Figure 3. (Below) The force on charge 1 due to several charges can be found by using Coulomb's law between charge 1 and each of the other charges in turn.

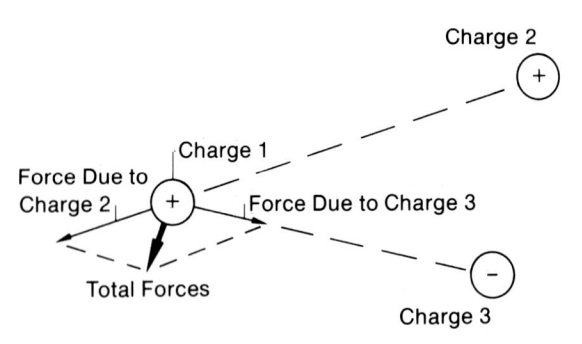

per second squared.) The number 9×10^9 is a numerical constant that has been determined experimentally. Now a number on the order of 9 billion is pretty large. If you were able to charge, say, two small rocks with 1 coulomb each and if you could bring them within 1 meter (about an arm's length) from each other, according to Coulomb's law the force between then would be something like 900,000 tons! If you count up all the positive charge even in a small piece of matter, it can be of the order of millions of coulombs. The catch is that there are also *negative* charges almost as great as the positive charge, and the net charge is close to zero.

Coulomb's law tells how to find the electric force between two charges. If there are more than two charges involved, experiment shows that the net force on one of the charges is the sum of the forces exerted by each of the other charges. The electric force between a given pair of charges is not changed by the presence of other charges.

Coulomb's law is accurate only when the size of the charged objects is small compared to the distance between them. If the objects are large, we can mentally subdivide each one into tiny bits of charge and then use Coulomb's law between the small bits.

Gravitational force and electric force have many similarities. Both vary with distance as the inverse square, and can act over long distances. However, on the scale of the solar system and the galaxy the gravitational force predominates and the electric force is unimportant. The reason is that all mass produces an attractive gravitational force, whereas electric forces are both attractive and repulsive and nearly cancel for large objects such as the earth and the sun. We think of gravitational force as being stronger than electric force simply because we haven't seen any "negative mass" canceling out the gravitational effect that masses exert on each other. Although it's possible that ordinary matter might exert a repulsive gravitational force on "antimatter," no one has been able to test this notion experimentally. The only antimatter available to us is certain subatomic elementary particles, and the gravitational force on them is too weak to measure.

One important difference between charge and mass is that mass can be destroyed by converting it into energy; the light and heat of the stars comes mainly from this process. However, there is no way to create or destroy net electric charge. Positive and negative charges can be made to combine to give zero net charge, but in every case exactly as much positive as negative charge disappears. The total charge in the universe is always the same, if it is correct to extrapolate our knowledge so far. In accelerators and in stars, charged particles can be created from the collision of neutral particles, but the creation of every positive charge is accompanied by the creation of equal negative charge. The net charge remains zero. You can say that electric charge is *conserved*.

ELECTRICITY AND THE STRUCTURE OF MATTER

Earlier chapters in this book presented a descriptive view of the atomic structure of matter and the many steps that led, finally, to an atom that is pictured as a positively charged nucleus surrounded by a cloud of electrons. Consider, now, the electric nature of the atom to bring that picture into focus.

Much of the practical importance of electricity stems from the electric nature of the electron, one of the fundamental units of matter. Electrons, the negatively charged particles in atoms, are bound by electric attraction to a central, massive nucleus that is positively charged. Electrons have a mass of 9×10^{-31} kilogram and carry an electric charge of -1.6×10^{-19} coulomb. The nucleus in its turn is made up of a number of

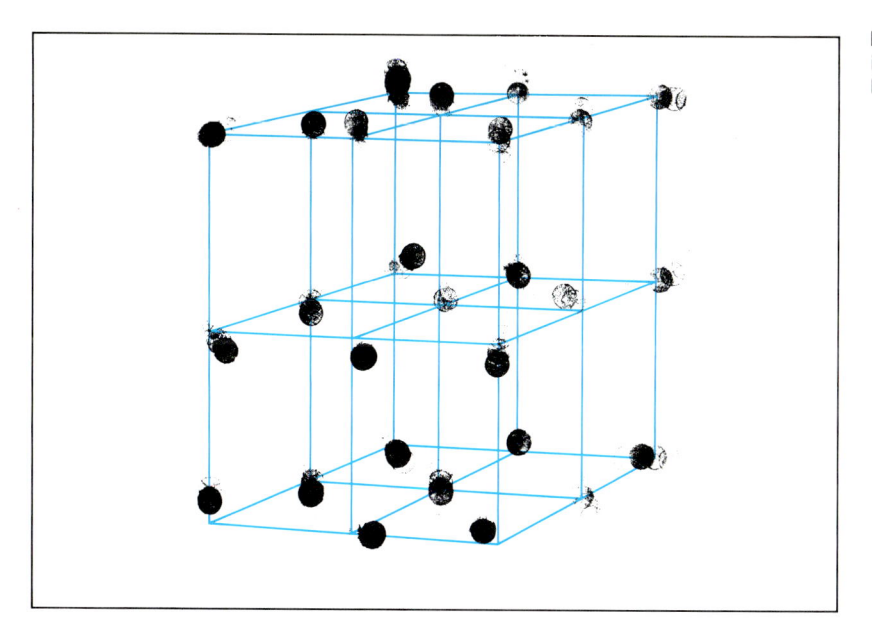

Figure 4. The atoms in a solid are fixed in position, but in conductors there are some electrons that are not bound to individual atoms.

protons and neutrons, each with a mass of about 1.6×10^{-27} kilogram. The mass of a proton or neutron is about 1,800 times greater than the mass of an electron. The protons carry an electric charge of $+1.6 \times 10^{-19}$ coulomb, which is exactly opposite that of the electron. The neutrons have zero charge.

A remarkable thing about all known elementary particles is that their charge either is zero or is equal in magnitude to the charge of the electron. There seems to be no smaller unit of electric charge than $\pm 1.6 \times 10^{-19}$ coulomb. Why charge comes in discrete bundles is unknown. Mass does not act this way at all—elementary particles come in a wide range of masses. You can see, however, that if charge could come in any amount it would be impossible to guarantee the exact equality of positive and negative charges and charge would not be strictly conserved.

The fact that electrons and protons have equal and opposite charges means that *the atom as a whole is electrically neutral if it has equal numbers of electrons and protons*. Two completely separated atoms exert negligible force on each other, which is why the electric force between two chunks of matter is effectively zero. (The electric neutrality of atoms has been checked experimentally to high accuracy by trying to exert electric forces on an atom isolated in a vacuum.)

If two atoms are pushed close enough together, however, the two nuclei are no longer shielded by the electrons and they repel each other strongly. When you push a book with your hand, the force is the electric repulsion of overlapping atoms in your hand and the book.

Electrons in Matter: Conductors and Insulators

Solid objects such as this book retain their shape and form because the atoms comprising them are bound together in a relatively rigid structure by electric forces. The massive atomic nuclei occupy fixed equilibrium positions, and most of the electrons are tightly bound to given nuclei (see Figure 4). But even though all the electrons remain bound to the material as a whole, some electrons are not bound to any particular atom and can move long distances through the material.

The number of unbound electrons in each atom depends on the detailed structure of the solid. In metals such as copper and aluminum, ap-

Figure 5. Charles Coulomb (1736–1806).

Figure 6. How can a comb that has been rubbed briskly pick up bits of paper?

proximately one free electron exists for every atom; in glass and plastic, the proportion of free electrons is much smaller. Materials having a large number of free electrons are called *conductors.* Those with a negligible number of free electrons are called *insulators.* But the distinction isn't always clear. Sea water, for instance, has a moderate number of free electrons and could be called a poor conductor *or* a poor insulator.

Although there are few free electrons in a good insulator such as glass, some of the electrons are only loosely bound to an atom. Very hot glass is a fairly good conductor because the additional thermal energy frees some of those loosely bound electrons. In fact, given enough thermal energy, some of the electrons in some kinds of matter can leave the material completely. This effect, known as *thermionic emission,* is used to produce the electron streams in television picture tubes and x-ray tubes.

Under normal conditions, then, matter is electrically neutral, with equal numbers of electrons and protons. The protons are bound and only electrons can move through a solid and from one solid to another. A *positively charged object* is one that has a deficiency of electrons, and a *negatively charged object* is one that has an excess of electrons. For electric phenomena on the human scale, the amount of excess charge involved is typically of the order of 10^{-6} coulomb, which is equal to the charge of about 10^{13} electrons. The granular, or bundle, nature of charge is not noticeable on this scale, and electric charge can be thought of as a continuous fluid when large numbers of electrons are involved.

Simple Electric Phenomena

Although the ancient Greeks are noted more for philosophical analysis than for experimental expertise, they did perform one of the earliest electric experiments. About 2,500 years ago they found that a piece of amber rubbed with fur or wool would attract lightweight objects. It's easy to repeat their experiment by rubbing a comb with a dry cloth. The comb becomes electrified and readily picks up small bits of paper. On the atomic scale, the rubbing breaks free some of the comb's loosely bound electrons, which transfer themselves to the cloth (or vice versa, depending on which of the two tends to bind electrons better). Plastic rubbed with fur generally gains electrons; glass rubbed with silk loses electrons. Metals, too, can be electrified by rubbing them, but unless the metal is on an insulated handle the excess charge rapidly flows away.

How is it possible for a charged object such as a comb to pick up uncharged bits of paper? Suppose a negatively charged comb is brought near a scrap of paper. The electrons in the paper will be repelled by the negative charge and will tend to move away from the comb, leaving the near end positively charged. Because *the force between two charges decreases with the square of the distance,* the attractive force the comb exerts on the positive charge outweighs the repulsive force it exerts on the more distant negative charges. The paper will leap up to the comb if the net electric force exceeds the weight of the paper.

You can see how a similar process could work on the molecular scale. Many neutral molecules have their electrons bunched more around one nucleus than the other. In a molecule of table salt (NaCl), for example, the valence electron of the sodium spends most of its time around the chlorine nucleus, leaving the sodium end positively charged and the chlorine end negatively charged. The molecule is called a *polar molecule.* If the NaCl molecule is near a negative charge, the attractive force will outweigh the repulsive force and the molecule will be urged toward the

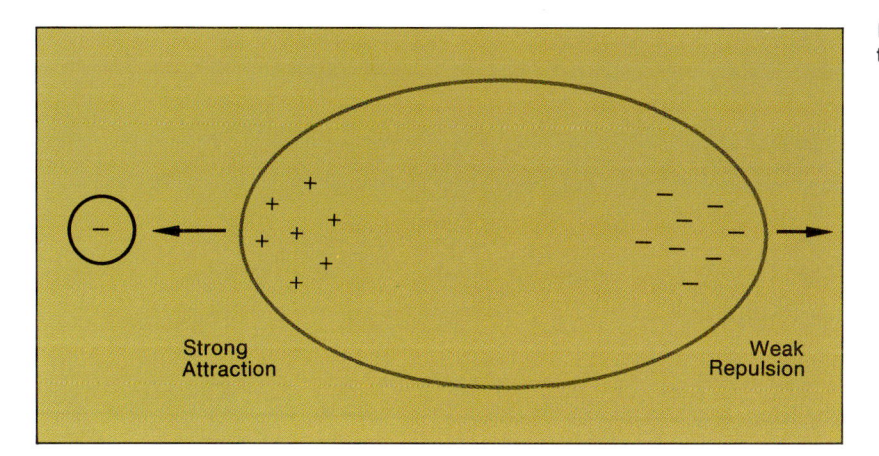

Figure 7. A charge can attract a neutral molecule if the charge in the molecule is distributed unevenly.

charge. The ability of charges to attract polar molecules helps in the formation of the complex molecules needed for life.

ELECTRIC CURRENT

In physics, as in other fields, definitions are made for their usefulness and convenience. For example, consider cars traveling down a freeway at about the same speed. Because the number of cars stays the same, all the cars that pass a certain point must pass a point down the road a little later. If the road forks, some of the cars will follow one fork and some the other, but all the cars can be accounted for. For traffic purposes, it might be useful to define a quantity that represents the number of cars passing a given point in unit time. This quantity is defined as the *current of the flow,* and the reason it is useful is that it obeys a kind of conservation law.

Suppose a road is carrying a steady flow of traffic with, say, ten cars passing a certain point every minute. The current at this point is ten cars per minute. Farther down the road, the current is also ten cars per minute because none of the cars is "lost" in traveling along the road. The current everywhere along the road is ten cars per minute. This is not a trivial observation. If the road has a speed limit of 30 miles per hour in one section and a speed limit of 60 miles per hour in another, the *speed* of flow can differ in the two sections but the *current* is the same. The cars are closer together in the slower section to maintain the current of ten cars per minute. Suppose, now, that the cars come to an exit and four cars a minute turn off the main road. Down the road from the exit, the current will be ten minus four, or six cars per minute.

To complete the highway example, suppose cars are driving into a beach parking lot at the average rate of five cars a minute and that two cars a minute leave the lot. You know that the lot is filling up at the rate of three cars a minute on the average. The current in equals the current out plus the number of cars accumulating in the lot per minute. This kind of relation is called an *equation of continuity.* Its physical basis in this example is the conservation of matter—all the cars have to be accounted for.

Electric current is the flow of electric charge. *Electric current is the amount of net charge passing a given point per second.* Electric current automatically satisfies an equation of continuity because net electric charge can be neither created nor destroyed.

The metric unit of electric current is the *ampere.* An electric current of 1 ampere corresponds to the flow of 1 coulomb of net charge per sec-

16,17,18!
SPEED LIMIT 60
NO STOPPING
START
FINISH
16,17,18!
Milk from COWS
16,17,18!
8,9,10!
SEE the ZOO!
16,17,18!
FOOD
6,7,8!
BACK TO freeway

ond. In a television picture tube, for instance, electrons are emitted from a hot filament at the small end of the tube and they fly through vacuum to the phosphor screen. The electrons moving from the filament to the screen represent an electric current of about 10^{-2} ampere. A lightning bolt is also an electric current. During its stroke a charge passes between thunderclouds and the earth, and the current may be of the order of many thousands of amperes.

The electric charge of an electron is -1.6×10^{-19} coulomb, so a current of 1 ampere corresponds to 6.3×10^{18} electrons per second. This number of electrons is far too large to count directly, and therefore the strength of currents is usually measured by the magnetic force it produces, as you shall see in the next chapter.

Many of the technical applications of electric current involve the flow of charge through a conductor such as a metal wire. An ordinary 100-watt light bulb requires a current of about 1 ampere, which is carried to and from the lamp by wires. How is it possible for a metal wire to carry an electric current? You know that some of the electrons in a metal are unbound. If they can be made to move in the same direction, the result will be a flow of net charge; in other words, an electric current.

To start the unbound electrons flowing, suppose you somehow manage to extract some of them from one end of the wire. Because the wire is electrically neutral at first, there will be a momentary excess of positive charge at the end. The unbound electrons still in the wire will be attracted and will start to move toward the end, creating an electric current. But as soon as the electrons start to move down the wire, an excess of positive charge is created at the other end as well, which tends to halt the flow. The only way to sustain a steady current is to extract electrons from one end and feed electrons back into the other. Current can flow steadily only in a complete circuit.

You can get the same result in another way by looking at the equation of continuity for electric current. Suppose you have a box with two wires coming out of it. If a current enters through one wire and another current leaves through the other wire, are the two currents equal or not? By the law of conservation of charge, the net charge entering equals the sum of the net charge leaving and the net charge accumulated in the box. According to the definition of electric current, this relation can be expressed as:

Current in = current out + net charge accumulated per second

Electric forces are so great, however, that if even a small amount of net charge accumulates in the box, the strong repulsive forces will prevent further charge from entering. So unless there is a special means for storing net charge, the incoming current is the same as the outgoing current. If a current of 1 ampere flows into a light bulb through one wire, a current of 1 ampere must leave through another. To stop the flow of current, the metallic contact in either wire can be broken by using a switch.

Applying this reasoning to a current-carrying wire leads to the conclusion that the current in a given wire is the same everywhere along its length, because there are no forces in a conductor that can hold together a great excess of charge. Although a wire may carry a charge flow of many coulombs per second, it remains electrically neutral. Electrons flowing into one section of the wire are balanced by the electrons flowing out, so there is no accumulation of net charge.

How fast do electrons travel in a current-carrying wire? You can find out the speed by doing some simple calculations. Consider a current-carrying copper wire with cross-sectional area of 8×10^{-3} square centimeter. A length of the wire 0.1 centimeter long has a volume of 8×10^{-4}

Figure 8. (Far left) In this analogy with electric current, cars move along a highway. The *current* of cars is the number of cars passing a given point per second. Moving charges in a wire are analogous to the moving cars:

Electric current is the amount of net charge passing a given point per second.

If the *number* of cars and their *speed* stays the same, the current is *steady*. When you speak of electric current, the same thing is true:

In steady flow, the electric current is the same everywhere in the system.

Suppose some of the cars leave or enter the highway at various ramps. All the cars can be accounted for—none of the cars are "lost." Electric charge is never lost, either. All the charge that flows into a device must, in steady flow, be balanced by an equal amount of charge flowing out. This important relation is called an *equation of continuity*.

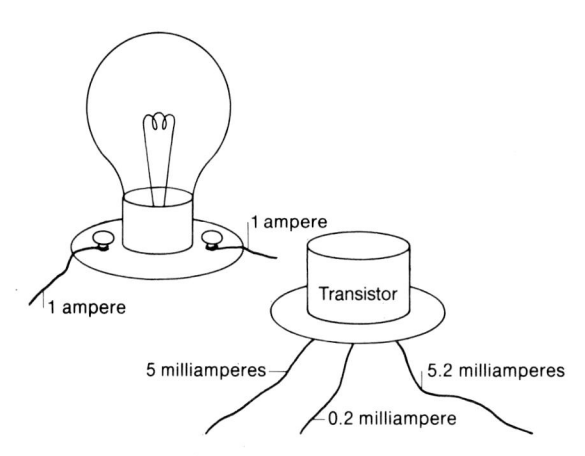

Figure 9. Electric current obeys the equation of continuity. In ordinary devices there is no continuing storage of net charge. The equation applies to devices with any number of wires. A transistor, for example, has three wire leads. If a current of 5 milliampere enters one lead and a current of 0.2 milliampere enters another, the outgoing current must be 5.2 milliamperes through the third lead.

cubic centimeter. The mass of this volume is 7×10^{-3} gram because the density of copper is 9 grams per cubic centimeter. Sixty-four grams of copper represent 1 mole of copper and, because 1 mole contains 6×10^{23} atoms, the length contains 7×10^{19} atoms. Now assume there is one unbound electron for every copper atom. If the unbound electrons are moving at a speed of 0.1 centimeter per second, all the 7×10^{19} free electrons originally present in the 0.1-centimeter length will pass into the next section of wire in 1 second. The corresponding current is $(1.6 \times 10^{-19}) \times (7 \times 10^{19}) = 11$ amperes.

Smaller currents evidently correspond to slower speeds. Nevertheless, current flow begins almost instantaneously when the switch in a circuit is closed. The electrons may have low speeds but they begin moving almost at once all along the wire, and it is this flow of charge that constitutes an electric current. Technically, it's possible to start and stop an electric current in a time interval as short as 10^{-9} second.

POTENTIAL ENERGY AND VOLTAGE

Work and energy are universal concepts in physical science and they have important application in the theory of electricity. You know that the work required to lift an object from one height to another against gravity is stored in the form of potential energy. At first glance, the concept of potential energy appears to be nothing more than a bookkeeping trick—a way of accounting for the work done on a system in order to obtain a conservation law for mechanical energy. But potential energy has a more fundamental meaning. Potential energy—the work we do—depends only on the initial and final positions and not on the path taken to get there. It is a fact of nature, and not a question of definition, that the work we do lifting a weight against gravity depends only on the initial and final heights and not on the particular path.

Potential energy is a way of storing mechanical work so that it can be returned, undiminished, at a later time. In some cases, however, the mechanical work done on a system is converted immediately to a nonmechanical form, and then the potential energy does not exist. The most notorious case is friction. You can spend a lot of time sanding a plank, but the mechanical energy you put out can't be retrieved undiminished; all you end up with is a hot board. In this case mechanical work has been changed to heat energy. The existence of potential energy is equivalent to the conservation of mechanical energy, and mechanical energy can't be conserved if friction converts part of it to heat.

Turning, now, to the electric force, you can see that it should have an associated potential energy. Suppose you pull a negative charge away from a fixed positive charge. You have to do work to pull it away because there is an electric force of attraction between the charges.

The electric potential energy of a charge acted on by other charges is written as qV. The symbol q always stands for electric charge, and the symbol V is a function of position called the voltage, or *potential*. The dimensions of voltage are energy/charge, so its metric unit is the joule/coulomb. Because of the physical importance of voltage, 1 joule/coulomb is given a name of its own: it is defined as 1 volt.

THE CONNECTION BETWEEN FORCE AND POTENTIAL

Where the electric force is strong, you have to do a lot of work to move a charge even a short distance, and the potential energy changes rapidly. So a rapid change of potential with distance implies the presence of a strong electric force. There is an exact analog in a gravitational case.

Voltage Drop in an Extension Cord

Imagine that a man runs an electric extension cord from a 120 volt outlet in his house to a barn 300 meters away. The cord, of the size used for table lamps, is made of rubber-covered copper wire and has a resistance of 6 ohms. In the barn, the man has a hot plate rated at 1,200 watts. How well will this appliance operate on the extension cord?

<u>Step 1.</u> To find the resistance of the hot plate, first determine the current by using the relation *power = voltage x current*.

A 1,200 watt hot plate operating on 120 volts draws a current given by

$$\text{Current} = \frac{\text{power}}{\text{voltage}} = \frac{1,200 \text{ watts}}{120 \text{ volts}} = 10 \text{ amperes}$$

Then the effective resistance of the hot plate is, by Ohm's law:

$$\text{Resistance} = \frac{\text{voltage}}{\text{current}} = \frac{120 \text{ volts}}{10 \text{ amperes}} = 12 \text{ ohms}$$

<u>Step 2.</u> When the hot plate is plugged into the extension cord, the total resistance of the circuit is the combined resistance of the cord and the hot plate:

$$\text{Total resistance} = 6 \text{ ohms} + 12 \text{ ohms} = 18 \text{ ohms}$$

With 120 volts applied, the current that flows is

$$\frac{120 \text{ volts}}{18 \text{ ohms}} = 6.7 \text{ amperes}$$

Then the drop in the extension cord is

$$6.7 \text{ amperes} \times 6 \text{ ohms} = 40 \text{ volts}$$

and only

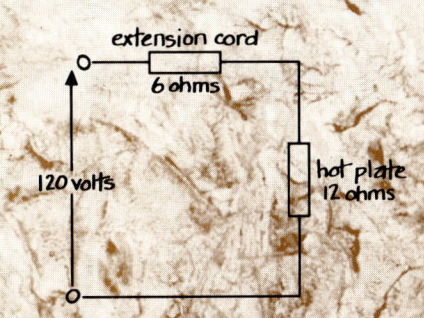

$$120 \text{ volts} - 40 \text{ volts} = 80 \text{ volts}$$

is available for the hot plate. The hot plate will not heat as well as usual; the power it dissipates is only

$$80 \text{ volts} \times 6.7 \text{ amperes} = 540 \text{ watts}$$

To allow the hot plate to operate properly, a thicker extension cord with less than .1 ohm resistance should be used.

Figure 10. Some charge configurations, their associated equipotentials, and a set of lines called *lines of force*. These lines are always drawn perpendicularly to the equipotentials. At each point, the line of force shows the direction of the electric force that would be exerted on a positive test charge placed at that point. The stength of the force is implied by the spacing of the equipotential lines, with relatively close spacing representing strong forces. This spacing is analogous to contour map spacings.

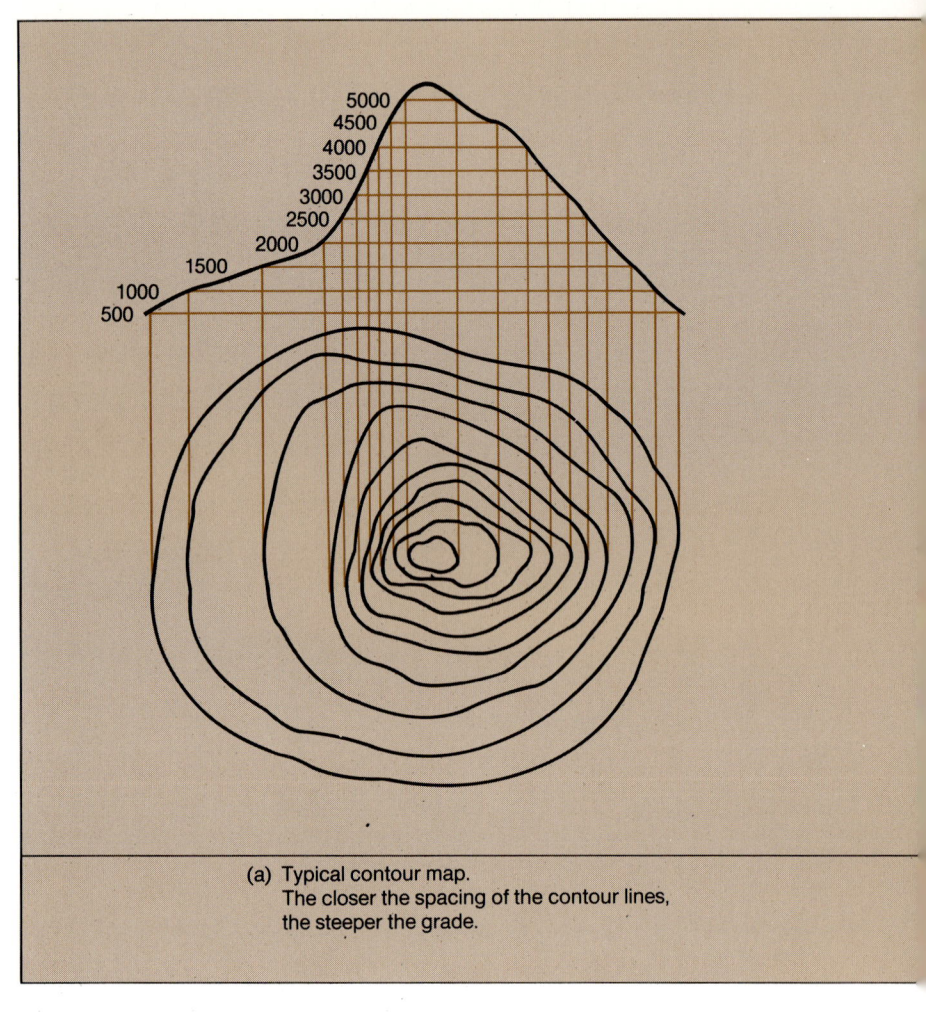

(a) Typical contour map. The closer the spacing of the contour lines, the steeper the grade.

Consider a contour map of a mountain, as shown in Figure 10. Hikers and mountain climbers know that where the contour lines are close together, the mountain is steep. Where the contour lines are relatively widely spaced, the grade is gentle.

The steepness of the grade on the mountain corresponds to the strength of electric force, and the values of the contour heights correspond to values of the electric potential. The analogy is a close one. Suppose, for instance, that a charge q is moved away from point A along a path so chosen that, at every point on the path, the potential is equal to its value at A. Because the potential does not change along this path, no work is done and the electric force on q must be zero. The analogous mountain situation corresponds to walking always on the same contour line; the mountain-climber would remain at the same altitude, neither climbing nor descending.

A path for which the electric potential has the same value everywhere is called an *equipontential*. Figure 10(b) shows equipotentials for several two-dimensional charge configurations. In a three-dimensional problem, equipotentials become surfaces rather than lines.

Work is needed to take a charge from one equipotential surface to another, but no work is necessary to go from one point of an equipotential surface to another point of the same surface. You can conclude that *the electric force is always perpendicular to equipotentials,* as illustrated in Figure 10(c). The lines with arrows in Figure 10(c) and 10(d) are called

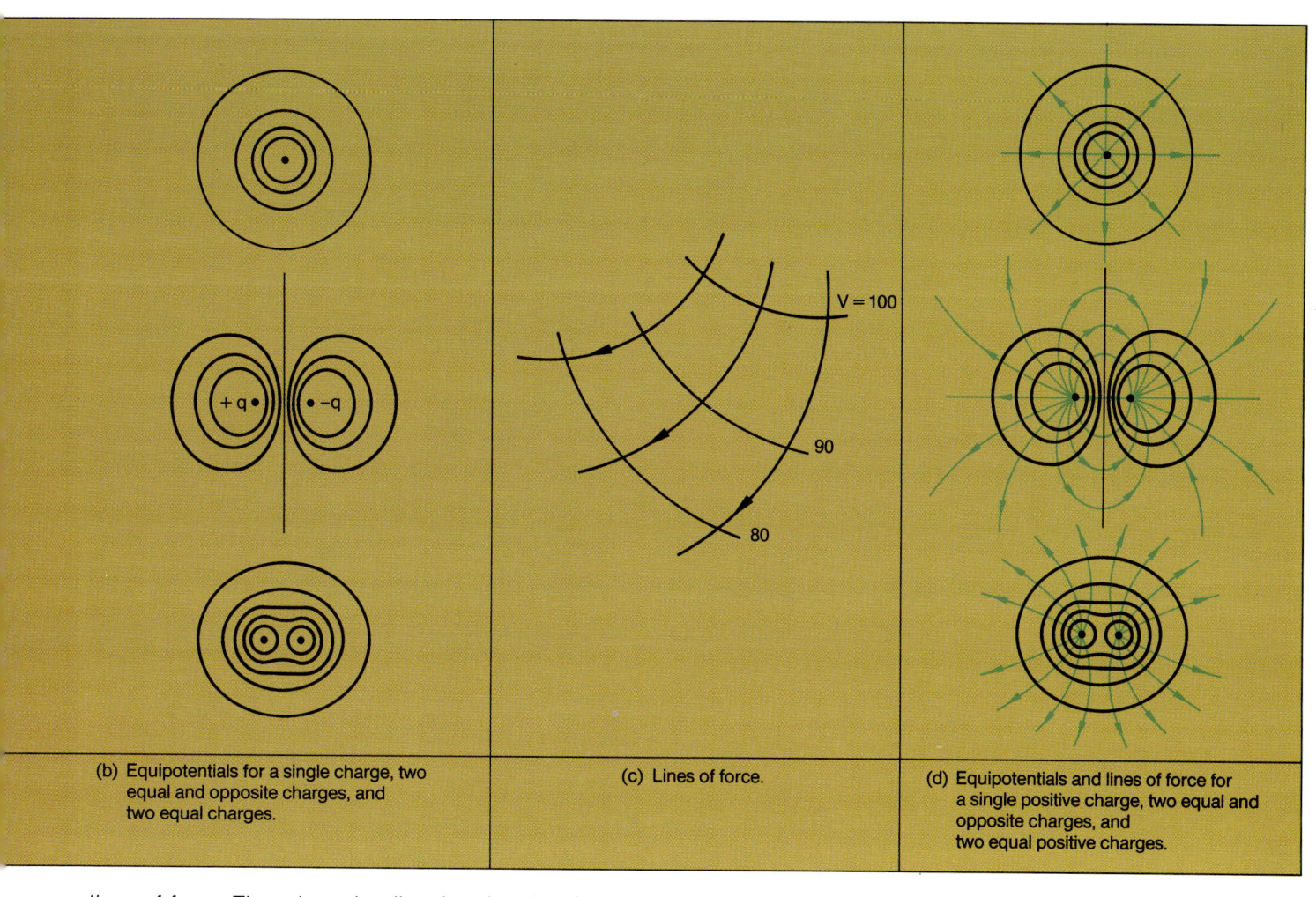

(b) Equipotentials for a single charge, two equal and opposite charges, and two equal charges.

(c) Lines of force.

(d) Equipotentials and lines of force for a single positive charge, two equal and opposite charges, and two equal positive charges.

lines of force. They show the direction the electric force would have on a positive test charge. Lines of force are always perpendicular to equipotentials.

CHARGES, CONDUCTORS, AND CAPACITORS

Given the concept of electric potential, several remarks can be made about charged conductors. Suppose, for example, there is momentarily a concentration of net charge inside a conductor. Such a concentration is not stable. By definition, a conductor has free electrons that are able to move long distances. Under the influence of the electric force produced by the charge concentration, the electrons will move through the conductor until a static or stable condition is reached. Under that condition, no electric force is working anywhere in the conductor. Three conclusions can be drawn:

1. There can be no net charge concentration anywhere in a conductor, in the static case, because such a concentration would certainly produce a strong electric force on electrons near it.

2. The only place for net charge to reside is on the surface of the conductor.

3. Because there is no electric force in a conductor in the static case, no

24 | The Original Charge Account

Figure 11. External static charges produce no effect in a region totally enclosed by a conductor.

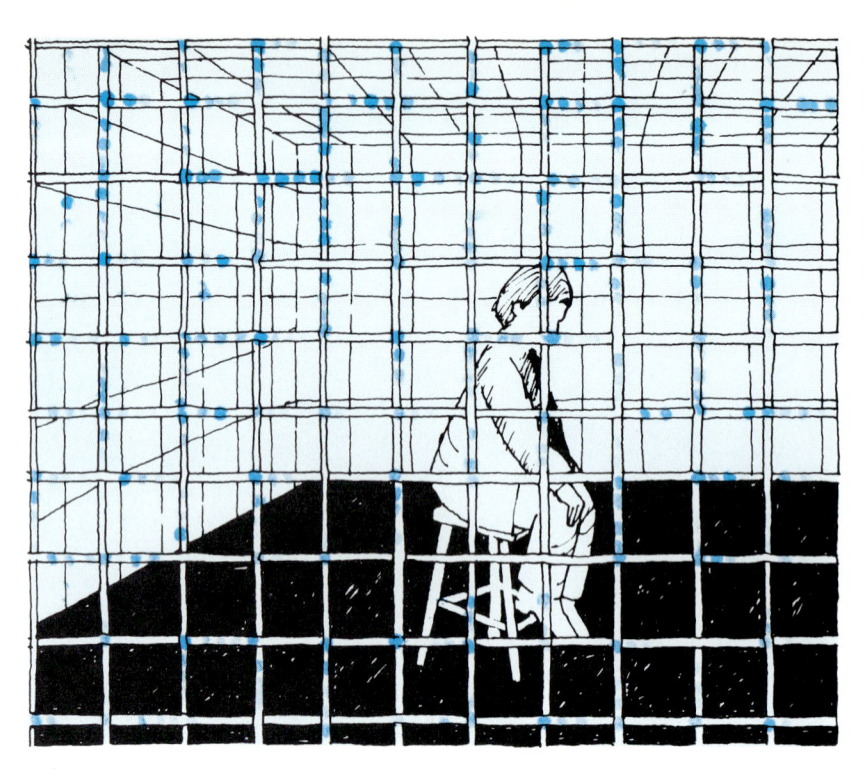

work is required to take a charge from one part of the conductor to any other part. So the electric potential must be the same everywhere in and on the conductor.

Figure 12. Michael Faraday (1791–1867).

These considerations apply only to isolated conductors; in current-carrying conductors, charges are always in motion and there is no static case. Electric forces act continuously in current-carrying conductors.

There are other conclusions concerning charges and conductors, but most are based explicitly on the inverse-square behavior of Coulomb's law and require mathematical treatment. One of the interesting results, however, is that there can be no electric force in a region entirely surrounded or shielded by a conductor even though there are charges outside. To demonstrate this, Michael Faraday built a cage of copper screen and sat inside it on an insulated stool with the screen connected to a high-voltage generator. Although sparks flew from the outside of the cage, Faraday was unable to measure any electric force inside. What happens is that the electrons in the conducting shield move about until the electric force they produce in the shielded region exactly compensates the electric force of the outside charges. Gravity is an inverse-square force like electric force, and in principle the same considerations apply. But there is no gravitational equivalent of a conductor, nor is there gravitationally neutral matter. And an antigravity shield analogous to the electrostatic shield can't be constructed.

Storing net charge in any quantity is difficult. If a net charge is placed on an isolated conductor the potential of the conductor must change, because work must be done to bring up a test charge against the electric force produced by the charge on the conductor. For instance, if you place a net charge of only 10^{-6} coulomb on an isolated conducting sphere with a radius of 10 centimeters, the potential of the sphere rises by 90,000 volts—and that wouldn't be a very convenient object to have around.

You will do much better to store equal amounts of positive and negative charge in a device called a capacitor. Suppose you place two isolated neutral conductors near each other and proceed to transfer electrons from one to the other. One conductor becomes positively charged, and the other, negatively charged. There is a potential difference between them, which becomes greater as more charge is transferred. This relationship may be written as:

Potential difference = $\dfrac{\text{magnitude of charge}}{\text{capacitance}}$

or

$V = Q/C$

Capacitance is a proportionality constant, and its numerical value depends on the size, shape, and separation of the conductors. It is always a positive number.

The metric unit of capacitance is the *farad,* named after Faraday. From the preceding equation,

1 farad = 1 coulomb per volt

A 1-farad capacitor is extremely large. The capacitors commonly found in electronic devices range from about 10^{-6} farad to 10^{-12} farad. A capacitor of 10^{-6} farad with a voltage difference of 100 volts has a charge of 10^{-4} coulomb stored on each of its conductors. That is a considerable improvement over the isolated sphere discussed earlier.

In a properly functioning capacitor, the two conductors are insulated from one another and electrons cannot cross the insulator. However, when being charged or discharged, a capacitor acts as if it carries a continuous current. As electrons flow into one conductor they flow out the other; no great amount of net charge can build up. Capacitors are advantageous when a large current of short duration is needed. In such applications, a capacitor is charged slowly with a small current until the magnitude of the stored charge reaches the desired value. If the terminals of the charged capacitor are then connected suddenly by a good conductor, a heavy current will flow through the conductor for a short time as the capacitor discharges.

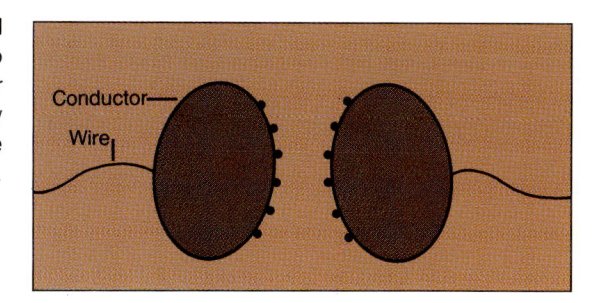

Figure 13. Charge can be held on two conductors by the mutual electric attraction.

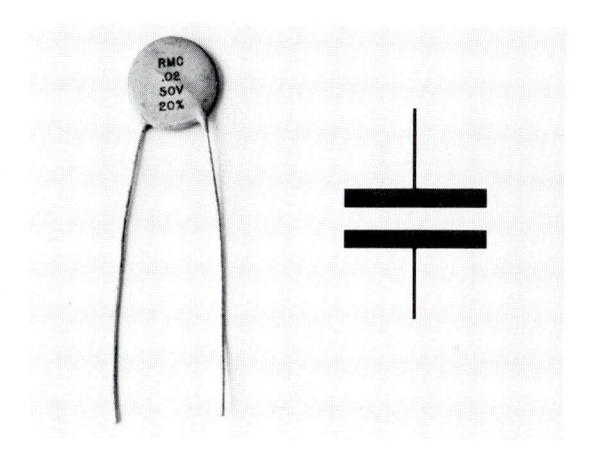

Figure 14. An example of a capacitor and the electric symbol for a capacitor.

ELECTRIC AND MECHANICAL POWER

Power is the rate at which physical work is done. You are already familiar with the term "horsepower," which is used when discussing automobile engines. By definition,

1 horsepower = 33,000 foot-pounds of work per minute

Suppose a 220-pound hiker climbs 5,000 feet up a mountain in 3 hours. The work he has done lifting his weight against gravity is $(220) \times (5{,}000) = 1.1 \times 10^6$ foot-pounds. Because it takes him 180 minutes to hike that far, the average power he "generates" is $(1.1 \times 10^6/180) = (1{,}100 \times 10^3/180) = 6{,}100$ foot-pounds per minute. From the definition of horsepower, the average power expended is 6,100/33,000, or 0.19 horsepower. Generally people can, with their muscles, generate only a fraction of 1 horsepower on a steady basis.

The electric power consumed by a light bulb, an amplifier, or some other electric device is usually rated in *watts.* But the watt is simply a unit of power in physics, and it can specify mechanical *or* electric power.

Figure 15. A charging capacitor acts as if it carries a current, because just as many electrons leave as arrive.

Figure 16. Georg Simon Ohm (1787–1854).

One watt is the same as 1 joule per second; 746 watts is about the same as 1 horsepower.

The 0.19 horsepower generated by the hiker was a mere 142 watts. A person working full time can barely generate enough power to light one or two light bulbs. By burning oil, gas, and coal he can generate much more power. The total amount of power generated in the United States for all purposes averages 6,000 watts per person on a steady basis; the corresponding figure for India is 300 watts.

What are the quantitative aspects of electric power? The general relation to consider is the one that exists between current, voltage, and electric power. Suppose a charged particle moves from one position to another. The work done on the particle by the electric force is the product of the charge and the potential difference, qV. The potential difference V is often called the "voltage drop," or the "voltage."

If a steady stream of charges moves through the potential difference, the total work done per second is the power expended:

$$\text{Electric power} = \frac{\text{charge}}{\text{second}} \times \text{voltage}$$
$$= \text{current} \times \text{voltage}$$

or, in a shorthand way,

$P = IV$

Consider, for example, a 100-watt light bulb. The 100-watt rating is the amount of electric power it *consumes,* not a measure of the light it produces. If a 100-watt light bulb is built to operate on a potential difference of 110 volts between the two ends of its filament, a current of about 0.9 ampere must be flowing.

$$\text{Current} = \frac{\text{power}}{\text{voltage}} = \frac{100 \text{ watts}}{110 \text{ watts}} \approx 0.9 \text{ ampere}$$

Only the product of current and voltage is important in determining electric power. A 100-watt light bulb, for example, could be designed that would require a voltage of 10 volts and a current of 10 amperes.

ELECTRIC RESISTANCE AND OHM'S LAW

A light bulb can be used to illustrate electric power, but a different sort of example is needed to explain how the flow of electric current heats a conductor white-hot in the first place. A return to the atomic level will show how electric energy is converted to heat energy.

The unbound conduction electrons in a current-carrying conductor do not move unimpeded along the wire. In fact, an electron moves along the wire like a person trying to make his way through a crowd. It frequently collides with fixed atoms, and its motion as it drifts down the wire under the influence of the electric force is quite irregular. In the space between the atoms, a conduction electron is accelerated by the electric force and its speed increases steadily. Before going past many atoms, however, it collides violently with an atom and loses the speed it gained during acceleration. But after the collision it accelerates again to repeat the process. At each collision, the electron transfers energy to the atoms in the wire, and the wire heats up. The collisions keep electrons in conductors from moving perfectly freely. They need an electric force to urge them on. If the electric force is large, the electrons will push their way more rapidly through the conductor and the current will be greater. Because increasing the voltage increases the electric force, you might expect that voltage and current are related in some way.

The nature of this relationship depends on the details of the collisions between the electrons and the atoms. For many conductors such as copper, iron, and sea water, experiments show that they are directly proportional:

Voltage = current × resistance

or

$V = IR$

The *electric resistance* is a proportionality constant and it is always considered a positive number. The metric unit for resistance is called the *ohm;* and the above equation is known as *Ohm's law,* named for the first experimenter to investigate the relationship between voltage and current. From this equation, 1 ohm is the same as 1 volt per ampere. For example, if a current of 0.5 ampere is flowing through a conductor with a resistance of 3 ohms, the voltage between the ends of the conductor must be 0.5 × 3, or 1.5 volts.

The resistance of a conductor depends on its shape and size as well as on the material of which it is made. A long, thin copper wire has a higher resistance than a short, thick copper bar. If you compare the resistance of a copper wire to the resistance of an iron wire of the same shape and size, you find that the resistance of the iron wire is about 6 times greater. Table 1 shows the resistance of wires of different materials 1 meter long and 1 millimeter in diameter. The resistance values span an enormous range. The resistance of a good insulator such as the quartz rod is 10^{24} times greater than the resistance of the silver wire. The extremely high resistance of the quartz reflects the almost total absence of conduction electrons—there are no unbound electrons to carry current.

The rate at which electric energy is expended in a current-carrying conductor is the electric power. Because the kinetic energy of the conduction electrons does not increase on the average, all the electric power goes into heating the conductor. The electrons act as intermediaries in transforming electric energy to heat energy.

The statement that electric power is the product of current and voltage is derived from general physical considerations and holds true for any conductor. If, however, the conductor happens to obey Ohm's law, it's possible to state that

Electric power = (current)² × resistance

or

Electric power = $\dfrac{\text{(voltage)}^2}{\text{resistance}}$

In a 3-ohm resistor carrying a current of 2 amperes, for example, heat is generated at the rate of $P = (2)^2 \times 3 = 12$ watts = 12 joules/second. The hotter an object, the greater is the rate it loses heat energy to its surroundings. The temperature of the resistor in this example will rise until the amount of heat energy lost per second equals the amount generated per second. If the rate of heat generation is too great, the temperature will rise until the resistor is burned out.

ELECTROMOTIVE "FORCE"

Suppose there is a current flowing in a continuous circuit. The conductors are heated by the current, and, by the conservation of energy, net

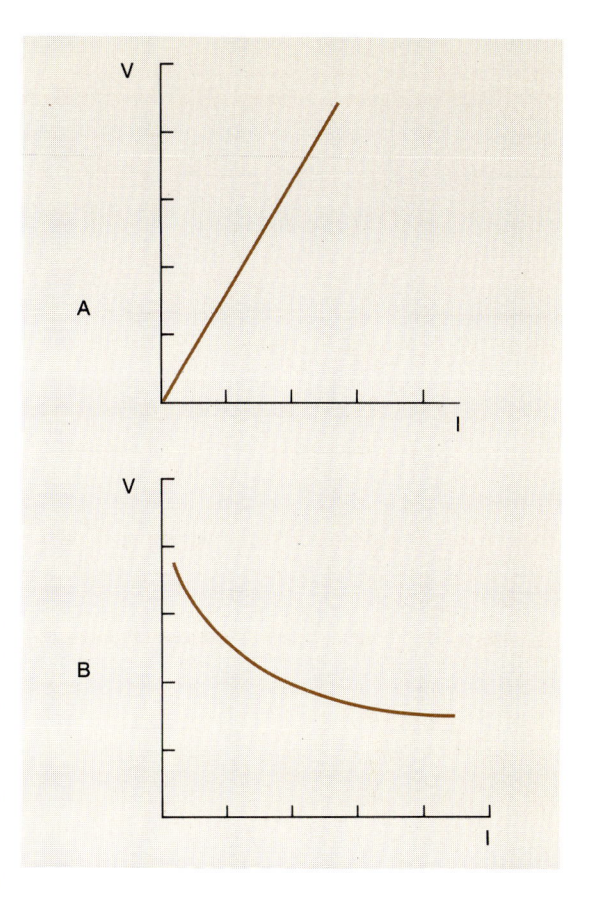

Figure 17. The relation between voltage and current for (a) a resistor obeying Ohm's law and (b) a neon lamp. Ohm's law, although useful and widely applicable, is not universally valid. It essentially is a statement concerning the conduction process in some materials. Other materials may behave differently. For example, the conduction of electricity through gases, as in neon lamps and fluorescent lights, does not obey Ohm's law: the voltage across a neon lamp decreases as the current increases.

Table 1.
Resistance of Wire 1 Meter Long and 1 Millimeter in Diameter

Material	Resistance (ohms)
Silver	1.9×10^{-2}
Copper	2.2×10^{-2}
Aluminum	3.7×10^{-2}
Iron	0.13
Graphite	12.7
Sea water	2.5×10^6
Fused quartz	$> 10^{22}$

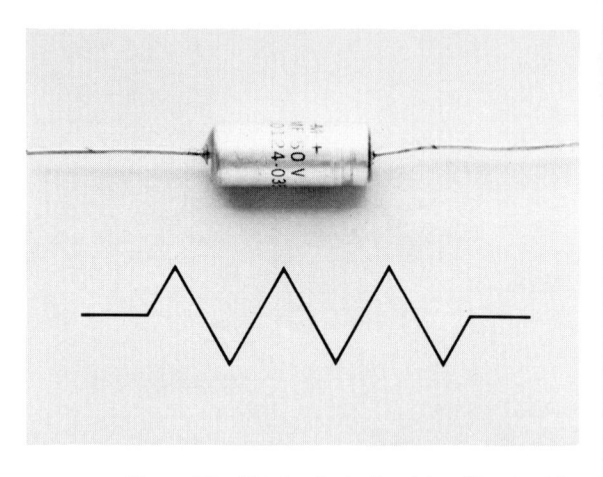

Figure 18. This is a typical resistor. The electric symbol used for resistors is shown above.

Figure 19. (Right) Photograph of an alternating current. The high-voltage arc was created by reversing the current 120 times in each second.

Figure 20. Neither gravity nor the electric force can do net work around a complete circuit. Some other agency must act.

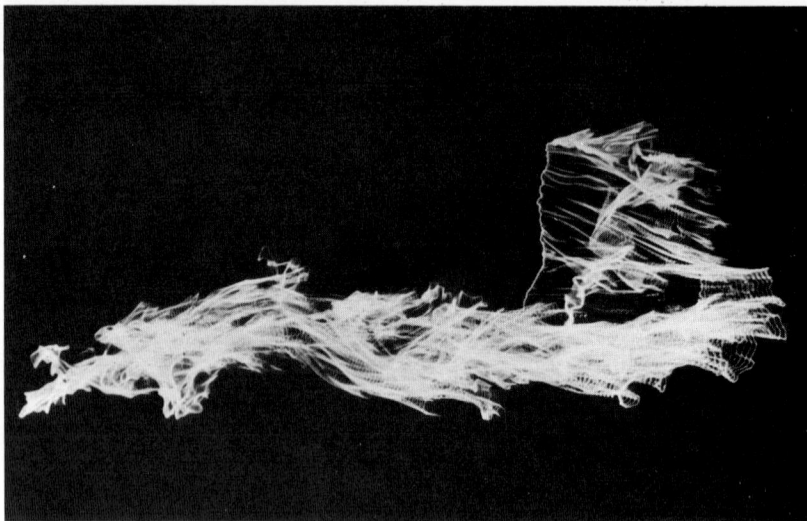

work must be done to supply this heat energy. But the electric force on the conduction electrons can't be responsible for this work. Consider an electron moving from a certain point of a circuit completely around the circuit and back to that point. Because the total charge in electric potential from start to finish is zero, the electric force does no net work on the electron.

Now the electric force *does* work on charges as it urges them from one end of a wire to the other. What is needed is some additional agency to carry the charges back to the starting point. The situation is similar to sledding down a hill in winter. Gravity does the work on the way down, but climbing back to the top requires work in the form of muscle power.

One way to get the electrons back to the starting point is to carry them back with mechanical forces. In a Van de Graaff electrostatic generator, the charges are transported on a moving belt made of an insulator: the work is done on the charges by the mechanical forces driving the belt. A Van de Graaff generator isn't suitable for generating large current, however. The next chapter discusses magnetism, which leads to a more practical way of converting mechanical energy to electric energy. Many other devices exist for doing nonelectric work on charges as they move around a circuit. Among the less familiar is the *solar cell,* which uses the energy of light to do work on charges. Solar cells are employed extensively on space missions because of their ruggedness and low weight. Another device that is useful for special applications is the *thermoelectric junction,* two dissimilar alloys joined together that convert heat energy to electric energy when they are heated. Thermoelectric junctions heated by a radioactive source make a sturdy power supply for untended locations, such as the bottom of the sea.

The most familiar device for doing work on charges is the *battery,* which converts chemical energy to electric energy. To maintain the flow of current in a conductor, electrons must be extracted continuously from the "positive" end of the conductor and taken around to the "negative" end. In a battery this journey is accomplished by the passage of charged atoms, or ions. An ordinary battery, such as the kind used for flashlights, contains two plates of different composition immersed in chemical solution. Neutral atoms in the solution near the negative plate, or *cathode,* give up electrons to the cathode and enter the solution as *positive ions.* The positive ions in the solution near the positive plate, the *anode,* tend to be repelled by the electric force but the force of attraction from the chemical binding forces is stronger. Positive ions at the anode become neutral by absorbing electrons from the anode. As positive ions are neutralized at the anode, the positive ions still in solution move closer toward the anode, maintaining the flow of current through the battery.

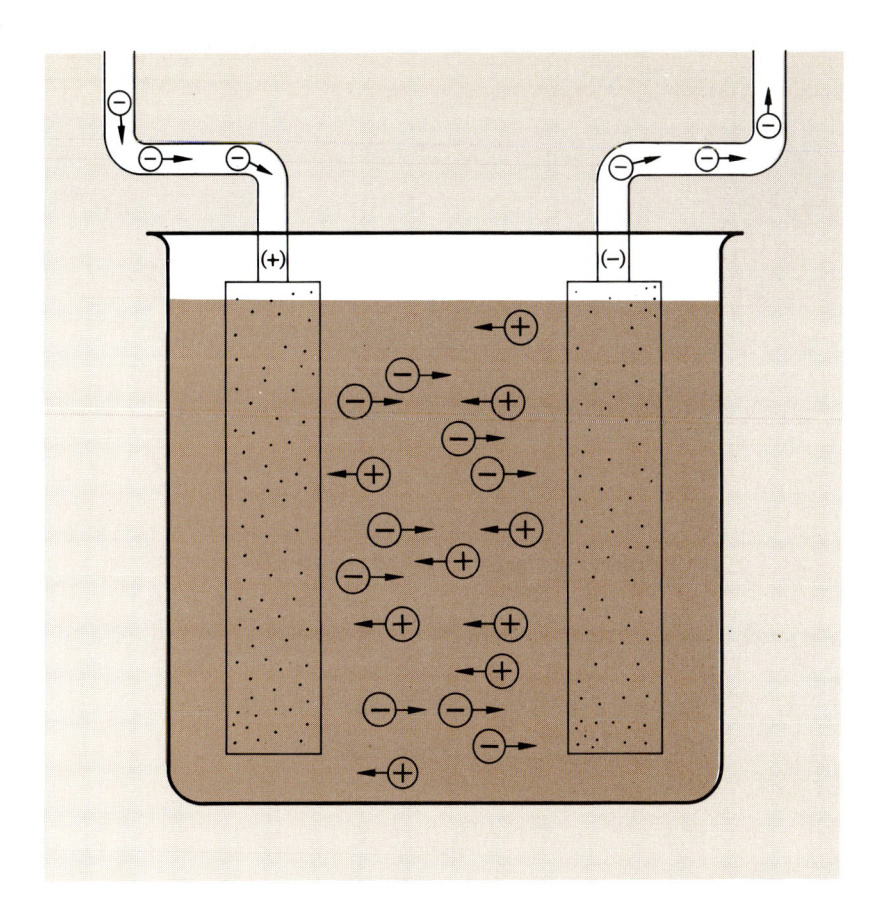

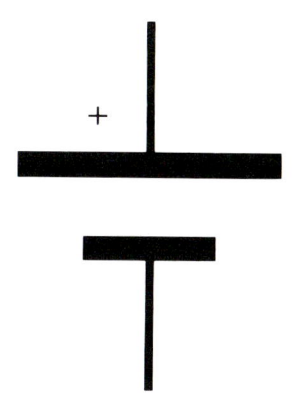

Generators and batteries do nonelectric work on charges in carrying them from one terminal to the other. The work per unit charge is called the "electromotive force," or *emf*. The name, although descriptive, is technically incorrect because emf is not a force. The metric unit of emf is the joule/coulomb. But 1 joule/coulomb equals 1 volt by definition, so emf is measured in volts.

The principle of conservation of energy can be applied to a complete circuit consisting of a source of emf and a resistor. The work done by emf per unit time is equal to the energy dissipated by the resistor in unit time:

Total emf = total potential drop

or

$V_0 = V$

This equation and the equation of continuity for electric current are both important in analyzing circuits.

Figure 21. In a battery, electrons are absorbed from the wire at the positive terminal (called the "anode"). These electrons are used to neutralize the positive ions traveling through the solution from the negative terminal (the "cathode") to the anode. The electric symbol for batteries is also shown.

ELECTRIC POWER TRANSMISSION

Suppose a generating station has to supply a certain amount of electric power to a city located a long distance away. Several forces determine

Figure 22. Power to the people. About 5 billion megawatts of energy go into the power plants that produce energy in the United States. Only about 35 percent is converted into electric power, but the percentage is increasing because of the ease with which electric energy is moved and used for thousands of different tasks. Only a small percentage of our electric power comes from atomic plants. But such plants are developing rapidly and by the year 2000 they may supply over one-half of the power we need. The growth of atomic power is a subject of active public debate.

Almost 70 percent of the energy going into power plants is lost as heat. A small percentage is lost during the distribution of electricity but most of the loss occurs in the generating station itself. The waste heat is usually disposed of by bringing in cooling water from lakes, rivers, or the sea. As we build more power plants we no longer will be able to dump heat into the earth's water systems without harming the environment. Large cooling towers, which transfer waste heat directly to the air, are being used more and more. The are expensive to build but they raise the actual cost of the power we use by only a small percentage.

This chart implies a continuous flow of energy from power plant to consumer. But we don't use the same amount of electric power throughout a 24-hour period. The "peak load" usually occurs during the evening hours. Late at night, during "off-peak" time, the load is very low. But power plants are efficient only if they generate power continuously. Utility companies therefore *store* the power generated during off-peak periods by using it to pump water uphill to a reservoir. Later, during peak load periods, the water is released and used to generate power. This method is called a "pumped storage system."

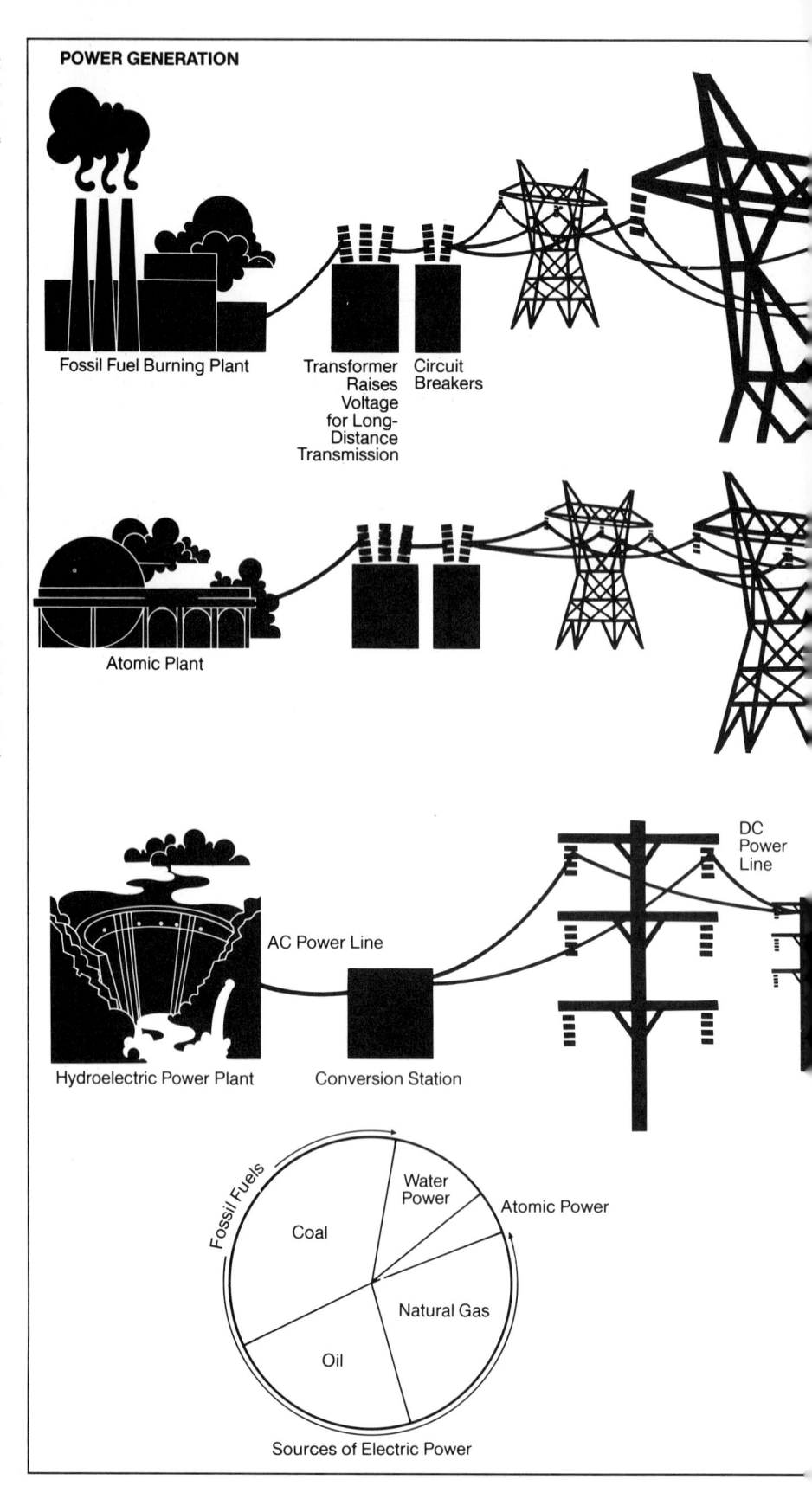

POWER GENERATION

Fossil Fuel Burning Plant — Transformer Raises Voltage for Long-Distance Transmission — Circuit Breakers

Atomic Plant

Hydroelectric Power Plant — AC Power Line — Conversion Station — DC Power Line

Sources of Electric Power: Fossil Fuels (Coal, Oil, Natural Gas), Water Power, Atomic Power

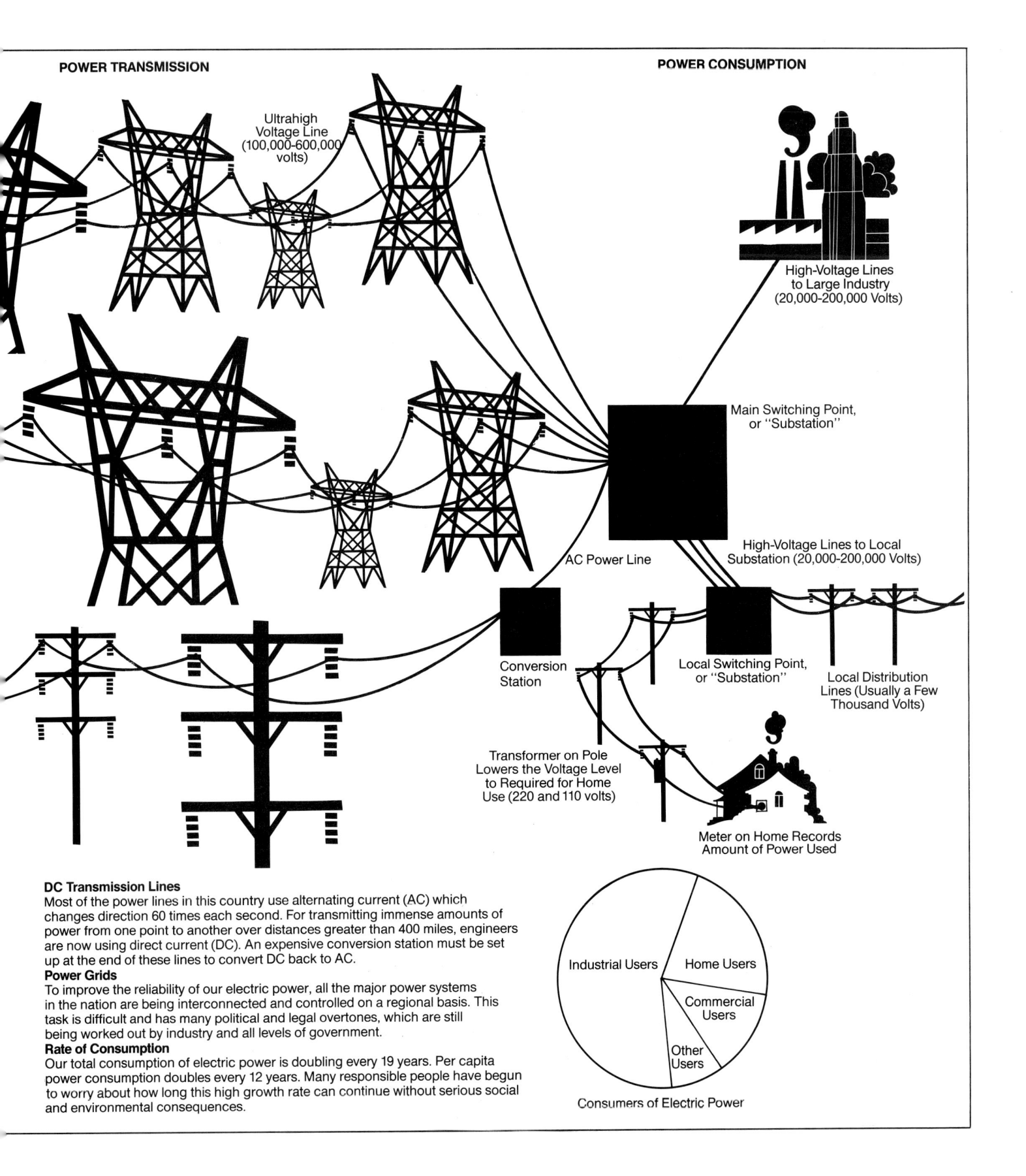

POWER TRANSMISSION

POWER CONSUMPTION

Ultrahigh Voltage Line (100,000-600,000 volts)

High-Voltage Lines to Large Industry (20,000-200,000 Volts)

Main Switching Point, or "Substation"

AC Power Line

High-Voltage Lines to Local Substation (20,000-200,000 Volts)

Conversion Station

Local Switching Point, or "Substation"

Local Distribution Lines (Usually a Few Thousand Volts)

Transformer on Pole Lowers the Voltage Level to Required for Home Use (220 and 110 volts)

Meter on Home Records Amount of Power Used

DC Transmission Lines
Most of the power lines in this country use alternating current (AC) which changes direction 60 times each second. For transmitting immense amounts of power from one point to another over distances greater than 400 miles, engineers are now using direct current (DC). An expensive conversion station must be set up at the end of these lines to convert DC back to AC.

Power Grids
To improve the reliability of our electric power, all the major power systems in the nation are being interconnected and controlled on a regional basis. This task is difficult and has many political and legal overtones, which are still being worked out by industry and all levels of government.

Rate of Consumption
Our total consumption of electric power is doubling every 19 years. Per capita power consumption doubles every 12 years. Many responsible people have begun to worry about how long this high growth rate can continue without serious social and environmental consequences.

Industrial Users | Home Users | Commercial Users | Other Users

Consumers of Electric Power

Figure 23. The principle of conservation of energy can be applied to a complete circuit consisting of a source of emf (here, a battery) and a resistor. The work done by emf per unit time is equal to the energy dissipated by the resistor in unit time, $V_o = V$.

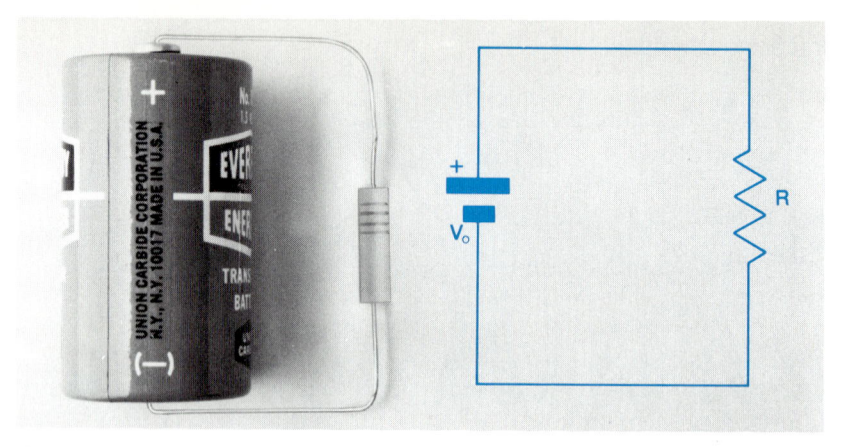

the efficiency of power transmission. The principle of the conservation of energy tells you that the power generated will be the sum of the power dissipated in the transmission lines and the power consumed by the city. To make the transmission process as efficient as possible, the power dissipated in the transmission lines must be minimized because that power only heats the conductors and does no useful work.

To minimize the power wasted, several alternatives are open. One possibility is to make the total resistance as small as possible. This can be done by using thick wires of a good conductor such as copper or aluminum and by locating the generating station as close to the city as possible to shorten the transmission lines. But other considerations limit the effectiveness of these remedies. Large amounts of copper for thick transmission lines would be expensive, for instance. Also, if power is being transmitted from the generating stations at, say, Niagara Falls to another city, the distance is fixed and can't be shortened.

By far, the most effective way to reduce the power loss is to reduce the current, because power loss equals the square of the current times the resistance of the power line. In order to keep the power delivered to the city constant, however, the voltage must be raised proportionately as the current is decreased, because power equals current times the voltage. The amount of power is the same, whether 10^6 amperes and 1,000 volts or 1,000 amperes and 10^6 volts are used. The advantage of using high voltage and low current is the reduction of wasted power in the transmission lines. For example, using 1,000 amperes instead of 10^6 reduces the wasted power by a factor of a million. The transmission of electric power over long distances is not feasible at low voltage and high current; the losses in the lines are prohibitively high. Of the power produced by the Four Corners power plant, 20 percent is wasted in its 500-mile transmission to Los Angeles.

Present-day power transmission is limited to voltages somewhat under 1 million volts for several technical reasons. Aside from the obvious problem of insulating the power lines, transmission at high voltages demands an efficient way of raising and lowering voltage. Local service in the home requires voltages of only 110 volts or 220 volts, so the transmission voltage must be lowered to these values with little waste of power once it reaches the city. The *transformer,* one of the devices discussed in the next chapter, plays a key role in power transmission systems because of its ability to raise and lower voltage with a power loss of only 1 percent or less.

Considering the human environment, it seems generally desirable to locate power plants far from the cities. However, the extent to which this can be accomplished is limited by the inefficiencies of power transmission. The problem will become even more severe if and when it proves feasible to generate power by a controlled hydrogen fusion process. It appears that such a power plant can operate only if it is immense, capable of producing sufficient power for many cities. Therefore, it will be necessary to transmit power over long distances. Controlled hydrogen fusion, if it can be made to work, will offer a nearly boundless supply of energy; and from the power standpoint, losses in transmission can then be tolerated. Nevertheless, the more power such plants generate, the more noxious by-products are produced, and concern over the environment would seem to preclude the generation of excess power.

More than 50 years ago, it was discovered that if some metals were made extremely cold, their electric resistance would drop to immeasurably small values. This phenomenon, called *superconductivity,* is receiving considerable attention as a means of developing a lossless transmission line. The technical problems are great: the superconductor must be held at a temperature less than −250 degrees Centigrade, and connecting the superconductor to a world at normal temperatures is a difficult problem. Moreover, there are many questions of safety and reliability. If the cooling system were to break down for some reason, the conductor would return to its normal resistance and would be vaporized unless the powerful currents were shut off in time.

Whatever the solution may turn out to be, more and more effort is being focused on the problem of power generation and transmission. Its solution clearly holds implications for the future integrity of the far-flung web of human activity. The fact that many nations are contemplating the construction of hydrogen fusion power plants and superconducting transmission lines is a measure of the universally felt need to mobilize the resources of society for the common good. Perhaps the people who organized to build Stonehenge felt the same way.

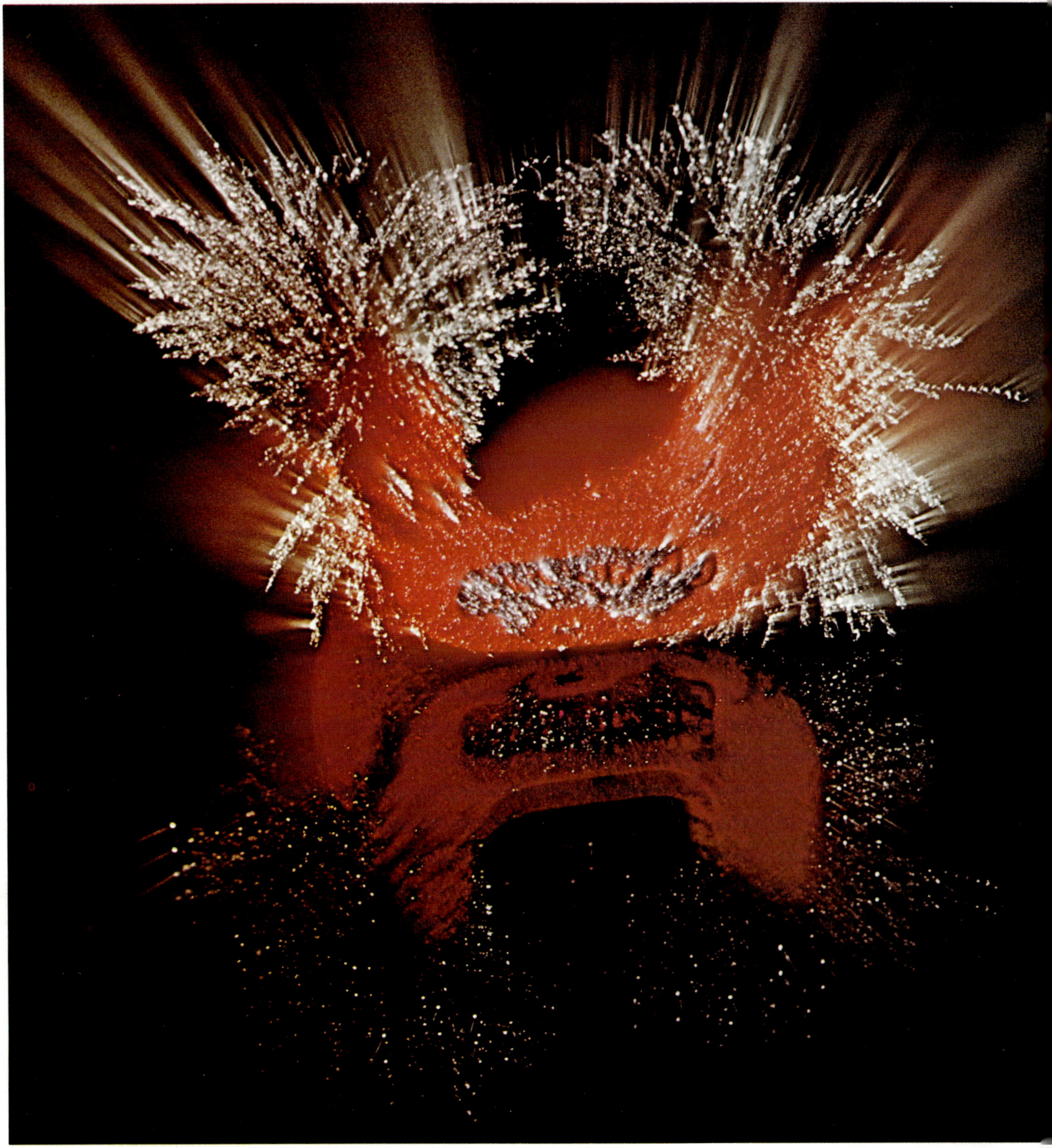

25

Magnetism and Electromagnetism

Slivers of metal fly through empty space in response to the field of force surrounding this bar magnet.

Have you ever played with a pair of magnets? The strong pulls and pushes through empty space, the rapid change of the force with distance, the unending flow of strength and apparent lack of motive power—all these things make magnetism the most interesting force in our everyday world. The gravitational force between ordinary objects is too small to notice. The electric force, although strong, only shows itself in special circumstances and dies away quickly in humid weather.

Magnetism is a fundamental property of matter, but it manifests itself differently in different materials. You know that a magnet will pick up a steel paper clip but not a copper penny or a piece of paper. Copper and paper are called "nonmagnetic" but, as you will see, every substance is magnetic to some degree. It's just that magnetic effects are thousands of times larger with steel or iron than with "nonmagnetic" materials such as copper and paper. Even an element such as oxygen is magnetic: liquid oxygen hangs like a rope from a strong magnet.

Some kinds of iron ore found in nature are strongly magnetic and the ability of pieces of this ore, called "lodestones," to attract iron was known to the ancient Greeks. Later it was found that a rod of lodestone pivoting about its middle turns until it points toward the north. It can be used to make a fairly accurate compass: the end that points toward the north is called the *north pole* of the magnet, and the other end is called the *south pole.* The magnetic compass was regularly used by European sailors in the thirteenth century, but it was known to the Chinese and Arabs much earlier. The reason for its action was explained in 1600 when William Gilbert, Queen Elizabeth's doctor, speculated that the earth itself is a magnet.

After the great success of Newton's inverse square law of gravitational attraction and Coulomb's discovery of the inverse square law of electric attraction, it was natural for scientists in the eighteenth century to look for an inverse square law for magnetic force. And they found that when two long magnets are close together, end to end, the force between them varies as the inverse square of their separation. Two north poles or two south poles repel, and a north pole and a south pole attract.

In general, the force between two magnets is not an inverse square force. Every magnet has two poles, a north pole and a south pole. When two magnets interact, the forces exerted by both poles of one magnet on both poles of the other have to be taken into account. For example, if two small bar magnets are far apart and oriented along the same straight line, the total force varies approximately as the inverse fourth power of the separation. That is why the force between two magnets seems to increase so rapidly as they are brought near one another.

The force between magnetic poles seems to be closely analogous to the force between electric charges. Like charges repel, unlike charges attract. Similarly, like poles repel, unlike poles attract. But there is one great difference between electric charges and magnetic poles: *poles never exist separately.* A magnet always has both a north pole and a south pole. Breaking the magnet to separate the poles won't work because new poles appear at the broken ends. Despite the superficial similarity, then, magnetic poles are basically different from electric charges. Electric charge is carried by particles like electrons and protons. As you saw in the preceding chapter, ordinary electric effects are due to the excess or

Figure 2. (Right) An iron filing map of the magnetic field of a current-carrying wire coil (above) is compared with an iron filing map of the magnetic field of a bar magnet.

Figure 3. (Below) William Gilbert, physician to Queen Elizabeth I, was the first to suggest that the earth acts as a giant magnet.

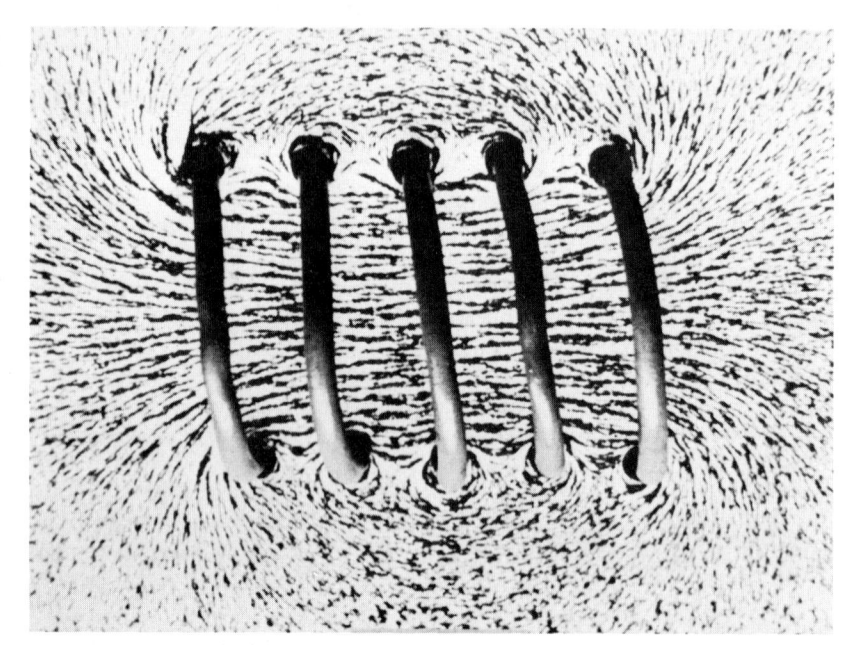

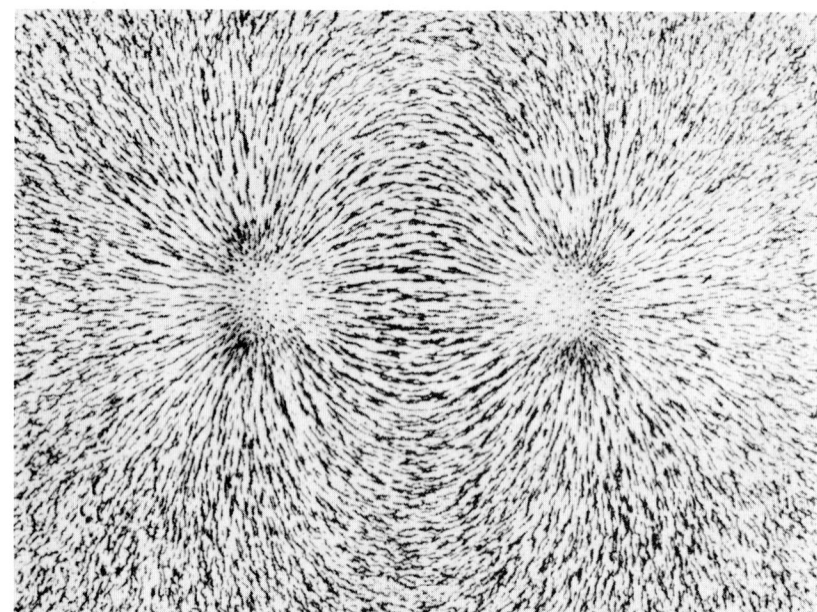

deficiency of electrons on a piece of matter. But there is no particle associated with a magnetic pole. In fact, the only evidence for the existence of magnetic poles is indirect—they are a way of describing the force between magnets. Perhaps a different model based on another picture could do just as well. When a certain model does a good job of accounting for a set of observations, it's tempting to think that we have an accurate picture of the nature of things. Sometimes, however, new observations can't be explained by the old model and a new picture must be developed.

The observation that challenged the pole picture of magnetism was made by Hans Oersted. Before 1819, electricity and magnetism seemed to be completely separate. Electric charges produce electric forces, and magnetic poles produce magnetic forces. Nevertheless, many experimenters including Oersted in Denmark felt there should be a connection between electricity and magnetism. How did such a belief arise without

Physical Science Today

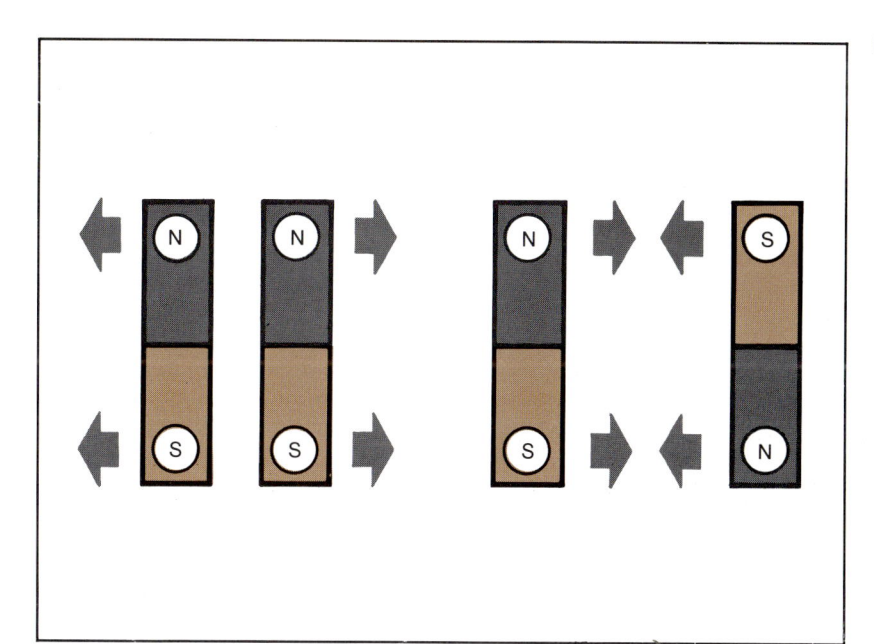

Figure 4. (Left) Like poles repel; unlike poles attract.

experimental evidence to support it? Part of the reason is philosophical and metaphysical. Many people believed that forces of different kinds had an underlying unity and that it should be possible to change one kind of force into any other kind. Such speculations and personal visions seldom appear in the final record of a scientific investigation, but they often play a valuable role in motivating the experimenters.

Oersted had sought a connection between electricity and magnetism for years. He found it in 1819 when he brought a magnetic compass needle near a wire carrying an electric current. He had done similar experiments in the past, always with the needle arranged at right angles to the wire, and he had noticed there was no effect when the current was turned on. But this time, while discussing his earlier attempts during a lecture, he started with the needle parallel to the wire. When the current was turned on, the needle swung *perpendicular* to the wire.

Oersted's experiment showed a new aspect of magnetism. A current-carrying wire had nothing to do with mangetic poles, yet it could exert forces on magnets. The next decisive step was accomplished within a few weeks by Andre Ampere, who eliminated poles altogether by showing that *two current-carrying wires exert forces on one another*. He was sure that the force was magnetic and not an ordinary electric force because of its direction. Two identical, parallel wires carrying current in the same direction *attract* each other. In the case of electric forces, however, like *repels* like.

Currents can exert magnetic forces on one another, and Ampere showed that the force between magnets may be accounted for by supposing that currents flow in magnets. Magnetic poles aren't needed at all. But there are no batteries in magnets—how can the currents keep flowing for years on end? Ampere thought that the currents in a magnet must be flowing around individual atoms, where there is no friction or resistance.

The magnetic pole picture of a magnet isn't really "wrong"—it's a good way of picturing the forces between magnets. But there is no way to use magnetic poles to describe the magnetic forces between currents. And single magnetic poles do not seem to exist. Ampere's theory, on the other hand, reduced all magnetic effects to a single cause: *currents*.

Figure 5. (Below) Magnetic poles never exist alone. If a magnet is broken, new poles appear at the break.

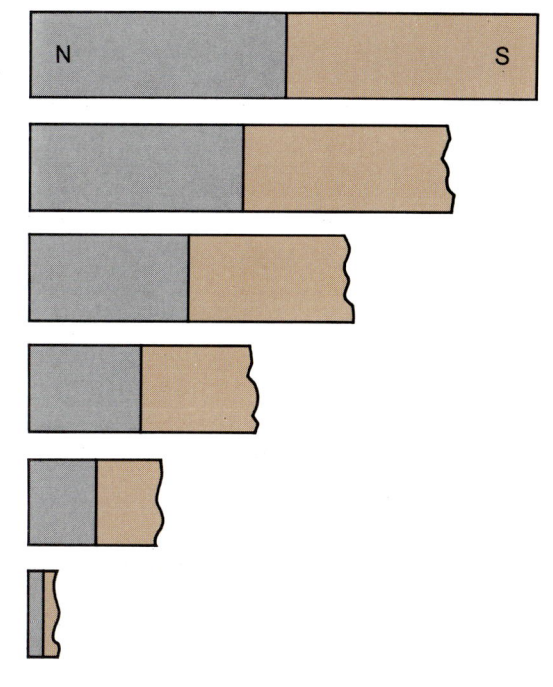

25 | Magnetism and Electromagnetism 479

Figure 6. (Left) Hans Oersted and (right) André Ampere.

Figure 7. The gravitational field of the sun (above left) and the earth (below left). The gravitational field of the earth near its surface is uniform and constant. The electric field of a positive charge (above right) and a negative charge (below right) points in the direction of the electric force that a positive charge would experience.

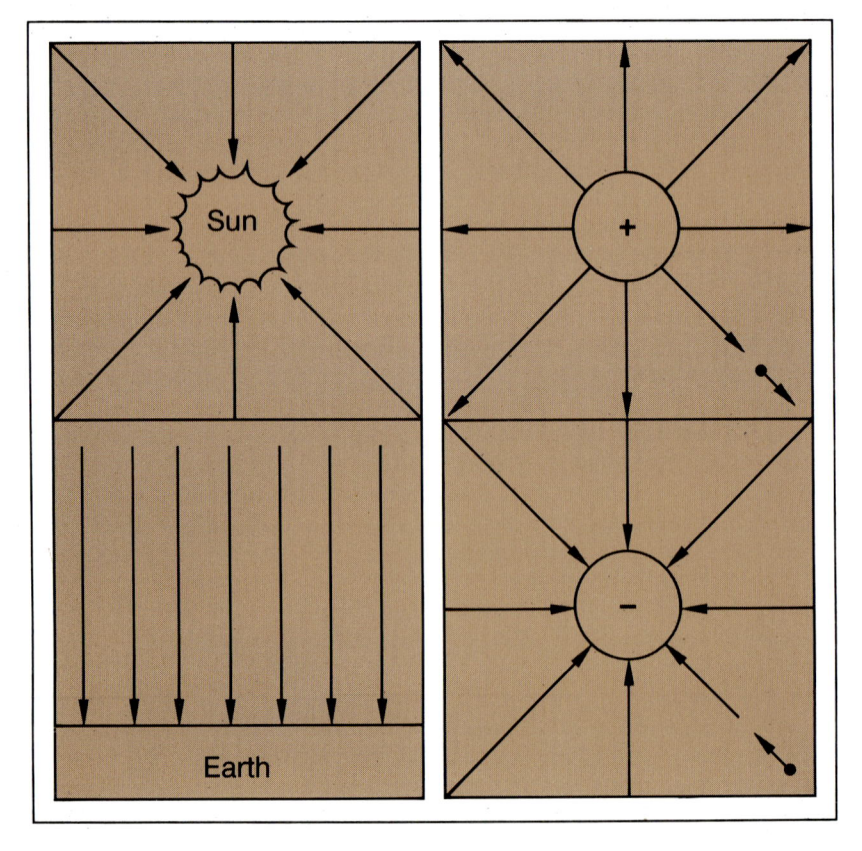

And eventually a physical basis for Ampere's atomic currents was found, as you will see. Before looking at magnetic forces in more detail, however, it's helpful to introduce another way of thinking about electric and magnetic effects.

ELECTRIC AND MAGNETIC FIELDS

All gravitational and electric and magnetic forces act through empty space. Two masses exert a gravitational force on each other even if they are separated by a considerable distance. Similarly, electric forces and magnetic forces are able to leap through space from one body to another. You've probably become accustomed to the way gravity pulls you down when you jump in the air, but it's puzzling to think of forces being exerted across empty space.

For one thing, how does the force know which way to go? The sun exerts a strong gravitational pull on the earth, but how does the sun know where the earth is? It's much more natural to suppose that the sun has the potentiality of exerting its gravitational attraction everywhere in space.

In Chapter 6 you were told of another way of looking at gravity. According to the *field* picture of gravity, the sun sets up a gravitational field through all of space. And the field exerts a force on the earth. The word "field" implies that the sun's gravity acts everywhere, over large regions. To describe a field completely, you have to know where it is everywhere in space.

Physically, the gravitational field is the force on a mass of 1 unit. Force has both magnitude and direction, so both the strength and the direction of a gravitational field have to be given at every point in space.

Figure 7 shows a way of portraying the gravitational field of the sun. The lines and arrows indicate the direction of the field. The field has the same direction as the force in a mass placed at the point. The sketch doesn't give the strength of the field directly, but the field is strongest where the lines are close together. Figure 7 shows the gravitational field of the earth near the ground. The field is straight down and has constant strength.

The field concept may be applied to electric and magnetic forces, too. The electric field produced by a single charge looks very much like the gravitational field produced by a single mass, as Figure 7 shows. The force on a charge due to the electric field is

Electric force = charge × electric field

The arrows on the electric field lines depict the direction of the electric force on a positive charge. You can interpret the electric force diagrams in the preceding chapter as electric field diagrams.

The magnetic field can be thought of in terms of the magnetic force that would be exerted on a single magnetic pole:

Magnetic force on a pole = strength of pole × magnetic field

Figure 8 shows the magnetic field of a straight current-carrying wire and the magnetic field of a loop. The direction of the field is the direction of the magnetic force that would act on a north magnetic pole placed at the point. You can see from the drawings that magnetic fields have a different nature from electric fields. The lines of an electric field start or stop on charges or go off to great distances. The magnetic field for a wire circles around the wire, never beginning or ending.

Suppose a compass needle is brought into a magnetic field. A compass needle is a long, thin magnet with a pole at each end. Figure 9 shows that the needle has a tendency to turn until it is lined up with the magnetic field at its location. A freely suspended compass needle shows the direction of the magnetic field. Near a current-carrying wire, a compass needle tends to line up at right angles to the wire, as Oersted found.

The magnetic field of the earth acts as though it was due primarily to a magnet near the center of the earth, lined up nearly along the earth's axis. The earth's magnetic field makes an angle with the ground in the Northern Hemisphere, as Figure 9 shows. A freely suspended compass needle not only points north but dips toward the ground as well. Near the North Pole, the needle points straight down.

But it isn't necessary to use a compass needle to map out magnetic fields: iron filings sprinkled on a sheet of glass or cardboard and held near the source of the field work even better. Even whiskers of iron become small magnetic needles in the field. If it is free to turn, each whisker lines itself up with the magnetic field at its location and the whole field is mapped out at once. Figure 2 shows iron-filing maps of the magnetic field of a bar magnet and the magnetic field of a coil of wire carrying a current. You can see the similarity between the magnetic field outside the magnet and the field outside the coil. If the coil is closely wound with many turns of fine wire, its field is indistinguishable from the field of a cylindrical magnet. Currents can account for the field of a magnet just as well as magnetic poles can.

MAGNETIC FORCES ON CHARGES

One of the most important properties of a magnetic field is its ability to exert forces on wires carrying electric current. But an electric current is electric charges in motion; therefore, the magnetic force on a current can be understood by looking at the magnetic force on an electric charge.

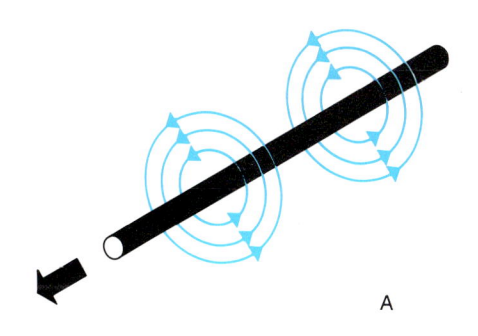

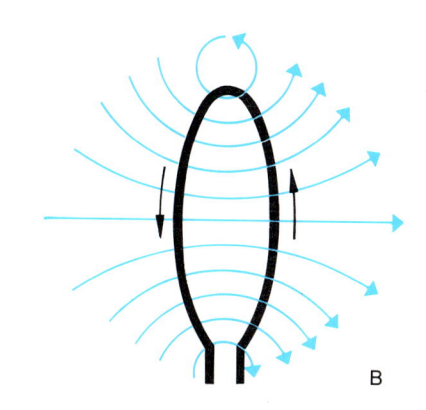

Figure 8. The lines of the magnetic field of a straight wire (A) and a loop (B) are perpendicular to the velocity of the moving charges and to the force driving the charges.

Figure 9. The magnetic forces on a compass needle tend to turn it so that the needle lines up with the direction of the field.

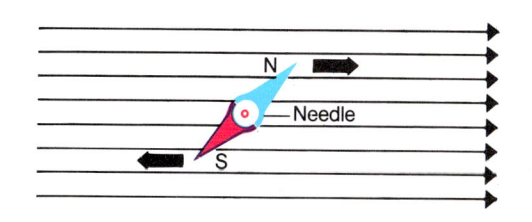

25 | Magnetism and Electromagnetism

Figure 10. The left-hand rule for the magnetic force on a moving positive charge: If the first finger points in the direction of the field and the second finger in the direction of the speed, the thumb gives the direction of the force. The force is *perpendicular* to the field and the velocity. The force on a negative charge is opposite in direction to the force on a positive charge moving the same way. (The right-hand rule is used for negative charges.)

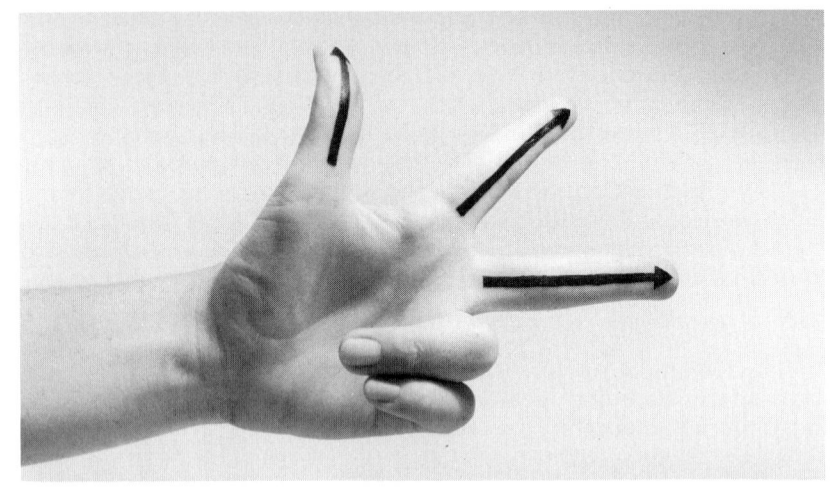

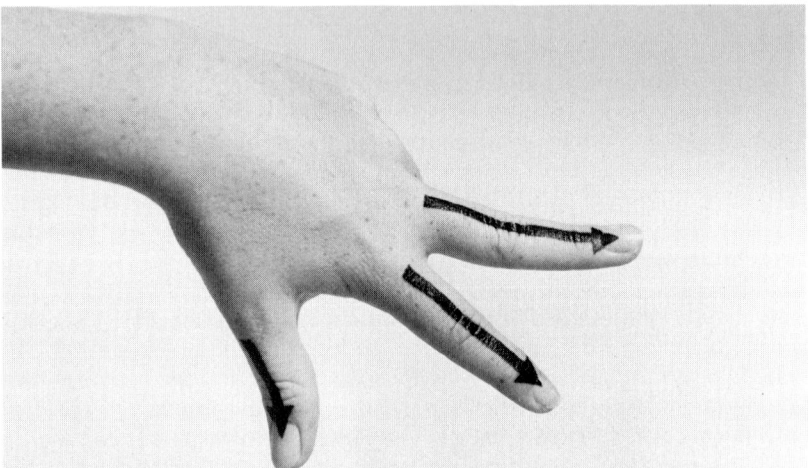

A wire in a magnetic field experiences a magnetic force if it is carrying current, but if the current is switched off, the magnetic force goes to zero. In other words, a magnetic field exerts a force on a charge only if the charge is moving. A charge at rest experiences no magnetic force.

The magnetic field acts only on moving charges.

Magnetic fields are used in television sets to deflect the stream of moving electrons and sweep it across the screen.

The strangest aspect of the magnetic force on a moving charge is the direction of the force. You have seen that the magnetic force on a pole is along the field direction. But the magnetic force on a *moving* charge is at *right angles* to the magnetic field. And the magnetic force is also at right angles to the instantaneous velocity vector of the charge.

The magnetic force on a moving charge is perpendicular both to the magnetic field at the position of the charge and to the charge's velocity vector.

This rule is illustrated in Figure 10.

The magnetic force on a moving charge is always perpendicular to the direction of motion of the charge. The magnetic force tends to bend the path of the charge. As the path bends, the magnetic force changes direction to stay always at right angles to the direction of motion. Here

482   Physical Science Today

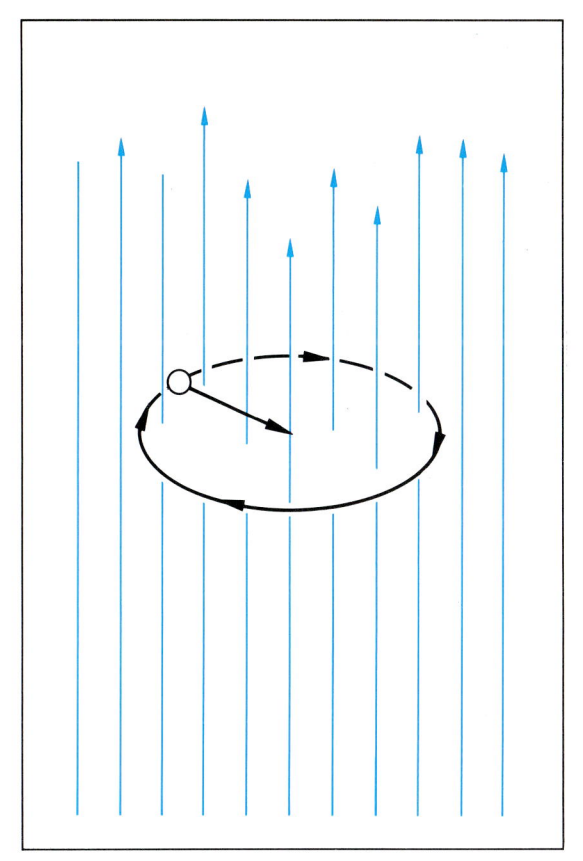

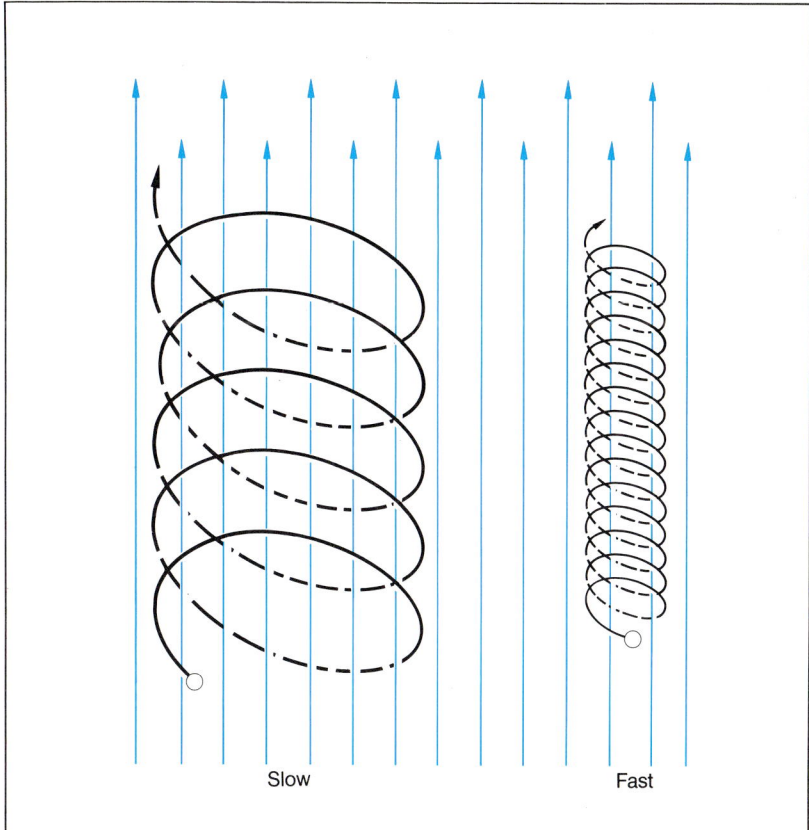

is a simple illustration of this idea. Suppose a charge is moving in a region where there is a constant, uniform magnetic field in the vertical direction (Figure 11). The magnetic force on a charge moving in this field is always horizontal, perpendicular to the magnetic field. If a charge is initially moving horizontally in the field, the magnetic force bends its path around and makes it move in a circle. This is the principle used to keep charged particles moving along the right path in ring-shaped high-energy accelerators.

The strength of the magnetic force on a moving charge depends on three factors: the size of the charge, the strength of the field, and the velocity of the charge. Only part of the velocity is important in determining the strength of the force, however. Velocity is a vector. Think of it resolved into two components: one along the magnetic field and the other at right angles to the field. The strength of the force depends on the part of the velocity perpendicular to the field.

Returning to the earlier example, suppose a charge is moving in a vertical magnetic field and starts off at an angle to the field. The charge has both vertical and horizontal velocity, and its motion is a spiral. The vertical motion is steady, but the horizontal motion is in circles.

Spiral motion is one way a magnetic field can trap a charged particle. A charge can move horizontally without limit where there is no field. If a magnetic field is present, however, the particle circles around the field lines. The faster the charge is going, the greater the magnetic force, and the tighter the circle. Very fast charged particles follow magnetic field lines closely. The sun gives off streams of energetic charged particles from time to time, and the particles that come near the earth can be trapped by the earth's magnetic field. As the particles follow the field lines

Figure 11. (Left) A charge moving in a horizontal plane in a uniform, vertical magnetic field must move in a circle. It is held in the circular path by the inward magnetic force.

Figure 12. (Above) Charges spiral about the magnetic field lines. The faster the charge, the tighter the spiral and the more limited the sideways motion of the charge.

25 | Magnetism and Electromagnetism

Figure 13. The aurora borealis that appears in northern latitudes. Auroras are formed in the earth's atmosphere at altitudes ranging from 100 to 1,000 kilometers.

and enter the atmosphere, they collide with atoms in the air and excite them to give off light. The result is an *aurora*. Frequently auroras seem to funnel down toward the magnetic poles of the earth, outlining the paths of trapped charges as they follow the magnetic field lines.

THE MAGNETIC FORCE ON CURRENTS

Ampere found that two long parallel wires attract each other when they carry current along the same direction. Consider the force on wire 1 in Figure 14. Wire 1 is immersed in the magnetic field produced by the current in wire 2. The field exerts a force on the charges moving through wire 1, and the direction of the force is toward wire 2. It doesn't even matter whether the current in wire 1 is caused by the motion of negative or positive charges. In wires, negative charges move from negative to positive, and positive charges move in the opposite direction. But the effect of the different direction is compensated for by the difference in the *sign* of the charge, and the force is always toward wire 2 when the currents in the two wires are parallel.

How is the force on the moving charges transferred to the wire itself? Remember from the preceding chapter that the current carriers in a wire make frequent collisions with the fixed atoms in the wire. When the magnetic field is present, the charge carriers are deflected to one side, and their collisions with the atoms push the wire sideways.

The direction of the magnetic force on a section of current-carrying wire can be found by thinking of the direction of the force on moving charges. The magnetic force on a section of wire is perpendicular both to the magnetic field and to the section.

The strength of the magnetic force on a piece of wire depends on the number of moving charges and on their speed. But these are just the

Physical Science Today

factors that determine the amount of current in the wire. In addition, the force depends on the length of the section: the force on a 1/2-centimeter of current-carrying wire is only half the force on a 1-centimeter length carrying the same current because there are only half the charges involved. And the force depends on the orientation of the section, just as it does for moving charges. The force is zero when the section is parallel to the field and maximum when the field and the section are at right angles.

As an example of these ideas, consider a loop of wire in the shape of rectangle, pivoted about a horizontal axis (Figure 15). Suppose there is a vertical magnetic field in the region of the loop and suppose further that current is flowing around the loop. Each side of the loop can experience a magnetic force. The magnetic forces on the ends are horizontal and do nothing except stretch the loop a bit. The forces on the sides of the loop are horizontal and act to turn the loop. The horizontal forces are constant in strength and direction, but as the loop rotates, their ability to turn the loop varies. There is no turning effect at all when the loop is perpendicular to the field. If the loop is left to itself, it will end up at rest with its plane perpendicular to the field. But suppose the current in the loop is switched off for a moment to let the loop coast by the neutral point. Then, if the current is switched on again but in the reverse direction, the forces will keep the loop rotating in the same direction. By repeating the process the loop can be kept turning. We have an electric motor—a means of converting electric energy to mechanical energy.

All practical electric motors operate on the principle that a current-carrying wire experiences a force when it is in a magnetic field. There is a subtle paradox involved, however. Magnetic force always acts sideways on the line of motion of a moving charge and can never do work. A mag-

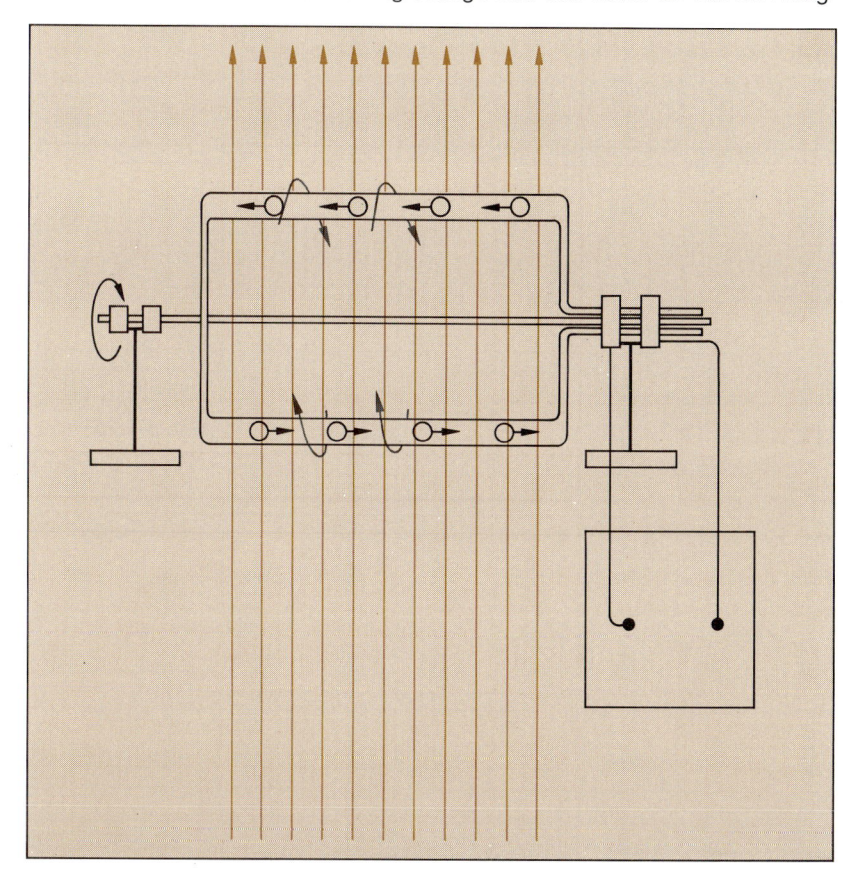

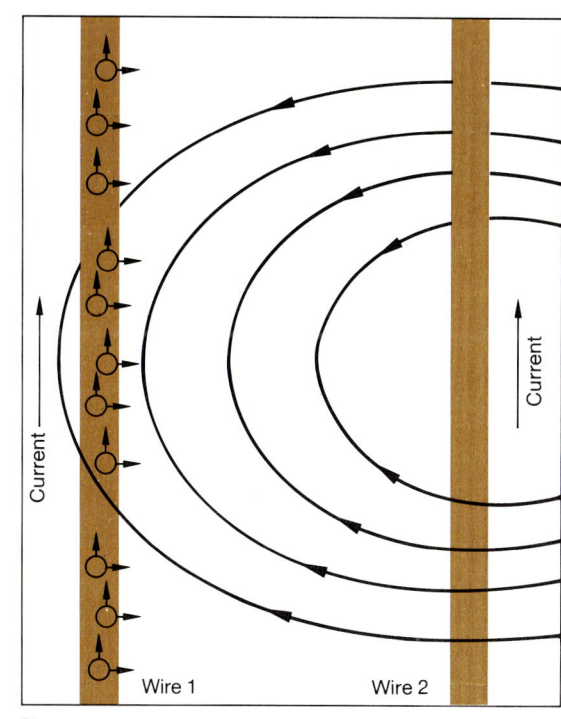

Figure 14. The moving charges in wire 1 experience a magnetic force toward wire 2.

Figure 15. (Left) The magnetic forces on the sides of the loop tend to turn the loop because the moving electrons in each side have a *component of velocity* perpendicular to the field.

25 | Magnetism and Electromagnetism

Calculating Electromotive Force

Suppose a rectangular loop of wire 20 centimeters long and 10 centimeters wide is immersed in a magnetic field of 3×10^3 gauss. Assume the field is perpendicular to the plane of the loop. If the loop is pulled steadily out of the field at a speed of 40 centimeters a second, how large an emf is developed in the loop?

Step 1. In the metric system, the emf in units of volts is numerically equal to the rate at which the magnetic flux through the loop is changing. When the field is perpendicular to the plane of the loop

$$\text{Magnetic flux} = \text{magnetic field strength} \times \text{area of loop}$$

The metric unit of magnetic field strength is equal to 10^4 gauss. Then the field strength for a field of 3×10^3 gauss is

$$\frac{3 \times 10^3 \text{ gauss}}{10^4 \text{ gauss}} = 3 \times 10^{-1} \text{ metric unit} = 0.3 \text{ metric unit}$$

To calculate the magnetic flux, the length and width of the loop must be expressed in meters so that metric units are used throughout:

$$\text{Magnetic flux} = (0.3) \times (0.2 \times 0.1) \text{ metric unit}$$
$$= 6 \times 10^{-3} \text{ metric unit}$$

After the loop has been pulled out of the field, the magnetic flux is zero.

Step 2. If the loop is withdrawn from the field at 40 centimeters a second, it requires 0.5 second to remove the loop completely. During this time the flux drops steadily from 6×10^{-3} to zero. The emf developed is then:

$$\text{emf developed} = \frac{\text{change in magnetic flux}}{\text{time required}}$$
$$= \frac{6 \times 10^{-3} - 0}{0.5} \text{ volt}$$
$$= 12 \times 10^{-3} \text{ volt}$$

If the coil had 1,000 turns instead of only one turn, an emf of 12 volts would be developed.

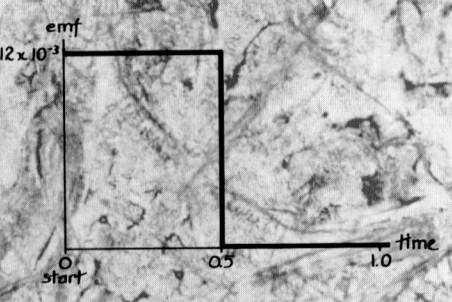

netic force cannot give a particle kinetic energy because all the force can do is change the particle's direction of motion, not its speed. So how can electric motors lift elevators, power automobiles, and run hoists and cranes? Think about what happens when a charge moves through a wire. It collides with atoms and loses energy. But it regains energy—not from the magnetic field, but from the electric forces produced by the battery or generator that keeps the current flowing. The battery or generator does all the work. The magnetic field is just a useful means of converting electric energy to mechanical energy.

Pointer types of electric meters are basically current-measuring devices. Going back to the pivoted loop in a magnetic field, suppose the loop is constrained by a spring. When current is flowing, the loop will turn until the effect of the magnetic forces is counterbalanced by the twist of the spring. If the current in the loop is increased, the magnetic forces become stronger and the loop turns a little farther. The angle that the loop turns through is a measure of the current. To make the meter more sensitive, many layers of wire can be wound around the loop to increase the total magnetic force for a given current.

UNITS OF CURRENT AND MAGNETIC FIELD

The unit of electric current, the ampere, is defined in terms of the force between two long, parallel wires carrying equal currents. If the wires are 1 meter apart, the current giving a force of 2×10^{-7} newton on each meter length of wire is defined to be 1 ampere. In practice, the force between cylindrical coils is used to check standards of current but the measurement is directly related to the fundamental definition involving long wires. The coulomb, in turn, is defined as the charge transported by a current of 1 ampere in 1 second.

The unit of magnetic field can be defined in terms of the force on a moving charge, or, equivalently, in terms of the force on a section of wire carrying a known current. The metric unit of magnetic field has not come into common usage. Magnetic fields are usually specified in *gauss*, which is 10^{-4} times smaller than the metric unit. The earth's magnetic field is about half a gauss. The field of a strong, permanent magnet is a few thousand gauss. The largest steady magnetic fields that have been produced are a few hundred thousand gauss.

The field 10^{-2} meter from a long wire carrying a current of 1 ampere is only 0.2 gauss. The field at the center of a circular loop 2×10^{-2} meter in diameter is about 0.6 gauss for a current of 1 ampere. The field can be increased by using many turns of wire, but it is clear that producing a field of several hundred thousand gauss requires immense currents. Moreover, the coil must be mechanically strong in order to resist the strong forces of repulsion between the parts of the coil.

MAGNETISM AND MATTER

All matter has magnetic properties of some kind because magnetism is built into the structure of the atom. You have seen that electric currents produce magnetic fields. Circulating electrons in atoms act like currents as they travel about the nucleus. This travel may be called the *orbital* part of the current. In addition, the electron itself acts like a tiny magnet, much as if it has internal circulating currents. That is the *spin* part.

The structure of the atom determines the degree of effectiveness of the atomic currents. An electron in an *s* state moves symmetrically with respect to all directions and produces no orbital current. An electron in a *p* state produces some orbital current, and an electron in a *d* state

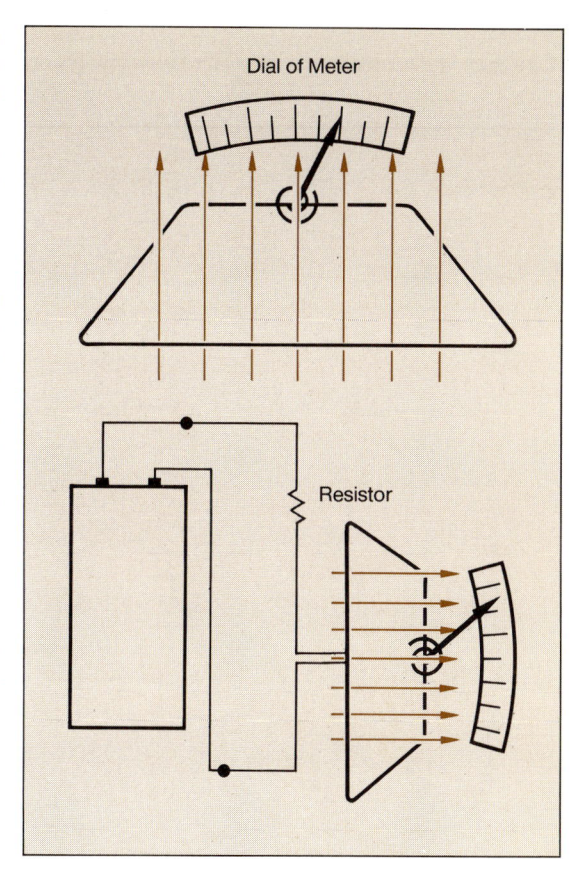

Figure 16. To make a meter that measures voltage, a resistor is put in series with the loop. The amount of current flowing through the loop then depends on the voltage at the terminals of the meter, according to Ohm's law. The meter dial may be calibrated in terms of voltage even though it actually measures current. Voltage applied to the terminals causes a current to flow through the loop, and the *angle* through which the loop turns can be used as a measure of that current.

Figure 17. Atoms with net currents can be thought of as tiny current loops or as tiny magnets.

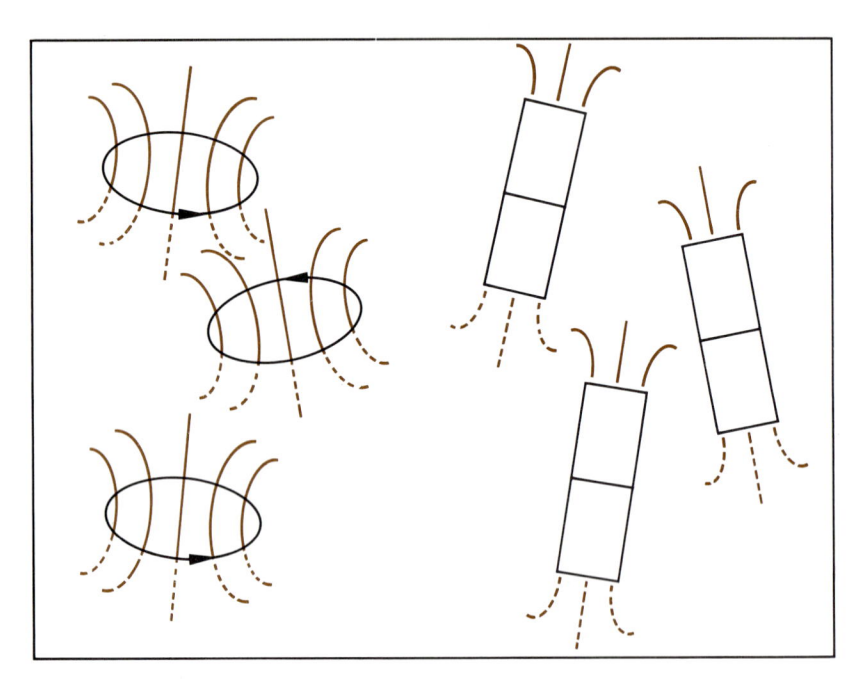

produces even more. If there are several electrons in an atom, however, the orbital parts may combine to add or subtract. The electrons in a closed shell always produce zero orbital current. If the shell is not closed, there is often considerable cancellation and the orbital current can even be zero.

The spin current depends on the number of unpaired electrons. If two electrons in the same shell are paired (in other words, if they spin in opposite directions), their spin current cancels. Closed shells always have zero spin current because the electrons in a closed shell are always paired. Generally, electrons in a partly filled shell don't start pairing up until the shell is at least half-filled. Atoms with half-filled or nearly half-filled shells tend to have strong magnetic properties because of the large total spin current. Oxygen is such an atom: it has four p state electrons in its outer shell. Two of the electrons are paired, and two are unpaired.

There is a simple model for thinking about the magnetism of atoms. In this model, *the total current in an atom is the combination of the orbital current and the spin current.* No matter what the structure of the atom, however, the atom acts magnetically like a tiny loop of current. Alternatively, then, the atom may be thought of as a *tiny bar magnet.* The two models are equivalent, according to Ampere. The details of atomic structure come into determining the strength of the current or the magnet.

The strength of the magnet in an atom can be measured by seeing how the atom responds to the pull or push of another magnet. If an atom is moving freely in vacuum and a magnet is brought near it, the atom will be deflected. The amount of deflection is a measure of the strength of the atomic magnet.

Otto Stern and W. Gerlach in 1922 were the first to accomplish the magnetic deflection of individual atoms. They worked with silver atoms moving in vacuum. The electronic structure of silver consists of a single unpaired electron outside closed shells. The magnetic properties of the silver atom are due entirely to the spin currents of the lone electron because it is in an s state and therefore has no orbital current. To understand the experiment, think of the forces between two bar magnets, one

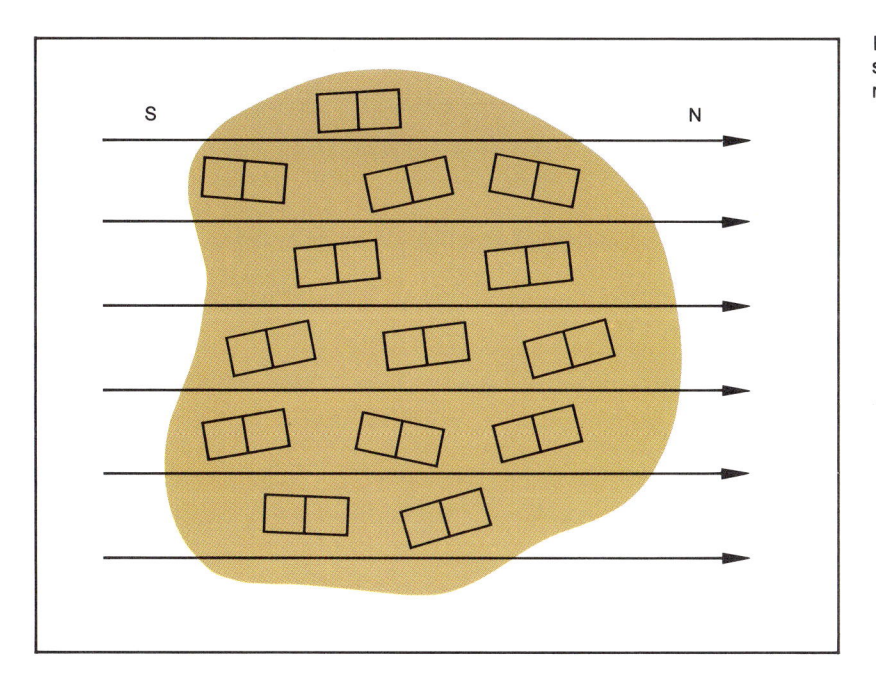

Figure 18. Atomic magnets in a paramagnetic substance orient themselves, with a slight excess of magnets pointing in the direction of the field.

representing the atomic magnet and the other representing the magnet applied by Stern and Gerlach. Two bar magnets attract or repel each other strongly if they are end-to-end. If the magnets are at an angle to each other, the force is intermediate. Taking the silver atoms at random, you might expect the force to vary from a strong push to a strong pull, depending on how the atomic magnet is oriented. But Stern and Gerlach found that about half the atoms were pulled as strongly as possible and the other half were pushed as strongly as possible. None of the atoms experienced an intermediate force. The silver atoms acted as if their atomic magnets were oriented in only two directions.

The orientation of atomic magnets in a magnetic field is *quantized*. For some atoms, such as silver and potassium, only two orientations are allowed. For a nitrogen atom, four orientations are possible. The important point is this: *For any atom, its atomic magnet can take only certain orientations.* This effect, called "space quantization," is also contained in the quantum theories of Erwin Schroedinger and Werner Heisenberg.

Paramagnetism

In some atoms, such as helium, the atomic magnet is zero because there is complete pairing of the electrons and a lack of orbital current. Moreover, even if an atom has an internal magnet, the strength may be altered drastically when the atom combines with other atoms in a molecule or in a solid because the electrons would move differently. A hydrogen *atom* has a strong internal magnet because of its unpaired electron. In a hydrogen *molecule,* however, the two electrons pair up and cancel their magnetic effects. A molecule or substance in which the currents do not cancel is called *paramagnetic*. Oxygen molecules, for example, are paramagnetic.

You can think of a paramagnetic substance as containing a multitude of tiny current loops or bar magnets. When a magnet is brought near paramagnetic material, the atomic magnets orient themselves. Suppose,

Figure 19. (Below) Paramagnetic materials are drawn into magnetic fields.

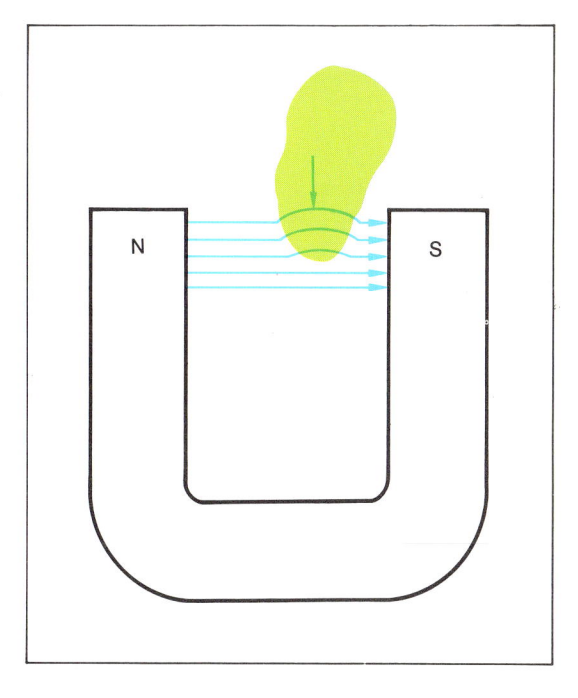

25 | Magnetism and Electromagnetism

Figure 20. (Far right) Iron crystals as seen through an electron microscope. These crystals have not ordered themselves into domains, either because they have not been exposed to a strong magnetic field or because they have been exposed to a weak magnetic field such as the earth's for a long period of time. Alignment of these domains would increase with an applied field.

for purposes of illustration, that only two orientations are permitted: with the field and against the field. Atomic magnets oriented in the direction with the field have somewhat lower energy, and ideally all the magnets would eventually end up pointing with the field. However, the energy of random thermal motion is comparable to the magnetic energy at ordinary temperatures, and the constant thermal jostling keeps the atoms about equally divided between the available orientations. The effect of the magnetic field is to produce, on the average, a slight excess of atomic magnets pointing with the field. In oxygen gas at room temperature in a field of 1,000 gauss, the excess amounts to a few molecules per thousand. The cooler the paramagnetic material and the stronger the field, the greater the excess.

The magnetic fields produced by the excess oriented atoms all add together and strengthen the applied field. Figure 19 shows the total magnetic field near a bar of paramagnetic material. Faraday pictured lines of magnetic field as acting like rubber bands. So paramagnetic materials are always drawn toward magnets.

Ferromagnetism

Iron is millions of times more magnetic than ordinary paramagnetic materials. However, an individual iron atom is not much more magnetic than other typical atoms. Iron metal has such strong magnetic properties because of the microscopic crystal structure of iron. The atoms in a microcrystalline domain of iron exert long-range forces on one another that make all the atoms in the domain orient themselves in the same direction (see Figure 20). The forces are much stronger than the disturbing effects of thermal motions. The magnetic field produced by each microcrystal is strong because all the atoms contribute. In ordinary unmagnetized iron, the crystals are all jumbled up and the net magnetic effect is zero. But in an applied magnetic field, the crystals orient themselves and grow in the direction that strengthens the structure. Materials that behave this way are called *ferromagnetic*.

A coil of wire with enough current to produce 10 or 20 gauss can give thousands of gauss when an iron core is inserted because of the contribution of the aligned crystals. Most electric magnets, relays, and motors use iron to enhance the magnetic field produced by ordinary current-carrying wires. At about 20,000 or 30,000 gauss, however, all the domains in iron are aligned and there is no further enhancement.

For some ferromagnetic alloys, there is a residual alignment of the domains remaining after the applied field is removed. These alloys are useful for making permanent magnets. Lodestones, or naturally occurring magnets, are ferromagnetic ores that cooled from liquid to solid under the aligning influence of the earth's magnetic field. Undisturbed bodies of magnetic ore are always found to be magnetically aligned along the north-south direction.

Diamagnetism

Atoms such as helium have no net internal currents, yet they show weak magnetic effects. The reason is that an applied magnetic field slightly influences the motion of the electrons so that symmetry is not complete. The effect of the applied field is to cause the atom to produce a weak magnetism in a counter-direction. Such behavior is called *diamagnetism*. It has nothing to do with orientation, and it occurs in all atoms. Normally the weak diamagnetic effects are swamped out if the substance is paramagnetic. Diamagnetic materials are *repelled* by magnets. Faraday was

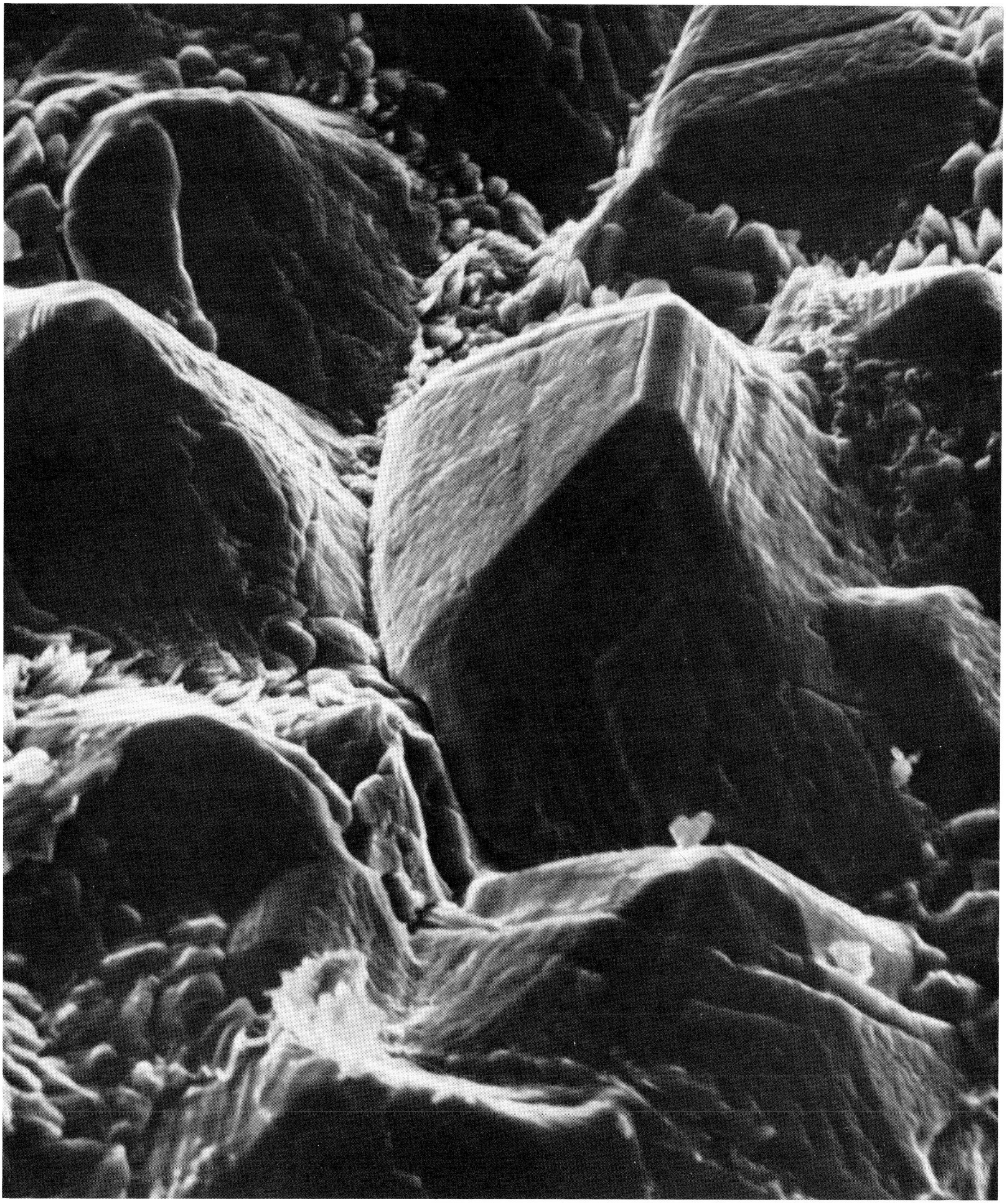

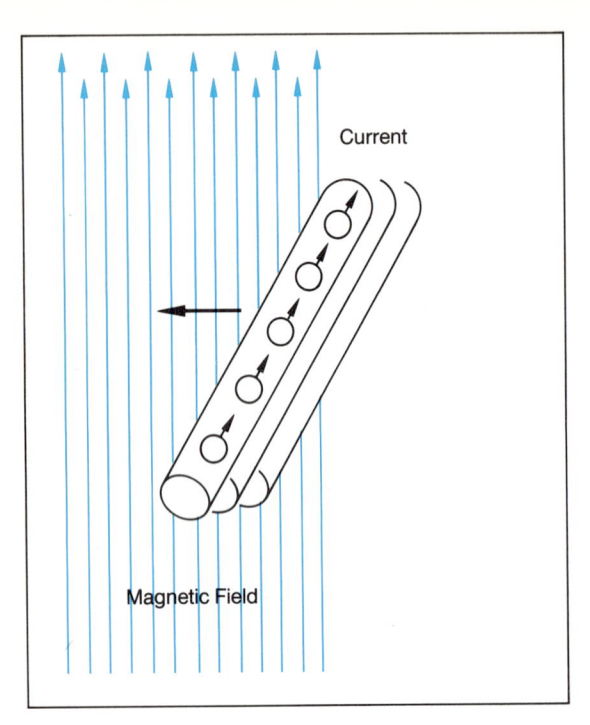

Figure 21. If a metal rod is moved through a magnetic field, the unbound charges experience a magnetic force.

Figure 22. If a complete loop is pushed into a magnetic field, current flows around the loop until the entire loop is submerged within the constant magnetic field.

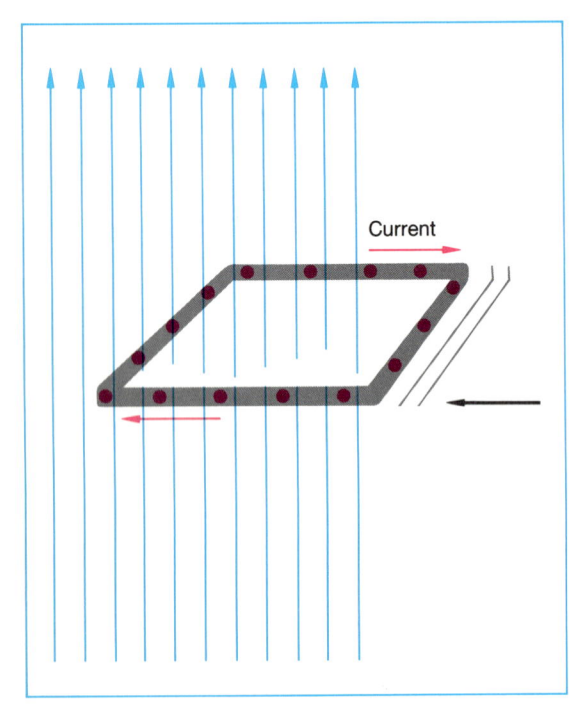

puzzled by the strange behavior of bismuth and was led to invent the concept of fields to explain his observations.

MAGNETISM AND ELECTRICITY

You have seen that electric currents produce magnetic fields. In 1831 Faraday in England (and, independently, Joseph Henry in the United States) found a way of producing electricity *from* magnetism.

Here is a modern explanation of what Faraday found. Suppose you move a metal rod in a magnetic field (Figure 21). The unbound charges in the rod experience a magnetic force because they are moving in a magnetic field. If the motion is at right angles to the rod, the charges will move along the length of the rod. A current will flow until charge builds up on the ends of the rod and counteracts the flow.

Suppose you use a complete loop instead of a rod (Figure 22). As you move the loop into the field, current will flow steadily around the loop. When the whole loop is moving in the field, however, the current produced in one end of the loop is opposed by the current produced in the other end, and the flow of charge stops. The current flows only when the field at one end of the loop is stronger than the field at the other end, and when the direction of motion lets the loop "cut" the magnetic field lines. Both conditions can be summarized as follows:

Current flows around a loop moving in a magnetic field only when the motion changes the number of lines of magnetic field through the area of the loop.

Generators and Transformers

As you move the loop into a magnetic field, current flows around the loop. The current can be used to light a light bulb or heat a resistor. Where does the energy come from? The energy comes from the work you do pushing the loop into the field. You have to push because there is current in the loop, and magnetic fields exert forces on current-carrying wires. If you check the direction of the magnetic force, you'll find it opposes your efforts to move the loop. The magnetic field does no work, as usual. But it makes possible the efficient conversion of mechanical energy into electric energy. A loop moved into or out of a magnetic field is a simple *generator*.

A more convenient form of generator is a loop turned steadily in a magnetic field, as in Figure 24. As the loop is turned, the number of lines of magnetic field through its area changes and a current flows. When the plane of the loop is momentarily horizontal, the current is momentarily zero. When the plane of the loop is vertical, the current generated is maximum. And when the loop has rotated by half a turn, the direction of the current reverses. The loop generates an *alternating current*. To bring the current out of the generator, the ends of the loop are connected to sliding contacts on the axle.

In a power plant, heat energy from nuclear or fossil fuels is converted to mechanical energy in a steam turbine. The steam turbine then produces electric energy by turning a generator. The overall efficiency is about 30 percent. A more efficient device, the *magneto-hydrodynamic generator,* uses no mechanical parts at all. In this type of generator, which is just beginning to appear in trial industrial service, heat energy is used to form a high-pressure jet of molten metal traveling 1,000 miles an hour. When the metal flows through a magnetic field, the conduction electrons experience a magnetic force that carries them to one face of the stream.

Figure 23. A loop turned in a magnetic field generates a current. The current can be led out through sliding contacts.

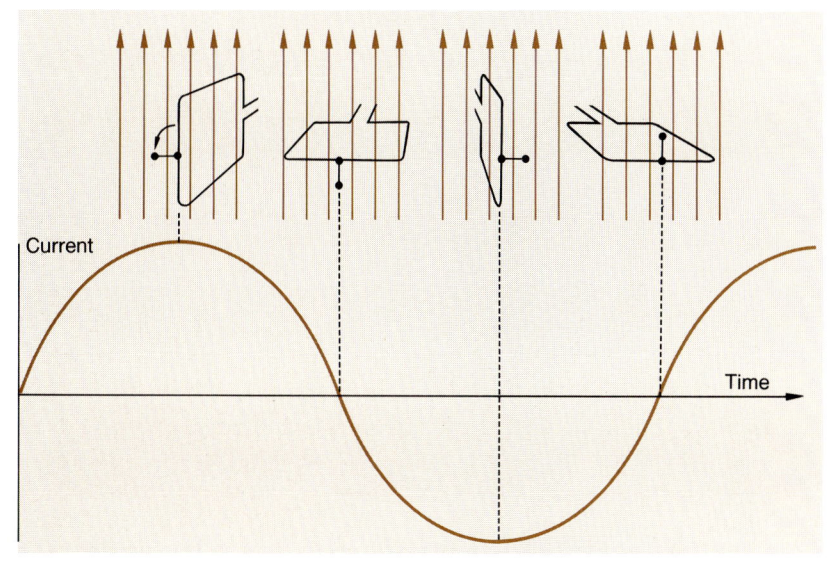

Figure 24. The current generated on the loop alternates as the loop is turned. The current is zero when the loop is perpendicular to the field.

If the circuit is completed through electric contacts, current flows. The efficiency of the process is about 50 percent, giving a significant saving in fuel over a conventional system.

With the help of magnetism, heat energy or mechanical energy can be converted to electric energy. And in an electric motor, magnetism allows electric energy to be converted to mechanical energy. Another element of an efficient power system is the *transformer,* which can change low voltages to high voltages and back again for efficient transmission of electric power. Here, too, magnetism plays a key role. But transformers have no moving wires, and to understand transformers you have to look at the relation of magnetism and electricity more carefully.

Electromagnetic Induction

Faraday wound two separate coils of insulated wire on an iron ring and connected the second coil to a current meter. When current from a

Figure 25. Electromagnetic induction can be demonstrated with two insulated coils wound on an iron ring. (a) One coil is hooked up to a current meter and the other can be connected to and disconnected from a battery. (b) When the battery is first connected, current flows briefly, then stops. (c) A steady current in one coil produces an absence of current in the other. (d) When the battery is disconnected, current again flows briefly in the second coil in the opposite direction.

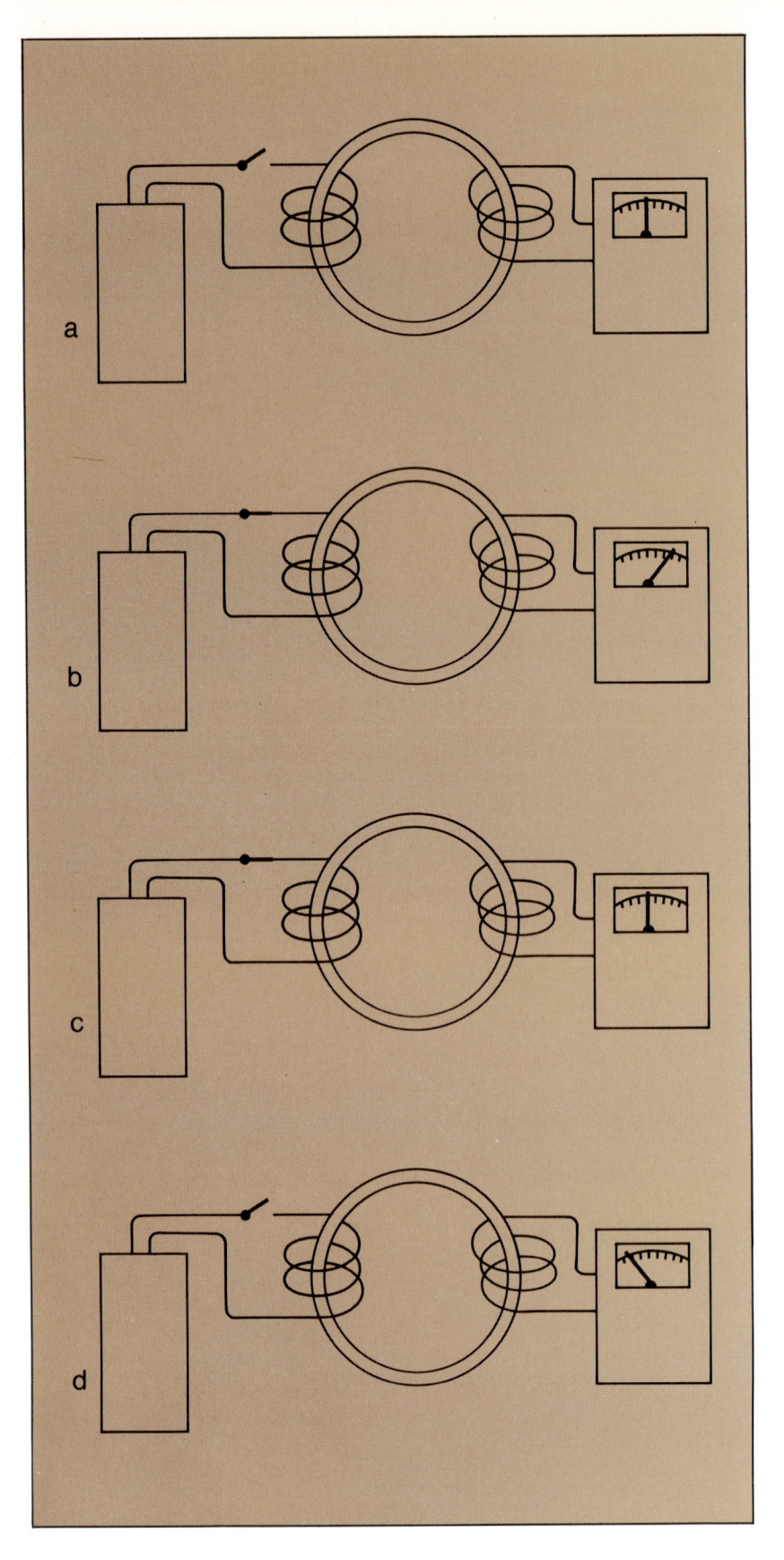

[Moving Loop   Stationary Magnet]   [Stationary Loop   Moving Magnet]

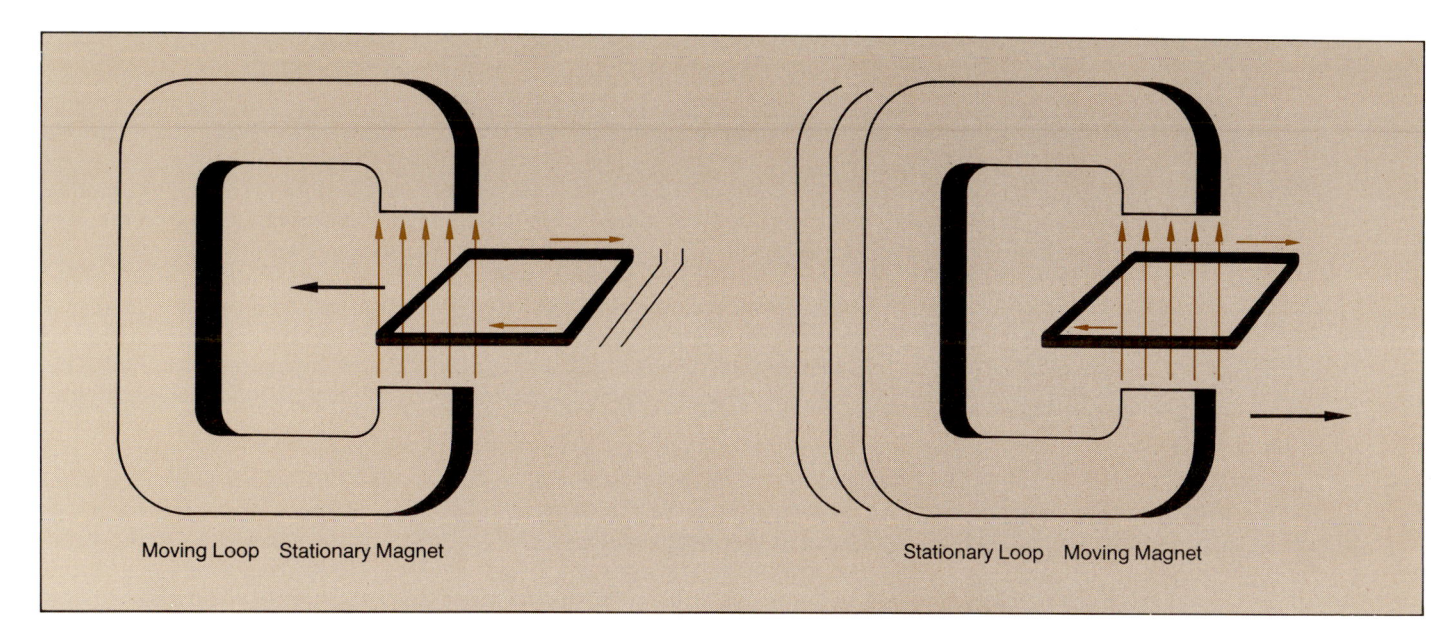

battery flowed through the first coil, current flowed in the second coil as well, but in an unexpected way. Faraday was looking for a steady current in the second coil. He found instead that when the battery was connected to the first coil, current flowed briefly in the second coil and then stopped. A steady current in the first coil produced no current in the second coil. And when the current in the first coil was turned off, current again flowed briefly in the second coil but in the opposite direction. Faraday called his discovery *electromagnetic induction*.

Here is an argument from relativity to help explain electromagnetic induction. You know that when a wire loop is pushed into a magnetic field, current flows because of the forces on the unbound charges in the metal. Suppose instead that the loop and the charges in it are stationary and that the *magnet* is moved toward the loop. What happens then? According to relativity, a current must still flow in the loop, for the principles of physical science depend only on relative motion and not on absolute velocity. But what makes the charges move in the wire?

Faraday said that a changing magnetic field produces an electric force on the charges in the wire. A changing magnetic field induces an electromotive force, measured in volts, in the wire. One way to make current flow in a circuit is to apply the electromotive force of a chemical battery. Faraday showed that a changing magnetic field does just what a battery does.

In the earlier discussion of generators, the emphasis was on the current that flowed in a moving loop. Now you see, however, that it is the induced *electromotive force* that is fundamental. How much electromotive force is induced depends on several factors that can be combined by introducing the idea of *magnetic flux*. The magnetic flux through a loop is found by multiplying the amount of area immersed in magnetic field by the component of magnetic field perpendicular to the plane of the loop. Then the electromotive force induced in the loop is equal to the rate at which the magnetic flux through the loop is changing. The amount of current that flows through the loop is incidental—it's determined by the induced electromotive force and the resistance of the loop, according to Ohm's law.

There are several ways in which the magnetic flux through a loop can be made to change. Rotating the loop in a magnetic field, for example,

Figure 26. A stationary magnet and moving loop is the physical equivalent of a stationary loop and moving magnet. A current must flow in each case.

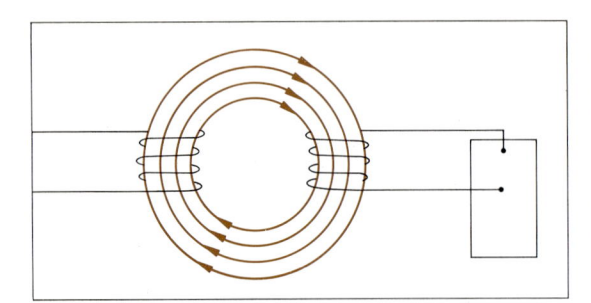

Figure 27. Current in a coil produces magnetic flux through the turns of the coil. If the current changes, the flux changes and an electromotive "force" is induced in the coil.

continually changes the component of magnetic field perpendicular to the loop. Here, then, is another way of looking at the simple generator we discussed before. In Faraday's experiment, neither coil was moving, but the flux in the second coil was made to change by turning the magnetic field on and off. The magnetic flux through the second coil is produced by the current in the first coil. As long as the current in the first, or primary, coil is changing, there will be an induced electromotive force in the secondary coil.

After Faraday had discovered electromagnetic induction, one of his friends, an amateur scientist, suggested that induction could occur in a single coil. Suppose you connect a coil to a battery. The current in the coil is initially zero, but as the current starts to flow a magnetic field builds up and the magnetic flux through each turn of the coil changes. An electromotive force is induced in the coil, and by conservation of energy, the electromotive force acts to oppose the increase of the current. So the battery has to do work even if the resistance of the wire is small. The electromotive force of the battery has to make the unbound electrons move against the electromotive force produced by the self-induction. Looking at it another way, there are two sources of electromotive force in the circuit—one due to chemical action in the battery and the other due to the changing magnetic flux. The total electromotive force in the circuit is the difference of the two, so by Ohm's law, the current in the circuit at any instant is:

$$\text{Current} = \frac{\text{total emf}}{\text{resistance}} = \frac{\text{emf of battery} - \text{induced emf}}{\text{resistance}}$$

At first, the current is small and the two electromotive forces nearly cancel. Later, when the current has built up and is changing less rapidly, the induced electromotive force is less than at the start. Eventually the current is steady and the induced electromotive force is zero.

A battery produces a steady electromotive force, but the rotating loop generator you looked at earlier produces an alternating electromotive force. Suppose an alternating electromotive force is applied to the coil. There is a continual change of magnetic flux and a continual self-induction. By Ohm's law,

Emf of generator − induced emf = current × resistance

If the coil is made of many turns of heavy wire and has an iron core to augment the magnetic field, the voltage drop across the resistance is negligible and the induced electromotive force is equal to that of the generator at every instant. A secondary coil wound on the same core experiences the same change of flux as the first coil, so if the coils are identical, the electromotive force induced in the secondary coil is also equal to the generator electromotive force. The total electromotive force in a coil is the total induced in each turn; the electromotive force's add like batteries in series.

For example, the electromotive force in the secondary coil is twice the generator electromotive force if the secondary coil has twice as many turns as the primary coil. The electromotive force in the secondary coil can be made higher or lower than the electromotive force applied to the primary by varying the relative number of turns.

$$\frac{\text{Emf in secondary coil}}{\text{Emf in primary coil}} = \frac{\text{number of turns in secondary coil}}{\text{number of turns in primary coil}}$$

Two coils coupled together by their magnetic fields constitute a transformer. In a transformer used for power transmission, current flows

Figure 28. Two moving charges exert both electric and magnetic forces on each other. In this analogy, if you were to move along with the footballs, they would seem to be at rest. If you were to move along with electric charges, they, too, would seem to be at rest and there would be no magnetic field. What seems to be the magnetic field of moving charges becomes an electric field if you move along with them.

in both the secondary and the primary coils, and the total magnetic flux depends on both currents. However, in a well-made transformer the electromotive force in the secondary coil and the primary coil are in the same ratio as the number of turns in each winding. Because electric power is current × voltage,

primary current × primary emf = secondary current × secondary emf

neglecting slight losses in the transformer. Hence a transformer that steps up voltage steps down current.

$$\frac{\text{Secondary current}}{\text{Primary current}} = \frac{\text{number of turns in primary coil}}{\text{number of turns in secondary coil}}$$

MAGNETISM AND RELATIVITY

Magnetism plays an essential role in the generation, transmission, and use of electric power. Magnetic forces do no work, but they allow energy to be converted from one form to another, and in a transformer they allow electric energy to be transferred from one circuit to another.

Yet in a sense, magnetic force is not a fundamental force of nature; it is only one aspect of electricity. Consider two positive charges moving along together at the same speed (Figure 28). The charges repel each other electrically, but they also exert an attractive magnetic force on each other. A single moving charge is not exactly like a steady current, but it produces a magnetic field that can act on a second moving charge in the usual way.

Suppose you are moving along with the charges. Then the charges would appear to be at rest. The force between them is purely electric. What happened to the magnetic force? A complete discussion requires the full use of the theory of relativity. But the point is that what seems like a magnetic field to one observer can seem like an electric field to an observer moving differently. What seems to us to be the magnetic field of moving charges becomes an electric field if you move along with charges.

And even the way an electron acts like a bar magnet is connected with relativity. In fact, the theory that combines relativity and quantum mechanics automatically predicts that the electron acts like a magnet.

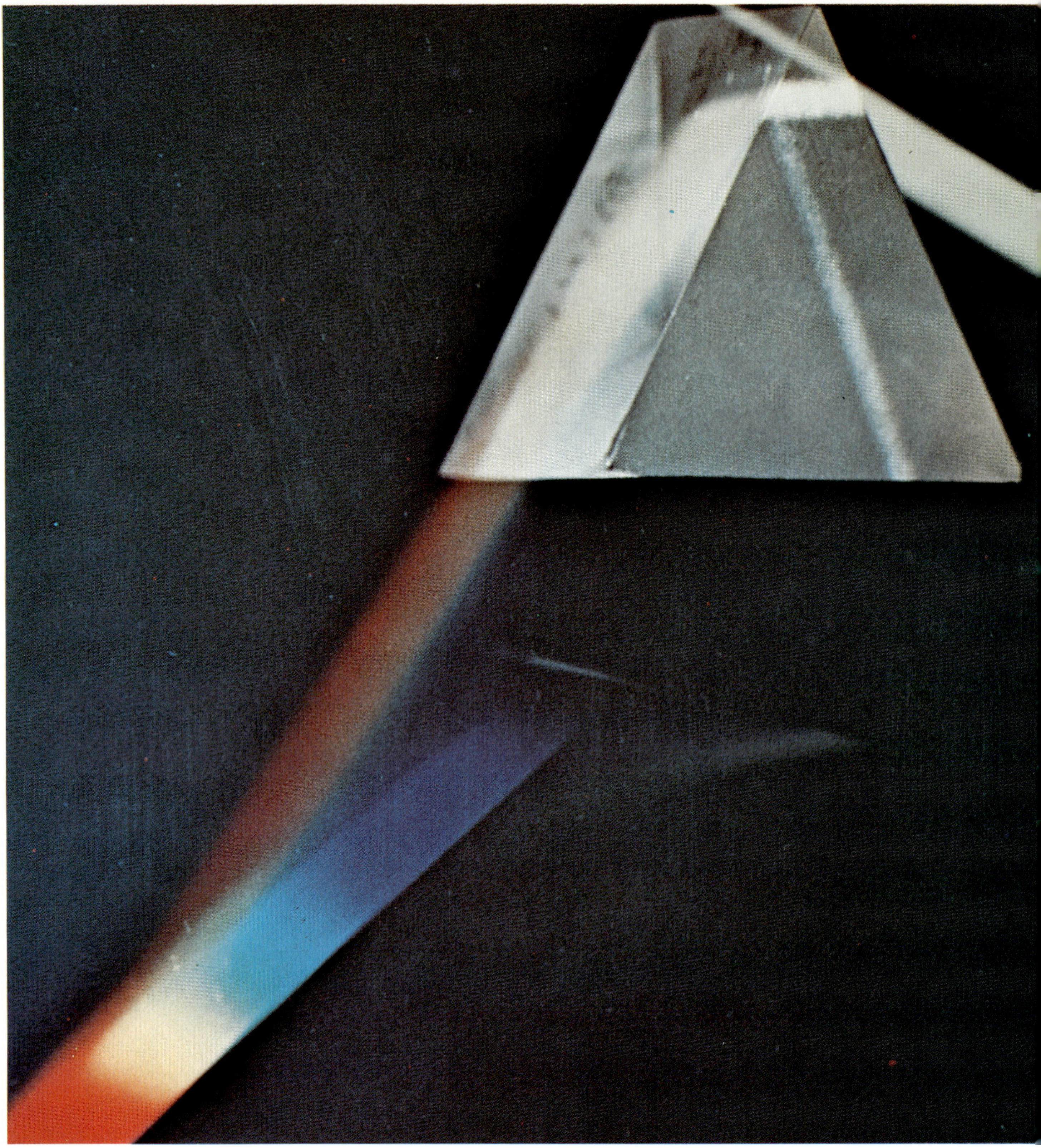

Fields are handy for visualizing electric and magnetic forces, as you saw in the preceding chapter. Michael Faraday's experiments with electric charges and with magnetic forces led him to believe that the concept of electric and magnetic fields was the best way to think about electricity and magnetism. The value of the field picture is even more convincing when time lags are involved. Suppose a radio transmitter on earth sends a brief radio signal to an instrument-carrying rocket in deep space, 10 million miles away. When the signal reaches the rocket, forces are exerted on electrons in the receiving antenna. But there is a time lag of about a minute between the time the transmitter is on and the time the electrons in the receiving antenna start to jiggle. Where is the energy stored during the time lag? It's hard to avoid the conclusion that something is happening, in the space between the earth and the rocket, to carry the energy along.

So far in the book the emphasis has been primarily on the photon picture. And it's true that a radio signal can be thought of as a bunch of photons traveling through space. But as useful as the photon picture is for discussing the interaction of light with individual atoms, it leaves many questions unanswered. What is the relationship of light to electric and magnetic fields? And what determines the speed of light?

These questions were answered by James Clerk Maxwell 40 years before the idea of the photon was introduced. By putting Faraday's conception of fields into mathematical form and by adding a new principle of his own, Maxwell in 1864 showed that light acts like a disturbance, or ripple, in electric and magnetic fields. His wave picture of light complements the photon model. The two pictures overlap to some extent. The average energy reaching the earth from the sun can be described either by waves or by photons. But the energy given to an individual electron in the photoelectric effect is best understood in terms of photons. And interference colors in a soap bubble are best understood in terms of waves. The wave picture is valid when many photons are involved at the same time.

This chapter is devoted to the wave nature of electromagnetic radiation. Particles, fields, and waves are fundamental concepts. Everything in physical science is described in terms of them and their interactions. You already know a lot about how particles behave from earlier chapters, and you learned about electric and magnetic fields in the chapter immediately preceding this one. Now look at the behavior of waves in general before going into the details of electromagnetic radiation.

WAVES

If you drop a stone into water, ripples spread out in all directions. And if you pluck a weak spring (such as a coiled telephone cord) you can see a kink whiz down the spring. Waves are *moving disturbances.*

There is something special about waves in water and waves on a spring that makes them different from the wavelike motion of a field of grain rippling under the wind. The stalks of grain merely respond to the external force of a puff of wind blowing across the field. But a wave in water or a wave on a spring is a *propagating* disturbance. A wave on a spring keeps moving along the spring without external forces. There is an inherent action that keeps the wave going.

A wave on a spring consists of an organized motion that is transmitted from coil to coil. Each coil moves only a slight distance. The elastic

26

Electro-magnetic Radiation

The wave picture of light complements the photon picture whenever photons act together, as they do when they are sent through a prism.

Figure 2. Water waves are analogous to the propagation of electromagnetic waves from a point source.

forces on each coil trying to bring the coil *back* to equilibrium interact with the inertia of the coil making it *overshoot* the equilibrium position. The motion is passed on from coil to coil, and the speed of the disturbance is governed only by the strength of the elastic forces and the mass of each coil. Similarly, the speed of the ripples in a puddle of water is determined by the elastic surface tension forces on the water and the inertia of the water. As each bit of water bobs up and down, it exerts forces on the next bit of water and passes on the motion.

A wave propagates because of the interplay of two or more contending effects. As you will see, electromagnetic radiation propagates because of the interaction of electric and magnetic fields. The strength of the contending effects determines the speed of propagation. For example, a sound wave propagates faster in a steel railroad track than in air because the elastic forces in steel per unit mass are stronger than they are in air.

One of the most important characteristics of a propagating wave is that it *carries energy* from one place to another. The sound waves from a thunderclap rattle dishes and windowpanes. And the energy carried by electromagnetic waves from the sun brightens and heats up the earth.

Figure 4 shows a disturbance propagating along a spring or an elastic cord. It also shows the disturbance a little while later, after it has moved farther along the cord. The vertical axis represents sideways displacement of the cord. But *all* waves propagate; Figure 4 may also be thought of as a representation of a pulse of sound traveling through air. For sound, the vertical axis represents pressure in the air, measured from equilibrium pressure. For an electromagnetic pulse, the vertical axis represents the strength of an electric or a magnetic field.

There is one form of propagating wave that is particularly important — the alternating wave shown in Figure 5. The alternating, or *sinusoidal,* waveform is the one associated with a pure *frequency.* If you imagine yourself standing at one spot watching a wave go by, the frequency is

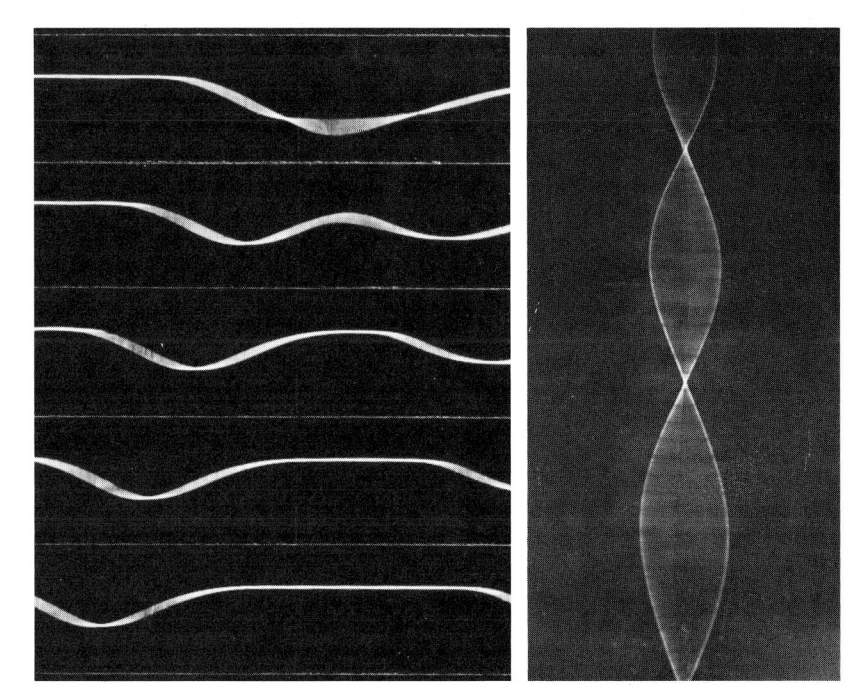

Figure 3. (Above left) If you stretch out this popular slinking toy and pluck it, nice transverse waves propagate down the coils.

Figure 4. (Above right) A propagating disturbance on a long elastic cord, photographed at two different times to illustrate the wave nature of the propagation.

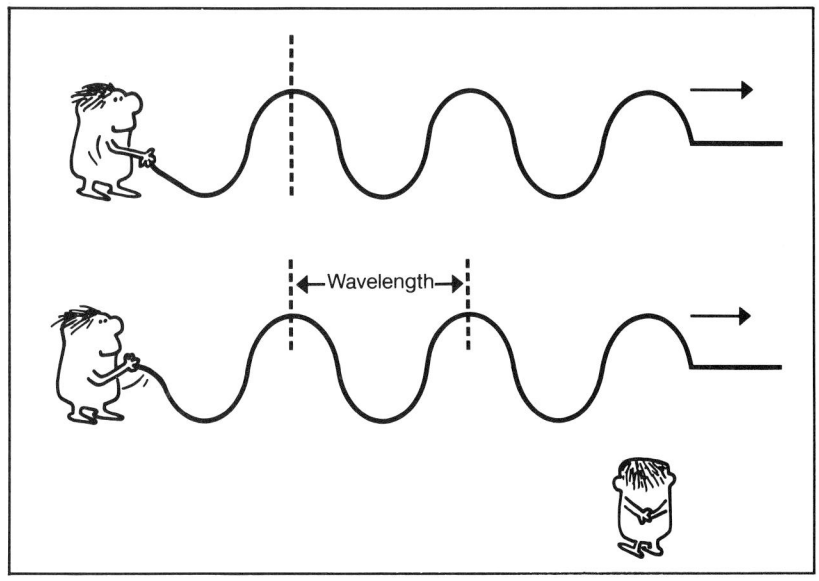

Figure 5. (Left) A sinusoidal wave is shown propagating on a long elastic cord. The frequency of the wave depends on the rate at which the man shakes the end of the rope. The *frequency* equals the number of crests that pass the stationary observer in each second. The *wavelength* is the distance between adjacent corresponding points on the wave.

the number of complete alternations, or oscillations, that go by every second. A definite frequency can't be associated with the pulse in Figure 4. The sound wave of a pure, sustained tone is represented by a sinusoidal waveform. A light wave of a pure color is also represented by a sinusoidal waveform.

In addition to its frequency, a sinusoidal waveform is characterized by its *wavelength* — the distance occupied by one complete oscillation of the wave. The number of oscillations per second times the length of one oscillation equals the distance the wave travels in a second. Therefore,

Frequency × wavelength = speed of propagation

The frequency of a wave is determined by its source. For example, the frequency of a pure tone from a loudspeaker depends only on how many

Figure 6. The amplitude of a wave is the maximum excursion to one side of equilibrium.

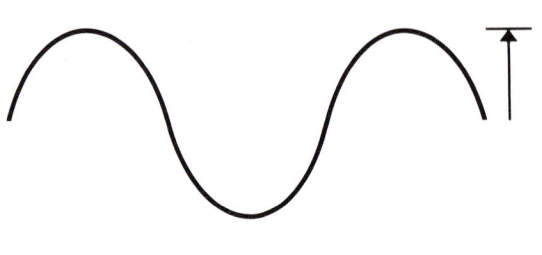

Figure 7. When two waves of the same frequency combine, the result is a wave of the same frequency. In (a), the two waves are of equal amplitude and are in phase. They combine to give a wave of twice their amplitude. In (b), the waves are of equal amplitude and are out of phase. They combine to give zero amplitude.

times a second the loudspeaker's cone is made to move in and out. So the wavelength varies with the speed of a wave. The wavelength of a sound wave completing 1,000 cycles per second is longer in steel than it is in air because sound travels faster in steel. The wavelength of red light is shorter in glass than in air because light travels more slowly in glass.

The *amplitude* of a sinusoidal wave is the maximum excursion along the vertical axis, as shown in Figure 6. For a wave on an elastic cord, the amplitude measures the maximum sideways displacement. For a sound wave, the amplitude represents the greatest pressure difference from equilibrium.

Particles, like waves, can also carry energy from one place to another. But waves have a property that isn't shared by particles: waves can *interfere.* Suppose two loudspeakers are each broadcasting a pure tone of the same frequency. What happens where the sound waves from the two speakers overlap? The total pressure change at each point is the sum of the pressure changes produced by each wave acting separately. But the waves can tend to reinforce or to cancel. Figure 7 shows a simple case of two sound waves of the same amplitude and frequency moving along the same direction. If the peaks of the waves coincide, the waves are said to be *in phase*. They add to give a sound wave of the same frequency but twice the amplitude. If the peak of one wave coincides with the trough of the other, the waves are completely *out of phase,* and they cancel each other. For intermediate phases between these extremes, the wave is of intermediate amplitude. This ability to interfere constructively or destructively is a versatile property that particles don't possess.

When sinusoidal waves of different frequencies are added together, the outcome is not another sinusoidal wave. Waves of different frequencies don't have a definite phase relationship—they reinforce part of the time and cancel part of the time. By adding together enough waves with the correct amplitudes it's possible to represent a wave of any shape, as Jean Baptiste Joseph Fourier proved at the beginning of the nineteenth century. It takes a wide range of frequencies to represent a wave in the form of a pulse—and the shorter the pulse, the wider the frequency range needed. If you hit a wooden table with the palm of your hand, the sound is much duller than if you rap the table with a wooden pencil. Because the pencil bounces away more quickly than your hand, the pulse is shorter and contains more high frequencies.

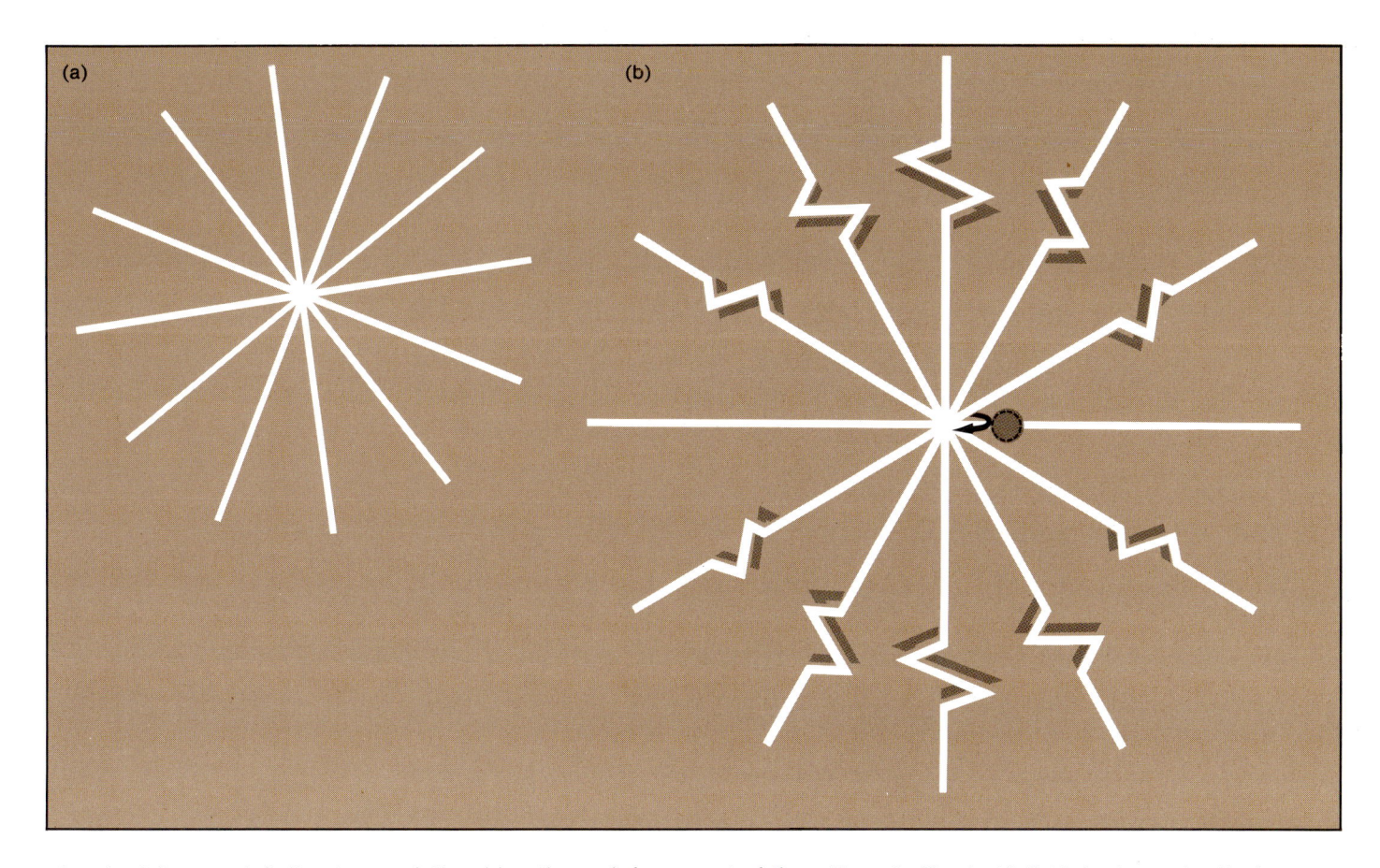

The constructive interference of electromagnetic waves will be described more fully later on. But first you need to know some of the special properties of electromagnetic radiation.

ELECTRICITY, MAGNETISM, AND ELECTROMAGNETIC WAVES

All electromagnetic waves are produced by the motion of electric charges. Without motion, there is nothing to start the disturbance and no way to supply the energy carried off by the waves. A charge at rest does not radiate electromagnetic energy. Furthermore, according to the principle of relativity there is no dynamic difference between uniform motion and a state of rest. So a charge moving at a steady speed along a straight line does not radiate electromagnetic energy, either.

Only accelerated electric charges radiate electromagnetic waves. Figure 8(a) represents the electric field of a charge at rest. Suppose an applied force accelerates the charge for a short time and then brings it to

Figure 8. The electric field of a charge at rest is shown to the left. If the charge is jerked back and forth, kinks appear in the electric field lines and *propagate outward.*

Figure 9. (Below) A signal that lasts for only a short time is the sum of many waves of high frequencies.

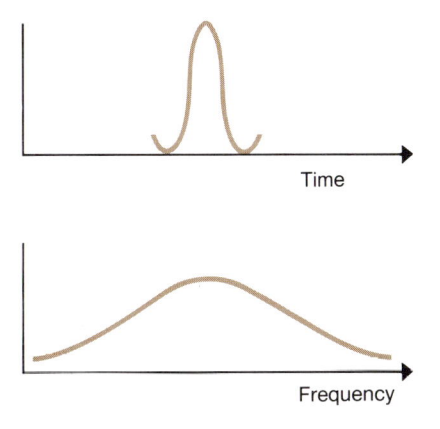

26 | Electromagnetic Radiation

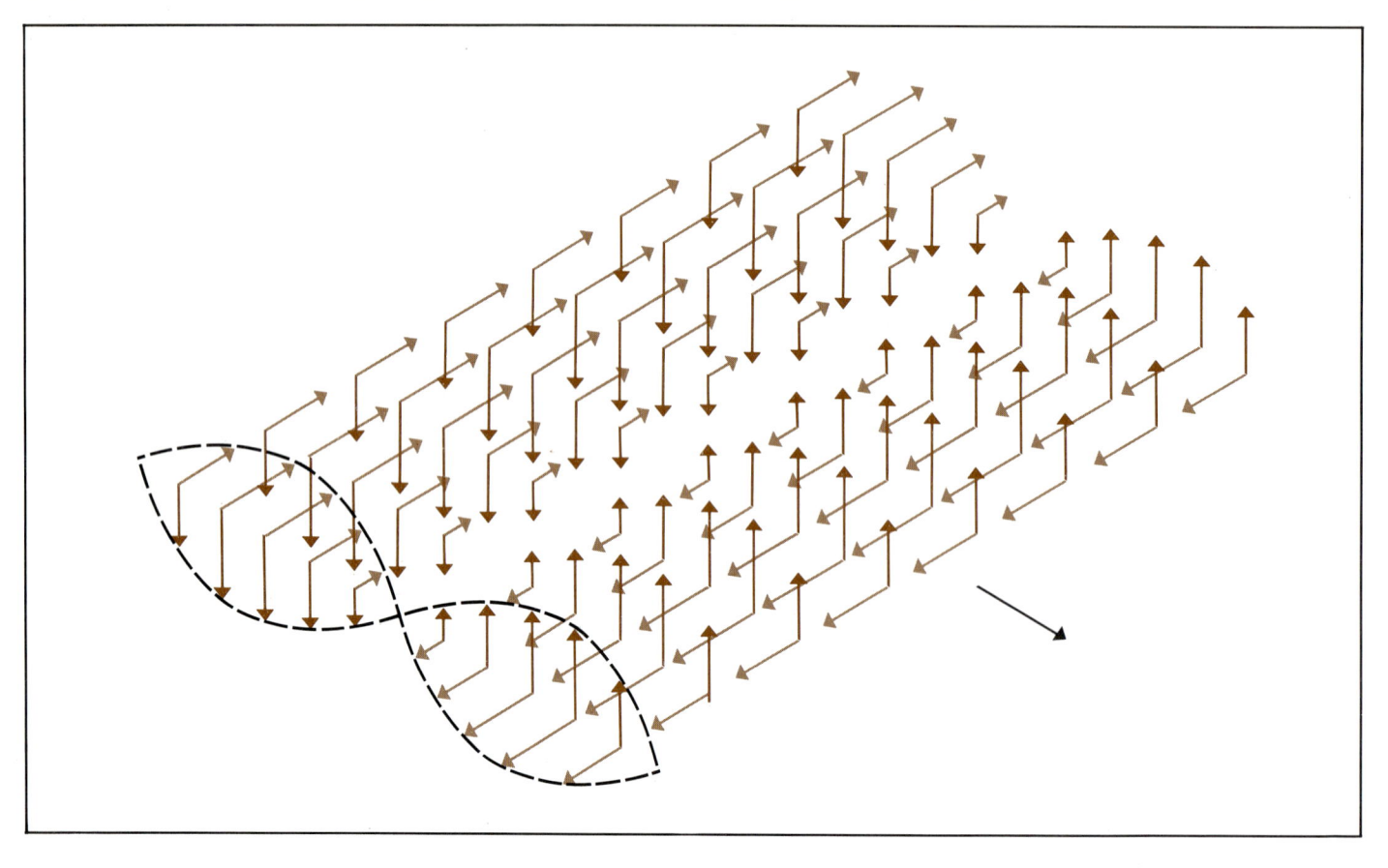

Figure 10. A schematic representation of a sinusoidal electromagnetic wave at one instant in time.

Figure 11. (Below) The propagating electric and magnetic fields are perpendicular to the direction of propagation and to each other.

a halt. A kink appears in the electric field lines, just as if the lines were elastic cords. The kink propagates away from the charge. The kink is strongest in the electric field lines at *right angles* to the acceleration; it is zero to the front and rear.

Why does the kink propagate outward, and what determines its speed? When you shake one end of a spring or a cord, the disturbance propagates along because of the interplay of inertia and elastic forces. An electric field line doesn't really have inertia or elasticity, however. The propagation of electromagnetic waves depends on the interplay of electric and magnetic fields.

As you recall from the preceding chapter, Faraday discovered that a changing magnetic field induces electric forces in a conductor. Even if a conductor is not present, a changing magnetic field still produces an electric field in space. Faraday was a self-taught scientist and expressed all of his discoveries verbally, without mathematics. Maxwell took up Faraday's physical insights and cast them into mathematical form. But the formulation seemed incomplete to Maxwell until he added a new principle: *a changing electric field can produce a magnetic field.*

Suppose you were stationed somewhere in space and the kink of electric field from an accelerated charge whizzed by. You would see a changing electric field. The changing electric field must produce a magnetic field—and the changing magnetic field must produce an electric field. Electromagnetic radiation consists of electric and magnetic fields in constant interaction, analogous to the interaction of inertia and elasticity for waves on a spring.

The speed of propagation depends on the facility with which electric fields can be changed into magnetic fields and vice versa. By using some coils and capacitors, it's easy to determine these factors in the laboratory without even looking at an electromagnetic wave. Maxwell's calculations predict that electromagnetic disturbances propagate with a speed of 3×10^8 meters per second—a speed equal to the measured speed of light and the first proof that it is related to electromagnetism.

Maxwell showed that the electric and magnetic fields associated with an electromagnetic wave in free space have certain properties:

1. The propagating electric field is always at right angles to the direction the wave is traveling. (You can see this from the kink in Figure 9.)

2. The propagating magnetic field is perpendicular both to the direction the wave is traveling and to the propagating electric field.

3. At any given point along the wave the strength of the magnetic field is directly proportional to the strength of the electric field.

4. The propagating electric and magnetic fields from a radiating charge fall off in strength inversely only as the *distance* from the charge, not as the inverse square of the distance.

If a charge is wiggled back and forth sinusoidally, a sinusoidal electromagnetic wave is generated. Figure 10 represents a sinusoidal wave a long distance from the charge. The fields are present over a large region of space, and to visualize them it's helpful to imagine yourself measuring electric and magnetic fields at a fixed spot. You would see the fields increase and decrease rhythmically as the wave travels by.

Like water waves or sound waves, mechanical waves travel in a material medium. And before the theory of relativity, it was thought that electromagnetic waves traveled in a medium, called the "ether," even though Maxwell's theory contains no reference to a medium. But you have seen that the medium for electromagnetic waves is the electric field itself: *electromagnetic waves represent disturbances propagating in the electric fields, which fill space.* There is no ether, nor is there any need for it.

ELECTROMAGNETIC WAVES AND ENERGY

All propagating waves carry energy. If an electric charge is in the path of an electromagnetic wave, it will be accelerated by the electric field of the wave and gain kinetic energy. *The amount of energy transported by an electromagnetic wave per second across a unit area is called the intensity.* The average intensity of a sinusoidal electromagnetic wave is proportional to the square of the electric field amplitude. So the intensity from a radiating charge falls off as the inverse square of the distance, because the field strength itself decreases with the distance. But as the wave spreads outward, it covers a larger and larger area; the total energy moving out from the charge is constant on the average. The energy is ultimately supplied by the force moving the charge back and forth.

To increase the intensity of a radiated wave, you would have to make the kink in the field lines stronger. One way to do this is to use a bigger electric charge to increase the strength of all the electric field lines. Another way is to give the charge a greater acceleration. The energy radiated from a charge depends on the square of the charge and the square of the acceleration.

THE ELECTROMAGNETIC SPECTRUM

Maxwell's theory spans an enormous range of applicability. Radio waves, heat waves, visible light, x-rays—all represent forms of electromagnetic

Figure 12. The *intensity* of electromagnetic waves falls of inversely as the square of the distance from the radiating charge because the same amount of energy is spread out over larger areas as the wave moves along.

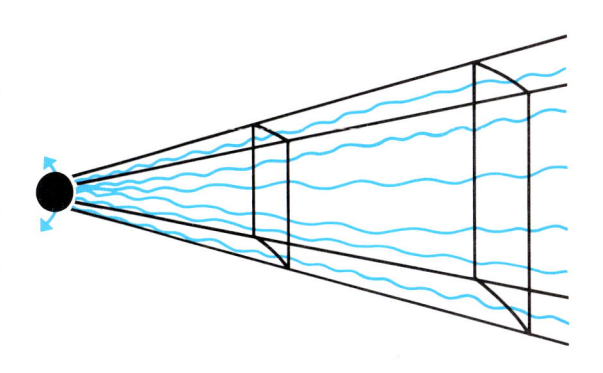

26 | Electromagnetic Radiation

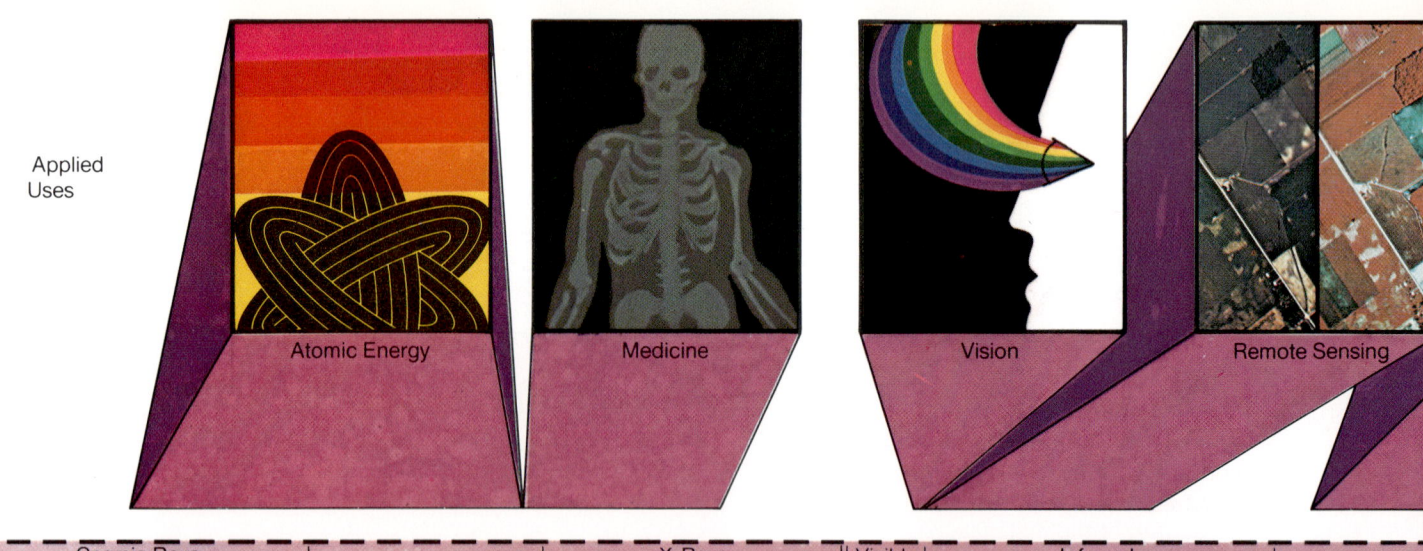

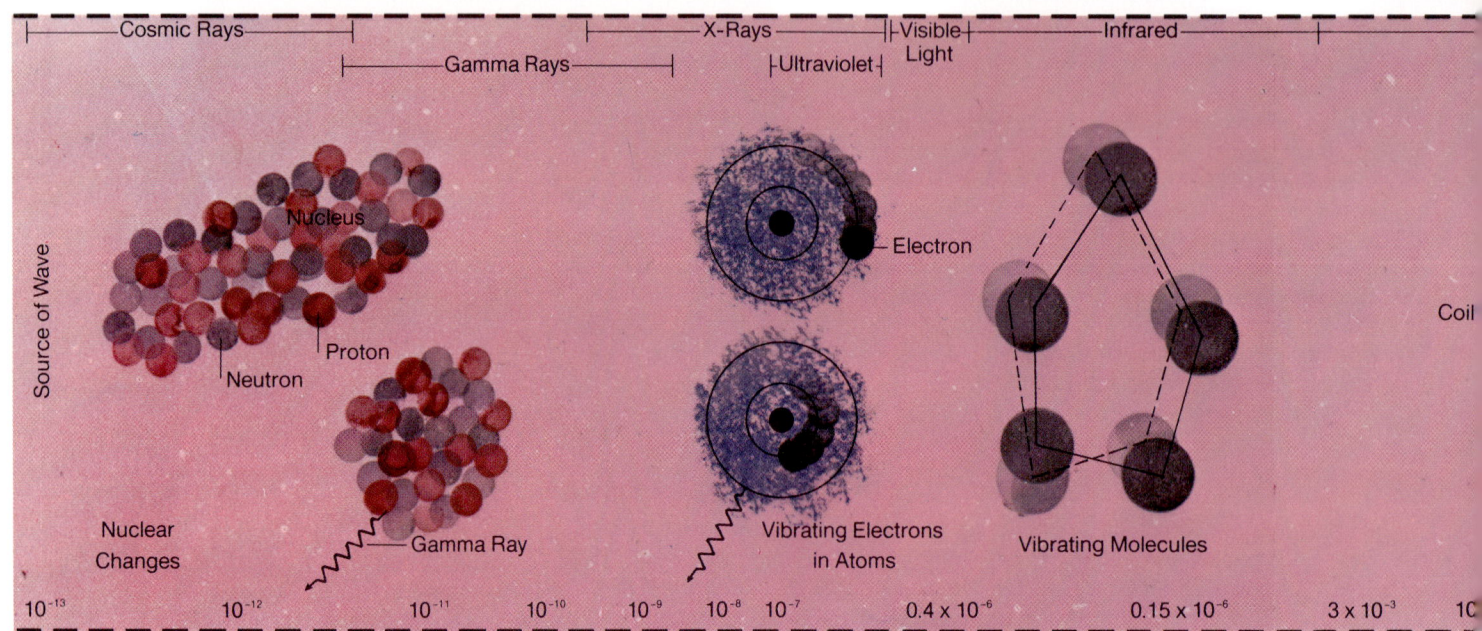

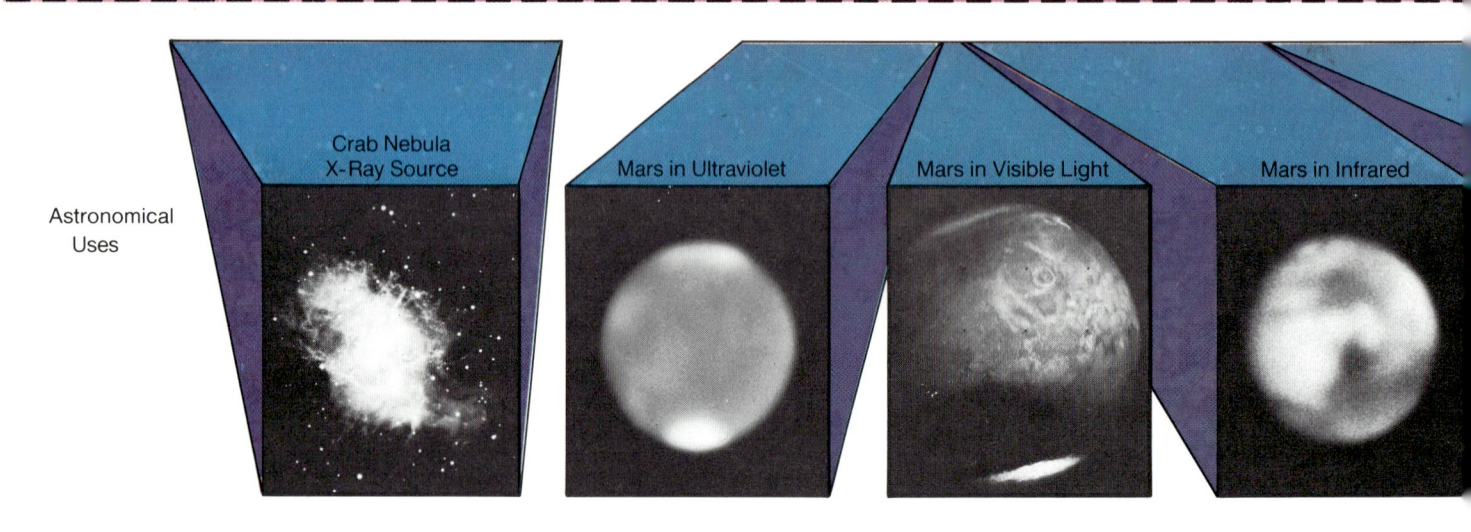

Applied Uses

Atomic Energy | Medicine | Vision | Remote Sensing

| Cosmic Rays | Gamma Rays | X-Rays | Ultraviolet | Visible Light | Infrared |

Source of Wave

Nucleus — Proton — Neutron

Nuclear Changes — Gamma Ray

Electron — Vibrating Electrons in Atoms

Vibrating Molecules

Coil

10^{-13} 10^{-12} 10^{-11} 10^{-10} 10^{-9} 10^{-8} 10^{-7} 0.4×10^{-6} 0.15×10^{-6} 3×10^{-3}

Astronomical Uses

Crab Nebula X-Ray Source | Mars in Ultraviolet | Mars in Visible Light | Mars in Infrared

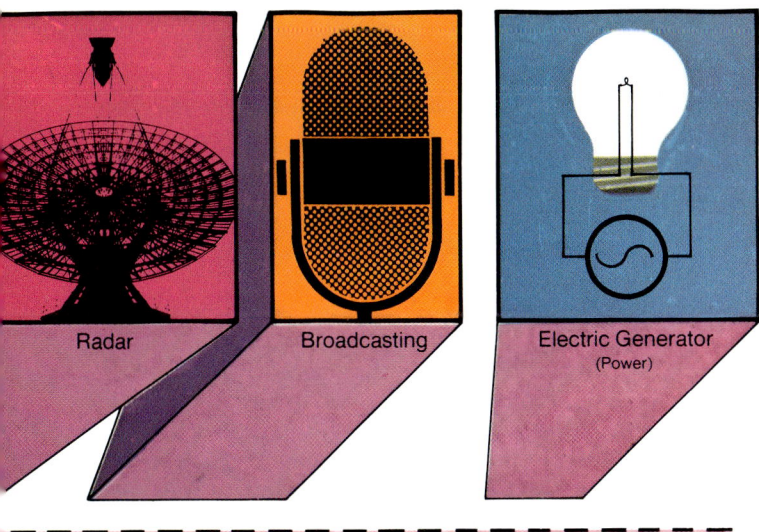

Figure 13. The electromagnetic spectrum. The center panel of this drawing shows the various forms of electromagnetic radiation in terms of wavelength. All such waves originate in the vibrations of atoms, but when we wish to *use* these waves we must look at them in different ways. For nuclear energy we work with the nuclear source; for vision we think of the changes that photons of different wavelengths cause in the atoms and molecules of objects, including the atoms and molecules of our eyes. And when we wish to use long waves, we require systems of electronic components that produce currents of radio frequencies. The upper panel of this drawing shows only a few of the ways we have learned to use electromagnetic radiation. The lower panel shows some of the remarkable astronomical images we receive from radiation of different wavelengths.

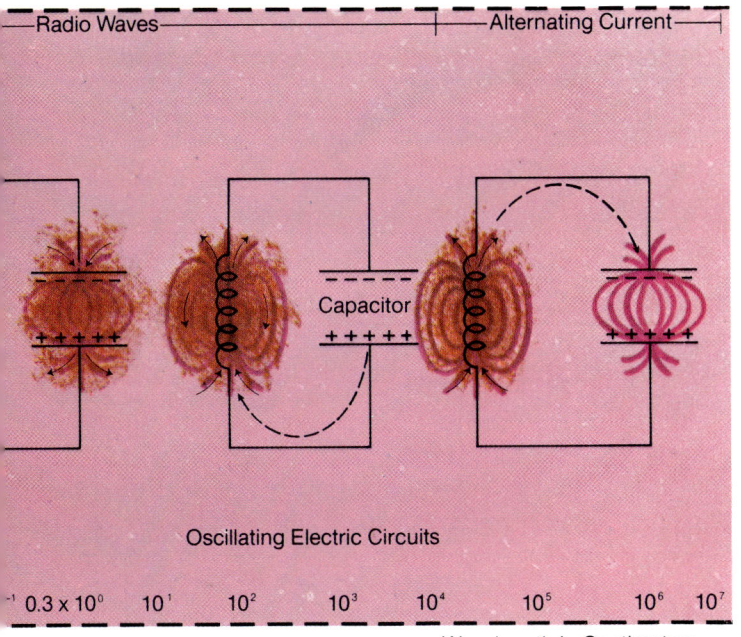

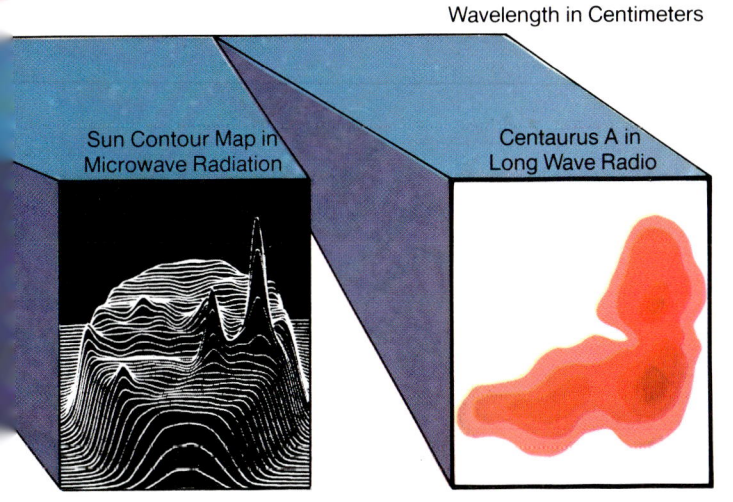

radiation. All travel in vacuum with the same speed—3×10^8 meters per second—and all represent undulations of electric and magnetic fields. Despite their underlying similarity, however, they differ greatly in the technical means for their generation and detection. It's convenient to list electromagnetic radiation in order of increasing frequency (decreasing wavelength), and to take up the various parts of the *electromagnetic spectrum* one by one (see Figure 13).

Radio Waves

Most radio and television transmissions are carried out at wavelengths greater than 20 or 30 centimeters, corresponding to frequencies below 10^9 cycles per second or so. In principle, you can generate a radio wave at as low a frequency as you wish. All you have to do is shake an electric charge back and forth at the desired rate. But the amount of power radiated varies as the square of the acceleration, which is small at low frequencies. Shaking a charge of 10^{-6} coulomb back and forth once a second only generates about 10^{-26} watt of electromagnetic radiation, which is far too small to be detected.

You know from an earlier chapter that electric currents represent the flow of electric charge, and that a current of 1 ampere is equivalent to the passage of 1 coulomb of charge per second. It's easier to generate electromagnetic waves by moving currents up and down wires than by shaking isolated electric charges. Suppose two wires are hooked up to

Figure 14. Two wires connected to an AC generator form an antenna. As the polarity of the generator voltage alternates, charges move back and forth along the wires and radiate electromagnetic waves.

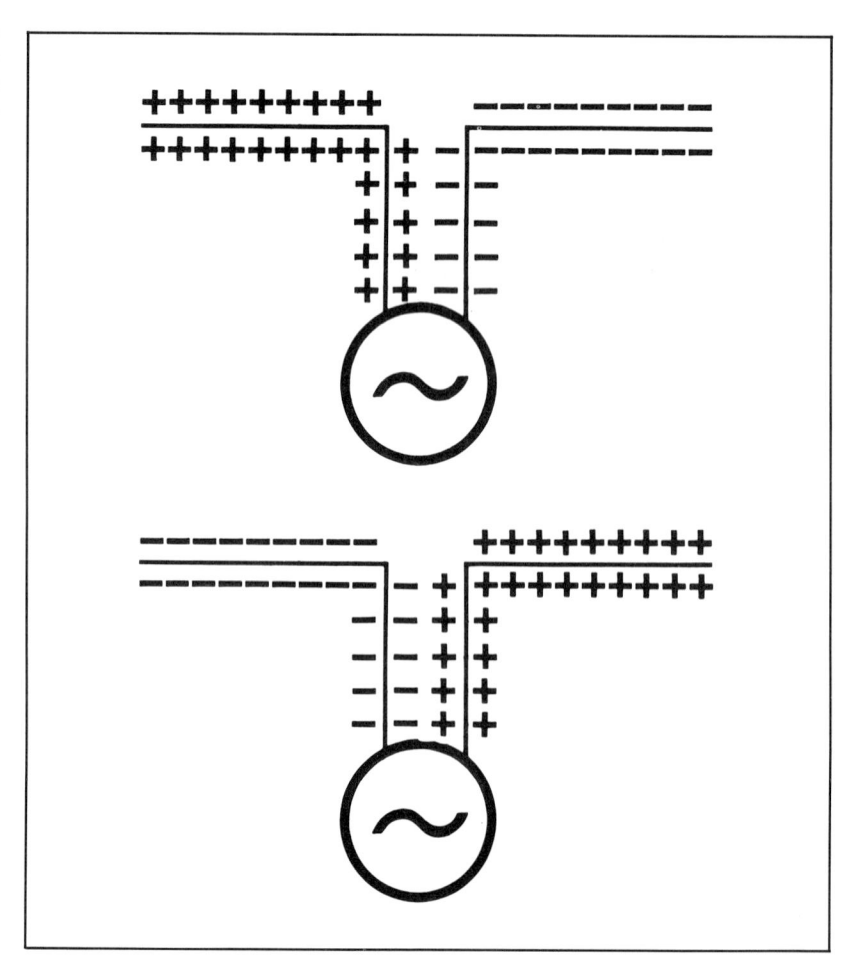

an alternating current generator. First, one wire is positive and the other negative, but after half a cycle the polarities reverse. Charge surges back and forth on each wire. And the accelerated charges radiate.

The two wires form a simple *antenna.* By raising the voltage applied to the wires, more charge flows and more power is radiated. But the antenna is most efficient if each arm is exactly one-quarter of a wavelength long. If you want to transmit a radio wave at a frequency of 10^6 cycles a second you would need two wires, each 75 meters long, to build the most efficient antenna. Such an antenna radiates about 35 watts for every ampere of current. Even a short antenna radiates some power, however. If each arm is one-tenth of a wavelength long, the radiated power is about 0.4 watt for each ampere of current.

Transmitting at high frequencies is convenient because the antennas can be small without losing efficiency. However, an alternating-current (AC) generator is not too useful for generating currents at high frequencies because it can't be turned fast enough. In addition, it is not flexible enough to transmit anything except a steady signal. The high-frequency currents needed in radio communications are generated electronically, which can be understood by considering a mechanical analogy.

Consider the swinging pendulum of a grandfather's clock in terms of *energy.* At the top of each swing, when the pendulum is instantaneously at rest, the pendulum's energy is entirely potential energy. At the bottom of the swing, its energy is in the form of kinetic energy. The energy flows back and forth between potential and kinetic forms. And the rate of energy flow is fixed by the strength of gravity and the size of the pendulum.

Suppose a charged capacitor is connected to a coil. At first the energy is purely electric energy in the charged capacitor. As the capacitor discharges, current flows and a magnetic field is built up in the coil. When the capacitor is completely discharged, all the energy resides in the magnetic field of the coil. Then, as the magnetic field collapses, an electromotive force is induced and charge again flows onto the capacitor. The charge surges back and forth, and the current alternates. The rate is set by the size of the capacitor and the size of the coil; the smaller they are, the higher the frequency.

Every circuit has resistance, and the energy of the coil and capacitor is eventually dissipated if they are left to themselves. Similarly, if left alone a pendulum finally comes to a stop because of friction. But the mechanism of the clock supplies a tiny kick to the pendulum to make up for lost energy. And the current in the oscillating circuit can be maintained by making up energy losses with an electronic amplifier.

The coil and capacitor circuit is also useful for tuning a radio receiver to a desired frequency. You can build up the motion of a pendulum with a feather if you hit it at just the right times. In the same way, a small signal at the natural frequency of the coil and capacitor can excite large currents, which can then be amplified further electronically. Different frequencies can be selected by varying the size of the coil or the capacitor.

The frequencies for radio communication are usually measured in two special units, the *kilohertz* and the *megahertz:*

1 kilohertz = 10^3 cycles per second

1 megahertz = 10^6 cycles per second

The coil and capacitor method for generating oscillations works well up to about 1,000 megahertz, a frequency which covers the range of most radio communication.

The radio spectrum is divided into bands for specified uses because of the practical importance of radio communication and the need for

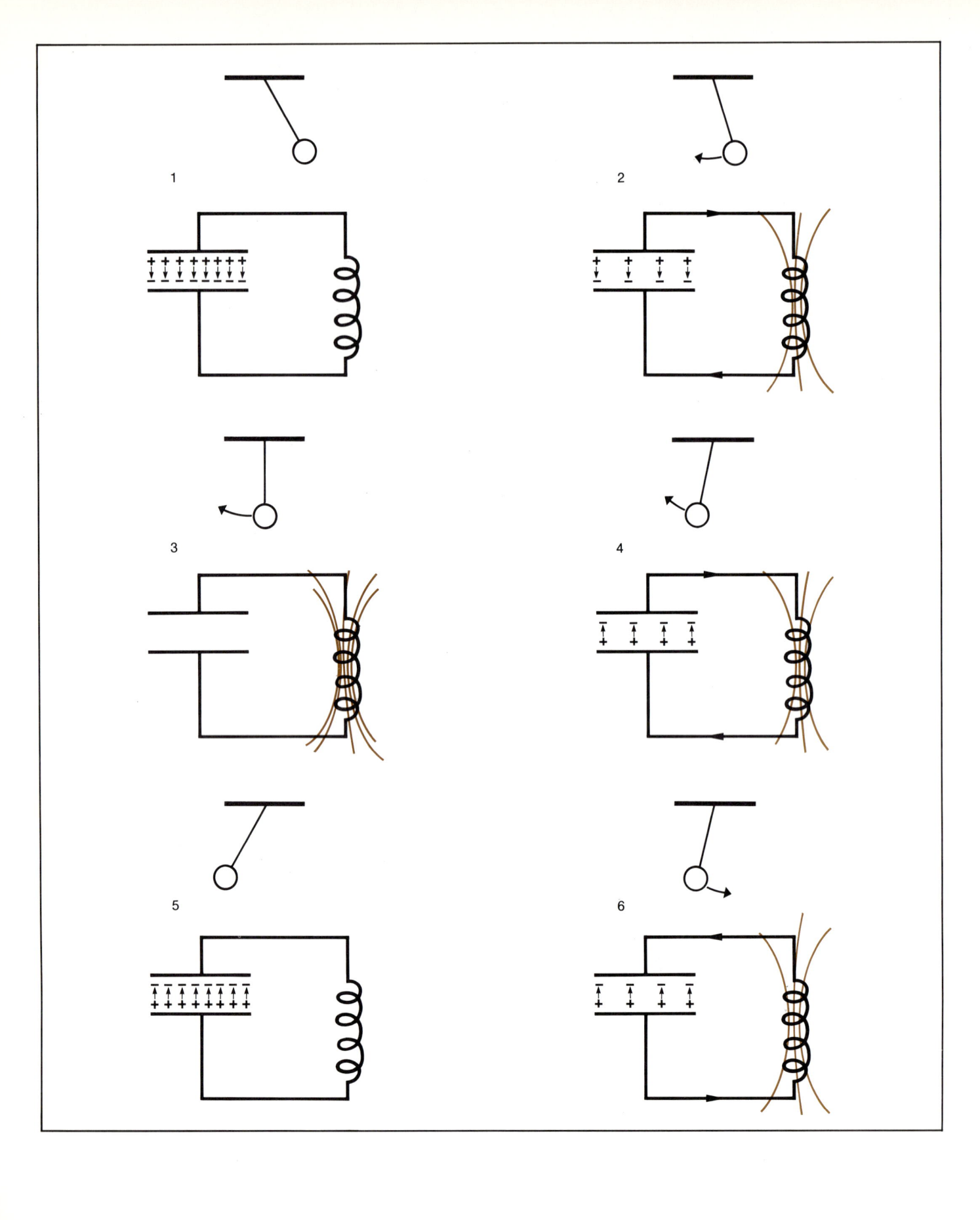

1

2

3

4

5

6

eliminating overlap and interference between signals from different transmitters. For instance, commercial AM broadcasting is confined to the frequency band between 535 and 1,605 kilohertz, and FM broadcasting is assigned the band from 88 to 108 megahertz. Television is assigned several frequency ranges:

Channels   2–6: 54–72, 76–88 megahertz

Channels   7–13: 174–216 megahertz

Channels 14–83: 470–890 megahertz

The remaining portions of the radio spectrum are reserved for amateur, maritime, business, and other types of communications.

The AM band occupies only about 1.1 megahertz of the electromagnetic spectrum, but the television bands are much broader. The AM broadcasting stations can be spaced as close together as 10 kilohertz, but each television station is allotted a bandwidth of 6 megahertz. A single television station needs more than 5 times as much of the radio spectrum as does the whole AM band. The reason is found in a fundamental principle of communications: *the greater the amount of information transmitted in a given time, the greater must be the bandwidth.*

For example, you can transmit the dots and dashes of Morse code by turning a transmitter on and off. The transmitted signal remains at nearly a constant frequency and little bandwidth is required. But it takes a long time to transmit a message by Morse code. By comparison, a television transmitter must send out enough information for a complete picture, with sound, 30 times every second. A television station requires a large bandwidth to send out that much information. An AM station, which transmits only sound, needs far less.

Furthermore, it's impossible to send television in a narrow band by using a rapidly transmitted Morse code system. A short signal is not a single frequency but is spread out over a wide range of frequencies. One example of a short electromagnetic signal is the pulse from a lightning stroke: the crackle from the flash can be heard at any frequency across the radio dial; it is not a single pure frequency.

The information-bandwidth relation imposes a restriction on the number of radio and television stations. Each station needs a certain amount of the radio spectrum. The stations have to be spaced in frequency because receivers can't distinguish between two overlapping stations. The crowding of the radio spectrum is alleviated somewhat by the tendency of high-frequency radio waves to travel out into space in straight lines. Two cities that are far enough apart can each have a television station on the same channel. The problem is more complicated for AM radio. Radio waves below 20 megahertz or so can bounce off layers of charge high in the atmosphere and return to earth quite a distance from the transmitter. This property is helpful for long distance communications, but it can also make the signal from a distant station overlap a local one.

One solution to the crowding of the spectrum is to send television signals by cable directly to receivers. Because the signal is not radiated into space, neighboring transmitters can't garble one another's signals and the whole radio spectrum is open for public and commercial use.

Figure 15. (Far left) In a coil-capacitor circuit, energy moves between electric and magnetic forms, just as energy in a swinging pendulum moves between potential and kinetic forms.

The Microwave Region

The microwave region of the electromagnetic spectrum extends from about 1,000 megahertz to 100,000 megahertz. The corresponding wavelengths range from 30 centimeters to 0.3 centimeter. Microwaves are the shortest wavelengths used for ordinary radio communication.

Radio waves with short wavelengths behave much like visible light. Microwaves tend to travel in straight lines, and they reflect from surfaces. Curved metal mirrors can focus microwaves into directed beams, which is a way that transmitter power can be economized. The transmitters on the Mariner spacecraft sent to Mars radiated only 10 watts of power at a frequency of about 2,300 megahertz. In order to concentrate the transmitter's power into a narrow bandwidth, television pictures of Mars were sent back at a low information rate—about three pictures a day.

Microwaves are ideal for radar because electromagnetic waves of such short wavelength are diffusely reflected by objects of ordinary size such as airplanes, ships, and buildings. In the operation of a typical radar set, the transmitter sends out a powerful signal at a microwave frequency for a period of about 10^{-6} second. Once the transmitter goes off, the antenna is reconnected from the transmitter to a sensitive receiver. The receiver picks up the feeble signal from the waves that have scattered off some object, and measurement of the time interval between transmission and reception gives the distance of the object. For example, the radar signal from a ship to a sandbar 1 kilometer away takes about 7×10^{-6} second for the round trip.

The technical means for generating microwaves is different from the coil and capacitor method suitable for frequencies below 1,000 megahertz. When the wavelength generated is smaller than the linear dimensions of the oscillatory circuit, all parts of the circuit store electric and magnetic energy, and ordinary circuit concepts are no longer valid. At microwave frequencies, the functions of the coil-capacitor network are performed by a single structure, called a *resonant cavity.* A resonant cavity for microwaves is simply a metal box, usually cylindrical in shape. Oscillating electric and magnetic wave patterns can be set up in a cavity. Like a capacitor and coil, a cavity can store electric and magnetic energy.

Only wave patterns of certain frequencies can be set up in a given cavity. The natural frequencies of a cavity depend on its shape and size. You know that the fundamental tone and overtones from a plucked guitar string depend on the length of the string. The lowest natural frequency of a cavity has a corresponding wavelength roughly comparable to the dimensions of the cavity. In other words, cavities are practical in size for the short wavelengths of microwaves but too large for the AM radio band, where the wavelengths are about 300 meters.

A sea shell acts like an acoustic resonant cavity. Perhaps you have held a sea shell to your ear to "hear the sea." Because of its small size, the sea shell cavity emphasizes stray sound of short wavelengths (high frequencies), and you hear a rushing noise.

To generate microwave radiation, fields from the cavity are allowed to interact through a hole in the cavity with a stream of fast electrons. The interaction is complicated, but if the fields mainly act to slow the electrons down, the kinetic energy of the electrons can be converted to electromagnetic field energy.

A microwave cooking oven is a resonant cavity and it usually operates at 2,450 megahertz. You read in an earlier chapter that polar molecules, such as water molecules, are positively charged at one end and negatively charged at the other. The oscillating electric field in a microwave oven works to turn water molecules first one way, then the other. But as the molecules turn, part of their energy is dissipated in collisions. In this way, microwave energy is converted to thermal energy and foods are heated almost uniformly throughout their entire volume. Cooking is

Figure 16. This x-ray negative of a chambered nautilus was formed because x-rays were absorbed by the calcium and did not go on to energize the photographic plate.

much faster because there is no wait for external heat to seep slowly into the food.

The Infrared Region

The infrared region extends from about 100,000 megahertz (10^{11} cycles per second) up to red visible light (4×10^{14} cycles per second). Electromagnetic waves in the infrared are difficult to work with. There is no electronic means for generating or detecting such high frequencies because the wavelengths are so short that resonant cavities are too small to be practical. Furthermore, infrared photons have too small an energy to be detected individually. Infrared radiation generally can be detected by allowing some object to absorb the radiation object and then measuring the resulting temperature change.

The most common source of infrared radiation is a hot object. The heating coils of an electric stove glow red at full power, but even before they start to glow you can feel heat coming from them. The heat energy is emitted in the form of electromagnetic waves, primarily in the infrared.

Charges are accelerated by thermal motion, so every object emits electromagnetic waves. Figure 17 shows the intensity of the electromagnetic radiation as a function of frequency for an object at room tempera-

Figure 17. The amount of electromagnetic radiation from an object increases rapidly with temperature, and the peak intensity moves toward higher frequencies as the temperature increases.

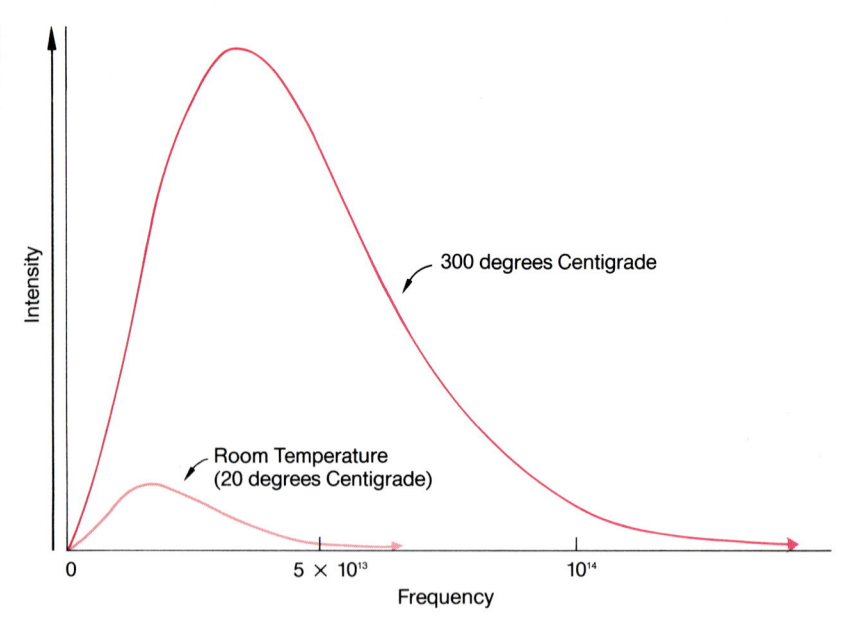

ture and at 300 degrees Centigrade. The radiation is spread out over a wide range of frequencies. There is little radiation at very low or very high frequencies. The frequency corresponding to maximum intensity increases with increasing temperature:

Frequency at maximum $\approx 6 \times 10^{10}$ absolute temperature

Figure 17 indicates that the emitted intensity increases rapidly with temperature. Taking all frequencies into account, the total energy emitted varies as the fourth power of the absolute temperature. An object at 300 degrees Centigrade (573 degrees absolute temperature) radiates 15 times more electromagnetic energy than when it is at room temperature (293 degrees absolute temperature).

The complete description of electromagnetic radiation from heated bodies was formulated by Max Planck in 1900. The curves in Figure 17 are calculated from his radiation formula. To explain the observed variation of intensity with frequency, Planck had to assume that the energy of the electromagnetic radiation was proportional to its frequency. A hot body has more heat energy than a cool one, and the energy it radiates can extend to higher frequencies. Planck's hypothesis marks the first appearance of the photon concept in physics—the idea that energy comes in bundles. In order to work with a clearly defined situation, Planck considered the problem of electromagnetic radiation as it occurs in a heated enclosure, such as an oven. When held at a steady temperature, such a system is in equilibrium and the amount of energy radiated from the interior surface into the cavity is equal to the amount of energy the surface absorbs. As Gustav Kirchhoff had stated earlier, the best absorber of radiation is also the best emitter. The best absorber is an object that looks perfectly black and reflects no light; so a black object is the best emitter and a shiny object, the worst. The interior of a Thermos® bottle is silvered because shiny surfaces radiate away the least energy for a given temperature. An unenclosed emitter, such as the hot tungsten filament in an incandescent light bulb, will not obey Planck's radiation formula exactly if its surface is shinier at some frequencies than at others.

An important phenomenon associated with the infrared region is *molecular vibrations.* The atomic nuclei in a molecule are bound to the molecule by electric forces, which act much like elastic springs. The nuclei can vibrate, and the rate of vibration is set by the strength of the forces and the masses of the nuclei. The vibration frequencies generally fall in the infrared region. Every molecule has its own characteristic frequencies of vibration, and this property is a useful tool for identifying molecules in chemical analysis. When a beam of infrared radiation is passed through a gas, the molecules can absorb energy from the beam if the frequency of the radiation matches one of the molecular vibration frequencies. The molecules reradiate the energy in all directions, weakening the beam. The method may also be useful in detecting pollutants such as carbon monoxide in the air over cities.

Visible Light

The short wavelengths associated with visible light are generally specified in *millimicrons*:

1 millimicron = 10^{-9} meter

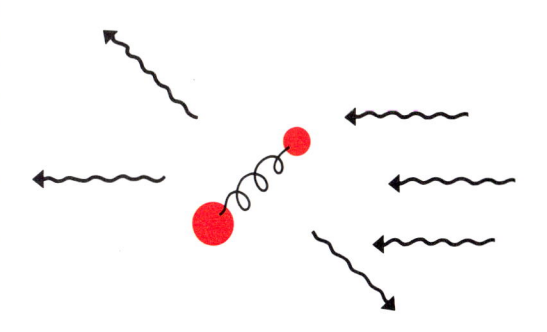

Figure 18. Molecules can weaken a beam of infrared radiation by absorbing and reradiating the energy in all directions if the frequency of the beam matches one of the natural vibration frequencies of the molecule.

The visible portion of the electromagnetic spectrum extends from about 700 millimicrons (deep red) to 400 millimicrons (deep indigo). The corresponding frequency range is 4.3×10^{14} cycles per second to 7.5×10^{14} cycles per second.

The principal source of visible light is the sun. The radiation from the sun is similar to that from an ideal black emitter at a temperature of 5,800 degrees absolute temperature. However, much of the sun's radiation is absorbed by gas molecules in the earth's atmosphere and never reaches the ground. The part that does reach the ground is concentrated primarily in the visible region. The human eye is adapted to detect only the solar radiation that actually reaches us.

Photons of visible light have enough energy to enable them to be detected one by one. A photon of purple light has an energy of 3 electron volts, enough to knock an electron out of some metals. By detecting the electron, which is a fairly simple task, the presence of the photon can be inferred. The eye itself is nearly sensitive enough to detect the energy deposited by a single photon.

Many of the phenomena associated with visible light can be understood by using simple wave ideas. You will look at some of these phenomena later in the chapter. For now, look at what happens when an electromagnetic wave interacts with matter. The oscillating electric field of the wave exerts forces on the electric charges in matter and jiggles them back and forth. The accelerated charges radiate away the energy they absorb and scatter the energy of the wave. When light from the sun passes into the earth's atmosphere, electrons in the air molecules vibrate and scatter the light in all directions. But the higher the frequency of the light, the greater the acceleration of the electrons and the more energy they radiate. Blue light, with its higher frequency, is scattered more than red light, and that is why the indirect light from the sky is blue. Outside the earth's atmosphere, the sky is black.

Ultraviolet Light, X-Rays, and Gamma Rays

For electromagnetic radiation above the visible region, frequencies are so high that the corresponding photons have energies of many electron volts. With this much energy, single photons are easy to detect. Furthermore, high-energy electromagnetic radiation is usually generated by individual atomic processes, so the wave picture becomes less useful in

Figure 19. The circles represent wavefronts traveling outward from a point of disturbance.

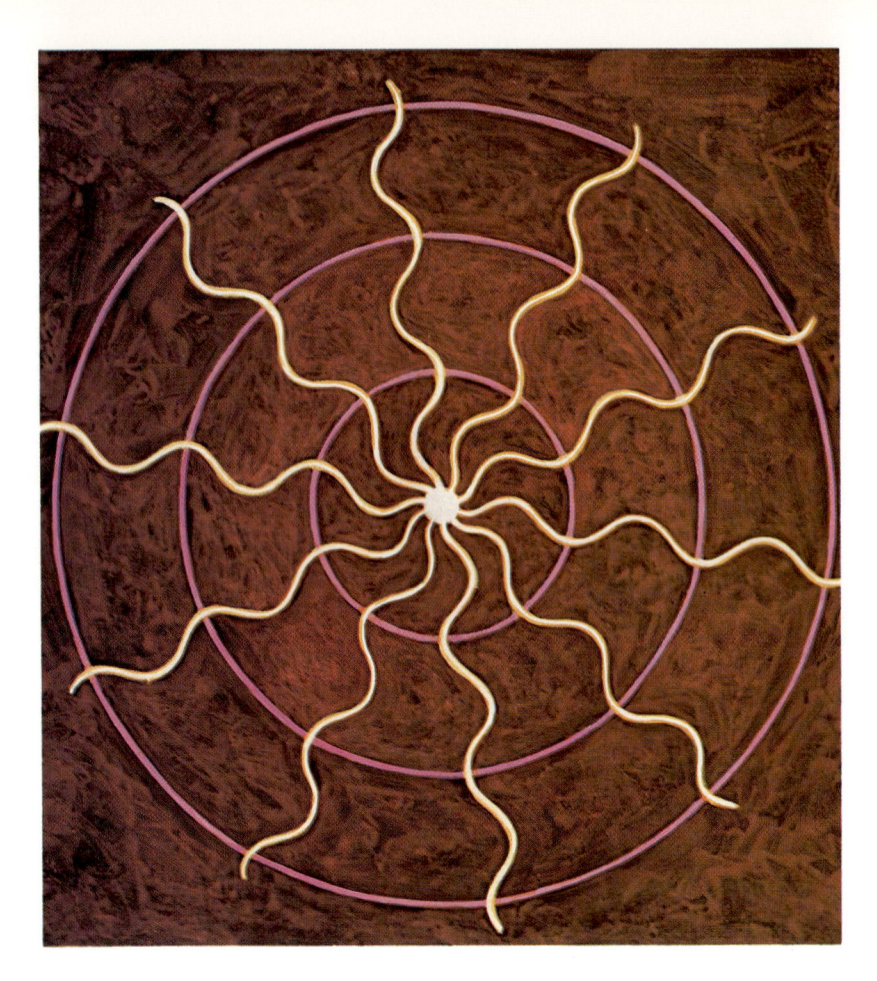

this region. Nevertheless, the wave properties of high-energy radiation can be important, as they are in the diffraction of x-rays by a crystal.

The ultraviolet region extends from the violet end of the visible spectrum down to wavelengths of 100 millimicrons or so. Ultraviolet radiation usually can be generated by exciting atoms to higher energy levels so that they emit photons in the ultraviolet region as they return to their ground state. For example, the ultraviolet light from a sun lamp can be produced by passing electric current through a vapor of mercury atoms so that the violent collisions excite them to higher energy levels.

The energy transitions that produce ultraviolet photons occur among the electrons in the outer shells of atoms. But what happens if a collision is so violent that an electron in one of the inner shells is entirely removed from an atom? The outer electrons cascade in to fill the vacancy and bring the atom to its lowest possible energy. The electric forces are so strong near the nucleus, particularly in atoms of high atomic number, that transitions involving inner electrons can have energy changes as high as 100,000 electron volts. In 1913, Henry Moseley used the energy of the characteristic x-rays from each element to rank the elements in order of atomic number. The higher the atomic number, the stronger the force on the inner electrons and the more energetic the x-rays.

How can x-rays be produced? A high voltage can be used to speed up electrons emitted from a heated filament, then the energetic electrons are allowed to crash into a target such as a plate of tungsten. In addition to producing x-rays by inner shell excitation, the energetic electrons radiate electromagnetic energy directly as they rapidly decelerate to a stop in the target. The x-rays produced this way cover a continuous spectrum, ranging up to the maximum initial energy of the energetic electrons. Therefore, photons of any desired energy can be produced, given electrons of sufficiently high energy.

High-energy photons are generally classified according to their origin rather than their energy. A gamma ray is a photon emitted by energy level changes in nuclei. The energy of gamma ray photons ranges from a few thousand electron volts to millions of electron volts.

The wavelength of very energetic photons is small compared to an atom. Such photons are relatively oblivious to the structure of matter, and it is not possible to focus them with lenses or mirrors. Energetic photons interact primarily with individual electrons or electrons bound in atoms.

LIGHT RAYS

The everyday behavior of light can be understood in terms of three laws of optics:

1. *In a uniform medium, light travels in straight lines.*

2. *The reflection of a light ray from a surface is symmetrical in the incoming and outgoing angles.*

3. *When a ray of light goes from one medium into another (for example, from air into glass), it is refracted into a new direction, as you read in Chapter 2.*

Toward the end of the seventeenth century, Christian Huygens showed that a simple wave picture could account for these three properties of light.

To understand Huygens' arguments, consider the sinusoidal electromagnetic wave generated by a vibrating electric charge. Corresponding crests at any instant form surfaces called *wavefronts:* all the waves along a given wavefront are in phase. The wavefronts are at right angles to the direction of propagation and they ripple outward, much like water waves when a stone is dropped into a pond. But once a wave has been generated, it propagates on its own with no further help from the source. Huygens incorporated the idea of propagation into his wave picture by assuming that every point on a wavefront acts as a generator for the later advance of the wave. Figure 20 shows how Huygens' construction implies the straight-line motion of light waves. According to Huygens, each point on a wavefront emits a spherical wavelet. In a uniform medium, the wave-

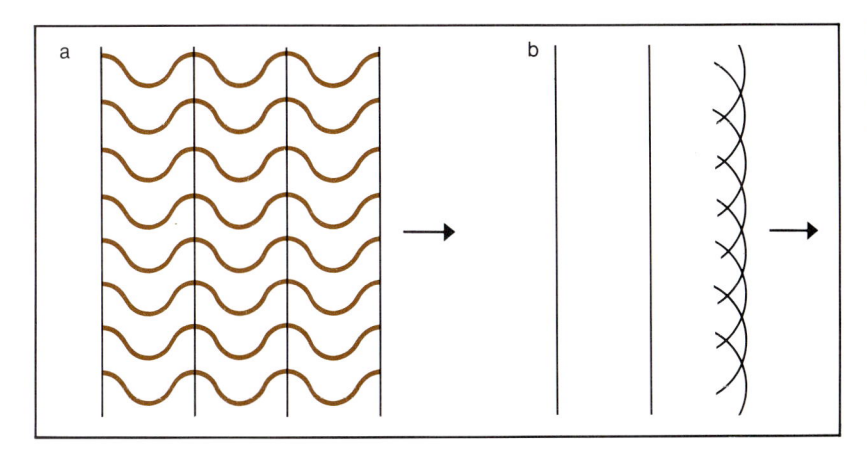

Figure 20. (a) When they are far from a radiating charge, wavefronts are approximately straight, parallel lines. (b) According to Christian Huygens' construction, each point on a wavefront acts as the source of a spherical wavelet.

The Doppler Shift

Oscillating electric charges in radio transmitters or in radiating atoms can produce electromagnetic radiation of a definite frequency. If the transmitter or the atom is moving, however, the frequency measured by someone standing still depends on the speed of the transmitter or of the atom. The apparent frequency change is the *Doppler shift*.

<u>Step 1.</u> To see how the Doppler shift arises, suppose that a transmitter generates 100 short electromagnetic pulses per second (measured when both the transmitter and the observer are at rest). The time interval between successive pulses is 0.01 second, and so the pulses arrive at the receiver once every 0.01 second. Then 100 pulses arrive at the receiver every second and the observed frequency is equivalent to the transmitted frequency.

<u>Step 2.</u> If the transmitter is moving away from the receiver at velocity v, then the observed time interval between pulses is longer than 0.01 second. Suppose the transmitter and the receiver are instantaneously a distance ℓ apart when pulse 1 is emitted. The time it will take for pulse 1 to reach the receiver is given by the relation

$$\text{Time 1} = \frac{\text{distance } \ell}{\text{speed of light}} \quad \text{or} \quad t = \frac{\ell}{c}$$

for a pulse traveling at the speed of light. When pulse 2 is emitted 0.01 second later, the transmitter and the receiver are farther apart than before:

$$\ell' = \ell + 0.01 \times v$$

Then the time at which pulse 2 reaches the receiver is

$$t' = 0.01 + \frac{\ell + 0.01 \times v}{c}$$

The time interval, or Δt, between pulses at the receiver is

$$\Delta t = t' - t = 0.01 + \frac{\ell + 0.01 \times v}{c} - \frac{\ell}{c} = 0.01 + \frac{0.01 \times v}{c}$$

The time interval between pulses is longer when the transmitter is moving with respect to the receiver, so the observed frequency ν' is lower:

$$\nu' = \frac{100}{1 + v/c}$$

In general,

$$\text{Observed frequency} = \frac{\text{transmitter frequency}}{1 + \text{velocity}/\text{speed of light}} \quad \text{or} \quad \nu' = \frac{\nu}{1 + v/c}$$

for a transmitter and receiver moving away from each other.

lets all travel the same distance in a given time. The new wavefront is made up of the surface defined by the wavelets.

Figure 21 illustrates Huygens' picture of reflection. When a light ray strikes a surface at an angle, one end of the wavefront reaches the surface first. The spherical wavelets from the surface start out at different times. The new wavefront, representing the reflected ray, leaves the surface symmetrically with respect to the incoming ray.

When a light ray goes from air into glass, a small part of the energy is reflected (about 4 percent). Most of the ray continues into the glass, but in a new direction. Huygens' construction to explain this refraction is shown in Figure 22. The speed of light is slower in glass than in air, so the wavelets move into the glass slower than the wavefront arrives at the surface. The net effect is to cause a bending of the ray. If the ray goes from glass into air, the bending is in the opposite direction.

INTERFERENCE

When two electromagnetic waves overlap in space, their electric fields combine. The waves can interfere constructively or destructively, depending on whether the crests of the two waves coincide or whether the crests of one coincide with the troughs of the other.

Light from ordinary sources, such as the sun or an incandescent lamp, is produced by the radiation from a large number of randomly accelerated electrons. A given electron only radiates for about 10^{-7} second at a time, but that interval is long enough for many millions of cycles of oscillation. Ordinary light consists of a collection of such short *wave trains.* For this reason, interference between light from two different lamps or from different parts of the same lamp is difficult to observe; even if two particular wave trains interfere constructively, the next two may not.

The interference effects you see in soap bubbles or in oil films come from splitting a wave train into two parts. One part is made to travel a different distance than the other, and when the two parts are recombined they interfere. Because every wave train is treated the same way, a steady interference pattern results. The interference is constructive if the effective path difference between the two parts of the wave train is a whole number of wavelengths.

The splitting of the wave train in a soap bubble takes place at the outer and inner surfaces of the soap film, as illustrated in Figure 24. Light reflected from the outer surface combines with part of the same wave

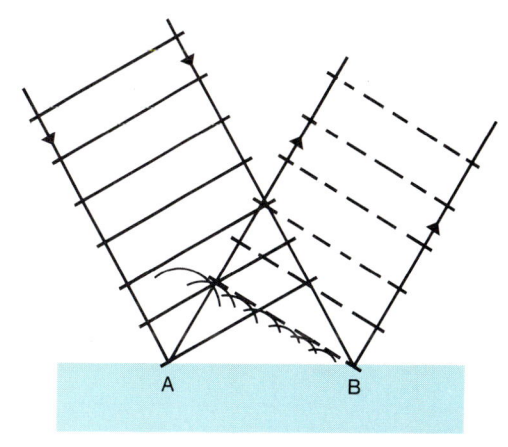

Figure 21. The incoming wavefront strikes the mirror at A before it reaches B, and the Huygens' wavelets leaving A have more time to travel (above). The reflected light ray leaves the mirror symmetrically.

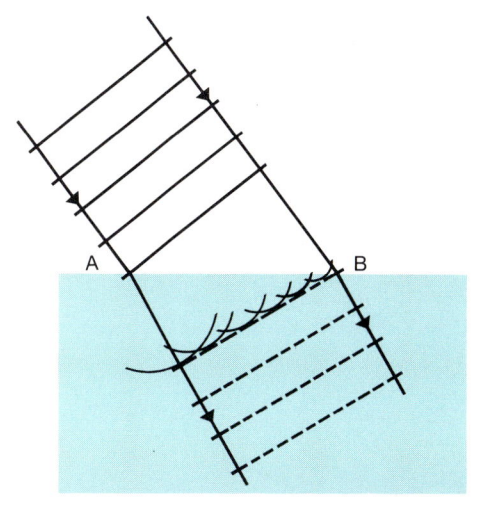

Figure 22. The Huygens' construction for a light ray is shown traveling from air into glass. Each wavefront reaches A before it reaches B. The ray of light is "bent" as it goes into the glass because the Huygens' wavelets travel slower in glass than in air.

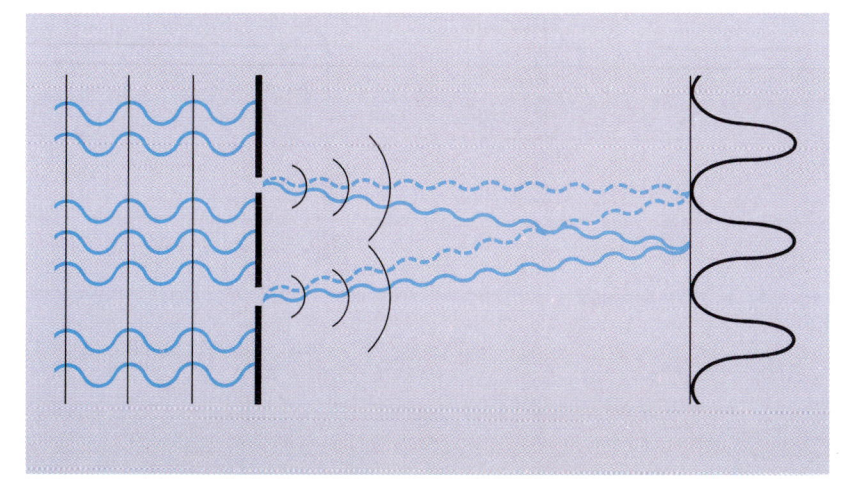

Figure 23. (Left) Interference of light can be demonstrated by using two slits to split a wave train (left). Each slit generates Huygens' wavelets. The two light rays arriving at A travel equal distances and arrive in phase. The light rays arriving at B travel distances that differ by one-half wavelength and they interfere destructively. If the interference pattern is projected on a screen placed a long distance L from the slits, the spacing between adjacent bright spots is approximately $\lambda L/d$, where λ is the wavelength of the light and d is the spacing of the slits. (L is much larger than d and λ.)

train that travels through the film and is reflected from the inner surface. Suppose that white light, which is a mixture of wave trains of all colors, shines on a soap bubble. Each part of the soap bubble reflects only the pure color whose wavelength is in the correct relation to the thickness of the film. Different thicknesses reflect different colors. And as the film evaporates, the color reflected from a particular section changes.

POLARIZED LIGHT

An electric field is represented by a vector and has direction as well as magnitude. The electric field of a propagating electromagnetic wave in free space is always perpendicular to the direction of propagation, according to Maxwell's theory. But imagine that a light ray is coming toward you. The electric field could be oscillating up and down; it could be oscillating right and left; or it could be oscillating in some other direction altogether. The plane of the electric field is called the plane of polarization of the wave.

The electromagnetic wave from an accelerating charge is inherently *polarized.* The kink of electric field lies in the plane determined by the direction of propagation and the direction of the acceleration. An individual wave train therefore has a definite polarization. However, a source

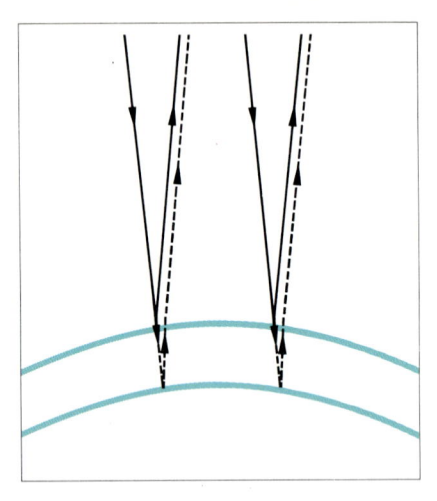

Figure 24. (Above) We see colors in a soap film because light rays reflected from the outer surface interfere with light rays reflected from the inner surface. (Other reflected and refracted rays are omitted from the sketch for clarity.) When light rays are effectively perpendicular to a soap film that is one-quarter wavelength in a thickness, constructive interference occurs. The light ray traveling into the film and out again goes one-half wavelength farther than the ray reflected directly from the outer surface. But the directly reflected ray suffers a change of phase equivalent to an additional one-half wavelength. The wavelength of red light in a soap film, is about 500×10^{-9} meter so a film 125×10^{-9} meter thick appears red.

Figure 25. The plane of the electric field in an electromagnetic wave is called the *plane of polarization.*

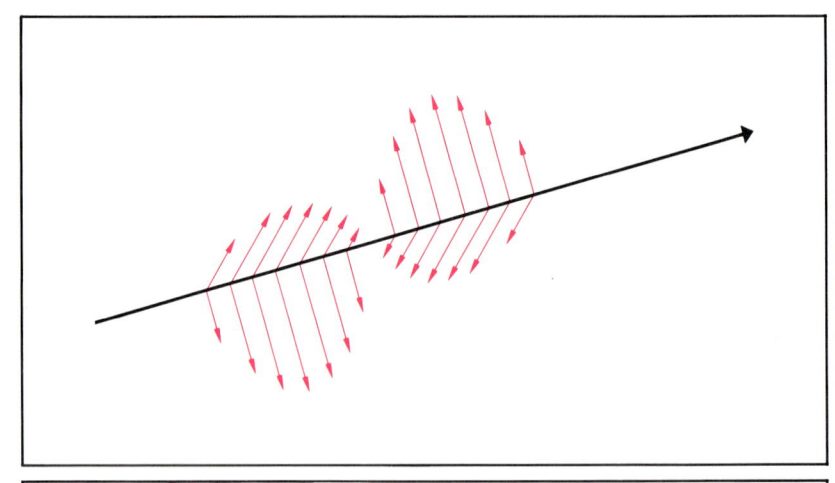

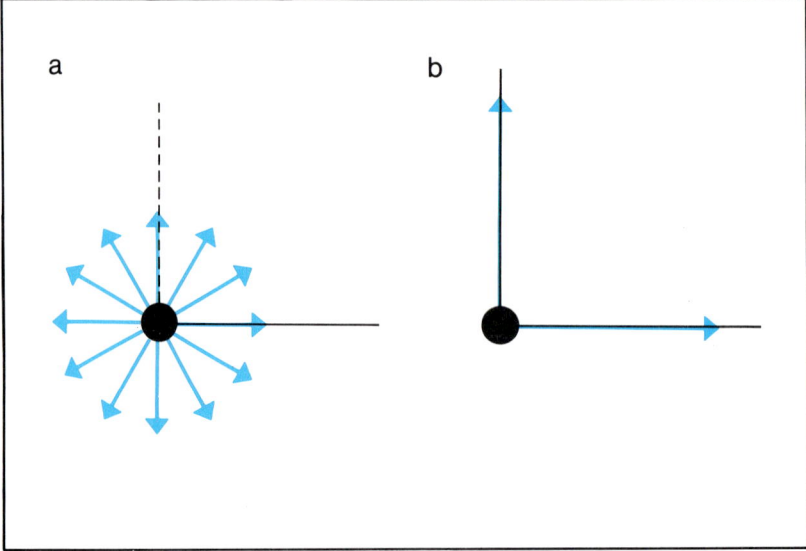

Figure 26. (a) An unpolarized beam of light consists of wave trains with random planes of polarization. (b) By decomposing each electric field vector along a particular set of axes, unpolarized light can also be represented as a mixture of two polarized beams at right angles with equal amplitudes, but with a constantly changing phase relationship.

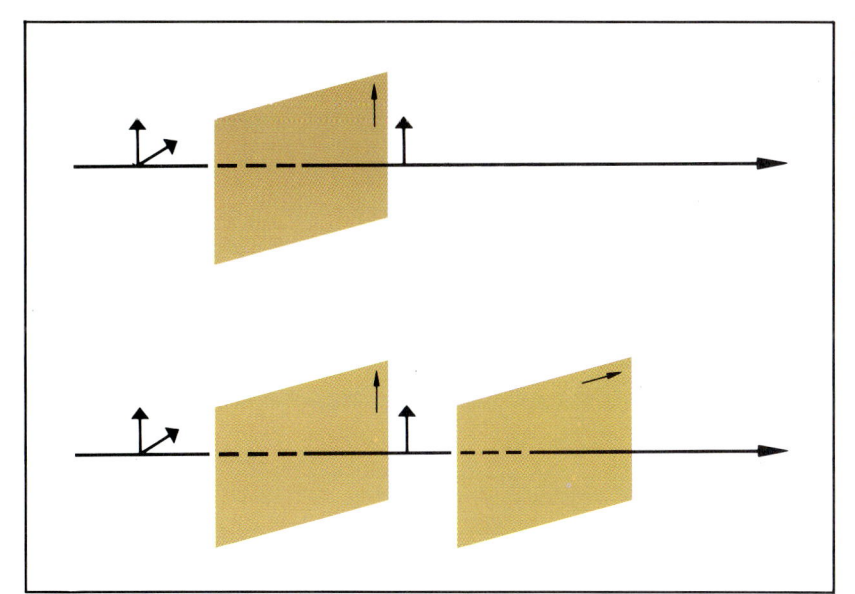

Figure 27. (a) A beam of unpolarized light is polarized after passing through a polarizer that absorbs one component of the electric field. (b) Two polarizers at right angles ideally do not transmit any unpolarized light.

such as a lamp emits wave trains at random, and on the average, light from a lamp has no definite polarization—it is *unpolarized.*

Consider again a beam of unpolarized light traveling toward you. The electric field vector for each wave train has a definite plane of polarization, but the plane is different from one wave train to the next. An electric field is a vector, and the electric field from each wave train can be resolved into a vertical and a horizontal component, as in Figure 26. You can think of unpolarized light as an equal mixture of two polarizations. A beam with a definite polarization can be produced by eliminating the unwanted component.

Unpolarized light passed through a piece of Polaroid® comes out polarized. Oriented crystals in the Polaroid® absorb one component of electric field strongly, but not the other. A piece of Polaroid® reduces the intensity of an unpolarized beam by half, neglecting absorption in the colored plastic. An additional thickness of Polaroid® has little effect on the intensity if its crystals are oriented the same way as in the first thickness. But if the crystals in the second Polaroid® are oriented at right angles to the crystals in the first, little or no light passes through.

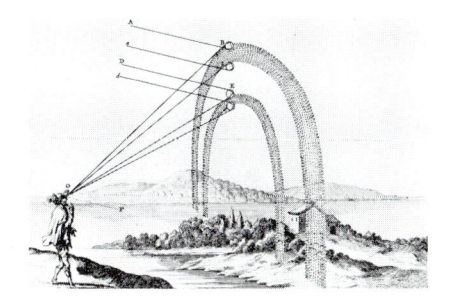

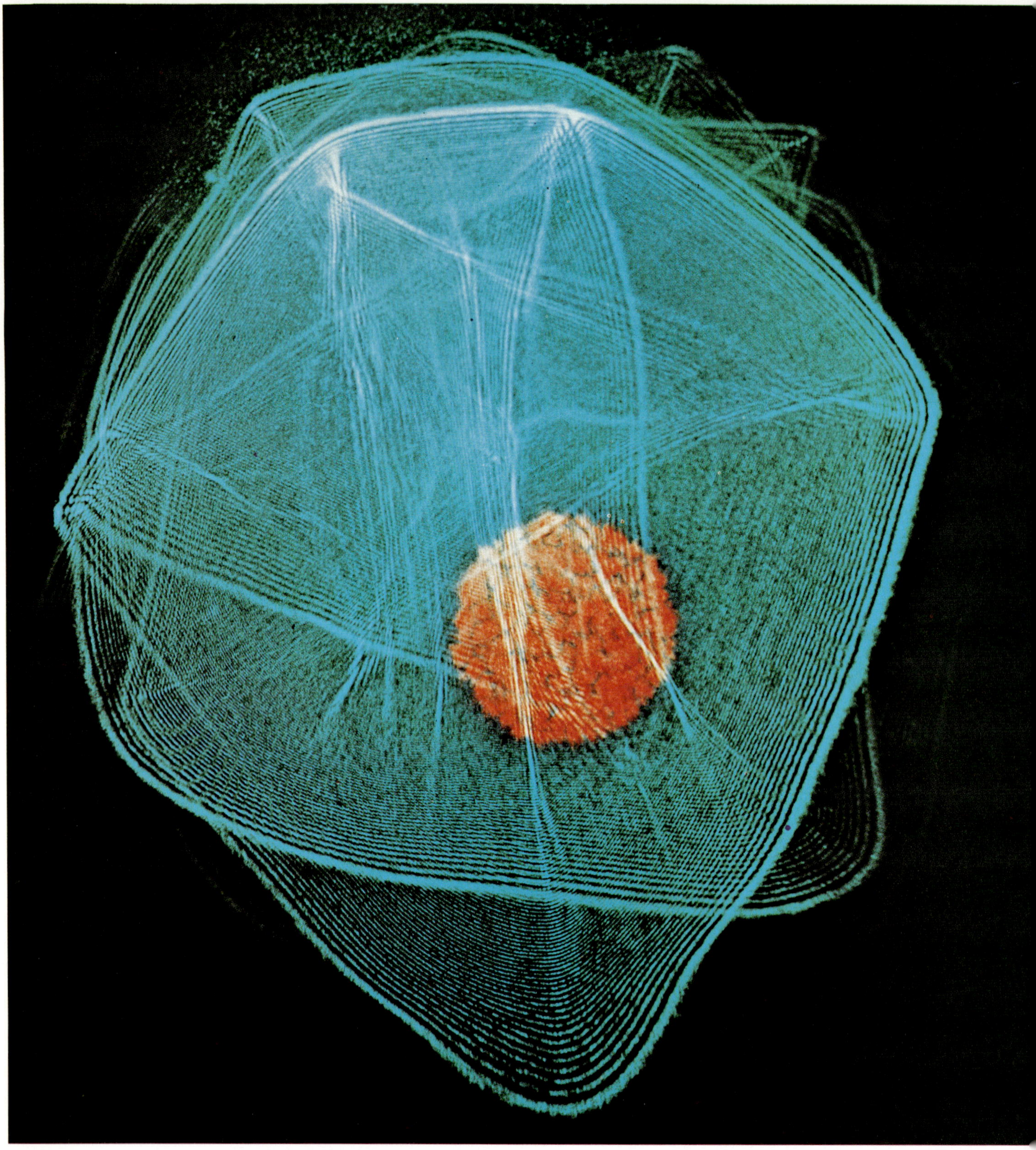

27

Lasers: The Light Fantastic

Laser images as an art form.

The science fiction stories of a light ray that reaches across great distances to "zap" things seem to be turning into fact with the advent of laser technology. Some lasers are weaker than a flashlight in power output, but others have beams that actually drill holes through diamonds and steel plates. Even though the old fantasies attributed a fearsome array of properties to the light ray, laser light has beneficial properties well worth developing. It holds tremendous potential for such wide-ranging fields as communications, medicine, chemical research, geophysics, and industry. As this potential is realized, you will be hearing more of this fantastic light and what it does. For now, consider this chapter a brief introduction to one of several products of twentieth-century science that will soon have a measurable effect on many levels of our society.

The term "laser" is an acronym for *light amplification by stimulated emission of radiation.* In simpler terms, lasers are devices that produce intense, highly directional light beams of an exceptionally pure color. It was not until 1960 that the first laser came into operation, and yet there is now a large family of lasers that come in almost invisibly small sizes as well as in lengths several hundred meters long.

Before lasers were invented, all light rays originated from atoms or molecules excited to higher energy levels by the somewhat random thermal jostling that goes on in hot bodies. When that happens, the energy stored in each atom is released as a burst of light—as a photon. The frequency of this emitted light wave depends on how much energy is stored in the atom.

Usually an excited atom spontaneously emits this light in a short time—perhaps a microsecond or less, depending on the kind of atom it is. Now a hot object contains great numbers of atoms, and each one is emitting bursts of light randomly. But consider what would happen if, during the short time that a group of atoms remains excited, a light wave of the same frequency were to reach it. The atoms would emit their stored energy simultaneously in a burst, and the light wave would be strengthened, or *amplified.*

By way of analogy, imagine a crowd of people streaming from one part of a city to another, each proceeding at his own pace and arriving at his own time. Then suddenly someone strikes up a band. Not only do the stragglers decide that this is the time to join the procession, they even fall into line and all march to the same step. Like a band beginning to play, a relatively weak light wave passing through a medium containing many excited atoms can be greatly intensified and laser action can result.

This process is called *stimulated emission,* in contrast to the *spontaneous emission* of ordinary light. It depends on having a medium of excited atoms, a column that contains those atoms, and a source of stimulation. A laser is a pencil-shaped device for amplifying the effect of such excited atoms. A small mirror at one end of its column faces a small mirror at the other end (Figure 3). One of the mirrors may be totally reflective, but the other must be at least partially transmissive; it must permit some light to pass through.

A light wave can originate anywhere in the excited region, and it usually does so by spontaneous emission. As it passes down the column it stimulates other atoms to emit energy, which amplifies the wave. In this way, more and more waves join the procession. Some light is lost through the transmitting mirror but most is reflected off the other mirror and be-

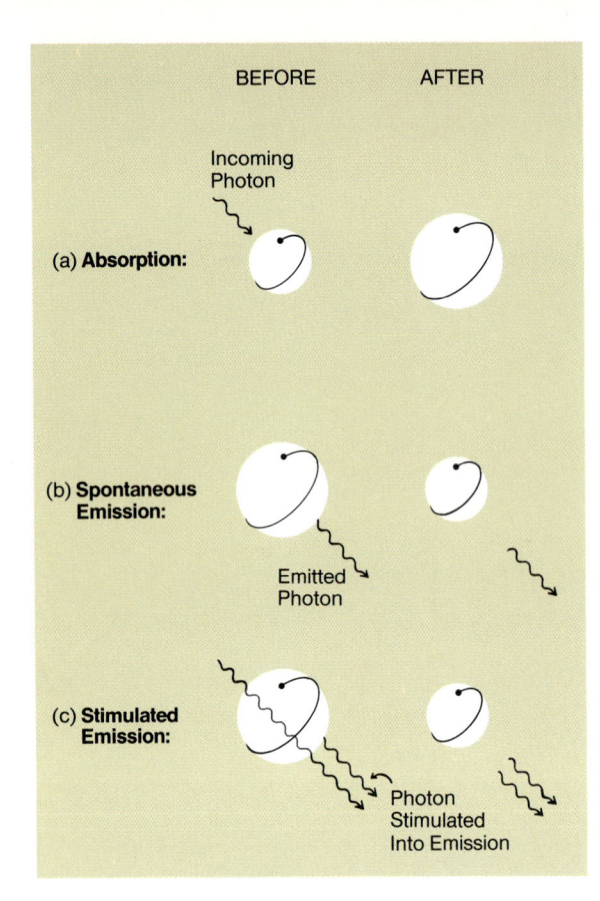

Figure 2. In *absorption* (a), a photon collides with an atom in the ground state and is absorbed. The increase in energy raises the atom to an excited state. In *spontaneous emission* (b), an atom that is in the excited state emits a photon, which lowers the energy and brings the atom to the ground state. In *stimulated emission* (c), photon energy in an atom is released when the atom is bombarded with a photon of the same frequency. The photon that is stimulated into emission is in phase, or "in step," with the incoming photon.

comes intensified as the wave returns along the amplifying column. As the wave passes back and forth in the column, its intensity increases until it is limited by the rate at which excited atoms are supplied by the medium. A wave that starts out in any direction *other* than forward or backward passes out the side of the column before its strength increases appreciably. But the light reflected back and forth between the mirrors attains such a high intensity that the portion finally transmitted through the end mirror is a powerful beam.

The properties of lasers can be understood by examining Figure 3. The beam that emerges from the partially transmissive end mirror is *highly directional* because only a wave traveling along the axis of the amplifying column will remain in the column long enough to become highly amplified. The light from a laser is relatively *powerful* because the atoms are stimulated to emit photons much faster than they would spontaneously. And laser light usually is of a single frequency, or *pure color*, because of the resonance nature of the stimulated emission process. (In other words, the atoms emit and respond to a frequency, or wavelength, of light only if they are in resonance with it. In a similar way, a child on a swing goes back and forth at the resonant, or characteristic, frequency of the swing.) Finally, laser light is *coherent*, which means that the photons are in phase (in step) with one another. Instead of radiating independently and randomly, the atoms are forced to contribute to the wave that is traveling back and forth between the mirrors.

In contrast to ordinary light sources, then, all lasers produce waves that are powerful, directional, monochromatic, and coherent. But the various individual types of lasers differ enormously in the extent to which they possess these properties.

What kind of medium has enough of the right kind of excited atoms to provide amplification of a light wave? Amplification by stimulated emission of radiation is the exact opposite of ordinary absorption of light. In other words, if an atom of some medium happens to be in a lower energy level, it can absorb a photon from a light wave, thereby acquiring the energy needed to rise to a higher level. This absorption weakens light because it removes photons from it. On the other hand, if the atom is already in the higher energy level, the same light wave can stimulate it to *emit* a photon and thus strengthen rather than weaken the wave.

A medium usually contains some atoms in the lower state and some in the upper state. If there are more in the lower state, absorption predominates. Only if there are more atoms in the upper state does amplification by stimulated emission exceed the absorption.

Now a medium in thermal equilibrium at any temperature always has more atoms in lower levels and progressively fewer in the higher, excited

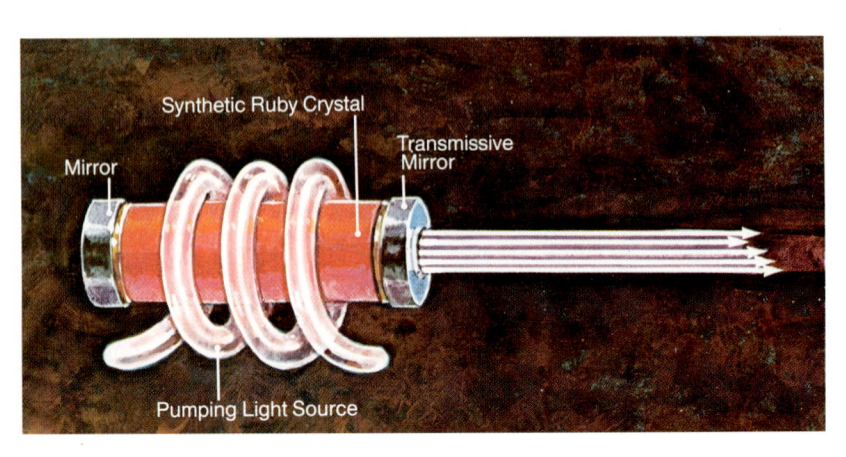

Figure 3. This drawing shows how a ruby laser is constructed.

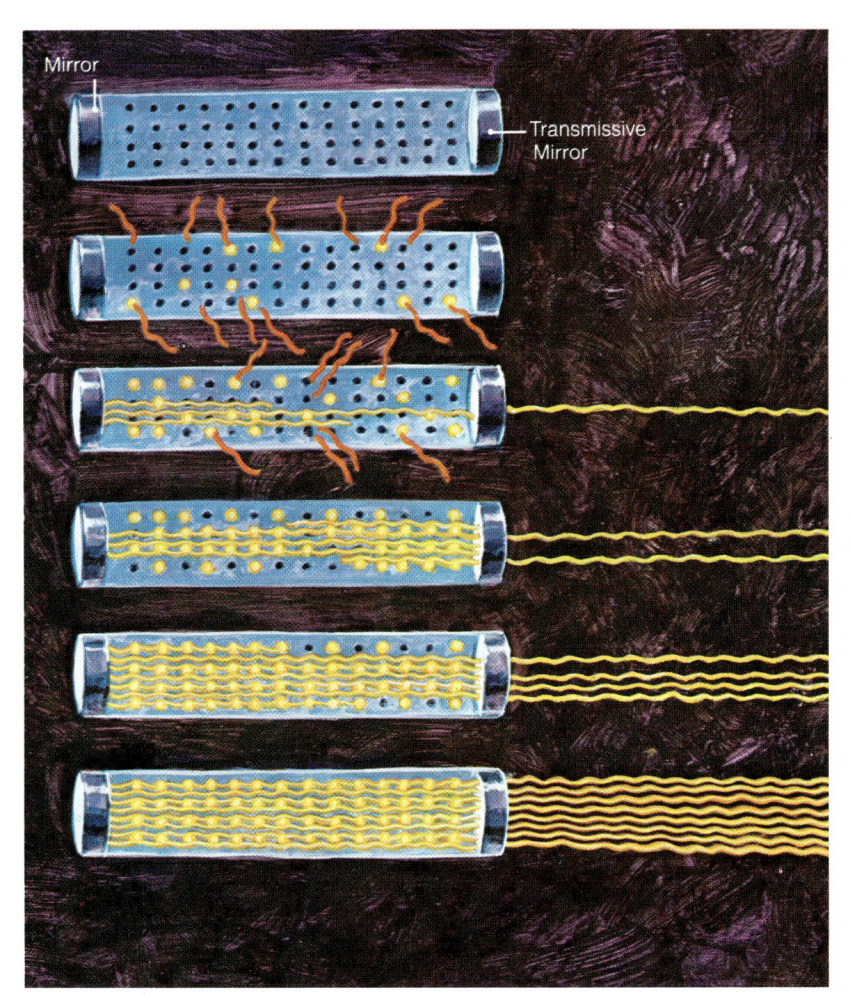

Figure 4. This diagram illustrates how a ruby laser (a) works. (b) The atoms making up the ruby crystal are excited to a particular energy level by bombardment with light photons of a particular wavelength. (The excited atoms are represented by the lighter circles.) (c) Photons of red light are emitted as the atoms return to their ground states. Some of the photons are emitted at angles that send them out of the crystal. But the photons traveling parallel to the axis of the crystal are reflected back into the crystal by mirrors. They interact with other atoms, triggering the release of still more photons of the same frequency and in the same direction of movement (c). A few of these photons "leak" out of the crystal through the transmissive mirror as the light builds up, but by now the amplitude is great enough so that the stimulated emission process is not affected. In (d) and (e), photons are flying back and forth between the two mirrors, triggering the emission of still more photons and increasing the intensity of the beam. (f) As the intensity of the red light in the crystal increases, the number of photons escaping through the transmissive mirror also increases, producing an intense laser beam.

levels. It was once assumed that emission would always be less than absorption because there would always be more atoms in the lower level of any pair of atomic states. During the years immediately following recognition of the phenomenon of stimulated emission, scientists were faced with this seemingly insurmountable barrier to the amplification of light.

It turns out that there are numerous methods of exciting atoms selectively into specific upper states. These methods provide the amplification by stimulated emission needed for laser action. Today many materials, including gaseous atoms, molecules and ions, solids, and even some liquids, can be used to create laser light.

In 1954, Charles Townes and his associates first produced short radio waves by stimulated emission of radiation from excited ammonia molecules. Nikolai Basov and Alexander Prokhorov independently proposed a similar device, and these three scientists were awarded the Nobel Prize in Physics in 1964. In 1958, Arthur Schawlow and Townes showed how to extend the stimulated emission principle to light and combine it with the two-mirror structure to produce what is now known as a laser. The first lasers were operated in 1960.

KINDS OF LASERS

Whether it is free to move about in a gas or is imbedded in a solid or liquid, every atom or molecule has many possible energy levels. An atom or

Figure 5. (Above) Laser beams were used for precision alignment in the construction of BART (Bay Area Rapid Transit System) in San Francisco.

Figure 6. (Above right) Laser machining is accomplished by focusing a parallel beam of light into a small spot to create temperatures in excess of 6,000 degrees Centigrade in 1/2 millisecond. The rapid temperature rise causes difficult-to-machine metals to vaporize in the high energy spot. The plume beneath the microscope is vaporized high carbon steel.

a molecule can be excited to a particular level by absorbing light of the right wavelength, or frequency, as determined by the Planck relation,

Change in stored energy $= h \times \nu$

where h is Planck's constant and ν is frequency.

A laser material can be energized, or *pumped,* by a flash of bright light that has been filtered to emit only those wavelengths capable of exciting the atoms to the desired upper level. Sometimes a broad band of wavelengths, such as that comprising white light, can be used for pumping. This form of light is useful with solids or liquids, particularly if the material has a band of energy levels suitable for absorbing the light. When an atom is excited to any of these levels, it then gives up part of its energy to the solid or liquid and ends up in one particular excited state. This pumping method is used with a number of important materials, including rubies, in what is called the *solid state laser.*

The ruby used in lasers is a synthetically grown material of large aluminum oxide crystals that contain a small amount of chromium oxide. To make a solid state laser, a rod of good optical quality is cut from this material and the ends are polished flat and parallel to each other. One end is coated to reflect nearly all the light; the coating on the other end reflects part of the light and transmits the rest. The reflective coatings act as mirrors at the two ends of the rod. A light source either is positioned right next to the rod or is focused on it by a reflector. The light excites the atoms of the rod, and the atoms in turn emit light that passes back and forth between the mirrored end surfaces. A portion of this light makes its way out of the partially transmissive end after each passage. Optically pumped lasers of this kind can be operated continuously or, more easily, in short pulses.

A *gas laser* operates in a manner similar to the familiar neon sign. When electric voltage is applied across the tube, electrons are knocked out of the gas atoms and then speeded up by the electric field. The charge is the means of "pumping"; the quickly moving electrons strike other atoms and give up some energy, leaving these atoms in excited states. It may happen that the excitation of one particular energy level of the atom is favored in these collisions between electrons. In that case, more atoms are excited to that particular level than to some lower level and stimulated emission can occur to produce laser action.

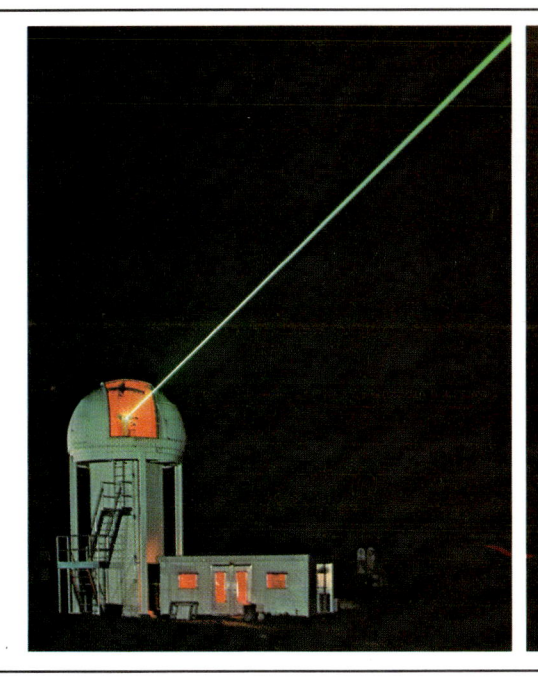

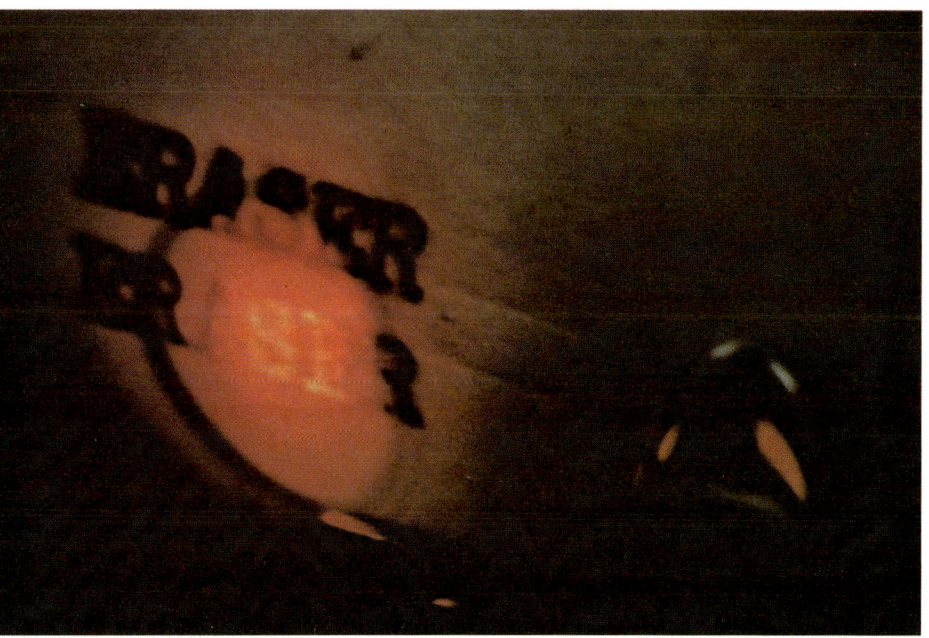

Sometimes the electric discharge is made to pass through a mixture of gases. Then the process may be more complicated. In the widely used helium-neon laser, the helium atoms attain a special "metastable" state. In other words, they achieve a relatively long-lived stability under a particular level of excitation. As a result, there are many more helium atoms in the metastable state during the discharge than in any other excited state.

The energy needed to raise the neon atom to an excited state is not supplied directly by an incoming photon; rather, energy becomes available when a neon atom collides with an excited helium atom. When that happens, the helium atoms deliver their stored energy to the neon atoms, which raises them to energy levels that are nearly the same as the metastable level of the helium atoms. During these collisions several lower energy levels of the neon atoms remain empty, so the conditions required for stimulated emission are satisfied. Helium-neon lasers generate a continuous beam of red light with a wavelength of 6.328×10^{-5} centimeter at a power level of about 1 milliwatt. They are commonly used for alignment of equipment and for laboratory experiments.

Electric discharges through many other gases produce laser action at some characteristic wavelengths. Carbon monoxide and especially carbon dioxide generate sustained infrared beams at power levels of hundreds and even thousands of watts. Their molecules also produce high-power laser action in a *gas dynamic laser,* which does not require an electric discharge. In such a device, gas heated to a high temperature is suddenly allowed to expand through a jet nozzle. It cools abruptly, and the result is that many molecules are excited to higher levels while lower energy levels are emptied. Continuous beams of many thousands of watts of infrared radiation have been produced by gas dynamic lasers.

Figure 7. (Left) The distance to the moon has been accurately measured with the use of a laser beam. This argon laser system is being tested at the Goddard Optical Test Range facility in Green Belt, Maryland for use as an earth beacon for orbiting spacecraft; it has been used as an earth beacon for the Surveyor spacecraft on the surface of the moon.

Figure 8. (Above) A laser eraser is fast and efficient but still too expensive for everyday use.

Figure 9. Laser beams are light waves, so no matter how close to parallel the waves may be when they start out, they eventually spread out. But in comparison with the beam of light sent out from, say, a searchlight, the angle of divergence of a laser beam is much smaller.

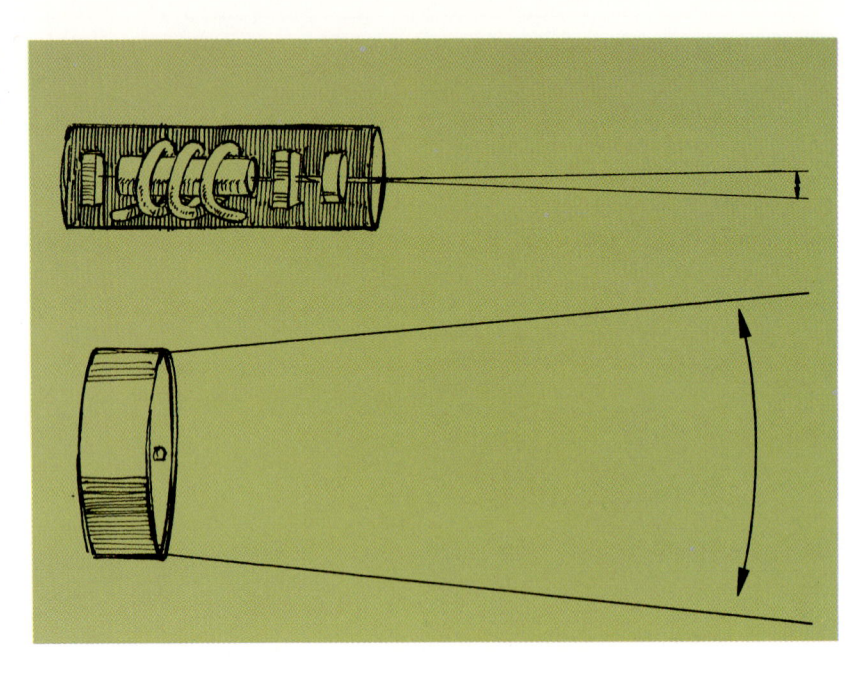

Electric discharges through semiconducting solids such as gallium arsenide also produce laser action. A small block of the material, typically smaller than a 1-millimeter cube, contains a flat junction between two regions with different kinds of impurities. The regions respectively donate positive or negative current carriers to the crystal. Laser action occurs along the junction layer when the large electric current is passed through the crystal. These *semiconductor lasers* are typically quite small and relatively efficient. They are particularly useful in portable, short-range optical communication devices.

In most lasers, the light produced by excited atoms or molecules emerges only in a few sharply defined wavelengths. But organic dyes in liquids or plastics can emit light over a broader range of wavelengths. When they are excited by a bright lamp or a powerful laser, these dyes can provide adjustable laser action. A prism inserted in the resonator between the amplifying dye and the end mirror can act as the tuning element in these *liquid lasers*.

PROPERTIES AND USES OF LASER LIGHT

The highly directional beam of laser light provides a nearly ideal straight line for all kinds of precision alignment. Lasers now guide huge earth-moving machines, for example, and can also be efficient surveying aids. Thus the laser becomes a sort of pathfinder, pointing the way.

Beam Divergence

But laser beams *are* light waves. And no matter how close to parallel these waves may be when they start out, they eventually spread out through diffraction. The divergence angle is small; its value (in radians, or the arc of a circle divided by the radius) is approximately the ratio of the wavelength of the light to the diameter of the beam. If, for example, the beam diameter is 0.3 centimeter and the wavelength is 0.00006 centimeter, an ideal laser beam will diverge by 1 meter, or about 3 feet, for every 5,000 meters of its

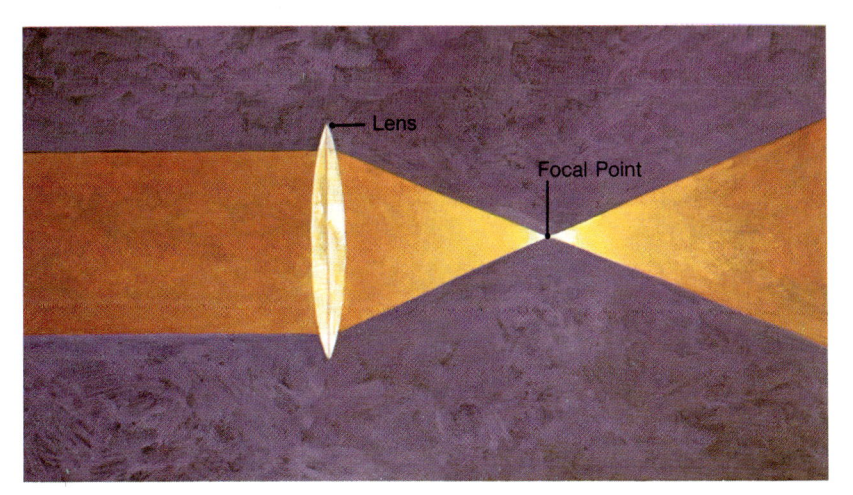

Figure 10. In order to reduce laser beam divergence even further, the light can be sent through a reversed telescope, which increases the beam's initial diameter but proportionally reduces the divergence angle. The diameter of the spot of a lens of a given focal length is approximately equal to the angle of divergence times the focal length.

path. Even though this divergence is within acceptable limits for many purposes, it can be reduced even further by passing the beam through a reversed telescope. The telescope increases the beam's initial diameter but at the same time it proportionally reduces the divergence angle. Ultimately, however, refinements are limited by fluctuations in the atmosphere, which impose a beam spread of at least 1 part in 100,000 for a beam traveling vertically through the atmosphere. This beam spread is equivalent to the diffraction produced by an aperture of 10 centimeters in diameter. Using a telescope with a larger aperture does not provide a gain in directionality.

Because of the spreading effect of the atmosphere, a beam of laser light sent from the earth to the moon, a distance of 384,400 kilometers (239,000 miles) would end up with a diameter of about 3–1/2 kilometers. This means that a normal-size mirror positioned on the moon would intercept only a small fraction of the light sent from even the best laser searchlight on earth. Curiously, the reverse is not true. A laser beam sent from the moon would be concentrated in a much smaller spot on earth. Do you know why?

A reflector actually has been positioned on the moon and it has intercepted small pulses of light from powerful, earthbound lasers. By measuring the time taken for the pulse to travel to the moon and back, scientists have measured the distance from a certain spot on the earth to a certain spot on the moon. These measurements are already accurate to within 1 foot. In fact, laser measurements are so precise that, if they are made over several years from stations on both sides of the Atlantic Ocean, it will be possible to determine the rate at which the continents are drifting apart.

Power Density

Because laser light is highly directional, it can be focused into a small area. And when a large amount of light energy is focused in a small area, the light intensity of the focal spot is correspondingly great (see Figure 10). For example, an argon ion gas discharge laser giving a continuous power output of about 1 watt, typically has a beam divergence of about 0.001 radian. Accordingly, the diameter of the spot produced by a lens of 1-centimeter focal length would be about 0.001 centimeter, and the corresponding area of the focal spot would be about 8×10^{-7} square

Calculating Continental Drift Using Lasers

Suppose that two stations on the earth's equator make simultaneous laser-ranging observations of the moon when it is over a point midway between them. One station is in South America and the other is in Africa. Each finds that it takes 2.5 seconds for a light pulse to travel from the earth station to a reflector on the moon and to return. They repeat their observations, under the same conditions, over a period of years. And each station finds that the time required for the laser to travel to the moon and back has increased by 8.0×10^{-9} second. How much has the distance between an observer and the reflector on the moon increased? If the moon is at an apparent angle of 60 degrees above the horizon for the observer on each continent, how far have the continents moved apart?

<u>Step 1.</u> The distance from each observer to the moon increases by

$$\tfrac{1}{2} \times \Delta t \times v$$

where Δt is the increase in the time it takes for the light pulse to travel to the moon and to return, and v is the velocity of light, or 300 million meters per second. Then the increase in distance is

$$\tfrac{1}{2} \times (8.0 \times 10^{-9} \text{ second}) \times (3.0 \times 10^{8} \text{ meters/second}) = 1.2 \text{ meters}$$

<u>Step 2.</u> The distance each continent has drifted is equal to the distance each station has moved. To determine this distance, draw a horizontal line and label one end of it position B. Then extend a line from B at an angle of 60 degrees. This line represents the increase in distance of the light path. If you use a scale of 1 inch = 1 meter, the length from B to C will be 1.2 inches.

To determine the position of A on the horizontal line, draw a line from C that is perpendicular to $\overline{BC}$. The point at which this line crosses the horizontal line is position A.

If you measure the length of the line from A to B on your diagram you find $\overline{AB}$ = 2.4 inches. Then the distance each station has moved is 2.4 meters, and the total distance of their separation is 2 × 2.4 meters, or 4.8 meters.

centimeter. The power density in such a focal spot would be more than *1 million watts per square centimeter.*

To understand what a fantastically high power density that is, consider the power emitted from the surface of the sun. The total power at all wavelengths is about 7,000 watts per square centimeter. Of course, the light emerging from the surface of a hot object such as the sun is emitted in all directions, so the intensity of focused sunlight can never exceed the intensity at the surface of the sun. In contrast, the light from a laser essentially comes out in only one direction and is therefore readily focused.

Even the earliest lasers were hotter than the surface of the sun because they could be focused into small spots with high power density. Of course, a laser cannot deliver light of this intensity to a large area because its total power is limited. But sometimes light energy concentrated in a small area is exactly what somebody wants. For example, small, pulsed lasers can drill holes in the tiny diamonds used for wire-drawing dies and rubies used in watch bearings. These lasers supply powerful bursts of heat to places that can be seen but not touched. Because they are so precise, such lasers are finding application in surgery on the retina of the eye. In still another application, clothing manufacturers employ computer-controlled carbon dioxide lasers to cut fabrics.

An object must absorb a substantial amount of laser light to become heated. If most of the light is reflected or transmitted, little heating will occur. For example, when dark ink on light-colored paper absorbs a flash of laser light, it becomes hot enough to vaporize completely. And yet the paper is unaffected—it merely reflects the light. Heating in this case occurs so swiftly that vaporization is nearly instantaneous; there is not enough time for heat from the ink to reach the paper by conduction. A laser eraser has been developed that can remove ink cleanly without damaging paper. Such devices are still prohibitively expensive, but laser erasers may soon be as common as electric pencil sharpeners and instant copying machines are today.

But the high density power of lasers has much greater implications. Extremely powerful lasers deliver brief pulses with peak powers of 10^{12} watts. With these devices, physicists are attempting to heat pellets of deuterium to a temperature high enough to produce thermonuclear fusion. So far they have created temperatures between 10 million and 100 million degrees Centigrade and have produced a few neutrons by thermonuclear processes. However, still larger lasers will be required to ignite a pellet that is large enough to release more energy than is put in. If controlled thermonuclear reactions can be achieved, they promise to provide nearly unlimited supplies of power while generating much less contaminating radiation than is produced by fission reactors.

Scattering and the Raman Effect

The well-defined, highly directional beam from a laser is easy to detect even when a small amount of its light is scattered as it passes through a medium. In fact, any light not traveling in the original direction will have been scattered by the medium. A laser radar can be used to detect and measure the amount of dust in the atmosphere at various distances and heights by observing how much light has been scattered out of the beam by the dust.

Because laser light is also monochromatic, even small shifts in wavelength from the original frequency can be detected and measured

(1) Incoming Light — Vibrating Molecule

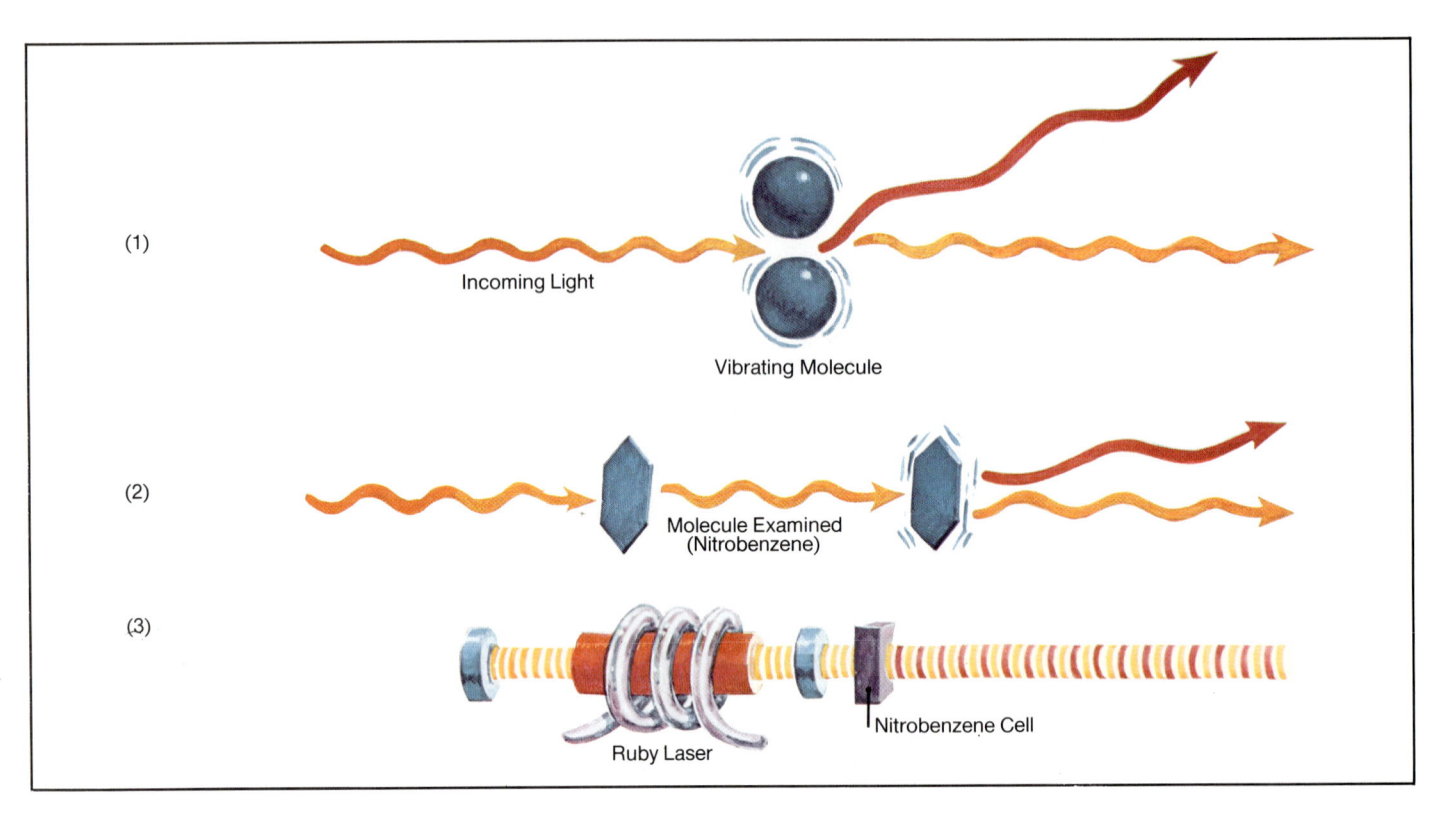

(2) Molecule Examined (Nitrobenzene)

(3) Ruby Laser — Nitrobenzene Cell

Figure 11. In Raman scattering, a substance that is to be examined is placed in the path of a laser beam. The change in frequency of the laser light is a direct measure of the vibrational frequency of the molecules that scatter it, and because each frequency identifies a particular kind of molecule, Raman scattering can be used to analyze samples of transparent substances to learn about their molecular structure.

when scattering occurs. In the chemical laboratory especially, lasers are now used as sources for what is known as "Raman scattering." Light beams from these lasers are sent pulsing through some substance, and the change in frequency of that light provides a direct measure of the vibrational frequency of the molecules that scatter it. Because each frequency is characteristic of a particular kind of molecule, the Raman effect can be used to analyze small samples of transparent substances, such as hydrocarbons, to learn about their molecular structure. Even the kinds of particles that are present in the atmosphere can be detected with moderately powerful pulsed lasers. Shifts in wavelength of the back-scattered light can be measured over considerable distances. For example, the percentage of water vapor present in the atmosphere has been measured at distances greater than 1 kilometer.

Raman scattering may also be used to detect small concentrations of harmful pollutants at distances of several hundred meters. Molecules of the major polluting gases—the nitrogen oxides, carbon dioxide, sulfur dioxide, and the hydrocarbons—absorb characteristic frequencies in the infrared portion of the spectrum. Because a laser can produce light of a single frequency, it can be tuned to different wavelengths to identify the infrared emissions of pollutants found in the atmosphere. When the laser beam encounters a pollutant molecule, the molecule absorbs infrared energy and weakens the light reaching a detector that has been positioned in the path of the beam.

One automobile manufacturer hopes to use diode lasers, tuned for each of the major pollutants, to test exhaust from cars on the production line. Some molecules absorb the laser light and become excited enough to emit radiation at their own characteristic wavelength. A spectrograph that is sensitive to radiation of those wavelengths is connected to a telescope that scans the path of the laser beam. The spectrographic device

Figure 12. Ordinary light is *incoherent*; its wave trains move out in all directions and interfere randomly with one another. Laser light is *coherent*; all the wave trains are in step.

picks up the induced radiation from the pollutant molecules in the path of the beam. When experimental systems are refined, it may be possible to pinpoint exactly where and in what concentrations polluting substances are being released into the atmosphere.

Coherence

The light produced by lasers not only is monochromatic, directional, and relatively powerful; it is also coherent. A coherent light wave is analogous to the orderly waves sent out in water when a plank is moved slowly up and down in it. Incoherent light can be likened more to the disorderly ripples produced by raindrops on the surface of a pond. In an ordinary light source, millions of atoms independently and randomly emit a wave as they are excited. At a given point not too far from the light source, the wave front fluctuates randomly and there is no persistent relationship between the vibrations at two different points. No wonder this random jumble of rapid fluctuations is called "incoherent."

In a laser, however, the excited atoms are not emitted randomly but are stimulated by the wave stored between the resonator mirrors. The atoms are forced to contribute to one highly directional wave that has well-defined wavefronts. The crests and troughs of this wave follow one another in regular succession.

If two waves have definite phases, they can add up in phase to provide a disturbance of twice the amplitude. But if coherent waves are brought together in such a way that their phases are opposite, they interfere destructively with one another. In the case of light waves, this means that light added to light can produce the absence of light — total darkness. It seems contrary to everyday experience that two light beams can come

27 | Lasers: The Light Fantastic

together to produce darkness. But all ordinary light is incoherent: the phases of its waves fluctuate relative to one another. These waves oscillate in and out of phase so rapidly that the eye sees only an average, uniform light intensity over an illuminated area.

Interference of light, particularly coherent light, can be used to measure small or large distances accurately. Consider what happens when a light beam is separated into two parts, as it can be by a partially transparent mirror. If the two parts travel different paths before being recombined, they will add up either constructively to produce a bright spot, or destructively to produce a dark one. If, then, the length of the path traversed by one of the two beams is subsequently changed by one-half wavelength of light (an amount ranging from 20 to 35 millionths of an inch, depending on the color), what was formerly a dark region will become a light one, and vice versa. Interference methods can be used to measure distances of much less than one-half wavelength.

LASERS AND LENSLESS PHOTOGRAPHY

A most fascinating application of the interference of light is the making and viewing of holograms. Dennis Gabor was awarded the 1971 Nobel Prize in Physics for his discovery of holography, which he first described in 1948. Holography is an ingenious method of making photographs that reproduce a truly three-dimensional view of an object. Sometimes called "lensless photography," holography produces *images,* not pictures. With the invention of the laser, the techniques of holography have improved rapidly in recent years.

A hologram looks like a half-exposed piece of film. When you illuminate it by reasonably coherent light from a laser or other source, the hologram becomes a magic window. Each of your eyes views the hologram from a different angle and, as a result, they reconstruct the subject of the hologram in three dimensions. When you move your head, the angle changes and so does the view of the holographic subject. You can even look around corners or behind obstacles that obscure other objects from one viewpoint.

Although the actual hologram looks like a nearly uniform blur, a powerful microscope would reveal a complex pattern of lines and spaces on the film. As coherent light passes through these spaces it bends in such complicated ways that it actually reconstructs what you would see if you were looking at the subject through a window. When light from a distant street lamp comes to your eyes after it passes through two narrow slits in a sheet of metal foil you see a row of dots; when it passes through a small circular hole you see a pattern of rings. But when light passes through the lines and spaces of the hologram, it produces the appearance of an actual object. In other words, in one case the original subject is used to produce an interference pattern, and in the other the interference pattern is used to recreate the object.

A hologram is the photographic record of an interference pattern obtained by the method shown in Figure 13. A lens spreads out the beam from a laser so that both a mirror and the subject of the hologram are illuminated. Scattered light from the subject falls directly on the photographic film without going through additional lenses, as it would in ordinary photography. A portion of the beam reflected by the mirror also falls on the film and uniformly illuminates it. The hologram records not only the light waves emanating from the subject but also the light waves, or the *reference beam,* from the mirror. The reference beam and the scattered

Figure 13. (Far right) Holograms are made in the following manner. Two lasers send out light of three distinct wavelengths—red, green, and blue—all of which are contained in both the reference beam sent directly to the hologram plate and in the signal beam diffracted by the object. The film emulsion of the plate is thick enough to behave as a three-dimensional volume, and the interface patterns generated by the three sets of wavefronts in both beams are recorded at different depths within the film. When the hologram is illuminated with a replica of the original reference beam, a full-color image appears with the "three-dimensional" properties characteristic of the object being photographed.

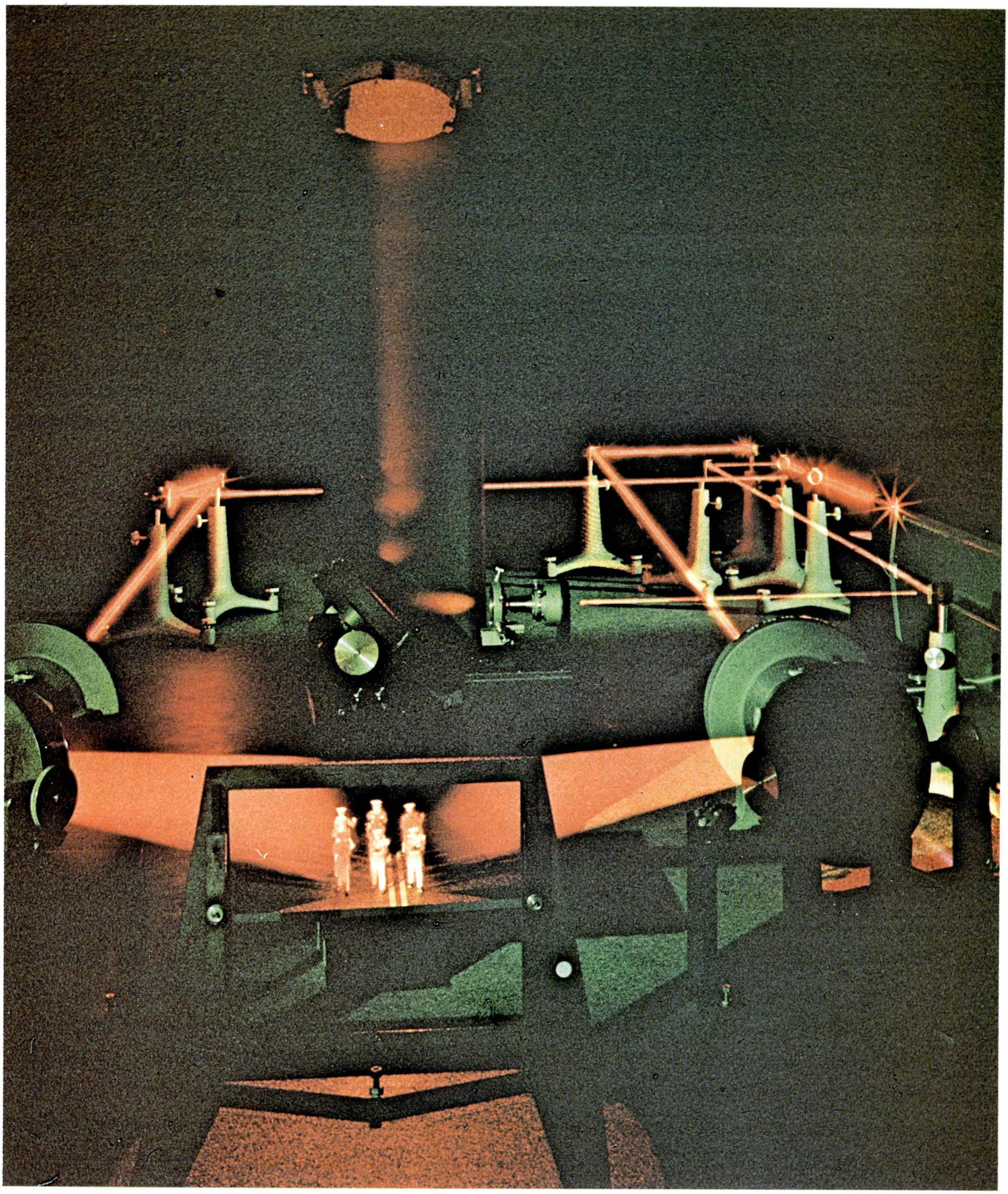

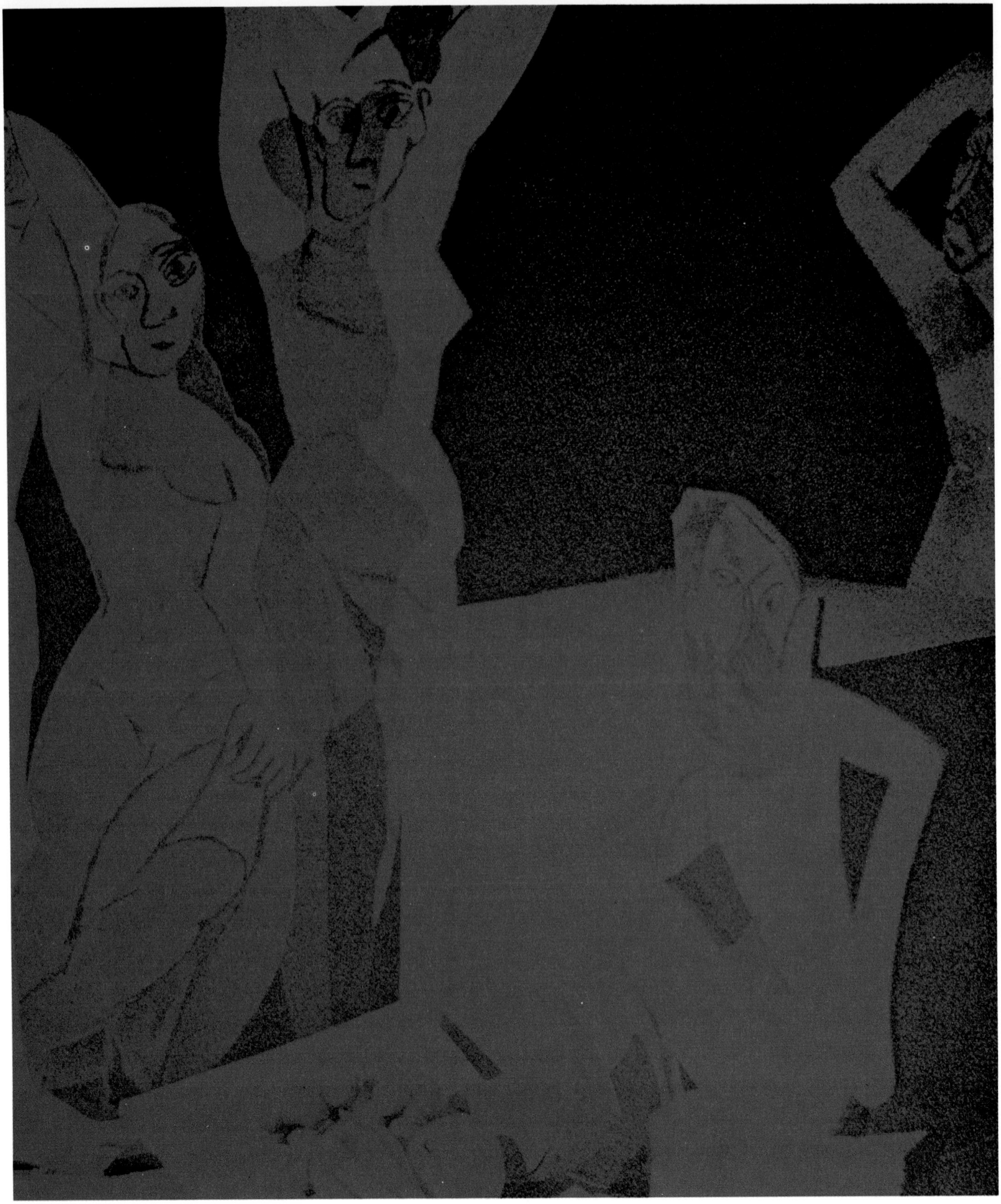

Figure 14. (Far left) Holography is also finding application in the arts: here a painting by Pablo Picasso is recreated in three dimensions.

Figure 15. (Left) A photograph of a double-exposed hologram shows the finger pressure deformation of the surface of a spacecraft antenna. Each fringe is a region of constant optical path length change; neighboring fringes correspond to a path difference of 0.35 micron.

light from the subject interfere with each other, producing the complex interference pattern that is recorded on the photographic film. When developed, the film becomes the hologram.

The hologram described above produces a *virtual image* — one that must be seen directly rather than by projection onto a screen. However, a hologram can also project a *real image* onto a screen when a small portion of it is illuminated by a beam of laser light. In viewing a hologram a virtual image appears to be located behind the film. In projecting a real image, the subject appears to be located in front of the film.

The number of uses for holograms grows steadily. A method has been proposed for storing large amounts of information on very small holograms. The information is not stored as an array of points, as in an ordinary photograph, but as an array of lines and spaces. On a tiny photograph, a small speck of dust can obliterate a large part of the picture. But on a hologram each small area contains a diffused encoding of the entire object, so that obliterating part of it only causes a slight blurring of the image. A potential application for holography lies in the realm of television. It may be possible to store television images for home use as a series of tiny holograms on a strip of film. In another application, images of, say, credit ratings are encoded on holograms as an array of dots to be projected onto a bank of small photocells for deciphering. Such information storage techniques will be useful for a computer memory of extraordinary capacity.

Other modifications in holography could make use of the thickness of the emulsion on the photographic film. The wavelength of light is so small that a sensitive layer on an ordinary photographic film is many wave-lengths thick. Like oil on puddles, there is room on the emulsion for a layer pattern on the scale of the wavelength that can selectively reflect certain colors while rejecting others. By taking advantage of this effect it is possible to make holograms in color.

One further use for holograms evolved because they can encode a three-dimensional image in considerable depth. When the hologram is reconstructed, your eyes can focus at various distances in the reconstructed scene. In other words, holograms offer a much greater depth of field than ordinary photography. Suppose you wanted to photograph a chamber full of fog droplets. With holography you would later be able to count and measure individual droplets at various depths. The depth of field is particularly valuable when holograms are made through a microscope.

Holograms also record information about the relative phases of waves from different parts of the subject when they reach the photographic plate. When the hologram is made, the beam from the mirror is a standard reference for the phase of the scattered light. If a double-exposed hologram is made when the subject is being moved slightly between the exposures, dark and light interference fringes can be detected in the reconstructed image. These fringes indicate in units of one-half wavelength of light how much the object has moved or deformed in the interval between the exposures. This procedure provides a sensitive means to test for small flaws in automobile tires.

Many more potential applications of holography are now being tested and will become practical when improved, less expensive lasers and recording media are available. One application often suggested is three-dimensional television broadcasting. This sort of broadcasting may some day be possible, although it seems likely that it will require an enormously wide band of radio frequencies to transmit all the information needed to reproduce the details in three dimensions. But perhaps some day you will be able to flip a switch to activate your home laser beam and television receiver, thereby producing three-dimensional "living" theater all around you.

LASER SAFETY

The truly powerful lasers could seriously injure someone standing in their path. The high-power carbon dioxide lasers and solid state lasers are especially dangerous and must be operated under conditions that prevent people from being struck by the laser beam. Inflammable materials, too, must be kept out of the path of the beam.

Most lasers in use, however, produce low-power beams. Helium-neon lasers, for instance, are widely used for aligning equipment and making precise measurements. The intensity of the light they produce is not high enough to cause injury directly, but it is a highly directional beam and it can be focused onto a small spot. The lens of the eye can focus this parallel laser light to a small, intense spot on the retina, which will damage this sensitive tissue. The laser light reflected from a mirror or even a shiny surface can be focused just as easily and is almost as dangerous as the direct beam. When working with such lasers it is essential to avoid looking directly into the beam or into its reflection. Pulsed lasers are even more hazardous than continuous wave lasers because of their high peak intensity. There is no time to respond to the sudden flash by closing your eyes or turning away.

There is little hazard, however, when the light of a low-power, continuous wave laser is used to illuminate a diffuse surface. Light scattered by the object is spread in all directions and the eye at any reasonable distance can intercept only a small fraction of the light. In fact, most lasers are far from being the death rays of science fiction. With reasonable precautions they can be used safely for many purposes. Safety standards and operating procedures are under study by various government public health and safety organizations, and high-power lasers are subject to regulations in many areas.

Lasers of the future may be used to stimulate nuclear reactions to provide a much-needed power source. Or they may assist a surgeon in delicate surgery impossible to perform with conventional instruments. Holograms may store information on a single piece of film that now fills a whole shelf of library books. And they may make it convenient to set up a central information storage center containing the vital statistics of an entire population. As in other areas of scientific research, the experiments or devices themselves are not intrinsically good or bad. Their applications must be a matter of interest and concern outside as well as within the scientific community.

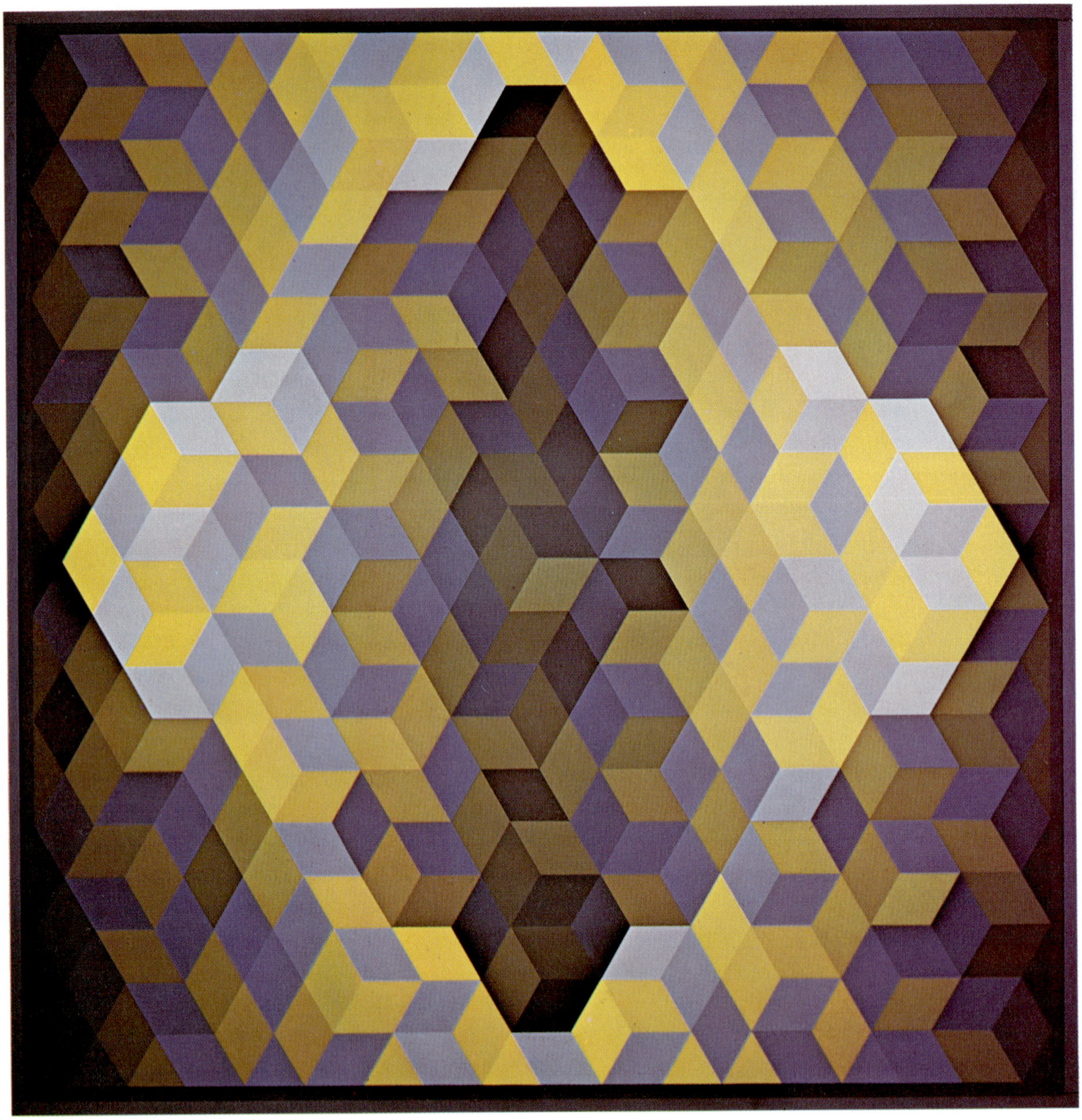

MEH-2
Victor Vaserely (1908–   )

If we were to say that reality resides only in matter, we would be hard pressed to explain the recognizable but intangible human awareness of the abstract as well as the material world. Even so, until recently the organization of physical phenomena was the undisputed domain of science, the fertile outpouring of the human consciousness was the domain of the arts and humanities, and it seemed that the two never would share the same philosophical framework. In the new quantum mechanics, however, they have come together to form a new philosophy. A reasonable interpretation of the quantum theory is that our understanding of the world is contained in the knowledge of the *probability* of space-time occurrences of mass and energy. These two "things" should not exist in the human mind as material objects according to this most recent scientific philosophy—a philosophy that is compatible with present observations.

The quantum theory blends our perceptions of the universe with the phenomenon we "observe" through measurements. The artist Vaserely uses positive and negative space to construct such perceptions of reality, both real and illusory. Each space of *MEH-2* has its own vitality, yet each complements the other to produce a new, integrated space. Even though the shapes of *MEH-2* seem familiar enough, the subtle interaction of these shapes causes us to ask if the "reality" of one is more probable than the "reality" of the other.

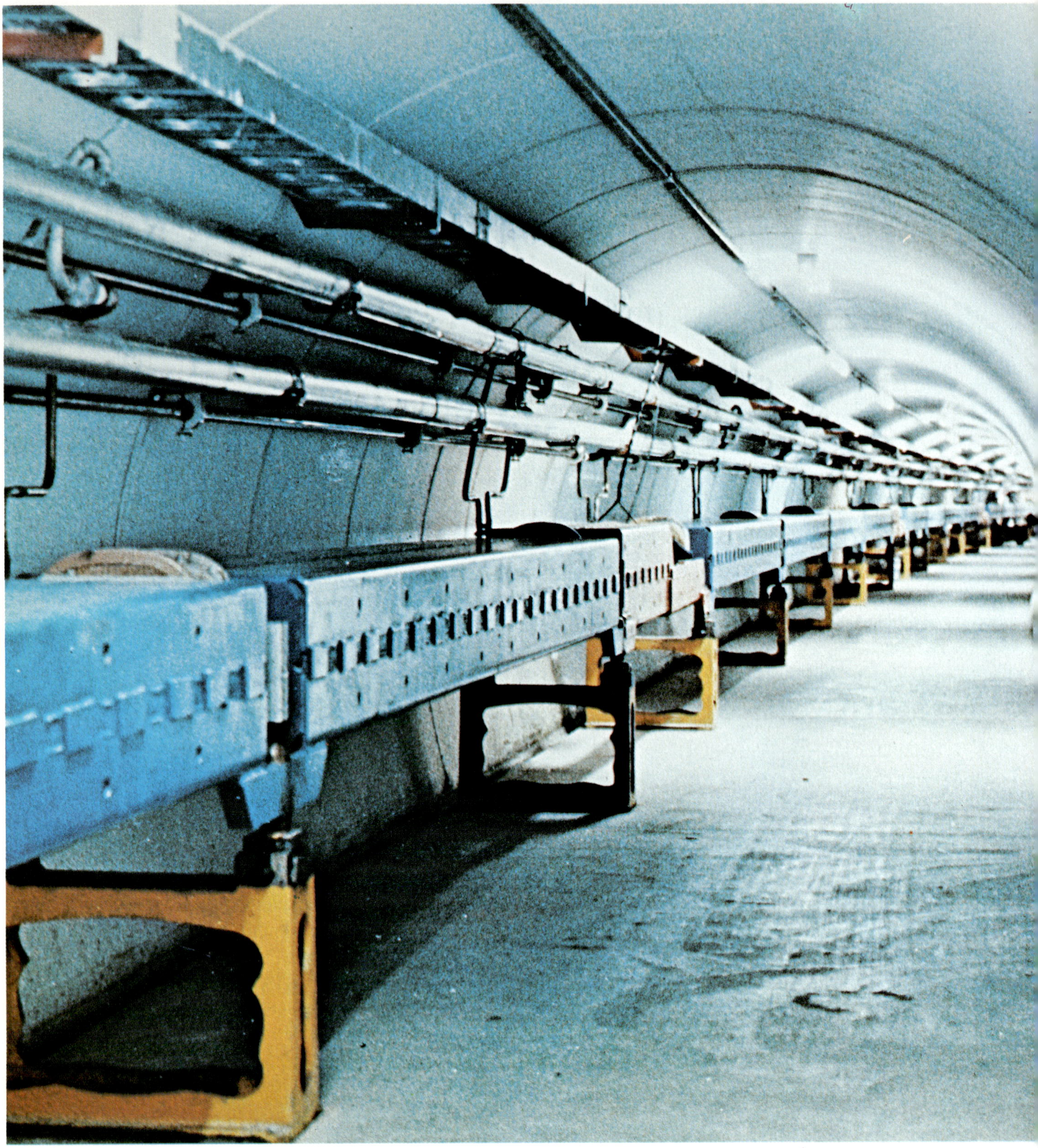

By now, you are familiar with the idea that an atom consists of a dense, small, positively charged nucleus and light, negatively charged electrons, and that the nature of each kind of atom is determined by the behavior of its electron cloud. But can anything be found in the nucleus itself, which is 10,000 times smaller than the atom? Remarkable as it may seem, we are in the process of unraveling the secrets of the internal structure of that incredibly small atomic core.

Today we know the nucleus is composed of two sorts of particles—neutrons and protons—of approximately the same mass. The main difference between them is that neutrons are electrically neutral and protons have a +1 charge. To emphasize their similarity, neutrons and protons in nuclei are often referred to collectively as *nucleons*. The nucleon number is merely the total number of protons *and* neutrons present in a nucleus. Moreover, both the neutron and the proton have a mass very close to 1 atomic mass unit, so the nucleon number is close to the total mass of a nucleus. For that reason the nucleon number is usually called the *mass number*. In nuclear science, the letter A stands for the mass number, Z for the number of protons, and N for the number of neutrons. A simple formula ties them together: $A = Z + N$.

Because the *atomic number* signifies the electric charge carried by a nucleus (or an element), it also signifies the number of protons in the nucleus. When you talk about a nucleus, both its atomic number and its mass number must be described because the two together indicate the number of protons and the number of neutrons. There is a convenient way to do this. A small subscript number may be written to the left of the name of an element to designate its atomic number, and a small superscript number may be written above that to designate its atomic mass. For example, the nitrogen nucleus with seven protons and seven neutrons is written $^{14}_{7}$nitrogen; the nitrogen nucleus with seven protons and eight neutrons is written $^{15}_{7}$nitrogen. Because the neutron and the proton both have masses close to 1 atomic mass unit, the mass number is numerically close to the atomic mass of the nucleus. For nitrogen 14 the mass number is 14 and the mass is 14.0032 atomic mass units.

The chemical nature of a nucleus is determined by how many protons it contains. Every nucleus that has seven and only seven protons in it can only be a nucleus of nitrogen; if it has eight protons it can only be oxygen; and if it has six protons it can only be carbon. But the number of *neutrons* has nothing to do with the chemical nature. In fact, the number of neutrons for a certain proton number has been found to vary. For example, nitrogen nuclei may have six neutrons, or seven neutrons, or eight neutrons, and so on. Nuclei that vary in this way are *isotopes*.

Frequently the atomic weight of many of the chemical elements is not close to a whole number. Consider chlorine, which has an atomic weight of 35.5 atomic mass units. Naturally occurring chlorine consists of two types of chlorine isotopes: 75 percent have a mass number of 35, and 25 percent have a mass number of 37, so there are 3 times as many atoms of mass 35 as of mass 37. Consequently,

$$\frac{(35 \times 3) + (37 \times 1)}{4} = \frac{142}{4} = 35.5$$

In those relative abundances the isotopes produce the observed mass.

28
Inside the Nucleus

Inside the world's largest particle accelerator in Batavia, Illinois.

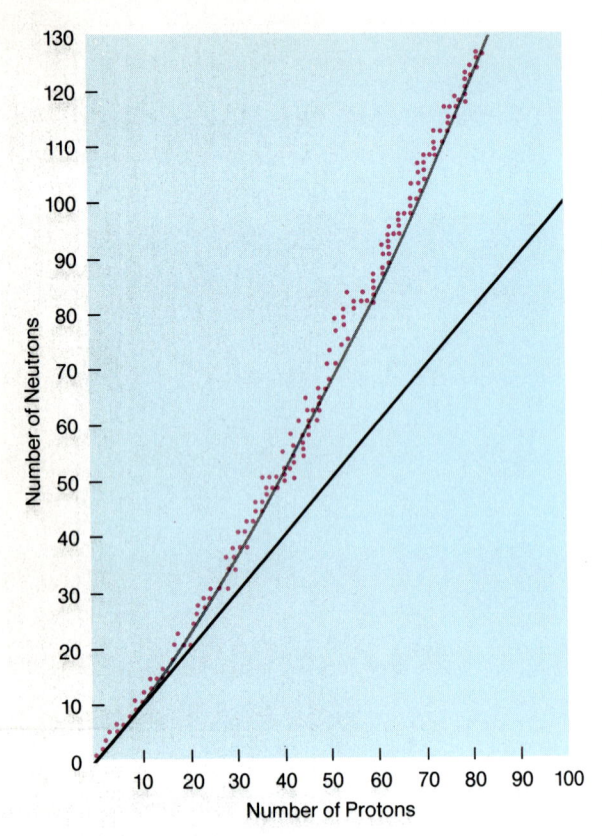

Figure 2. In this graph, the number of neutrons is plotted against the number of protons for stable nuclei (greater than 10 percent abundance). The solid line corresponds to N = Z, or equal numbers of protons and neutrons. The excess neutrons provide attraction without coulomb repulsion and therefore stabilize the heavier nuclei.

NUCLEAR STABILITY AND THE NEUTRON/PROTON RATIO

Some nuclei transform spontaneously into other nuclei. For example, the carbon isotope with mass number 14, written as $^{14}_{6}$carbon, transforms into the nitrogen isotope of mass number 14, or $^{14}_{7}$nitrogen. And the isotope $^{238}_{92}$uranium changes into the isotope $^{234}_{90}$thorium. In such spontaneous transformations, the nuclei are *radioactive* and we say that they undergo *radioactive decay.* The rate at which radioactive atoms decay varies greatly. Some decay so rapidly that any total sample is transformed in less than 1 second, and yet others decay so slowly that even after millions of years only a fraction of a sample has been transformed. There are about 300 different nuclei which, within the limits of present measuring methods, are not observed to decay at all. They are known as the *stable nuclei.* Nuclei having measurable decay rates are called *unstable.*

Of the 300 stable nuclei, about 200 have an even number of protons and an even number of neutrons. They include $^{4}_{2}$helium, $^{16}_{8}$oxygen, $^{120}_{50}$tin, and $^{208}_{82}$lead. Far fewer have an even atomic number and an odd neutron number (such as $^{17}_{8}$oxygen and $^{115}_{50}$tin), or an odd atomic number and an even neutron number (such as $^{39}_{19}$potassium or $^{121}_{51}$antimony). There are between 60 and 70 of each of those two kinds of nuclei. Very few stable nuclei have both an odd atomic number and an even atomic number. Only four are known—$^{2}_{1}$hydrogen, $^{6}_{3}$lithium, $^{10}_{5}$boron, and $^{14}_{7}$nitrogen—and they are very light. When you compare all these numbers you can see at once that even numbers of nucleons are related to the stability of nuclei, and that instability to radioactive decay is promoted when either or both protons and neutrons are odd numbers.

The preference of nature for even numbers rather than odd numbers is even more marked if you consider the number of stable isotopes for the elements. Those elements of even atomic numbers frequently have between four and eight stable isotopes. The element tin, with fifty protons, has ten stable isotopes of mass numbers 112, 114, 115, 116, 117, 118, 119, 120, 122, and 124. On the other hand, elements with odd values of the atomic number have only one and at most two stable isotopes. Indium, with forty-nine protons, has stable isotopes of mass numbers 113 and 115; for antimony, with fifty-one protons, the stable mass numbers are 121 and 123.

Figure 2 shows the ratio of the number of neutrons in the stable nuclei as a function of atomic number. For the lighter elements, nuclear stability is favored not only when there are even numbers of neutrons and protons but when there is an equal number of each. As the atomic number increases, the ratio of the neutrons to protons increases above a value of 1, reaching a value of about 1.6 for uranium. Why does the ratio increase? The answer lies in the fact that each (positively charged) proton exerts a repulsive electric force on all the other protons in the nucleus. As the number of protons in the nucleus increases, this total repulsive electric force increases. But because the nucleus holds together despite this repulsive force, there must also be an *attractive* force present—a force whose origin is independent of the proton's electric charge. Nuclei also contain neutrons, so this attractive nuclear force must exist not only between protons, but also between neutrons and protons, and between neutrons and other neutrons. The addition of neutrons, then, serves to increase the attractive force and thereby stabilize the nucleus.

NUCLEAR SIZE

How are neutrons and protons arranged in a nucleus? The size of the nucleus provides a clue to the answer. The simplest model you can think

544 Physical Science Today

Figure 3. A packing model of incompressible nucleons shows that the volume of a nucleus is proportional to the total number of nucleons.

about is one that has the neutrons and protons simply "packed together" like marbles or steel balls in the smallest possible volume. In other words, the nucleons are not compressible but they pack together so that the void space is minimal. If this model is valid, the volume of the nucleus should be proportional to the total number of nucleons (assuming protons and neutrons are the same volume). The nuclear volume, then, is related to the number of nucleons. And if you assume the nucleus has a spherical shape and remember that for a sphere the volume is proportional to the cube of the radius, you know the cube of the radius is *also* proportional to the number of nucleons. This simple relationship, based on a model of close packing of incompressible nucleons, agrees fairly well with experimental data on nuclear sizes.

We have learned that the total nuclear size is on the order of 10^{-12} centimeter, or about 10,000 times smaller than the atom. When nuclear sizes are studied in even finer detail, it turns out that not all nuclei are truly spherical. Some are spheroidal, like the shape of a football. The major axis in those nuclei can be 20 percent to 30 percent longer than the minor axis.

NUCLEAR ENERGY

In a typical chemical system, the energies involved in a reaction that forms a compound from its elements are on the order of 100 kilocalories per mole of product. For example, the reaction carbon + oxygen → carbon dioxide + energy releases 94.03 kilocalories per mole of carbon dioxide formed. A similar reaction for nuclei is measured not in kilocalories but in energy units of electron volts or of million electron volts. We could express the formation of carbon dioxide in electron volts, in which case it would correspond to approximately 4 electron volts per atom. In contrast, the formation of $^{4}_{2}$helium from two protons and two neutrons would liberate an energy of 28.3 million electron volts. Clearly, nuclear reactions are approximately a million times more energetic than chemical reactions.

Albert Einstein expressed the relationship between the change of energy and mass in a system as $E = mc^2$, where E is the energy change, m is the mass change, and c^2 is the square of the velocity of light in a vacuum. This equation may be used to calculate the change in mass (or the change in energy) in chemical and nuclear equations. For the formation of a mole of carbon dioxide (44.009 grams) from a mole of graphite (12.011 grams) and a mole of oxygen gas (31.998 grams), it is impossible with our present techniques to detect any mass change (mass of 1 mole carbon dioxide = mass of 1 mole carbon + mass of 1 mole

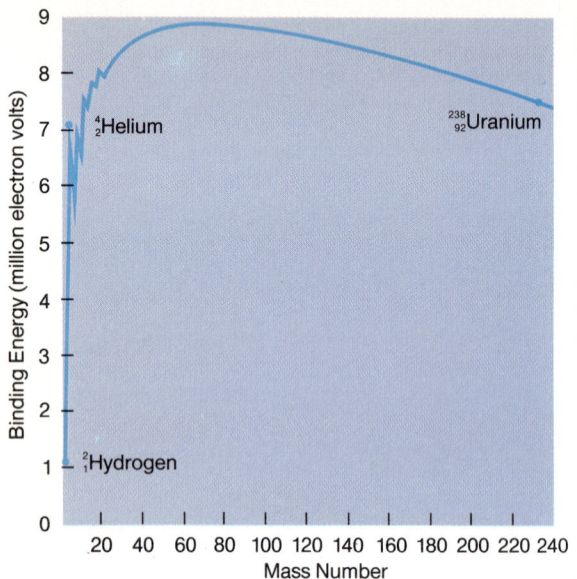

Figure 4. The graph of binding energy per nucleon as a function of mass number is rather erratic for small mass number. When the mass number becomes large enough to produce statistical effects, the graph becomes smooth. The most stable nuclei are those at about mass number 60, such as iron and nickel.

Figure 5. (Far right) In the *fusion* process, energy is released when light nuclei combine into heavier nuclei, as shown in the above drawings. In the *fission* process, there is an increase in the value of binding energy per nucleon when a heavy nucleus is decomposed into lighter nuclei, and this dissociation releases energy.

R. B. Leachman, "Nuclear Fission." Copyright 1965 by Scientific American, Inc. All rights reserved.

oxygen). The amount of energy released in this reaction is 94.0 kilocalories per mole of carbon dioxide formed. And from Einstein's equation, it is associated with a mass decrease of 4.4×10^{-9} gram, which is much too small to be detected at present. In contrast, the energy changes and, hence, the mass changes in nuclear reactions are often a million times larger per mole of product than in chemical reactions. The formation of a mole of $^4_2$helium (4.0026 grams) from 2 moles of neutrons (2×1.0087) and 2 moles of $^1_1$helium (2×1.0078) would have a mass change of $4.0026 - (2.0174 + 2.0156)$. In other words, there would be a decrease in mass of 0.0304, which can be measured easily on an ordinary analytical balance. This value corresponds to 28.3 million electron volts of energy released per nucleus of $^4_2$helium formed.

BINDING ENERGY

The energy released when a mole of a chemical compound is formed from its constituent elements is known as the *heat of formation*. The same energy is the minimum required to decompose the compound back to the elements. The energy released when a nucleus forms from its component nucleons is known as the *binding energy*. Conversely, the binding energy is the amount of energy required to make a nucleus break apart into individual neutrons and protons. Referring to the preceding paragraph, the binding energy of $^4_2$helium is 28.3 million electron volts.

Just as the heat of formation measures the chemical stability of the molecule, so the binding energy measures the stability of a nucleus. It tells how difficult it would be to decompose that nucleus in terms of energy requirements. Rather than using the total binding energy (*BE*), nuclear scientists often use the average binding energy per nucleon (*BE/A*) as a convenient measure of the stability of the nucleus. For $^4_2$helium, the average binding energy per nucleon is 28.3/4, or 7.1 million electron volts; for $^2_1$hydrogen it is 2.2/2, or 1.1 million electron volts; for $^{238}_{92}$uranium it is 18.017/238, or 7.6 million electron volts. Figure 4 shows the values of binding energy per nucleon for the stable (nonradioactive) nuclei as a function of their mass number.

The curve in Figure 4 ascends to a broad maximum at a mass number of about 60, which implies that the most stable nuclei are those having a mass number close to that value, such as iron and nickel. Because the most abundant elements in the universe are hydrogen and helium and *not* iron and nickel, the universe obviously has not reached an energy condition of maximum stability. This observation provided important clues for theories about the origin of the elements.

The curve in Figure 4 also tells us another significant thing about nuclear stability. When the very light nuclei are combined to form a heavier nucleus, the value of binding energy per nucleon increases. The heavier nucleus is more stable than the two light, separate nuclei. For light nuclei, then, energy is released by combining or fusing nuclei into heavier ones. Such *exothermic fusion* reactions of light elements are the basis of the hydrogen bomb and are under study as a source of energy in thermonuclear reactors. For heavy nuclei there is an increase in the value of binding energy per nucleon when the heavy nucleus is decomposed into lighter nuclei. Therefore, in the region of the heavy nuclei, dissociation or *fission* of the heavier nucleus releases energy. These energy-releasing fission reactions of the heaviest elements are the source of energy in atomic bombs and in nuclear reactors.

A final feature of this curve is that, except for the very lightest stable nuclei, the values of binding energy per nucleon vary only between

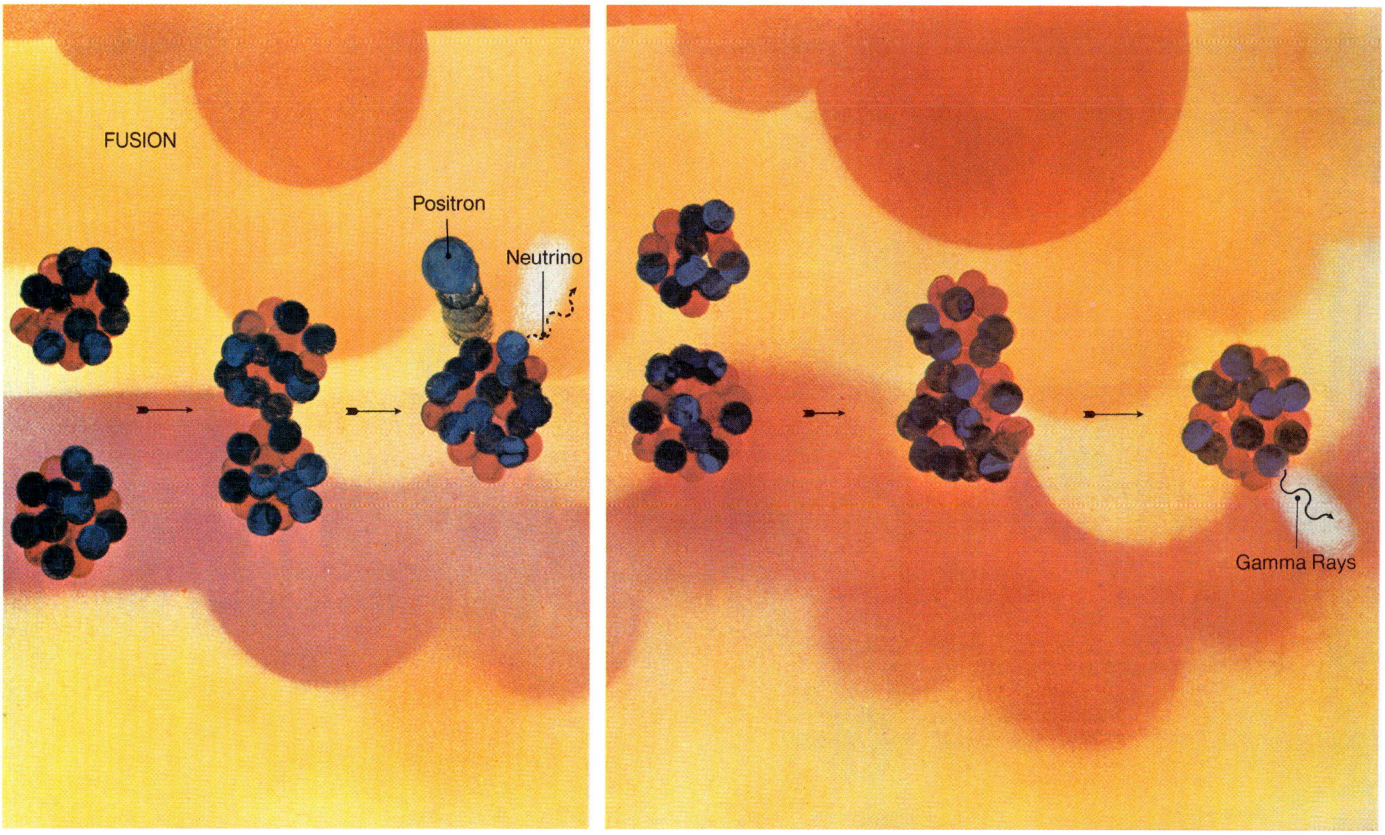
FUSION
Positron
Neutrino
Gamma Rays

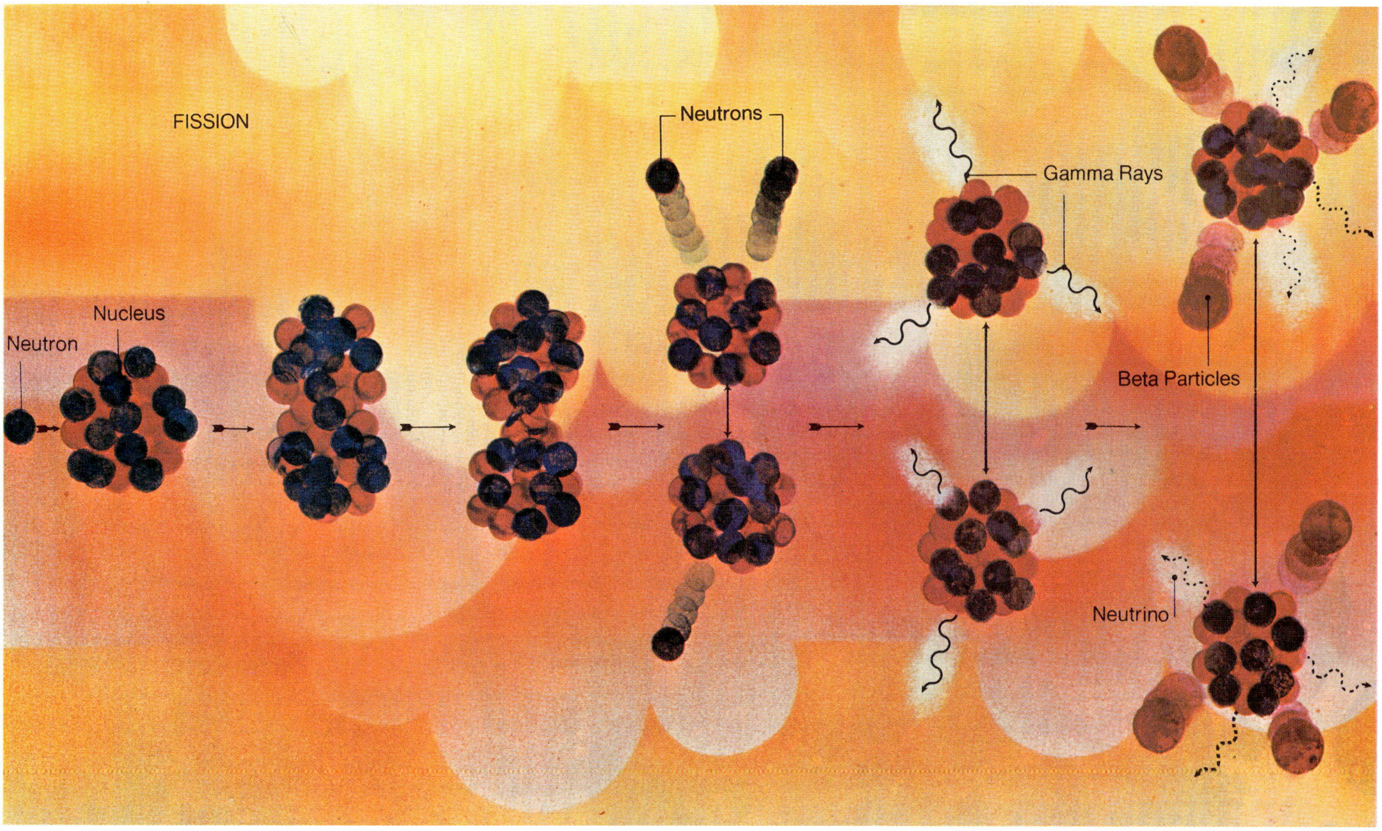
FISSION
Nucleus
Neutron
Neutrons
Gamma Rays
Beta Particles
Neutrino

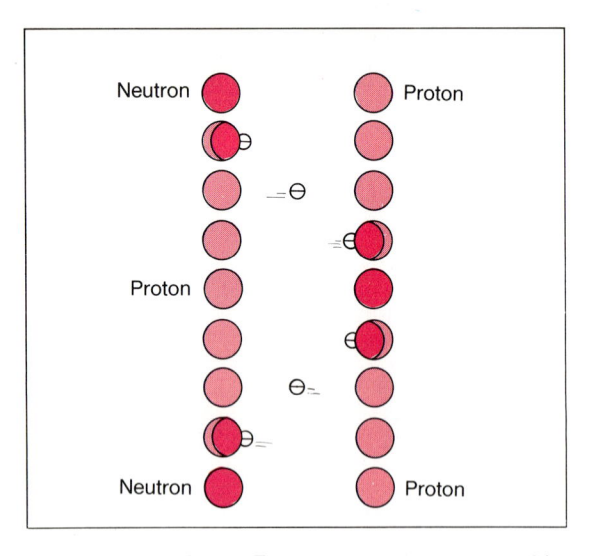

Figure 6. (Above) The exchange of mesons provides the attractive force between nucleons. Here a neutron and proton share a meson, and in the process they switch back and forth to proton and neutron. (Below) Hideki Yukawa of Kyoto University gained world-wide attention in 1935 with his startling prediction of the existence of an elementary particle that is now known to be the meson.

7 million and 9 million electron volts. Therefore, to a first approximation it is almost constant. This means that the total binding energy is roughly proportional to the total number of nucleons. As you shall see in the next section, that is a basic piece of information about the nature of nuclear attractive forces.

NUCLEAR FORCES

The attractive forces in stable nuclei must be stronger than the proton repulsive forces; after all, the nucleons do stick together to form these nuclei. So far, a complete understanding of the attractive nuclear forces has eluded nuclear scientists, but many characteristics of those forces have been observed in a wide variety of experiments.

Nuclear forces operate over much shorter distances than electric or gravitational forces. The electric repulsion between nuclear protons exist between each proton and all the others in the nucleus. For a nucleus with Z protons, the total energy of electric repulsion would be proportional to $Z \times (Z - 1)$, or approximately Z^2.

Precise measurements on the interaction between two protons or a proton and a neutron indicate that the distance over which the nuclear force is felt is of the order of 10^{-13} centimeter. In other words, in a nucleus containing many nucleons, nuclear attractive forces exist only between those nucleons that are immediately adjacent. Now you can understand why the values of binding energy per nucleon are roughly constant. If each nucleon in a nucleus shared nuclear forces with all the other nucleons present, the total binding energy would be expected to be proportional to the number of nucleons A times the number of other nucleons, $A - 1$, to which each was attracted, or $A \times (A - 1)$. This is approximately A^2. So the total binding energy would be proportional to A^2 and BE/A^2 should equal a constant value. If nuclear forces are shared only by adjacent nucleons, however, the total binding energy will be proportional to the total number of nucleons, giving a nearly constant value of BE/A. Because the latter is the true case, you can assume nuclear forces are shared *only* between adjacent nucleons.

The second major characteristic of nuclear forces is that they are charge-independent. The nuclear force of attraction is the same between two protons, two neutrons, or a neutron and a proton. However, between two protons the electric repulsive force is also present and it slightly decreases the net attractive force. Experimentally, this charge-independent character is demonstrated by bombarding hydrogen gas with neutrons and protons and observing the scattering, just as Ernest Rutherford studied the scattering of alpha particles. When the scattering of the protons is corrected for electric repulsion, the scattering is almost identical for both neutrons and protons, which shows that the two have the same nuclear force.

In 1937 a Japanese physicist, H. Yukawa, proposed a theory that attempted to explain nuclear forces as the result of an exchange of an unknown particle between the adjacent nucleons. In his theory, the two nucleons experience an attractive force because they share the exchanging particle, in somewhat the same way that atoms are held together by a covalent chemical bond in which electrons are shared by the two atoms. Yukawa calculated that the "exchanging" particle must have a mass approximately 200 times that of an electron. In 1947 the *pion* (pi meson), with a mass equal to 273 electron masses, was discovered. These pions, as well as other, more recently discovered mesons, are now believed to be the Yukawa exchange particles.

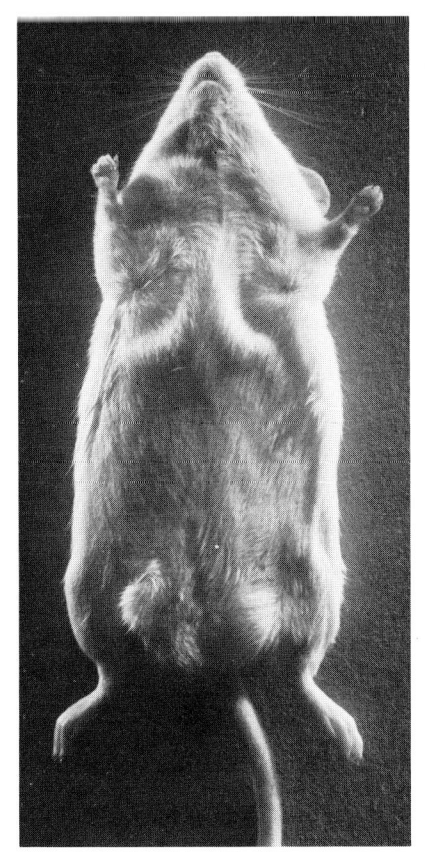

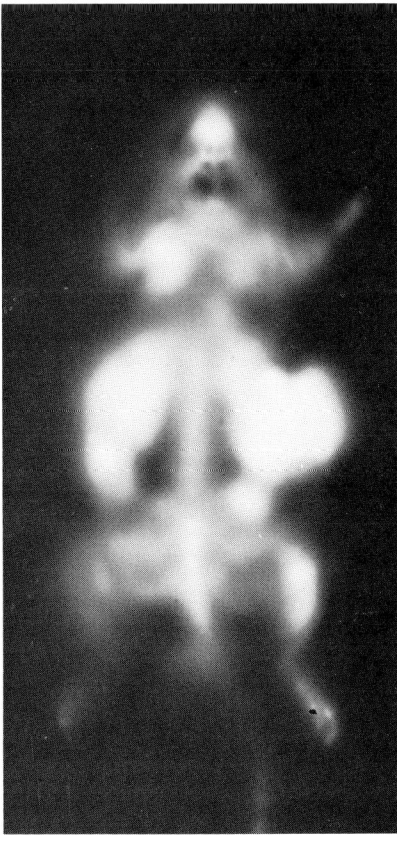

Figure 7. Henri Becquerel made his experiments in 1896 by exposing a photographic plate to uranium salt. The same principle is used today to obtain x-ray photographs by injecting an organism with a radioactive substance.

TYPES OF RADIOACTIVE DECAY

The radioactive decay of unstable atoms is immensely valuable in such fields as chemistry, physics, biology, geology, oceanography, agriculture, and environmental studies. Following Henri Becquerel's discovery of radioactivity (Figure 7), it was soon learned that three distinct types of emissions could be discerned in radioactive decay. These emissions were assigned the names "alpha," "beta," and "gamma." In 1940 a fourth mode of decay, spontaneous fission, was discovered.

All nuclei heavier than bismuth are energetically unstable to decay by emission of alpha particles. An alpha particle is identical to the nucleus of helium, and its emission causes a nucleus to lose two protons and two neutrons. When that happens, the atomic number decreases by 2 units and the mass number decreases by 4 units. The nuclear reaction for alpha decay of the isotope of radium of mass number 226 is written as:

$$^{226}_{88}\text{radium} \longrightarrow {}^{222}_{86}\text{radon} + {}^{4}_{2}\text{helium}$$

Figure 8. Henri Becquerel (1788–1878).

The general term *beta decay* actually refers to three different radioactive decay processes. In nuclei having an excess of neutrons for the number of protons (their value of A is larger than that of the nonradioactive isotopes), the nucleus undergoes decay whereby the nuclear charge is increased by 1 unit but the mass number does not change. Effectively, a neutron is converted to a proton. Simultaneously a negative electron is created and emitted from the atomic system. Electrons, as such, do not exist in nuclei but nuclear changes may result in the creation or destruc-

Figure 9. Beta decay processes involve the emission or absorption of an electron (positron) and a neutrino. The three basic reactions are β^- decay, β^+ decay, and electron capture.

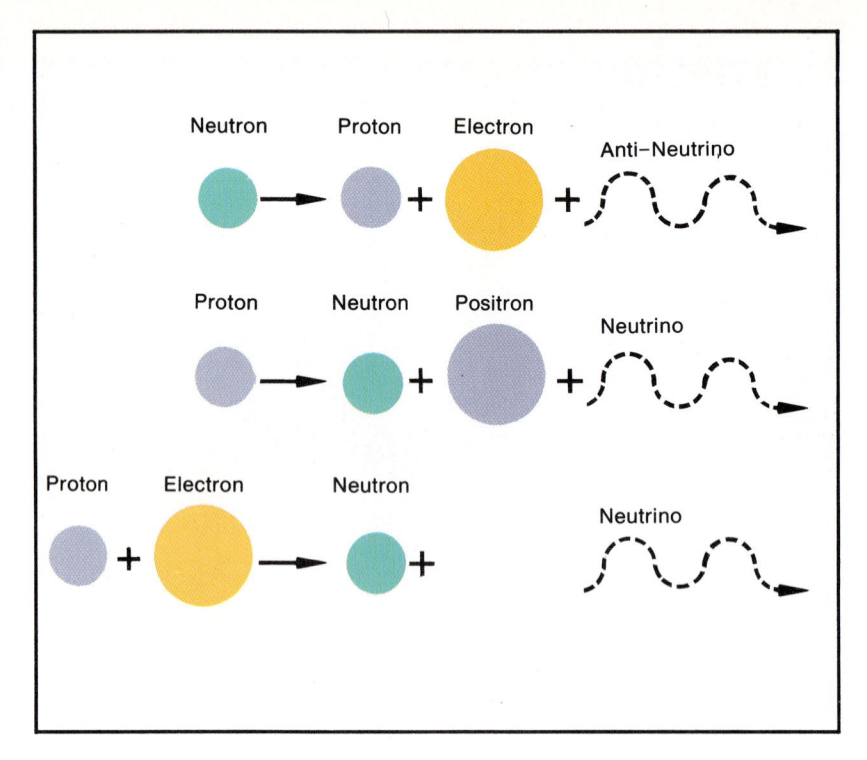

tion of electrons. The emitted electron is designated by β^-, as in the following reaction:

$${}^{14}_{6}\text{carbon} \longrightarrow {}^{14}_{7}\text{nitrogen} + \beta^-$$

If the nuclide has a value of A less than that of its stable isotopes, it has too few neutrons for that atomic number. A beta decay process may occur which removes a positive charge from the nucleus; again, there is no change in the mass number. Effectively, a proton is converted to a neutron. The positive charge is removed by creation and emission of a positive electron, or a *positron*, designated by β^+. A sample reaction is

$${}^{22}_{11}\text{sodium} \longrightarrow {}^{22}_{10}\text{neon} + \beta^-$$

An alternate process to positron emission involves absorption by the nucleus of an extranuclear, orbital electron. This is known as electron capture. A sample reaction is

$${}^{195}_{79}\text{silver} \xrightarrow{\text{electron capture}} {}^{195}_{78}\text{platinum}$$

The absorption of the negative charge into the nucleus effectively changes a proton to a neutron with no change in mass number, so electron capture changes the nucleus in the same fashion as positron emission. In the lighter elements, β^+ decay occurs whereas in the elements heavier than platinum, electron capture is predominant. In the elements of middle weight, both processes occur.

In all the beta decay processes, *no change in the mass number occurs.* Decay by β^- emission increases the atomic number by 1 unit; positron decay and electron capture decrease it by 1 unit. All three modes of beta decay involve emission of a *neutrino,* which is a particle of no charge and of a mass corresponding only to its velocity.

Another type of radioactive decay is observed for the very heaviest elements such as californium. In this decay, known as *spontaneous fission,* the nucleus splits into two roughly equal parts and simultaneously releases a great amount of energy. For $^{252}_{98}$californium, of every 100 nuclei that decay approximately 3 do so by spontaneous fission. The rest undergo alpha decay.

Gamma ray emission is observed also in radioactive decay. These are a form of high-energy electromagnetic radiation similar to x-rays and light rays. Neither the atomic nor the mass number of the emitting atom are changed by gamma ray emission. Emission of gamma rays results from the rearrangement of the neutrons and protons in a nucleus into lower energy states.

HALF-LIFE OF RADIOACTIVE DECAY

The rate of radioactive decay is directly proportional to the amount of radioactive species present. In other words, it follows the equation $A = \lambda N$, where A is the decay rate expressed as the number of disintegrations in some unit of time, N is the number of radioactive atoms in the sample, and λ is called the decay constant. Radioactive decay can be classified by a *half-life.* The half-life of a radioactive species is the time period for the amount of material initially present to be reduced by one-half. For example, the half-life of iodine 131 is 8 days. If a 1-gram sample of iodine 131 were isolated, in 8 days there would be only 0.5 gram of it left; in another 8 days only 0.25 gram would remain, and so on. Because the half-life is a nuclear property, it is unaffected by differences in temperature, pressure, or chemical state. This property of a constant half-life is valuable in dating archaeological and geologic material.

Carbon 14 decay was one of the earliest nuclear time clocks used. Now $^{14}_{6}$carbon is continuously produced in the atmosphere by the reaction of neutrons from cosmic rays with nitrogen atoms:

$^{14}_{7}$nitrogen + $^{1}_{0}$neutron $\longrightarrow$ $^{14}_{6}$carbon + $^{1}_{1}$hydrogen

The $^{14}_{6}$carbon decays to $^{14}_{7}$nitrogen:

$^{14}_{6}$carbon $\longrightarrow$ $^{14}_{7}$nitrogen + β^-

with a half-life of 5,800 years. Because all evidence indicates that the rate of formation of $^{14}_{6}$carbon and its rate of decay have been in balance for over 50,000 years, the proportion of $^{14}_{6}$carbon in the carbon found in carbon dioxide in the air—and in live organisms that use carbon dioxide—is constant.

When the organism dies, it no longer adds new carbon 14, but the old carbon 14 continues to decay. By comparing the proportion of carbon 14 of the total carbon content in a "dead" object such as a piece of wood (or charcoal) with the corresponding proportion of carbon 14 in a piece of wood from a living tree, it is possible to calculate the time that has elapsed since the "death" of the old material. $^{14}_{6}$Carbon dating can be used for objects up to about 50,000 years old. If the age exceeds 50,000 years, decay of the $^{14}_{6}$carbon has proceeded so far (almost nine half-lives) that the decay rate is too small to detect accurately. Linen wrappings from the Dead Sea Scrolls were found to be 1,917 ±200 years old, which definitely ruled out theories that attempted to date them much later than some historians had hoped. By dating charcoal, rope sandals, and other organic material in prehistoric sites around the United States, it is possible to show that man inhabited this country at the very least 11,500 years ago. In Africa, some caves were inhabited over 43,000 years ago (Table 2).

In some nuclear dating methods, there is no constant production of radioactive species so it is impossible to know how much radioactive isotope was present in the sample when it assumed its present form. In

Table 1.
Radioactive Decay Rates

Radioactive Atom	Half-Life
$^{238}_{92}$Uranium	Uranium I
	$\alpha \downarrow$ 4.5 × 10^9 years
$^{234}_{90}$Thorium	Uranium X$_1$
	$\beta \downarrow$ 24 days
$^{234}_{91}$Protactinium	Uranium X$_2$
	$\beta \downarrow$ 1.1 minutes
$^{234}_{92}$Uranium	Uranium II
	$\alpha \downarrow$ 250,000 years
$^{230}_{90}$Thorium	Ionium
	$\alpha \downarrow$ 80,000 years
$^{226}_{88}$Radium	Radium
	$\alpha \downarrow$ 1,620 years
$^{222}_{86}$Radon	Radon
	$\alpha \downarrow$ 3.8 days
$^{218}_{84}$Polonium	Radium A
	$\alpha \downarrow$ 3.0 minutes
$^{214}_{82}$Lead	Radium
	$\beta \downarrow$ 26.8 minutes
$^{214}_{83}$Bismuth	Radium C
	$\beta \downarrow$ 19.7 minutes
$^{214}_{84}$Polonium	Radium C
	$\alpha \downarrow$ 1.6 × 10^{-4} second
$^{210}_{82}$Lead	Radium D
	$\beta \downarrow$ 22 years
$^{210}_{83}$Bismuth	Radium E
	$\beta \downarrow$ 5.0 days
$^{210}_{84}$Polonium	Polonium
	$\alpha \downarrow$ 138 days
$^{206}_{82}$Lead	Lead

Figure 10. By comparing the amount of carbon 14 remaining in a creature that no longer is animate with the amount present in its living counterpart, you can calculate the time since its death. The amount of carbon 14 is determined by the rate of radioactive decay (beta disintegration) per gram of carbon. Many significant archaeological findings have been dated in this manner.

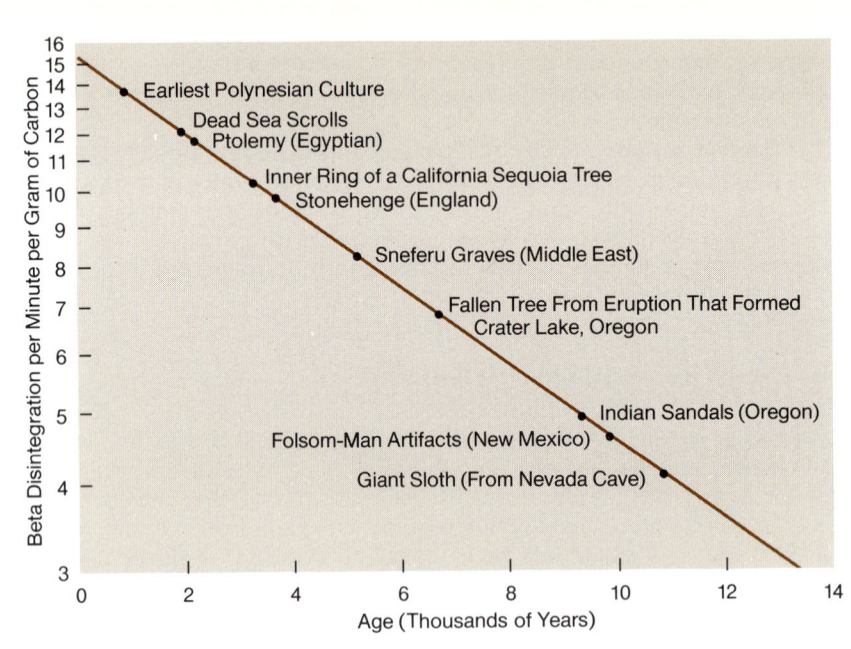

such cases, however, it is often possible to measure both how much radioactive isotope is still present and how much of the decay product is present. We may assume there was no decay product present when the sample was formed; its behavior would be chemically different from that of the element from which it forms by radioactive decay. For example, $^{87}_{37}$rubidium undergoes radioactive decay to $^{87}_{38}$strontium with a half-life of 4.7×10^{10} years. Therefore, the amount of $^{87}_{38}$strontium present is a measure of how much of the original $^{87}_{37}$rubidium has decayed. From this information and the half-life of the decay of the $^{87}_{37}$rubidium, the age of the rubidium-strontium containing material can be determined. It is because it has such a long half-life that rubidium-strontium dating is useful for geologic samples taken from the earth or the moon.

DETECTION OF RADIATION

As the emissions from radioactive decay travel rapidly through matter, they transfer energy to the atoms and molecules. Chemical bonds in the molecules break because of this energy transfer. In addition to the bond breakage, electrons are ionized from the atoms. In biological systems, the bond breakage from the passage of the radioactive particles as well as secondary bond breakage from the energetic electrons produced in the ionization cause damage and disruption. Exposure to high levels of radiation results in sickness, even death. On the other hand, some foods can be preserved for long periods without refrigeration by killing bacteria with radiation. Radiation can also be used to change the chemical bonding in some chemical systems to produce new, more desirable products.

Alpha particles, which travel 10,000–20,000 miles per second, have the least penetrating ability of the radioactive emissions. About 5 centimeters of air, a sheet of paper, or a thin sheet of aluminum will stop them. Beta particles, electrons, liberated at speeds more than six times that of alpha particles, require several meters of air or several millimeters of aluminum to absorb them. Gamma rays have still greater penetrating power—tens of centimeters of aluminum are often required to stop them (Figure 11).

If the molecules that are ionized by the passage of alpha, beta, or gamma rays are part of a gas between two electrodes, the ions and the

Figure 11. Relative penetration power of alpha particles, beta particles, and gamma rays. (Alpha particles are *helium nuclei*, beta particles are *positrons*, and gamma rays are *photons*.)

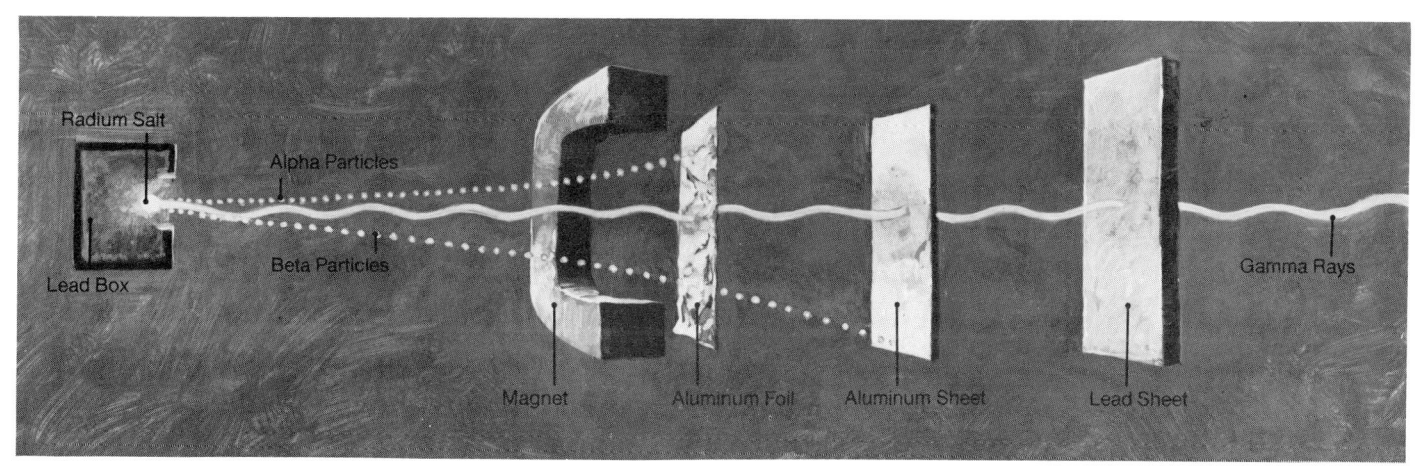

Table 2.
Basic Measurement Methods

Method	Material	Time Dated	Useful Time Span (years)
Carbon-14	Wood, peat, charcoal	When plant died	1,000–50,000
	Bone, shell	Slightly before animal died	2,000–35,000
Potassium-argon	Mica, some whole rocks	When rock last cooled to about 300° Centigrade	100,000 and up
	Hornblende Sanidine	When rock last cooled to about 500° Centigrade	10,000,000 and up
Rubidium-strontium	Mica	When rock last cooled to about 300° Centigrade	5,000,000 and up
	Potash feldspar	When rock last cooled to about 500° Centigrade	50,000,000 and up
	Whole rock	Time of separation of the rock as a closed unit	100,000,000 and up
Uranium-lead	Zircon	When crystals formed	200,000,000 and up
Uranium-238 fission	Many	When rock last cooled	100–1,000,000,000 (Depending on material)

electrons can be moved apart before they recombine. The electrons are accelerated to a positively charged electrode, where their collection causes a minute flow of current. The current is amplified and registered in auxiliary equipment. This process is the basic principle of operation of Geiger counters.

One type of Geiger counter consists of an argon-filled tube containing a metal cylinder and a thin wire. A high voltage is maintained between the two electrodes, and when an electron enters the tube and collides with the gas molecules, the collisions produce many ions and free elec-

Radioactive Dating

One of the most reliable methods for dating organic archaeological remains such as charcoal or wood is using the isotope carbon 14. Its half-life is 5,568 years. The farthest back in time we can date with this technique is about 40,000 years, for beyond this date, the amount of carbon 14 left is too small to work with. You could apply this technique to dating an early Indian circular campsite in the Southwest desert.

<u>Step 1.</u> Early man cleared these circular campsites on the rocky desert floor and built his sage and pine firepits. Now an archaeologist, digging in the sand, finds there are three distinct layers of charcoal remains, each directly above the other. He proceeds to date the times of the campsite usage.

<u>Step 2.</u> In his carbon decay counter, which measures the ratio C^{12}/C^{14}, he finds the rates for the three samples:

Sample number	Ratio C^{12}/C^{14}
1	½/1
2	3/1
3	5/1

From these ratios you can determine the dates.

<u>Step 3.</u> When one atom of carbon 14 decays, it forms one of carbon 12. Now, assuming there was no carbon 12 to start with in the burnt wood, all carbon 12 has to be from the decay of carbon 14. Using this assumption, you can get the ages by simple ratios. For sample 1, the ratio C^{12}/C^{14} is ½/1, or ½. You know that when half of the sample has decayed, or $C^{12}/C^{14} = 1/1$, 5,570 years have passed. So the date of Sample 1 would be:

$$\frac{1}{5,570 \text{ years}} = \frac{½}{x \text{ years}}$$

$$x = ½ \times 5,570 = 2,785 \text{ years}$$

For Sample 2:

$$\frac{1}{5,570 \text{ years}} = \frac{3}{x \text{ years}}$$

$$x = 3 \times 5,570 = 16,710 \text{ years}$$

For Sample 3:

$$\frac{1}{5,570 \text{ years}} = \frac{5}{x \text{ years}}$$

$$x = 5 \times 5,570 = 27,850 \text{ years}$$

So you can see that the site was occupied intermittently by hunters, and that the earliest level perhaps represents indications of the first groups of men to migrate down the American continent from the north.

Table 3.
Reactions Used to Synthesize Transuranium Elements

Target	Projectile	Product	Emitted Particles
$^{238}_{92}$uranium	+ $^{2}_{1}$helium	⟶ $^{238}_{93}$neptunium	+ 2 neutrons
$^{238}_{92}$uranium	+ $^{4}_{2}$helium	⟶ $^{240}_{94}$plutonium	+ 2 neutrons
$^{239}_{94}$plutonium	+ $^{4}_{2}$helium	⟶ $^{241}_{95}$americium	+ proton, neutron
$^{239}_{94}$plutonium	+ $^{4}_{2}$helium	⟶ $^{240}_{96}$curium	+ 3 neutrons
$^{244}_{96}$curium	+ $^{4}_{2}$helium	⟶ $^{254}_{97}$berkelium	+ proton, 2 neutrons
$^{238}_{92}$uranium	+ $^{12}_{6}$carbon	⟶ $^{245}_{98}$californium	+ 5 neutrons
$^{238}_{92}$uranium	+ $^{14}_{7}$nitrogen	⟶ $^{247}_{99}$einsteinium	+ 5 neutrons
$^{238}_{92}$uranium	+ $^{16}_{8}$oxygen	⟶ $^{250}_{100}$fermium	+ 4 neutrons
$^{253}_{99}$einsteinium	+ $^{4}_{2}$helium	⟶ $^{256}_{101}$mendelevium	+ neutron
$^{246}_{96}$curium	+ $^{12}_{6}$carbon	⟶ $^{254}_{102}$nobelium	+ 4 neutrons
$^{252}_{98}$californium	+ $^{10}_{5}$boron	⟶ $^{257}_{103}$lawrencium	+ 5 neutrons
$^{249}_{98}$californium	+ $^{12}_{6}$carbon	⟶ $^{257}_{104}$rutherfordium	+ 4 neutrons
$^{249}_{98}$californium	+ $^{15}_{7}$nitrogen	⟶ $^{260}_{105}$hafnium	+ 4 neutrons

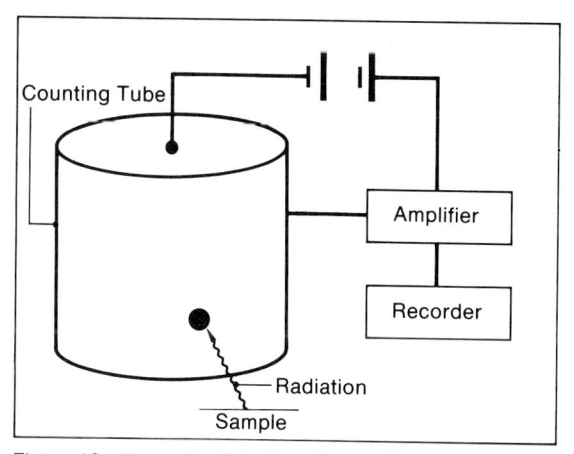

Figure 12. A counter detects and records radioactive decay that causes ionization in the gas of the counting tube.

trons. The electrons move rapidly to the positively charged wire, where they cause a pulse of current. The small pulse of current is amplified in the electronic circuit and is recorded as a count on a register, as a movement of a needle on a dial or as a click in a headphone (Figure 12).

Semiconductor materials such as silicon and germanium are also used for radiation detection and measurements. These solid state devices resemble gas counters because the ionization produced by the passage of radiation results in a pulse of current. Other solid and liquid detectors operate in a different manner. The passage of radiation in these crystals—for example, sodium iodide, or anthracene—causes small flashes, or scintillation, of light. The scintillations are detected by sensitive photomultiplier tubes and converted to electronic pulses, which may be recorded. The counters using this type of detection system are known as scintillation counters.

NUCLEAR REACTIONS

Although a number of radioactive nuclides are found in nature, many more have been made in nuclear reactions. Well over 1,000 new radioactive nuclides have been produced and studied. Target nuclei have been bombarded with neutrons from nuclear reactors or with high-energy protons, deuterons, alpha particles, carbon ions, oxygen ions, and heavier ions produced in accelerators. Table 3, which lists some reactions that have been used to produce the synthetic elements heavier than uranium, illustrates the rich variety of nuclear reactions studied by nuclear chemists and physicists.

If a beam of protons is shot at a target to cause a nuclear reaction, no reaction occurs unless the protons have a certain minimum

Figure 13. An accelerator tube gives a proton an energy boost of 5 million electron volts by the use of two oppositely charged rings.

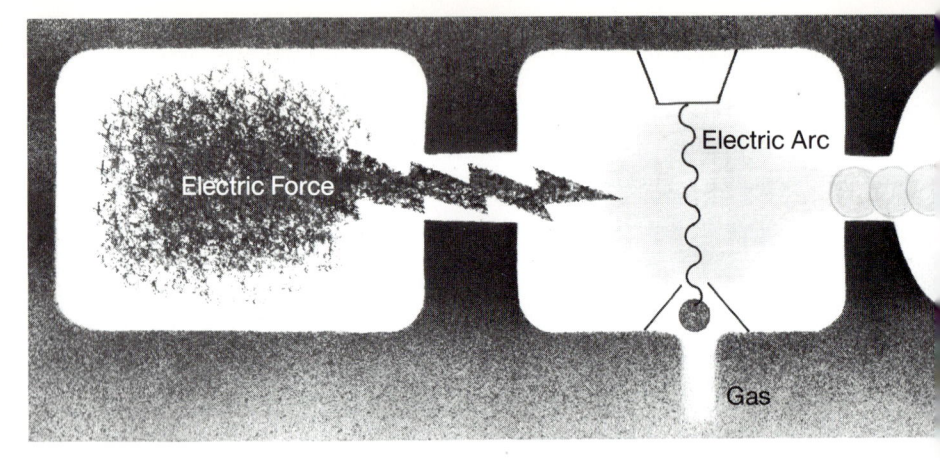

kinetic energy. This *activation energy* is approximately twice as large for alpha particles as it is for protons for the same target. In addition, it increases as the atomic number of the target increases. This behavior is consistent with a model in which the minimum kinetic energy is the amount needed to overcome the repulsion to close approach that exists between a positively charged projectile (such as a proton) and the positively charged target nucleus. An alpha particle with a +2 charge has approximately twice as much repulsion to overcome as a proton with a +1 charge. For the target, the greater the atomic number (the number of protons), the greater the repulsive effect on an approaching positive particle. This hypothesis of a repulsive barrier to reaction by charged particles is strengthened by the observation that neutrons, bearing no charge, react with target nuclei even when the kinetic energy of the neutrons is essentially zero.

ACCELERATORS

In order to impart sufficient kinetic energy for reaction to take place, physicists have developed various types of charged-particle accelerators. The basic principle is similar in most types; the particle is accelerated by being attracted to an electrode of opposite charge. If protons are placed at the entrance of a hollow tube, which has 0 voltage at the entrance and 5 million volts at the exit end, the protons are attracted to the negative end and acquire 5 million electron volts of kinetic energy in moving from the entrance to the exit end of the tube.

In cyclotrons and synchrotons, the particles are made to go in circles by imposing a magnetic field, which causes charged particles to bend in their path. Again, acceleration is achieved by a difference of electrostatic potential at an "accelerating gap." The magnetic field does *not* increase the energy of the particles, but by bending them in circles it causes them to pass repeatedly through the same voltage gap, where they do gain energy. Thus, very high energies on the order of 500 billion electron volts can be obtained.

THE NUCLEAR SHELL MODEL

The success of the atomic shell model led to attempts to develop a description of the energy levels for nuclei using a shell model. The basis of the atomic shell model is that the energy levels can be calculated from a quantum mechanical wave equation in which the interaction between

Figure 14. The nuclear shell model predicts the possible energy levels of nucleons in shells similar to the electrons around the nucleus. The medium and heavy nuclei are best described with separate proton and neutron shells; filled shells correspond to neutron or proton numbers of 2, 8, 20, 28, 50, 82, and 126.

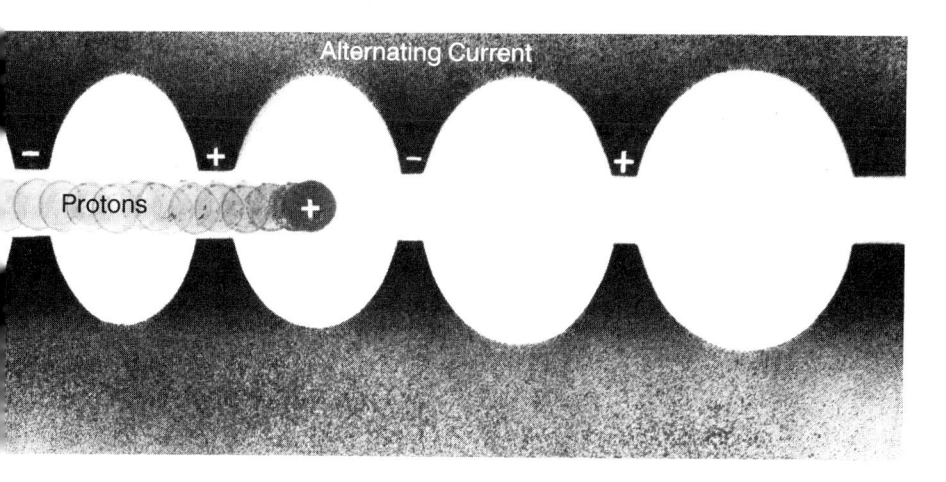

an electron and the nucleus is described by Coulomb's law. The application of such a model to the nucleus is faced with a number of complications. First, the neutrons and protons must be treated separately, so there would be a set of neutron levels and an independent set of proton levels. As was discussed earlier in this chapter, a repulsive electric force exists between the protons, whereas the attractive nuclear forces operate for both neutrons and protons. The total interaction, then, must include both repulsive and attractive terms. But we are not sure of the exact equation to use for the latter. Finally, as in the case of the atoms of more than a single electron, the mathematical complexity prevents direct solutions of the quantum wave equation for nuclei.

As a result of these complexities, approximations are used in calculating nuclear energy level patterns. It is assumed that any individual neutron (or proton) interacts with a potential energy field generated collectively by all the other neutrons (or protons). In the case of the electron in the atom, the potential field is created by the nucleus, which may be considered to be a point because its diameter is so small (10^{-13} centimeter) compared with the size of the atom (10^{-8} centimeter). For each individual particle within a nucleus, the potential field is created by all the other similar particles throughout the nucleus. The field does not vary with distance from a central point; rather, it is constant over the whole nuclear volume. Because the value of the potential field is constant, the energy level pattern is different than that for electrons in atoms.

In atoms, the electronic "shells" of energy levels are filled when the number of electrons is 2, 10, 18, 36, 54, and 86. This filling accounts for the chemical inactivity of the rare gases such as helium and neon and of ions such as sodium$^+$, chlorine$^-$, and barium^{+2}. In nuclei, the filled shells correspond to neutron or proton numbers of 2, 8, 20, 28, 50, 82, and 126. The nuclear stability for the proton number of 50 explains why tin has ten stable isotopes whereas cadmium, with a proton number of 48, has only eight and indium, with a proton number of forty-nine, has 2.

From this brief discussion you have learned that a given energy level pattern is a consequence of the assumption of the interaction potential. The more detailed our knowledge of the interactions between particles, the better becomes the ability of theory to explain experimental observations of atomic and nuclear systems. Because our understanding of the interactions between nucleons is relatively poor, it comes as no surprise that the nuclear shell model can explain much less than the atomic shell model. The next chapter discusses a different model that is useful in explaining nuclear fission.

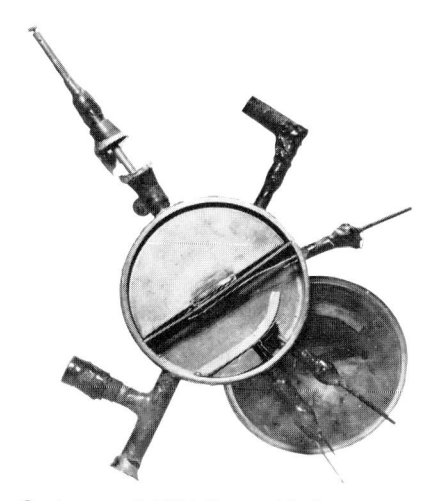

On January 2, 1931, the world's first cyclotron produced 80,000-volt hydrogen molecular ions. The power source was a 4.5-inch-diameter chamber placed between the poles of a 4-inch-diameter magnet.

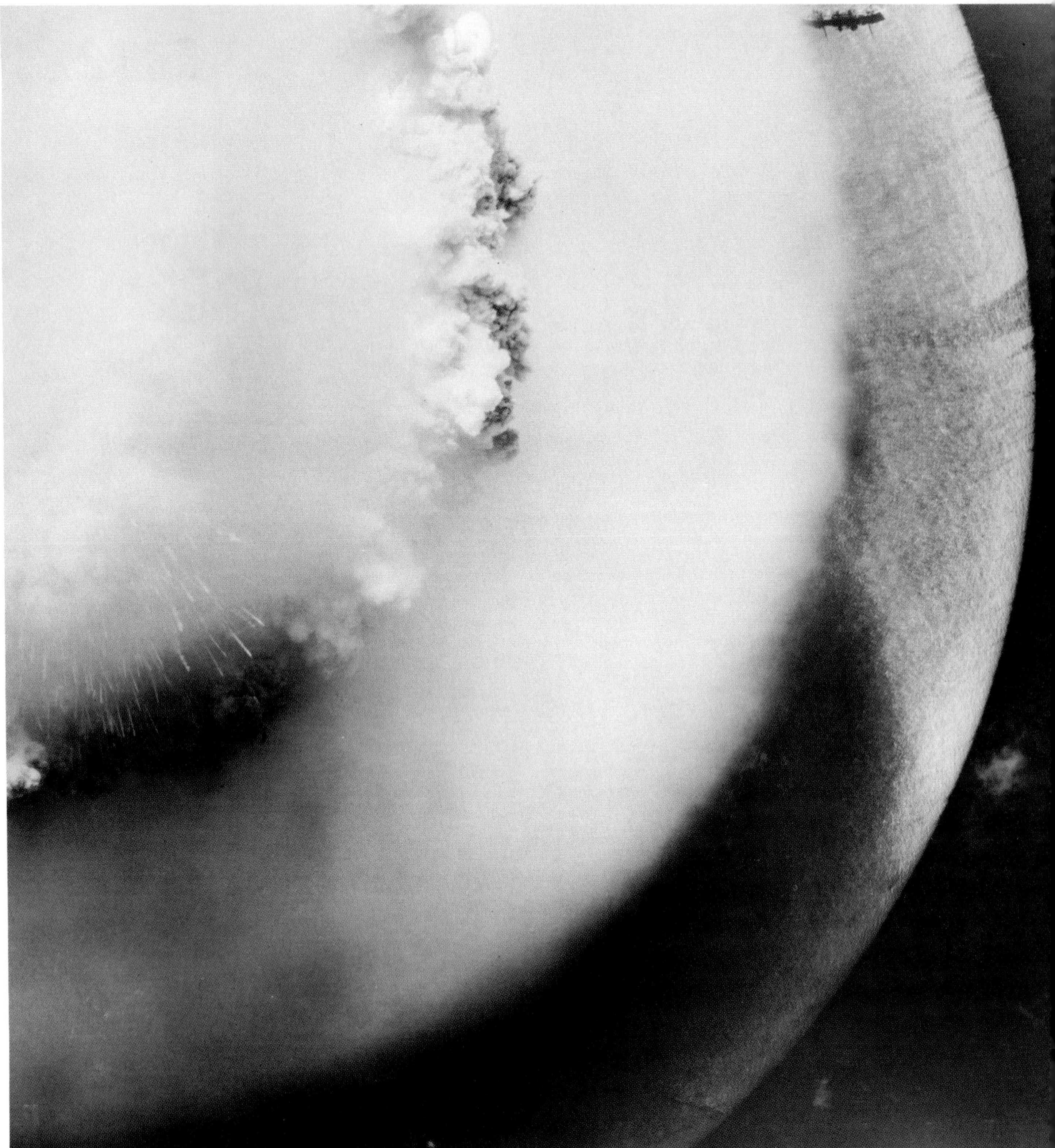

29

Fission, Fusion, and Nuclear Energy

Energy released by a nuclear explosion.

A few million years ago the first human creature walked the earth. He was a forager, for the most part, but occasionally he gathered up rocks and threw them at creatures smaller than himself. His aim must have been pretty good; he managed to survive, and in any event he was the only creature to see the value of rocks. Half a million years ago he was working away at them to get a nice sharp edge on an assortment of tools. In a few hundred thousand years he would learn to shape not only rocks but bones and wood into fishhooks and axes, scrapers and spears. Fifty thousand years ago he fell into the category *Homo sapiens,* the thinking animal, the manipulator *par excellence.* And the sorts of materials he now began to think about knew no bounds. In the space of a few million years this creature managed to unravel more and more rapidly the intricately woven mysteries of more than just rocks. He finally began to expose the nuclear core of *all* matter and the incredible energy underlying it. Now that he has done it he has suddenly paused, with some confusion, to ask what he will do next.

And pause he must. Today none of us can afford to be unaware of the raw potential of nuclear energy. We can use it to help shape a better world or we can, with its gigantic force, manipulate ourselves out of existence. We must learn to understand the two words that underwrite nuclear energy: fission and fusion. Atomic bombs are based on fission—but so are the nuclear reactors that supply an ever increasing amount of our electric power. Hydrogen bombs are based on fusion—but so are the fusion reactors that hold promise as the long-term solution to the energy needs of the world.

In nuclear *fission,* a heavy nucleus divides into two fragments of approximately equal mass; in nuclear *fusion,* two light nuclei are combined into a single, heavier nucleus. Both processes release large amounts of energy, popularly known as "atomic energy." Unlike the truly "atomic" energy in a candle flame, however, which arises from changes in the electron energy levels of the atom, the energy released in fission and in fusion results from changes in the proton and neutron configuration of the nucleus. It is better expressed as "nuclear energy."

DISCOVERY OF FISSION

In 1932, the discovery of the neutron provided scientists with a valuable tool for producing nuclear reactions. Because the neutron is uncharged, the positive charge of the nucleus does not repel it when it approaches. For that reason neutrons of low kinetic energies easily strike nuclei and cause nuclear reactions. Enrico Fermi and his coworkers in Rome recognized what the neutron could do and bombarded practically every known element with neutrons to produce many new radioactive species. A characteristic feature of the radioactive nuclides produced by capture of a neutron is β^- decay. For example, the capture of a neutron by $^{75}_{33}$arsenic produces $^{76}_{33}$arsenic, which decays to $^{76}_{34}$selenium. First,

$$^{75}_{33}\text{arsenic} + ^{1}_{0}\text{neutron} \longrightarrow ^{76}_{33}\text{arsenic}$$

then

$$^{76}_{33}\text{arsenic} \longrightarrow ^{76}_{34}\text{selenium} + \beta^-$$

Because the β^- decay causes the atomic number to increase by 1 unit,

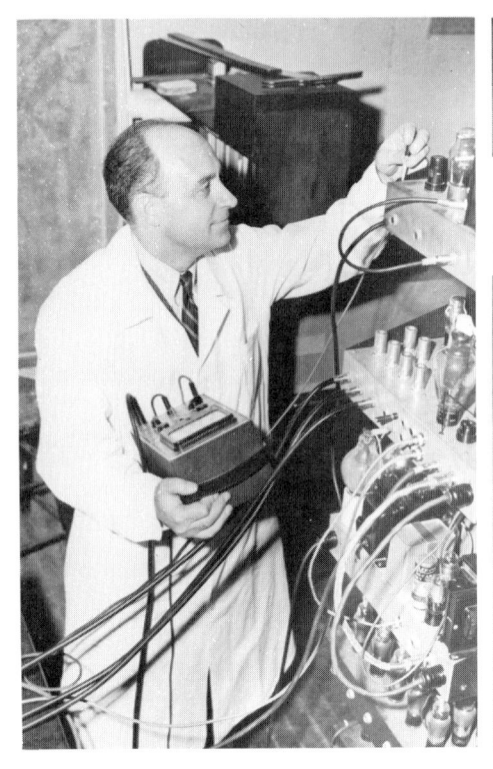

Figure 2. Enrico Fermi (left) was the leader of the group of scientists who succeeded in initiating the first man-made nuclear chain reaction.

Figure 3. Otto Hahn and Lise Meitner (center) were among the first to explore the phenomenon of fission.

Figure 4. Irene Joliot-Curie (right), the daughter of the discoverers of radium, investigated the radioactivity of uranium irradiated with neutrons. Together with Peter Savitch, she found chemical properties that differed from those expected for a transuranium element.

Fermi realized in 1934 that if uranium (element 92) captured a neutron, it might undergo β^- decay to become element 93. At that time, uranium was the heaviest element known and the search was on to find elements of atomic numbers greater than 92. Soon Fermi's group found in a bombarded sample of uranium a radioactivity with a 13-minute half-life and β^- decay mode that could be separated from all the known elements heavier than lead, including uranium. They concluded the radioactivity must be from a new element heavier than uranium.

Otto Hahn, Lise Meitner, and Fritz Strassman in Germany confirmed that Fermi's radioactivity could not arise from any element between radium and uranium. Using the same logic, these workers and others in Europe found within the next few years a number of other radioactivities which they attributed to new transuranium elements. In fact, it seemed that *four* such elements, presumably with atomic numbers 93, 94, 95, and 96, had been made that had chemical properties corresponding with those of homologs of rhenium, osmium, gold, and platinum, respectively.

The intense interest in this area of research can be measured by the fact that an article in 1938 reviewing this research cited over ninety references in scientific journals between 1934 and 1938 supporting the discovery of the four new elements. Within a short time, however, the conclusions were shown to be almost completely wrong. Indeed, the early history of the neutron bombardment of uranium is convincing evidence that even the best scientists may make errors in the interpretation of their data.

In 1938, Irene Joliot-Curie, the daughter of Pierre and Marie Curie, and Peter Savitch investigated a radioactivity of 3.5-hour half-life produced when uranium was irradiated with neutrons. The substance was puzzling. They thought it was a transuranium element, yet its chemical properties were different from those reported for substances that were

Figure 5. (Left) Niels Bohr (1885–1962) received the Nobel Prize in Physics in 1922.

Figure 6. John Wheeler (right) and Niels Bohr proposed the liquid drop model in 1939. Their model is still useful in explaining many nuclear reactions.

believed to be transuranium elements. Hahn and Strassman, intrigued by this report, reinvestigated the Joliot-Curie and Savitch research. Because they found the radioactivity could be precipitated with barium sulfate, they believed it arose from an isotope of radium that is chemically similar to barium. To their consternation, *their* "radium" could not be separated from barium. On the other hand, in their experiments they were able to show definitely that real radium could be separated from barium.

Many similar experiments performed with great patience and skill led Hahn and Strassman to conclude that their radioactivity, produced as a result of the reaction of neutrons with uranium, was not radium at all but, in fact, must be barium. As chemists they were quite convinced of their results. Nevertheless, when they published their findings in January 1939, they spoke cautiously of the "bursting" of uranium (atomic number 92) that had yielded an element such as barium (atomic number 56), much lighter than uranium. The idea of atoms splitting into small fragments was almost too much to even suggest! It was obvious, nevertheless, that all the so-called "transuranium" elements of the 1934–1938 researches were, in fact, isotopes of lighter elements formed by splitting the uranium nucleus.

Lise Meitner, who had been forced by the Nazi government to leave Germany, was informed of those experiments. Immediately she and Otto Frisch published a paper with calculations based on BE/A values (discussed in the preceding chapter), indicating that if a heavy atom were split, or fissioned, into two lighter ones a large amount of energy would be released. Very shortly, confirming reports came out of several laboratories. A few months later Niels Bohr and John Wheeler published a paper explaining many features of fission by using a model of nuclear behavior based on the analogy to a droplet of liquid.

Within months the attention of various governments had been drawn to the potential power of a bomb that released fission energy. From then

29 | Fission, Fusion, and Nuclear Energy

on, until 1945, research on fission was cloaked in secrecy. Nevertheless, achievements of the United States Manhattan Project in studying nuclear fission were almost beyond comparison. Fission has since been studied intensively, and today many thick volumes could be filled with descriptions of the phenomenon. Even so, more volumes will have to be filled before sufficient theory is developed to describe satisfactorily the division of mass, nuclear charge, and energy and other details of the fission process.

THE LIQUID DROP MODEL OF THE NUCLEUS

In the preceding chapter you read about the application of a shell model to the nucleus and how it compares to the atomic shell model. The shell model is unsuccessful in explaining nuclear fission, and a quite different model of the nucleus must be used. It is known as the *liquid drop model*.

In this model, each nucleon is interacting individually with the other nucleons in the nucleus. In other words, the neutrons and protons resemble the molecules in a droplet of liquid. The nuclear particles undergo constant motion and collisions in a manner similar to the thermal motions of molecules in a liquid. The molecules in a liquid droplet are held together by attractive forces, but if the temperature is increased sufficiently their energy exceeds the attraction and evaporation occurs. Similarly, nuclear reactions in which neutrons and protons are emitted from nuclei can be described with the liquid drop model by a process of evaporation. The liquid drop model of the nucleus, originally proposed by Bohr and Wheeler in 1939, has been useful in explaining the principal features of nuclear reactions.

When a droplet of liquid is given more energy, the increased molecular motion causes distortions in the shape of the droplet. The surface tension energy of the droplet causes it to assume a spherical shape while the thermal energy causes it to assume distorted shapes. As a result, the droplet oscillates between spherical and deformed shapes. The deformation increases with the temperature until the surface tension is inadequate to cause restoration to a spherical shape (Figure 7). The droplet then separates into two smaller droplets. This pattern of oscillation, along with the resultant deformation and separation, explains the major features of the fission of the nucleus (see Figure 8).

MASS DISTRIBUTION IN FISSION

How is mass distributed among fragments formed in fission? No matter how nuclei are made to undergo fission, fragments of various masses are formed that result in the production of chemical elements as light as gallium (atomic number 31) and as heavy as gadolinium (atomic number 64).

When the isotope $^{235}_{92}$uranium fissions from bombardment with low-energy neutrons, the amount of different masses formed may be plotted as a curve with two maxima—one near mass number 97 and another near mass number 137 (see Figure 9). These two masses are formed together in the most probable split. Because the mass number of uranium is 235 and that of the neutron is 1, the fissioning system has a total mass of 236. But the sum of the fission products is only 234. The difference is accounted for by the emission of two neutrons during fission. An average of 2.5 neutrons are emitted in neutron fission of uranium 235. The reaction can be shown as

$^{235}_{92}$uranium + neutron $\longrightarrow$ $^{236}_{92}$uranium $\xrightarrow{\text{fission}}$

fission products + 2–3 neutrons

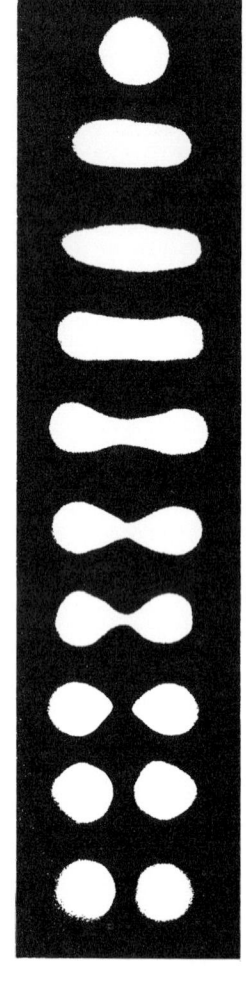

Figure 7. An illustrative sequence showing the variation in shapes of an ordinary drop of water suspended in oil when deformation is induced by voltage applied across the oil. The deformation is sufficiently large that the drop will fission rather than return to a spherical shape. Bohr and Wheeler used the analogy of the splitting of such a drop of liquid to explain nuclear fission.

R. B. Leachman, "Nuclear Fission." Copyright 1965 by Scientific American, Inc. All rights reserved.

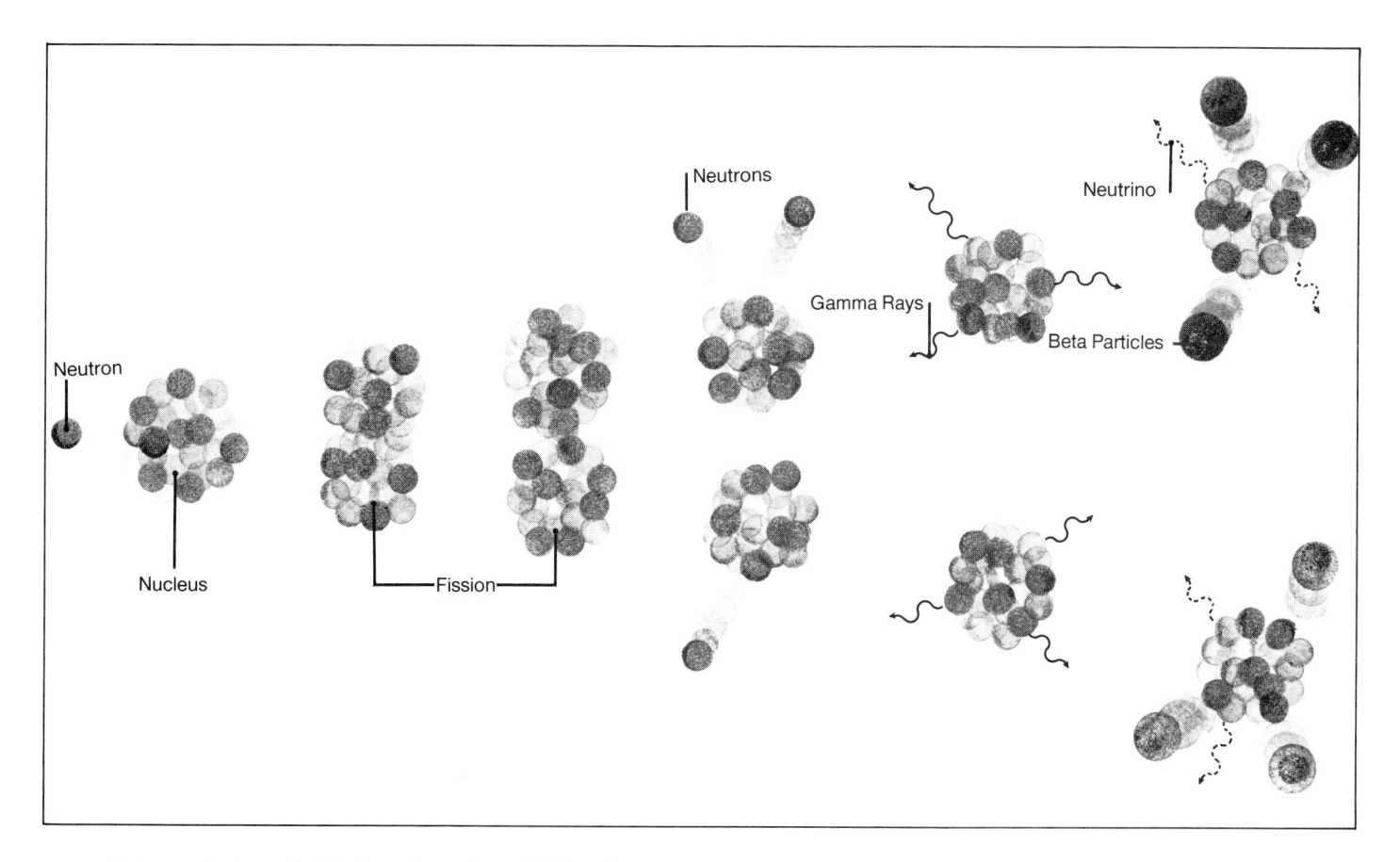

Only rarely does the fission of uranium 235 by a low-energy neutron result in two fragments of equal masses because mass number 117 is the minimum in the curve of Figure 9. Division into fragments of unequal mass numbers is termed *asymmetric* fission and division into fragments of equal mass number is termed *symmetric* fission.

Bombardment can cause fission of heavy elements other than uranium to occur, particularly if charged particles such as protons accelerated to high energies are used. For example, 10 million electron volts can be used for thorium and 30 million electron volts can be used for gold. As the energy increases, the mass distribution becomes symmetric.

ENERGY OF FISSION AND THE CHAIN PROCESS

The large amount of energy released when a heavy nucleus fissions is easily explained. In nuclear science, the energy released when a nucleus is formed from its constituent neutrons and protons is called the *binding energy*. Now the binding energy of $^{236}_{92}$uranium is 7.6 × 236, or 1,790 million electron volts. The total binding energy of a nucleus having a mass number 97 is 8.6 × 97, or 830 million electron volts; and that of one of mass number 137 is 8.3 × 137, or 1,140 million electron volts. The total binding energy of the two product nuclei is 1,970 million electron volts. Using these data, an approximate value for the energy released can be obtained by assuming the fission reaction to be

$$^{235}_{92}\text{uranium} + \text{neutron} \longrightarrow [^{236}_{92}\text{uranium}] \xrightarrow{\text{fission}} {}^{97}x + {}^{137}y$$

The energy released in fission would be the difference between the binding energy of the two product nuclei (1,970 million electron volts) and that

Figure 8. Sequence of events in the fission of a uranium nucleus by a neutron. In (a), the neutron strikes the nucleus and is absorbed, causing the nucleus to undergo deformation. (b) In about a hundredth of a trillionth of a second (10^{-14} second) one of the deformations, (c), is so drastic that the nucleus cannot recover and it fissions (d), releasing two or three neutrons. In about a trillionth of a second, the fission fragments have lost their kinetic energy to the surrounding matter and have come to rest (e), emitting a number of gamma rays as their excited states decay. In the final stage (f), the excess nuclear energy is removed from the fission fragments by the emission of beta particles and gamma rays over a period of time from seconds to years, and stable nuclei result.

R. B. Leachman, "Nuclear Fission." Copyright 1965 by Scientific American, Inc. All rights reserved.

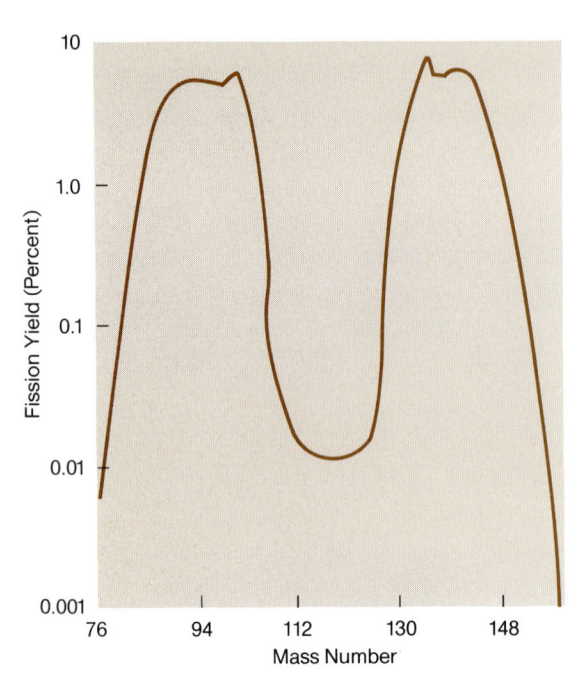

Figure 9. The mass yield curve for fission of uranium 235 with neutrons of very low energy shows symmetric peaking.

Table 1.
Approximate Fission Energy Balance

Distribution	Fission Energy (millions of electron volts)
Kinetic energy of fission fragments	170
Energy of gamma rays	16
Energy of beta decay	9
Kinetic energy of emitted neutrons	5
	200 total

of the $^{236}_{92}$uranium nucleus (1,790 million electron volts). In other words, about 200 million electron volts of energy are released in fission.

How is this energy distributed in fission? Given the large energy release in a reaction that tears the nucleus apart, it is not surprising to learn that more goes on than the simple division of a nucleus into two fragments. You know that stability to radioactive decay is associated with certain values of the neutron/proton ratio. In the region of the masses of the fission products, the stable ratios of neutron/proton vary from 1.2 to 1.4. The neutron/proton ratio for $^{236}_{92}$uranium is 1.6, so fission fragments always have too large a value for the ratio to be stable. To adjust partially to a value close to stability, several neutrons are emitted in the act of fission. But the number of neutrons emitted is not enough to lower the neutron/proton ratios all the way to stable values. To achieve stability, the fission fragments undergo a series of radioactive β^- decay steps after neutron emission which lowers the values of the neutron/proton ratio.

Assume two neutrons are emitted in the fission of uranium 235 by a neutron. As an example of the fission, you may use:

$$^{235}_{92}\text{uranium} + \text{neutron} \longrightarrow [^{236}_{92} \text{uranium}] \xrightarrow{\text{fission}}$$

$$^{94}_{38}\text{strontium} + ^{140}_{54}\text{xenon} + 2 \text{ neutrons}$$

The $^{140}_{54}$xenon then decays by the following sequence (with the half-life shown over the arrow):

1. $^{140}_{54}$xenon $\xrightarrow{16 \text{ seconds}}$ $^{140}_{55}$cesium $+ \beta^-$
2. $^{140}_{55}$cesium $\xrightarrow{66 \text{ seconds}}$ $^{140}_{56}$barium $+ \beta^-$
3. $^{140}_{56}$barium $\xrightarrow{12.8 \text{ days}}$ $^{140}_{57}$lanthanum $+ \beta^-$
4. $^{140}_{57}$lanthanum $\xrightarrow{40.2 \text{ hours}}$ $^{140}_{58}$cerium $+ \beta^-$

The final product of the decay sequence is $^{140}_{58}$cerium, which is stable. The decay chain starting with $^{94}_{38}$strontium ends with stable $^{94}_{40}$zirconium.

In addition to neutrons, several gamma rays are emitted in fission. Therefore, the fission energy appears as the energy of gamma rays, as the kinetic energy of the emitted neutrons and of the fission fragments, and as the energy of beta decay of the fragments. Table 1 shows how the fission energy is distributed for a total fission energy of 200 million electron volts.

When scientists learned that more neutrons are emitted by a fissioning nucleus than are absorbed, they realized it might be possible to have a self-sustaining *chain reaction.* In such a reaction, an initial reaction is multiplied in succeeding generations without further external help. The first fission absorbs one neutron and releases two. The result is two fissions in the next step, four in the third, eight in the fourth, sixteen in the fifth, thirty-two in the sixth, and so on. Because fission occurs rapidly — on the order of 10^{-17} second — this multiplication is incredibly fast. If

How Can Fission Take Place?

One of the primary problems with thermonuclear reactions is the high temperature requirement caused by the small size of the nuclei involved and the electric repulsion forces between them. If you wished to make a fission reactor, how hot would it need to be for two deuterons to be able to touch (that is, their centers would be about 10^{-12} centimeter apart) and for fission to occur?

Step 1. At the separation given, the repulsive electric potential energy of particles would be given by

$$PE = \frac{(charge)^2}{separation} = \frac{(4.80 \times 10^{-10} \text{ electrostatic unit})^2}{10^{-12} \text{ centimeter}}$$

$$= 2.30 \times 10^{-7} \text{ erg}$$

$$= 1.44 \times 10^{5} \text{ electron volts}$$

$$= 144 \text{ kilo-electron volts}$$

Although this is a small amount of energy compared to cyclotron energies, it is enormous compared to ordinary thermal energy. This energy, then, is a potential barrier to the interaction of the deuterons.

Step 2. To determine the temperature necessary to go over this potential barrier and come within 10^{-12} centimeter of each other, let us suppose that each deuteron has an energy equal to half of the total required, or 72 kilo-electron volts. From the ideal gas law equations, it is possible to show that the energy of an atom or molecule can be given as

$$E = \tfrac{3}{2} kT \quad \text{where } k = \text{Boltzmann constant}$$

Step 3. Now you have

$$E = \tfrac{3}{2} kT = 72 \text{ kilo-electron volts} = 1.15 \times 10^{-7} \text{ erg}$$

If you use $k = 1.38 \times 10^{-6}$ erg/degree Kelvin, you find

$$T = \frac{\tfrac{2}{3}(1.15 \times 10^{-7} \text{ erg})}{1.38 \times 10^{-6} \text{ erg/degree Kelvin}}$$

$$= 560 \text{ million degrees Kelvin}$$

The actual temperature requirement isn't quite this high because some nuclei have much more energy than the average. In practice, thermonuclear explosions proceed at temperatures of about 60 million degrees Kelvin.

allowed to proceed unimpeded, it leads to an explosion as in an atomic bomb. If the rate is controlled once it reaches a certain level, however, the energy released can be maintained at a controllable level for power use. Nuclear reactors operate by control of the chain reaction of fission.

THE MANHATTAN PROJECT

At the request of several prominent nuclear scientists, Albert Einstein wrote a letter to President Franklin Roosevelt in August 1939, pointing out the possibility that nuclear fission might "lead to the construction of bombs and it is conceivable—though much less certain—that extremely powerful bombs of a new type may thus be constructed."

As a result of that letter, the United States government initiated the Manhattan Project. The world has never known a comparable project in science. Thousands of the most capable scientists, many of whom came from Europe, worked in the greatest secrecy and haste to construct the atomic bomb. Eventually whole new cities were constructed in Tennessee (Oak Ridge), New Mexico (Los Alamos), and Washington (Hanford) for the laboratories and facilities needed to accomplish the task.

The fission of uranium and of the element plutonium, which had been discovered only the year before, was studied intensively. Completely new techniques were developed to investigate the physical, chemical, and nuclear properties of these radioactive elements. And scientists learned that low-energy neutrons could cause fission in uranium 235 but not in uranium 238 so methods of separating uranium 235, present in amounts of only 0.7 percent in uranium, had to be devised and a plant built for the separation. Huge nuclear reactors were designed and built to produce plutonium from uranium.

Finally, in July 1945, the first atomic bomb was exploded in the New Mexico desert. After watching the explosion J. Robert Oppenheimer, the director of the Los Alamos Laboratory, reflected on a quotation from the sacred Hindu epic *Bhagavad-Gita:*

If the radiance of a thousand suns
Were to burst at once into the sky
That would be like the splendor of the Mighty One . . .
I am become Death,
The shatterer of worlds.

In August two bombs, one made of uranium and the other of plutonium, were dropped on Japan and World War II ended. Mankind had entered a new atomic era.

NUCLEAR REACTORS

Fermi, as his part in the Manhattan Project, had undertaken the task of controlling the fission chain reaction. Fermi knew that the neutrons must be slowed to extremely low speeds in order to increase the probability of their being captured by a uranium nucleus, thereby causing fission. Collision with light atoms such as carbon does an effective job of slowing down, or "moderating," the neutrons. Under the west stands of the football stadium at the University of Chicago, Fermi's team began to assemble an array of graphite blocks and uranium. On December 2, 1942, they achieved a self-sustaining chain reaction, which marked the first controlled release of nuclear energy.

Figure 10. It was not too long ago that weary men spoke of "the war to end all wars." Even more recently one of our greatest scientists spoke to Franklin Delano Roosevelt of a bomb that would end still another terrible war. It did that; and in the aftermath mankind finds itself on an even more precarious level of conflict, increasingly weary and still seeking a lasting peace.

Figure 11. J. Robert Oppenheimer, Director of the Los Alamos Laboratory when the United States exploded the first "atomic" bomb in July 1945 in the New Mexico desert.

29 | Fission, Fusion, and Nuclear Energy

Figure 12. The operation of a nuclear reactor. The uranium *fuel* undergoes fission reactions, giving off neutrons. The *moderator* slows down these neutrons to speeds at which they can be easily captured by more uranium to aid in further fission; the *control rods* keep the concentration of these neutrons at a reasonable level in order to pace the entire reactor. An outside *pressure vessel* contains a *pumped*, flowing liquid *coolant* that carries heat produced by the reactor to the *heat exchanger* where water is turned to *steam*. The steam operates a *turbine* which, in turn, operates a *generator* that sends electricity out to the consumer over *power lines*. The steam is turned back to water in the process, and is recycled by a *pump*. The reactor itself is enclosed in a *shielding* to protect living organisms.

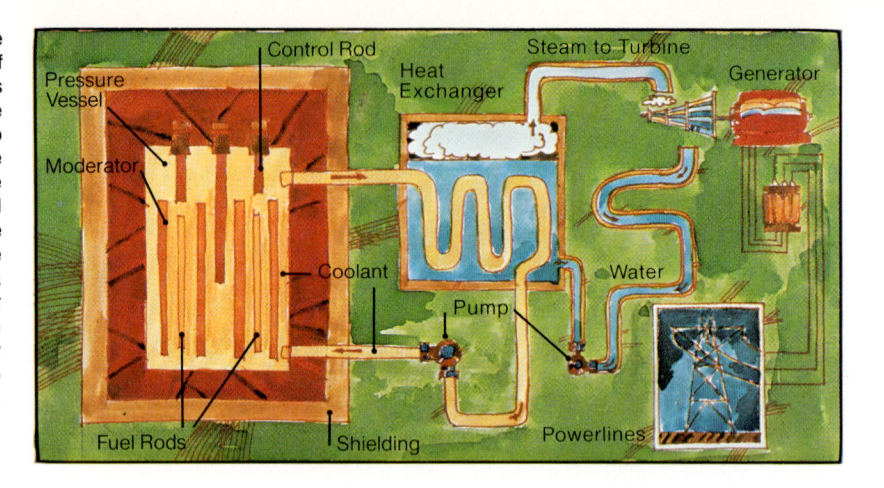

Although a great variety of reactors are in operation around the world, all have five basic components:

1. Fuel. This is either natural uranium or uranium enriched in uranium 235. In a few cases plutonium 239 is being used as the fuel. Usually the fuel is used in the form of metal plates alloyed with aluminum.

2. Moderator. Graphite, ordinary water (H_2O), and deuterium oxide (D_2O) are the most common materials used to slow the fission neutrons down to the extremely low energies at which they are most effective in causing further fissioning.

3. Control Rods. Cadmium and boron are used to control the concentration of neutrons present in the reactor because they have a great capacity for absorbing neutrons.

4. Coolant. The coolant conducts the heat away in order to keep the temperature at a reasonable level. In power reactors the coolant is the vehicle for transferring the heat derived from the fission energy to external use. Water, deuterium oxide, air, even molten sodium, or a mixture of sodium and potassium are used as coolants. In many reactors the same water or deuterium oxide serve as both moderator and coolant.

5. Shielding. Fission is accompanied by gamma ray emission as well as neutron emission. This emission as well as the intense radioactivity of the fission products requires the presence of thick layers of absorbing shielding. Usually water, concrete, or both are used as shields.

Figure 12 is a schematic drawing of a nuclear reactor that uses a heat exchanger to provide steam for a turbine that generates electricity.

NUCLEAR ENERGY

In nuclear reactors, the large amount of energy that fission releases can be used to generate electricity. And in a few countries, at least, nuclear power supplies a growing percentage of energy needs. Unfortunately, nuclear power comes with a price tag: it also produces large quantities of intensely radioactive fission products. These radioactive elements must be stored in some fashion for many thousands of years. Only then can we be certain that radioactivity will not escape into the environment.

And yet, even now the world faces a serious shortage of energy. Even if the population growth rate *were* to level off — and that is an unlikely prospect for the immediate future — the per capita consumption of energy still would increase for a considerable length of time as the standard of living rises in less-developed countries. We may expect that the per capita energy consumption of the more developed countries will increase for a brief period of time and then begin to level off; and perhaps it might even decrease as we achieve more efficient utilization of our energy resources. If an eightfold increase in total energy consumption by the year 2200 is projected for the United States alone, and even if that projection is high by a factor of 2, we still must consider how we will find energy sources that will quadruple the energy available now.

Today oil and gas supply about 75 percent of the nation's demand for energy, with most of the rest supplied by coal. Unfortunately, both natural gas and oil will be of diminishing importance after the next 50 to 100 years. The supply of coal available to the world is extremely large and it will continue to play an important role in the energy problem. Unfortunately, not only does coal burning result in a fair amount of atmospheric pollution, but the cost of mining and transportation raises its price considerably. Improved mining technology will alleviate some of the problems, as will the location of coal-burning plants near the mines themselves, in areas remote from cities.

By the year 2200, nuclear energy is expected to supply approximately one-quarter of the total energy needs of the United States. But even with the availability of economical nuclear power, after the year 2050 there will be an increasing *energy deficit* unless we develop a readily available source. One potential new source is nuclear fusion.

The only material found in nature that can be used as a nuclear fuel is uranium; and of uranium only the isotope uranium 235 provides the fission power used in reactors. That isotope constitutes 0.7 percent of the total abundance of natural uranium, and unless we can figure out some way to use the more abundant isotope uranium 239 in power production, we will soon exhaust the availability of nuclear power from uranium.

Fortunately, the nonfissionable isotope uranium 238 can be converted by the capture of neutrons to the fissionable species uranium 239, an isotope of element 94, plutonium. The first step is

$$^{238}_{92}\text{uranium} + \text{neutron} \longrightarrow {}^{239}_{92}\text{uranium}$$

which is quickly followed by:

$$^{239}_{92}\text{uranium} \longrightarrow {}^{239}_{93}\text{neptunium} + \beta^-$$

and

$$^{239}_{93}\text{neptunium} \longrightarrow {}^{239}_{94}\text{plutonium} + \beta^-$$

The plutonium 239 can then be used in reactors as a fuel. An additional source of nuclear power may become available through the use of thorium. There are no naturally occurring isotopes of thorium that undergo fission in a reactor; however, the thorium 232 that is found in nature can be made to capture neutrons and be converted into a fissionable isotope of uranium 233. The reactions are

$$^{232}_{90}\text{thorium} + \text{neutron} \longrightarrow {}^{233}_{90}\text{thorium}$$

$$^{233}_{90}\text{thorium} \longrightarrow {}^{233}_{91}\text{palladium} + \beta^-$$

$$^{233}_{91}\text{palladium} \longrightarrow {}^{233}_{92}\text{uranium} + \beta^-$$

This conversion of the nonfissionable uranium and thorium isotopes into fissionable plutonium 239 and uranium 233 is achieved in *breeder reactors*.

You have learned that more neutrons are released in fission than

Figure 13. Plutonium (left) and neptunium (right), two elements whose isotopes are used in the production of nuclear fuel.

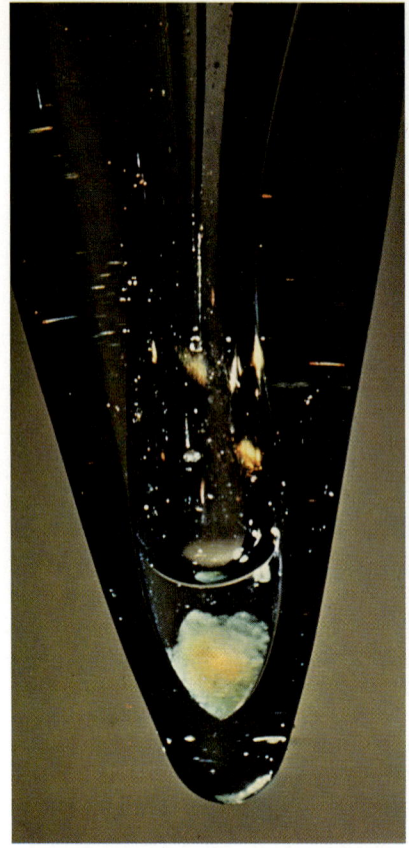

are required to cause fission. The excess of neutrons is responsible for the propagation of the chain reaction. Theoretically, however, in a reactor operating at a constant level only one of the emitted neutrons is required for further fission. In other words, if an average of 2.5 neutrons are emitted in fission, the excess of 1.5 neutrons can be used for capture by non-fissioning material. Those neutrons then can be used to convert the uranium 238 and thorium 232 to fissionable plutonium 239 and uranium 233. In fact, because more neutrons are available for capture than are involved in fission, more fissionable material can possibly be made in this process than is used in the reactor for power production. That is the basic concept of a breeder reactor. In principle, after we have exhausted all the available uranium 235 in the world, there will still be a large amount of available fission material in the form of uranium 233 and plutonium 239 that has been produced in the breeder reactors. By such a breeding process, we may assume that all the thorium and all the uranium in nature is available for the production of power in nuclear reactors. On this basis, sufficient nuclear fuel exists to provide power for several hundred years.

Figure 14 is a diagram of a breeder reactor being developed in the United States. This particular reactor is known as the liquid-metal-cooled-fast-breeder reactor, or "LMFBR." Liquid metals such as sodium or sodium-potassium alloys are used to cool the fuel core and to transfer the heat to a heat exchanger for the subsequent generation of electric power. At the temperature of operation, the average energy of the neutrons is greater than the energy in a normal reactor; hence the name "fast breeder." Breeder reactors have the advantage of producing more fissionable material per unit fission of uranium 235. Moreover, the high boiling point of the sodium coolant permits the system to operate at a higher temperature, which results in some advantages in the heat transfer process. The operation of large, reliable commercial breeders of this type is at least a decade away, perhaps longer. A number of experimental breeder reactors are in operation around the world.

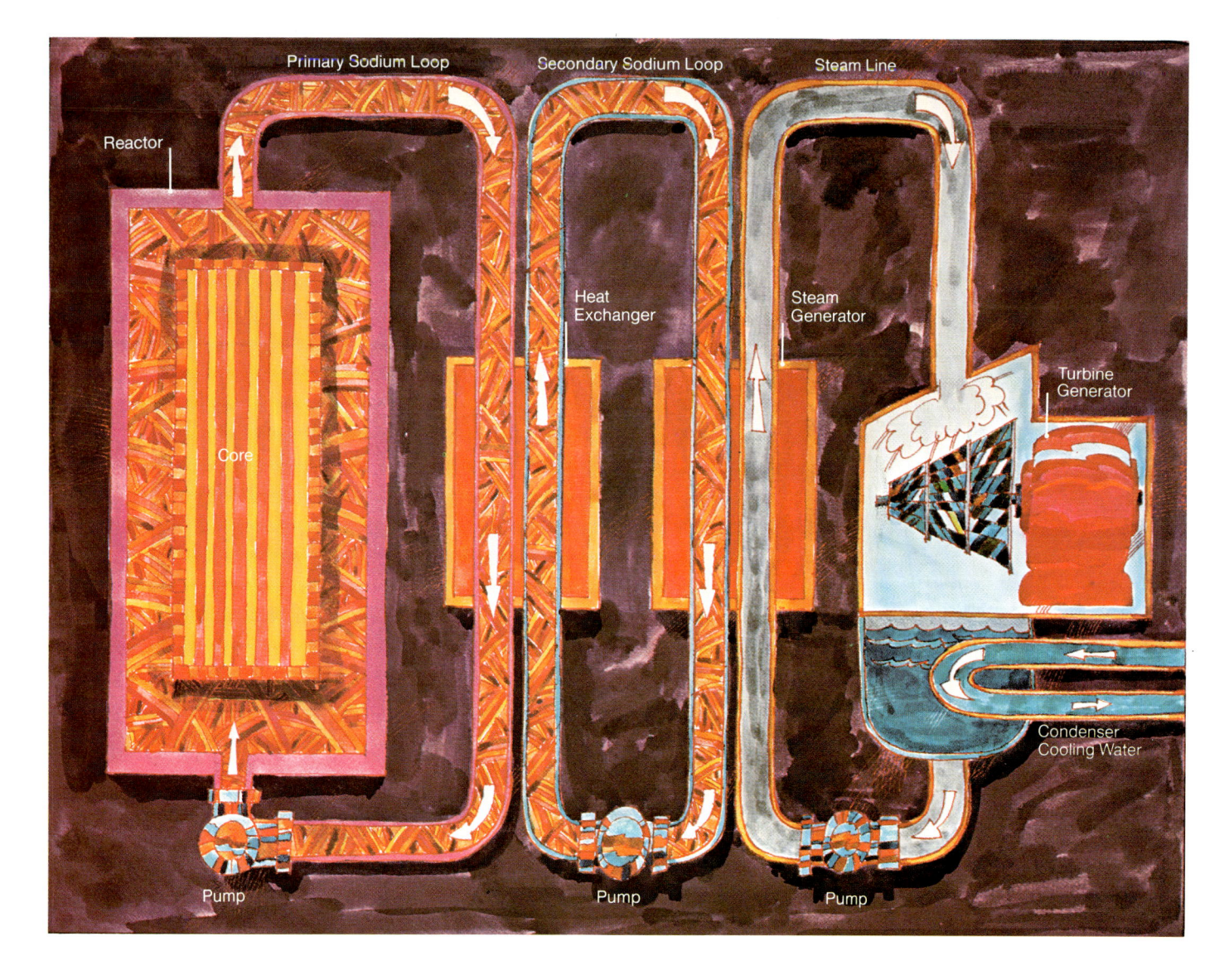

NUCLEAR FUSION

In the past two decades several nations have developed hydrogen bombs, weapons of almost unbelievable destructive power. Unlike the atomic bomb and nuclear reactors, where energy release comes about when a large atom splits into two smaller parts, the reaction in the hydrogen bomb comes from the fusion of two light nuclei into a heavier one.

Nuclear fusion involves the collision of two positively charged and, therefore, mutually repulsive nuclei. In order to overcome repulsion and cause the nuclei to combine, extremely high energies must be imparted to the nuclei. For fusion to occur, atoms must be initially heated to about 2 million degrees Centigrade. In a hydrogen bomb, a small atomic bomb

Figure 14. The liquid metal fast breeder operates at a higher temperature than a normal nuclear reactor. Liquid sodium is used to cool the fuel core and to absorb heat in the secondary loop.

29 | Fission, Fusion, and Nuclear Energy

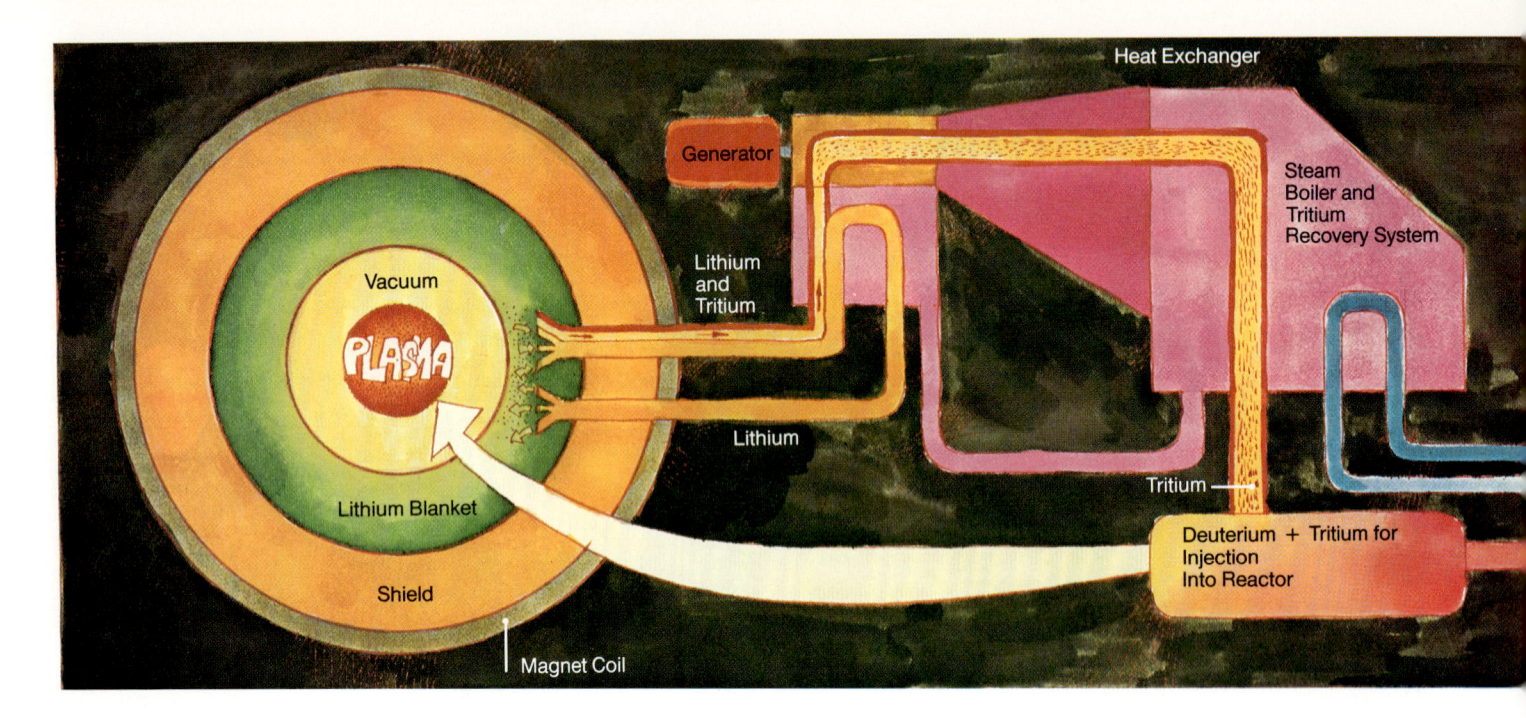

Figure 15. The schemes for fusion power plants differ from those for fission power plants primarily in how the energy is produced, not in how it is exchanged and used to generate power. In this plan, deuterium and tritium are injected into the plasma core to provide fuel for fusion into helium. Liquid lithium is used as a heat-exchange bath. The tritium produced in the lithium by interaction with neutrons given off by the fusion process can be recycled into the fuel.

is exploded which triggers the fusion reaction. Under proper conditions, once nuclear fusion begins the heat from the thermonuclear reaction is more than enough to ensure the reaction will be self-sustaining. Nearly 100 million calories would be released for each gram of hydrogen nuclei fused in reactions, as in the combination of the isotopes deuterium and tritium:

$$_1^2\text{hydrogen} + {_1^3}\text{hydrogen} \longrightarrow {_2^4}\text{helium} + \text{neutron}$$

DEUTERIUM   TRITIUM

Deuterium is available for fusion reactions and it seems to be the answer to the long-range energy needs of the world. There is about 1 atom of deuterium for every 6,500 atoms of ordinary light hydrogen in water. The enormous amounts of water available on earth thus represent a virtually inexhaustible source of fusion energy.

But there are obstacles to achieving practical fusion power. What, for example, will contain the fuel gases at the fantastically high temperatures required? Obviously, if the atoms of deuterium and tritium were to be striking the container walls they would soon impart enough energy to the walls to melt and even vaporize them. In order to avoid this dilemma scientists have been studying the use of magnetic walls to contain the fusing gases. At extremely high temperatures gases exist as ions in the plasma state. Because those gases have an electric charge, they will react to magnetic fields. If a magnetic field is shaped properly, then, it can act as a container or bottle, holding the fusing gases away from the walls of the reactor vessel. It can even cause the gases to "pinch together" to increase their concentration and thereby increase the probability of reaction.

Figure 15 depicts a possible fusion power plant operating on the reaction of deuterium plus tritium. In this scheme, deuterium is obtained from sea water and injected, along with tritium, into the fusion reactor. A blanket of lithium metal surrounds the reactor vessel, absorbing the heat of the fusion reaction. The liquid lithium metal will also be acted upon by the neutrons released in the fusion process to cause the formation of tritium. The lithium can be pumped off to a heat exchanger, where the fusion power can be used to generate electricity. The tritium in the lithium can also be removed at this stage and sent off to mix with the deuterium being fed into the reactor. Of course, the fusion power plants that are finally developed may bear little resemblance to this possible scheme, but some of the elements will probably be incorporated.

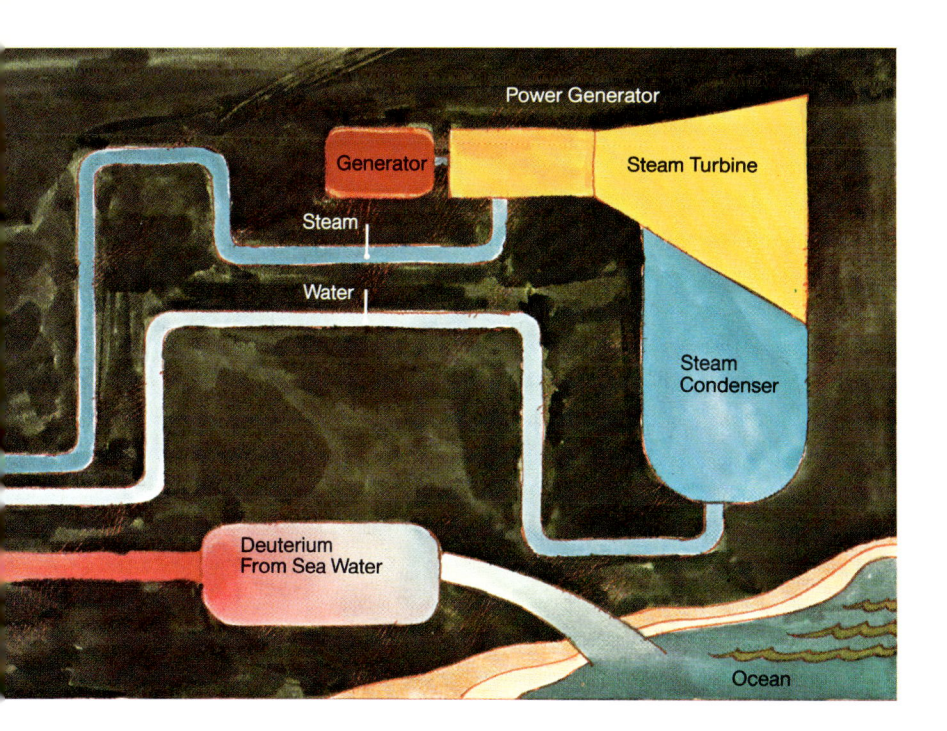

The technology for practical fusion power plants will probably require 20 or 30 more years of development. Once they are developed, however, we can expect fusion to supply great amounts of energy indefinitely. Fusion may replace nuclear fission power long before the depletion of fissioning energy sources because it does *not* produce radioactive isotopes. Unlike fissioning energy, the fusion process creates no residual by-products that will require processing and storing for thousands of years in order to preserve our environment even as we use its resources. Utilization of fusion power in reactors may well be the next technological step we take to keep pace with the increasing needs of mankind.

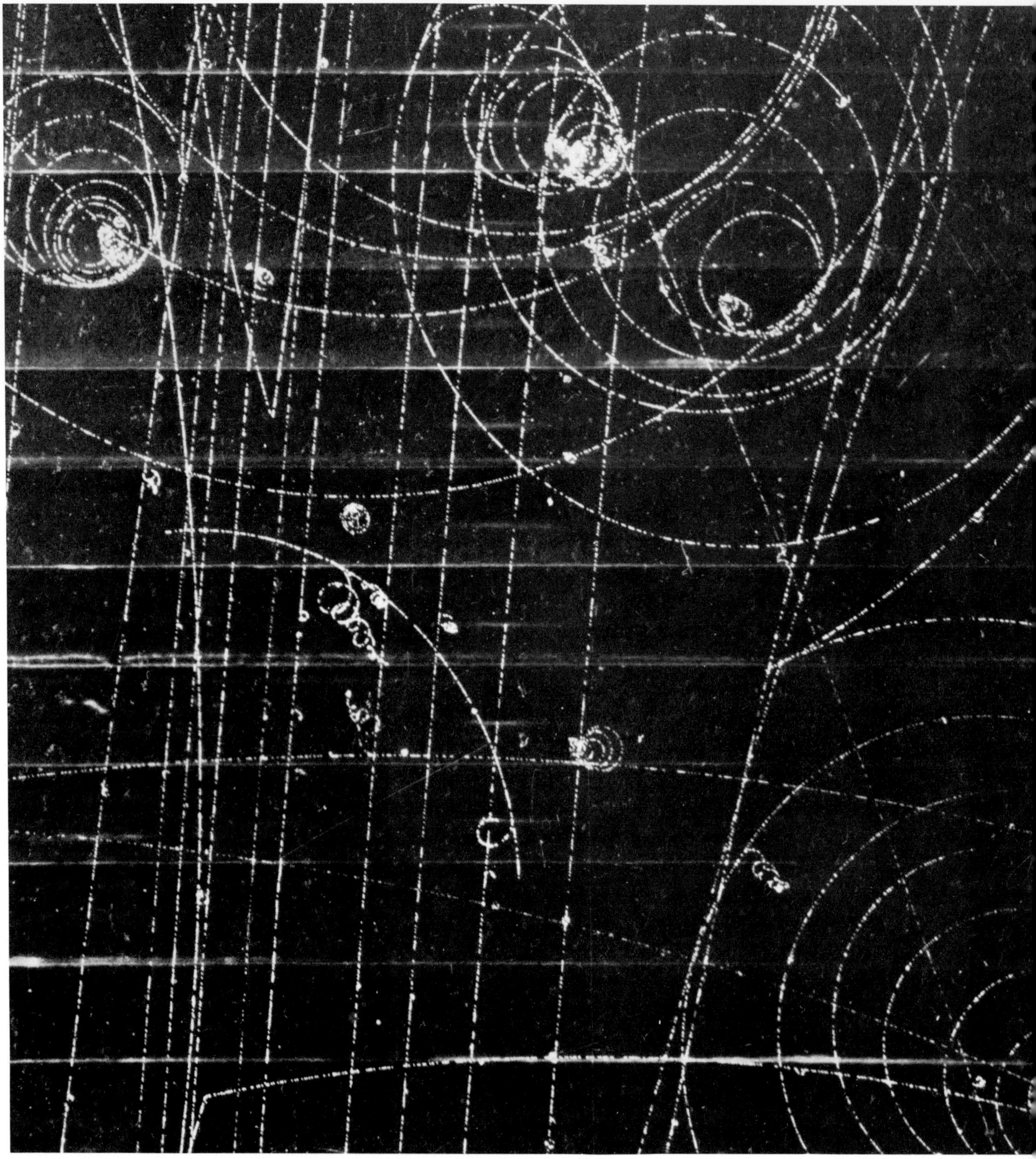

30

Elementary (?) Particles and Symmetry

Elementary particles interacting with matter in the presence of electromagnetic fields leave a bubbly record of their motion through the liquid hydrogen of a bubble chamber.

James Chadwick's discovery of the neutron in 1932 seemed to complete the picture of matter. Every kind of matter appeared to be the manifestation of only four fundamental particles: electrons, protons, neutrons, and photons. But in the same year, another particle was discovered: the positive electron, or *positron.* Positrons were found originally in cosmic rays, but it was soon shown that gamma rays from radioactive substances produce positrons if the gamma ray has sufficiently high energy. And a few years later the mu meson, or *muon,* was detected in cosmic rays. The new particles didn't fit into the simple electron-proton-neutron picture of matter. And more particles continued to be discovered. The advent of high-energy accelerators in the 1940s made possible the controlled study of particle interactions, and a host of new particles has been identified in such reactions. The count of distinct particles or particle-like entities now stands at well over a hundred.

The properties of the known particles differ widely. The electron and the proton appear to be completely stable if left alone, whereas some of the other particles decay spontaneously to new forms. The lifetimes span a wide range; free neutrons decay in about 1,000 seconds on the average, but a neutral pi meson, or *pion,* exists for only 10^{-16} second or so. The masses of the particles range from zero for the photon to over 3 times the proton mass for certain short-lived particles. There are intriguing similarities as well. Positrons are identical to ordinary electrons in every way except for the sign of their electric charge. And although protons and neutrons differ in their electric and magnetic properties, they otherwise seem to act similarly when bound in nuclei.

How can we find ways of classifying and understanding the multitude of different particles? Partial answers have been found, but knowledge is by no means complete. Some believe that the situation is analogous to the state of chemistry in the middle of the nineteenth century. Dmitri Mendeleev's periodic table showed there was order in the properties of the chemical elements even though the reason underlying the order was not understood at the time. Later development of atomic models and the corresponding discovery of new principles eventually explained the physical basis for the periodic table. Many hope that the same thing will happen in particle physics—that new principles will be found to account for the order we can glimpse today.

Atoms and their chemical behavior were eventually found to be understandable in terms of a few simple building blocks. Perhaps all the elementary particles known today will turn out to be aspects of something even more fundamental, as well. Many believe that the physical world is based on simplicity, and that there must be a simple way to understand the vast array of known particles. One of the purposes of studying elementary particles is the search for new principles to reduce the complexity apparent today. You will see some of the partial successes later in this chapter.

The study of elementary particles is hampered because their behavior is so far removed from everyday experience. When Ernest Rutherford and others at the turn of the century were attempting to understand the structure of the atom they could at least rely on the well-known Coulomb law of force between electric charges. But for elementary particles, the electric force is only one of several that can act. Here the other important forces—the nuclear force and the weak interaction force—are both short-range and not well understood. Where does the search for order begin?

CONSERVATION LAWS

Fortunately, some of the principles of physical science seem to be valid over wide ranges of experience and may be applicable on the subnuclear level. Among the most fundamental concepts of physical science are the *conservation laws,* many of which you have read about earlier in this book. Conservation laws frequently place stringent limitations on the kinds of events that can occur. For instance, one of the important conservation laws is that the total electric charge in the universe is constant. An example of charge conservation occurs in the annihilation reaction of a positron with an electron:

Positron + electron ⟶ photons

Here charge is conserved. The total charge at the beginning is zero because the positron and the electron are *equally but oppositely charged;* and the total charge after the annihilation is zero, because photons carry no charge.

Usually more than one conservation law applies to a reaction. For example, energy, as well as electric charge, must be conserved in every reaction. You know from Chapter 8 that energy can exist in a variety of forms, and that total energy is conserved. In reactions between elementary particles, energy takes on only a few simple forms. First, it can be the mass-energy mc^2 of each particle involved in the reaction. Second, it can be kinetic energy of the particles. And finally, it can be photon energy if any photons take part in the reaction. Suppose, for example, that the reaction

Positron + electron ⟶ photons

takes place when the positron and the electron are moving slowly, with negligible kinetic energy. Because the mass-energy of a positron or an electron is 0.51 million electron volts, you can conclude that the total energy of the photons produced is 1.02 million electron volts. Alternatively, if the energy of the photons was measured to be 1.02 million electron volts and the mass-energy of an electron was known to be 0.51 million electron volts, you could conclude something about the mass-energy of the positron.

The important point about a general conservation law, such as the law of conservation of energy, is that *it holds regardless of the details of the interactions or how the reaction actually takes place.* Conservation laws are valuable in two ways. First, they allow unknown quantities such as the positron mass-energy in this example to be found from experiment. And second, they tell us that any reaction that violates a conservation law cannot take place. The boxed figure on this page is an example of how conservation laws limit reactions.

The conservation laws for charge, energy, and linear momentum apply to everyday situations and were known long before the discovery of elementary particles. You saw in Chapter 8 that the idea of energy started in classical mechanics with kinetic energy and potential energy. Gradually the concept was widened to include nonmechanical forms of energy, such as light and heat. But suppose Newton's laws of motion had never been devised. Could the law of conservation of energy have been inferred from the study of nature? After all, conservation of charge was discovered by experiment. One of the tactics of elementary particle physics is to guess at new conservation laws by studying reactions. Con-

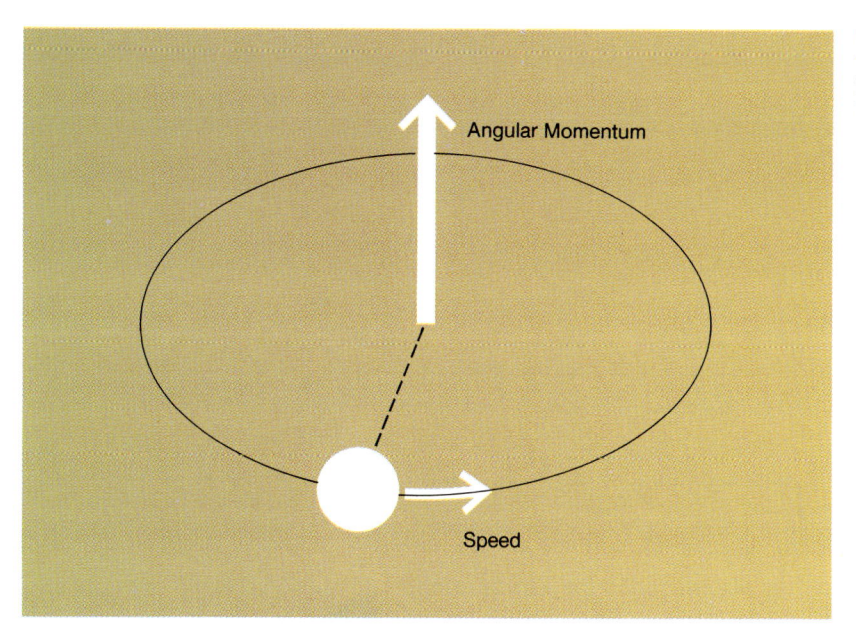

Figure 2. Angular momentum can be assigned a vector direction according to the right-hand rule. Let your fingers point along the direction of motion, and this vector points along the direction of your thumb.

servation laws manifest themselves most clearly when they *forbid* a reaction to occur. For example, the reaction

Proton ⟶ electron + two positrons

has never been observed, even though it would conserve charge, energy, and momentum. Proton decay is forbidden by an additional conservation law, which you will read about later in this chapter.

ANGULAR MOMENTUM

Suppose a particle is traveling in a circular path. For such motion, the *angular momentum* of the particle is defined to have a magnitude of

Mass × speed × distance from the axis of rotation

In addition, it is conventional to assign a direction to the angular momentum, as illustrated in Figure 2. Angular momentum can be thought of as a *vector,* having both magnitude and direction.

The concept of angular momentum can be applied to a massive object by visualizing the object as a collection of particles. A door swinging on its hinges has angular momentum because every part of the door is moving in a circular path. Furthermore, all the contributions to the total angular momentum vector lie in the same direction, so they add together.

There is no reason to invent a vector such as angular momentum unless it has some special properties that make it useful. To see why angular momentum is useful, consider an analogy between linear momentum and angular momentum. Linear momentum depends on the mass of the object and the speed at which it is moving. The only way to change the linear momentum of a body is to exert a force—this book, for example, won't start moving unless you push it. But a force won't necessarily make the book rotate and gain angular momentum. To make a body rotate, forces have to be applied that produce a twisting effect. Suppose you want to swing a door open. You know if you push near the hinges, it is difficult to get the door moving; it is much easier to push on the doorknob side, as far away from the hinges as possible. *The twisting effect of a force depends*

30 | Elementary (?) Particles and Symmetry

Figure 3. The torque on a door depends on the distance from the axis and on the perpendicular component of force.

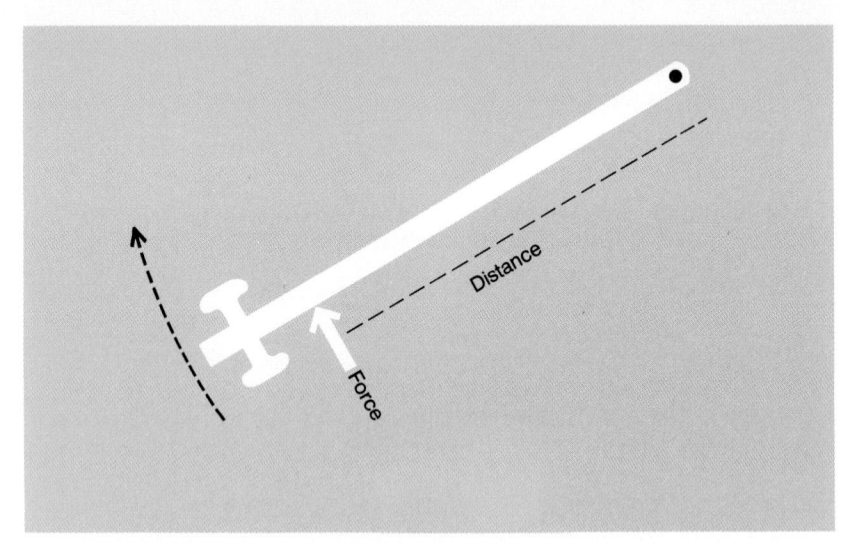

on the distance from the axis of rotation. Torque, you recall from Chapter 6, is a measure of the twisting effect of a force. For the simple case shown in Figure 3,

Torque = force × distance from the axis of rotation

Change of linear momentum depends on the applied force: in the absence of applied force, linear momentum remains constant—it is conserved. Similarly, angular momentum changes only when torque is applied. That is why angular momentum is important—it obeys a conservation law. *In the absence of torque, angular momentum remains constant.*

Any quantity that obeys a conservation law is helpful in the effort to understand elementary particles. But on the submicroscopic scale, angular momentum does not act the same way it does on the everyday scale. You know that Planck's constant determines the fundamental properties of the quantum world. Not only does it establish the relationship between the energy of a photon and its frequency, it also establishes the relationship between the momentum of a particle and its wavelength. In the metric system,

$h = 6.62 \times 10^{-34}$ joule-second

Now 1 joule equals 1 newton-meter, and 1 newton equals 1 kilogram-meter per second per second. So it is equally correct to write

$h = 6.62 \times 10^{-34}$ kilogram-meter-meter/second

In other words, Planck's constant has the *same units* as angular momentum. Both are expressed in terms of mass times speed times distance. Surprisingly, nature seems to have provided us with a fundamental unit of angular momentum.

The fundamental unit of angular momentum turns out to be Planck's constant divided by 2π. This quantity is so important it is given a special symbol of its own: $\hbar$ (called "h-bar"). The angular momentum of a spinning bicycle wheel is about 10^{35} units. But the angular momentum of an electron in an atom is typically of the order of only 1 unit in magnitude because an electron has such a small mass and because the radius of its orbit is so tiny.

Consider the angular momentum produced by the motion of an electron circling about a nucleus. If a component of this orbital angular momentum vector is measured, *the result is always a whole number times the unit of angular momentum:*

The *largest* component value of the orbital angular momentum (in units of ℏ) for a quantum state is equal to the orbital quantum number of the state.

In a *p* state, for example, the electron moves with a maximum value of 1 unit of angular momentum. The minimum component value for a *p* state must be −1, corresponding to the angular momentum vector pointing in the opposite direction. And an intermediate value of 0 is also possible, but no others, because 0 is the only whole number between 1 and −1. Similarly, for a *d* state, the maximum value is 2 and the others must be 1, 0, −1, −2, giving five possible values in all. The states with a given orbital angular momentum but with *different* component values are the explanation for the so-called "hidden states." The component quantum numbers such as 1, 0, −1 are often called *magnetic quantum numbers* because in a magnetic field, states with different magnetic quantum numbers have different magnetic energies.

Orbital motion is not the only source of angular momentum in an atom. As an analogy, consider the earth orbiting about the sun. In addition to orbital angular momentum, the earth also possesses *spin angular momentum* because it rotates about its polar axis. Similarly, the electron possesses spin angular momentum as if it were rotating.

There are puzzling aspects to electron spin. If the spin of the electron were 1, for example, an applied magnetic field would reveal the presence of the three states corresponding to 1, 0, and −1. But in actual experiments only *two* states are found. According to our understanding of angular momentum, the spin of the electron must be 1/2. The states correspond to 1/2 and −1/2, separated by one unit as we require. So even though orbital angular momentum is always a whole number of units, *spin angular momentum can come in half units or whole units.* Furthermore, all other evidence seems to show that the electron has no size whatsoever. How can a particle with no size have spin angular momentum? All you can say is that as far as can be measured, it *acts* as if it is spinning.

The spin of a particle is one of its fundamental properties. An electron can have different orbital angular momentum in different orbits, but its spin is always 1/2. The spins of protons and neutrons are also 1/2. Some particles, such as pions, have spin 0. The photon has spin 1.

The spin of a particle is only one of the quantum numbers that describe it. So one way of trying to understand elementary particles is to search for new quantum numbers. The search is intimately related to the search for new conservation laws discussed in the preceding section. You have seen that the spin quantum number is related to an important physical quantity, angular momentum. And the reason angular momentum is important is that it obeys a conservation law. The search for new conservation laws is the same as the search for new quantum numbers.

HADRONS AND LEPTONS

One of the complications in dealing with elementary particles is that three distinct interactions are involved: the strong nuclear force, the electromagnetic force, and the weak interaction force. (In the case of elementary particles, the fundamental gravitational force seems to be too weak to be of importance.) Elementary particles can be grouped into two broad classes, depending on how they interact. Particles that exert strong nuclear forces are called *hadrons* ("the strong ones"). Protons and neutrons are hadrons. Other important hadrons and some of their properties are

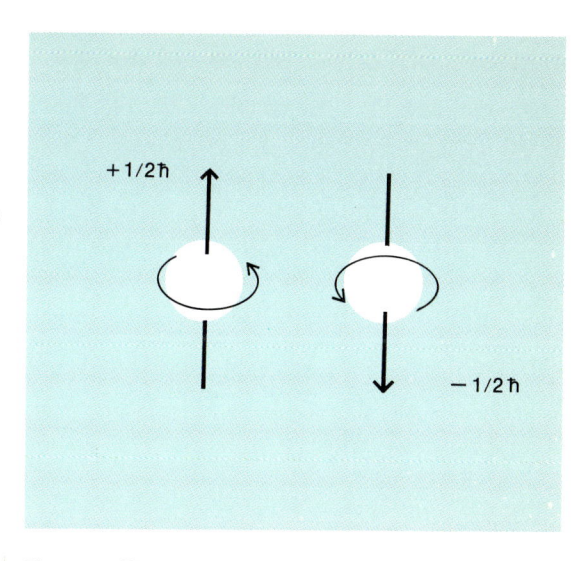

Figure 4. The spin angular momentum of an electron can only be components + ½ or −½.

listed in Table 1. Particles that are associated with the weak interaction, but cannot exert strong nuclear forces, are called *leptons* ("the weak ones"). The only known leptons are electrons and positrons, muons, and various kinds of neutrinos (Table 2). The photon is in a separate class by itself. Photons are produced by radiating electric charges and interact only electromagnetically.

Table 1.
Some Important Hadrons

Name	Symbol	Spin	Mass (million electron volts)	Mean Lifetime (seconds)
Baryons				
Proton	p	1/2	938.3	stable
Neutron	n	1/2	939.6	932
Lambda	Λ	1/2	1,115	2.5×10^{-10}
Sigma	Σ^+	1/2	1,189	0.8×10^{-10}
	Σ^0	1/2	1,192	$<10^{-14}$
	Σ^-	1/2	1,197	1.5×10^{-10}
Xi	Ξ^0	1/2	1,315	3.0×10^{-10}
	Ξ^-	1/2	1,321	1.7×10^{-10}
Omega	Ω^-	3/2	1,673	1.3×10^{-10}
Mesons				
Pion	π^+	0	140	2.6×10^{-8}
	π^0	0	135	0.9×10^{-16}
	π^-	0	140	2.6×10^{-8}
Kaon	K^+	0	494	1.2×10^{-8}
	K^0	0	498	$0.9 \times 10^{-10}, 5 \times 10^{-8}$*
	K^-	0	494	1.2×10^{-8}
Eta	η	0	549	0.3×10^{-18}

*Fifty percent have lifetimes of 0.9×10^{-10} second; fifty percent have lifetimes of 5×10^{-8} second.

Table 2.
The Leptons

Name	Symbol	Spin	Mass (million electron volts)	Mean Lifetime (seconds)
e-neutrino	ν_e	1/2	0	stable
μ-neutrino	ν_μ	1/2	0	stable
positron	e^+	1/2	0.5	stable
electron	e^-	1/2	0.5	stable
muon	μ^+	1/2	106	2×10^{-6}
	μ^-	1/2	106	2×10^{-6}

Some particles, such as protons, can interact according to all three forces. Similarly, electrons can interact electromagnetically as well as through the weak interaction. Nevertheless, there are some useful guides for telling which interaction will be the main one in a particular reaction. If the nuclear force is assigned the relative value 1.0, then the electromagnetic force has relative strength 10^{-2}, and the weak force is about

10^{-14} or so. These relative strengths imply that in general the strong interaction dominates. But sometimes a conservation law forbids reactions via the strong interaction, and in this case only reactions consistent with the electromagnetic interaction can occur. Finally, if additional conservation laws bar the electromagnetic interaction as well, the weak interaction will manifest itself.

Each type of interaction has a characteristic time scale based on its strength. The strong interaction is so strong that a reaction almost certainly will occur when two strongly interacting particles come into contact with each other. The time scale for the strong interaction is roughly the time required for a particle moving at nearly the speed of light to travel across the diameter of a proton—about 10^{-23} second. Correspondingly, reactions proceeding according to the weak interaction take about 10^{-10} second. Considerably longer times are not unusual, however, because the reaction times with weak interactions depend on the amount of energy available and other details.

QUANTUM THEORY OF FORCES

The chapter on electromagnetic radiation presented the classical picture of electromagnetic interactions. Charges produce electric and magnetic fields, and the fields exert forces on other charges. Although this picture works reasonably well even on the atomic scale, an alternative and more accurate model has been developed that takes into account the quantum "graininess" that the fields themselves exhibit. A charged particle such as an electron is considered to be surrounded by a cloud of photons. The photons are constantly being emitted and absorbed by the charge. According to this model, charges exert electric forces on one another by exchanging photons. The photons carry momentum from one charge to the other, like a beach ball being thrown back and forth by two children.

In 1935, Hideki Yukawa in Japan proposed that the nuclear force was also due to an exchange of momentum carriers between strongly interacting particles. These momentum carriers, called *mesons,* must be massive because of the short range of the nuclear force. To see why, suppose a proton emits a meson and then reabsorbs it. Classically, such a process does not conserve either energy or momentum. But according to the distance-momentum uncertainty principle, which was presented in the chapter on the quantum atom,

$$\begin{pmatrix}\text{Uncertainty}\\\text{in position}\end{pmatrix} \times \begin{pmatrix}\text{uncertainty}\\\text{in momentum}\end{pmatrix} > \text{Planck's constant}$$

The mesons must stay within about 10^{-13} centimeter of the proton, the range of the nuclear force. This makes the uncertainty in position small, so that the uncertainty in momentum must be large. And a large uncertainty in momentum means that you can't tell whether momentum is exactly conserved or not. By equating the position uncertainty to the range of the nuclear force, and by assuming that the meson is traveling at nearly the speed of light, you get an estimated meson mass of about 200 times the mass of an electron.

The mesons around a strongly interacting particle are called *virtual mesons* because they cannot be observed directly. In a collision between two particles, however, enough energy may be transferred to a meson to allow it to leave completely. Two years after Yukawa's prediction, a particle with about the expected meson mass was detected in cosmic rays. Unfortunately, later experiments showed that the new particle, the *muon,* interacted only weakly with matter. The strongly interacting meson expected by Yukawa should exist for only 10^{-18} second or so in matter before it reacts with a nucleus. The muon, however, lives for 10^{-6} second. The discrepancy was resolved in 1947 with the discovery of another meson, the *pion,* which interacts strongly and fits the Yukawa prediction. Most pions produced in the upper atmosphere by cosmic rays react almost immediately with other matter and never reach the earth. Muons are produced in these reactions, however, and because of their weak interaction, many muons pass through the atmosphere to the earth. Pions were discovered by searching at altitudes above the atmosphere. Today accelerators produce copious quantities of pions and muons for study; pions are even being used in cancer therapy.

PARTICLES AND ANTIPARTICLES

In our world, electrons have negative electric charge and are bound in atoms by their attraction to a positively-charged nucleus. Why does nature seem to favor negative electrons and positive nuclei? Positive electrons and negative nuclei would work just as well, but ordinary matter doesn't contain atoms with such reversed construction.

Nevertheless, one of the remarkable discoveries of particle physics is that every kind of particle has a corresponding *antiparticle.* A particle and its antiparticle always have opposite electric charge. But other properties, such as mass or interaction strengths, are the same for both. One of the first new particles to be discovered, the *positron,* is the antiparticle to the electron. (You could also say that the electron is the antiparticle to the positron.) Similarly, the antiproton, first produced in 1955 with the help of a large accelerator, has a negative charge but is otherwise much the same as an ordinary proton in its outward properties. Antineutrons are electrically neutral, like the neutron itself.

Table 3.
Some Particles and Their Antiparticles

Particle	Symbol	Antiparticle	Symbol
Proton	p	antiproton	$\bar{p}$
Neutron	n	antineutron	$\bar{n}$
Sigma	Σ^+	antisigma	Σ^-
	Σ^0		$\bar{\Sigma}^0$
	Σ^-		Σ^+
Pion	π^+	pion	π^-
	π^0		π^0
	π^-		π^+
Kaon	K^+	kaon	K^-
	K^0	antikaon	$\bar{K}^0$
	K^-	kaon	K^+
Muon	μ^+	muon	μ^-
	μ^-		μ^+
Positron	e^+	electron	e^-
Electron	e^-	positron	e^+
e-neutrino	ν_e	anti-e-neutrino	$\bar{\nu}_e$
μ-neutrino	ν_μ	anti-μ-neutrino	$\bar{\nu}_\mu$

You can imagine an antihydrogen atom made up of a positron and an antiproton. By using antineutrons and antimesons, you could build up the antiversion of any kind of ordinary matter. Unfortunately, antiparticles have a disconcerting property: a particle and its antiparticle always annihilate each other when they come into contact. An electron and a positron annihilate to give photons. Protons and antiprotons annihilate in a variety of ways to give pions and photons. Whatever the internal quantum numbers of particles and antiparticles may be, they are artfully arranged so that no conservation law ever forbids particle-antiparticle annihilation.

Antimatter would not last long in our world, and we wouldn't last long in a world of antimatter. The question remains why all matter seems preponderantly to be the kind with negative electrons and positive protons. One possibility is that distant galaxies may be made of antimatter. If such a galaxy ever came into contact with a galaxy made of normal matter, vast amounts of energy would be emitted in the ensuing annihilations. Such energy release has been looked for, but at present the existence of large amounts of antimatter in the universe remains a speculation.

REACTIONS OF ELEMENTARY PARTICLES

One of the first reactions to be studied was beta decay, in which a neutron seems to disintegrate spontaneously to a proton and an electron:

Neutron $\longrightarrow$ proton + electron

The mass-energy of a free neutron is greater than the combined mass-energies of a proton and an electron, so the proton and electron should carry away the additional energy in the form of kinetic energy. But when the energy was measured, some was missing and could not be accounted for.

Wolfgang Pauli in the early 1930s suggested what he called a "desperate remedy." He postulated that a *third* particle was produced in the decay and carried off some of the energy. The third particle would not be detected, Pauli said, if it had no electric charge and otherwise interacted only weakly with matter.

Pauli's suggestions received complete confirmation in 1956 with the detection of the third particle in the intense neutron flux near a nuclear reactor. The correct beta decay reaction was proved to be:

Neutron $\longrightarrow$ proton + electron + $\bar{\nu}_e$

The ordinary beta decay reaction produces the antiparticle to the neutrino. The neutrino itself is produced in a reaction such as

Proton $\longrightarrow$ neutron + positron + ν_e

which can occur in nuclei or in collisions of a proton with other particles. The sun is a copious emitter of neutrinos because of reactions of this type, and about 5 percent of the energy emitted by the sun is in the form of neutrinos.

Like photons, neutrinos appear to have zero mass and must move at the speed of light. But neutrinos do not interact electromagnetically. They interact only through the weak interaction, which is so weak that neutrinos pass through matter almost like ghosts. A neutrino striking the earth is almost sure to pass completely through: the chance of its interacting with matter is only 1 in 10^{10}. About 10^{15} or so neutrinos from the sun pass through our bodies every second. Of these, perhaps one is absorbed each minute.

One useful property of particle reaction equations is that given enough energy, another reaction is obtained by replacing a particle on one side of the reaction by its antiparticle on the other. For example, the reaction

Proton $\longrightarrow$ neutron + positron + ν_e

implies the reaction

$\bar{\nu}_e$ + proton $\longrightarrow$ neutron + positron

This reaction is the one used to detect the antineutrino. An antineutrino reacting with a proton in hydrogen produces a free neutron and a positron, and the detection of both products provides unambiguous evidence that an antineutrino was present.

Neutrinos and antineutrinos are truly distinct particles. From the reaction

Neutron $\longrightarrow$ proton + electron + $\bar{\nu}_e$

you may derive the reaction

ν_e + neutron $\longrightarrow$ proton + electron

A detector based on this reaction gives no counts when placed near a nuclear reactor because reactors produce antineutrinos, not neutrinos.

The beta decay reaction

Neutron $\longrightarrow$ proton + electron + $\bar{\nu}_e$

is a typical example of a reaction occurring by way of the weak interaction. It involves two leptons, the electron and the antineutrino. The muon is also a lepton and reacts in a similar way. For example,

Neutron $\longrightarrow$ proton + μ^- + $\bar{\nu}_\mu$

It was discovered in the early 1960s that the neutrinos in such reactions are not the same as the neutrinos associated with electrons. The muon has its own kind of neutrino, and a reaction such as

ν_μ + neutron $\longrightarrow$ proton + electron

has never been observed. Muon decay is interesting because it involves both kinds of neutrinos:

$\mu^+ \longrightarrow$ positron + ν_e + $\bar{\nu}_\mu$

The long average lifetime of the muon, 2×10^6 second, shows in another way that the decay proceeds according to the weak interaction.

The hadrons, or strongly interacting particles, can undergo a wide variety of reactions by way of the strong interaction, given sufficient energy. Here is a sample:

π^- + proton $\longrightarrow \pi^0$ + neutron

π^+ + neutron $\longrightarrow \Lambda + K^+$

K^- + proton $\longrightarrow \Lambda + K^+ + K^-$

π^+ + proton $\longrightarrow$ proton + $\pi^+ + \pi^+ + \pi^- + \pi^0$

There is a striking fact about these and all other known reactions. Suppose the hadrons are divided into two classes on the basis of their spins. All hadrons with whole number spins, such as the pion (spin 0), are called *mesons*. All hadrons with fractional spins, such as the proton (spin 1/2), are called *baryons* ("the heavy ones"). It turns out that in any reaction, any number of mesons can be produced given enough energy, but the number of baryons is strictly conserved.

The law of conservation of baryons is similar to the law of conservation of charge. To bring out the analogy, assign a baryon "charge" of +1 to all baryons, a baryon charge of −1 to all baryon antiparticles, and a baryon charge of 0 to mesons. An example of how this works is the reaction

$K^- + \text{proton} \longrightarrow \Lambda + K^+ + K^-$

$0 \;\;+1 \qquad\qquad 1 + 0 \;\;+0$

The total baryon number is 1 on both sides of the reaction; baryon number is conserved. Conservation of baryon number prevents protons from decaying into pions and electrons, even though there is enough energy for the decay.

Lepton number obeys a strict conservation law as well, apparently because all known leptons have fractional spin 1/2. Again assign leptons the quantum number +1 and antileptons the number −1. However, baryon numbers and lepton numbers must be totaled separately; baryons have zero lepton number and vice versa. A good example is the beta decay reaction

Neutron $\longrightarrow$ proton + electron + $\bar{\nu}_e$

Baryon number: $1 \longrightarrow 1 + 0 + 0$

Lepton number: $0 \longrightarrow 0 + 1 - 1$

ISOTOPIC SPIN AND HYPERCHARGE

Telling the difference between a proton and a neutron is easy—protons have electric charge and neutrons do not. But from the standpoint of the strong interaction, protons and neutrons seem to be identical. Modern accelerator equivalents of Rutherford's scattering experiment show that after the electromagnetic interaction is subtracted, the force between two protons is the same as the force between a proton and a neutron or the force between two neutrons. Furthermore, the mass of the proton is nearly the same as the mass of the neutron. To help organize ideas about the strong interaction, a model has been developed in which the proton and the neutron are viewed not as distinct particles but as two aspects of the same particle. That particle is called the *nucleon*.

An analogy based on electron spin is useful here. An electron has spin 1/2 and therefore possesses two hidden states with angular momentum components +1/2 and −1/2. Similarly, the nucleon is assigned a new quantum number 1/2; the proton corresponds to the component +1/2 and the neutron to the component −1/2. The new quantum number is called *isotopic spin* because the proton and neutron are like "isotopes" of the nucleon. You can say that the proton and the neutron form an isotopic spin *doublet*. However, isotopic spin has nothing at all to do with ordinary spins and rotations. It is based on the recognition that the kind of mathematics used in dealing with angular momentum can be applied to new situations.

The three pions—positive, neutral and negative—have the *same* strong interactions and nearly the *same* masses. It is natural to view the pions as three states +1, 0, −1 of *one particle with isotopic spin 1*. The pions form an isotopic spin *triplet*. The neutral lambda baryon, on the other hand, seems to have no natural partners and forms an isotopic spin *singlet* with isotopic spin 0. The other hadrons can be classified into sets, or *multiplets,* according to isotopic spin as well. The concept of isotopic spin does not apply to leptons but only to particles that interact strongly.

You have seen that quantum numbers are connected with conservation laws. Assigning new quantum numbers such as baryon number or isotopic spin has no purpose unless there is an associated conservation law. To see one aspect of the conservation law for isotopic spin, consider the reaction

$\pi^- +$ proton $\to \Sigma^- + K^+$

$-1 + 1/2 \quad -1 + 1/2$

Table 4.
Isotopic Spin and Hypercharge for Some Important Hadrons

Name	Symbol	Isotopic Spin Component	Hypercharge
Proton	p	+1/2	1
Neutron	n	−1/2	1
Lambda	Λ	0	0
Sigma	Σ^+	+1	0
	Σ^0	0	0
	Σ^-	−1	0
Xi	Ξ^0	+1/2	−1
	Ξ^-	−1/2	−1
Pion	π^+	+1	0
	π^0	0	0
	π^-	−1	0
Kaon	K^+	+1/2	1
	K^-	−1/2	−1

However, the similar reaction

$\pi^- +$ proton $\to \Sigma^+ + K^-$

$-1 + 1/2 \quad 1 - 1/2$

does not conserve isotopic spin and experimentally is found not to proceed by way of the strong interaction.

Each particle is an isotopic spin multiplet has a different value of isotopic spin component, depending on its electric charge. Because electric charge is conserved in all reactions, it is possible to combine isotopic spin and charge to form another quantum number, *hypercharge:*

Hypercharge = 2 × (electric charge) − 2 × (isotopic spin component)

The advantage of hypercharge is that every particle in a multiplet has the same value of hypercharge. For example, all three pions have hypercharge 0, and both the proton and the neutron have hypercharge +1. Hypercharge adds simply, like electric charge. For the reaction

$\pi^- +$ proton $\Sigma^- + K^+$

the hypercharge is $0 + 1 \longrightarrow 0 + 1$.

Unlike the conservation laws for energy or electric charge, the conservation law for hypercharge is not universally valid. Hypercharge is not necessarily conserved in weak interactions. A reaction forbidden by hy-

Figure 5. A ball possesses *rotational symmetry*; a decorative frieze possesses *translational symmetry*.

percharge conservation may in fact occur by way of the weak interaction. This insight explains a puzzling feature of baryon production and decay. The negative sigma (Σ^-) baryon, for example, can be produced in the reaction

π^- + proton $\longrightarrow$ Σ^- + K^+

It is clear that the Σ^- is a strongly interacting particle. Yet the Σ^- decays according to the scheme

$\Sigma^- \longrightarrow$ neutron + π^-

with a mean lifetime of 10^{-10} second or so, a lifetime on the time scale of the weak interaction. Furthermore, the hypercharge (0, and 1 + 0) is not conserved in the decay, another indication that the weak interaction is involved.

Why does the Σ^- decay according to the weak interaction if it is a strongly interacting particle? The answer is that if the Σ^- produced is associated with another hadron such as the positive kaon (K^+), hypercharge can be conserved and the strong interaction dominates. However, after the Σ^- flies off alone, there is no decay reaction available to it that conserves hypercharge and only the weak interaction can act.

CONSERVATION LAWS AND SYMMETRY

Conservation laws have been exceptionally helpful in the study of elementary particles. One reason is that a conservation law represents a fundamental and general way of stating a physical principle. But there is

Figure 6. The distance between these attracting spheres is the same in both coordinate systems.

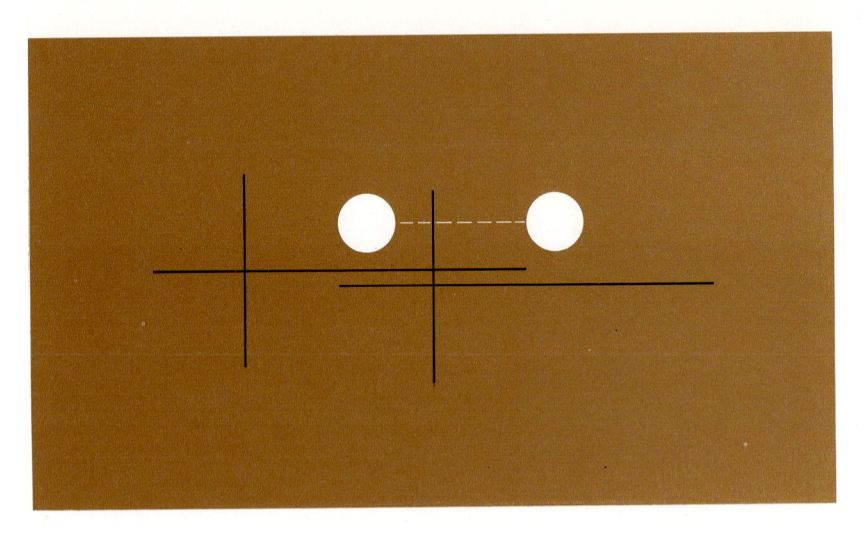

something perhaps even more fundamental than conservation laws: symmetry. You know that a ball or a block or a decorative pattern or your body seem to be "symmetrical." Here is a definition of symmetry that sharpens these ideas: an object possesses symmetry if some operation can in principle be performed that leaves the object unchanged in appearance. For example, rotating a ball through any angle leaves it looking the same. But only certain rotations leave a cube unchanged—rotations by 90 degrees, for instance. And a decorative frieze with a repeating pattern looks unchanged if you move along it the distance of one pattern. There are many kinds of symmetry: *rotational* symmetry for the ball, *translational* symmetry for the decorative frieze.

No real object ever has perfect symmetry. Balls aren't perfectly round, blocks may have nicks in the edges. So representations of symmetry occur most often in the ideal world of art and design. Mathematics is an ideal world as well, and the mathematical formulation of physical laws may also possess true symmetry of various kinds. Some of the symmetry operations that can be applied to mathematical equations cannot be carried out with actual objects. One example is the mirror symmetry operation, which is equivalent to changing a left hand to a right hand. So the symmetry operations useful in physics are often abstract and more difficult to visualize than the ordinary geometric symmetries you are most familiar with.

Our experience indicates that in empty space one position is as good as another, and one direction is as good as another. So physical laws have to be formulated in a way that implies translational symmetry and rotational symmetry. As a simple example, take the gravitational force between two bodies. Gravitational force depends on the distance between bodies, not on their absolute positions. Using either of the coordinate systems shown in the sketch gives the same distance, hence, the same calculated value for the force.

By a mathematical argument too complicated to develop here, it turns out that if the laws of motion have translational symmetry, linear momentum must automatically obey a conservation law. And if the laws of motion have rotational symmetry, angular momentum automatically must be conserved.

Another general requirement for the laws of motion is that they be independent of the direction of time flow. To see what this means, suppose you throw a ball in the air and make a movie film of its flight. Someone seeing the film running backward through a projector would not

Figure 7. The laws of motion must be independent of the direction of time flow. Can you tell from this drawing the direction in which the ball is moving?

notice anything peculiar about the motion. The motion itself contains no clue about the direction of time and looks physically reasonable when it is run either forward or backward in time. More complicated events seem at first sight to define a particular direction for time. If you drop a glass and it breaks, the pieces will not come back together and leap back up to your hand. But if every bit of glass was started off properly, the reverse event *could occur.* It is the difficulty of obtaining the exact initial conditions that makes the reverse of complicated events improbable. Simple events such as the collision of two marbles are clearly reversible. The symmetry of the laws of motion with respect to the direction of time leads to the law of conservation of energy.

Conservation laws may hold for some interactions but not for others, as you saw in the case of hypercharge. In the 1950s, the law of conservation of *parity* was shown to be violated by the weak interactions. Parity is not a familiar quantity in the everyday world, and it is easier to understand parity conservation by discussing its equivalent symmetry operation, mirror reflection. When you look into a mirror, the mirror world you see appears to be perfectly reasonable from the standpoint of physical laws. In other words, there seems to be no way to distinguish the mirror world from this world. There is an alternative way of describing mirror symmetry. If you hold your right hand in front of a mirror, the reflection looks like a left hand. As long as mirror symmetry holds, right-handedness and left-handedness are equally valid. Under mirror symmetry, no physical law can distinguish right hand from left, or clockwise from counterclockwise, except by arbitrary conventions without physical significance.

Everyone was so sure that mirror symmetry had to hold in all physical laws that any theory not taking into account this notion was

30 | Elementary (?) Particles and Symmetry

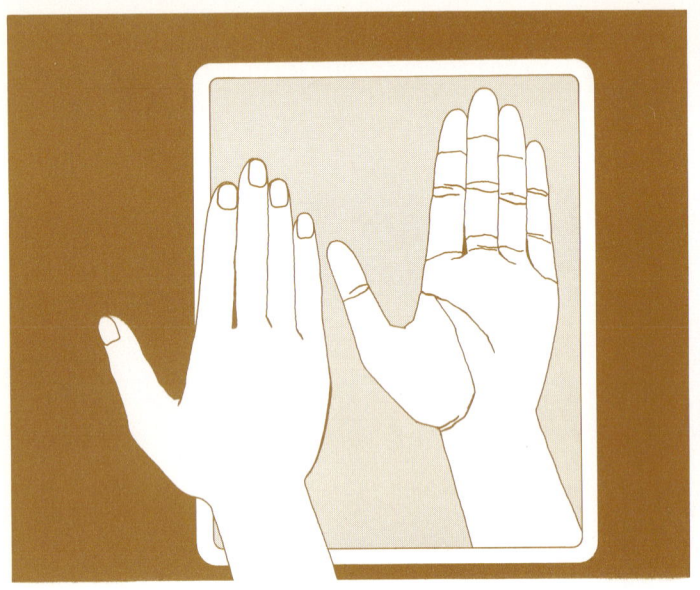

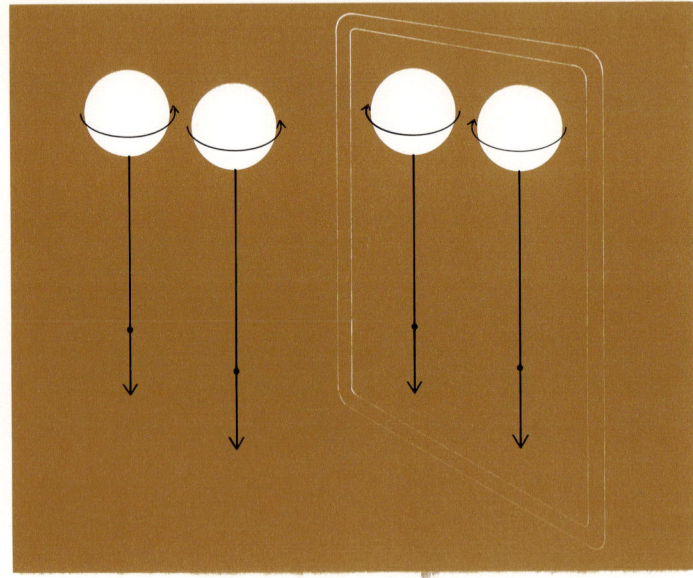

Figure 8. A right hand reflected in a mirror looks like a left hand.

Figure 9. In the real world, beta rays from cobalt 60 are given off in a certain direction with respect to the direction of nuclear rotation. The situation in the mirror does not represent a possible event in the real world.

rejected immediately. But only experimentation can tell us what the physical world is actually like; preconceived ideas cannot be relied upon. In 1956, T. D. Lee and C. N. Yang examined the existing experimental evidence for mirror symmetry and concluded that it had not been effectively checked for the weak interaction. Within a few months, C. S. Wu and her colleagues had performed a suitable experiment with cobalt-60 nuclei. And she found that nature *does* make a fundamental distinction between right hand and left.

Cobalt-60 is radioactive, undergoing beta decay. The cobalt-60 nucleus is spinning and can be lined up with its spin along a particular direction. The surprising result of the experiment is that the beta particles are always emitted along the direction opposite to the spin of the nucleus. The experiment is not correct in the mirror world, as Figure 10 shows. The nuclei are spinning in the wrong direction.

In order for mirror symmetry to hold, the beta rays would have to be given off equally in both directions. Instead, nature goes wholeheartedly to one extreme in the weak interaction. Furthermore, extensions of the alignment experiment showed that an antineutrino always has its spin along its direction of flight. Now you can explain to anyone in the universe what people on the earth mean by counterclockwise—it is the direction an antineutrino rotates, as seen from in front. But for someone living in an antimatter world, beta decay would emit neutrinos, and your directions would be wrong. An antimatter mirror world looks physically reasonable, even with respect to the weak interaction. Symmetry is restored.

Even time reversal may be violated by weak interactions, however. It may be that only an antimatter mirror world with reversed time looks like our world. The comparatively insignificant weak interaction has told us many strange things so far. Perhaps there are more surprises to come.

Symmetry principles and conservation laws appear to be closely connected. However, there is no known symmetry principle that leads to conservation of electric charge or conservation of baryon number. On the other hand, symmetry principles are valuable in the search for new conservation laws and quantum numbers. Even though a detailed theory for the elementary particles does not exist, guesses can be made as to the symmetries that such a theory might exhibit. Most of the symmetries that have been tried cannot be described in terms of symmetry operations in ordinary space and time. One scheme, suggested by Murray Gell-Mann and others, is based on generalizations of the mathematics of angular momentum and isotopic spin. Just as sets of particles are grouped into multiplets by isotopic spin, the new symmetries group multiplets together into even larger sets. Two of the sets, with eight particles in each, are shown in Figure 11. There is more to the model than the order suggested by the snowflake patterns. Some of the groupings were initially incomplete. But when the missing particles were discovered experimentally they proved to have the predicted properties. This achievement was comparable to Mendeleev's prediction of new elements. And like Mendeleev's accomplishment, it has stimulated the search for the source of this order.

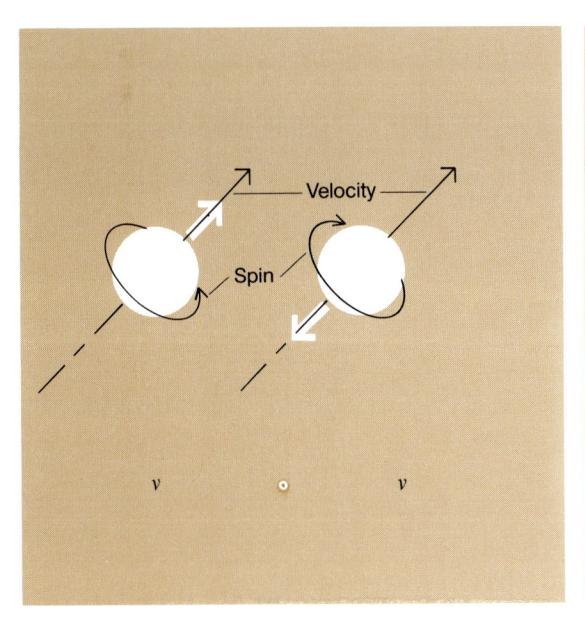

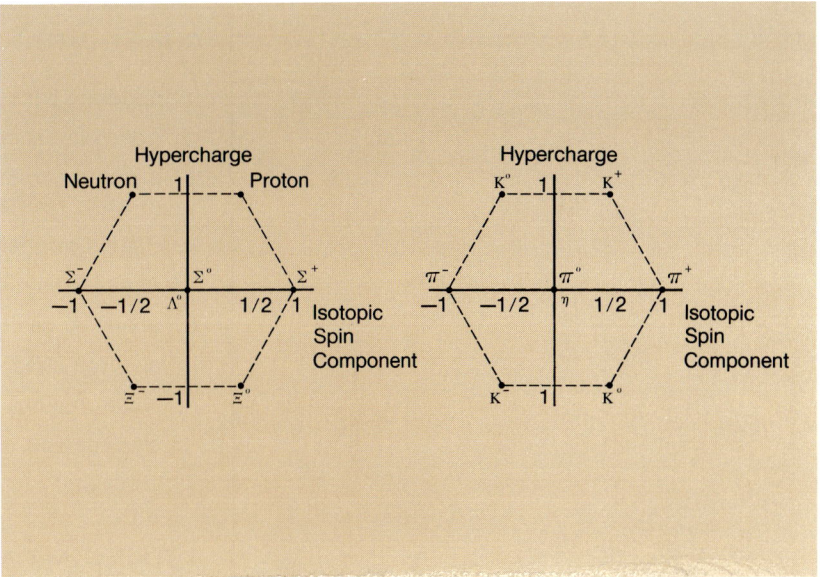

Figure 10. The spins of both antineutrinos and neutrinos bear a fixed direction to the respective directions of motion.

Figure 11. Here, two sets of "multiplets" are grouped into symmetric sets of eight particles each. Such classifications aid physicists in establishing some order in the little-understood world of elementary particles.

Appendix 1. Constants, Conversions, and Astronomical Data

CONSTANTS

c	speed of light (in vacuum)	3×10^8 meters/second
G	gravitational constant	6.7×10^{-11} newton-meter^2/kilogram^2
g	acceleration due to gravity at earth's surface	9.8 newtons/kilogram
h	Planck's constant	6.6×10^{-34} joule-second
N_0	Avogadro's number	6×10^{23} atoms/gram-mole
R	universal gas constant	8.31 joules/gram-mole-°K
	electron charge	1.6×10^{-19} coulomb
	electron mass	9×10^{-31} kilogram
	electron volt	1.6×10^{-19} joule

ASTRONOMICAL DATA (Average Values)

earth mass	5.98×10^{24} kilograms
earth radius	6,400 kilometers
moon radius	1,740 kilometers
sun mass	1.99×10^{30} kilograms
sun radius	696,000 kilometers
distance from earth to sun	149×10^6 kilometers
light-year	9.46×10^{15} meters

CONVERSIONS

Energy Units

    1 joule = 0.239 gram-calorie at 4°C
        = 0.738 foot-pound
        = 3.725×10^{-7} horsepower hours
1 gram-calorie = 4.18 joule
    1 kilowatt = 1,000 watts
        = 738 foot-pound/second
        = 1.34 horsepower

Length

    1 centimeter = 0.394 inch
        1 meter = 100 centimeters
            = 39.4 inches
            = 3.28 feet
    1 kilometer = 1,000 meters
        = 0.621 mile
1×10^{-10} meter = 1 angstrom

Pressure Units

1 gram/centimeter^2 = 1.42 × 10^{-2} pound/inch^2
= 7.36 × 10^{-1} mm mercury
= 9.68 × 10^{-4} atmosphere

Speed

1 meter/second = 3.60 kilometers/hour
= 0.37 mile/minute
= 196.8 feet/minute
= 3.28 feet/second

Temperature

degree Centigrade = 5/9(°F−32°)

degree Fahrenheit = 9/5(°C) + 32°

degree Kelvin = °C + 273°

Weight or Mass

1 gram = 0.035 ounce (avoirdupois)

1 kilogram = 2.20 pounds (avoirdupois)

1 atomic mass unit = 1.66 × 10^{-24} gram
= 1.49 × 10^{-10} joule
= 931 million electron volts

Appendix 2. Star Charts

You can use the first four charts for the evenings and mornings on the dates indicated if you live between the latitudes 30 degrees and 50 degrees. To use the charts, you must know your approximate latitude, which you can determine from the small map of the earth. Draw a line across the star map at your latitude position and use that portion of the map above the horizontal line. If you look north, use the maps on the right; the horizon will be the portion of the circle above your latitude line. If you look south, use the maps on the left. Looking north, Polaris stays relatively fixed throughout the seasons; it can be found simply by following the "handle" of Ursa Minor. If you were on the equator, Polaris would be on the northern horizon. The last maps illustrate some of the stars visible from the Southern Hemisphere. Looking south, Crux serves as a point of reference, although it is not as "stationary" as Polaris is in the Northern Hemisphere. In both hemispheres, you can trace the flow of the Milky Way—the light, cloudy area running through the sky.

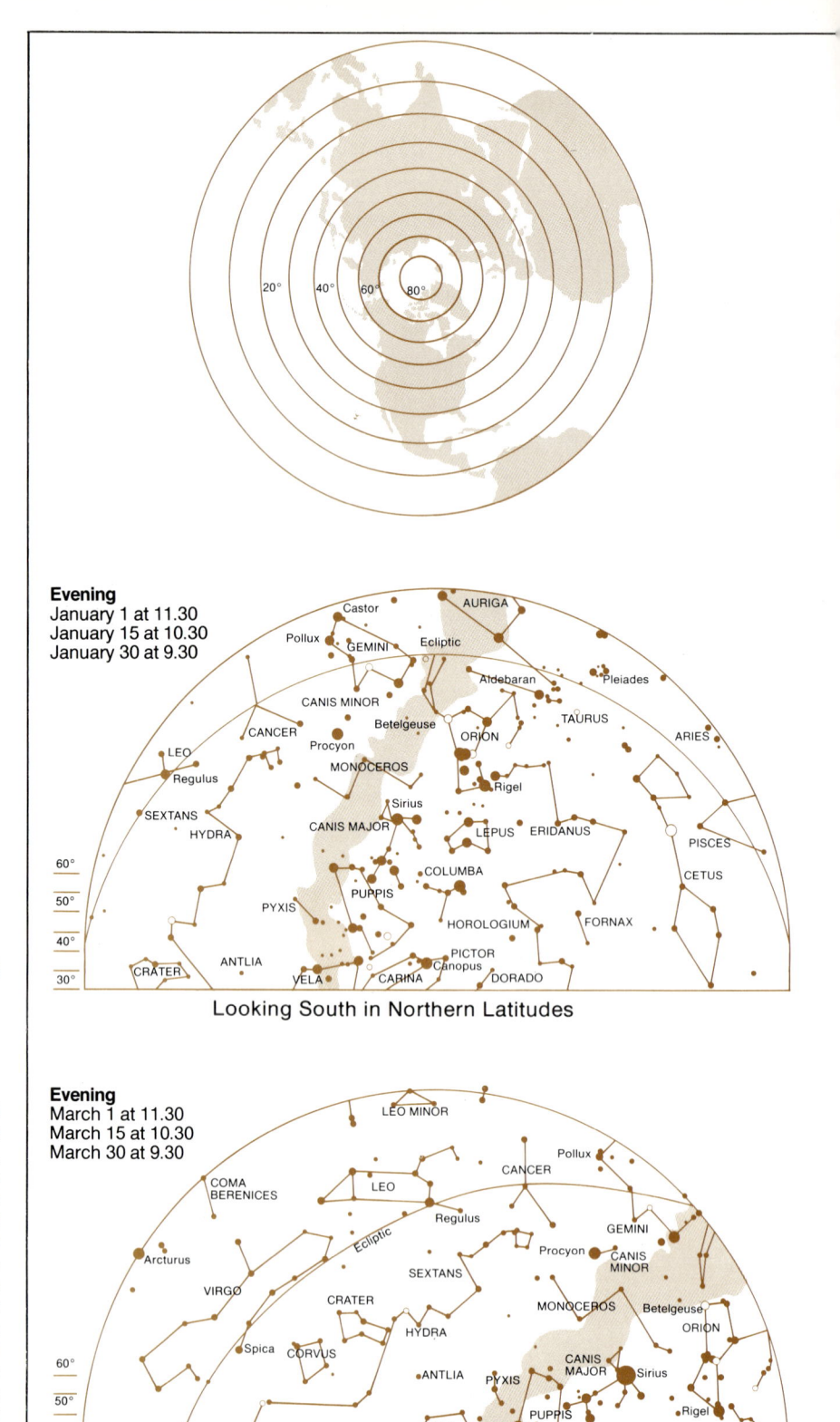

Evening
January 1 at 11.30
January 15 at 10.30
January 30 at 9.30

Looking South in Northern Latitudes

Evening
March 1 at 11.30
March 15 at 10.30
March 30 at 9.30

Looking South in Northern Latitudes

594

Morning
October 1 at 5.30
October 15 at 4:30
October 30 at 3:30

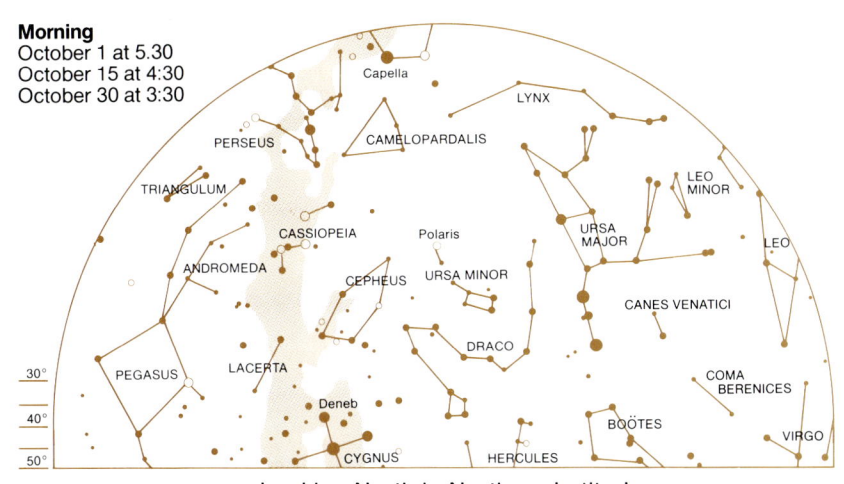

Looking North in Northern Latitudes

Morning
November 15 at 6.30
December 1 at 5.30
December 15 at 4.30

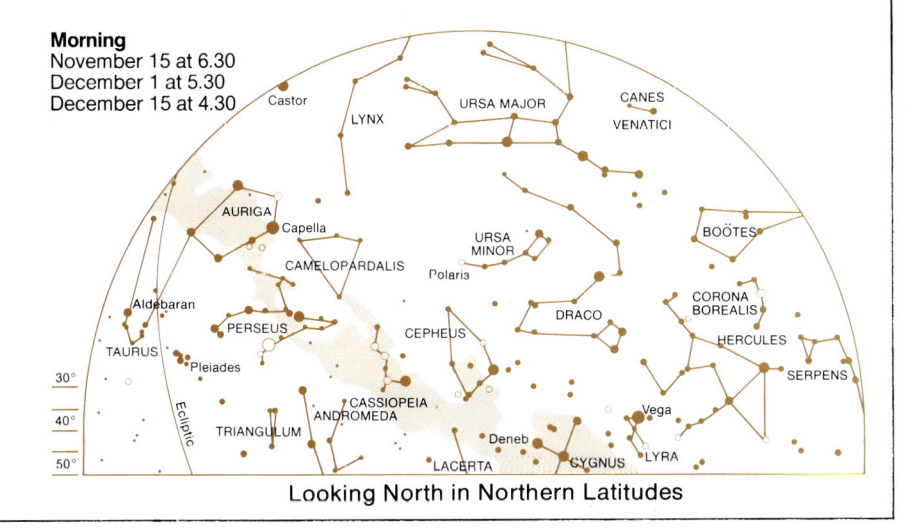

Looking North in Northern Latitudes

Evening
May 1 at 11.30
May 15 at 10.30
May 30 at 9.30

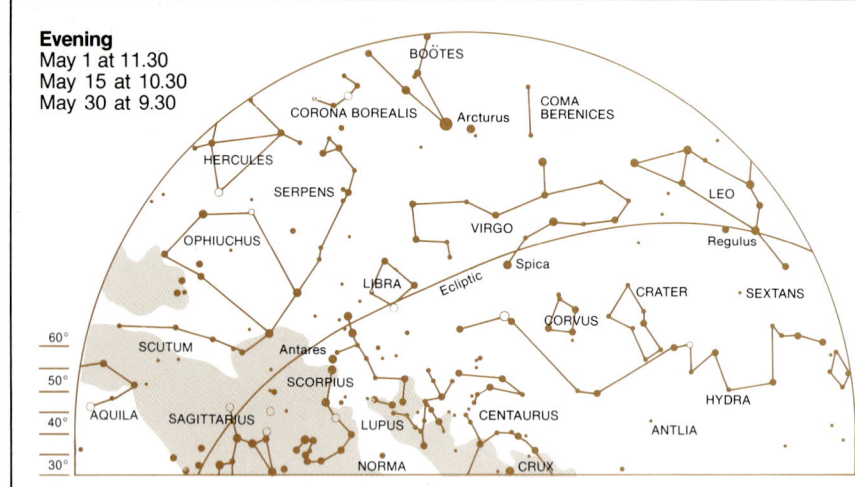

Looking South in Northern Latitudes

Evening
July 1 at 11.30
July 15 at 10.30
July 30 at 9.30

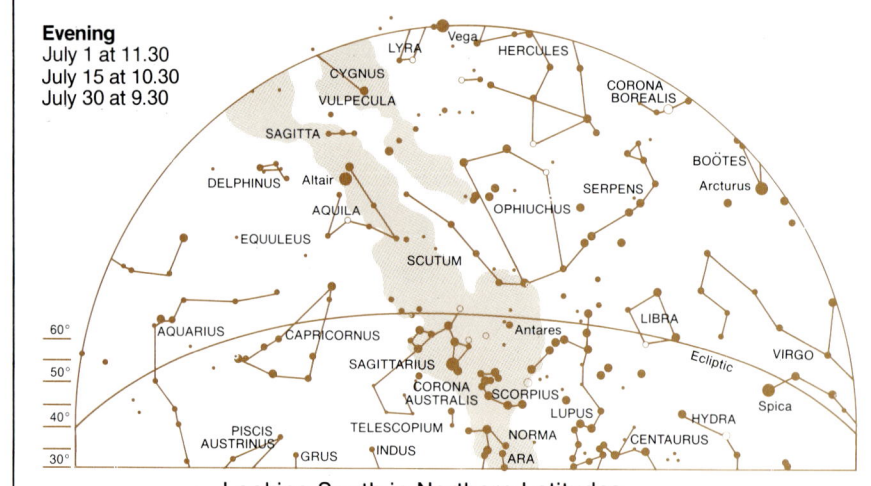

Looking South in Northern Latitudes

Morning
October 1 at 5.30
October 15 at 4.30
October 30 at 3.30

Southern Hemisphere

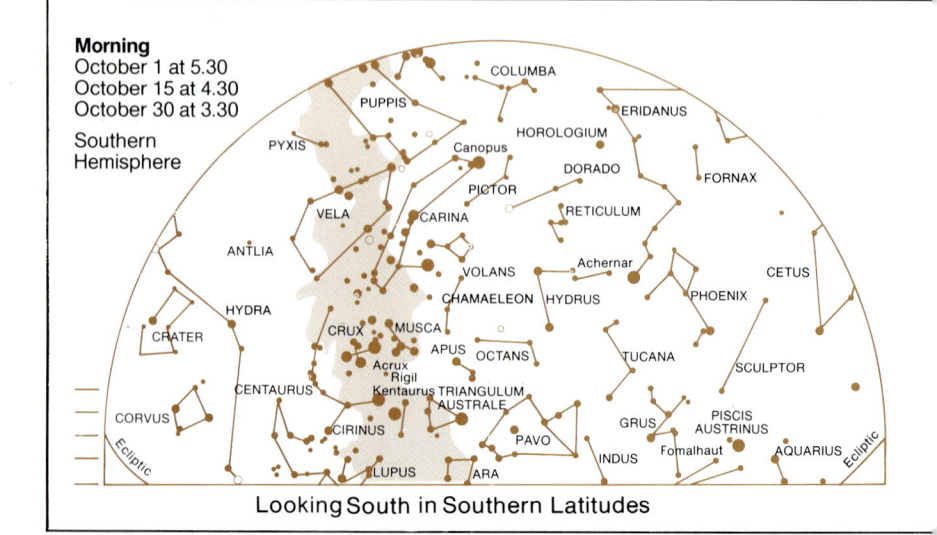

Looking South in Southern Latitudes

Morning
January 15 at 6.30
February 1 at 5.30
February 14 at 4.30

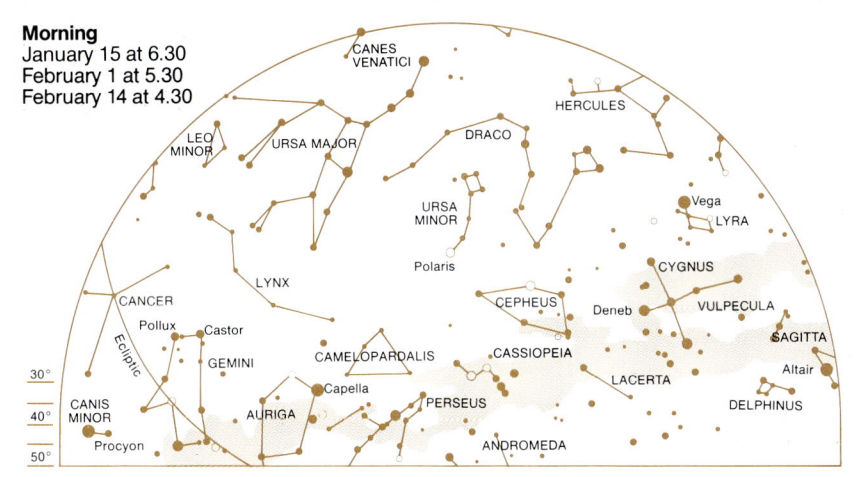

Looking North in Northern Latitudes

Morning
April 1 at 5.30
April 15 at 4.30
April 30 at 3.30

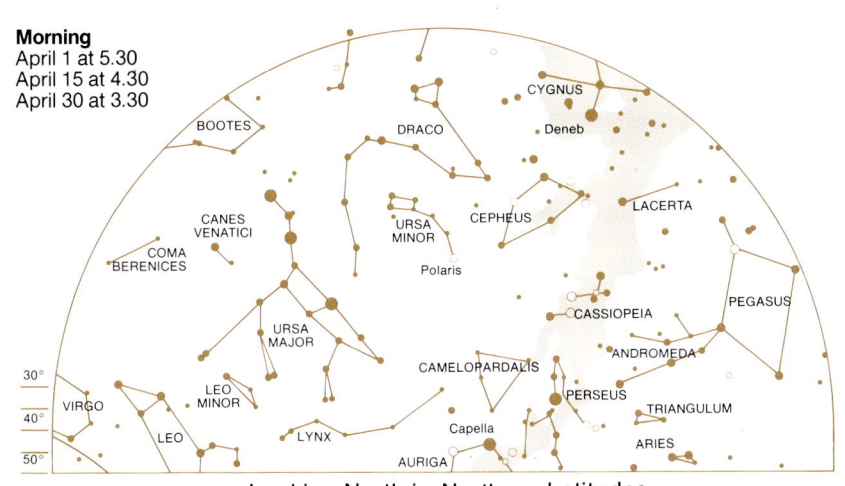

Looking North in Northern Latitudes

Evening
January 1 at 11.30
January 15 at 10.30
January 30 at 9.30

Southern Hemisphere

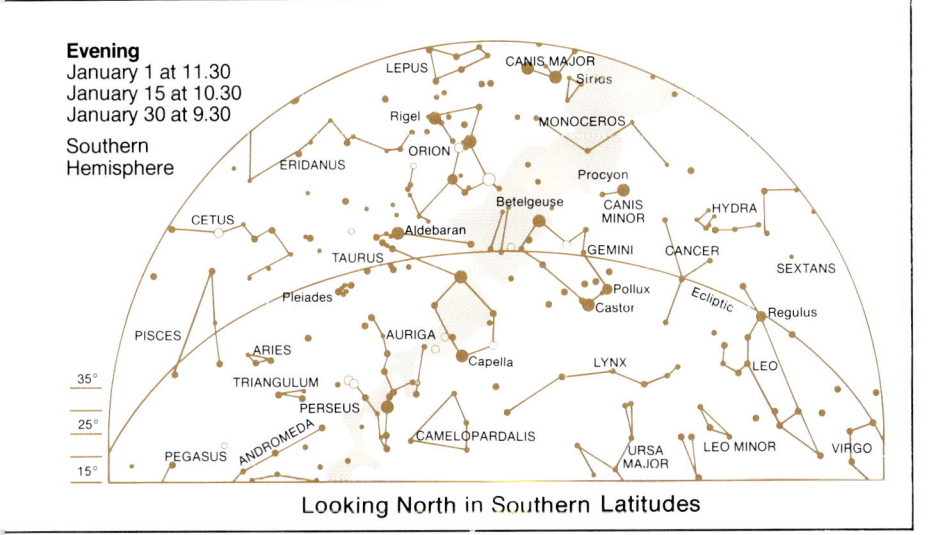

Looking North in Southern Latitudes

Contributing Consultants

Isaac Asimov received his doctorate in chemistry in 1948 from Columbia University, where he also earned his bachelor's and master's degrees. He joined the faculty of the Boston University School of Medicine in 1949 and is presently associate professor of biochemistry there. Dr. Asimov began writing professionally for magazines at the age of 18 and published his first book in 1950; in July 1972, his one-hundred-and-twenty-third book was published. His imaginative essay in this book reflects his widely acclaimed ability for provocative writing in both science and science fiction. Dr. Asimov's interests span the fields of anatomy, biology, chemistry, mathematics, astronomy, geography, and history. Two of his published works related to his essay for this book are *The Universe* and *Understanding Physics* (three volumes).

Adolph Baker, currently professor of physics at Lowell Technological Institute in Massachusetts, was awarded a doctorate in physics from Brandeis University after obtaining four previous degrees in the humanities as well as in science. He has worked in both electronics and theoretical physics and has conducted research in nuclear physics, scattering theory, optics, and geophysics. One of his special interests is the communication between scientists and humanists. He is the author of *Modern Physics and Antiphysics*, a book addressed primarily to nonscientists. His papers have appeared in *Physical Review* and other physics journals. In addition to contributing an essay to this book, Dr. Baker served as a writing consultant for the project.

Marion Bickford received his doctorate in 1960 from the University of Illinois, and is now professor of geology at the University of Kansas. He is a Fellow of the Geological Society of America and of the Mineralogical Society of America. Dr. Bickford's principal research interest lies in the study of long-lived radionuclides for geochemical purposes. His most recent research efforts have been directed toward the measurement of uranium-lead radiometric ages from the common accessory mineral zircon. In most of his published research, Dr. Bickford has dealt with measurements of the age of rock units from ancient terrains, attempting to reconstruct geologic history using the ages of rocks and other geologic and petrologic data. He contributed to the conceptualization of the earth science material for this book and was major writer for that unit.

Geoffrey Burbidge received his doctorate from the University of London in 1951. He is presently professor of physics at the University of California, San Diego. His previous appointments have taken him to the Enrico Fermi Institute for Nuclear Studies, the California Institute of Technology, and the Cavendish Laboratory at Cambridge University. He is a participant in numerous committees on astronomy and astrophysics. In 1959 Dr. Burbidge was awarded the Warner Prize jointly with E. M. Burbidge. He was elected Fellow of the Royal Society of London in 1968 and was made Fellow of University College in London in 1970. Dr. Burbidge contributed to the planetary science material in this book.

Gregory Choppin was awarded his doctorate from the University of Texas in 1953, and is presently professor and chairman of the Department of Chemistry at Florida State University. Dr. Choppin previously worked in the Lawrence Radiation Laboratory at the University of California, Berkeley, as a member of the team that in 1955 discovered the chemical element mendelevium. He is a member of the Committee on Nuclear Science and chairman of the Subcommittee on Radiochemistry of the National Research Council of the National Academy of Sciences. His research interest is in inorganic and nuclear chemistry. He is the author of four textbooks and approximately one hundred research papers. Dr. Choppin was a major advisor and writer for this project on the topics of the chemical atom, the nucleus, and nuclear energy.

Michael Chriss, a general contributor to this project, is planetarium director at the College of San Mateo and is an instructor in astronomy and physical sciences there. He received both his bachelor's and master's degrees in astronomy from the University of Arizona and pursued graduate work in the history of science at the University of California, Berkeley. He worked on the dynamics of satellite orbits at Lockheed and set up a satellite tracking station in Spain for the Smithsonian Institute. In addition, he has taught an extension course called "Spaceship Earth" and has presented a television series entitled "Dimensions, the Universe."

David Eisenberg is a member of the Molecular Biology Institute and the Department of Chemistry at the University of California, Los Angeles. He studied biochemical sciences as an undergraduate at Harvard University and theoretical chemistry as a Rhodes Scholar at Oxford University, obtaining his doctorate degree from the latter institution. He then went to Princeton University to study the structures of ice and water and later to the California Institute of Technology to investigate methods of protein structure analysis. He is currently studying the structures of enzymes. Dr. Eisenberg combined his research and writing interests to produce this book's material on giant molecules.

John Fowler, who contributed material on the laws of energy, received his doctorate in physics from The Johns Hopkins University in 1954. He is presently a visiting professor at the University of Maryland, where he served as project director of the Commission on College Physics from 1965 to 1972. He received the 1969 Award in the Teaching of Science from the Washington Academy of

Sciences and the Millikan Lecture Award in the same year. Dr. Fowler has written and lectured on nuclear reactions; angular distributions and polarization of neutrons, protons, and deuterons; and polarization of cosmic-ray muons.

Harry Gray, professor of chemistry at the California Institute of Technology, obtained his doctorate in 1960 from Northwestern University. He is a member of the National Academy of Sciences and of the National Research Council in the Division of Chemistry and Chemical Technology. Dr. Gray received the 1970 American Chemical Society Award in Pure Chemistry and a 1972 Guggenheim Fellowship Award to study the spectroscopy of metalloproteins at the University of Copenhagen and the Biochemical Institute of the University of Rome. Dr. Gray was recently named one of four notable college chemistry teachers in the United States by the Manufacturing Chemists Association. His current research interests are in the area of transition metal chemistry. He has published seven books and more than one hundred papers. Also well known as "Harry the Horse" after the horse costume he sometimes wears to liven up his classes, Dr. Gray contributed the material on chemical change for this book. William Beranek, a graduate student at the California Institute of Technology, assisted Dr. Gray in the preparation of the material on photochemical smog.

Alan Holden was graduated from Harvard University in 1925 and joined Bell Telephone Laboratories the same year. He has been a visiting professor of physics at the Massachusetts Institute of Technology and at Cambridge University, and has participated in curriculum revision projects in high school and college science. Currently he is serving as a trustee of William Paterson College in New Jersey. In 1968 he was honored with the Robert Andrews Millikan Award of the American Association of Physics Teachers. He has written a number of well-known books including *Crystals and Crystal Growing, The Nature of Solids,* and *Shapes, Space, and Symmetry.* He contributed to this book in the areas of molecular bonding, states of matter, and the development of atomic theory.

Lester Ingber, who served as a writing consultant for this book, was awarded his bachelor's degree in 1962 from the California Institute of Technology and his doctorate in 1966 from the University of California, San Diego. Presently a research physicist at UCSD, he previously taught at the University of California, Berkeley, at the State University of New York, Stony Brook, and was a National Science Foundation Postdoctoral Fellow. His research interests center on elementary particles, nuclear physics, and astrophysics. The first Westerner to receive the Instructor's license and third Dan black belt from the Japan Karate Association, Dr. Ingber teaches an extension course at UCSD entitled "Application of Karate to the Studies of Attention and Physics." He is president of the Institute for the Studies of Attention, which runs a private school in California that emphasizes "Learning to Learn."

Robert Kolenkow, a major advisor and major writer for this book, is currently a visiting scientist at the Massachusetts Institute of Technology, where he was a member of the Research Staff of the Radioactivity Center, and, later, associate professor in the Department of Physics. Before receiving his doctorate in physics from Harvard University in 1959, he studied at MIT and at the University of Göttingen, Germany, as a Fulbright exchange student. At Harvard, he was the recipient of a Danforth Fellowship and a National Science Foundation Fellowship. In 1968 he was awarded the Everett Moore Baker award for outstanding undergraduate teaching at MIT. He has recently coauthored a freshman textbook in mechanics. Dr. Kolenkow's research interests are focused on the interaction of nuclear radiation with matter, medical physics, atomic and molecular beam studies, and atomic and molecular interactions. In addition to assisting in unifying the conceptual framework for this book, Dr. Kolenkow contributed the material on electromagnetism, elementary particles, quantum mechanics, and the intellectual history of atomism.

Cindy Lee received bachelor's degrees in both chemistry and chemical engineering from Arizona State University in 1970. She is currently working toward a doctorate in oceanography at Scripps Institution of Oceanography at the University of California, San Diego. She contributed to the chapter on water systems of the earth for the earth science unit of this book.

Peter Lonsdale, who contributed to the material on oceanography and fresh water systems, is a doctoral candidate at Scripps Institution of Oceanography in La Jolla, California. He received a First Class Honours Degree from Trinity College at Cambridge University in geography, with a specialization in geomorphology. While a student there he conducted research in Yugoslavia, southern France, and East Africa. His research at Scripps concerns the disturbance of the deep ocean floor by erosive currents of cold bottom water sweeping north from Antarctica. He has participated in several research cruises throughout the Pacific Ocean.

M. Granger Morgan is acting assistant professor in the Department of Applied Physics and Information Science at the University of California, San Diego, where he obtained his doctorate in applied physics in 1969. He is presently studying the impact of computers on society for the National Science Foundation in Washington, D.C. Dr. Morgan has specialized in radio physics and is skilled in techniques of remote probing, wave propagation, signal processing, and computer science. He is currently attempting to determine cost benefit ratios for new fossil fuel and nuclear power capacity. His teaching objectives are to relate technology to modern society using a problem-oriented approach. Dr. Morgan's contributions include the material on the earth's atmosphere and on man and the environment.

Manuel Rotenberg received both his undergraduate and doctorate degrees from the Massachusetts Institute of Technology, and is now professor of applied physics at the University of California, San Diego. Before coming to San Diego, he conducted research at the Los Alamos Scientific Laboratory and taught at Princeton University and the University of Chicago. Dr. Rotenberg served as an initial blueprint advisor for the book.

Matthew Sands, the overall advisor and a major writer for this project, is presently professor of physics and a Fellow of Kresge College at the University of California, Santa Cruz. After receiving his doctorate from the Massachusetts Institute of Technology he held positions on the faculties of MIT, the California Institute of Technology, and Stanford University. Dr. Sands has served as Vice-Chancellor, Division of Natural Sciences, at Santa Cruz, and recently has become involved in the planning and development of a student-centered education experiment at Kresge College. Dr. Sands was one of the authors of *The Feynman Lectures on Physics*. Many of the innovative ways of presenting physics in this book are the outcome of Dr. Sands' imagination and deep interest in undergraduate education.

Arthur Schawlow, professor and former chairman of the Department of Physics at Stanford University, was awarded his doctorate from the University of Toronto in 1949. He is a member of the National Academy of Sciences and is a fellow of several physics societies and associations. Dr. Schawlow coauthored with C. H. Townes the first paper describing optical masers, now called lasers. For this work, Schawlow and Townes were awarded the Stuart Ballantine Medal by the Franklin Institute in 1962 and the Thomas Young Medal and Prize of the American Physical Society and the Institute of Physics in 1963. Schawlow was also honored with the Morris N. Liebmann Memorial Prize Award by the Institute of Electrical and Electronics Engineers in 1964. Dr. Schawlow's research interests include optical and microwave spectroscopy, nuclear quadrupole resonance, superconductivity, and lasers. He contributed the material on lasers for this book.

Victor Weisskopf, presently chairman of the Department of Physics at the Massachusetts Institute of Technology, was educated in Europe and received his doctorate from the University of Göttingen, Germany. He came to the United States in 1937 after working with Werner Heisenberg, Erwin Schroedinger, Niels Bohr, and Wolfgang Pauli. He taught at the University of Rochester and then joined the faculty of MIT in 1945. In 1961 he returned to Europe for five years as Director-General of the Center of European Nuclear Research in Geneva, Switzerland. Dr. Weisskopf is a member of the National Academy of Sciences and a Fellow of the American Physical Society and the American Academy of Arts and Sciences. He was the recipient of the Planck Medal in 1956 and the George Gamow Memorial Lectureship Award in 1971. Dr. Weisskopf has written more than one hundred published papers and books, including the popular introduction to science, *Knowledge and Wonder*.

George Wetherill, a major advisor and contributor of the space science material for this book, was awarded his doctorate in physics from the University of Chicago in 1953. He is presently professor of geophysics and geology and chairman of the Department of Planetary and Space Science at the University of California, Los Angeles. He is a Fellow of the American Academy of Arts and Sciences and the American Geophysical Union. In 1969-70, he was a member of the National Aeronautics and Space Administration Lunar and Planetary Mission Board and participated in the National Academy of Sciences studies on Lunar Exploration, the Viking Mission to Mars, the Exploration of Venus, and the Summer Study on Space Priorities. He is currently chairman of the NASA working group on the post Viking Exploration of Mars. Dr. Wetherill's research activities have centered on geochronology and the evolution of continents; lunar and planetary geology; meteorites; and the origin, evolution, and dynamics of the solar system.

Selected Bibliography

1 Thinking About Things

*Achinstein, Peter. 1968. *Concepts of Science.* Baltimore: Johns Hopkins Press.

*Ackoff, Russell. 1962. *Scientific Method.* New York: Wiley.

Arons, Arnold, and Alfred Bork. 1964. *Science and Ideas.* Englewood Cliffs, N.J.: Prentice-Hall.

*Ashford, Theodore. 1960. *From Atoms to Stars.* New York: Holt, Rinehart and Winston.

*Asimov, Isaac. 1970. *The Universe.* New York: Walker.

*Gamow, George. 1965. *Matter, Earth, and Sky.* 2nd ed. Englewood Cliffs, N.J.: Prentice-Hall.

*———. 1962. *One, Two, Three... Infinity.* New York: Viking.

Grayson-Smith, Hugh. 1967. *Changing Concepts of Science.* Englewood Cliffs, N.J.: Prentice-Hall.

*Hecht, Selig. 1960. *Explaining the Atom.* 2nd ed. New York: Viking.

*Weisskopf, Victor F. 1962. *Knowledge and Wonder: The Natural World as Man Knows It.* Garden City, N.Y.: Anchor Books.

2 Seeing Things

*Baker, Robert H. 1971. *Astronomy.* 9th ed. New York: Van Nostrand.

*Bragg, William. 1959. *The Universe of Light.* New York: Dover.

*Cassidy, Harold G. 1970. *Science Restated.* San Francisco: Freeman, Cooper & Co.

Eisenbud, Leonard. 1971. *The Conceptual Foundations of Quantum Mechanics.* New York: Van Nostrand.

Gamow, George, and J. Cleveland. 1960. *Physics: Foundations and Frontiers.* Englewood Cliffs, N.J.: Prentice-Hall.

Klein, Miles V. 1970. *Optics.* New York: Wiley.

Orear, Jay. 1967. *Fundamental Physics.* 2nd ed. New York: Wiley.

*Physical Science for Non-Science Students Staff. 1969. *An Approach to Physical Science.* New York: Wiley.

*Physical Science Study Committee. 1960. *Physics.* Boston: Heath.

Robertson, John K. 1954. *Introduction to Optics.* 4th ed. Princeton, N.J.: Van Nostrand.

3 Measurements and Numbers

*Belser, Arthur. 1961. *Basic Concepts of Physics.* Reading, Mass.: Addison-Wesley.

Eddington, Arthur. 1958. *The Nature of the Physical World.* Ann Arbor, Mich.: Ann Arbor Books.

*Gamow, George. 1957. *One, Two, Three... Infinity.* New York: Mentor Books.

Grayson-Smith, Hugh. 1967. *The Changing Concepts of Science.* Englewood Cliffs, N.J.: Prentice-Hall.

*Kemble, Edwin C. 1966. *Physical Science, Its Structure and Development: From Geometric Astronomy to the Mechanical Theory of Heat.* Cambridge, Mass.: MIT Press.

*Lee, O. 1950. *Measuring Our Universe.* New York: Ronald Press.

4 The Atom Hunters

*Born, Max. 1969. *Atomic Physics.* John Dougle (tr.). 8th ed. New York: Hafner.

*Crombie, Alistair C. 1959. *Medieval and Early Modern Science.* 2 vols. Garden City, N.Y.: Anchor Books.

*Duveen, Dennis I. 1956. "Lavoisier," *Scientific American,* 194 (May): 84-94.

*Gillispie, Charles C. 1960. *The Edge of Objectivity: An Essay in the History of Scientific Ideas.* Princeton, N.J.: Princeton University Press.

*Hall, Marie Boas. 1967. "Robert Boyle," *Scientific American,* 217 (August): 96-102.

*Holliday, Leslie. 1970. "Early Views on Forces Between Atoms," *Scientific American,* 222 (May): 116-122.

*Read, John. 1952. "Alchemy and Alchemists," *Scientific American,* 187 (October): 72-76.

*Schonland, Basil. 1968. *The Atomists (1805-1933).* Oxford, England: Clarendon.

5 Stuff and Substance

*Alder, B. J., and T. E. Wainwright. 1959. "Molecular Motion," *Scientific American,* 201 (October): 113-126. Also Offprint No. 265.

Chalmers, Bruce, 1959. "How Water Freezes," *Scientific American,* 200 (February): 114-122.

Crewe, Albert. 1971. "A High-Resolution Scanning Electron Microscope," *Scientific American,* 224 (April): 26-35.

*Parsonage, Neville G. 1966. *The Gaseous State.* New York: Pergamon.

Richtmyer, Floyd K. et al. 1955. *Introduction to Modern Physics.* 5th ed. New York: McGraw-Hill.

*Romer, Alfred. 1960. *The Restless Atom.* Garden City, N.Y.: Anchor Books.

6 Pushes and Pulls

*Brandwein, P. 1968. *Energy: Its Forms and Changes.* New York: Harcourt, Brace & World.

*Cohen, I. Bernard. 1960. *The Birth of a New Physics.* Garden City, N.Y.: Anchor Books.

*Cooper, L. 1968. *An Introduction to the Meaning and Structure of Physics.* New York: Harper & Row.

*De Broglie, L. 1962. *New Perspectives in Physics.* New York: Basic Books.

Hesse, Mary B. 1961. *Forces and Fields.* Totowa, N.J.: Littlefield.

Holton, Gerald J. 1952. *Introduction to Concepts and Theories in Physical Science.* Reading, Mass.: Addison-Wesley.

*Jammer, Max. 1957. *Concepts of Force: A Study in the Foundations of Dynamics.* Cambridge, Mass.: Harvard University Press.

*Kemble, Edwin C. 1966. *Physical Science, Its Structure and Development: From Geometric Astronomy to the Mechanical Theory of Heat.* Cambridge, Mass.: MIT Press.

*Rogers, E. M. 1960. *Physics for the Inquiring Mind: The Methods, Nature, and Philosophy of Physical Science.* Princeton, N.J.: Princeton University Press.

7 Motion

*Andrade, Edward N. 1958. *Sir Isaac Newton: His Life and Work.* Garden City, N.Y.: Anchor Books.

Goldstein, Herbert. 1951. *Classical Mechanics.* Reading, Mass.: Addison-Wesley.

*Hogben, Lancelot. 1937. *Mathematics for the Millions.* New York: Norton.

Holton, Gerald J. 1952. *Introduction to Concepts and Theories in Physical Science.* Reading, Mass.: Addison-Wesley.

*Kemble, Edwin C. 1966. *Physical Science, Its Structure and Development: From Geometric Astronomy to the Mechanical Theory of Heat.* Cambridge, Mass.: MIT Press.

Mach, E. *The Science of Mechanics.* T. J. McCormack (tr.). 6th ed. LaSalle, Ill.: Open Court.

*Miles, Vadem W. et al. 1969. *College Physical Science.* 2nd ed. New York: Harper College Books.

*Rogers, E. M. 1960. *Physics for the Inquiring Mind.* Princeton, N.J.: Princeton University Press.

Sears, Francis W., and Mark W. Zemansky. 1970. *University Physics.* 4th ed. Reading, Mass.: Addison-Wesley.

Triffet, Terry. 1968. *Mechanics: Point Objects and Particles.* New York: Wiley.

Whittaker, E. 1944. *Analytical Dynamics.* 4th ed. New York: Dover.

8 Energy

*Born, Max. 1951. *The Restless Universe.* 2nd ed. New York: Dover.

*Brandwein, P. 1968. *Energy: Its Forms and Changes.* New York: Harcourt, Brace & World.

*Emmerich, Werner et al. 1964. *Energy Does Matter.* New York: Walker.

*Harrison, George R. 1968. *Conquest of Energy.* New York: Morrow.

Holton, Gerald J. 1952. *Introduction to Concepts and Theories in Physical Science.* Reading, Mass.: Addison-Wesley.

*Kemble, Edwin C. 1966. *Physical Science, Its Structure and Development: From Geometric Astronomy to the Mechanical Theory of Heat.* Cambridge, Mass.: MIT Press.

Pollack, H. 1971. *Applied Physics.* Englewood Cliffs, N.J.: Prentice-Hall.

Sears, Francis W., and Mark W. Zemansky. 1970. *University Physics.* 4th ed. Reading, Mass.: Addison-Wesley.

Theobald, David W. 1966. *The Concept of Energy.* New York: Barnes & Noble.

9 Relativity

*Bergmann, Peter G. 1968. *The Riddle of Gravitation.* New York: Scribner Library.

*Bronowski, J. 1963. "The Clock Paradox," *Scientific American*, 208 (February): 134-144.

*Cohen, I. Bernard. 1955. "An Interview with Einstein," *Scientific American*, 193 (July): 68-73.

Dicke, R. H. 1961. "The Eötvös Experiment," *Scientific American*, 205 (December): 84-94.

Einstein, Albert et al. 1924. *The Principle of Relativity.* W. Pennett and G. B. Jeffery (trs.). New York: Dover.

French, Anthony P. 1968. *Special Relativity.* New York: Norton.

*LeCorbeiller, Phillipe. 1954. "The Curvature of Space," *Scientific American*, 191 (November): 80-86.

*Rothman, Milton A. 1960. "Things that Go Faster than Light," *Scientific American*, 203 (July): 142-152.

*Shankland, R. S. 1964. "The Michelson-Morley Experiment," *Scientific American*, 211 (November): 107-114. Also Offprint No. 321.

10 Exploring Outer Space

*Abell, G. O. 1964. *Explorations of the Universe.* New York: Holt, Rinehart and Winston.

*Asimov, Isaac. 1970. *The Universe.* New York: Walker.

*Baker, Robert H. 1971. *Astronomy.* 9th ed. New York: Van Nostrand.

*Bok, B. 1958. *The Astronomer's Universe.* New York: Cambridge University Press.

*De Vancouleurs, Gerald H. 1957. *Discovery of the Universe.* New York: Macmillan.

Feynman, Richard P. et al. 1964. *Feynman Lectures on Physics.* 3 vols. Reading, Mass.: Addison-Wesley.

*Flammarion. 1964. *The Flammarion Book of Astronomy.* New York: Simon and Schuster.

*Hoyle, F. 1962. *Astronomy.* Garden City, N.Y.: Doubleday.

*Struve, Otto, and Velta Zebergs. 1962. *Astronomy of the Twentieth Century.* New York: Macmillan.

11 Origin and Evolution of the Universe

*Bondi, Hermann. 1968. *Cosmology.* London: Cambridge University Press.

*———. 1960. *Rival Theories of Cosmology.* New York: Oxford University Press.

*Charon, J. 1970. *Cosmology.* New York: McGraw-Hill.

Graham-Smith, F. 1960. *Radio Astronomy.* London: Penguin Books.

*Lovell, B. 1962. *The Exploration of Outer Space.* New York: Harper & Row.

McVittie, G. C. 1962. *Fact and Theory in Cosmology.* New York: Macmillan.

*Page, Thorton, and W. Lou (eds.). 1968. *Evolution of Stars.* Vol. 6, Sky and Telescope Library of Astronomy. New York: Macmillan.

Peebles, P. J. E. 1971. *Physical Cosmology.* Princeton, N.J.: Princeton University Press.

Reddish, V. 1967. *Evolution of the Galaxies.* Edinburgh, Scotland: Oliver & Boyd.

Schatzman, E. L. 1965. *Origin and Evolution of the Universe.* New York: Basic Books.

———. 1968. *Structure of the Universe.* New York: McGraw-Hill.

12 Early History of the Solar System

*Abell, G.O. 1964. *Exploration of the Universe.* New York: Holt, Rinehart and Winston.

*Aller, L. H. 1961. *The Abundance of the Elements.* New York: Interscience.

*Heide, Fritz. 1964. *Meteorites.* Edward Anders and Eugene Du Fresne (trs.). Chicago, Ill.: University of Chicago Press.

*Jastrow, Robert. 1967. *Red Giants and White Dwarfs.* New York: Harper & Row.

Mason, Brian H., and William G. Nelson. 1970. *Lunar Rocks.* New York: Interscience.

---

*Introductory level books.

*Nininger, H. H. 1952. *Out of the Sky*. New York: Dover.

Page, Thorton, and W. Lou (eds.). 1968. *The Evolution of Stars*. Vol. 6, Sky and Telescope Library of Astronomy. New York: Macmillan.

*Whipple, Fred L. 1968. *Earth, Moon, and Planets*. 3rd ed. Books on Astronomy. Cambridge, Mass.: Harvard University Press.

*Wood, J. A. 1968. *Meteorites and the Origin of the Solar System*. New York: McGraw-Hill.

13 The Third Planet

*Bates, David R. (ed.). 1957. *The Planet Earth*. New York: Pergamon.

*Gamow, George. 1965. *Matter, Earth, and Sky*. 2nd ed. Englewood Cliffs, N.J.: Prentice-Hall.

Gass, I. (ed.). 1971. *Understanding the Earth*. Cambridge, Mass.: MIT Press.

*Holmes, Arthur. 1965. *Principles of Physical Geology*. New York: Ronald Press.

*Mehlin, Theodore G. 1959. *Astronomy*. New York: Wiley.

*Rogers, John J. and J. A. Adams. 1966. *Fundamentals of Geology*. New York: Harper College Books.

*Strahler, Arthur N. 1971. *The Earth Sciences*. New York: Harper & Row.

*Stumpff, Karl. 1959. *The Planet Earth*. Ann Arbor Science Library. Ann Arbor, Mich.: Ann Arbor Books.

14 Materials of the Earth

*Dunbar, Carl O., and John Rodgers. 1957. *Principles of Stratigraphy*. New York: Wiley.

Gass, I. (ed.). 1971. *Understanding the Earth*. Cambridge, Mass.: MIT Press.

*Harker, Alfred. 1960. *Petrology for Students*. London: Cambridge University Press.

*Holmes, Arthur. 1965. *Principles of Physical Geology*. New York: Ronald Press.

Hurlbut, Cornelius S., Jr. 1970. *Dana's Manual of Mineralogy*. 18th ed. New York: Wiley.

Mason, Brian. 1966. *Principles of Geochemistry*. 3rd ed. New York: Wiley.

*Simpson, Brian. 1966. *Rocks and Minerals*. New York: Pergamon.

*Spock, Leslie E. 1962. *Guide to the Study of Rocks*. 2nd ed. New York: Harper College Books.

15 The Changing Atmosphere

*Hidore, John. 1968. *Geography of the Atmosphere*. 2nd ed. Dubuque, Iowa: William C. Brown.

Humphreys, W. J. 1964. *Physics of the Air*. New York: Dover.

*Oort, Abraham H. 1970. "The Energy Cycle of the Earth," *Scientific American*, 223 (September): 54-64.

*Orr, Clyde. 1959. *Between Earth and Sky*. New York: Macmillan.

*Riehl, Herbert. 1965. *Introduction to the Atmosphere*. New York: McGraw-Hill.

Singer, S. F. (ed.). 1970. "Global Effects of Atmospheric Pollution," in *Proceedings of an American Association for the Advancement of Science Symposium*. New York: Springer-Verlag.

Wise, William. 1970. *Killer Smog*. Walden Editions. New York: Ballantine.

16 Waters of the Earth

*Chorley, Richard J. (ed.). 1969. *Water, Earth, and Man*. London: Methuen.

*Fleming, Richard H. et al. 1942. *The Oceans: Their Physics, Chemistry, and General Biology*. New York: Prentice-Hall.

Hill, Maurice N. 1962-1970. *The Sea*. 4 vols. New York: Interscience.

*Menard, Henry W. 1969. *Anatomy of an Expedition*. New York: McGraw-Hill.

*Tureklan, Karl K. 1968. *Oceans*. Foundations of Earth Science. Englewood Cliffs, N.J.: Prentice-Hall.

*Weyl, Peter K. 1969. *Oceanography: An Introduction to the Marine Environment*. New York: Wiley.

17 The Restless Earth

Billings, Marland P. 1954. *Structural Geology*. 2nd ed. Englewood Cliffs, N.J.: Prentice-Hall.

*Eicher, Don L. 1968. *Geologic Time*. Englewood Cliffs, N.J.: Prentice-Hall.

*Holmes, Arthur. 1965. *Principles of Physical Geology*. 2nd ed. New York: Ronald Press.

Jacobs, John A. et al. 1959. *Physics and Geology*. New York: McGraw-Hill.

Phinney, Robert A. (ed.). 1968. *History of the Earth's Crust: A Symposium*. Princeton, N.J.: Princeton University Press.

*Rittmann, Alfred. 1965. *Volcanoes and Their Activity*. New York: Interscience.

*Scientific American Editors. 1969. *The Ocean, A Scientific American Book*. San Francisco: W. H. Freeman.

*Strahler, Arthur N. 1971. *The Earth Sciences*. New York: Harper & Row.

*Takeuchi, H. et al. 1970. *Debate About the Earth*. 2nd ed. San Francisco: Freeman, Cooper & Co.

18 A Nice Place to Visit...

*Alden, D., and Marjorie Meinel. 1972. "Physics Looks at Solar Power," *Physics Today*, 25 (February): 44.

*Barnea, Joseph. 1972. "Geothermal Power," *Scientific American*, 226 (January): 70.

*De Nevers, Noel. (ed.) 1972. *Technology and Society*. Reading, Mass.: Addison-Wesley.

Hogerton, John F. et al. 1971. *Nuclear Power Waste Management*. New York: Atomic Industrial Forum.

Meadows, Donella H. et al. 1972. *The Limits to Growth: A Report of the Club of Rome's Project on the Predicament of Mankind*. New York: Universe Books.

*Teick, Albert H. (ed.). 1972. *Technology and Man's Future*. New York: St. Martin's Press.

19 The Quantum Atom

*Andrade, E. 1956. "The Birth of the Nuclear Atom," *Scientific American*, 195 (November): 93-104.

*D'Abro, A. 1951. *The Rise of the New Physics*. 2nd ed. 2 vols. New York: Dover.

*Gamow, George. 1958. "The Principle of Uncertainty," *Scientific American*, 198 (June): 51-57. Also Offprint No. 212.

*Hecht, Selig. 1960. *Explaining the Atom*. 2nd ed. New York: Viking.

*Hermann, Armin. 1971. *The Genesis of Quantum Theory (1899-1913)*. Cambridge, Mass.: MIT Press.

*Holden, Alan. 1971. *The Nature of Atoms*. New York: Oxford University Press.

*Nash, Leonard K. 1966. *Stoichiometry*. Chemistry Series. Reading, Mass.: Addison-Wesley.

*Rich, R. 1965. *Periodic Correlations*. New York: W. A. Benjamin.

*Schonland, Basil. 1968. *The Atomists (1805-1933)*. New York: Oxford University Press.

20 Bonds Between Atoms

*Asimov, Isaac. 1965. *A Short History of Chemistry*. Garden City, N.Y.: Anchor Books.

*Companion, Audrey. 1964. *Chemical Bonding*. New York: McGraw-Hill.

Coulson, Charles A. 1961. *Valence*. 2nd ed. New York: Oxford University Press.

*Dubos, Rene. 1960. *Pasteur and Modern Science.* Garden City, N.Y.: Anchor Books.

*Holden, Alan. 1971. *Bonds Between Atoms.* New York: Oxford University Press.

Jolly, William L. 1966. *The Chemistry of the Nonmetals.* Englewood Cliffs, N.J.: Prentice-Hall.

Krauskopf, Konrad B. 1967. *Introduction to Geochemistry.* New York: McGraw-Hill.

Larson, E. M. 1965. *Transitional Elements.* New York: W.A. Benjamin.

*Light, R. J. 1968. *A Brief Introduction to Biochemistry.* New York: W.A. Benjamin.

21 States of Matter

Barker, J. A. 1963. *Lattice Theories of the Liquid State.* New York: Pergamon.

*Bragg, William. 1948. *Concerning the Nature of Things.* New York: Dover.

*Gottlieb, Milton et al. 1966. *Seven States of Matter.* New York: Walker.

Holden, Alan. 1965. *The Nature of Solids.* New York: Columbia Books.

*Kepler, Johannes. 1966. *A New Year's Gift, or On the Six-Cornered Snowflake.* Colin Hardie (tr.). Oxford, England: Clarendon.

Kittel, Charles. 1966. *Introduction to Solid State Physics.* 3rd ed. New York: Wiley.

*Stewart, Alec T. 1965. *Perpetual Motion: Electrons and Atoms in Crystals.* Garden City, N.Y.: Anchor Books.

Uman, Martin A. 1964. *Introduction to Plasma Physics.* New York: McGraw-Hill.

22 Chemical Change

*Angrist, Stanley W., and Loren G. Hepler. 1967. *Order and Chaos: Laws of Energy and Entropy.* Science and Discovery. New York: Basic Books.

*Beranek, W., Jr. (ed.). 1972. *Science, Scientists, and Society.* Tarrytown-on-Hudson, N.Y.: Bogden and Quigley.

Dence, J. B. et al. 1968. *Chemical Dynamics.* New York: W. A. Benjamin.

Firth, D. C. 1969. *Elementary Chemical Thermodynamics.* New York: Oxford University Press.

*Hammond, G. S. et al. 1971. *Models in Chemical Science.* New York: W. A. Benjamin.

*Herz, W. 1964. *The Shape of Carbon Compounds.* New York: W. A. Benjamin.

*King, E. L. 1963. *How Chemical Reactions Occur.* New York: W. A. Benjamin.

*Vanderwerf, C. A. 1961. *Oxidation-Reduction.* New York: Van Nostrand.

23 Giant Molecules

Caspar, D. L. D., and A. Klug. 1962. *Cold Spring Harbor Symposia on Quantitative Biology,* 27:1.

Dickerson, Richard E., and Irving Geis. 1969. *The Structure and Action of Proteins.* New York: Harper & Row.

Eisenberg, D. 1971. "X-ray Crystallography and Enzyme Structure," in Paul D. Boyer (ed.), *Enzyme Structure, Control.* 3rd ed. Vol. 1, The Enzymes. New York: Academic Press.

Eiserling, F. A., and R. C. Dickson. 1972. "Assembly of Viruses," *Annual Review of Biochemistry,* vol. 41.

*Kaufman, Morris. 1968. *Giant Molecules: The Technology of Plastics, Fibers, and Rubber.* Garden City, N.Y.: Doubleday.

Klug, A. 1972. "Assembly of Tobacco Mosaic Virus," *Federation Proceedings,* 31 (January-February): 30.

Lehninger, A. 1970. *Biochemistry.* New York: Worth.

Watson, J. D. 1970. *Molecular Biology of the Gene.* 2nd ed. New York: W. A. Benjamin.

24 The Original Charge Account

*Anderson, David. 1964. *Discovery of the Electron.* Princeton, N.J.: Van Nostrand.

Barthold, L. O., and H. G. Pfeiffer. 1964. "High Voltage Power Transmission," *Scientific American,* 210 (May): 38-47.

Duckworth, Henry. 1960. *Electricity and Magnetism.* New York: Holt, Rinehart and Winston.

*Ehrenreich, Henry. 1967. "The Electrical Properties of Materials," *Scientific American,* 217 (September): 194-204.

*Halliday, David, and Robert Resnick. 1962. *Physics.* Part 2. New York: Wiley.

*Purcell, Edward M. 1965. *Electricity and Magnetism.* Vol. 2, Berkeley Physics Course. New York: McGraw-Hill.

*Rainey, G. L. 1966. *Basic Electricity.* New York: Holt, Rinehart and Winston.

*Sands, Leo. 1971. *One Hundred and One Questions and Answers About Electricity.* Indianapolis, Ind.: Howard W. Sams.

25 Magnetism and Electromagnetism

*Akasofu, Syun-Ichi. 1965. "The Aurora," *Scientific American,* 213 (December): 54-62.

*Bates, Leslie F. 1961. *Modern Magnetism.* New York: Cambridge University Press.

*Bitter, Francis. 1959. *Magnets: The Education of a Physicist.* Garden City, N.Y.: Anchor Books.

Ford, Kenneth W. 1963. "Magnetic Monopoles," *Scientific American,* 209 (December): 122-131.

*Freeman, Arthur J., and Henry H. Kolm. 1965. "Intense Magnetic Fields," *Scientific American,* 212 (April): 66-78.

*Keffer, Frederic. 1967. "The Magnetic Properties of Materials," *Scientific American,* 217 (September): 222-234.

*Kondo, Herbert. 1953. "Michael Faraday," *Scientific American,* 189 (October): 90-98.

*Purcell, Edward M. 1965. *Electricity and Magnetism.* Vol. 2, Berkeley Physics Course. New York: McGraw-Hill.

*Sharlin, Harold I. 1961. "From Faraday to the Dynamo," *Scientific American,* 204 (May): 107-116.

*Williams, L. Pearce. 1971. *Michael Faraday.* New York: Clarion Books.

*Wilson, Mitchell. 1954. "Joseph Henry," *Scientific American,* 191 (July): 72-77.

26 Electromagnetic Radiation

*Battan, Louis J. 1962. *Radar Observes the Weather.* Garden City, N.Y.: Anchor Books.

Baumeister, Philip, and Gerald Pincus. 1970. "Optical Interference Coatings," *Scientific American,* 223 (December): 58-75.

*Connes, Pierre. 1968. "How Light is Analyzed," *Scientific American,* 219 (September): 72-82.

*Crawford, Bryce, Jr. 1953. "Chemical Analysis by Infrared," *Scientific American,* 189 (October): 42-48. Also Offprint No. 257.

*Ditchburn, R. W. 1963. *Light.* 2nd ed. 2 vols. New York: Wiley.

*Feinberg, Gerald. 1968. "Light," *Scientific American,* 219 (September): 50-59.

Fowles, Grant R. 1968. *Introduction to Modern Optics.* New York: Holt, Rinehart and Winston.

Gordy, Walter. 1957. "The Shortest Radio Waves," *Scientific American,* 196 (May): 46-53.

*Halliday, David, and Robert Resnick. 1962. *Physics.* Part 2. New York: Wiley.

*Heirtzler, James R. 1962. "The Longest Electromagnetic Waves," *Scientific American,* 206 (March): 128-137.

*Sears, Francis W. 1958. *Mechanics, Wave Motion and Heat.* Reading, Mass.: Addison-Wesley.

27 Lasers: The Light Fantastic

*Brown, Ronald. 1969. *Lasers: Tools of Modern Technology.* Science Study Series. Garden City, N.Y.: Doubleday.

Collier, Robert J. et al. 1971. *Optical Holography.* New York: Academic Press.

*Fishlock, David. 1967. *A Guide to the Laser.* New York: American Elsevier.

---

*Introductory level books.

Harvey, A. F. 1970. *Coherent Light.* New York: Interscience.

*Klein, H. Arthur. 1963. *Masers and Lasers.* Introducing Modern Science Books. Philadelphia: Lippincott.

*Kock, Winston E. 1969. *Lasers and Holography.* Garden City, N.Y.: Doubleday.

Maitland, A., and M. H. Dunn. 1970. *Laser Physics.* New York: American Elsevier.

*Schawlow, Arthur L. (intro. by). 1969. *Lasers and Light: Readings from Scientific American.* San Francisco: W. H. Freeman.

Siegmann, Anthony E. 1971. *An Introduction to Masers and Lasers.* New York: McGraw-Hill.

28 Inside the Nucleus

*Choppin, Gregory R. 1964. *Nuclei and Radioactivity.* New York: W. A. Benjamin.

*Cuninghame, J. G. 1965. *Introduction to the Atomic Nucleus.* Vol. 3, Topics in Inorganic and General Chemistry. New York: American Elsevier.

Friedlander, Gerhart *et al.* 1964. *Nuclear and Radiochemistry.* 2nd ed. New York: Wiley.

*Glasstone, Samuel. 1958. *Sourcebook on Atomic Energy.* 2nd ed. New York: Van Nostrand.

Harvey, Bernard G. 1962. *Introduction to Nuclear Physics and Chemistry.* Englewood Cliffs, N.J.: Prentice-Hall.

*_____. 1965. *Nuclear Clemistry.* Englewood Cliffs, N.J.: Prentice-Hall.

*Johnson, N. R. *et al.* 1963. *Nuclear Chemistry.* New York: Interscience.

*Overman, Ralph T. 1963. *Basic Concepts of Nuclear Chemistry.* New York: Van Nostrand.

*United States Atomic Energy Commission, Division of Technical Information. n. d. *World of the Atom.* 58-booklet series. Oak Ridge, Tenn.: USAEC.

29 Fission, Fusion, and Nuclear Energy

Choppin, Gregory R. 1964. *Nuclei and Radioactivity.* New York: W. A. Benjamin.

*Compton, Arthur H. 1956. *Atomic Quest: A Personal Narrative.* New York: Oxford University Press.

Cuninghame, J. G. 1965. *Introduction to the Atomic Nucleus.* Vol. 3, Topics in Inorganic and General Chemistry. New York: American Elsevier.

*Fermi, Laura. 1961. *Atoms in the Family: My Life with Enrico Fermi.* Chicago: Phoenix Books.

Friedlander, Gerhart *et al.* 1964. *Nuclear and Radiochemistry.* 2nd ed. New York: Wiley.

Glasstone, Samuel. 1958. *Sourcebook on Atomic Energy.* 2nd ed. New York: Van Nostrand.

*Hahn, Otto. 1966. *Otto Hahn: A Scientific Autobiography.* W. Ley (ed.). New York: Scribner.

Harvey, Bernard G. 1962. *Introduction to Nuclear Physics and Chemistry.* Englewood Cliffs, N.J.: Prentice-Hall.

_____. 1965. *Nuclear Chemistry.* Englewood Cliffs, N.J.: Prentice-Hall.

Johnson, N. R. *et al.* 1963. *Nuclear Chemistry.* New York: Interscience.

*Overman, Ralph T. 1963. *Basic Concepts of Nuclear Chemistry.* New York: Van Nostrand.

*United States Atomic Energy Commission, Division of Technical Information. n. d. *World of the Atom.* 58-booklet series. Oak Ridge, Tenn.: USAEC.

30 Elementary (?) Particles and Symmetry

Baker, Adolph. 1970. *Modern Physics and Antiphysics.* Reading, Mass.: Addison-Wesley.

*Feinberg, Gerald, and Maurice Goldhaber. 1963. "The Conservation Laws of Physics," *Scientific American,* 209 (October): 36-45.

Fonda, L., and G. C. Chirardi. *Symmetry Principles in Quantum Physics.* New York: Marcel Dekker.

*Gell-Mann, Murray, and E. P. Rosenbaum. 1957. "Elementary Particles," *Scientific American,* 197 (July): 72-88. Also Offprint No. 213.

Lewis, G. M. 1970. *Neutrinos.* New York: Springer-Verlag.

*Sachs, Robert G. 1972. "Time Reversal," *Science,* 176 (May 12): 587-597.

*Swartz, Clifford E. 1965. *The Fundamental Particles.* Reading, Mass.: Addison-Wesley.

*Weisskopf, Victor F. 1968. "The Three Spectroscopies," *Scientific American,* 218 (May): 15-29.

*Wigner, Eugene P. 1965. "Violations of Symmetry in Physics," *Scientific American,* 213 (December): 28-36. Also Offprint No. 301.

Wu, C. S., and S. A. Moszkowski. 1965. *Beta Decay.* New York: Interscience.

Introductory level books.

Glossary of Scientific Terms

a

absolute humidity: the concentration of water vapor in the air; it is expressed in grams per cubic centimeter.

absolute motion: motion with respect to a fixed coordinate system. See *relativity*.

absolute zero: the temperature at which the volume of an ideal gas disappears; also, the temperature at which the average kinetic energy of a gas molecule is zero; it is equal to -460 degrees Fahrenheit or -273 degrees Centigrade.

absorption: in physics, the process by which radiant energy is taken in and transformed into other forms of energy.

abyssal plain: the flat, relatively featureless sea bottom on either side of a mid-oceanic rise.

acceleration: the rate of change of velocity; a body is accelerated when there is change in its speed, direction of motion, or both.

acceleration due to gravity: acceleration with which bodies fall when acted on only by gravity; it is equal to 9.8 meters per second per second near the earth's surface.

accreting plate boundary: the edge along which two crustal plates are pulling apart.

achondrite: a stony meteorite possessing an internal structure of coarse-grained material resembling plutonic igneous rocks.

acid: a substance whose water solution contains a high concentration of hydrogen ions; a solution whose pH is less than 7.

acoustics: the study of sound, including its production, transmission, and effects.

activation energy: the energy barrier that must be overcome to allow a chemical reaction to proceed.

active site: the region on an enzyme to which the substrate attaches, activating it for a chemical reaction.

activity series: a listing of elements in a sequence of decreasing reactivity with other elements.

adiabatic: occurring without a loss or gain of heat energy.

adiabatic process: a change in the thermodynamic state of a system in which there is no transfer of heat across the boundaries of the system.

adsorption: the adhesion of a thin film of molecules to the surface of a solid substance, the surface not combining chemically with the adsorbed substance.

affinity: the preference of one type of atom or molecule to interact with another type.

air masses: parcels of air of relatively uniform temperature and humidity that project into another region of air of different temperature and humidity.

air pollution: the fouling of the atmosphere by noxious gases and solid particles. See *smog*.

alchemy: the practice of chemistry in the Middle Ages, involved with the transmutation of metals into gold and the search for a universal remedy for diseases; alchemy was a mixture of magical doctrines and scientific principles marking the beginnings of modern chemistry.

alkali metals: a family of soft, light, extremely reactive metals with similar chemical properties; the alkalis are lithium, sodium, potassium, rubidium, cesium, and francium.

alkaline earth metals: the metals that immediately follow each of the alkali metals in the periodic table; they include beryllium, magnesium, calcium, strontium, barium, and radium.

allotropy: an ability of certain elements to exist in several forms with different physical and chemical properties.

alloy: a metallic substance composed of two or more chemical elements, at least one of which is an elemental metal.

alluvial fan: a thick wedge of water-deposited coarse sedimentary debris along the base of a mountain range.

alpha helix: a polypeptide chain folded into a certain spiral structure and held together by hydrogen bonds.

alpha particle: historically, a heavy positively charged particle emitted during radioactive decay; now known to be the nucleus of a helium atom.

alpha ray: an alpha particle.

alternating current: an electric current that periodically reverses its direction.

AM broadcasting: the broadcasting of radio waves with the amplitude changed, or modulated, in accordance with the sound being transmitted.

amino acid: a carboxylic acid containing an amino group; twenty different amino acids are polymerized in various combinations by living organisms to make proteins.

amorphous: without form; having no definite crystalline structure.

ampere: a unit of electric current corresponding to the passage of one coulomb of electric charge per second.

amphiboles: a group of silicate minerals containing iron and magnesium; a common example is hornblende.

amplitude: the range in variability, such as in the light from a variable star; also, the maximum displacement or strength of a wave.

angstrom: a unit of length used mainly to express short wavelengths; an angstrom is equal to 10^{-8} centimeter.

angular momentum: a rotational quantity analogous to linear momentum.

anode: an electrode to which electrons flow; in an electrolytic cell, oxidation occurs at the anode.

anomaly: a deviation of an observed value from an expected or theoretical value.

antenna: in radio and television, a wire or set of wires used in sending or receiving electromagnetic waves.

anticyclone: a system of air circulation centering on an area of high barometric pressure; it rotates clockwise in the Northern Hemisphere and counter-clockwise in the Southern Hemisphere.

antimatter: matter made of antiparticles; it can undergo complete annihilation upon contact with ordinary matter.

antiparticle: either of two particles, one of matter and one of antimatter, that can undergo complete annihilation upon contact with the other; a positron and an electron are examples.

aquifer: a porous bed of rock capable of transmitting or storing ground water.

aromatic compounds: carbon compounds containing benzene-like rings.

asteroids: minor planets located between the orbits of Mars and Jupiter.

astronomical unit: a unit of length equal to the distance from the earth to the sun, or 149,599,000 kilometers.

astronomy: a branch of science treating the physics and morphology of that part of the universe which lies beyond the earth's atmosphere.

atmosphere: the gaseous envelope surrounding the earth, made up of layers of varying pressure and temperature.

atmospheric pressure: the pressure at any point in an atmosphere due to bombardment by the gas molecules; commonly the pressure measured at sea level.

atom: the smallest particle of an element that retains the properties characterizing the element; it is visualized as a positively charged nucleus surrounded by a negatively charged electron cloud.

atomic mass number: the number of nucleons, both protons and neutrons, in the nucleus.

atomic nucleus: the small dense core of an atom containing all of the atom's positive charge and most of its mass.

atomic number: the number of positive charges in the nucleus; it is equal to the number of electrons in an uncharged (neutral) atom.

atomic pile: a nuclear reactor in which a controlled and self-sustaining nuclear fission reaction can occur, releasing a great amount of energy.

atomic spectrum: a series of distinct colors emitted by the atoms of an element when they are excited to energies above the ground level energy.

atomic weight: the relative weight of an atom; now expressed on a scale in which the atomic weight of carbon 12 is exactly 12.00.

atomism: the concept that matter consists of vast numbers of tiny individual particles called atoms.

aurora: light radiated by atoms and ions in the ionosphere; it is induced by the influx of particles from the sun and is seen mostly in the polar regions.

autumnal equinox: the time when the sun crosses the equator moving from north to south about September 21.

Avogadro's hypothesis: states that equal volumes of all gases and vapors at the same pressure and temperature contain the same number of molecules.

Avogadro's number: the number of atoms in one gram-atomic weight of an element, or approximately 6.023×10^{23} atoms.

b

Balmer series: a family of lines in the visible and near-ultraviolet atomic spectrum of hydrogen.

barometer: a device for measuring atmospheric pressure.

barred spiral galaxy: a spiral galaxy in which the spiral arms begin from the ends of a "bar" running through the nucleus rather than from the nucleus itself.

baryon: an elementary particle of spin 1/2, 3/2, and so on, that is capable of interacting via the strong interaction.

basalt: fine-grained, black to medium gray igneous rock containing a high percentage of ferromagnesian minerals.

base: a substance capable of neutralizing an acid, yielding a salt; a solution whose pH is greater than 7.

basin of sedimentation: a locality of deposition of clastic debris.

battery: a collection of chemical cells for the production or storage of electric energy.

beam: a collection of focused rays of radiant energy.

beta decay: a radioactive transformation of a nucleus in which the atomic number changes by ±1 with emission of a negative electron or a positron.

beta particle: historically, a particle emitted from a nucleus during radioactive decay; now known to be an electron, or positron.

beta ray: a beta particle.

betatron: a particle accelerator in which electromagnetic induction is used to accelerate electrons.

big bang theory: the view, based in part on Hubble's law, that all matter in the universe was once gathered in one place, then exploded outward; the recession of galaxies now observed is thought to result from this tremendous explosion marking the beginning of time.

black body: an ideal emitter that radiates energy at each wavelength at the maximum possible rate per unit area for any given temperature, while also absorbing all the radiant energy falling on it.

black body radiation: the electromagnetic waves radiating from light originating in the vibrations of atoms at the inner surface of an empty, opaque cavity when it is heated; it is the theoretical maximum amount of radiant energy that can be emitted by a body at a given temperature.

blueschist: rock that is metamorphosed under extremely high pressures but relatively low temperatures.

Bohr theory: a theory of the structure of an atom in which electrons are envisioned as circling a nucleus in orbits of definite size, emitting no energy unless they jump from one orbit to another.

boiling point: the temperature at which the vapor pressure of a liquid is equal to the atmospheric pressure, causing the liquid to boil.

bond, chemical: the mechanism that joins atoms together to form molecules.

bond, covalent: a chemical bond in which atoms are held together by electron pairs shared more or less equally.

bond, ionic: a chemical bond in which ions are held together by strong electrostatic forces of attraction between oppositely charged particles.

bond, polar: a chemical bond in which atoms are held together by electron pairs shared unequally.

bonding: the joining together of atoms or molecules by means of electron sharing, ion formation, or other processes.

bond length: the separation distance between bound nuclei in a molecule.

bore: an incoming tidal wave that is steepened into a turbulent wall of water as it moves upriver.

Bose-Einstein statistics: formulas relating to the statistical behavior of particles with whole-number spin.

Boyle's law: states that the volume of a gas is inversely proportional to its pressure provided the temperature is held constant.

breeder reactor: a nuclear reactor in which another nuclear fuel, usually plutonium, is generated.

bremsstrahlung (braking radiation): electromagnetic radiation produced by a moving electron or other charged particle as it is accelerated or decelerated.

Brownian motion: the random movement of microscopic particles suspended in a fluid; caused by the bombardment of molecules.

c

calcite: a mineral composed of calcium carbonate; the principal constituent of limestone.

calorie: a unit of heat energy, originally defined as the amount of heat needed to raise the temperature of 1 gram of water 1 degree Centigrade; one calorie equals 4.18 joules.

Calorie: a unit equal to 1,000 calories; it is used to measure potential heat energy present in the chemical bonds in food; one Calorie equals 4,180 joules.

Cambrian: the earliest subdivision of the Paleozoic era, beginning approximately 600 million years ago, during which time invertebrates first appeared.

capacitance: the ratio of the electric charge given to a body to the resultant change of potential.

capacitor: a system of two or more conducting bodies, separated by layers of dielectrics, making possible accumulations of large charges at comparatively small voltages.

carbonaceous chondrite: a stony meteorite containing large amounts of carbon material.

carbonate: a salt or mineral containing the radical CO_3^-.

carbon cycle: a sequence of nuclear reactions and spontaneous radioactive decays that serves to convert matter into energy.

Cartesian coordinates: a coordinate system in which the location of a point in space is expressed by reference to three mutually perpendicular planes.

catalysis: the process of using a substance to speed up or slow down a chemical reaction, with no permanent change occurring in the catalytic agent.

catalyst: a substance that can alter the rate of a chemical reaction without being permanently changed by the reaction.

catalytic site: the location on a molecule where a catalyst acts to aid in chemical reaction.

catenane: a chained hydrocarbon compound.

cathode: an electrode from which electrons flow; in an electrolytic cell, reduction occurs at the cathode.

cathode ray: a stream of electrons usually emitted from a hot cathode.

cementation: the filling of rock pores with a cementing mineral precipitated from solutions passing through the rock.

centimeter: a unit of distance in the metric system; it is equal to 1/100 of a meter.

central force: a force directed toward or away from a central point; its strength depends only on the distance to the point.

central force field: a spatial distribution of the influence of a central force.

centrifugal force: an apparent outward force experienced by a body moving in a circle.

centripetal acceleration: the acceleration of a particle moving in a curved path; it is directed toward the instantaneous center of the path's curvature.

cepheid variable star: a star whose luminosity varies in a regular way; a typical example is Cephei in the constellation Cepheus.

chain reaction: a reaction in which one of the compounds that is used up in forming the products is simultaneously being produced by a side reaction.

charge, electric: a measure of the ability to produce or experience electric force.

charged body: an object with an excess or deficiency of electrons.

Charles's law: states that a volume of a gas is proportional to its absolute temperature provided the pressure is held constant.

chemical change: the formation of new kinds of molecules from the atoms of other molecules.

chemical element: the class of all atoms having the same atomic number; there are 92 naturally occurring elements.

chemical equilibrium: the state occurring when both a chemical reaction and its reverse reaction take place at the same rate.

chemical reaction: the interactions of chemical substances to produce new chemical substances.

chemical weathering: the destruction of rock by aqueous solutions, which convert it into soil and sediments.

chondrite: a relatively primitive and undifferentiated stony meteorite formed early in the history of the solar system but later than the highly differentiated meteorites.

clastic: pertaining to sediments that are mechanically derived, transported, and deposited.

clay minerals: a group of silicate minerals rich in aluminum and hydrogen and having laminated sheet structures; they typically comprise the finest-grained materials in soils and sediments.

cleavage: the tendency of a substance to split along certain planes, as determined by the atomic arrangement in its crystal lattice.

climate: the average weather for any region.

closed foliation: a type of rock formation in which platy minerals are tightly packed in a parallel manner and are not separated by alternating bands of some other mineral.

cloud: an airborne body of condensed water droplets resulting from cooling as a column or mass of air moves upward.

coherent light: light whose wave trains have all the same wavelength and phase.

cohesion: the force by which molecules of a substance hold together in a solid or liquid.

cold front: the forward boundary of a cold air mass invading a region of warm air, pushing the warm air upward.

color: the sensation resulting from stimulation of the retina of the eye by light waves; different colors correspond to different wavelengths as determined by the energy of the photons.

color temperature: a temperature assigned to an incandescent object to characterize its radiation; the peak intensity from an ideal black body at this temperature falls at the same wavelength as the peak intensity from the object.

combining weight: the weight of an element that will combine with or displace one gram-atomic weight of hydrogen.

combustion: the rapid oxidation of a substance that is marked by noticeable heat and light.

comet: a conglomerate of solid particles and gases moving about the sun, usually on an orbit of high eccentricity.

compaction: a decrease in the volume of sediments due to compression by the weight of overlying strata.

compound: a homogeneous chemical combination of elements in a definite proportion.

compound, covalent: a compound made up of atoms bound to one another by electron pair sharing.

compressional deformation: changes in structure and shape of an object due to compressional forces acting on two or more sides.

condensation: a physical process by which a vapor becomes a liquid or solid; condensation is the opposite of evaporation.

condensation nuclei: the particles or droplets needed to initiate the process by which liquids or vapors condense to form solid or liquid particles.

conduction of heat: the transmission of heat by means of a medium without any movement of the medium itself.

conductor: a metal or body through which an electric charge can easily travel.

conservation of angular momentum: the principle that the total angular momentum of an isolated system never changes.

conservation of energy: the principle that the total energy of an isolated system remains constant; thus, energy can be neither created nor destroyed—it is only changed from one form to another.

conservation of momentum: the principle that the total linear momentum of an isolated system never changes.

constellation: a configuration of stars in the night sky.

constructive interference: interference occurring when two or more wave trains are exactly in phase with each other, crest to crest, so that the resultant wave amplitude is the sum of the individual amplitudes.

consuming plate boundaries: the edges where two crustal plates converge.

contact: the surface between adjacent rock bodies.

continent: a large body of land elevated by its buoyancy relative to the surrounding denser oceanic crust.

continental drift: the hypothesis that continents move about at a rate of a few centimeters per year, sliding on a layer of deformable mantle rocks.

continental shelf: a land mass covered by shallow, coastal waters, gently sloping seaward to an average depth of about 200 meters.

continental shield: the ancient "core" areas of the continents that have been uplifted and eroded over a long period of time.

continental slope: a seaward continuum of the continental shelves having an inclination of 3 to 6 degrees and dropping to depths in the range of 1,300 to 3,200 meters.

continuous spectrum: a spectrum with all colors present.

convection: mass motions within a fluid resulting in transport and mixing of the fluid's properties; in meteorology, convection refers to atmospheric motions that are mainly vertical.

core: the center of the earth's interior of about 3,470 kilometers radius, possibly composed of liquid iron and nickel.

Coriolis effect: the apparent force due to the earth's rotation that causes all particles moving on the earth's surface to be deflected toward the right in the Northern Hemisphere and toward the left in the Southern Hemisphere.

cosmic rays: extremely high-energy particles that enter the earth's atmosphere from outer space.

coulomb: a unit of electric charge equal to the amount of charge provided by a current of 1 ampere flowing for 1 second.

Coulomb's law: states that the force between two electric charges is directly proportional to the product of their charges and inversely proportional to the square of the distance between them.

Crab Nebula: a remnant of supernova explosion that occurred in AD 1054; a powerful sky emitter of radio waves and light.

cracking: the process of breaking up long chain molecules, such as hydrocarbons, into smaller pieces by heat.

craton: a stable continental mass contributing sediments to portions of the earth's crust that warp downward.

critical mass: the minimal amount of concentrated fissionable material necessary to support a self-sustaining fission reaction.

crust: the relatively cool and solid outer layer of the earth; its thickness ranges from 5 to 60 kilometers.

crystal: a solid having the structure of a regular crystal lattice.

crystal face: a symmetrical, planar surface formed during the growth of a crystal lattice.

crystal imperfection: any form of irregularity in the structure of a crystal, such as holes or dislocations.

crystal lattice: one of the limited number of ways that atoms may be arranged to form a symmetrical, repetitious pattern.

crystalline: showing orderly internal arrangement that is also expressed in external form.

crystallography: the science of the form, structure, properties, and classification of crystals.

cubic meter: a volume equal to the volume of a cube one meter on each side.

cumuliform: towering, internally turbulent cloud structures developed in an unstable atmosphere.

curie: a unit of the rate of radioactive decay; it is equal to 3.7×10^{10} disintegrations per second.

Curie point: the temperature in a ferromagnetic material above which the material becomes largely nonmagnetic.

current: the flow of a substance, such as of electrons.

cyclone: a weather system of low barometric pressure; it rotates counterclockwise in the Northern Hemisphere and clockwise in the Southern Hemisphere.

cyclotron: a device for accelerating charged particles to high energies by giving successive increments of energy boosts from an alternating electric field to particles held in a spiral path by a magnet.

d

de Broglie equation: states that the motion of an electron or of any other particle is associated with a wave motion of a certain wavelength inversely proportional to the particle's momentum.

decay, radioactive: the spontaneous transmutation of one atomic nucleus into another, usually by the ejection of subatomic particles.

decomposition: the chemical alterations of rock materials.

definite proportions, law of: states that proportions by weight of the different elements that make up a chemical compound are the same in every sample of the compound.

degassing: the emission of gases from a planetary interior; it is also known as "outgassing."

dehydration: the removal of water from a mineral or a chemical compound.

delta: the accumulation of sediments where a stream empties into a body of quiet water, resulting in building out of the shore line.

denaturation: a physical change in the structure of a protein resulting in the loss of its characteristic biological activity.

density: the ratio of the mass of an object to its volume.

deposition: the laying down of eroded materials into a basin of sedimentation.

destructive interference: interference occurring when two or more waves are out of phase with each other (crests not lined up); destructive interference produces partial or complete cancellation.

deuterium: an isotope of hydrogen in which the nucleus contains one proton and one neutron; the most common isotope of hydrogen has no neutrons.

devitrification: the breaking of the internal bonds of a glassy structure.

dewpoint: the temperature at which a given parcel of air must be cooled at constant pressure and constant vapor content in order for saturation to occur.

diagenesis: the chemical changes taking place in sediments previous to and during their accumulation and consolidation.

diamagnetic: a substance in which the induced magnetism is opposite in direction to the applied magnetic field; diamagnetic materials tend to move from stronger to weaker magnetic fields.

diastrophism: the process of large-scale deformation, metamorphism, and intrusion that occurs in orogenic (mountain-building) belts.

diffraction: the process by which the direction of radiation is changed so that it spreads into the shadow region of an opaque or refractive object that lies in a radiation field; the spreading out of light in passing the edge of an opaque body due to the wave nature of light.

diffraction gratings: a system of closely spaced equidistant slits or reflecting strips which, by diffraction and interference, can separate light into its constituent colors.

diffusion: the process of spreading of objects into regions not already occupied by the objects.

dipole moment: the product of one of the charges of a dipole and the distance separating the two charges.

discharge tube: a tube containing a gas or vapor at low pressure whose atoms can be excited to radiate by the passage of electric current.

dislocation: a linear defect in crystals due to an offset of the crystal lattice.

disordered state: a random arrangement, as in an amorphous material.

dispersion force: a force between atoms or molecules due to the interaction of instantaneous electric dipole moments.

dissociation energy: the energy needed to break apart a molecule.

disturbance (moving): a force or field that alters the surrounding space and the already-present fields and objects.

dolomite: a mineral composed of equal quantities of calcium and magnesium carbonates in a regular array.

donor-acceptor bond: a bond between two elements, one of which donates electrons, the other receiving them.

Doppler effect: the apparent change in wavelength of the radiation from a source caused by its motion relative to the observer or, as in the case of sound, relative to the medium. See also *red shift*.

drag: the resistance to movement due to the medium through which an object is trying to move.

dry adiabatic lapse rate: the rate at which temperature drops as unsaturated air rises; it is equal to about 1 degree Centigrade per 100 meters.

dyne: the force required to accelerate a mass of 1 gram by 1 centimeter per second per second.

e

earth core: see *core*.

earth crust: see *crust*.

earth mantle: see *mantle*.

earthquake: an abrupt shifting of giant blocks of the earth's crust along fracture surfaces, or faults; it triggers a group of elastic shock waves that radiate outward in all directions.

eclipse: the apparent cutting off, wholly or partially, of the light from a luminous body by a dark body coming between it and the observer; in a lunar eclipse the earth causes a shadow on the moon; in a solar eclipse the moon causes a shadow on the sun.

edge dislocation: a movement of position resulting from a plane of atoms or ions terminating inside a crystal's interior instead of going completely through the crystal.

elastic collision: a collision between particles in which no change occurs in the total kinetic energy.

elasticity: a measure of the ability of a body that has been deformed by applied force to return to its original shape when the force is removed.

electrical potential energy: the energy stored in electric or magnetic fields.

electric conductivity: a measure of the quantity of electricity that will flow through a unit cube of a given substance in unit time with unit applied potential difference.

electric current, alternating: a current that reverses its direction of flow at regular intervals.

electric current, direct: a current that flows in one direction only; the flow is steady and free of pulsations.

electric dipole: any assemblage of charge that produces an electric field similar to the field produced by two equal and opposite electric charges separated by a small distance.

electric field: a field in which an electric charge would experience an electric force.

electric force: the force of attraction or repulsion exerted by electrically charged bodies.

electricity: the sum of all phenomena associated with stationary or moving charges.

electric power: the rate at which an electric current does work.

electric resistance: in conductors, the resistance to the flow of electrons because of collisions; generally, the ratio of potential difference to current.

electrode: a terminal at which electric charge passes from one medium to another.

electrolysis: a process of decomposing a substance by the passage of an electric current.

electrolyte: a substance whose molten form or whose solution conducts an electric current.

electromagnetic induction: the phenomenon in which a changing magnetic flux sets up an electric field.

electromagnetic radiation: radiation consisting of waves propagated through the interaction of electric and magnetic fields; it includes radio, infrared, light, ultraviolet, x-rays, and gamma rays.

electromagnetic spectrum: the range of electromagnetic radiation from very low to very high frequencies.

electromagnetism: the study of electric and magnetic forces between currents and moving charges.

electromotive force: the positive work on electric charges carried around a complete circuit; it is also known as *emf*.

electron: a subatomic particle, the least massive electrically charged particle, which carries one unit of negative charge, 1.6×10^{-19} coulomb; electrons constitute the outer part of atoms, and their interactions determine chemical bonding.

electron affinity: the energy released when an atom gains one extra electron.

electron avalanche: a process in which a relatively small number of free electrons in a gas subjected to a strong electric field accelerate and ionize gas atoms by collision, thus forming many new free electrons.

electron "cloud": the average distribution of electrons in an atom around the nucleus.

electronegativity: a measure of the ability of an atom to attract electrons from atoms to which it is bonded.

electron microscope: a microscope analogous to an optical microscope but which uses streams of electrons to magnify objects.

electron orbit: the path of an electron about an atomic nucleus; it is appropriate to semiclassical atomic models.

electron sharing: the sharing of a pair of electrons between two elements forming a chemical bond.

electron shell: one of the major groupings of energy levels that form the electronic structure of the atom.

electron spin: the intrinsic angular momentum of an electron, independent of any orbital motion.

electron volt: a unit of energy equal to the energy gained by an electron accelerated through a potential difference of one volt; it is equal to 1.60×10^{-19} joule.

electrostatic force: the force exhibited by an electric charge at rest.

element, chemical: see *chemical element*.

element, radioactive: an element that spontaneously changes to another element, usually by emission of a particle from the nucleus.

elementary particle: one of the various indivisible subatomic particles in nature.

elimination: a reaction in which a part of a molecule is removed, and replaced by the formation of a new internal bond.

elliptical orbit: an oval path characteristic of the planets and of bound particles under inverse square central force.

emission: the process by which electromagnetic radiation escapes a body; also, the sending out of particles from a surface.

emission spectrum: a broad term denoting any spectrum from a source; it may be continuous, line, or band, depending on the source.

endothermic reaction: a reaction during which energy is absorbed.

energy: the ability to do work; some forms of energy are kinetic, potential, heat, chemical, electric, magnetic, and radiant.

energy, atomic: commonly, the energy generated by nuclear fission or fusion.

energy, chemical: the energy involved in bonding atoms to each other; it is released or absorbed in chemical reactions.

energy, electric: the energy stored in electric and magnetic fields.

energy, heat: the energy of a body in virtue of the random motion of its molecules.

energy, mechanical: the energy of a body in virtue of its kinetic and potential energy.

energy, thermal: heat energy.

energy level: a stationary or long-lived state of definite energy of an electron in an atom.

entering group: a group joining to a molecule, either causing expulsion of another part of the molecule or adding on to a multiple bond.

enthalpy: the sum of the internal energies of a system plus the pressure-volume work done by or on the system.

entropy: a measure of the randomness of a system.

enzyme: a protein that catalyzes a chemical reaction in a living organism.

enzyme-substrate complex: the state in an enzyme-catalyzed reaction in which the substrate is bound to the enzyme.

epicenter: the focal point of energy release during an earthquake.

equation of continuity: a mathematical means of expressing conservation of a quantity.

equinoxes: two points of intersection of the ecliptic and the celestial equator.

equipotential: a path or surface on which the electric potential (or other potential) has the same value everywhere.

erg: the work done by a constant force of one dyne when the body on which the force is exerted moves a distance of 1 centimeter in the direction of the force; one erg equals 10^{-7} joule.

erosion: the physical removal of material mainly by fluid transport, but also by processes such as slumping and sliding.

error: the difference between the computed or measured value of any quantity and the true value.

escape velocity: the initial speed a body requires to leave a given gravitational field permanently.

ether: an imaginary invisible substance postulated as pervading space and serving as the medium for transmission of radiant energy.

evaporation: the process by which a liquid is transformed into the gaseous state.

excited state: a state in which an electron is at an energy level higher than the lowest one available.

exclusion principle: see *Pauli exclusion principle*.

exothermic reaction: a reaction during which energy is released.

exponent: the power to which a number is raised.

exponential notation: the use of powers of 10 to indicate large and small numbers.

extrusive rocks: molten rocks that flow out and solidify on the surface of the earth.

f

farad: a unit of electric capacitance equivalent to the storage of one coulomb of charge with one volt potential difference.

fault: a fracture in a rock or a geologic structure along which movement has occurred.

feldspars: the most abundant family of minerals in the earth's crust; they are light colored minerals classed as aluminosilicates and rich in sodium, calcium, or potassium.

Fermi-Dirac statistics: formulas relating to the statistical behavior of particles with spin of 1/2, 3/2, and so on.

ferromagnetism: a phenomenon in which microscopic magnetic domains of certain materials can be aligned by an applied magnetic field to produce a strong total magnetism.

field: a physical quantity having a value at every point of space.

fission: see *nuclear fission*.

FM broadcasting: broadcasting in which the frequency of the radio waves is changed in accord with the frequency of the sounds being transmitted.

foliated: exhibiting parallel orientation of platy minerals; it is characteristic of many metamorphic rocks.

foot-pound: the work done by a constant force of one pound when the body on which the force is exerted moves a distance of 1 foot in the direction of the force.

force: any influence capable of producing a change in the speed or direction of motion of a body.

formula, chemical: a representation of molecules using letters and numbers of the constituent elements.

fractional distillation: the separation of a mixture of chemical substances by means of their different boiling points.

fracture zone: a continuous series of steep cliffs that may extend for thousands of kilometers along the ocean floor; it is commonly associated with offset magnetic anomalies.

Fraunhofer lines: a set of dark lines in the spectrum of the sun or of a star due to absorption of light by the cooler outer layers of gas.

freezing point: the temperature at which a liquid freezes and becomes solid.

frequency: the number of vibrations per unit of time, or the number of waves passing a given point per unit of time.

friction: the resistance to motion of an object moving through or on another object.

friction force: the tangential force exerted at the contact surface of two bodies.

front: the surface of contact between adjacent air masses that typically differ in temperature and relative humidity.

fusion: see *nuclear fusion*.

fusion crust: the crust formed by the superficial heating and melting of a meteor during its fall through the atmosphere; the interiors of such stones have been observed to remain cold to the touch even after impact on the earth's surface.

g

g: a measure of the acceleration due to gravity, or about 9.8 meters per second per second at sea level.

G: the universal constant of gravitation; it is equal to 6.67×10^{-11} metric unit.

galaxy: a vast assemblage of stars, nebulas, and other objects separated from other galaxies by great distances.

Galilean transformations: simple vector addition of velocities to relate motion relative to moving coordinates to motion relative to stationary coordinates.

gamma rays: historically, a form of penetrating radiation from radioactive decay; photons produced by the change of a nucleus from one energy level to another; also, any photons of high energy.

garnet: any of a group of dense silicate minerals that typify certain classes of metamorphic rocks; garnets are sometimes used as gems and abrasives.

gas: a state of matter in which the molecules are practically unrestricted by intermolecular forces, thus being free to occupy any space within an enclosure.

gas laser: a laser using a gas or mixture of gases to form the active medium.

gauss: a unit of magnetic field intensity.

Geiger counter: an instrument for detecting and counting ionizing radiations.

generator: any device used to convert mechanical to electric energy.

genetic mutation: a permanent change in a gene of an organism which, if not lethal, will be passed on to subsequent generations.

geocentric viewpoint: one that regards the earth as the center of the solar system.

geologic time: the expanse of time since the formation of a solid crust on the earth; the earliest known rocks have measured ages of over 3.5 billion years.

geology: the study of the minerals and rocks that make up the earth, the processes that change the surface and interior of the earth, and the history of the planet as deciphered from the rocks.

geomagnetism: the study of the magnetic properties of the earth.

geostrophic wind: wind blowing with steady speed along lines of constant pressure (isobars).

geothermal energy: a means of generating electric power from the steep thermal gradients around cooling bodies of rock.

geothermal gradient: the rate of increase in the temperature of the earth with depth.

glacial period: a period of lowered temperatures and lowered sea level during widespread glaciation of continental surfaces.

glaciology: the study of the physics and chemistry of glaciers and ice caps.

globular cluster: one of a number of large star groups that form a system of clusters centered on the nucleus of the galaxy.

gneiss: a coarse-grained, imperfectly foliated feldspathic rock, commonly of granitic composition, representing a high grade of metamorphism.

graben: a valley formed by a subsidence of a block bounded on each side by faults.

gram: a unit of weight in the metric system; one pound is equal to 454 grams.

granite: a coarse-grained, intrusive igneous rock composed of feldspar, quartz, and a ferromagnesian mineral.

gravitation, law of: states that every body attracts every other body in the universe with a force depending on the masses of the bodies and the inverse square distance between them.

gravitational field: a field in which a particle is subject to a gravitational force.

gravitational force: the force exerted between two bodies by virtue of their masses.

gravitational heating: the heating of a body due to energy released by gravitational contraction.

gravitational instability: a lack of stability in a mass due to gravitational disruptive forces acting on the body.

gravitational mass: a measure of the ability of an object to produce or respond to gravitational force.

gravitational red shift: a displacement toward the red of the spectral lines of an atom in a gravitational field; also, the slowing down of a clock in a gravitational field.

gravity: the tendency of masses to attract one another.

gravity anomalies: the differences in the force of gravity in certain areas from that normally found for the earth.

graywacke: a hard, dark-colored coarse sandstone or grit having a clay-rich matrix.

greenhouse effect: a principle of heating in which visible light easily penetrates the atmosphere to reach the earth, but where water vapor and carbon dioxide block the escape of heat radiation, causing a rise in temperature.

ground state: the lowest stable energy level of an electron.

ground water: water that moves at varying depths below the surface of the earth.

group: in the periodic table, a column of elements with similar chemical characteristics.

gyres: large-scale horizontal motions taking the form of closed loops of more or less circular patterns.

h

Hadley cell: a model of the atmosphere in which a cell of circulating air consists of horizontal and vertical motions that form a complete circuit.

hadron: an elementary particle capable of interacting via the strong interaction.

half-life: the time it takes for half the number of nuclei in a radioactive sample to disintegrate.

halogen: a member of the seventh group of elements, lacking just one electron to make a closed outer shell; the common halogens are fluorine, chlorine, bromine, and iodine.

hardness: the resistance of minerals and rocks to abrasion or scratching.

heat: a form of energy produced by the random motion of atoms and molecules.

heat, radiant: the heat radiated from an object in the form of electromagnetic waves.

heat island: a localized area of a temperature higher than the temperature of surrounding areas.

heat of formation: the change in enthalpy accompanying the formation of a molecule; it is an indication of the stability of a molecule.

heliocentric viewpoint: one that regards the sun as the center of the solar system.

heterogeneous: composed of diverse elements or constituents.

holography: a technique using laser beams for making three dimensional displays.

homogeneous matter: matter whose properties do not vary from one part to another.

horsepower: a unit of power equal to the rate of work done in raising 33,000 pounds at the rate of one foot per minute, or approximately 746 watts.

Hubble's law: states that the velocity of a receding galaxy increases with its distance according to the equation $v = Hr$, where v is the velocity of recession, H is a proportionality constant (75 kilometers per second per 10^6 parsecs), and r is the distance between the galaxy and the observer.

humidity, relative: the ratio of the actual amount of water vapor present in the atmosphere to the amount needed to completely saturate air at the given temperature.

hurricane: an intense tropical cyclone formed near the east or west coasts of the southern part of North America.

hydrocarbon: a compound containing only hydrogen and carbon; gasoline is an example.

hydrogen bond: a weak electrostatic chemical bond in which two atoms are joined to a hydrogen atom.

hydrologic cycle: the earth's water recycling system in which all fresh waters are derived from evaporation from land and water surfaces and return through precipitation and runoff, with a multitude of possible side journeys.

hydrology: the study of water, especially in reference to its occurrence in streams, lakes, underground structures, and as snow.

hydronium ion: an ion formed by the addition of a proton to a water molecule; it is written as H_3O^+.

hydrosphere: a term that broadly refers to the earth's waters, including the bodies of salt water, fresh water, ice, and, in some cases, water vapor.

hygroscopic nuclei: a nuclei having a chemical affinity for water.

hypercharge: a quantity conserved in strong interactions; it is analogous to electric charge and equal to twice the difference of the electric charge and the isotopic spin component.

hypothesis: a scientific generalization presented without complete evidence.

i

ideal gas: a gas that obeys the ideal gas law, or the equation $PV = nRT$, where P is the pressure, V is the volume, n is the number of grams, R is the universal gas constant, and T is the absolute temperature.

igneous rock: a rock formed by the crystallization of molten magma.

incoherent light: light whose wave trains have no definite phase relationship.

index of refraction: a measure of the refracting power of a transparent substance; specifically, the ratio of the speed of light in a vacuum to its speed in the substance.

inertia: the resistance a material body offers to any change in its state of motion.

infiltration capacity: the maximum rate at which a soil can transmit water.

infrared radiation: electromagnetic radiation of a wavelength longer than the longest red wavelength that can be perceived by the eye but shorter than the wavelength of radio waves.

insulator: a substance through which electric charges can move only with difficulty.

intensity: a measure of the amount of energy falling on a unit area.

interference: in light, the phenomenon that is observed when two wave trains having a definite phase relationship follow the same path at the same time. See *constructive interference* and *destructive interference*.

interglacial: pertaining to a period of warm climate that separates glaciations.

intrusive rock: an igneous rock formed by cooling and crystallization of magma beneath the earth's surface.

inverse square law: states that force depends on the reciprocal of the distance squared.

ion: an atom, or a chemically combined group of atoms, having either a surplus or a deficiency of electrons.

ionization: the process of adding or removing an electron from an atom or molecule to create a charged species.

ionization energy: the amount of energy required to remove an electron from an atom.

ionization potential: ionization energy expressed in units of electron volts.

ionosphere: an upper region of the atmosphere characterized by the presence of a large ion density, beginning at the top of the mesosphere at about 50 kilometers and continuing to an altitude of several thousand kilometers.

irregular galaxy: a galaxy without rotational symmetry, neither a spiral or elliptical galaxy.

island arc: a curved group of oceanic islands at a converging plate junction; the islands are usually volcanic with an associated deep sea trench.

island universe: a synonym used by Immanuel Kant for the term "galaxy."

isobar: a line of constant pressure.

isomerization: the process of conversion from one isomer to another.

isomers: compounds having the same composition but different structural arrangements of their atoms.

isostasy: general equilibrium of the earth's crust, maintained by equalization of pressure by the slow flow of material.

isothermal: of constant temperature.

isotopes: atoms of the same element that have different atomic mass numbers.

isotopic spin: a quantity mathematically analogous to quantized angular momentum, but which is not conserved by the weak interaction; the components of isotopic spin specify the different elementary particles in a charge multiplet.

j

jet stream: a thin meandering ribbon of high-speed flowing air in the upper atmosphere.

joule: a metric unit of energy equal to the work done by a force of 1 newton moved through a distance of 1 meter; equal to 10^7 ergs.

k

kilocalorie: a unit of energy equal to 1,000 calories.

kilogram: a unit of mass equal to 1,000 grams.

kilohertz: a unit of frequency equal to 1,000 hertz.

kilometer: a unit of distance equal to 1,000 meters.

kilowatt: a unit of work equal to 1,000 watts.

kilowatt-hour: a unit of work equal to the work done in 1 hour by a device working at a rate of 1 kilowatt.

kinetic energy: the energy associated with motion; classically, it is equal to one-half the product of the body's mass and the square of its speed.

kinetic theory: a science that treats the properties of gases in terms of the motion and behavior of their molecules.

l

land bridge: a passage of land that connects two land masses across a large area of water.

laser: an instrument for generating intense beams of light by the process of stimulated emission.

lateral slippage boundaries: edges where plates slip and slide laterally against one another.

leaving group: the part of a molecule that is replaced in a chemical reaction.

lepton: an elementary particle not capable of interacting via the strong interaction.

Lewis acid and base: a theory in which acids are considered compounds able to accept an electron pair, and bases are compounds able to give an electron pair.

light: electromagnetic radiation that is visible to the eye.

light, dispersion of: the phenomenon of splitting white light into the spectral colors.

light rays: electromagnetic radiation visible as beams of light.

light year: the distance traveled by light in one year, traveling at a speed of 3×10^8 meters per second; a light year is approximately 10^{16} meters.

limestone: a sedimentary rock consisting mostly of calcium carbonate.

limiting nutrients: nutrients essential to the food chain, which by their availability set a limit on the amount of life forms able to survive.

limnology: the study of fresh water environments of living organisms, both plants and animals.

linear momentum: a physical quantity associated with motion, equal to the product of a particle's mass and velocity; the momentum of a body can be changed only by an applied force.

line defect: a defect occurring along a line in a crystal.

lines of force: imaginary lines representing the direction of a vector field; the density of the lines indicate the strength of the field at a point.

liquid: a substance which, unlike a solid, flows readily, but which, unlike a gas, doesn't tend to expand indefinitely.

liquid laser: a laser using a liquid as the active element of the system.

liter: a unit of capacity in the metric system equal to the volume of a kilogram of distilled water at 4 degrees Centigrade; it is equal to approximately 1.06 quarts.

lithification: a complex process that converts unconsolidated sediments into solid rock.

lithosphere: the relatively cool and solid upper layers of the earth.

lodestone: a naturally occurring stone containing oriented magnetic particles that tend to align with the earth's magnetic field; the first compasses and magnets.

Lorentz-Fitzgerald contraction: the apparent contraction or shrinkage in the direction of motion of a moving body as seen by a stationary observer.

luminosity: the rate of radiation of electromagnetic energy into space by a star or other body.

m

magma: molten rock of various composition, called "lava" if extruded on the earth's surface; its temperature may range from 1,200 degrees Centigrade to as low as 500 degrees Centigrade for certain intrusions; it cools to form igneous rocks.

magnet: any object that has the property of attracting iron magnetically.

magnetic anomalies: variations in the strength and direction of the earth's magnetic field.

magnetic field: a field within which magnetic forces can be detected.

magnetic flux: the total number of magnetic lines of force through an area.

magnetic pole: one of two points on a magnet at which the greatest density of lines of force emerge.

magnetic potential energy: a measure of the energy necessary to carry a unit north pole from infinity to some point.

magnetism: a branch of physics dealing with magnets and the forces associated with them.

magnetometer: an instrument used to measure magnetic fields.

magnetosphere: an area within the earth's external magnetic field, extending out to perhaps 1.3×10^5 kilometers.

mantle: the intermediate layer of the earth between the crust and the core; it extends to a depth of about 3,000 kilometers.

marble: a metamorphic rock composed mainly of granular calcite or dolomite; marble is the product of the recrystallization of limestone at elevated temperature and pressure.

mare: a sea-like area on the Moon or on Mars.

mass: a measure of the total amount of material in a body; also, the measure of a body's inertia.

mass number: the number of protons and neutrons in an atomic nucleus.

mass spectrograph: a machine that is used to separate atoms, nuclei, or molecules of different masses.

Maxwell-Boltzmann statistics: formulas relating to the statistical behavior of classical particles, neglecting quantum theory.

mechanical weathering: the physical processes resulting in the disintegration of rocks.

mechanics: a branch of physics dealing with motion and with the action of forces on bodies.

medium: a surrounding or pervading substance in which a body exists or moves.

megahertz: a unit of frequency equal to 10^6 hertz.

melt: a pool of molten rocks and minerals.

melting point: the temperature at which a solid is transformed into a liquid.

meson: an elementary particle with whole number spin capable of interacting via the strong interaction.

mesosphere: an atmospheric zone of diminishing temperature located above the stratopause and extending to about 85 kilometers.

metals: a class of electropositive elements that characteristically give up electrons in chemical reactions and that bond together with nonsaturating bonds to form extensive structures.

metamorphic facies: an association of rock types all formed under approximately the same conditions of temperature, pressure, and fluid medium.

metamorphic rock: rock formed by alteration of sedimentary or igneous rock under heat and pressure inside the earth.

metamorphism: the processes of change in the structure and composition of rock materials that take place in the solid state within the earth, under conditions of elevated temperature and pressure; it yields metamorphic rocks.

metastable: in atoms, a quasi-stationary state of comparatively long lifetime.

meteorite: a solid object from outer space that falls through the earth's atmosphere as a meteor and survives to land on the earth.

meteorology: the study of atmospheric phenomena at every scale from the microenvironment around leaves to global circulation.

meter: the basic unit of length in the metric system; it is approximately equal to 39.38 inches.

metric system: a system of measurement based on powers of ten, using the meter, kilogram, and second as basic units.

mica: a soft mineral with conspicuous cleavage in one plane, composed of sheet-structured aluminosilicates.

microwaves: short radio waves of frequency above 1,000 megahertz or wavelengths below about 30 centimeters.

mid-oceanic ridges: the 40,000-mile-long system of diverging plate junctions or "rifts" that encircle the globe; they were first discovered as a chain of mountains in mid-ocean.

Milky Way: the name for our Galaxy, a band of stars crossing the sky in a diffuse pattern.

millibar: a unit equal to 1/1,000 of a bar of atmospheric pressure.

mineral: a naturally occurring, inorganic substance of homogeneous structure and composition; minerals are crystalline in most areas.

mineralogy: the study of minerals, including their structure, form, chemistry, and occurrence.

mixture: a substance containing two or more elements which are not in fixed proportions and which maintain their individuality.

model: a plan, pattern, or conceptual idea for the way something is thought to work.

moderator: a material in a reactor used to slow neutrons, increasing their effectiveness in inducing fission.

Mohole: a proposed hole drilled through the earth's crust; the proposal was not carried out.

Mohorovičić discontinuity: a seismic discontinuity defined as the base of the earth's crust; it is often abbreviated as "moho."

moist adiabatic lapse rate: the rate at which temperature changes in a rising saturated air mass.

mole: an amount of substance containing Avogadro's number of molecules; it is equal to 1 gram molecular weight of the material.

molecular vibrations: vibrations of the atomic nuclei in a molecule.

molecule: the smallest particle of any substance that can exist freely and still exhibit all the chemical properties of the substance.

momentum: usually, the linear momentum of a body.

monochromatic: containing a single color.

monomer: a single unit of a substance or chain compound, which can join together in long chains or polymers.

motion, first law of: states that every body in motion or at rest remains so unless acted on by an outside force.

motion, second law of: states that the Change in a body's momentum in a short time interval is proportional to the total force acting on the body, multiplied by the change in time. Also, for objects not moving too fast, that the acceleration produced on a body by a force is inversely proportional to its mass.

motion, third law of: states that forces always exist in pairs; the mutual forces between two bodies are equal and opposite.

mountain: a large mass of rock and earth rising more or less abruptly above the surrounding level land.

multiplet: a grouping having several quantum numbers in common but differing in a quantum number such as the component of isotopic spin.

muon: an elementary particle intermediate in mass between an electron and a proton; it behaves similarly to a heavy electron.

n

natural law: a statement of order or relation between phenomena; it is presumed always to hold given the specified conditions.

nebula: a large cloud of dust and gases in outer space that may glow when near bright stars; otherwise it appears as a dark cloud.

negatively charged: having an excess of electrons.

negentropy: a measure of the disorder involved in changing from one form of energy to another; the greater the disorder introduced, the greater the degradation of the energy to less available forms.

neutral equilibrium: a state in which the lapse rate for an atmosphere of unsaturated air is the same as the adiabatic lapse rate for a parcel of air within it; in physics, a state of equilibrium unchanged by small perturbations.

neutrino: a particle having no mass or charge but carrying away energy in the course of certain nuclear transformations; it interacts only via the weak interaction.

neutron: a subatomic particle of no charge and of mass approximately equal to that of the proton.

neutron star: an extremely small, dense star of very intense magnetic fields; neutron stars are formed in the final stages of stellar evolution.

newton: a metric unit of force equal to the force required to accelerate a kilogram of mass one meter per second per second.

noble gases: any of the elements helium, neon, argon, krypton, xenon, and radon, which are characterized by closed major shells of electrons.

nonmagnetic: lacking the ability to attract or be attracted to a magnetic pole with noticeable strength.

nonmetal: any of the class of electronegative elements that in chemical reactions take on or share electrons.

north pole (magnetic): the pole of a magnet that is attracted to the north magnetic pole of the earth.

nuclear energy: the energy generated by the combining or disruption of atomic nuclei.

nuclear fission: the splitting of a heavy nucleus into nuclear fragments of lesser mass, accompanied by a release of energy.

nuclear force: a strong nonelectromagnetic force between particles, responsible for the binding of nucleons in atomic nuclei.

nuclear fusion: the joining of light nuclei to form a heavier nucleus, accompanied by the release of energy.

nuclear reactor: a device in which nuclear fission may be sustained in a self-supporting chain reaction.

nucleic acid: a long chain complex molecule; living organisms use two types of nucleic acids, DNA and RNA, to carry genetic information.

nucleon: a proton or a neutron.

nucleotide: a molecule consisting of phosphate, a five-carbon sugar, and a purine or pyrimidine base; also, the chemical unit comprising DNA and RNA.

nucleus: the heavy core of an atom, composed of protons and neutrons.

O

obsidian: volcanic glass, characterized by glassy luster.

occluded front: a front in which cold air has pushed into a warm air mass, lifting the warm air out of contact with the ground.

ocean: a great body of salt water that covers more than two-thirds of the surface of the earth.

ocean basin: a bowl sunk into the depths of the ocean, to 4,000 to 6,000 meters deep; it occupies 30 percent of the earth's surface.

oceanic trenches: long, narrow depressions containing the greatest depths in the oceans; they are associated with island arcs and with a sinking oceanic crustal plate.

oceanography: the study of the physical, chemical, and biological workings of the oceans.

ohm: a unit of electrical resistance.

Ohm's law: states that the current through many conductors is equal to the potential difference across the conductor divided by the resistance of the conductor: $I = V/R$.

olivine: a green, rock-forming silicate mineral rich in magnesium and iron.

ooze: mud found on the deep ocean floor; the dominant content can be detrital clay minerals, or the calcareous or siliceous shells of minute surface organisms.

opaque substance: material through which light can't penetrate.

optically active: in reference to molecules, having the ability to rotate the plane of polarized light.

optical window: transparency of the earth's atmosphere to a band of wavelengths that allows visible light to pass through without being greatly absorbed.

orbital: the probabilistic description of an electron's position in an atom or molecule.

orbital quantum number: a quantum number characterizing the angular momentum of an electron.

ordered state: nonrandomness.

orogenesis: the process of mountain building.

orthorhombic: pertaining to a set of crystals in which the axes are of unequal lengths and at right angles to each other.

oscillator: an apparatus producing an alternating voltage or current, generally of a definite frequency.

oxidation: the chemical process of giving up electrons.

oxidation number: a positive or negative number indicating the ionic charge of an element or ion.

oxidation-reduction: a chemical reaction in which oxidation numbers of atoms or molecules are changed; the reaction is also called "redox."

oxidizing agent: a substance that causes an element or compound to be oxidized.

ozone: a reactive unstable gas whose molecule consists of three oxygen atoms it is formed when oxygen is subjected to electric discharge.

p

paleomagnetism: the study of the orientation of the earth's magnetic field as it was in past geologic ages as recorded by the natural magnetism of rocks formed at those times.

paleontology: a science dealing with the evolution of life forms of past geologic ages and with the use of those forms as fossil geologic records.

Paleozoic: pertaining to an era of ancient life forms in which the fossil record is dominated by invertebrates; it began about 600 million years ago and lasted for nearly 400 million years; life forms prior to the Paleozoic are only faintly known.

paramagnetic: pertaining to material in which the magnetic effects of the electrons do not cancel, so that the internal magnetism can be aligned by an applied magnetic field to give a stronger field.

particle: see *elementary particle*.

Pauli exclusion principle: states that no more than one electron in an atom may have the same set of quantum numbers.

percolation: the process of the passing of an aqueous solution through small interstices such as in sand or porous stone.

perfect gas law: states that the product of pressure and volume is porportional to temperature.

peridotite: a distinctive rock type composed of olivine and pyroxene minerals.

period: the time required to execute one complete cycle; in chemistry, a sequence of elements in the periodic table, arranged by atomic number.

period, geologic: a period of geologic time distinguished from other periods on the basis of episodes of more or less continuous sedimentary deposition.

periodic table: an arrangement of the elements in the form of a table that illustrates the periodicity in the chemical and physical properties of the elements.

petrology: the study of the structure, composition, and natural history of rocks.

pH: a measure of the acidity of solutions; a solution below 7 on the pH scale is acidic and a solution above 7 is basic.

phase: a homogeneous bounded part of a system, such as a liquid phase; in a periodic quantity, the interval between some point in the period and a standard starting point, often expressed as an angle between 0 and 360 degrees.

photoelectric effect: the emission of electrons from a substance when light hits it.

photoelectron: an electron that has been liberated from a substance by a photon.

photon: a discrete unit of electromagnetic energy; the particle of light and the mediator of electromagnetic interactions.

photosynthesis: the process by which green plants, when exposed to sunlight and water, use carbon dioxide from the atmosphere and give off oxygen in the production of carbohydrates.

physical science: the study of the material world; it encompasses physics, chemistry, geology, and astronomy.

physics: the study of matter and of the different forms of energy.

phytoplankton: microscopic plants living in the shallow oxygenated ocean zone.

pion: a meson having a mass between that of electrons and protons; it is a mediator of the strong nuclear force.

Planck's constant: a constant of proportionality relating the energy of a photon to its frequency; it is equal to 6.62×10^{-34} joule-second.

Planck's radiation law: describes the intensity of radiation from a black body at a given temperature as a function of wavelength; it implies that radiant energy is gained or lost by vibrating atoms in discrete bundles.

planet: a satellite of the sun; from the earth it appears in the sky as a bright object whose position changes relative to the stars.

planetary albedo: the fraction of incident radiant energy reflected by a planet.

plasma: a high-temperature ionized gas composed of electrons and positive ions in such numbers that the gaseous medium is essentially electrically neutral.

plate tectonics: the hypothesis that the earth's crust is made of approximately nine plates moving relative to each other over a yielding mantle; most of the tectonic behavior of the earth is accounted for by interaction of the plates.

plutonic: derived from or emplaced deep within the earth; the term is particularly used in conjunction with igneous rocks and processes.

point defects: defects occurring at particular points in a crystal's structure.

polarized light: light in which polarization, or a preferred plane of vibration of the electric field, is present.

polar molecule: a molecule that possesses an electric dipole moment.

pollutant: a harmful chemical or waste material discharged into water or into the atmosphere, causing them to become unclean and contaminated.

polymer: a chain of repeating single units known as monomers.

polymerization: the process by which long chains, or polymers, are formed from single chain, or monomeric, units.

polymorphism: the ability of certain compounds to exist in more than one form.

polypeptide: a polymer of amino acid molecules.

porphyries: igneous rocks, generally extrusive types, in which large phenocrysts (early-formed crystals) are set in a finer groundmass.

positively charged: having a deficiency of electrons.

positron: a positively charged elementary particle otherwise similar to an electron; the antiparticle of an electron.

postulate: a position or idea assumed without proof at the beginning of an argument.

potential difference: the amount of potential energy between two points gained by one coulomb of charge moved from one point to the other.

potential energy: the stored energy that can be converted into other forms without loss.

pound: the basic unit of weight in the English system; it is equal to approximately 0.454 kilogram.

power: the rate at which work is done.

power density: the rate at which energy is delivered to a unit area.

precession of equinoxes: slow, conical motion of the earth's axis caused by the gravitational torque of the moon and sun on the earth's equatorial bulge; a complete period requires approximately 26,000 years.

precipitate: a solid that separates from a solution in a chemical reaction.

pressure, atmospheric: see *atmospheric pressure.*

pressure gradient: a change of barometric pressure with distance.

primary waves: longitudinal seismic waves of a compressional type, like sound waves, in which the particle motions are back and forth in the direction of propagation; they are the first waves to arrive after a quake, and can be transitted through solid, liquid, or gaseous matter.

principal quantum number: a quantum number that identifies the shell occupied by an electron in an atom.

probability: for events of equal likelihood, the ratio of the number of favorable outcomes to the total number of outcomes.

products: the resultants of a chemical reaction.

propagating disturbance: a motion, such as a wave-like motion, that maintains its movement by the interchange of various forms of energy.

proportionality: a relationship between two quantities in which one quantity is equal to a constant times the other.

protein: any one of an important class of molecules in living organisms consisting of one or more intertwining polypeptide chains.

proton: a subatomic particle, the nucleus of an ordinary hydrogen atom; stable and positively charged, it interacts via the strong interaction and is a principal constituent of nuclei.

proton-proton scattering: a means for studying the interaction of protons by allowing a moving proton to be deflected by another proton.

pulsar: a star whose light and radio waves vary rapidly and periodically with a definite rhythm.

pulsating star: a variable star that physically pulsates in size and luminosity.

pump: in a laser, to raise atoms to higher levels of energy by a means such as a gas discharge.

pyroxenes: a group of ferromagnesian silicate minerals having a chain structure linking the silica tetrahedra.

q

quantization: the restricting of the magnitude of an observable quantity to a discrete set of values.

quantized: to be divided into discrete units.

quantum: a "bundle," or discrete amount, of a quantity.

quantum jump: a jump of an electron between levels of an atom due to the absorption or emission of a quantum of energy.

quantum mechanics: a description of the behavior of matter that takes into account its wave nature.

quantum numbers: numbers, usually whole numbers or half whole numbers, that label the state of a system, such as an electron in an atom.

quartz: a mineral composed of pure silicon dioxide in a network structure; it forms six-sided crystals tapering to pyramids at the end.

quartzite: an extremely hard quartz rock formed by metamorphic recrystallization of sandstone, or by aqueous deposition of silica cement between sand grains in the sediment.

quasi-stellar galaxy: extremely small sources thought to be galaxies emitting intensely powerful radio emissions, with their spectral lines showing a large red shift.

r

radiant energy: energy emitted as electromagnetic radiation.

radioactive decay: the spontaneous transmutation of one atomic nucleus into another, usually by the ejection of subatomic particles.

radioactivity: the emission of energetic particles and radiant energy by certain atomic nuclei.

radio astronomy: the technique of making astronomical observations at radio wavelengths.

radio waves: a part of the electromagnetic spectrum extending up to about 10^{11} hertz.

Raman effect: a phenomenon in which monochromatic light scattered from a molecule is accompanied by scattered light at longer and shorter wavelengths that depend on the energy levels of the molecule.

Raman scattering: a means of studying molecules by observing Raman effect in light scattered from them.

randomness: lacking order or rule, such as in random thermal motion of molecules.

reaction mechanism: the postulated means or pathway a reaction follows as it goes from reactants to products.

real image: an image formed by the rays proceeding from an object.

recrystallization: the growth of certain minerals or grain fragments at the expense of others after crystallization has already occurred once.

red giant: a large, cool star of high luminosity.

red shift: a displacement of spectral lines of a star or galaxy toward the red end of the spectrum; it is generally ascribed to the Doppler effect.

reduction: the process of removing oxygen from its chemical combination; generally, a gain of electrons in a chemical reaction.

reference beam: in holography, the beam that sets the standard phase in the generation of the interference pattern.

reference point: a point from which all subsequent directions and measurements are based.

reflection: the turning back of a wave or a part of a wave at a surface.

refraction: the bending of light rays passing from one transparent medium into another.

relative humidity: the ratio between the amount of moisture in a volume of air and the maximum amount of moisture that volume of air can hold when saturated at the given temperature.

relativity, theory of: a theory, formulated essentially by Albert Einstein, that all motion is relative to the frame of reference from which it is measured and that space and time are relative; the special theory of relativity deals with uniform motion and holds only in special frames of reference that move with uniform, nonaccelerated motion, and the general theory deals with gravity and its effects on motion.

remanent magnetism: magnetism acquired by the cooling of a rock through the Curie point in the presence of a magnetic field such as the earth's.

residue, amino acid: an amino acid in a polypeptide chain.

resonant cavity: a cavity in which standing waves are set up; generally resonance is possible only at certain wavelengths.

rest mass: the mass of an object measured in the frame in which the object is at rest.

resultant vector: the vector that is the vector sum of two or more vectors.

retrograde motion: the apparent backward motion of a planet.

rift: a trenchlike depression in the earth; the term commonly refers to features of diverging, or accreting, plate boundaries, such as the mid-oceanic rift and the East African rift.

rock: an aggregate of one or more mineral species.

rubidium 87: an isotope of the element rubidium that decays to strontium isotopes; it is used in dating rock materials.

S

salinity: a measure of the total amount of dissolved salt present in one kilogram of water.

salt: the product of reaction between an acid and a base; an ionic solid containing metallic and nonmetallic ions.

satellite: a secondary body that revolves about a larger one; the moon is a satellite of the earth.

scalar: having a magnitude but no direction in space, such as volume or temperature.

scale height: the vertical distance marking a change by a factor of 10 in atmospheric density.

schist: metamorphic rock exhibiting a foliated structure with prominent mica minerals.

scintillation: a process producing a flash of light due to the interaction of an energetic particle with a phosphor screen or crystal.

screw dislocation: in crystallography, a dislocation that is a structure defect caused by the introduction of a screw axis not normally present.

sea: a body of salt water, either contiguous with the world ocean or enclosed by land.

sea floor spreading: the process by which oceanic crust forms beneath the sea at ridge crests and moves laterally away in both directions.

second: a metric unit of time.

secondary waves: seismic waves that travel in a transverse, or crosswise motion, forcing the transmitting medium to vibrate at right angles to the direction of wave propagation; secondary waves are propagated only through solids.

sedimentary rock: rock formed from the disintegrated matter of preexisting rocks.

sedimentation: the process of deposition of material transported by erosional agents.

seismic wave: a means by which energy released in earthquakes is transmitted to all parts of the world.

seismograph: an instrument used to sense and measure seismic waves.

seismology: a branch of geophysics that studies earthquakes and the shock waves associated with them.

self-luminous: shining by the production of its own light; the sun is an example of a self-luminous body.

semiconductor: a substance whose electrons must be given a moderate amount of energy to become conduction electrons.

sequence: the order of amino acids in a protein, or of monomeric units in a polymer.

shadow zone: a zone on earth in which no secondary waves are received from an earthquake; the principal basis for believing the earth has a liquid core.

shell, electron: a major grouping of energy levels of nearly the same energy in an atom.

shielding: the barriers protecting against particles released during radioactive decay or reactions.

side chain: a secondary group attached to the main sequence of an amino acid.

silicate: a mineral in which various metallic ions are combined with a silicate ionic group.

sinusoidal: a quantity whose variation with time or distance has the form of the sine trigonometric function.

slate: a fine-grained metamorphic rock possessing a well-developed secondary cleavage which allows it to break into sheets having smooth surfaces.

smog: a brownish contaminant in urban air, caused by the interaction of pollutants in the atmosphere in the presence of sunlight; a major cause of reduced visibility and of respiratory problems.

solar cell: a device, often made of silicon, that converts solar energy to electric energy.

solar flare: a sudden and temporary outburst of light and matter from an extended region of the sun's surface.

solar spectrum: the spectrum of light from the sun.

solar system: a system of the sun and the planets, their satellites, the minor planets, comets, meteoroids, and other objects revolving around the sun.

solar wind: a steady flow of plasma outward from the sun.

solid; solid state: a state of matter in which a substance possesses definite volume, definite shape, and elasticity of shape and bulk, resisting any force that tends to alter its volume or form; solids are characterized by very stable surfaces of distinct outline on all sides.

solid state physics: a branch of physics dealing with the structure and properties of solids; it encompasses crystallography, conductivity, cohesive forces, band structure, plasticity, impurity, lattice defects, and dislocation theory.

south pole (magnetic): the pole of a magnet that is attracted to the south magnetic pole of the earth.

space quantization: the axis of rotation of a particle appears to point only in certain directions due to the quantization of angular momentum.

specific gravity: the ratio of the density of a substance to that of water.

spectral analysis: the determination of the elements present by an identification of their spectral lines.

spectral lines: pure colors arising from electron transitions in an atom or molecule.

spectroscope: an instrument using a prism or diffraction grating to disperse light into a spectrum.

spectrum: an array of colors or wavelengths obtained when light from a source is dispersed, as in passing it through a prism or grating.

speed: the rate at which an object moves without regard to its direction of motion; also, the numerical magnitude of velocity.

speed, light: the rate at which a photon of light of any wavelength moves; it is equal to 186,000 miles per second, or 3×10^8 meters per second.

spin: the rotation of a body about its own axis through its center.

spiral galaxy: a flattened, rotating galaxy with pinwheel-like arms of interstellar matter and young stars winding out from the nucleus.

spring: a seepage of ground water in large amounts through rock and unconsolidated material.

spring scale: a balance scale based on the stretching of a calibrated spring.

square meter: an area of one meter length on each side.

stable equilibrium: an equilibrium at minimum energy that is not upset by small perturbations.

star: a self-luminous sphere of gas.

stars, apparent brightness of: a measure of the observed light flux received from a star or other object at the earth.

state: the condition of a substance, such as "gas"; the condition of a quantum system, such as "electron state."

statics: the physics of bodies and forces at rest or in equilibrium.

stationary front: a contact surface between adjacent air masses that are not moving.

statistical theory: a mathematical theory based on probability that interprets the action of large numbers of objects.

stimulated emission: the enhancement of the probability that an atom in an excited level will emit radiation; stimulation can be produced by light of the same frequency.

stoichiometry: the study of the heat, energy, and material balance of a chemical system.

stratiform cloud: a cloud formed of horizontal layers.

stratosphere: the atmospheric layer lying above the troposphere; it has a steady temperature and little weather activity.

strong interaction: one of the fundamental forces in nature, a strong short-range force between certain elementary particles; the "glue" that holds atomic nuclei together.

strontium 87: an isotope of the element strontium involved in the decay of rubidium 87; it is useful for radioactive dating.

structural trends: parallel orientations of folds or structures along a continuous or discontinuous line.

substrate: a substance used as a base for the deposition of other substances; in biochemistry, the compound acted on by an enzyme.

sun: the star about which the earth and other planets revolve.

superconductivity: the phenomenon in which the resistance of many metals drops entirely to zero at a temperature above absolute zero.

supercool: to cool, as a liquid, below its freezing point without solidification.

supernova: a star that suddenly flares into extreme brightness, perhaps 200,000,000 times its previous state.

supersaturated solution: a solution in which more solute is dissolved than is normally possible at that temperature.

surface current: a current in the ocean of predominantly horizontal water movement near the upper surface.

surface tension: a force exerted by a surface, much as if it were an actual membrane.

swell: a wave moving through a region of weak winds or calms and decreasing in height and steepness.

symmetry: the correspondence in size, form, and arrangement of parts on opposite sides of a plane, line, or point; a form that can be turned into itself by an operation such as reflection or rotation.

synchrotron: a ring-shaped particle accelerator in which the particles are held to a fixed path.

t

tectonism: a process of deformation of the earth's crust, producing the many structurally dominant features of the crust, such as mountains, rifts, and basins.

telescope: an optical instrument used to aid the eye by producing a magnified image of a distant object.

temperature: a quantity proportional to the average kinetic energy of an ideal gas.

temperature, absolute: temperature measured in Centigrade degrees from absolute zero.

temperature, Centigrade: temperature measured on a scale where water freezes at 0 degrees and boils at 100 degrees.

temperature, Fahrenheit: temperature measured on a scale where water freezes at 32 degrees and boils at 212 degrees.

temperature, Kelvin: absolute temperature measured in Kelvin degrees; it is approximately equal to Centigrade temperature minus 273 degrees.

template: a basic pattern used in the reproduction of the same unit over and over; for example, the model of coded information stored in the DNA molecule from which the other DNA or RNA molecules are replicated.

tetragonal: in reference to crystals, having three axes at right angles, two of the axes being of equal lengths.

tetrahedral bonding: bonding in which four bonds are directed toward the center of a tetradedron.

theory: a proposed explanation of a wide range of phenomena associated with a particular subject using a set of scientifically acceptable hypotheses and laws.

thermionic emission: electron emission due to the temperature of the emitter.

thermoelectric junction: a contact between two metals; an emf is generated if the junction is at a different temperature from the rest of the metal.

thermohaline circulation: deep water circulation based on water becoming denser with increased salinity, decreased temperature, or both.

thermometer: a device for measuring temperature.

thermonuclear reaction: a nuclear reaction or transformation, usually fusion, that results from encounters between nuclear particles that are given high velocities by heating them.

thermosphere: a zone of rapid temperature increase at an altitude of 200 kilometers above the mesopause.

threshold frequency: the frequency which light falling on a surface must exceed before any electrons will be emitted.

tide: a deformation of a body by the gravitational force exerted on it by another body.

time: a fundamental dimension of physical science, measuring a period of duration.

torque: a measure of the ability of a force to produce a twisting or rotating motion.

transformer: an electric device having no moving parts that enables an efficient increase or decrease of AC voltage; it is used in the long-distance transmission of electric power.

translucent: a material through which light can penetrate but through which no definite shapes are discernible because the light is diffused.

transparent: a material through which light shines clearly.

transportation: movement of eroded material via running water, wind, or ice.

transuranic element: an element of atomic number greater than 92; all transuranium elements are artificially made and do not exist in nature.

transverse wave: a wave in which the physical quantity constituting the wave oscillates at right angles to the direction of wave travel.

tropopause: a region between the troposphere and stratosphere.

troposphere: the lowermost layer of the atmosphere, through which the temperature decreases; it contains nearly all of the atmospheric clouds, rain, and storms.

typhoon: a nearly circular cyclonic storm with extremely low pressure at the center, accompanied by high winds, dense clouds, and heavy precipitation; it forms in the western Pacific as a form of tropical cyclone.

u

ultraviolet light: electromagnetic radiation of wavelength shorter than the shortest (violet) wavelengths to which the eye is sensitive; its range is from 100 to 4,000 angstroms.

uncertainty principle: states that it is impossible to simultaneously determine accurate values for position and velocity or for energy and time interval.

uniform motion: motion in a straight line with steady speed.

unpolarized light: light in which there is a lack of any preferred direction of polarization. See *polarized light*.

uplift: a process of raising land above its current position through tectonic or isostatic forces.

upwelling: the raising of a liquid from below when the upper layers are displaced laterally; the principal source of nutrients in many of the fertile areas of the world ocean.

v

vacuum: the absence of matter in a region of space; in practice, the near-absence of matter compared to ordinary conditions.

valence: the capacity of an atom to combine with other atoms to form a molecule; it is specified as the number of hydrogen atoms with which one atom of the element will combine or displace.

van der Waal's force: a weak force of attraction between atoms caused by the interaction of instantaneous charge distributions.

variable star: a star that varies in luminosity.

vector: a quantity that has both magnitude and direction.

vector direction: the location of a point determined vectorially from a reference point.

vector magnitude: the numerical size of a vector without reference to the direction.

vector position: a position in space, assigned vectorially, with a designated direction from a reference point and at a designated distance away.

vector sum: the vector resulting from the addition of two or more vectors.

velocity: strictly, a vector that denotes both the speed and direction a body is moving.

vernal equinox: the time when the sun crosses the equator of the earth from south to north about March 21.

virtual image: an image formed by the prolongation of the rays that proceed from an object; no light actually passes through the apparent location of the image.

virus: a submicroscopic particle that consists of either DNA or RNA and a protein coat; it is parasitic and reproduces by means of synthetic processes in the host cell.

viscosity: an internal property of a fluid offering resistance to flow.

volatile: the ability to evaporate; a highly volatile liquid is one that evaporates quickly, such as methanol in air.

volcanic rock: rock produced by volcanism.

voltage: an electromotive force or a difference in potential expressed in volts.

voltage drop: a potential difference, particularly across a circuit element.

w

warm front: the boundary of a relatively warm air mass moving into a region occupied by colder air, with the warm air sliding up over it on a broad, gently sloping front.

water table: the upper surface of the water-saturated zone of a soil and rock profile.

wave: a propagating disturbance varying with time and position; examples are water waves and sound waves.

wave front: a term describing the position of corresponding points on a travelling wave disturbance at a given time.

wavelength: the spacing of the crests or troughs in a wave train.

wave train: a set of waves of the same wavelength and frequency following each other in sequence.

weak interaction: one of the fundamental forces in nature, acting prominently in beta decay; all particles except the photon seem capable of exerting this force.

weather: atmospheric phenomena near the ground that take place over a relatively short time interval.

weathering: chemical and mechanical wearing away of rock material.

white dwarf: a star that has exhausted most or all of its nuclear fuel and has collapsed to a very small size.

wind: air motion in a dominantly horizontal motion relative to the surface of the earth.

wind-driven circulation: circulation of the atmosphere due to large-scale rising and sinking motions of the winds.

wind shear: a characteristic tendency of a moving mass of air to cause an almost adjacent medium, such as the ocean surface to slide.

wind wave: a wave that is formed or sustained by winds.

work: the product of displacement and the force component in the direction of displacement.

work units: derived force-distance units, because work is done by a force acting through a distance; examples are the newton-meter, or 1 joule, and the dyne-centimeter, or 1 erg.

x

x-ray: a penetrating high-frequency electromagnetic wave produced whenever fast electrons are brought to rest quickly or when electrons in atoms make transitions to inner shells.

x-ray diffraction: the diffraction of x-rays by crystals to produce regular patterns indicative of the crystal structure.

z

zodiac: a band on the celestial sphere divided into twelve sections, each one containing one of the constellations known as the signs of the zodiac.

zooplankton: very small floating animals growing in vast numbers in the shallow, well-oxygenated surface layers of the ocean.

Index

a

absolute humidity, 280
absorption, of photon energy, 20, *21*, 25–26, 367
abyssal plain, 247, *246–247*
acceleration, 127, 153
   See also momentum, change in
accelerator, *543*, 556, *556*, *557*
acid, 428
acid-base reaction, 428, *431*
action and reaction
   See reaction force
activation energy, 556
adamantane, bonding of, 391, *391*
addition reaction, 430, *431*
adhesion, 89, 90, 108
adiabatic lapse rate, 281
air
   and color of sky, 31
   density of, 71
   molecules in, 12–13, *13*
   pressure, 75–76
   propagation of light through, 27–28
air pollution, 275, 331, 419, 432
   detecting pollutants, 515
   during rush-hour traffic, 428
   nuclear, 342
   and photochemical reaction, 435, *436*
   smog, 275, *276*, 435–437, *436*
   thermal, 336, *337*
air resistance, theory of, 360, *360*
albedo, planetary, 282, *282*, 338
alchemy, 57
   secret signs in, *67*
alkali metals, 80
allotropy, 405–407, *406*, *407*
alluvial fan, 264
Alpha Centauri, 186
alpha helix, 445, *445*
alpha rays (helium nuclei)
   penetration power, *553*
   and radioactive decay, 67, 549
   and Rutherford's experiment, 361, *362*
alternating current, *470*, *473*, *492*, *492*
aluminum, 69–70
amino acid, structure of, *394*, 440, *441*
amorphous solid, 251
   atomic arrangement of, 253, 254
   behavior of, 253
ammonia, bonding geometry of, 387, *388*
Ampere, André, 479, *480*
amplitude, 502, *502*
Andes Mountains, *245*, 324
Andromeda, 191
angular momentum, 353, 577, *577*
   and electron orbital motion, 578
   and electron spin, 579
   unit of, 578
annihilation of matter, 352
anode, 470, *471*
anticyclone, 287, *288*
antimatter, 456, 582–583
antiparticle, 352, 582–583
Apollo manned lunar landings, 224, *226*
Appalachian Mountains, *245*, 325, *326*
apparent photographic magnitude of nebulas, 193
aquifer, *303*
Arecibo Ionosphere Laboratory, *173*
Aristotelian view of matter, 56–57, *57*
Aristotle, 56–57, *56*, 170

asthenosphere, *317*, 319
asteroid belt, *182*, 223
Aswan Dam project, *332–333*
atmosphere
   and behavior of gases, 71, 74, 75
   composition of, 280, *280*
   early, 234, *234*, 241
   heat balance of, 282–286, *282*, *284*, *285*
   modeling of, 288
   particulates in, 334
   regions of, 279, *279*
   scale height of, 277
   structure of, *279*
atmospheric circulation
   and climate, 338
   models of, 284–286, *284–285*
atmospheric pressure, 277
   unit of, 277
atom, 9–17, 55, 357
atom
   configuration of, 16–17
   currents in, 487
   electron photomicrograph of, 66, *60*
   energy levels of, 365
   internal structure of, 15–17, *16*, 85, 487
   kinetic energy of, 12, 19, 73–74
   minimum energy value of, 367
   selectivity for photons, 367
   size of, 10, 16–17, 65, 375
atomic clock, 47
atomic elements
   See elements
atomic models
   Bohr's quantum atom, 368
   de Broglie's steady-wave model, 373, *373*
   Dalton's atom, *61*, 357
   early pictures of, 55–59, 357
   Rutherford's atom, 361, *361*, *362*, 363
   shell model, 371, *371*
   solar system model, 363
   Thomson's raisin cake, 361, *361*, 362
   modern picture of, 9–13, 15–17, *16*, 69, 85, 376
atomic nucleus
   See nucleus
atomic number, 16–17
   and number of electrons, 16–17, 85
   and periodic table, 362
atomic radiation, measurement of, 37–38
atomic spectra
   as clue to atomic structure, 84, 357
   and combination principle, 366
   examples of, *83*
atomic structure, internal, 15–17
atomic weight, 16–17
   and Avogadro's number, 63
   and combining weight, 60–62
   and density, 70
atomism, 55–67, 72, 377
augite, *254*
aurora, 484, *484*
Avogadro, Amadeo, 62, *62*
Avogadro's hypothesis, 62
Avogadro's number, 63–65, *63*

b

Bahama Banks, *266*
balancing chemical equations, 419
Balmer, J. J., 366
Balmer's formula, 366
Balmer spectrum for hydrogen, *365*
Barnard's star, 186, *186*
baryon, 584

624

basalt
    and age of ocean floor, 313
    in earth's crust, 243
    fossil rift zone of in United States, 324
    and paleomagnetism, 317
bases, 428
    of DNA, 440
basic electric charge, 16
Basov, Nikolai, 525
Batavia accelerator, *543*
battery, 470, *471*, 487
Bay of Fundy, 299
Becquerel, Henri, 549, *549*
benzene, structure of, 395, *395*
Bernal, J. D., 401
Bernoulli, Daniel, 399, *400*
beta decay, 549, *550*
beta rays (electrons)
    penetration power of, *553*
    and radioactive decay, 67, 549
big bang, 194-195
    evidence for, 193-195, *194*, *195*
    and theories of origin of universe, 348
binding energy, 546, 563
black body radiation, 514, *514*
black hole, 353
blueschist, 319
Bohr, Niels, 369, 561, *561*
Boltzmann, Ludwig, 72, *72*, 78
bond angle
    in carbonates, 254
    and molecular geometry, 387-389
bonding, atomic
    covalent, 382, 383, *384*, 387
    delocalized, 395, 396
    geometry of, 387
    ionic, 379, *384*
    metallic, 397
    mirror-image structures, 392, *393*
    polar molecules, 384
    rotation in, 389, *391*
bore, tidal, 299
Born, Max, 399
Boyle, Robert, 57-59, *58*, 379
Bragg, Laurence, 401, *401*
Bragg, William, 401, *401*
Bragg's law, *402*
Brahe, Tycho, *174*
Bravais lattices, 252
breeder reactor, 570, *571*
de Broglie, Louis, 372, *372*, 413
Brown, Robert, 67
Brownian motion, 67, *67*, 357
Bryce Canyon, 263
bubble chamber, *573*
Bunsen, Robert, 84, 357

C

calcite, structure of, *256*
calcium carbonate, aragonite form of, 407, *407*
calibration, 91
Calorie, 135, 136, 144
camera, *28-29*
Cameron, A., 230
capacitance, 467
capacitor, 467, *467*, 468
carbon
    and allotropy, 406, *406*
    bonding, 389
    carbon-oxygen groups, 254, *254*
    in combustion, 14, *15*
    compounds of, 389, *391*, 405
    cycle in thermonuclear reaction, 208
    isomers of, 389
    in meteorites, 407-408
    and organic chemistry, 389

carbonates, 254, *255*
carbon cycle, 208-209
carbon dioxide, 14, *14*, 334
    in atmosphere, 280
    as combustion by-product, 331
    in ocean, 334
carbon monoxide, 14, *14*
catalyst, 426
catenanes, 389
cathode, 470, *471*
Cavendish, Henry, 60
Centigrade scale of temperature, 73, *73*
Chadwick, James, 575
chambered nautilus, *513*
change of state, 11-13, 76, *77*
    and disposition of heat energy, 76
charcoal, burning of, 14
charge, electric
    See electric charge
chemical change, 14-15, 57-60, 419
    combustion, 14-15, *15*, 431
    energy transfer in, 422
    in environment, 432
    and reactivity, 421
    regulation of, 424
chemical compound, 59
    combining weights of, 60
    formulas for, 420
    and law of definite proportions, 419
    nomenclature, 14
    relative weights of, 60
chemical element, 13-14, 59
    and atomic theory, 57-60
    See also elements
chemical reaction, 57, 419
    acid-base, 428, *431*
    addition, 430, *431*
    chain-terminating, *437*
    and chemical structure, 421
    and conservation of matter, 419
    elimination, 430, *431*
    endothermic, 422, *422*
    energy changes in, 422
    exothermic, 422, *422*
    isomerization, 430
    photochemical, 435-437, *436*
    rates of, 424, 425
    redox, 431
    substitution, 427, *431*
chlorine gas, and covalent bonding, 382
chondritic (stony) meteorite, 215, *216*, 219
    abundance of elements in, 222-224, *223*, *224*
    and age of solar system, 218-221, *221*
    carbonaceous, 223
chymotrypsin enzyme, 447, *447*
circuit, electric, 461
    and electromotive "force," 469
    and equation of continuity, 471
circular motion, 131-133, *132*
Clausius, Rudolf, 72, *72*
cleavage, in crystalline solid, 253, *253*
clocks, 45-47, *45*, *46*
    atomic, 47
    and relativity, 161-163, *162*
cloud structures
    cumuliform, 281, *282*
    stratiform, 281, *282*
coil-capacitor circuit, 509, *510*

color
    and atomic spectra, 357
    and photon energy, 31-34, *32*, *33*
    of rainbow, 31-32, *33*
    of sky, 31
    of visible spectrum, 31-33, *33*
collision
    of atoms, 75-76, *75*, 367, 468, 484
    and conservation of energy, 137-138
    and conservation of momentum, 122-125
    elastic, *136*
    between equal masses, 122, *122*
    between unequal masses, 122, *123*
comb, electrified, 458, *458*
combining weights, 37, 60
combustion, 14-15, *15*, 60, 331
    by-products, 330-334
    defined, 421
    and oxidation, 431
compound, chemical
    See chemical compound
compound, stable, 80
condensation nuclei, 335
condensation, of water vapor, 335
conduction, 76-78
    in crystals, 409
    of electricity, 77, 409
    by electrons, 77
    in gases, 77
    in liquids, 77
    in metals, 77, 397
conductor
    current-carrying, 461, 466, 468
    defined, 458
    and energy bands, 416
    speed of electrons in, 461, 468
    in static condition, 465
    and thermionic emission, 458
    and unbound electrons, 416, 457, *457*
conservation laws
    importance of, 576
    and symmetry, 587
conservation of electric charge, 456, 461, 576
conservation of energy, 137-138, 145, 576
    and circuits, 471, *472*
    and first law of thermodynamics, 145
    in relativity theory, 164
conservation of matter, 37, 59, *59*
    in chemical reaction, 419
    in combustion, *59*, 60
conservation of momentum, 122-125, *122*, *123*, *124*, 353, 576, 578
constructive interference, 502, *502*
continental drift, 312, *313*, 319, *320-321*, *322*, 339
continental shield, 246
Copernican planetary system, *7*, *174*
core of earth, 238, 241, *241*
    and shadow zone, *236*, 238
Coriolis effect, 285, *285*, 286
cosmic ray photons
    in electromagnetic spectrum, 33, *33*
    and mesons, 579
coulomb, 107, 362
Coulomb, Charles, *457*
Coulomb's constant, 107
Coulomb's law, 107, 359, 362, 456, *456*
covalent bonding, 382, 383, *384*
    charge distribution in, 387, *387*
    in polyatomic molecules, 386, *386*
Crab Nebula, 1, 2, 210, *210*
cracking, 426
craton, 247
Crick, Francis
    and DNA replication, 443
    use of diffraction techniques, 404

crust of earth, 238, 239, *241*
    composition of, 243, *243*, 251
    continental, 243
    oceanic, 243,
    relation to mantle, 244
    thickness of, 314
crystal
    diffracting properties of, 65, 253, 375, 400
    electrons in, 414, *415*
    faces, 251, *252*, 400
    grain boundaries in, 400, *400*
    growth, *253*, 400
    of iron, *491*
    lattice, *252*
crystal domains, 490, *490*
crystal, energy bands in, 415, *415*
crystalline solid, 251, 399
    atomic arrangement of, *252*, 253
    behavior of, 253, 399
    cleavage in, 253, *253*
    crystal faces of, 253
    differential hardness of, 253
    and electron diffraction, 375
    and x-ray diffraction, 65, 253
current
    See electric current
cyclone, *286*, 287, *288*

d

Dalton, John, 60–62, *60*, *61*, 379
Davisson, Clinton, 375, 401
Davy, Humphry, 59, 405, *405*
Delta, as change, 114
delta, river, 264, *264*
Democritus, 55, *56*
denaturation, 446
density
    of atmosphere, 71, 75, *75*
    definition of, 69
    and sound waves, 78
    and specific gravity, 69
    ultimate, 70
    of various materials, *70*
deoxyribonucleic acid (DNA), *10*, 443
    backbone of, 440
    double helix of, *443*
    and diffraction techniques, 404
    and DNA polymerase, 443
    encoding of information, 443–444
    function of, *440*
    molecular structure of, *442*
    as optically active molecule, 394
    size of, 441–442
deposition, of sedimentary rock, 259
    and alluvial fans, 264
    and deltas, 264, *264*
destructive interference, 502, *502*
detritus, 265, 300
deuterium ("heavy hydrogen"), 202, *205*, 572
devitrification, 251
diamagnetism, 490
diamond
    as allotropic substance, 406–407
    bonding of, 411
    natural state of, *408*
    in polarized light, *408*
    structure of, *406*
differential hardness, of crystalline solid, 253

diffraction, 25
    in crystal gratings, 65, *65*
    in diffraction gratings, 34
    of electrons, 375, 401, *403*
    and wave nature of light, 23–25, *24*, *25*, 34
    x-ray, 65, *65*, 400–404, *402*, 446, *446*
dipole moment, 384, *385*
direct current, *470*, 473
dislocation, edge, 409,*410*
dislocation, screw, 410, *411*
dispersion force, 385
DNA
    See deoxyribonucleic acid
dolomite, structure of, *256*
Doppler effect, 193–194, *195*, 518
drag force, 118, 119, *119*
drainage basin, 301, *301*
Duke, Charles, Jr., 226

e

earth, *233*
    abundance of elements in, 238–239
    average density of, 182
    composition of, 238–239
    density at core, 70
    formation of, 233
    geologic time scale of, 234, *234*
    interior of, 238, *240*
    mass of, 236
    model of, 239, *240*,*241*
    outer surface of, 238
    size of, 236
earthquake
    defined, 236
    epicenter, *236*
    and fault planes, 314
    focus, 236, *242*, 314, *317*
eclipse, solar, *213*
ecliptic, *178*
E. coli bacteriophage, *439*, 442
Einstein, Albert, 151, *156*, 170, 545, 566, *567*
    on Brownian motion, 67
    and energy-frequency relation, 35, 365
    and energy-mass relation, 164, 545
    and general theory of relativity, 165–167
    life of, 156–157
    on relativity principles, 157–167
    and special theory of relativity, 157–165, 170
eka-silicon (germanium), prediction of, 83
Ekman "spiral," *297*
Ekman, V. W., 297
electric charge, 16, 106, 455
    of atom, 16, 456
    on capacitor, 467, *467*
    conservation of, 456, 461
    in magnetic field, *480*, 481–484, *483*
    negative, 16, 107–108, 455, *456*, 458
    neutral, 16, 107–108, 455
    positive, 16, 107–108, 455, *456*, 458
    and potential difference, 467
    size of, 455
    unit of, 107, 362, 455
    and voltage, 462
electric current
    alternating, *470*, 473
    defined, 459, *461*
    direct, *473*
    and electromotive "force," 469
    and equation of continuity, 459, *461*
    unit of, 459, 487
electric energy
    See energy, electric

electric field, and electric force, *480*, 481
electric force
    See force, electric
electric motor, principle of, 485
electric power
    consumption of, 143, 331, *472–473*
    and current and voltage, 468
    generation of, *472–473*
    loss, 474
    pumped storage systems, 472
    sources of, 338, 339, 342, *472–473*
    transmission of, 471, *472–473*, 474
    units of, 467
electric resistance, 469, *469*
    unit of, 469
electrolysis, 423, *423*
electromagnetic induction, 493, *494*
electromotive "force," 469, *470*, 495, *496*
electromagnetic radiation
    applied uses of, *506–507*, 508–517
    forms of, 33, 505–517
    interference of, 502
    and polarization, 518
    principles of, 499–505
    propagation of, 33–35, 151, 500
    speed of, 33
electromagnetic spectrum, 33, *506–507*
    energy forms in, 32–33, 508
electromagnetic waves
    amplitude of, 502
    analogous to water waves, 33–34, *34*, *500*
    applied uses of, *506–507*, 508–517
    and atmospheric heat balance, 282
    constructive interference of, 502, *502*
    defined, 505
    destructive interference of, 502, *502*
    energy carried by, 505
    intensity of, 505
    interaction with matter, 515
    as probability waves, 33
    propagation of, 33–35, 151, 500, 503, *503*
    and quantum theory, 33–35
    sinusoidal, 500, *501*, *504*
    source of, 503, *503*
    speed of, 151
electron, 16–17, 359
    building up of energy states, 371, *371*
    charge, 456
    charge/mass ratio of, 359
    configuration of orbitals, 16, *16*, 380, 487
    in a crystal, 414, *415*
    distance from proton, 366
    emission, 359, 460
    emission from television filament, 461
    energy of, and color, 365
    flow in conductor, 461
    mass, 456
    in a metal, 413, *413*, 414
    motion, and angular momentum, 578
    motion, and magnetism, 490
electron affinity energy, 380
electron cloud, 16, *16*, 17, 376, *376*, 385, *385*
electron density, contour maps of, 404, *404*
electron diffraction, 375, 401, *403*
    compared with x-ray diffraction, *402*
electron microscope, 66, *66*
electron shells, *371*
    examples of, 85
    groupings of, 85
    and hidden energy levels, 370, 371
    and Pauli exclusion principle, 371
    and periodic table, 371
    and valence, 85

electron spin, 376, 413, 487–488, 579
electrostatics, law of, 107, 109
elementary particles, 575, 579–591
  and antiparticles, 582, 582
  charge of, 457
  classes of, 579, 580
  model for, 589, 589
  mass of, 457
  properties of, 575
  reactions of, 583
elements
  chemical nature of, 13–14
  chondritic abundance of, 222–224, 223
  definition of, 13
  formation of, 196–198
  origin of, 197–198
  solar abundance of, 222–224, 223
  transuranium, 555, 560
elimination reaction, 430, 431
endothermic reaction, 422, 422
energy
  forms of, 135, 136, 364
  minimum value for atoms, 367
  runs downhill, 146–147, 348
  quantization of, 368
energy, chemical, 145, 424
energy, conservation of, 137–138, 145, 364, 576
  and energy levels of atoms, 365
  and first law of thermodynamics, 145
  and relativity theory, 164, 203
energy, consumption of
  current United States level of, 330, 330
  in prehistory, 135
  rate increase, 136, 136, 568
energy-frequency relation, 35, 365, 526
energy, electric
  and electromagnetic waves, 505
  as photons, 19, 31–33
  as potential energy, 145, 462
energy, elastic potential, 138, 140
energy, kinetic, 137–138, 137, 139
  and "electromagnetic waves," 505
  and elastic collision, 137–138
  and gravitational "accretion," 227
  and gravity, 138–140
  as heat, 12, 19, 143–145, 227
  and molecular motion, 73–75
energy levels, atomic, 365
  ground level, 367
  excited level, 367
  hidden level, 370
  of hydrogen, 366, 367
  and ionization potential, 367
  and life, 367
  minimum value for, 368, 369
  notation for, 368, 369
  and wavelength, 374
energy-mass relation, 164, 545, 576
energy, mechanical, 137–138
  and heat energy, 143, 144–145
energy, nuclear, 545, 559
  and energy-mass relation, 545
  and energy resource crisis, 568
  fuel source for, 569, 560
energy, potential, 137–138, 137, 139
  and gravity, 138–140, 141
  and position, 139, 140, 462
  and voltage, 462
energy state letters, 369, 369, 370
entropy, 348

enzyme, 394
  active site, 446, 446
  as catalyst, 426, 445–446
  denaturing of, 446
  as optically active molecule, 394
  size of, 442, 442
  x-ray diffraction of, 446, 446
Epicurus, 55–57, 56
equation of continuity, 459, 461
equinoxes, 177
equipotential, 464, 464
erosion, 259, 261
  and resulting landforms, 262–263
"ether," 31, 152, 505
evaporation, 12, 76, 77, 282, 284, 304, 309, 338
exchange particles, 548, 548
expanding universe, 193–195, 194, 195, 346
exponential notation, 48, 50–51
exothermic reaction, 422, 422, 546
extinct radioactivities, 217, 221
extrusive (volcanic) rock, 257, 259, 325

f

Fahrenheit scale of temperature, 73, 73
Faraday, Michael, 466, 499, 504
Faraday's copper cage, 466, 466
faulting, 236, 314, 323, 323
feldspar, 254
  and metamorphism, 268
  in meteorites, 220
  structure of, 254
Fermi, Enrico, 559, 560, 566
ferromagnetism, 490
Fischer, Emil, 394
fission, nuclear, 559, 565
  chain reaction in, 563, 563
  discovery of, 559
  energy release in, 546, 547
  mass distribution in, 562
  mass yield curve, 564
Fitzgerald, George, 154
Fizeau, Armand, 22–23, 23
folding, 246, 259, 260
force
  definition of, 89, 89
  unit of, 91
force, components of, 96, 97
force, electric, 89–90, 105–109, 105
  between matter, 457
  and electron deflection, 359
  and equipotentials, 464, 464
  and gravitational force, 106
  fundamental law of, 107
  unit of, 107
force, gravitational, 89, 102
  at earth's surface, 103, 103
  and electric force, 106
  and formation of stars, 198–200, 201
  and geometry of space-time, 167
  and mass, 100–104, 101, 165–166
  and planetary motion, 182, 187
  universal law of, 100, 165
force, magnetic, 109
  and electric charge, 481, 482
  and electric current, 484
  and left-hand rule, 482, 482
  and right-hand rule, 482, 482
force, nuclear, 89, 90, 109, 544
  charge independence of, 548
  range of, 109, 548
force, as vector, 91–93, 92, 93
forces, combining, 93–94
  law of, 91
  and resultant vector, 94
  and rule of vector addition, 94, 94

fossil fuel, 331, 472–473
Foster, Carey, 81
Fourier, Jean Baptiste Joseph, 503
von Fraunhofer, J., 222
Fraunhofer lines, 221, 222
  original count of, 377
  source of, 222
free fall, 128–129, 129
  and gravitational collapse, 199–200
  velocity of, 129
frequency, 35, 500–501
  and energy relation, 34–35
friction, 89, 90, 108, 119, 462
  See also drag
Friedrich, W., 400
fronts, 286, 287
fusion, nuclear, 559, 571
  conditions for, 571
  energy release in, 203, 546, 547
  as energy supply, 475
  in formation of elements, 200, 205, 207–209
  power plant for, 572, 572

g

galaxy
  changing configurations of, 176
  configurations of, 190, 191, 208
  Milky Way, 189–191, 188–189
  M104, 208
  quasi-stellar, 191, 193
Galilean transformations, 151
Galileo, 118, 151, 170–171, 210
gamma rays (photons), 517
  and electromagnetic spectrum, 33, 506–507
  penetration power, 553
  and radioactive decay, 67, 549
  and red shift, 195
garnet, 254, 268
gas, average density of, 71, 71
gas giant, 184, 230
gas laws, 62, 79, 405
gas pressure
  and molecular energy, 144–145
  and perfect gas law, 405
  in pistons, 144–145, 145
gas volumes, 62, 405
Gay-Lussac, Joseph, 59
Geiger counter, 37, 553, 555
Gell-Mann, Murray, 591
generator
  and alternating current, 492
  and electromotive "force," 495–497
  electrostatic, 470
  magneto-hydrodynamic, 492
  principle of, 492, 492
  solar cell, 470
  thermoelectric junction, 470
geocentric frame of reference, 175
geologic time scale, 234, 234
  and radioactive decay, 551, 552, 553
geostrophic wind, 286
geothermal gradient, 256
geothermal power, 339, 340
Gerlach, W., 488
Germer, Lester, 375, 401
Gilbert, William, 477, 478
glacier, 305, 305
glaciology, 293, 306

glass
  atomic arrangement of, 252
  as conductor, 458
  example of, 252
  index of refraction for, 27
  as "supercooled liquid," 254
Glossopteris flora, 312, *312*
glutamine synthetase, 442, *442*
Gondwanaland, *313*, *320–321*
Goudsmit, Samuel, 377
graben, 314
Grand Canyon, *260*
granite, 258, *261*
  chemical equivalent in extrusive rock, *259*
  in earth's crust, 243
graphite
  as allotropic substance, 406
  imperfections in, 408
  structure of, *406*
gravimetric survey, *338*, 340
gravitation, and general theory of relativity, 102, 165–167
gravitation, Newton's universal law of, 100,
gravitational collapse, 198–200, *201*, 210
gravitational constant, 101
  measurement of, *102*
gravitational force
  See force, gravitational
gravitational heating, and melting of moon, 227
gravitational instability, 198–199
gravity anomalies, 245
  in earth's crust, 245
  negative, 314, 319
  in prospecting, *338*
gravity, effect on light, 166, *166*, *167*
gravity field
  of earth, 104–105, *105*, 480, *480*
  and potential energy, 138
  of sun, 480, *480*
graywacke, 324
Great Nebula in Orion, *193*
greenhouse effect, 284
ground water, 302, *302*
  effect on limestone, 303, *303*
  and springs, 303
  and water table, *302*, 303

h

Hadley cell, 284, *285*
Hadley, George, 285
hadron, 579, *580*
Hahn, Otto, 560, *560*
Half Dome, *262*
half-life, 163, 196–197
  and age of universe, 197, 217
  of gold, 192, *196*
  and rate of radioactive decay, 551, *551*
halides, and electron attraction, 384
halite, 253, *253*
halogens, in periodic table, 80, *80*, 81
heat energy
  of atmosphere, 281–286, *282*, *284*
  in changes of state, 76
  in combustion, 14–15
  and electric energy, 468
  as energy of motion, 12, 19, 73–74, 143–145
  and mechanical energy, 144–145, *145*, 462
  and second law of thermodynamics, 146–147
  waste, 331, *337*, 338
  See also energy, kinetic

heat engines, 144–145, *145*
"heat island," 336, *336*
"heavy hydrogen" (deuterium) 202, *205*
Heisenberg, Werner, 375, *375*, 489
Heisenberg's uncertainty principle, 376
heliocentric frame of reference, 175
heterogeneous substance, 419
Hodgkin, Dorothy, 404
holography, 534, *534*
homogeneous substance, 419
horsepower, 143, 467
Hubble, E., 193
Hubble's law, 194, 195
Humasen, M., 193
humidity
  absolute, 280
  relative, 280, *281*, 335
hurricane, *275*
Huygens, Christian
  and kinetic energy, 137
  and wave picture of light, 517, *519*
hybrid orbital, 388
hydrocarbons, 389
  forms of, 389
  and photochemical reactions, 435–437, *436*
  unsaturated, 395
hydrogen atom
  combining weight of, 60
  density of, 70
  radius of, 70
  spectral lines of, 190
hydrogen bomb, 200
  energy release in, 200, *202*
hydrogen-helium cycle, 200–205, *205*
  "bottleneck" in stars, 203
hydrogen sulfide, bonding geometry of, 387, *388*
hydrologic cycle, 293, *295*
hydrology, 293
hydronium ion, 428
hydrosphere, 241, 293, 306
hypercharge, 585, 586

i

ice
  and change of state, 12, *12*, 76
  density of sea ice, 305
  forms of, 408
  as fresh water supply, 306
  and glaciations, 306
  movement of ice sheets, 305, *305*
Ice Ages, 306, *306*
igneous rock, 246, 254
imperfections, in solids, 409
  edge dislocations, 409, *410*
  screw dislocations, 410, *411*
Indarch meteorite, 220, *221*
index of refraction, 27, *27*, 517
infiltration capacity, 302
infrared radiation
  and astronomy, 190
  and black body radiation, 514
  and earth's atmosphere, 282
  and electromagnetic spectrum, 32, 506–507
  and molecular vibrations, 514, *515*
  and photographing "heat islands," 336
insulator
  defined, 458
  and energy bands, 416
  and unbound electrons, 416, 457

interference
  bands of light waves, 24–25, *24*
  constructive, 502, *502*
  destructive, 502, *502*
  effects on soap film, 519, *520*
intrusive (plutonic) rock, 258, *258*, *259*, 325
inverse square law
  for electricity, 106, 363, 458
  and electromagnetic waves, 505, *505*
  and magnetism, 477
  for gravitation, 101, *101*, 102, 363, 466
  for light, *20*, 195
ion
  defined, 379
  and electromotive "force," 470, *471*
  in mineral groups, 254
  radii of, 382, *382*
ionic bonding, 379, 380, *384*
ionization potential, 367
ionosphere, 279
iron filing maps, *478*
iron meteorite, 215, *216*
Isaacs, John, 306
island arcs, 248, *246–247*, 313, 319
isobars, 286, *287*
isomer, 430
isomerization, 430
isostatic balance, of earth's crust, 244
isotope, 543
isotope, stable
  defined, 195
  source of, 217–218
  See also extinct radioactivities
isotope, unstable
  and age of universe, 197, 219–221
  defined, 196
  and half-life, 196–197
  and radioactive decay process, 196–197
  source of, 217–218
isotopic spin, 585, 586

j

jet-stream, 286, 287
Joliot-Curie, Irene, 560, *560*
joule, 47
Jupiter
  composition of, 184
  density of, 182
  formation of, 230
  as gas giant, 184
  mass of, 182
  orbit in galactic plane, *177*
  as star, 187
  size of, 182, *182*
  volume of, 184

k

Kant, Immanuel, and island universe, 191
Kelly meteorite, *216*
Kelvin scale of temperature, *73*, 74
Kepler, Johann, *174*, 210
kinetic energy
  See energy, kinetic
kinetic theory of gases, 72–76, 357, 405
  and density, 75–76, *75*
  and temperature, 73–75, *74*
  and molecular weight, 74
Kirchhoff, Gustav, 84, 357, 514
Knipping, P., 400

l

lake, 301
  basins, 304, *304*
  interior, *293*
Langmuir, Irving, 399, 416
Larderello geothermal field, 339
laser, 523
  applications, *526*, *527*, 528–533, *534*
  and stimulated emission, 523–525, *524*, *525*
  types of, 526
von Laue, Max, 400, *401*
lava, 256, *257*
Lavoisier, Antoine, *58*, 59–60
law of definite proportions, 419
Lee, T. D., 588
lenses, 28–29, *30*, 529, *529*
leptons, 580, *580*
light
  absorption of, 21, *21*, 25–26
  behavior of, 19–36, 517
  coherent, 524, 533, *533*
  color of, 31–35, *32*, *33*, *34*
  dual nature of, 19, 24, 33–35, 372, *373*, 499
  and electromagnetic spectrum, 33, *506–507*
  incoherent, 533, *533*
  model of, 35
  polarized, 520, *520*, *521*
  propagation of, 21, 26, 151, 517, *519*
  speed of, 22
  and vision, 19, *28–29*, 31–32, *32*, 515
  wavelength of, 34–35, *34*, 65
light beam, 22, *23*
  laser, 528
  polarization of, 392
light rays, 22, 65, 517
  and gravity, 166, *166*, *167*
  and lenses, 28–29, 529, *529*
limestone
  Bahama banks, *266*   beds, *251*
  beds, 251
  caverns, 303, *303*
limnology, 293
lines of force, *464*, 465, *480*, 481
liquid state
  atomic vibrations in, 399
  behavior of, 11, *11*, 404
  explanation of flow in, 405
  and kinetic theory of gases, 73
Lisbon earthquake, *236*
lithification, 259, 265, *265*
lithosphere, 241, 293, 319
  accreting plate boundaries of, 319
  consuming plate boundaries of, 323
  lateral slippage boundaries of, 323
living systems
  and diffraction techniques, 404
  distinction from giant molecules, 451
  origin of, 234, 395
  simplest organisms of, 448
lodestone, 477, 490
Lorentz, Hendrick, 155, *156*
Lorentz transformations, 155–156, 161
luminosity, 19, 22, 67

m

Magellanic Clouds, *190*, 191
magma, 254, 256, *257*, 258
magnet
  bar, 477
  coil, *478*
magnetic compass, 479, *479*, 481, *481*
magnetic field, *478*
  in current-carrying loop, 481, *481*
  in current-carrying wire, 481, *481*
  of earth, 317
  effect on moving charge, 482
  iron-filing maps of, *478*
  and magnetic force, 481
  unit of, 487
magnetic flux, 495, *496*
magnetic force
  See force, magnetic
magnetic poles, 477
  of earth, 315, 477
  indirect evidence for, 478
magnetic potential energy, 370
magnetic quantum numbers, 577
magnetic survey, *338*
magnetism, 477
  and Ampere's theory, 479
  and current in atom, 488
  and electricity, 492, *492*
  and relativity, 497, *497*
  and structure of atom, 487
magnetometer, 225
magnetosphere, *279*, 280, *280*
magnification, 28–29
Malaspina Glacier, *305*
Manhattan Project, 566
mantle of earth, 238, 241, *241*
  relation to crust, 244
Mariner 9, *184*
Mars
  apparent retrograde motion of, 180, *180*
  density of, 182
  in infrared light, *506*
  Mariner 9 probe of, *184*
  orbit in galactic plane, *177*
  size of, 182, *182*
  in ultraviolet light, *506*
  in visible light, *506*
mass
  definition of, 100
  equality of inertial and gravitational, 165–166
  gravitational, 102, 165
  inertial, 165, 166, *238*
  measurement of, 100
  relativistic, 164
  unit of, 100
matter, conservation of, 37, 59, *59*
matter, dual nature of, 373
matter, states of, 399
  allotropic, 405
  atomic vibrations in, 399
  changes in, 11–13, 76, 77, 399, 404, 411
  orderliness in, 402
  polymorphic, 406
  plasma, 416
  structure of solids, 399
  temperature and restlessness, 404
Maxwell, James Clerk, *72*
  and kinetic theory of gases, 72, 78
  on electromagnetic wave propagation, 151, 152, 499, 504
measurement
  approximation and error, 8, 40, 49
  averaging procedure, 8–9, *9*, 71, 72
  counting as, 37
  derived units of, 47
  of distance and length, 41–42, *41*
  of forces, 90–91, *90*
  indirect, 41, *41*
  and probability, 37, 72
  and proportionality, 41, *41*, 45, 91
  and ratios, 38
  and significant figures, 49
  of speed, 47
  of speed of light, 22–23, *23*
  and statistics, 72
  of surfaces, 42–43, *42*, *43*
  of temperature, 73–74, *73*
  of time, 45–47
  of volume, 44, *44*
  units of, 38–40
Meitner, Lise, 560, *560*
melting point, 83
Mendeleev, Dmitri, 81–82, *81*, 362
mercuric iodide, as allotropic substance, 406
Mercury
  density of, 182
  motion through Zodiac, 177
  orbit in galactic plane, *177*
  size of, 182, *182*
mercury
  atomic spectrum of, 83
  changing states of, 399
meson, 548, *548*, 584
  as attractive force between nucleons, 548
  and uncertainty principle, 581
mesosphere, 279
Messier, Charles, *187*
metamorphic rock, 254, 268, *269*
  classification of, 269, *269*
  and rock cycle, 270–271, *272*
metamorphism, 268
  of blueschists, 319
  and dehydration, 268
  and foliation, 268, *269*, 272
  and recrystallization, 268
  in orogenesis, 324
  in rock cycle, 270–271, *272*
meteor, defined, *214*
meteorite
  and age of solar system, 217, 219–221
  defined, 213, *214*
  examples of, *216*
  formation of, 215
  recovery of, 213–215
  types of, 215, *216*
methane, bonding geometry of, 387, *388*
metric system, 39
Meyer, Lothar, 81
mica, *254*
  and metamorphism, 268, *269*
  structure of, *254*
Michelson, Albert, 152, *153*
Michelson-Morley experiment, 152–154, *152*, *154*
microscope
  electron, 66, *66*
  principles of, *28–29*
microwaves, 511–513
  in electromagnetic spectrum, 33, *506–507*
  and radar, 512
mid-oceanic ridge, 247, *246–247*, 295
  and rifting, 248, 314, *315*
Milky Way Galaxy, 189–191, *188–189*
  diameter of, 189
  evolution of, 198, 211
  and galactic plane, *188–189*, 189
  mass of, 189
Millikan, Robert, 359, *360*
  and electron charge/mass ratio, 359
  oil drop experiment, 359, *360*
mineral
  defined, 251
  major groups of, 254
  mining, *267*
  prospecting for, *338*

629

mirrors, *28–29*, *30*
mirror symmetry, 586
model building, 5–7, 37, 55, 377, 478, 575
Mohorovičić discontinuity, 238, 241, *241*, 244
mole, 63, 423
molecule, 11, 379
    dipole moment in, 385
    model of, 11
    orbitals in, 388
    paramagnetic, 489
    polar, 384
    polyatomic, 386, *386*, 387
    shapes of, 387
    size of, 65
    speed distribution in air, 73, *73*
momentum, 119–125, *120*
    angular, 353, 575, *575*
    change in, 125–126, *126*, 359
    conservation of, 122–125, *122*, *123*, *124*, 574, 576
    definition of, 120, *121*
    law of (second law of motion), 125
    unit of, 120
momentum, change in, 125–126, *126*
    and second law of motion, 126–127
    unit of, 127
monomer, 440
monsoon, *288*, *289*
moon, 227
    age of, 224
    craters of, *226*, *227*
    distance from earth, *182*
    electric conductivity in, 225
    formation of, 224–231
    interior of, 225
    magnetic field of, 225, 226
    melting of, 225–226
    and history of solar system, 224–231
    volcanism of, 226–227
    water vapor of, 226–227
moonquakes, 226
Morley, Edward, 152, *153*
Moseley, Henry, 516
motion
    and change in position, 111–114, *113*, *114*
    and "curved universe," 350
    and change in time 113, *113*
    circular, 131–133, *132*, *133*
    with constant force, 127–129
    and free fall, 128–129, *129*
    and kinetic energy, 140
    spiral, in magnetic field, 483
    and time flow, 350, 586
    and trajectory, 130
motion, contraction of, 154–155, *155*, 163
motion, laws of
    and dependence on velocity, 127, 153, 157, 164
    first law, 118, 149
    and planetary motion, 187
    and relativistic mass, 164
    and relativity principles, 151, 153, 155–156
    second law, 125, 126–127
    third law
        see reaction force
motion, uniform, 116–119, 149
    "absolute," 149–150
    and constant driving force, 118
    definition of, 116
    and Galileo, 118
mountain, 245, *245*
    orogenesis, 324–327, *326*
muon, 575, 581
mutation, genetic, 444

n

nebula
    Crab Nebula in Taurus 2, 210, *210*
    and formation of stars, 198–200, *199*
    Great Nebula in Orion, *193*, 199
    solar, 218, 229
negentropy, 146–147
neon
    atomic spectrum of, *83*
    discharge tube, 359, *360*
Neptune, 177
    composition of, 184
    density of, 182
    as gas giant, 184
    orbit of, *177*, *182*
    size of, 182, *182*
neutrino
    defined, 202
    in elementary particle reaction, 583
    in thermonuclear reaction, 202–203, 205
neutron
    and atomic weight, 195
    charge of, 457
    mass of, 457
    as nuclear component, 457, 543
neutron star, *207*, 209, 211, 348
    and pulsars, 211
Newlands, John, 81
Newton, as unit of force, 91, 455
Newton, Isaac, 48, 63, *63*, 69, 150
    and invention of calculus, 103
    and laws of motion, 125, 149, 182
    and relativity principle, 150,.153, 155–156, 157
    and universal law of gravitation, 100
Newtonian physics
    and atomic spectra, 84
    and electron orbit, 363
    and molecular motion, 72
    and laws of motion, 118, 125, 149, 157, 160, 171, 363
    and relativity principle, 151, 153, 155–156, 364
    and relativistic mass, 164
    universal law of gravitation, 100, 165
nitric oxide, and air pollution, 331, 336
nitrogen molecules
    structure of, *10*
    speed distribution in air, 73, *73*
noble gases, 80, *80*, 81
North Slope, *342*
nova, 188
nuclear chain reaction, 563, *563*, 566
nuclear force
    See force, nuclear
nuclear models
    liquid drop model, 562, *562*
    packing model, *545*
    shell model, 556, *556*
nuclear power, 342
    by-products of, 342
    plant, 572, *572*
nuclear radiation
    detection of, 552–555
    effects of, 552–555
nuclear reaction, 555, *555*
nuclear reactor, 566, 568, 570, *571*
nuclei, formation of, 197–198
    and carbon cycle, 208–209
    and energy release, 203
    and hydrogen-helium cycle, 200–205, *205*
    in stars, 200

nucleon, 543, 583
    and meson exchange, 548, *548*
nucleotides, 440
nucleus, atomic, 15–16, *16*, 543
    components of, 195, 456
    configurations of, 545
    density of, 70
    liquid drop model, 562, *562*
    packing model, *545*
    radius of, 363
    Rutherford's discovery of, 362, *362*
    shell model, 556, *556*
    size of, 544, *545*
    stability of, 544, *544*
    volume of, 545
nylon 6, 441

o

obsidian, *252*, 257
ocean basin, 247, *246–247*
    composition of, 295, *295*
    fracture zones in, 248, *317*
    island arc-oceanic trench system, 248, *246–247*, 313, 323
    topographic features of, 247–248
ocean floor
    age of, 313
    composition of, 295
    Pacific, *246–247*
ocean, world
    composition of basin, 295, *295*
    dimensions of, 293
    and inundation of coastal cities, *308*
    man's use of, 300
    zones of, 293–295, *295*
oceanic currents, 295–298, *296*, *297*
    deep, 296, *296*
    gyres, 297
    surface, 296, *296*
    thermohaline circulation, 297
    warm, *296*
    wind-driven circulation, 296
oceanic trench, 247, *246–247*, 313, 315, *317*
octane (normal), rotation in bonding, *391*
Oersted, Hans, 478, *480*
Ohm, Georg Simon, *468*
Ohm's law, 469, *469*
oil drop experiment, 360, *360*
Olbers, Heinrich, 195
Olbers' paradox, 195, *195*
olivine, *254*
    in meteorites, 220
    structure of, *254*
ooze, *295*, 300
opacity, 26
operational definition, 171
Oppenheimer, J. Robert, 566, *567*
optical window, 282, *342*
orbital quantum number, 369, *369*
organic chemistry, 389
organic compounds, 389
    examples of bonding in, *390*
    functional groups in, *392*
    nomenclature, 391, *391*
orogenesis, 324–327, *326*
oscillator, 45
oxidation, 424, 431
oxidation-reduction reaction, 431, *432*
oxygen
    combining weight of, 60
    molecular structure of, *10*
ozone, in atmosphere, 280, 435, *436*

TELL US WHAT YOU THINK...

Students today are taking an active role in determining the curricula and materials that shape their education. Because we want to be sure *Physical Science Today* is meeting student needs and concerns, we would like your opinion of it. We invite you to tell us what you like about the text—as well as where you think improvements can be made. Your opinions will be taken into consideration in the preparation of future editions. Thank you for your help.

Your name_____ Institution_____

City and State_____ Course title_____

How does this text compare with texts you are currently using in other courses?

☐ Excellent   ☐ Poor
☐ Good   ☐ Very poor
☐ Adequate

Name other texts you consider good and why._____
_____
_____

Do you plan on selling the text back to the bookstore, or will you keep it for your library?   ☐ Sell it_____   ☐ Keep it_____

Circle the number of each chapter you read because it was covered by your instructor.

1 2 3 4 5 6 7 8 9 10 11 12 13 14 15 16 17 18
19 20 21 22 23 24 25 26 27 28 29 30

What chapters did you read that were not assigned by your instructor? (Give chapter number.)_____
_____

Please tell us your overall impression of the text.

	Excellent	Good	Adequate	Poor	Very poor
1. Did you find the text to be logically organized?	_____	_____	_____	_____	_____
2. Was it written in a clear and understandable style?	_____	_____	_____	_____	_____
3. Did the graphics enhance readability and understanding of topics?	_____	_____	_____	_____	_____
4. Did the captions contribute to a further understanding of the material?	_____	_____	_____	_____	_____
5. Were difficult concepts well explained?	_____	_____	_____	_____	_____

Can you give examples which illustrate any of your above comments? _____
_____
_____
_____

Which chapters did you particularly like and why? (Give chapter number.)_____
_____
_____
_____

Which chapters did you dislike and why?_____
_____
_____
_____

After taking this course, are you now more interested in the physical sciences? ☐ Yes ☐ No

Do you feel this text had any influence on your decision? ☐ Yes ☐ No

Further comments or suggestions: _____

_____

_____

_____

_____

BUSINESS REPLY MAIL
No Postage Stamp Necessary if Mailed in the United States

Postage will be paid by

CRM BOOKS

Del Mar, California 92014

FIRST CLASS
Permit No. 59
Del Mar, Calif.

Just fold, staple and mail. No stamp necessary.

p

Pacific Ocean basin, *246-247*
paleomagnetism, *314, 317, 319, 320-321*
Pangaea, *313*
parallelogram, *93*, 94, *114*
paramagnetism, 489, *489*
parity, downfall of, 589
particulates, in atmosphere, 334, *335*
Pauli exclusion principle, 371
Pauli, Wolfgang, 371, *375*, 583
percolation, through soil, 302
perfect gas law, 405
periodic table of the elements, 80
   and atomic structure, 368
   and electron shells, 371, *371*
   explanation of groupings in, 81-83
   history of, 81-83
perpetual motion machine, 145
a philosophy of science, 168-171
pH scale, 428, *429*
photoelectric effect, 364
photon
   absorption of, 20, *21*, 25-26, 367
   and color, 31-33, 34, 365
   counting, 364
   as electric energy, 19, 31-35, *32, 33, 34*
   and electron selectivity, 367, *367, 368*
   emission of, 19, 364, 523, *524*
   and energy-frequency relation, 35, 365
   and particle nature of light, 19-35, 372
   and Planck's radiation formula, 514
   propagation of, *22*, 22
   and rest mass, 164
photosynthesis
   energy changes in, 424, *424*
   in geologic time, 234, *234*
   and oceanic distribution of nutrients, 299
   and solar energy, 341
pion, 548, 573, 580
piston, 144-145, *145*
Planck, Max, 35, *364*, 514
Planck's constant, 35, 365, 413, 526, 578
planets
   and Barnard's star, 186, *186*
   distance from sun, 177, 180, *181*
   major, 184
   orbits of, 176, *177*
   terrestrial, 182
plasma
   confinement of, 417
   in ionosphere, 279, 416
   state of matter, 416-417
plate tectonics, theory of, 319, *320-321*
   See also tectonism
Pleiades, 187, *187*
Pluto, 177
   density of, 182
   orbit in galactic plane, *177*
   size of, 182, *182*
polar bonding, 384
   and dipole moment, 384
   in sodium chloride, 458
polar ice caps, 306, *306*
Polaris, 188
polarized light, 520, *520, 521*
pollen grain, motion in water, 66-67, *67*
pollution
   air, 275, 336, *419*, 428, 515
   global, 329
   ocean, 301, 340
   mercury, 432
   nuclear, 342
   thermal, 331, 336, *337*
polyatomic molecules, 386, *386*
   bonding geometry of, 387
polymer
   defined, 439
   types of, *440*
porphyries, *257*, 258
positron, 550, 573
potential energy, *135*, 138, 462
   See also energy, potential
potential (voltage), 462
   defined, 462
   and equipotential, 464, *464*
   and force, 462
   unit of, 462
power
   defined, 142, 467
   units of, 143
power, electric
   See electric power
powers of ten, 49
precipitation
   and aquifers, 303
   and drainage basins, 301
   in monsoon season, *289*
pressure
   atmospheric, 76
   definition of, 76
pressure gradient, 286
Priestley, John, 60
principal quantum number, 368, *368*
prism, effect on light, *33*, 83, 151
Problems Pages
   Are the Seacoast Cities in Trouble?, *308*
   Calculating Continental Drift Using Lasers, *530*
   Calculating the Distance to the Focus of an Earthquake, *242*
   Calculating Electromotive Force, *486*
   Calculating Gravitational Collapse of a Star, *201*
   Calculating Planetary Distance, *181*
   Calculating Pressure Changes of a Gas, *79*
   Continental Drift, *322*
   The Cost of Mining Earth Materials, *267*
   Diffracted Electrons, *403*
   The Doppler Shift, *518*
   Finding the Vector Sum of Several Forces, *95*
   Fuels and Bonds, *381*
   How Can Fission Take Place?, *565*
   Images in Mirrors and Light Reflection, *30*
   Is the Waste Heat of the U. S. at the Danger Level?, *337*
   Measuring a Film of Oil on Water, *64*
   Radioactive Dating, *554*
   Understanding the Chemistry of Chocolate Chip Cookies, *433*
   Voltage Drop in an Extension Cord, *463*
   Working Out the Motion When the Force Is Different for Each Position, *132*
Prokhorov, Alexander, 525
protein
   backbone of, 440
   fibrous, 444
   molecular bonding of, 412
proton
   and atomic number, 543
   charge, 457
   mass of, 457
   as nuclear component, 195, 457, 543
proton-proton scattering, 202, 203
Ptolemaic planetary system, *174*
pulsar, 211
   and neutron star, 211

q

quantization
   of energy, 368
   of numbers, 577
   of space, 489
quantum electrodynamics, and energy-frequency relation, 34-35, 365, 526, 578
quantum theory of forces, 581
quantum theory of light
   See quantum electrodynamics
quantum wave mechanics, 33, 372-377, *373, 375, 376*, 413, *413*
quartz, *252*
   atomic arrangement of, *252*
   nature of, *252*
   structure of, *254*

r

radiation, atomic
   and isotopes, 195-197
   measurement of, 37-38
radical, 435
radioactive decay, 67, 163, 196
   and age of universe, 197, 217, 219-221
   and geologic time scale, *234*, 551, *552, 553, 554*
   modes of, 549
   process in radioactive isotopes, 195-197
   rate of, 551, *551*
   and stability of nucleus, 544, *544*
radioactive element
   isotopes of, 195-197
   nuclear source of, 109, 544
radioactive waste storage, 343
radio astronomy, 190, 195
radio spectrum, 509-511
radio telescope, *173*, 191, 195
radio transmitter, 363, 499
radio waves, 508-511
   antenna for, *508*, 509
   and astronomy, 190, 195
   in electromagnetic spectrum, 33, *506-507*
   frequencies for, 509
   in ionosphere, 279
   transmission of, 509
Raman scattering, 531, *532*
reaction force, 96-98, *97, 98*
red giant
   Betelgeuse, 186, 207
   evolution of, 207, *207*, 208
   our sun as, 207
redox reaction
   See oxidation-reduction reaction
red shift, 166, 193-194, *194, 195*, 346
reflection
   and Huygen's constructions, 517, *519*
   of light energy, 20, *21*, 28-29, *30*, 517
refraction
   double, in crystalline material, 253, *254*
   index of, 27, *27*, 517
   of light energy, 26-31, *27, 28-29, 30*
   in prisms, *33*
   of seismic waves, 236
   Huygen's explanation of, 519
Rejkanes Ridge, 317, *319*
relative humidity, 280, *281*, 335
relativistic mass, 164
relativity, general theory of, 165-167

relativity principle, 149, 150, 160
    Einstein on, 157
    and equality of inertial and gravitational mass, 165–166
    and magnetism, 497, *497*, 503
    in Newtonian mechanics, 155–156, 161
relativity, special theory of, 157–165
remanent magnetism, 315
resistor, 469, *469*, *470*
    and conservation of energy, 471, *474*
resonant cavity, 512
resource crisis, and technological innovation, 330
"rest energy," 164–165, 203
retrograde motion, apparent, 180, *180*
Reynolds, John, 217
Richardton meteorite, 217, *218*
    and age of solar system, 217
rifting, 248, *315*, 317, 323, 339
Rigel, 186
river systems
    drainage basins of, 301, *301*
    headwaters, 304
    underground, 303
rock
    defined, 243, 251
    igneous, 246, 254, *259*, 325
    metamorphic, 246, 256, 268, 272, 319, 324
    plutonic (intrusive), 258, *258*, *259*, 325
    sedimentary, 254, 313, 324
    volcanic (extrusive), 257, *259*, 325
rock cycle, *270–271*, 272, 324
Roentgen, W. C., 400, *401*
Rousseau, Jean Jacques, on earthquakes, 236
rubidium 87
    and age of moon, 224
    and age of solar system, 218–221
rusting, *432*
Rutherford, Ernest, 361, *362*, 548
Rydberg formula, 366
Rydberg, J. R., 366

s

salinity, 295, 309
salt, table
    See sodium chloride
San Andreas fault, 323, *323*
Santa Catharina meteorite, *216*
Saturn, *177*
    composition of, 184
    density of, 182
    formation of, 230
    as gas giant, 184
    orbit in galactic plane, *177*
    size of, 182, *182*
Savitch, Peter, 560
scalar quantity, 93
scaling, 40, *41*, 43, *43*, 93
Schawlow, Arthur, 525
schist, 272, 319
scientific process, 7–9, 37, 55, 59, 63, 156–157, 173–175, 168–171, 377, 478, 479, 560, 575–577
sea floor spreading, 317, *317*, 319
sedimentary rock, 254, *260*, *263*
    classification of, 266
    detrital, 265
    diagenesis in, 273
    formation of, 259
    in orogenesis, 324
seismic surveys, *338*

seismic wave
    primary, 236, *236*, 238, *238*
    refraction of, 236
    secondary, 236, *236*, 238, *238*
    velocity/depth curve for, 236, *239*
seismograph, 236, *238*
seismology, 236–238, *236*, *238*
shadows, 21–25, *22*
ship coring, 313, *314*
Shroedinger, Erwin, 375, *375*, 489
silicate, 254, *254*
silicon
    atomic arrangement of, 254
    bonding of, 412
    molecular structure of, *10*
    silicon-oxygen groups, 254, *254*, 412
silk, β-pleated sheets of, 444, *444*
silver, 65
    as conductor, 77
Sirius, 184
smog, 275, *276*, 435–437, *436*
snow crystal, 399, *417*
sodium
    atomic spectrum of, 83
    reactivity of, *420*, 421
sodium chloride, 14, 379
    bonding of, 379, 411
    as insulator, 416
    molecular structure of, *380*
    packing model of, *383*
solar cell, 470
solar eclipse, *213*
solar flare, 225, *225*
solar nebula, 218, 221
    chemical composition of, 221–224
    and formation of planets, 229–230
solar power, 340, 341, 470
solar prominence, *182*, *213*
solar system, 175, 213
    age of, 176, 218
    composition of, 221–224, *223*, *224*
    formation of, 198–199, 213, 221, 229, *229*
solar wind, 279, 280, *280*
solid state
    allotropic forms, 405, 406, *406*
    amorphous forms, 251
    atomic vibrations in, 399
    behavior of, 12
    crystalline, 251
    imperfections in, 409
    and kinetic theory of gases, 73
    orderliness in, 404–405
    polymorphic forms, 406
soltices, *178*
Sommerfeld, Arnold, *412*, 413
Sommerfeld's box, 413
sound waves, 78, 151–152, 166, 373
specific gravity
    definition of, 69, 143–144
    of earth, 236
    water density as standard for, 69, 236
specific heat capacity, 144, *144*
spectra, atomic
    See atomic spectra
spectral analysis, 222
spectral lines, 190, 193, 221, *222*
    analysis of, 222
    and combination principle, 366
    of galaxies, 346
    and red shift, 193
    source of, 221
    See also Fraunhofer lines
spectroheliogram, *210*
speed
    definition of, 47, 114
    unit of, 116
    and velocity, 114

speed of light, 22
    calculating the, 22, *23*
    in energy-frequency relation, 35
    and laws of motion, 157
    principle of constancy of, 160
speed of sound, 78
spin angular momentum, 579
spring scale, 90–91, *90*
spring, water, 303
stable star, 205
star
    cluster, 187, *187*
    evolution of, 205–211, *207*
    first-generation, 202
    formation of, 198–200, *201*, 205, *207*
    nova, 188
    neutron, *207*, 209, 211, 348
    populations, 186–188
    red giants, 186, *207*, *207*, 208, *208*
    stable, 205
    sun as, 175
    supernova, 188, 210
    variable, 188
    white dwarf, *207*, 210, 348
statics, first law of, 98, *98*, *99*
statics, second law of, 99, *99*
steady-state universe, theory of, 349
Stern, Otto, 488
stoichiometry, 419
stony meteorite
    See chondritic meteorite
Strassman, Fritz, 560
stratosphere, 279, *279*
substitution reaction, 427, *431*
sulfur
    and air pollution, 336
    as combustion by-product, 331
sun
    abundance of elements in, 222–224, *223*, *224*
    age of, 176
    average density of, 176
    chemical composition of, 176, 222–224
    contour map of, *507*
    density at center of, 70
    energy consumption in, 205
    formation of, 205
    as G2V star, 175
    motion through Zodiac, *177*
    temperature of, 175–176
sunspot, *225*
superconductivity, 83, 416, 475
supernova, 188, 210, 217
    energy release in, 210
    occurrence of, 210, 217
    as source of new isotopes, 217, 218
Surtsey, *311*
symmetry, and conservation laws, 587

t

tartaric acid, as optically active molecule, *392*, *393*
technetium, and formation of elements, 197, 209
tectonism, 311
    and compressional features of crust, *315*
    oceanographic evidence for, 313
    and orogenesis, 324–327, *326*
    and paleomagnetism, 314
    and sea floor spreading, 317, *317*, 319
    and tensional features of crust, *315*

telescope
   Galileo's, *191*
   optical, 191, *191*
   radio, *173*, 191
   reflecting, *28–29*
temperature
   and heat energy, 143
   and kinetic theory of gases, 73–74
   mean, 8
   scaling of, 73, *73*
   versus time, 7–9, *9*
temperature, measurement of, 73–74, *73*
terrestrial planets, 182, 238
thermal pollution, 336, *336*
thermionic emission, 458
thermodynamics, first law of, 145, 348
thermodynamics, second law of, 146–147, 348, 349
thermoelectric junction, 470
thermometer, *73*
thermonuclear reaction
   carbon cycle, 208–209
   and formation of elements, 197–198, 207–209
   hydrogen-helium cycle, 200–205, *205*
thermosphere, 279
Thomson, George, 401
Thomson, J. J., 357, *359*
Thomson's electric discharge tube, 359, *360*
thorium, electron photomicrograph of, *66*
thought experiments, 149, 157, 164, 415
tidal bore, 299
tides, 298, 299
time, direction of, 350–351, 587
time, measurement of, 45–47
time transformation
   See Lorentz transformations
tin, as polymorphic substance, 406
tobacco mosaic virus, *448*, 450
du Toit, Alexander, 312
torque, 99, 577–578, *578*
Townes, Charles, 525
transformer, 474, 493
translucence, 26
transparency, 26
transportation, of sedimentary rock, 259, 261–264
transuranium elements, *555*, 560
troposphere, 279
tsunami, 298, *299*

u

Uhlenbeck, George, 377
ultraviolet light, 515
   and earth's atmosphere, 282
   and electromagnetic spectrum, 32, *506–507*
uniform motion, 116–119
   See also motion, uniform
universe
   age of, 197
   evolution of, 198–199, 203, 211
   formation of, 173, 194, 198
   time scale of, 199
universe, "curved," 350
universe, expanding
   See expanding universe
unpolarized light, 521, *521*
upwelling, *297*, 299–300
Uranus, 177
   composition of, 184
   density of, 182
   as gas giant, 184
   orbit in galactic plane, *177*
   size of, 182, *182*

v

vacuum pump, 70
valence, 78–80, *81*
   change in periodic table, 80–82
   definition of, 78
Van de Kamp, Peter, 186
variable star, 188, 205
vector, 93
   force as, 91–93
   position as, 111
   properties in crystalline material, 251, 253
vector addition
   examples of, *94*
   rule of, *93*
vector position, 111, *112*
vector sum, 94, *95*, 99
velocity
   definition of, 115
   of recession in red shift, 194, *194*
   as vector quantity, 115–116, *115*, *116*, 483
Venus, 177
   brightness of, 180
   density of, 182
   motion through Zodiac, 177
   orbit in galactic plane, *177*
   size of, 182, *182*
Vine, F. J., 317
virus
   as giant molecule, 450, *451*
   self-assembly of, 450
   as simplest living organism, 448, 451
   weight of, 65
viscosity, 254
visible spectrum, 33, *33*
   absorption lines in, 221
   interaction with matter, 515
vision, 19, *28–29*, 515
   and color, 31–32, *32*
vis viva, 137
volcanism, 256, *270–271*, 273, *311*, 317, *317*, 323, 324
voltage, 462
   See also potential
voltage meter, 487, *487*

w

water
   allotropic forms of, 408
   balancing equation for, 420
   ground, 302, *302*
   and hydrologic cycle, 293, *295*
   as liquid, 11–12, *11*
   ocean, 285
   as solid, 12, *13*, 305, *305*
   as standard for specific gravity, 69
   molecular structure of, *10*
water table, *302*, 303
Watson, James, 404
   and DNA replication, 443
   use of diffraction techniques, 404
Watt, James, 143
waveform, sinusoidal, 500, *501*, *504*
wavefront, 517
wavelength
   definition of, 34, 501
   of electron, 373
   of light, 34–35, *34*, 65
   "location" of, 376
   of sinusoidal waveform, 501, *501*

waves
   amplitude of, 502
   constructive interference of, 502, *502*, 533
   defined, 499
   destructive interference of, 502, *502*, 533
   and dual nature of light, 19, 24, 502
   propagation of, 33–35, *34*, 499
waves, seismic, 236–238, *236*
waves, sound, 78, 500, 501
waves, tidal, 298
wave train, 519
waves, water, *25*, 152
   motion of, 298, *298*
   propagation of, 499, *500*
weak interaction, 109, 579, 590
weather
   control of, 289
   prediction of, 288
weathering, 259, *261*
Wegener, Alfred, 312
Wegener hypothesis of continental drift, 312, *312*
Wheeler, John, 561, *561*
white dwarf, *207*
winds
   defined, 286
   and oceanic circulation, 296
   and wave height, 298, *298*
wind wave, velocity of, 298
Winkler, Clemens, 83
Worden gravimeter, *338*
work, 135, 138–143, *140*, 462
   and circular motion, 141, *142*
   and constant velocity, 141, *141*
   defined, 140, *140*
   as scalar quantity, 142
   unit of, 142

x

x-ray diffraction, 65, *65*, 400–404
   and Bragg's law, *402*
   compared with electron diffraction, *402*
   of enzyme crystals, 446, *446*
x-rays, 516
   and astronomy, 190
   and electromagnetic spectrum, 33, *506–507*
   of organisms, *513*, *549*
   producing, 516
   and red shift, 195

y

Yang, C. N., 588
Yukawa, Hideki, 548, *548*, 581

z

zinc sulfide
   sphalerite form of, 407, *407*
   wurtzite form of, 407, *407*
Zodiac, constellations of, 176, *178–179*
   and Milky Way, 189
   motion of moon through, 177
   motion of planets through, 176, 177
   motion of sun through, *177*
zone of subduction, 317, *320–321*, 323

Credits and Acknowledgments

Cover Design by Tom Lewis, *Photography by* Stephen McCarroll

The World Around Us
3—Photograph courtesy of the Hale Observatories. Copyright by the California Institute of Technology and the Carnegie Institution of Washington.

Chapter 1
4—Lisa Starr; 6-7—Illustrations by Howard Saunders, 7—(right) Photograph courtesy of the Hale Observatories. Copyright by the California Institute of Technology and the Carnegie Institution of Washington; 8—Doug Armstrong after U.S. Department of Commerce, Weather Bureau, Phoenix, Arizona; 10—Photo by Phil Stotts, Diagram by Marie Feallock; 11-14—Tom Lewis; 15—(left) Photo by Fritz Goro, (right) Tom Lewis; 16—Donald H. Andrews and Richard J. Kokes, *Fundamental Chemistry,* © 1962, John Wiley & Sons, Inc.; 17—Medieval Concept of the Sky, courtesy Deutsches Museum, Munich.

Chapter 2
19—Ernst Haas/Magnum Photos; 20—(top) Phil Kirkland, (bottom) John Dawson; 21—(left) Walter Chandoha, (right) Tom Lewis; 22—Tom Lewis; 23—(top) John Dawson, (bottom) Dennis Stock/Magnum Photos; 24—Fundamental Photographs; 25—United States Navy; 27—Fundamental Photographs; 28-29—Doug Armstrong; 30—Bonnie Weber; 32—Doug Armstrong; 33—(left and right) Doug Armstrong, (center) Joe Thein; 34—(top right) John Dawson, (center) Doug Armstrong, (left) Dennis LaBerge; 35—The Bettmann Archive.

Chapter 3
36—David Miller; 38—John Dawson; 40—Doug Armstrong; 41—Tom Lewis; 42-43—Doug Armstrong; 44—Fred Hartson; 45—Doug Armstrong; 46—Eldon P. Slick; 50-51—Fourteen aquatint etchings in black and white by Patricia Collins and line drawing, p. 159, *The Scale of Nature* by John Tyler Bonner; © 1969 by John Tyler Bonner and Patricia Collins; by permission of Harper and Row Publishers; 53—The Granger Collection.

Chapter 4
54—Harold Cohen; 56—(left and center) The Granger Collection, (right) The Bettmann Archive; 57—Don Fujimoto; 58—(top left) Prints Division, The New York Public Library, Astor, Lenox and Tilden Foundations, (top right, center, and bottom) The Crerar Library; 59—Photos by Fritz Goro; 60—The Granger Collection; 61—(top) Don Fujimoto; 62—The Granger Collection; 63—The Crerar Library; 64—Bonnie Weber; 65—G. E. Research and Development Center; 66—A. V. Crewe; 67—(top) Tom Lewis, (bottom) Alchemist manuscript from the late Middle Ages, courtesy Germanisches National Museum.

Chapter 5
68—Photo by Fritz Goro; 70—Jell-O 1-2-3 Dessert Mix; 71—Tom Lewis; 72—(left) Culver Pictures, (center) The Bettmann Archive, (right) The Granger Collection; 73—(top) Doug Armstrong, (bottom) Tom Lewis; 74—Tom Lewis; 77—David Miller; 79—Bonnie Weber; 80—Howard Saunders; 81—(left) Doug Armstrong, (top right) The Granger Collection, (bottom right) Novosti Press Agency; 82—Albert Fenn; 83—Bausch & Lomb; 84—Marie Feallock; 85—"Mirror of Consolation," by Petrarch, 1532, courtesy Zentralbibliothek, Zürich.

Graphic Pause
86—René Magritte, "The Castle of the Pyrenees," Collection Harry Torczyner, New York. Photo by Geoffrey Clements.

Chapter 6
88—*Son of the Sheik*, United Artists; 90—John Dawson; 92—(top) *Sunnyside*, United Artists, (bottom) *One Good Turn*, Hal Roach Studios; 93—(top) Tom Lewis, (bottom) Doug Armstrong; 94—Doug Armstrong; 95—Bonnie Weber; 97—(top) *Gold Rush*, United Artists, (center) Ron Estrine, (bottom) *Girl Shy*, Paramount Pictures, courtesy Museum of Modern Art Film Library; 98—(top right) The Bettmann Archive, (bottom right) Don Fujimoto, Bonnie Weber, (bottom left) Werner Kalber/PPS; 99—(left) *The Speakeasy*, Mack Sennet, (right) *The General*, United Artists, courtesy Raymond Rohauer; 100—*Dracula*, Universal, courtesy The Museum of Modern Art/Film Stills Archive; 101—(top) Marie Feallock, (bottom) Ron Estrine; 102—John Dawson; 103—Ron Estrine; 104—*Frankenstein*, Universal Studios; 105-106—Tom Lewis; 109—"Natuurkunde Uit Ondervindingen Opgemacht," John Theophilus Desaguliers, 1751, courtesy Zentralbibliothek, Zürich.

Chapter 7
110—*Two Tars*, Hal Roach Studios; 112—Ron Estrine; 113—Marie Feallock; 114—Doug Armstrong; 115—*Modern Times*, United Artists; 116—(left) Doug Armstrong, (right) *Scarface*, United Artists; 117—(top) Ron Estrine, (bottom) Marie Feallock; 118-119—*Magnificent Men In Their Flying Machines*, Twentieth Century-Fox; 120—*City Lights*, United Artists; 121—*Magnificent Men In Their Flying Machines*, Twentieth Century-Fox; 122—*Three Jumps Ahead*, Universal Studios, courtesy The Museum of Modern Art/Film Stills Archive; 123-124—John Dawson; 125—*Early To Bed*, Hal Roach Studios; 126—Marie Feallock; 129—(top) Ron Estrine, (bottom) *Sundown Jim*, Twentieth Century-Fox; 130—Doug Armstrong; 132—Bonnie Weber; 133—(top) Marie Feallock, (bottom) Baroque Variant of Villard de Honnecourt's Perpetuum Mobile.

Chapter 8

136—(left) Doug Armstrong after Earl Cook, "The Flow of Energy in an Industrial Society," Copyright © 1971 by Scientific American, Inc. All Rights Reserved, (right) John Dawson; 137—Shirley Dethloff; 139—(top) *Liberty,* Hal Roach Studios, (bottom left) *The Greatest Show On Earth,* Paramount Pictures, (bottom right) Doug Armstrong; 141—Shirley Dethloff; 142—(left) Shirley Dethloff, (right) Marie Feallock; 145—John Dawson; 146—*Sunnyside,* United Artists; 147—"Monuments of Nineveh, II," Austen Henry Layard, 1853, courtesy Zentralbibliothek, Zürich.

Chapter 9

148—National Aeronautics and Space Administration; 150—Ron Estrine; 151—Marie Feallock; 152—Shirley Dethloff; 153—Brown Brothers; 154-155—Doug Armstrong; 156—(top) The Nobel Foundation, (bottom) Courtesy California Institute of Technology Archives; 157—Ron Estrine; 158-159—Howard Saunders after Payson Stevens; 160—Shirley Dethloff; 162—(top) Ron Estrine, (bottom) Doug Armstrong; 165—Doug Armstrong; 166—Ron Estrine; 167—(top) Shirley Dethloff, (bottom) Escher Foundation, Haags Gementemuseum, The Hague.

Wishing Won't Make It So

169—Terry Lamb

Chapter 10

172—Lawrence Lowry/Rapho-Guillumette; 174—(top left) The Bettmann Archive, (top right) Zentralbibliothek, Zürich, A. Cellarius "Harmonia Macrocosmica," 1661; 175—(left) Zentralbibliothek, Zürich, A. Cellarius "Harmonia Macrocosmica," 1661, (right) Musse de l'Homme, Paris; 176—Marie Feallock, (caption) Kenneth Rexroth, *Collected Shorter Poems.* Copyright 1940 by Kenneth Rexroth. Reprinted by permission of New Directions Publishing Corporation; 177—Doug Armstrong; 178-179—Karl Nicholason; 180—Ignacio Gomez; 181—Bonnie Weber; 182-183—From *Atlas of The Earth,* page 15, © Mitchell Beazley, Ltd., 1971; 185—(top and bottom) JPL/NASA, (center) Steve McCarroll; 186—(top) Yerkes Observatory Photograph, (bottom) Marie Feallock after Peter Van de Kamp, "Alternate Dynamical Analysis of Barnard's Star," *Astronomical Journal,* August 1969; 187—(left) Photograph courtesy of the Hale Observatories. Copyright by the California Institute of Technology and the Carnegie Institution of Washington, (right) Official U.S. Naval Observatory Photograph; 188-189—Lund Observatory, Sweden; 190—(left) Photograph courtesy of the Hale Observatories. Copyright by the California Institute of Technology and the Carnegie Institution of Washington, (top right) Lick Observatory Photograph, (bottom right) Official U.S. Naval Observatory Photograph; 191—Courtesy of the American Museum of Natural History.

Chapter 11

192—Official U.S. Naval Observatory Photograph; 194—(top) Marie Feallock after Edwin Hubble, *Red Shifts in the Spectra of Nebulae,* Halley Lecture, May 8, 1934, Clarendon Press, Oxford, (bottom) Howard Saunders; 195—Doug Armstrong; 196—Doug Armstrong after Don L. Eicher, *Geologic Time,* © 1968, reprinted by permission of Prentice-Hall, Inc., Englewood Cliffs, N.J., and *Physics for the Modern Mind* by Walter R. Fuchs. Copyright © 1967 by Weidenfeld and Nicolson, Ltd. and The Macmillan Company; 199—Photograph courtesy of the Hale Observatories. Copyright by the California Institute of Technology and the Carnegie Institution of Washington; 201—Bonnie Weber; 202-204—Howard Saunders; 206—Phil Kirkland; 207—Ron Estrine; 208—Photograph courtesy of the Hale Observatories. Copyright by the California Institute of Technology and the Carnegie Institution of Washington; 210—(top) NASA Photograph, (bottom) Photograph courtesy of the Hale Observatories. Copyright by the California Institute of Technology and the Carnegie Institution of Washington; 211—"Creation of the Universe" from *Histoire Naturelle* by Buffon, Vol. 1, Deux Ponts, 1785, courtesy Bibliothek der Eidgenossische Technische Hochschule, Zürich.

Chapter 12

212—Royal Greenwich Observatory; 214—(bottom) P. Ahnert/Sonneberg Observatory; 216—Photograph by Julius Weber; 218—Doug Armstrong, adapted from J. H. Reynolds "Determination of the Age of the Elements," *Physical Review Letters,* January 1, 1960, Vol. 4, p. 8; 220-221—Doug Armstrong after K. Gopalan and G. W. Wetherill, *Journal of Geophysical Research,* Vol. 75, pp. 3464 and 3465; 222—From *Atlas of the Universe,* page 142, © Mitchell Beazley, Ltd., 1971; 224—Doug Armstrong after Brön Mason, *Handbook of Elemental Abundances,* 1971, Gordon and Breach Science Publishers, Inc.; 225—Photograph courtesy of the Hale Observatories. Copyright by the California Institute of Technology and the Carnegie Institution of Washington; 226-227—NASA Photograph; 228—Phil Kirkland; 231—Courtesy of Offentliche Bibliothek der Universität Basel from Julius Obsequens "De Prodigiis," Basel, 1552.

Chapter 13

232—NASA Photograph; 235—Joe Garcia; 236—Mansell Collection; 237—John Dawson; 238—(top) John Dawson after *Atlas of the Earth,* © Mitchell Beazley, Ltd., 1971, (center) From *Atlas of the Earth, page 39,* © Mitchell Beazley, Ltd., 1971; 239—Doug Armstrong after B. Gutenberg, *Internal Constitution of the Earth,* Dover Publications, Inc., New York; 240—From *Atlas of the Universe,* page 64, © Mitchell Beazley, Ltd., 1971; 241—Don Fujimoto; 242—Bonnie Weber; 243—Doug Armstrong; 244—Dan Morrill; 245—Reflejo; 246-247—Courtesy of the National Geographic Society; 249—Library of Congress.

Chapter 14

250—Darrell Grehan/Photo Researchers, Inc.; 252—(top left) John Sinkankas Collection, photo by Werner Kalber/PPS, (top right) Doug Armstrong, (center) Doug Armstrong after W. G. Ernst, *Earth Materials,* © 1969. By permission of Prentice-Hall, Inc., Englewood Cliffs, N.J., (bottom) Doug Armstrong after F. C. Phillips, *An Introduction to Crystallography,* New York, John Wiley and Sons, Inc.; 253—John Sinkankas Collection, photo by Werner Kalber/PPS; 254—By permission of Ward's Natural Science Establishment, Inc., Rochester, New York; 255-256—John Sinkankas Collection, photographs by Werner Kalber/PPS, diagrams by Doug Armstrong; 257—(top) By permission of Ward's Natural Science Establishment, Inc., Rochester, New York, (center) M. Bickford; 258—M. Bickford; 260—(top) Bob and Ira Spring, (bottom) G. R. Roberts, Nelson, New Zealand; 261—By permission of Ward's Natural Science Establishment, Inc., Rochester, New York; 262—(top and bottom left) G. R. Roberts, Nelson, New Zealand, (center left) Don Fujimoto, (bottom) David Miller; 263—(top) David Miller, (center left) G. R. Roberts, Nelson, New Zealand, (bottom) Don Fujimoto; 264—G. R. Roberts, Nelson, New Zealand; 265—Rob Ratkowski; 266—Photo by Fritz Goro; 267—Bonnie Weber; 269—John Sinkankas Collection, Werner Kalber/PPS; 270-271—Shirley Dethloff; 273—*Traite de Geognosie,* Jean d'Aubuisson de Voisons, 1828-1835, courtesy Zentralbibliothek, Zürich.

Chapter 15

274—NASA Photograph; 276—UPI Compix; 278—Nona Remoz; 280—Ignacio Gomez; 282—Claire Steinberg; 283—Ignacio Gomez; 284—(bottom) John Dawson, (top) Tom Lewis; 285—Tom Lewis; 287—(left) NOAA, (right) Tom Lewis; 288—(left) Tom Lewis, (right) NOAA; 289—National Oceanic and Atmospheric Administration, Geophysical Fluid Dynamics Laboratory; 290—(top) Brian Brake/Rapho-Guillumette, (bottom) UPI Compix; 291—"An Introduction to a General System of Hydrostaticks and Hydraulicks," Vol. I, 1729, courtesy Science Museum, London.

Chapter 16

292—Tom Hollyman/Photo Researchers, Inc.; 294—Ignacio Gomez; 295—Doug Armstrong after *Galathea Report,* Vol. 1, Fig. 6, 1959; 296—Tom Lewis; 297—(top) U.S. Naval Oceanographic Office, (bottom) Doug Armstrong after H. U. Sverdrup, 1942, *Oceanography for Meteorologists,* New York, Prentice-Hall; 298—(top) Doug Armstrong after Sverdrup and Munk, (bottom) Tom Lewis; 299—NOAA; 301—Doug Armstrong; 302—John Dawson; 303—Harvey Caplin; 304—John Dawson; 305—Austin Post, U.S. Geological Survey; 307—Tom Lewis; 308—Bonnie Weber; 309—"Mundus Subterraneus," by Athanasius Kircher, 1678, courtesy Zentralbibliothek, Zürich.

Chapter 17

310—Solarfilma; 312—(bottom) Photo by Werner Wetzel, courtesy of Professor A. Seilacher; 312-313—(top) *Atlas of the Earth,* p. 36, © Mitchell Beazley, Ltd., 1971; 314—(bottom) Calvin Woo after *Oceans,* June 1969, p. 19; 314-315—(top) Tom Lewis; 316—Calvin Woo after F. J. Vine, "Magnetic Anomalies Associated with Mid-Ocean Ridges," in Robert A. Phinney, *The History of the Earth's Crust.* Copyright © 1968 by Princeton University Press, figure facing p. 82. Reprinted by permission of Princeton University Press; 318-321—John Dawson; 322—Bonnie Weber; 323—(top) R. J. P. Lyon, Stanford University, (bottom) NASA; 325—Don Fujimoto; 326—(top) John Dawson, (bottom) Grant Heilman; 327—"Systema Ideale Pyrophylaciorum Subterraneorum Quorum Montes," 1676, by Kirchner.

Chapter 18

328—Ernest Braun; 330—Howard Saunders; 332—(top left) Photographer William MacQuitty; 332-333—(bottom) *Atlas of the Earth,* page 94-95, © Mitchell Beazley, Ltd., 1971; 333—(top left and top center) Photographer William MacQuitty, (top right) NASA, (center left) *Atlas of the Earth,* p. 110, © Mitchell Beazley, Ltd., 1971, (center) George S. Nelson, London School of Hygiene and Tropical Medicine; 335—(left) British Columbia Government Photograph, (right) Boone Morrison; 336—Howard Sochurek; 337—Bonnie Weber; 338-339—*Atlas of the Earth,* pp. 96-97, © Mitchell Beazley, Ltd., 1971; 340—Pacific Gas & Electric Company; 341—Don Cowen, artist concept of the "Solar Power Farm" proposed by Aden and Marjorie Meinel of the University of Arizona; 342—Joe Rychetnik/Photo Researchers, Inc.; 344—(top) Reproduced from *Engineering and Science,* March 1971, published at the California Institute of Technology, (bottom) John Sinkankas Collection, photo by Werner Kalber/PPS.

The Beginning and the End

347—Peter Lloyd; 348-354—Doug Armstrong.

Chapter 19

356—Raman effect, Jon Brenneis; 358—John Dawson; 359—The Granger Collection; 360—(top) John Dawson, (bottom) Brown Brothers; 361—Doug Armstrong; 362—(top) Doug Armstrong, (bottom) The Granger Collection; 364—(top) Brown Brothers, (bottom) Doug Armstrong; 365—Dennis LaBerge; 366—Doug Armstrong; 367—(top) Doug Armstrong, (bottom) Howard Saunders; 368-372—Doug Armstrong; 373—(top) The Granger Collection, (bottom) Doug Armstrong; 374—Bonnie Weber; 375—(top left) UPI Compix, (top right) Segrè Collection, Niels Bohr Library, (bottom) Ron Etrine; 376—(top), Don Fujimoto after Harvey-Porter, *Introduction to Physical Inorganic Chemistry,* 1963, Addison-Wesley, Reading Massachusetts, (bottom) Doug Armstrong; 377—(top) Doug Armstrong, (bottom) Fraunhofer Solar Spectrum colored by himself, 1814, courtesy Deutsches Museum.

Chapter 20

378—John Oldenkamp; 380—Jane Daram; 381—Bonnie Weber; 383—Alan Holden and Phyllis Singer, *Crystals and Crystal Growing,* Doubleday, 1960; 385—Miriam Wohlgemuth; 387—(top) Miriam Wohlgemuth; 391-394 and 396—Jane Daram; 397—(bottom) Don Fujimoto.

Chapter 21

398—E. R. LaChapelle; 400—(top left) The Granger Collection, (top center) Brown Brothers, (top right) The Nobel Foundation, (bottom) Courtesy of Bell Laboratories; 401—Brown Brothers; 402—(top) Doug Armstrong, (bottom left) Diffraction pattern by J. Vander Sande, M.I.T., Cambridge, Massachusetts, (bottom right) Eastman Kodak Company; 403—Bonnie Weber; 404—Ray Saleme; 405—The Granger Collection; 406—Jane Daram; 407—(top) Jane Daram, (bottom) Lee Bolton; 408—(top) Photographs by Julius Weber, (bottom) Doug Armstrong; 409—Doug Armstrong after C. F. Elam, *The Distortion of Metal Crystals,* 1936, The Clarendon Press, Oxford; 410—Doug Armstrong after Alan Holden, *The Nature of Solids,* Columbia University Press, 1965, p. 47, by permission of the Publisher; 411—Doug Armstrong after W. T. Read, Jr., *Dislocations in Crystals.* Copyright 1953 by McGraw-Hill, Inc. Used with permission of McGraw-Hill Book Company; 412—(top) Jane Daram, (bottom) Niels Bohr Library, American Institute of Physics; 413—Doug Armstrong after Alan Holden, *The Nature of Solids,* Columbia University Press, 1965, p. 167, by permission of the Publisher; 414-415—Doug Armstrong; 417—Vermont Life Magazine/The Bentley Collection.

Chapter 22

418—Dan Morrill; 420—Ron Estrine; 421—Marie Feallock; 422—UPI Compix; 423—Marie Feallock; 425—Tom Lewis; 426—Doug Armstrong after Edward L. King, *How Chemical Reactions Occur,* W. A. Benjamin, 1963, Fig. 9.2, p. 115; 429—Calvin Woo; 431—Doug Armstrong; 433—Bonnie Weber; 434—Don Fujimoto; 436—(left) Bonnie Weber, (right) Los Angeles County Air Pollution Control District; 437—Germanisches National Museum.

Chapter 23

438—Courtesy of A. K. Kleinschmidt; 442—(left) Doug Armstrong, (right) R. C. Valentine *et al, The Journal of Biochemistry,* no. 7, p. 2143, 1968; 443—Doug Armstrong; 445-446—Doug Armstrong; 447—Doug R. Henderson and D. M. Blow, *Nature,* May 13, 1967; 449—Robert Kinyon/Millsap and Kinyon; 451—"Human Foetus in the Fourth Month," Michael Pius Erdl, *Die Entwickelung des Menschen un des Hühnchens im Ei,* Leipzig, 1845.

Graphic Pause

452—Vincent Van Gogh, "The Sower," June 1888, courtesy Kröller-Müller Foundation.

Chapter 24

454—Robert Gomel/Life Magazine, © 1965 by Time, Inc.; 456—(bottom) Marie Feallock; 457—(top) Doug Armstrong, (bottom) Brown Brothers; 458—Tom Lewis; 459—Doug Armstrong; 460—Ron Estrine; 461—Marie Feallock; 463—Bonnie Weber; 464-465—Doug Armstrong; 466—(right) Don Fujimoto, (bottom) The Granger Collection; 467—Diagrams by Doug Armstrong, Photo by Werner Kalber/PPS; 468—(top) Doug Armstrong, (bottom) The Granger Collection; 469—Doug Armstrong; 470—(top left) Diagram by Doug Armstrong, Photo by Werner Kalber/PPS, (top right) Photo by Fritz Goro, (bottom left) Ron Estrine; 471—Diagrams by Doug Armstrong, Photo by Werner Kalber/PPS; 472-473—Tom Lewis; 474—Diagram by Doug Armstrong, Photo by Werner Kalber/PPS; 475—"Essai Sur L'Electricite des Corps," by Nollet, Paris, 1746, courtesy Eidgenossische Technische, Zürich.

Chapter 25

476—Dan Morrill; 478—(top and center) *Photographs of Physical Phenomena,* Kodansha Ltd., (bottom) The Bettmann Archive; 479—(top) Fred Hartson, (bottom) Richard Benson; 480—(top) The Bettmann Archive, (bottom) Doug Armstrong; 481—Fred Hartson; 482—Werner Kalber/PPS; 483—Doug Armstrong; 484—NOAA; 485—Doug Armstrong; 486—Bonnie Weber; 487—Doug Armstrong; 488—Doug Armstrong; 489—Doug Armstrong; 491—Herbert Ohlmeyer; 492-496—Doug Armstrong; 497—(top) Ron Estrine, (bottom) "De Magnete" by William Gilbert, 1600, Zentralbibliothek, Zürich.

Chapter 26

498—American Optical Corporation; 500—Hugh Wilkerson; 501—(top) From *PSSC Physics,* D. C. Heath and Company, Lexington, Massachusetts, 1965, (center) Alice Harmon; 502—Alice Harmon; 503—(top) Dennis LaBerge, (bottom) Alice Harmon; 504-505—Alice Harmon; 506—(top right) Photographs courtesy of Purdue University, Laboratory for Application of Remote Sensing, (bottom center left) Lowell International Planetary Patrol Photograph, (bottom right) New Mexico State University Observatory Photograph, (bottom center right) JPL/NASA, (bottom left) Photograph courtesy of the Hale Observatories. Copyright by the California Institute of Technology and the Carnegie Institution of Washington; 507—(bottom left and bottom right) *Atlas of the Universe,* p.19, © Mitchell Beazley, Ltd., 1971; 506-507—Illustrations by Tom Lewis after Payson Stevens; 508-510—Alice Harmon; 513—G. E. Medical Systems Division; 514—Marie Feallock; 515—Alice Harmon; 516—John Dawson; 517—Alice Harmon; 518—Bonnie Weber; 519-520—Alice Harmon; 521—(top) Alice Harmon, (bottom) Rainbow of Cartesian Physics.

Chapter 27

522—Michael Heller; 524—(top) Calvin Woo, (bottom) John Dawson; 525—John Dawson; 526—(left) Bay Area Rapid Transit District Photo, (right) Western Electric Company, photo by Dave Thomas; 527—(left) NASA Photograph, (right) A. L. Schawlow; 528-529—John Dawson; 530—Lettering by Bonnie Weber, Illustration by John Dawson; 532-533—John Dawson; 535—Laser Studio set-up at McDonnell Douglas Facilities in producing Master Hologram plate for Reflection Film Holograms; 536—Margaret Benyon; 537—T. R. W., Inc.; 539—By permission of National Newspaper Syndicate, Chicago.

Graphic Pause

540—Victor Vasarely, "Meh-2," 1967-68.

Chapter 28

542—Tony Frelo, NAL; 544—Doug Armstrong; 545—Doug Armstrong after Walter R. Fuchs, *Physics for the Modern Mind,* Illustrations by Klaus Burglz. Copyright © 1967 by Wiedenfeld and Nicolson, (Educational) Ltd. and the Macmillan Company; 546—Doug Armstrong; 547—(top) Tom Lewis, (bottom) Tom Lewis after R. B. Leachman, "Nuclear Fission," Copyright © 1965 by Scientific American, Inc. All rights reserved; 548—(top) Doug Armstrong after Walter R. Fuchs, *Physics for the Modern Mind,* Copyright © 1967 by Weidenfeld and Nicolson, (Educational) Ltd. and the Macmillan Company, (bottom) UPI Compix; 549—(top) Photos by Fritz Goro, (bottom) Brown Brothers; 550—Dennis LaBerge; 552—Doug Armstrong after Gregory R. Choppin, *Nuclei and Radioactivity,* Copyright © 1964, W. A. Benjamin, Inc., Menlo Park, California; 553—John Dawson; 554—Bonnie Weber; 555—Doug Armstrong; 556-557—(top) Tom Lewis; 556—(bottom) Doug Armstrong; 557—(bottom) Lawrence Berkeley Laboratory.

Chapter 29

558—Photo by Fritz Goro, courtesy Army & Navy Air Task Force; 560—(top left) Photograph courtesy of the University of Chicago, (top center) American Institute of Physics, Niels Bohr Library, (top right) Archives de l'Institut du Radium; 561—(top left) The Nobel Foundation, (top center) Photo by Heka, Niels Bohr Library; 562-563—Tom Lewis after F. B. Leachman, "Nuclear Fission," © 1965 by Scientific American, Inc.; 564—Doug Armstrong after Gregory R. Choppin, "Fission," *Chemistry,* Vol. 40, 1967; 565—Bonnie Weber; 567—(top) Ernst Haas/Magnum Photos, (bottom) Photo World/F.P.G.; 568—Ron Estrine; 570—Photos by Fritz Goro/Time-Life Syndication; 571—Ron Estrine; 572-573—(top) Ron Estrine; 573—(bottom) The Granger Collection.

Chapter 30

574—Brookhaven National Laboratory; 577-590—Calvin Woo; 591—(top) Calvin Woo, (bottom) Don Fujimoto.

Appendix 2

594-597—Doug Armstrong after *Atlas of the Universe,* Endpapers, © Mitchell Beazley, Ltd., 1971.

Contributing Consultants

598-602—Robert Kinyon/Millsap and Kinyon.

PHYSICAL SCIENCE TODAY Book Team
John H. Painter, Jr. • *Publisher*
Cecie Starr • *Associate Publisher and Editor*
JoAn Rice • *Associate Editor*
Susan Harter • *Editorial Assistant*
Donald Fujimoto • *Senior Designer*
Linda Higgins • *Associate Designer*
Bonnie Weber • *Art Assistant*
Payson R. Stevens • *Scientific Graphics Consultant*
Linda Rill • *Graphics Research*
Shelagh Dalton • *Graphics Research Assistant*
Donald Umnus • *Production Manager*
Sandie Marcus • *Production Assistant*
Bobbye Hammond • *Book Team Coordinator*

John Osche • *Sales Manager*
William Bryden • *Science Marketing Manager*
Debby Green, Leslie Gilbert • *Sales Coordinators*
Nancy Sjoberg • *Rights and Permissions Supervisor*

CRM BOOKS
Richard Holme • *President and Publisher*
John H. Painter, Jr. • *Publisher, Life and Physical Sciences*
Roger Emblen • *Publisher, Social Sciences*
Tom Suzuki • *Director of Design*
Charles Jackson • *Managing Editor*
Henry Ratz • *Director of Production*

Officers of Communications Research Machines, Inc.
Charles C. Tillinghast III • *President*
Richard Holme • *Vice-President*
James B. Horton • *Vice-President*
Paul N. Lazarus III • *Vice-President*
Wayne Sheppard • *Vice-President*

The Periodic Table of the Elements

The periodic table condenses a great store of information about physics and chemistry that enables us to understand various relationships among different elements. This information is based on the periodic law, which states that the properties of elements are periodic functions of their atomic numbers. It is this periodicity that provides a basis for organizing the chemical and physical properties of the elements.

When elements are listed together on the basis of increasing atomic number, elements having similar chemical and physical properties appear at regular intervals in the listing. More than a century ago Dmitri Mendeleev worked out a scheme for showing the periodicity of the elements, but his scheme was based on atomic mass, not on atomic number. The large foldout table presented here is a modified version of Mendeleev's original layout.

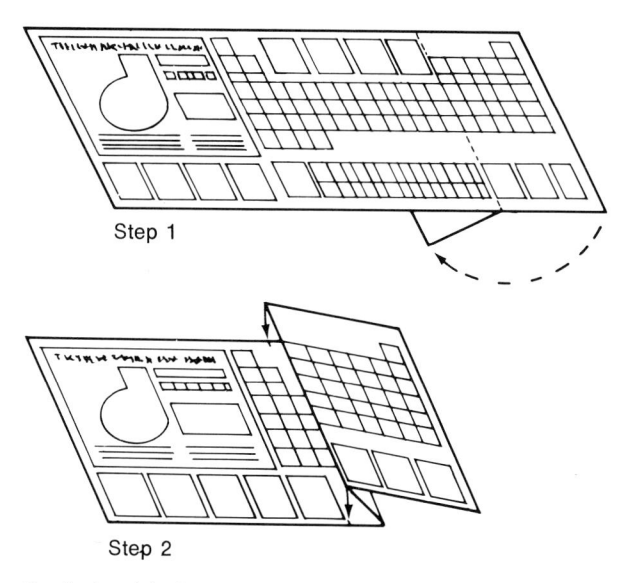

Step 1

Step 2

To distinguish the transition elements from the representative families, fold along the leading edge of Group IIIA (Step 1). After the fold has been made, bring the two short line marks together (Step 2).

Conceptual Design: Payson R. Stevens
Copyright © 1973 Communications Research Machines, Inc.
All rights reserved.

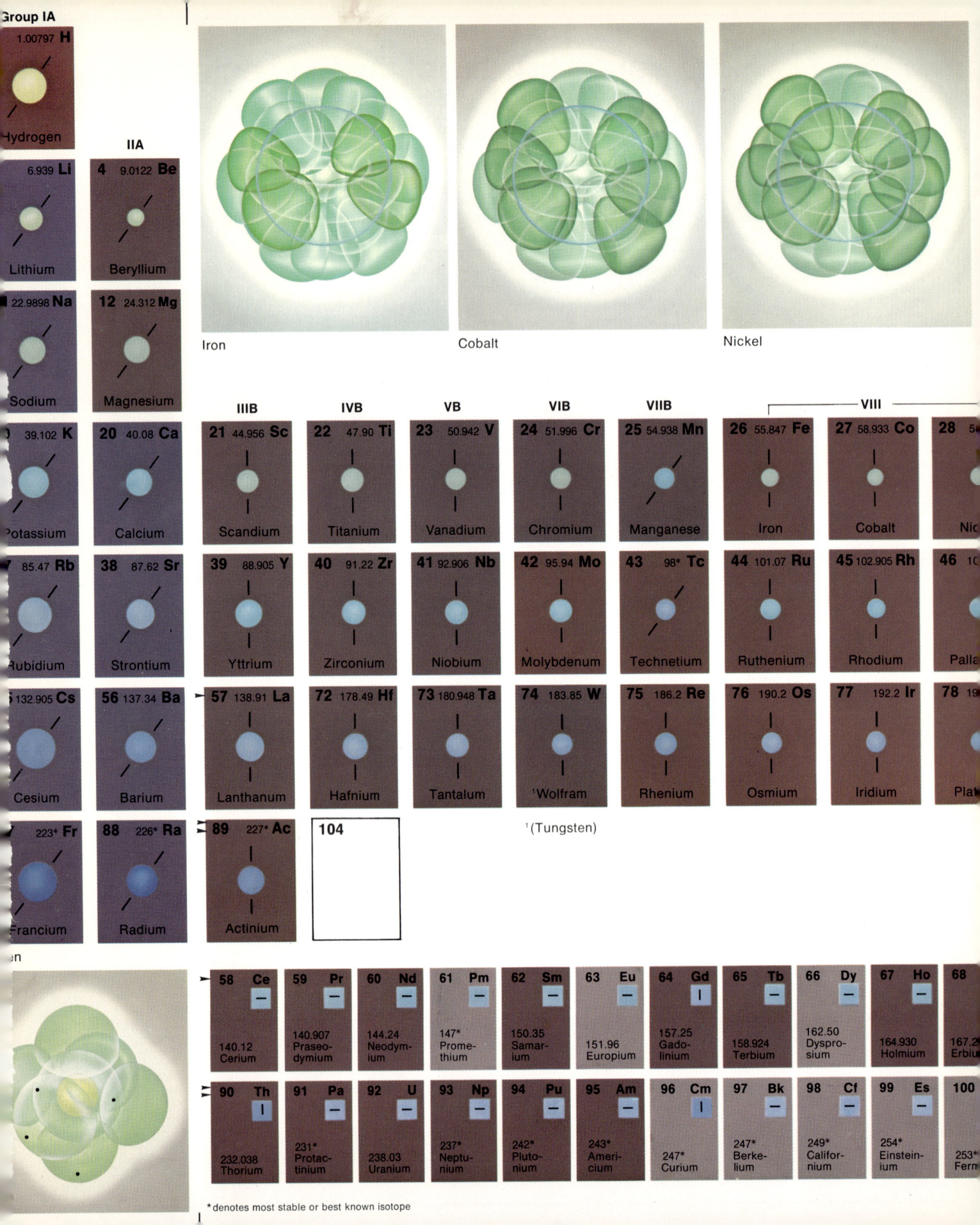

Group IA											
1.00797 H Hydrogen	IIA										
6.939 Li Lithium	4 9.0122 Be Beryllium										
22.9898 Na Sodium	12 24.312 Mg Magnesium										
		IIIB	IVB	VB	VIB	VIIB		VIII			
39.102 K Potassium	20 40.08 Ca Calcium	21 44.956 Sc Scandium	22 47.90 Ti Titanium	23 50.942 V Vanadium	24 51.996 Cr Chromium	25 54.938 Mn Manganese	26 55.847 Fe Iron	27 58.933 Co Cobalt	28 Ni
85.47 Rb Rubidium	38 87.62 Sr Strontium	39 88.905 Y Yttrium	40 91.22 Zr Zirconium	41 92.906 Nb Niobium	42 95.94 Mo Molybdenum	43 98* Tc Technetium	44 101.07 Ru Ruthenium	45 102.905 Rh Rhodium	46 Palla
132.905 Cs Cesium	56 137.34 Ba Barium	57 138.91 La Lanthanum	72 178.49 Hf Hafnium	73 180.948 Ta Tantalum	74 183.85 W ¹Wolfram	75 186.2 Re Rhenium	76 190.2 Os Osmium	77 192.2 Ir Iridium	78 Pla
223* Fr Francium	88 226* Ra Radium	89 227* Ac Actinium	104						

¹(Tungsten)

58 Ce 140.12 Cerium	59 Pr 140.907 Praseo-dymium	60 Nd 144.24 Neodym-ium	61 Pm 147* Prome-thium	62 Sm 150.35 Samar-ium	63 Eu 151.96 Europium	64 Gd 157.25 Gado-linium	65 Tb 158.924 Terbium	66 Dy 162.50 Dyspro-sium	67 Ho 164.930 Holmium	68 167. Erbiu
90 Th 232.038 Thorium	91 Pa 231* Protac-tinium	92 U 238.03 Uranium	93 Np 237* Neptu-nium	94 Pu 242* Pluto-nium	95 Am 243* Ameri-cium	96 Cm 247* Curium	97 Bk 247* Berke-lium	98 Cf 249* Califor-nium	99 Es 254* Einstein-ium	100 253* Ferm

*denotes most stable or best known isotope

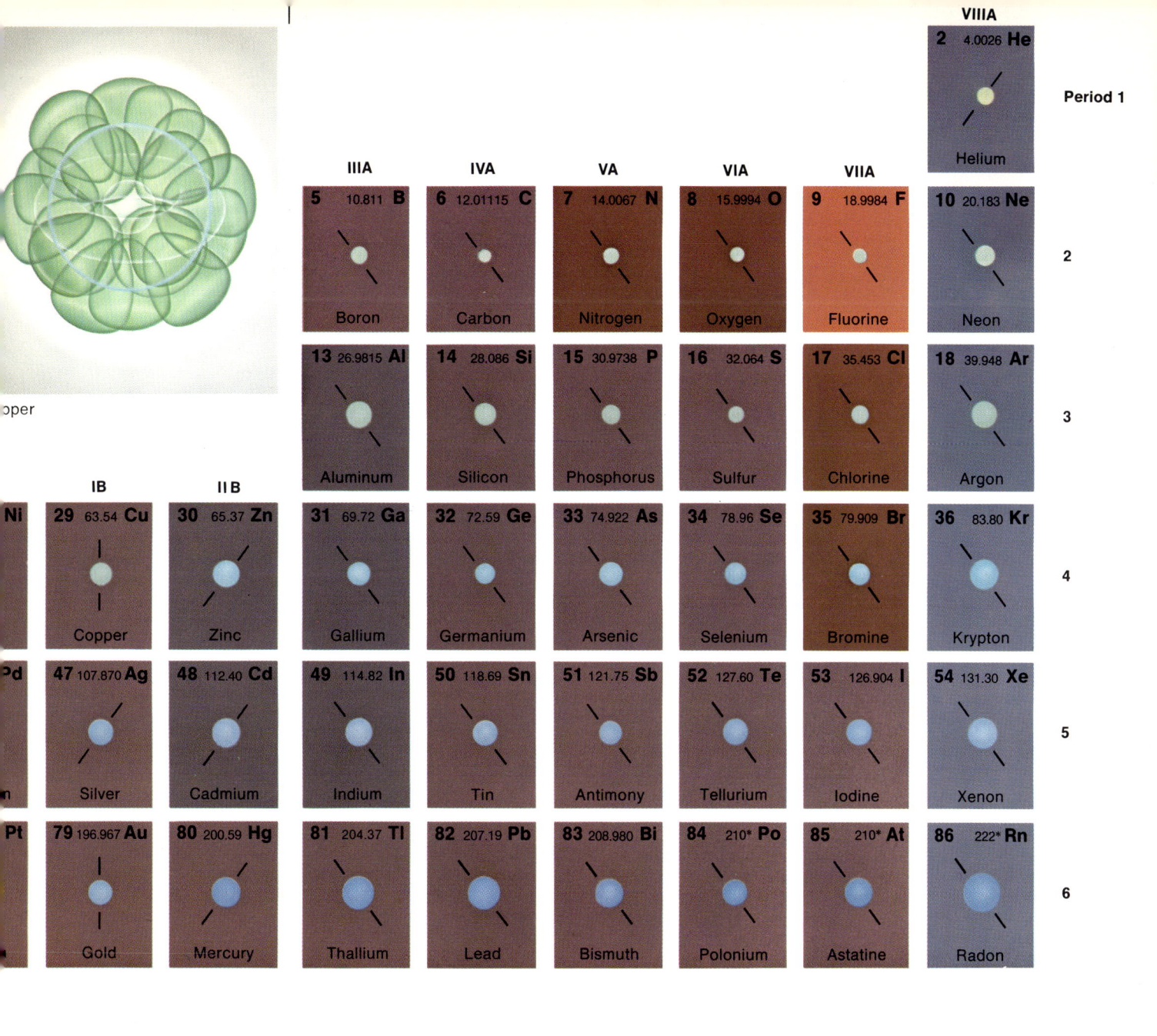

		IIIA	IVA	VA	VIA	VIIA	VIIIA	
							2 4.0026 **He** Helium	Period 1
		5 10.811 **B** Boron	6 12.01115 **C** Carbon	7 14.0067 **N** Nitrogen	8 15.9994 **O** Oxygen	9 18.9984 **F** Fluorine	10 20.183 **Ne** Neon	2
		13 26.9815 **Al** Aluminum	14 28.086 **Si** Silicon	15 30.9738 **P** Phosphorus	16 32.064 **S** Sulfur	17 35.453 **Cl** Chlorine	18 39.948 **Ar** Argon	3
IB	IIB							
29 63.54 **Cu** Copper	30 65.37 **Zn** Zinc	31 69.72 **Ga** Gallium	32 72.59 **Ge** Germanium	33 74.922 **As** Arsenic	34 78.96 **Se** Selenium	35 79.909 **Br** Bromine	36 83.80 **Kr** Krypton	4
47 107.870 **Ag** Silver	48 112.40 **Cd** Cadmium	49 114.82 **In** Indium	50 118.69 **Sn** Tin	51 121.75 **Sb** Antimony	52 127.60 **Te** Tellurium	53 126.904 **I** Iodine	54 131.30 **Xe** Xenon	5
79 196.967 **Au** Gold	80 200.59 **Hg** Mercury	81 204.37 **Tl** Thallium	82 207.19 **Pb** Lead	83 208.980 **Bi** Bismuth	84 210* **Po** Polonium	85 210* **At** Astatine	86 222* **Rn** Radon	6
								7

69 **Tm** — 168.934 Thulium	70 **Yb** — 173.04 Ytterbium	71 **Lu** I 174.97 Lutetium
101 **Md** — 256* Mendelevium	102 **No** — 254* Nobelium	103 **Lw** I 257* Lawrencium

Oxygen

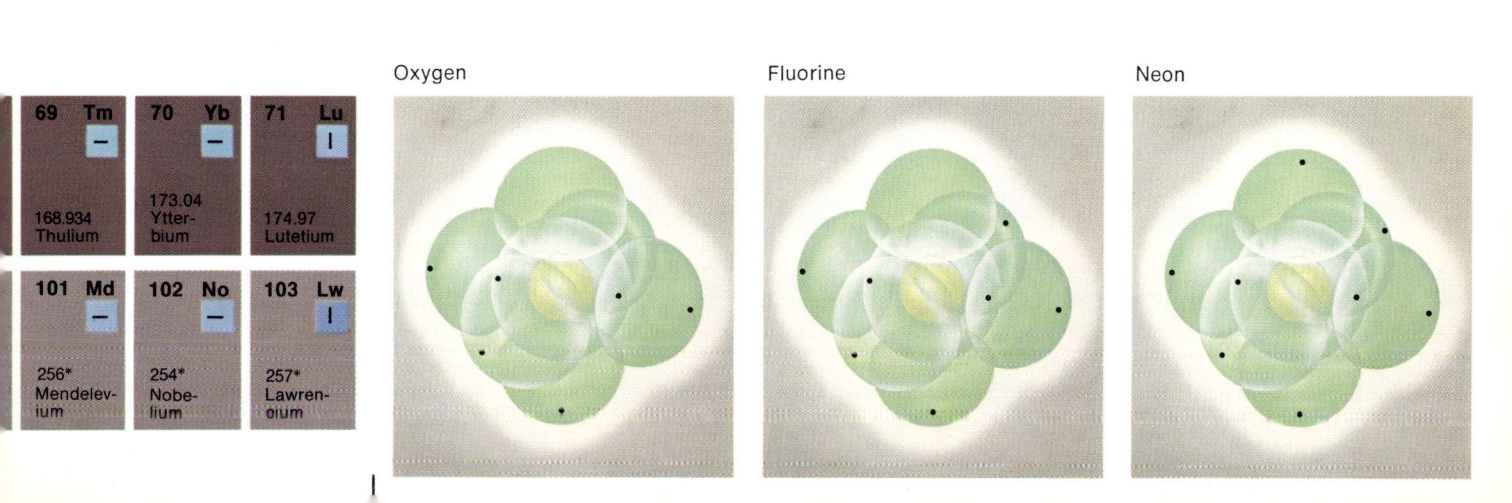

Fluorine

Neon

Traditional tables have relied on the printed word to describe periodic relationships. In contrast, this version provides a series of visual reference points for the main elemental properties in order to emphasize this periodicity. The large table lists elements with similar properties in vertical columns, or *groups*; elements having dissimilar properties are listed in horizontal rows, or *periods*. The arrangement shows the order in which atoms lose or gain electrons, which is called their *order of activity*. It is on the basis of this ordering that we classify elements as "metals," "non-metals" (including inert gases), and "semi-metals" (or *transition elements*). When the large table is folded as indicated, the transition elements are separated from the other, more representative elements.

The more active the element, the more intense the color used to shade the square for that element in the table. From this shading it becomes evident that the most active metal (the one with the strongest tendency to lose electrons in chemical reactions) falls in the lower left-hand corner of the table, and that the most active nonmetal (the one with the strongest tendency to acquire electrons) falls in the upper right-hand corner. This gradation is referred to as an "electronegativity" scale, and it provides a qualitative way of describing the strength with which atoms bond to each other. Generally, the greater the difference between electronegativities of two elements, the greater the strength of the bond between them.

What is the physical basis for the periodicity seen in this table? Each atom has only certain allowed energy levels, which are designated by n—a whole number (or *principal quantum number*) in a set starting with $n=1$. But within each set of n are further divisions. For example, 1s stands for the lowest energy level $n=1$; and 2s and 2p stands for the two energy levels corresponding to $n=2$. These divisions contain "hidden" energy states. For example, every s level contains two states; every p level contains six states; every d level, ten states, and so on. According to the Pauli exclusion principle, there can be no more than one electron in each of these states. On the average, *all* of the electrons with the same principal quantum number n are roughly the same distance away from the nucleus, and all have similar energy values. Such electrons are said to belong to the same electron *shell*.

The schematic color wheel is a visual key to the successive ordering of these shells. Each shell is given a letter (K, L, M, N, O, ...) to correspond with a principal quantum number (1, 2, 3, 4, 5, ...). The table above the wheel shows the order of filling for the orbitals, based on quantum mechanics. Each *s, p, f,* and *d* subshell is indicated by a slash mark. Using this slash mark, the table can be examined in an orderly way. The group A elements (or representative families) can be divided into two *s* and *p* blocks on the left- and right-hand side of the table. The transition elements show *d*-orbital filling, and the inner transition elements, *f*-subshell filling. Usually in these last two groups, inner *d* and *f* subshells are being filled even though electrons already occupy orbital positions in the next outer subshell. Some of the outer subshells are associated with lower energies, and the atom can achieve a more stable configuration by filling these shells first.

The energy an electron has in a given shell also depends somewhat on the *orbital quantum number* ℓ, which describes the *magnitude* of the electron's angular (or rotational) momentum. The smaller the orbital quantum number, the closer to the nucleus the electron is likely to be, and the lower its total energy. Like linear momentum, angular momentum is a vector quantity and to fully describe it we must give its *direction* as well. The orientation of these orbitals affects the way atoms combine into molecules. An electron subshell consists of all the electrons having the same principal quantum number n and the same orbital quantum number ℓ.

When a given shell or subshell contains all the electrons it can hold, it is said to be *closed*. And whether a shell or subshell is "open" or closed relates to the order of activity found in the periodic table. When an atom has only closed shells or subshells, its electric charge is distributed uniformly. It does not attract additional electrons, nor can its electrons be detached very easily. (This behavior is characteristic of the "passive" inert gases.) When an atom has only one electron in its outermost shell, it tends to lose that electron because the negative electron is far away from the attractive pull of the positive charge of the nucleus. (Alkali metals fall in this category.) And when an atom lacks only one electron in an otherwise filled shell or subshell, it tends to acquire that electron because its nuclear attraction is not completely "sealed off" by the innermost electrons. (This behavior is characteristic of the voracious halogens.)

These patterns are quite important, because the electron configuration of an element determines its properties. Interpretation of these configurations enables, for example, an understanding of crystal structure, electric properties or conductivity, oxidation state, relative atomic size and electronegativity. One example is the relationship between the valence number of an element (a measure of the combining power of elements with each other) and the position of the element within the table. There is a strong similarity between elements in the same vertical group: they all have the same valence and exhibit similar chemical properties even though they have different quantum numbers. There is even some resemblance between the properties of elements in different groups but having the same number of valence electrons.

When you look at the relative atomic *size* of the elements, another pattern emerges. As atomic number increases within a group, there is an increase in atomic size. But within a given horizontal row, or period, atomic size decreases with an increase in atomic number. This behavior is also explainable in terms of the changes that take place in electronic structure. Within a group, an increase in atomic number means there is an increase in the principal quantum number and a corresponding increase in the distance between the outermost electrons and the nucleus. Within a given period, the principal quantum number remains the same, but the increased nuclear charge exerts a stronger force on the outermost electrons and keeps them closer to the nucleus.

These are just a few examples that show how the electronic structure of the elements is related to their chemical and physical properties. Once these basic relationships are understood, the table may also be used to explore the properties that matter possesses, in both elemental and compound forms.

The Periodic Table of the Elements

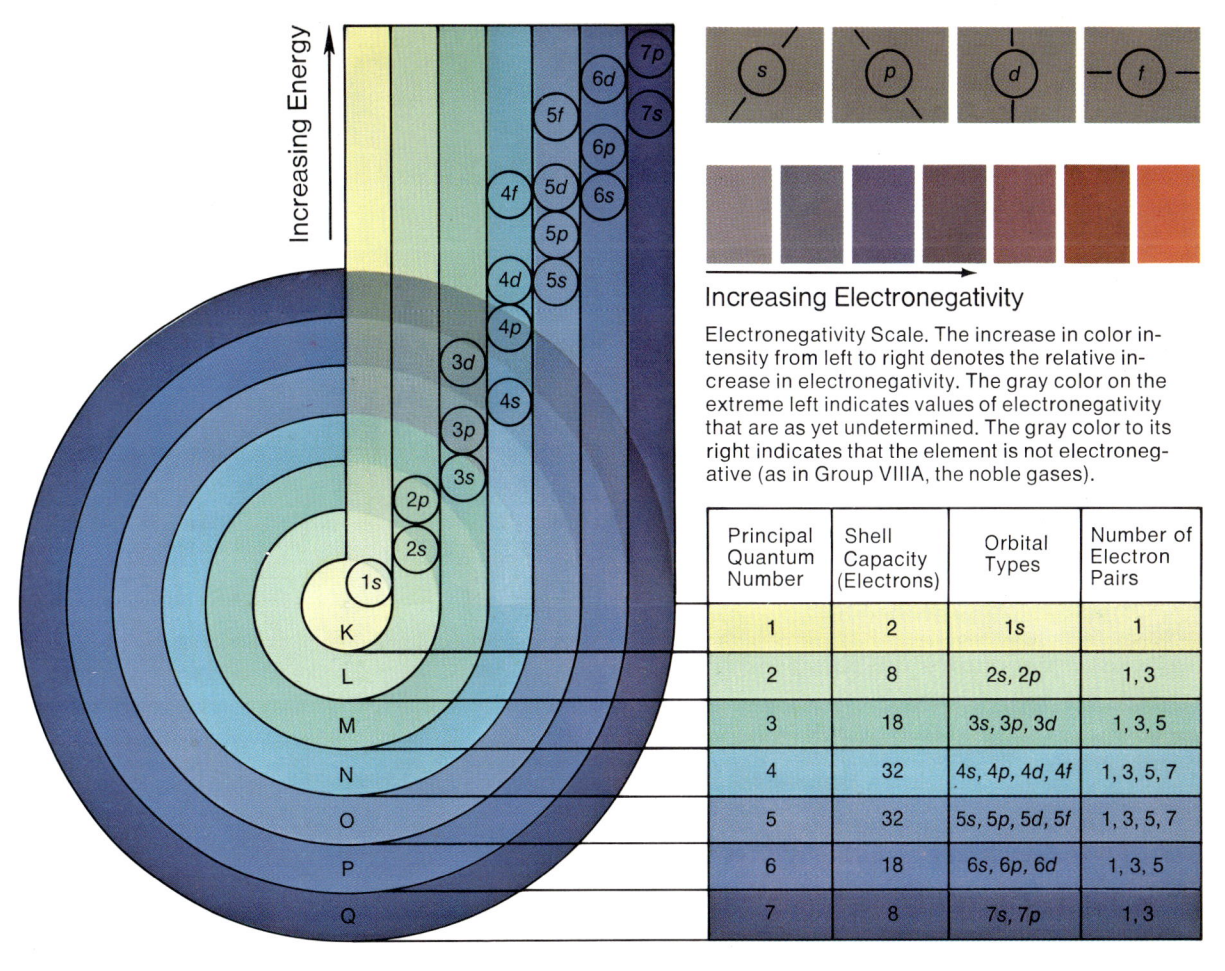

Increasing Electronegativity

Electronegativity Scale. The increase in color intensity from left to right denotes the relative increase in electronegativity. The gray color on the extreme left indicates values of electronegativity that are as yet undetermined. The gray color to its right indicates that the element is not electronegative (as in Group VIIIA, the noble gases).

Principal Quantum Number	Shell Capacity (Electrons)	Orbital Types	Number of Electron Pairs
1	2	1s	1
2	8	2s, 2p	1, 3
3	18	3s, 3p, 3d	1, 3, 5
4	32	4s, 4p, 4d, 4f	1, 3, 5, 7
5	32	5s, 5p, 5d, 5f	1, 3, 5, 7
6	18	6s, 6p, 6d	1, 3, 5
7	8	7s, 7p	1, 3

This color wheel is the visual key to the table. Successive shells are shown from K to Q. These shells are related to principal quantum numbers and to other characteristics shown in the table to the right. The spheres in the periodic table that depict relative atomic sizes are color-keyed back to this wheel. The sphere color indicates which shell is being filled by electrons. The graph above the wheel is an energy level diagram, which shows the order of orbital filling. The various orbitals within each shell are indicated.

The sequence for the filling of the s and p orbitals from lithium to neon is shown below. Here, electrons are indicated symbolically. The p orbitals are considerably less complex than the d orbitals.

The sequence of electron filling for the transition elements is shown above the main table. Orbitals that possess an electron pair are depicted as a darker green, whereas those that possess only one electron per orbital are depicted as a lighter green. The sequence of iron, cobalt, and nickel shows that there is an outer 4s orbital (blue) which already has two electrons in it while the complex inner 3d orbitals are being filled. In copper, all the 3d orbitals are filled while there is only one electron in the 4s orbital. It should be noted that these images represent the *average* probability of finding an electron occupying the volume in space for a specific orbital.

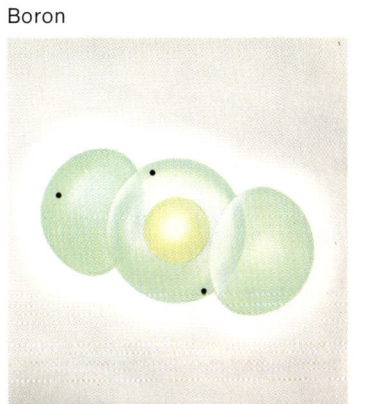

Lithium

Beryllium

Boron

Carbon

Ni